Electronics and Computer Math

SEVENTH EDITION

Bill Deem
Tony Zannini

Prentice
Hall

Upper Saddle River, New Jersey
Columbus, Ohio

Library of Congress Cataloging-in-Publication Data

Deem, Bill R.
    Electronics and computer math / Bill Deem, Tony Zannini.-- 7th ed.
      p. cm.
    Previous ed. has title: Electronics math.
    Includes index.
    ISBN 0-13-091127-5
      1. Electronics--Mathematics. 2. Computer engineering--Mathematics. I. Zannini,
Tony. II. Deem, Bill R. Electronics math. III. Title.
TK7864 .D44 2003
510′.246213--dc21

          2002070095

Editor in Chief: Stephen Helba
Editor: Frank I. Mortimer, Jr.
Media Development Editor: Michelle Churma
Production Editor: Alexandrina Benedicto Wolf
Production Coordination: Clarinda Publication Services
Cover Design Coordinator: Diane Ernsberger
Cover Designer: Linda Sorrells-Smith
Cover art: Digital Stock
Production Manager: Brian Fox

This book was set in Times Roman by The Clarinda Company. It was printed and bound by R.R. Donnelley & Sons Company. The cover was printed by The Lehigh Press, Inc.

Pearson Education Ltd.
Pearson Education Australia Pty. Limited
Pearson Education Singapore, Pte. Ltd.
Pearson Education North Asia Ltd.
Pearson Education Canada, Ltd.
Pearson Educación de Mexico, S.A. de C.V.
Pearson Education—Japan
Pearson Education Malaysia, Pte. Ltd.
Pearson Education, *Upper Saddle River, New Jersey*

10 9 8 7 6 5 4 3 2 1
ISBN 0-13-091127-5

As a top-selling text in its field, *Electronics and Computer Math* has been used in several hundred classrooms over the last six editions. The book is written for students in high schools, community colleges, and technical institutes and for technicians in the field of electronics. There are no course prerequisites for this text. It is intended to be used as a separate text in electronics math or as a text that could be used as a reference throughout the study of electronics.

Electronics and Computer Math provides a thorough, complete, and practical study of electronics math and its relationship to the world of electronics. The mathematical topics chosen are those that the authors feel are most useful in solving electronics problems. As such, this book places greater emphasis on certain areas of the discipline than does abstract math. The grouping and sequencing of topics are designed to support various configurations of related courses in DC, AC, and digital electronics.

Because the calculator is an integral part of all technical students' classroom tools, the use of the calculator in problem solving is introduced in the text as the need arises. Algorithms are presented when appropriate.

To reinforce new concepts and to help students test their understanding of the material, *Electronics and Computer Math* features:

- Key Concept highlights
- over 300 examples
- over 1300 practice problems within chapters
- over 2600 end-of-chapter problems
- chapter summary tables
- self-tests at the end of topics

If the text is being used for self-study, the self-tests can be used to determine whether or not the student already possesses that skill. To ensure accuracy, technical reviewers have worked every example, practice problem, self-test, and end-of-chapter problem. In addition, each supplement has been technically checked by additional technical reviewers.

CHANGES IN THIS EDITION

Although the title of the seventh edition has been changed to *Electronics and Computer Math*, it retains all of the material from the sixth edition. In addition, it incorporates numerous ideas suggested by instructors who have used previous editions for many years.

New Title

The most obvious change in this edition is the new title, which reflects changes in technology. Today's electronics technicians must not only know how to work with computers, they must also know how computers work. Section six combines three chapters from the previous edition into one section, "Math for Digital Electronics," which is important for the understanding of how computers work. In addition, some of the new example and end-of-chapter problems are taken from the computer field.

New Co-author Tony Zannini

A less obvious change is the addition of a co-author. Tony Zannini has joined Bill Deem in the development of this edition. Tony brings his experience as an electronics design engineer (over twenty years) and his experience in the education field (over ten years in electronics, computers, and math) to help students understand the math principles upon which all electronics and computers are built.

Arrangement of Chapters

We have, in general, retained the previous order of the chapters. Twenty-six of the twenty-eight chapters remain in the same order as in the previous edition. The biggest change is that the chapter "Computer Number Systems" has been moved from Chapter 4 in the sixth edition to Chapter 25 in this edition, so that it directly precedes the chapter on Boolean algebra. The other change is that the chapter "Fractions, Decimals, and Percents," which does not include any algebraic concepts, now precedes "Algebraic Terms: Roots and Powers."

Section Organization

Upon examining the table of contents, you will notice that the chapters have now been divided into seven sections:

- Section 1—Review of Arithmetic
- Section 2—Algebra Fundamentals
- Section 3—Math for DC Electronics
- Section 4—Math for AC Electronics
- Section 5—Logarithms in Electronics
- Section 6—Math for Digital Electronics
- Section 7—Introduction to Statistics in Electronics

This helps students see the relationship between math topics, other courses in their curriculum, and applications in technology. The grouping also helps guide instructors adjust the order in which math topics are taught because of changes in course offerings from term to term.

Sections 1 and 2 should be taught in sequence, and can be covered quickly with advanced math students. Section 3, "Math for DC Electronics," would normally follow Section 2; however Section 5, 6, or 7 could follow Section 2 if those math topics were needed to support other courses.

Section 3 (Chapters 10 through 14) is designed to be taught in a program where a course in basic electronics is taught concurrently. Chapters 10, 11, and 12 provide support for the principles usually taught in a DC electronics class. Chapters 13, "Graphing," and 14, "Simultaneous Linear Equations," can be used to solve some of the problems in Chapters 10, 11, and 12 but are not dependent on them so they can be taught any time after Section 2.

Section 4, "Math for AC Electronics," (Chapters 15 through 21) is normally taught after Section 3, "Math for DC Electronics." Chapter 13 is a prerequisite for Chapters 15, "Complex Numbers," and 16, "The Right Triangle." The three chapters on AC circuits, like the three on DC circuits, provide support for the principles taught in an electronics class.

Sections 5, 6, and 7 can be taught in any order after Section 2. Usually the sequence is dictated by the math support required by other courses.

Estimating

New to this edition is a discussion on an estimating technique that uses scientific notation and rounding in Chapter 2. Making quick mental mathematical estimates is helpful when troubleshooting electronic circuits, verifying calculator answers, and taking timed pre-employment tests when calculators are not allowed.

Calculator Usage

The instructions for using a calculator have been expanded. We continue to provide the instructions for the Texas Instruments TI-36X and we have added instructions for the Casio fx-115W. Both are low-cost, popular, scientific calculators and most keystrokes are the same for both calculators. We point out the instructions for the Casio when they are different from the TI. Students who are already calculator proficient can easily skip these instructions.

New Word Problems

We have added some math problems that are stated with words rather than numeric symbols. Technicians often have to translate written or verbal descriptions of problems into math symbols before they can begin a solution.

CHAPTER ORGANIZATION

Chapters 1, 2, and 3 deal with decimal numbers, powers of ten, and prefixes. These chapters introduce the student to the calculator and to problem solving involving electrical units.

Chapter 4 reviews addition, subtraction, multiplication, and division of fractions. Conversions between mixed numbers and decimal fractions and finding common denominators are covered.

Chapter 5 introduces the student to algebraic terms, roots, and powers. Typical electrical problems involving literal numbers, squares, and square roots are solved using the calculator. Chapters 6 through 9 contain topics in algebra, including linear equations, second-degree equations, fractional equations, and factoring. Throughout these chapters, problem solving applications using the calculator are presented.

In Chapters 10 through 12, student use the algebraic skills they developed in Chapters 6 through 9 to solve dc circuit problems using Kirchhoff's and Ohm's law, and Thévenin's, Norton's, and the superposition circuit theorems. Graphical and algebraic solutions to circuit problems and linear equations are presented in Chapters 13 and 14. Practical applications are presented for each technique discussed.

Chapters 15 through 18 introduce algebra and trigonometry elements needed to solve ac circuit problems. Angular velocity and the sine wave are introduced in these chapters. Problem solving using trigonometric functions and the calculator is presented. In Chapters 19 through 21, ac series, parallel, and complex circuit problems are solved. In Chapter 19, the student learns how to express phasors in either polar or rectangular form. In Chapter 20, circuit theorems are again presented as an aid in solving complex circuit problems. These problem-solving techniques are used in Chapter 21 in determining the parameters for several types of filter circuits.

Chapters 22, 23, and 24 cover both common and natural logarithms and their applications. Logarithmic equations are covered in Chapter 23. Applications including the Bode plot are found in Chapter 24.

Chapter 25 presents the various number systems (binary, octal, and hexadecimal) that are used in the study of computers. Conversions between the number systems and addition and subtraction in these systems are covered. Chapter 26 discusses the basic logic functions inherent in all logic circuits and presents those theorems, laws, and postulates used in the simplification of logic expressions. Chapter 27, Karnaugh Maps, offers an alternative method of logic circuit simplification.

In Chapter 28, Introduction to Statistics, we introduce the student to frequency distribution tables, histograms, measures of central tendency, and the normal curve.

EXTENSIVE SUPPLEMENTS PACKAGE

Electronics and Computer Math comes with a wide variety of optimal supplements for both the instructor and student.

- A **Student Study Guide with Selected Solutions** (ISBN 0-13-048782-1) contains chapter overviews and additional study questions for each chapter. It includes fully worked-out solutions to selected end-of-chapter problems.
- An **Instructor's Solutions Manual with PowerPoint slides** (ISBN 0-13-091128-3) contains fully worked-out solutions to end-of-chapter problems and provides the instructor with over 120 illustrations to use.
- An **Instructor's Test Item File** (ISBN 0-13-091131-3) contains 1000 additional test questions. It is also available in computerized format.
- A **Companion Website** (ISBN 0-13-091120-8) can be accessed at www. prenhall.com/deem. It includes an online study guide with practice problems, Syllabus Manager™, and links to other resources on the Web.
- A Study Wizard continuing multiple-choice questions is found on the CD-ROM packaged with the text.

ACKNOWLEDGMENTS

We wish to acknowledge the assistance given by the editorial and production staffs of Prentice Hall and Clarinda Publication Services and the many students and teachers who aided and assisted us in preparing this edition. Also, we wish to acknowledge the time and effort given by Ginger Deem in preparing the artwork for the Instructor's Solutions Manual and Nikki Zannini for helping with data and equation entry.

We would also like to thank Professors Bob Derby and David Greiser, DeVry University, Pomona, CA, and Professor Ted Wu, DeVry University, Long Beach, CA for their valuable suggestions.

Finally, a great big thank you to our technical reviewers: Alan Krause, DeVry University, Addison, IL; Timothy Staley, DeVry University, Irving, TX; Terry O'Laughlin, Madison Area Technical College, WI; and Vicky Saling, Heald College, Hayward, CA.

Bill Deem
Tony Zannini

BRIEF CONTENTS

19   AC Circuit Analysis: Series Circuits   531

20   AC Circuit Analysis: Parallel Circuits   568

21   Filters   599

Section 5   Logarithms in Electronics   617

22   Logarithms   619

Review of Arithmetic Fundamentals

Topics in this section include a review of arithmetic fundamentals, scientific notation, electrical units, and terms related to electronics and computers. Many of the concepts in this section are usually first taught in grade or high school; consequently, detailed explanations are not always provided. It is important to master the concepts in this section because they form the foundation for the next section on algebra. Adult learners who have been out of school for several years or recent high school graduates who want to review their math skills need to practice solving these chapter problems until they can do them quickly with no errors.

Chapter 1 reviews the decimal number system, whole numbers, fractions, rounding, addition, subtraction, multiplication, division, and mathematical expressions and terms.

Chapter 2 covers powers of ten, scientific notation, reciprocals, and exponent operations. Calculator usage is described in detail and a method of estimating calculations is presented.

Chapter 3 introduces engineering notation, the International System of Units, prefixes of electrical units, and coversions between the metric and English systems of measurements. Scientific units are introduced here so that students will become familiar with the terminology of typical electronics and computer problems. Students will learn the physical characteristics of units such as "volts" and "watts" in their Electronics courses.

In Chapter 4 we will study fractions again, learn how to reduce them to lowest terms, and convert them to decimal numbers and percentage points.

"How many . . .?" is a common question in technology. Technologists use algebra (and sometime calculus) to model electronic and computer systems so that mathematical techniques can be used to find solutions to problems. At some point in the problem solving process the topics covered in this section are used to calculate the answer to the question, "How many . . .?" Students who master the topics in this section will improve their problem solving skills and find the next section challenging and fun.

The Decimal Number System

Introduction

The decimal number system is the number system that we have always used. We all grew up using this system for all computations. Many of the concepts we will discuss in this chapter we have applied for years without really thinking about them. We are going to spend some time discussing the decimal numbering system in order to help us better understand working with numbers and to pave the way for a quicker understanding of other number systems. These other number systems, such as the *binary number system* and the *hexadecimal number system,* are important number systems used in the computer world and will be discussed in Chapter 25.

The chapter is primarily a review of basic arithmetic. We will review the process of adding, subtracting, multiplying, and dividing mathematical terms and expressions. In this chapter we will convert between decimal numbers and decimal fractions, and we will learn how to round numbers. Rounding numbers is a common practice in electronics because we seldom need accuracies beyond three decimal digits. The concepts learned or relearned in this chapter are used throughout the text.

Chapter Objectives

In this chapter you will:

1. Develop an understanding of the decimal number system and the concept of place value.
2. Learn how to read whole numbers and decimal fractions.
3. Learn how to convert from decimal fractions to decimal numbers.
4. Learn the rules for rounding both whole and nonwhole numbers.
5. Learn how to work with signed numbers.
6. Learn how to add, subtract, multiply, and divide decimal numbers.
7. Learn how to recognize and simplify various mathematical expressions and terms.

Our number system is called the decimal number system. *Decimal* means ten. In the decimal number system there are ten symbols. These symbols are called *digits*. The ten digits are 0, 1, 2, 3, 4, 5, 6, 7, 8, 9. Zero is the digit having the least value. In counting, when the count reaches 9, a limit has been reached because 9 is the digit of greatest value. An additional count produces a *carry*. This carry is equal to 10 and is said to occupy the tens position; the 0 occupies the units position. This process continues each time 9 is reached in the units position. When a 9 appears in the tens position, the next carry to the tens position produces a carry to the hundreds position. In Table 1–1 we show the names for places up to millions.

It is important to note that a 1 in the tens position possesses a value *(weight)* 10 times that of a 1 in the units position. The same is true for each position of greater weight. Because of this relationship, the decimal numbering system is called a *place value system* and its base is 10.

Table 1-1   Names for Places from Millions to Millionths.

Units							Tenths
Tens						Hundredths	
Hundreds					Thousandths		
Thousands				Ten-thousandths			
Ten-thousands			Hundred-thousandths				
Hundred-thousands		Millionths					
Millions							

The decimal number system not only includes the ten numerical symbols, it also includes several symbols that are called mathematical operators. Examples of these operator symbols are the addition sign $(+)$ and the multiplication sign $(\times)$. These operator symbols are similar to punctuation marks in English, or any written language. Mathematics is the international language of numbers. It is a written and spoken language. To be an effective communicator in this language you must understand the meaning of the mathematical operator symbols. You can increase your math vocabulary, and become a better math communicator, by learning more of the math operators.

Math operator symbols usually tell us what to do, or what mathematical operations we should perform on numbers; for example the addition sign in $6 + 2$ tells us to add six and two. Sometimes the symbols are used just to make our understanding of numbers easier, for example we use a comma to separate thousands in the number 3,247. The table below describes some common math operators:

Table 1-2 Common Math Operators.

NAME	SYMBOL	EXAMPLE	USAGE
Decimal Point	.	3.7	The reference point for the place value of each numerical symbol
Addition Sign	+	3 + 2	Shows that 3 must be added to 2
Plus Sign	+	+3	Indicates a positive number and is usually omitted unless clarification is necessary
Subtraction Sign	−	3 − 2	Shows that 2 must be subtracted from 3
Negative Sign	−	−3	Indicates a negative number
Multiplication Sign	×	3 × 2	Shows that 3 must be multiplied by 2
Division Sign	÷ or /	3 ÷ 2 or 3/2	Shows that 3 must be divided by 2
Equals Sign	=	3 × 2 = 6	Shows that the quantity on the left is the same as the quantity on the right

EXAMPLE 1-1

(a) In the number 24, what weight does 2 have?

SOLUTION 2 is in the tens position so its weight is (2 times 10) which equals 20.

(b) In the number 5,264, what weight does 2 have?

SOLUTION 2 is in the hundreds position so its weight is (2 times 100) which equals 200

(c) In the number 375, how many tens are in the tens position?

SOLUTION The tens position is the second from the right so there are 7 tens.

PRACTICE PROBLEMS 1-1

1. In the number 893,462:
   (a) What weight does the 9 have?
   (b) What weight does the 4 have?
   (c) How many thousands are there?

2. In the number 7,603,418:
   (a) How many ten-thousands are there?
   (b) How many tens are there?
   (c) How many hundred-thousands are there?
   (d) How many units are there?

1. (a) 9 ten-thousands or 90,000
   (b) 4 hundreds or 400   (c) 3

2. (a) 0   (b) 1   (c) 6   (d) 8

Additional practice problems are at the end of the chapter.[1]

<table>
<tr><td>**1-2**</td><td>**DECIMAL FRACTIONS**</td></tr>
</table>

To discuss decimal digits, let's review some names used in dealing with fractions. In the fraction $\frac{7}{10}$ the number above the line (7) is called the *numerator*. The number below the line (10) is called the *denominator*. The line itself is called the *vinculum*.

A decimal fraction is a fraction whose denominator is 10 or a multiple of 10 (100, 1000, 10,000, and so on). If the denominator is 10, the fraction is read one-tenth if the number is $\frac{1}{10}$, two-tenths if the number is $\frac{2}{10}$, and so on.

Here are some decimal fractions and how they are read:

$$\frac{3}{100} \qquad \text{3 hundred}ths$$

$$\frac{1}{1000} \qquad \text{1 thousand}th$$

$$\frac{12}{10,000} \qquad \text{12 ten-thousand}ths$$

Notice that we read the denominator just as we read a whole number except that we add *th* or *ths*. Similarly, if we write "4 thousandths," we know this is a fraction (the *ths* at the end of thousand tells us this); so we know the fraction is $\frac{4}{1000}$.

Table 1–3 gives the names of places from millions to millionths. The tenths position is the first position to the right of the decimal point. The hundredths position is the next position, and so on, to the millionths position, which is the sixth position to the right of the decimal point.

EXAMPLE 1–2

Read the following fractions:

(a) $\frac{5}{10}$ (b) $\frac{27}{100}$ (c) $\frac{27}{1000}$ (d) $\frac{17}{10,000}$

---

[1] At the end of each chapter there are additional problems labeled **End of Chapter Problems.** These problems are additional practice problems. For easy reference these *End of Chapter Problems* are numbered by section just like the Practice Problems. Thus, Practice Problems *1–1* and *End of Chapter Problems 1–1* relate to the material covered in section 1–1. Practice Problems *1–4* and *End of Chapter Problems 1–4* relate to the material covered in section 1–4, and so on.

Table 1–3   Names of Places from Millions to Millionths.

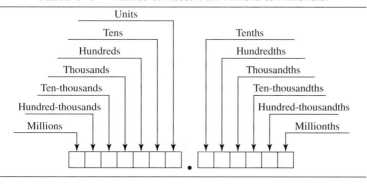

(a) 5 tenths   (b) 27 hundredths   (c) 27 thousandths
(d) 17 ten-thousandths

To express a written number as a decimal fraction, we determine the number in the denominator from Table 1–3. The numerator is the decimal number part. For example, if a number is written as 143 thousandths (one hundred forty-three thousandths), the "thousandths" tells us that the number in the denominator is 1000. The numerator is 143.

$$143 \text{ thousandths } = \frac{143}{1000}$$

The number 9 ten-thousandths is $\frac{9}{10000}$. The "ten-thousandths" tells us that the denominator is 10,000. 9 is the value of the numerator.

EXAMPLE 1–3

Change the following numbers to fractions:

(a) 73 hundredths   (b) two hundred eleven thousandths
(c) 33 hundred-thousandths

(a) $\frac{73}{100}$ (b) $\frac{211}{1000}$ (c) $\frac{33}{100,000}$

These same numbers can be written as decimal numbers. In the number 143 thousandths, the "thousandths" tells us that there are three digits to the right of the decimal point as shown in Table 1–3. Once we determine how many digits there must be, then we place the rightmost digit of the number, (3), in this position. The number would be written 0.143. (We placed a 0 to the left of the decimal point to remind us that there is no whole number part, and to help us keep track of where the decimal point belongs.)

The number 6 ten-thousandths has four digits to the right of the decimal point. 6 would be placed in the ten-thousandths position and zeros would be placed in the thousandths, hundredths, and tenths positions signifying that there are no thousandths, no hundredths, and no tenths. The number would be written 0.0006.

EXAMPLE 1–4

Convert the following numbers to decimal numbers:

(a) 47 thousandths (b) 7 hundredths
(c) six hundred forty-nine ten-thousandths

SOLUTIONS

(a) 0.047 (b) 0.07 (c) 0.0649

To write a decimal number, determine the position of the rightmost digit (refer to Table 1–3). Then write the value of the number. Let's look at the number 0.003. The rightmost digit is 3 and occupies the thousandths position. Therefore, the number is 3 thousandths. In the number 0.0567, the 7 is in the ten-thousandths position so the number is 567 ten-thousandths.

EXAMPLE 1–5

Write the following decimal numbers:

(a) 0.432 (b) 0.0077 (c) 0.01047

SOLUTIONS

(a) 432 thousandths (b) 77 ten-thousandths
(c) 1047 hundred-thousandths

1-2-1 Converting Between Decimal Numbers and Decimal Fractions

Usually quantities between zero and one are represented as decimal numbers or as decimal fractions. In this section we will convert from one to the other. First, let's convert from decimal fractions to decimal numbers. Consider the fraction $\frac{27}{100}$. To convert decimal fractions to decimal numbers, first determine the value of the denominator. In this fraction, the denominator is 100, making the fraction 27 hundredths. Place the rightmost digit of the numerator, 7, in the hundredths position. The 2 then must occupy the tenths position, so the answer is 0.27.

EXAMPLE 1–6

Change the decimal fraction $\frac{17}{100}$ to a decimal number.

SOLUTION   The value of the denominator is 100. The rightmost digit in the numerator is 7; therefore, the 7 is placed in the hundredths position. This puts the 1 in the tenths position and the number is 0.17.

EXAMPLE 1–7

Change the decimal fraction $\frac{17}{1000}$ to a decimal number.

SOLUTION   The denominator is 1000 so the 7 must appear in the thousandths position. That puts the 1 in the hundredths position. As with whole numbers, zeros must be used as place holders when necessary. Therefore, a zero goes in the tenths position. The decimal number is written 0.017 (17 thousandths).

Notice that in each case a zero was placed to the *left* of the decimal point. This is done to help us remember where the decimal point belongs. The zero also tells us that there is no whole-number part.

☞ **RULE 1–1**   To convert decimal fractions to decimal numbers, determine the value of the denominator. Place the rightmost digit of the numerator in this position.

Now let's convert from decimal numbers to decimal fractions. Consider the number 0.039. Determine the position of the rightmost digit (9). This tells us the value of the denominator. The 9 is in the thousandths position, so the denominator is 1000. The numerator is the number with the decimal point and leftmost zeros removed. Stated numerically:

$$0.039 = \frac{39}{1000}$$

EXAMPLE 1–8

Convert 0.007 to a fraction.

SOLUTION   The rightmost digit, 7, appears in the thousandths position so the answer is $\frac{7}{1000}$ (seven-thousandths).

EXAMPLE 1–9

Convert 0.0023 to a fraction.

SOLUTION   The rightmost digit, 3, is in the ten-thousandths position so the answer is $\frac{23}{10,000}$ (twenty-three ten-thousandths).

☞ **RULE 1–2**   To convert decimal numbers to decimal fractions, determine the position of the rightmost digit. This tells us the value of the denominator. The numerator is the decimal number with the decimal point removed.

1. Convert the following fractions to decimal numbers:

    (a) $\dfrac{4}{100}$ (b) $\dfrac{23}{1000}$ (c) $\dfrac{203}{100,000}$

2. Write the following numbers as decimal fractions:

    (a) 0.05 (b) 0.00073 (c) 0.00009

3. Write each of the following first as a decimal fraction and then as a decimal number:

    (a) 7 tenths (b) 37 hundredths
    (c) 7 ten-thousandths
    (d) 417 hundred-thousandths
    (e) 6 millionths

4. In the number 0.00246, in which place does the 2 appear? The 6?

5. Write the names of the following numbers:

    (a) 0.03 (b) 0.0005 (c) 0.00073

SOLUTIONS

1. (a) 0.04 (b) 0.023 (c) 0.00203

2. (a) $\dfrac{5}{100}$ (b) $\dfrac{73}{100,000}$ (c) $\dfrac{9}{100,000}$

3. (a) $\dfrac{7}{10} = 0.7$ (b) $\dfrac{37}{100} = 0.37$

    (c) $\dfrac{7}{10,000} = 0.0007$

    (d) $\dfrac{417}{100,000} = 0.00417$

    (e) $\dfrac{6}{1,000,000} = 0.000006$

4. The 2 appears in the thousandths position. The 6 appears in the hundred-thousandths position.

5. (a) 3 hundredths
   (b) 5 ten-thousandths
   (c) 73 hundred-thousandths

Additional practice problems are at the end of the chapter.

1–3 WHOLE NUMBERS AND FRACTIONS

Now let's look at numbers in which there is a whole-number part and a fractional part. For convenience, Table 1–4 (a repeat of Table 1–1) is shown below.

Consider the number 13.36. This would be read "thirteen *and* thirty-six hundredths." The word *and* is used in a number to define the position of the decimal point. We read the whole-number part, add the word *and,* and then read the

Table 1-4   Names for Places from Millions to Millionths.

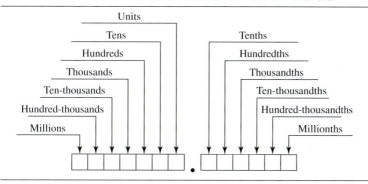

decimal-number part. We could also write the number as a *mixed* number: $13\frac{36}{100}$. Mixed numbers indicate the addition of two numbers: a whole number and a fraction. $13.36 = 13\frac{36}{100} = 13 + \frac{36}{100}$. Typically, though, we wouldn't use the + sign but would simply write the number as a decimal number (13.36) or as a mixed number ($13\frac{36}{100}$).

Consider the number 432.243. This number is read "four hundred thirty-two *and* two hundred forty-three thousandths." Remember, the word *and* is used to connect the whole-number and decimal-number parts. As a mixed number we would write

$$432\frac{243}{1000}$$

EXAMPLE 1-10

Convert the following decimal number to a mixed number.

$$57.045$$

SOLUTION   Change the fractional part of the number (.045) to a decimal fraction, as we did in Example 1–9 and 1–10. This yields $\frac{45}{1000}$. Then add the decimal fraction and the whole number to get the answer, which is $57\frac{45}{1000}$.

$$57.045 = 57\frac{45}{1000}$$

EXAMPLE 1-11

Convert the following mixed number to a decimal number.

$$27\frac{7}{100}$$

SOLUTION Change the decimal fraction to a decimal number as we did in Examples 1–7 and 1–8. This yields 0.07. Then add the whole-number part and the fractional part to get the answer, which is 27.07.

$$27\frac{7}{100} = 27.07$$

EXAMPLE 1–12

Express the following number as a mixed number and then as a decimal number.

Three hundred two and seventy-three thousandths

SOLUTION The whole-number part is 302. The fractional part expressed as a decimal number is 0.073 (see Example 1–5), or expressed as a fraction is $\frac{73}{1000}$ (see Example 1–4). Combining the whole number and fractional parts, we get $302\frac{73}{1000} = 302.073$.

☞ **KEY POINT** Mixed numbers are numbers that are made up of a whole-number part and a fraction. When reading the number, the word *and* is used to connect the whole number and decimal number parts.

PRACTICE PROBLEMS 1–3

1. Convert each of the following decimal numbers to mixed numbers:
   (a) 24.007   (b) 706.024
   (c) 4.00017   (d) 7.400   (e) 370.006
   (f) 47.113

2. Convert each of the following mixed numbers to decimal numbers:
   (a) $1076\frac{7}{100}$ (b) $8\frac{23}{10,000}$
   (c) $76\frac{14}{1000}$ (d) $35\frac{270}{100,000}$
   (e) $713\frac{7}{1000}$ (f) $8\frac{87}{10,000}$

3. Express the following numbers as decimal fractions and as mixed numbers:
   (a) Fifty-six and seventy-eight hundredths
   (b) Fifteen and thirty-five thousandths
   (c) One hundred six and eight ten-thousandths
   (d) Seven hundred four and two hundred fourteen hundred-thousandths
   (e) Four thousand seventy-five and fourteen ten-thousandths
   (f) Eighty and forty-three thousandths

1. (a) $24\dfrac{7}{1000}$ (b) $706\dfrac{24}{1000}$

    (c) $4\dfrac{17}{100,000}$ (d) $7\dfrac{400}{1000}$

    (e) $370\dfrac{6}{1000}$ (f) $47\dfrac{113}{1000}$

2. (a) 1076.07   (b) 8.0023   (c) 76.014
    (d) 35.00270   (e) 713.007
    (f) 8.0087

3. (a) $56.78 = 56\dfrac{78}{100}$

    (b) $15.035 = 15\dfrac{35}{1000}$

    (c) $106.0008 = 106\dfrac{8}{10,000}$

    (d) $704.00214 = 704\dfrac{214}{100,000}$

    (e) $4075.0014 = 4075\dfrac{14}{10,000}$

    (f) $80.043 = 80\dfrac{43}{1000}$

Additional practice problems are at the end of the chapter.

ROUNDING WHOLE NUMBERS   1-4

In problem solving it is common practice to simplify multidigit numbers. This simplification is done by replacing the rightmost nonzero digits with zeros. For example, the number 72,348 could be simplified to 72,350 to the nearest ten; 72,300 to the nearest hundred; 72,000 to the nearest thousand; or 70,000 to the nearest ten-thousand. This simplification is called *rounding*.

Rounding numbers is done to make problem solving easier. Some accuracy is lost as a result of the rounding. The more rounding we do, the less accurate the number. In electronics we work with components that typically have 5% or 10% tolerances. We normally make measurements with instruments whose tolerances are within the 5% or 10% range (or worse). For these reasons, great accuracies in problem solving are not necessary.

Let's look at the number 72,348 again. When we rounded to the nearest ten, we rounded to 72,350. We had the choice of rounding *down* to 72,340 or rounding *up* to 72,350. Since 48 is closer to 50 than to 40, we rounded *up* to 72,350. When we rounded to the nearest hundred, we rounded to 72,300. We had the choice of rounding *down* to 72,300 or rounding *up* to 72,400. Since 348 is closer to 300 than to 400, we rounded *down* to 72,300. The same logic was used in rounding to the nearest thousand. Since 2348 is closer to 2000 than to 3000, we rounded to 72,000. Finally, 72,000 was rounded to 70,000 because 72,000 is closer to 70,000 than to 80,000.

Although the digit 5 could indicate either up or down, it is typically rounded *up*.

EXAMPLE 1-13

Round the number 2735 to the nearest ten, hundred, and thousand.

SOLUTION   To round to the nearest ten, we examine the digit in the units position. Since the digit is 5, we round *up* to 2740 by replacing 5 with 0 and adding 1 to the next significant digit. To round to the nearest hundred, we note that the digit in the tens position is 3; so we round *down* to 2700 by replacing 3 with 0. (The digit in the units position is also changed to 0 since we are rounding to the nearest hundred.) Since 7 is the digit in the hundreds position, we round *up* to the nearest thousand and the answer is 3000.

☞ **RULE 1-3**   When rounding numbers, round *up* if the digit is 5, 6, 7, 8, or 9. To round up, the digit is replaced with 0, and 1 is added to the digit to the left. Round *down* if the digit is 1, 2, 3, or 4. To round down, the digit is replaced with zero.

PRACTICE PROBLEMS 1-4

Round the following numbers to the nearest ten, hundred, and thousand:

1. 2714
2. 6526
3. 43,258
4. 76,547
5. 82,803
6. 26,764
7. 78,226
8. 18,999
9. 4050
10. 243,671

SOLUTIONS

	TO NEAREST TENS	TO NEAREST HUNDREDS	TO NEAREST THOUSANDS
1.	2710	2700	3000
2.	6530	6500	7000
3.	43,260	43,300	43,000
4.	76,550	76,500	77,000
5.	82,800	82,800	83,000
6.	26,760	26,800	27,000
7.	78,230	78,200	78,000
8.	19,000	19,000	19,000
9.	4050	4100	4000
10.	243,670	243,700	244,000

In problem 5, since a 0 already exists in the tens position, rounding to the nearest hundred is not necessary. In problem 8, when we round to the nearest ten, we round up because the units digit is 9. Remember that we round up by adding 1 to the digit to the left (the tens position). Adding 1 to 9 in the tens position produces a 0 and a carry into the hundreds position. A carry into the hundreds position results in 0 and a carry into the thousands position, resulting in the answer 19,000.

In problem 9, since a 0 already exists in the units position, rounding to the nearest ten is not necessary. The 5 in the tens position causes a carry to the hundreds, resulting in the answer 4100 to the nearest hundred.

Additional practice problems are at the end of the chapter.

ROUNDING NONWHOLE NUMBERS   1-5

Nonwhole numbers are rounded by using the same rules as for whole numbers. As with whole numbers, we work from the rightmost digit to the left. Consider the number 12.736. We could round to the nearest hundredth, tenth, unit, or ten. To round to the nearest hundredth, we would note the digit in the thousandths position. Since the digit is 6, we change the digit to 0 and add 1 to the next digit, making the number 12.74. We don't write the number as 12.740. The zero is not accurate since we rounded to the nearest hundredth. Therefore, we just drop the zero. To round to the nearest tenth, we note the digit in the hundredths position. Since the number is 3, we simply drop it and the answer is 12.7. To round to the nearest unit, we note the digit in the tenths position. Because it is a 7, a number that is 5 or greater, we drop it and add 1 to the number in the units position. The answer is 13. To convert to the nearest ten, we change the 2 to a 0. The answer is 10. Remember that in rounding whole numbers the zero is a place holder and cannot be dropped. In a decimal fraction, if a zero exists in the rightmost position as a result of rounding, the zero is dropped because it is not accurate.

PRACTICE PROBLEMS 1–5

Round the following numbers to the nearest hundredth, tenth, unit, and ten:

1. 73.647
2. 26.401
3. 17.0419
4. 30.6908
5. 50.4736
6. 48.047
7. 33.781
8. 68.147
9. 78.6671
10. 15.5554

	HUNDREDTH	TENTH	UNIT	TEN
1.	73.65	73.6	74	70
2.	26.40	26.4	26	30
3.	17.04	17.0	17	20
4.	30.69	30.7	31	30
5.	50.47	50.5	50	50
6.	48.05	48.0	48	50
7.	33.78	33.8	34	30
8.	68.15	68.1	68	70
9.	78.67	78.7	79	80
10.	15.56	15.6	16	20

Additional practice problems are at the end of the chapter.

1-6 SIGNIFICANT DIGITS

The term "Significant Digits" is used in conjunction with the process of rounding numbers and with Scientific Notation which will be covered in the next chapter. When we say we are going to round a number to "two significant digits", we mean the same as rounding to two places.

> ☞ **KEY POINT** Significant Digits, also called Significant Figures, are the digits in a number (not counting leading zeros) that are known to be accurate.

Leading zeros are the zeros to the left of the leftmost non-zero digit. When writing whole numbers, leading zeros are meaningless and are usually not written. For example, we would not write the number 52 as 052 because the leading zero adds no meaning. When writing numbers between zero and one, the leading zeros are important as placeholders, but are not customarily counted as Significant Digits. For example: 0.04 has one significant digit, which is four. The zero to the left of the decimal point tells us that there is no whole number part. The second zero is important because it states that there are no tenths in the number. But neither of these two zeros are considered Significant Digits.

Trailing zeros (the rightmost zeros) are considered Significant Digits when to the right of the decimal point, because their presence indicates a known accuracy. For example: the number 52.00 tells us that there are zero tenths and zero hundredths associated with this number. Their presence indicates that no tenths or hundredths quantities were measured. The absence of any digit in the thousandths position indicates an uncertainty as to the quantity of thousandths in the measurement.

There is an ambiguous case; when a whole number has trailing zeros we can not be sure of their accuracy. For example: 5,000 has 4 significant digits according to the above definition but it may be a number that has been rounded to the nearest thousand. If it has been rounded then the trailing zeros are not accurate.

The ambiguity problem is eliminated by the use of Scientific Notation covered in the next chapter. For all problems in this text, we will assume that trailing zeros in whole numbers are accurate and therefore they should be considered Significant Digits.

EXAMPLE 1–14

How many significant digits are in the following numbers?

1. 593   2. 8000   3. 27,061   4. 0905

SOLUTIONS

1. 3   2. 4   3. 5   4. 3

Significant digits are independent of the decimal point. The following numbers each have four significant digits:

5432   543.2   54.32   5.432   0.5432   0.05432   0.005432

Let's examine the number 687. The digit 7 is in the units position so its value is 7. The eight is in the tens position so its value is 80 (10 × 8). The six is in the hundreds position so its value is 600 (100 × 6). In this number the 7 has the least value because it is in the units position. A digit in the units position, the rightmost position, is referred to as the *least significant digit* (LSD).

The digit with greatest value will always be the leftmost nonzero number. In this number the digit 6 has the greatest value or weight. A nonzero digit in the leftmost position is called the *most significant digit* (MSD).

EXAMPLE 1–15

In the following numbers, (a) which digit is the most significant digit? (b) Which digit is the least significant digit? (c) List each digit and its value.

1. 79,630   2. 485,607

SOLUTION

1. (a) 7 is the most significant digit (MSD). (b) 0 is the least significant digit (LSD). (c) 7 occupies the ten-thousands position so it has a value of 70,000 (7 × 10,000). 9 occupies the thousands position so it has a value of 9000 (9 × 1000). 6 is in the hundreds position, making its value 600 (6 × 100). 3 is in the tens position so its value is 30 (3 × 10), and 0 is in the units position and its value is 0 (0 × 1).

2. (a) 4 is the most significant digit (MSD). (b) 7 is the least significant digit (LSD). (c) 4 occupies the hundred-thousands position, making its value 400,000 (4 × 100,000). 8 occupies the ten-thousands position so its value is 80,000 (8 × 10,000). 5 is in the thousands position, giving it a value of 5000 (5 × 1000). 6 is in the hundreds position so its value is 600 (6 × 100). 0 is in the tens position so its value is 0 (0 × 10), and 7 is in the units position, making its value 7 (7 × 1). See Table 1–1.

☞ **KEY POINT** The most significant digit (MSD) is the leftmost nonzero digit. The least significant digit (LSD) is the rightmost digit.

PRACTICE PROBLEMS 1–6

1. How many significant digits are in the following number?
   (a) 489.0   (b) 0.489   (c) 40,089
   (d) 4809.0   (e) 0.00489

2. In the number 4762:
   (a) Which digit is the most significant digit (MSD)?
   (b) Which digit is the least significant digit (LSD)?
   (c) Which digit occupies the hundreds position?

SOLUTIONS

1. (a) 4   (b) 3   (c) 5   (d) 5   (e) 3

2. (a) 4   (b) 2   (c) 7

SELF-TEST 1–1

1. In the number 20,378:
   (a) Which digit is the MSD?
   (b) Which digit is the LSD?
   (c) Which digit occupies the thousands position?
   (d) Which digit occupies the tens position
   (e) What weight does the digit 3 have?
   (f) How many units are there?
3. Write each of the following first as a decimal fraction and then as a decimal number:
   (a) Three-tenths
   (b) Eighty-five hundredths
   (c) Eighteen ten-thousandths

2. Convert the following fractions to decimal numbers:
   (a) $\dfrac{41}{100}$ (b) $\dfrac{9}{1000}$ (c) $\dfrac{1783}{100,000}$

4. Write the names of the following numbers:
   (a) 0.33   (b) 0.004

5. Convert each of the following decimal numbers to mixed numbers:
   (a) 7.46 (b) 18.006

6. Convert each of the following mixed numbers to decimal numbers:
   (a) $76\frac{14}{100}$ (b) $6\frac{23}{1000}$

7. Express the following numbers as decimal fractions and as mixed numbers:
   (a) Three and seven-hundredths
   (b) Twenty-eight and sixty-three thousandths

8. Round to the nearest ten, hundred, and thousand:
   (a) 4765 (b) 9705 (c) 15,789
   (d) 37,046

9. Round to the nearest tenth, hundredth, and thousandth:
   (a) 0.0746 (b) 0.4605 (c) 0.4056
   (d) 0.3748

10. Round to the nearest ten, unit, tenth, and hundredth:
    (a) 17.486 (b) 23.462 (c) 36.547
    (d) 20.706

Answers to Self-test 1–1 are found at the end of the chapter.

ADDITION AND SUBTRACTION OF SIGNED NUMBERS   1-7

1-7-1 Signed Numbers Defined

In mathematics, numbers may be either positive ($+$) or negative ($-$). The plus sign is not usually written. If we saw the numbers 6, -4, 15, and -9 written, we would recognize that the numbers 6 and 15 are positive and the numbers 4 and 9 are negative.

In this section we introduce the symbols $<$ and $>$. The symbol $<$ means *less than* and the symbol $>$ means *greater than*.

In Figure 1–1 we have drawn a straight line that starts at zero and extends both right and left. Notice that numbers starting at zero and extending to the right are positive. Numbers starting at zero and extending to the left are negative. In comparing signed numbers, we see that the rightmost number is the greater and the leftmost number is the lesser. For example, compare 4 and -3. We could say that 4 is greater than -3 (expressed mathematically $4 > -3$), or we could say that -3 is less than 4 (expressed mathematically $-3 < 4$).

Consider the numbers -7, -2, and 6. We could say that $-7 < -2 < 6$. Or we could say that $6 > -2 > -7$. Notice that in comparing -2 and -7 we said

FIGURE 1–1
Scale showing positive and negative numbers.

that $-2 > -7$. Remember, on the number scale -2 is to the right of -7; therefore, it is greater. -2 is more positive than -7. -7 is more negative than -2. If we compare 3 and 11, we would say that $3 < 11$. We could also say that 3 is more negative than 11.

The *absolute* value of a number is that number without regard to sign. The absolute value of $+5$ and -5 is 5 and is symbolized as $|5|$. This tells us that no matter whether 5 is $-$ or $+$, the distance from zero is 5.

1-7-2 Addition and Subtraction

In addition to using the symbols $+$ and $-$ to denote the sign of a number, we also use these symbols to indicate the operations of addition and subtraction.

Anyone who has trouble visualizing adding signed numbers should refer to Figure 1-1. Adding two or more positive numbers does not present any problem with signs. $2 + 3 + 5 = 10$. Let's add two negative numbers: $(-2) + (-3)$. -2 and -3 are enclosed in parentheses to indicate that the $-$ is the sign of the number and not a sign of subtraction.

$$(-2) + (-3) = -5$$

$$(-6) + 7 + (-9) = -8$$

We could simplify the problem before finding the sum by adding the negative numbers together first:

$$-6 + 7 + (-9) = -6 - 9 + 7 = -15 + 7 = -8$$

In each case we move left on the number scale if the number is $-$ and we move right if the number is $+$.

> ☞ **RULE 1-4** To add numbers with like signs, add the numbers and affix the sign to the answer.
> To add numbers with unlike signs, subtract the smaller from the larger and affix the sign of the larger to the answer.

In subtraction we call the first number the *minuend*. The second number is the *subtrahend*. The answer is the *difference*.

$$
\begin{array}{rl}
6 & \text{minuend} \\
-3 & \text{subtrahend} \\
\hline
3 & \text{difference}
\end{array}
$$

> ☞ **RULE 1-5** To subtract, change the sign of the subtrahend and then add, following the rule for addition.

$$6 - (+3) = 6 + (-3) = 3$$

$$
\begin{array}{rcr}
6 & & 6 \\
-3 & = & + -3 \\
\hline
3 & & 3
\end{array}
$$

When more than two numbers are subtracted, the same rule applies. Whenever subtracting is indicated, we change the sign of the following numbers and add. $17 - (+6) - (+3) = 17 + (-6) + (-3) = 17 + (-9) = 8$. When addition and subtraction are both indicated, we follow the rules for both.

$$25 - (+6) + 4 = 25 + (-6) + 4 = 23$$

$$-7 + 16 - (-4) = -7 + 16 + 4 = 13$$

PRACTICE PROBLEMS 1–7

Perform the indicated operations:

1. $6 + (-4)$
2. $-14 + (-6)$
3. $16 - (+4)$
4. $9 - (-3)$
5. $-10 - (-6)$
6. $-15 - (+4)$
7. $25 + (-3) - (+4)$
8. $4 - (-6) + (-15)$
9. $-6 - (-3) - (-10)$
10. $40 + (-30) - (-6)$

SOLUTIONS

1. 2
2. -20
3. 12
4. 12
5. -4
6. -19
7. 18
8. -5
9. 7
10. 16

Additional practice problems are at the end of the chapter.

MULTIPLICATION AND DIVISION OF SIGNED NUMBERS   1–8

Both $\times$ and $\cdot$ are symbols used to indicate multiplication. The symbols $\div$ and / are used to indicate division. Each part of a multiplication or division has a name as shown in Figure 1–2 on the following page.

When the multiplicand and multiplier are whole numbers then they are also called the factors of the product.

Signed numbers are multiplied and divided by using the same methods learned in arithmetic. The rules are:

> ☞ **RULE 1–6** When multiplying or dividing signed numbers, if all numbers are positive, the answer is positive. If there is an even number of negative signs, the answer is positive. If there is an odd number of negative signs, the answer is negative.

FIGURE 1–2
Parts of
multiplication
and division
identified.

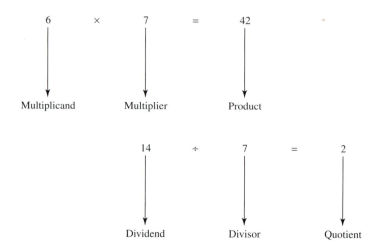

The general equations are:

$$(+) \cdot (+) = +$$
$$(+) \cdot (-) = -$$
$$(-) \cdot (-) = +$$

Examples of Rule 1–6

$$9 \times 8 = 72 \qquad (+) \cdot (+) = +$$
$$7 \times (-6) = -42 \qquad (+) \cdot (-) = -$$
$$-8 \times (-6) = 48 \qquad (-) \cdot (-) = +$$
$$16 \div 4 = 4 \qquad (+) \div (+) = +$$
$$12 \div (-4) = -3 \qquad (+) \div (-) = -$$
$$-20 \div 5 = -4 \qquad (-) \div (+) = -$$
$$-35 \div -7 = 5 \qquad (-) \div (-) = +$$

PRACTICE PROBLEMS 1–8

Perform the indicated operations:

1. $17 \times (-5)$
2. $-14 \times (-16)$
3. -9×12
4. -12×14
5. $-14 \div 2$
6. $-42 \div 6$
7. $172 \div (-4)$
8. $98 \div (-14)$
9. $-121 \div (-11)$
10. $-144 \div (-12)$

1. -85
3. -108
5. -7
7. -43
9. 11

2. 224
4. -168
6. -7
8. -7
10. 12

Additional practice problems are at the end of the chapter.

MATHEMATICAL EXPRESSIONS AND TERMS   1–9

Many problem-solving situations in electronics involve all four operations: addition, subtraction, multiplication, and division. Therefore, it is important that we understand the order in which these operations are performed.

Before we talk about these problem-solving situations, we must define some new words. Here and in future chapters we are going to be talking about mathematical *terms* and *expressions*.

☞ KEY POINTS

· A mathematical *term* is a number preceded by a $+$ or $-$ sign.
· A mathematical *expression* is made up of one or more terms.

Thus, $4 + 7$ is an expression containing two terms. This expression can be reduced to *one* term (11).

$4 + (7 \times 3)$ is a two-term expression. 4 is one term and 7×3 is the other. We have used parentheses to help you identify the term. Remember, *terms* are separated by $+$ and $-$ signs. Signs of multiplication and division are signs within a term. In problem solving, all operations within a term are performed first and then the terms are combined. The rule is:

> ☞ **RULE 1–7** Unless directed by other mathematical operators, such as parentheses, a mathematical expression can be simplified by performing the multiplications and divisions first and then the additions and subtractions.

Here are some examples:

$$(7 \times 6) + (4 \times 2) = 42 + 8 = 50$$

$$(-14 \div 7) - (-6 \times 3) = -2 - (-18) = -2 + 18 = 16$$

$$4 + (-3) - (-4 \times 6) = 4 - 3 - (-24) = 4 - 3 + 24 = 25$$

Notice in each example that the operations within the terms were performed *first,* and then the terms were combined.

My colleague, Barbara Snyder, uses the phrase "**My D**ear **A**unt **S**ally" to help her students remember which function is performed first—**M**ultiply, **D**ivide, **A**dd, and **S**ubtract.

PRACTICE PROBLEMS 1–9

Simplify the following expressions to one term:

1. $8 + 6 \times 3$
2. $-3 + 12 \div (-2)$
3. $6 \times 3 - 2 \times 4$
4. $16 \div (-2) - (-4) + 6 \times (-6)$
5. $(-6) \times (-4) + 7 \times (-3)$
6. $6 \times (-4) \div 2 - 4 + (-6)$
7. $15 \times 3 \div 3 + 4$
8. $8 \times 4 - (-6) \times (-4)$

SOLUTIONS

1. 26
2. -9
3. 10
4. -40
5. 3
6. -22
7. 19
8. 8

Additional practice problems are at the end of the chapter.

1-10   PARENTHESES ( ) AND OTHER SIGNS OF GROUPING

Parentheses are mathematical symbols that can be used in a mathematical expression to direct the sequence of performing calculations. For example when an expression is written as $3 \times 4 - 5$ we should first multiply 3 times 4, which is 12, and then subtract 5 to get 7. $3 \times 4 - 5 = 12 - 5 = 7$

If the expression was written as $3 \times (4 - 5)$ we should first perform the mathematical operation inside the parentheses. The sequence of operations, as directed by the parentheses, is:

$$3 \times (4 - 5)$$

$$3 \times (-1)$$

$$-3$$

In complex technical problems, we often have terms within terms. For example in the expression $3 + 9(5 + 2)$, the first term is the 3. The second term is $9(5 + 2)$ but it contains another term, $5 + 2$, which must be evaluated before being multiplied by 9. This example can be further complicated by dividing it by 5, which we could write as $(3 + 9(5 + 2))/5$. The order of operation, as directed by the parentheses, is:

(3 + 9(5 + 2))/5

(3 + 9(7))	/5	Perform the operation within the innermost parentheses (5 + 2).
(3 + 63)	/5	Remove the innermost parentheses by performing the indicated multiplication.
(66)	/5	Perform the operation within the parentheses (3 + 63).
13.2		Divide 66 by 5.

When we have terms within terms, we refer to them as *nested* terms. The order of operation is to remove the innermost parentheses by performing its operation first, like the (5 + 2) above. The next operation is to remove the innermost parentheses which enclose the seven. We would continue to remove parentheses until the outermost parentheses are gone. Then we would use "My Dear Aunt Sally" to remove the remaining operators.

When we have nested terms, we can use brackets [ ] and braces { } along with parentheses ( ) to identify the beginning and end of terms. If a division is involved, like the example above, then the *vinculum* (the horizontal bar that separates the numerator from the denominator) can be used to dictate the order of operations. For example, if the above problem is written as:

$$\frac{3 + 9(5 + 2)}{5}$$

then the order of operations is the same as before. The vinculum tells us that the numerator must be reduced to a single term before the division can be performed. The vinculum is doing the same job as the outermost parentheses in (3 + 9(5 + 2))

Parentheses ( ), brackets [ ], braces { }, and the vinculum are often called *signs of grouping* because they are used to group terms together to show the order in which mathematical operations should be performed. For example in the expression (2 + 4 + 6 + 8) × (3 + 5 + 7 + 9) the two groups of terms in the parentheses must be evaluated before the multiplication can be performed.

If all four signs of grouping are used, for example in the problem {(3 + 9 [5 + 2])/5} + 9, we still work from the innermost group toward the outermost group. Using different signs of grouping often make it easier to see the order of operations.

> ☞ **RULE 1-8** When there are nested terms in a mathematical expression, simplify (evaluate) the innermost term first and then work toward the outermost term.

EXAMPLE 1–16

Perform the indicated operations:

(a) 7(6 + 3)   (b) 4 − (−6 + 3) × 2   (c) 3[4(3 + 2) − 4]
(d) −3 − [2 + 4 × (−6)]

SOLUTIONS

(a) $7(6 + 3) = 7 \times 9 = 63$
(b) $4 - (-6 + 3) \times 2 = 4 - (-3) \times 2 = 4 - (-6) = 4 + 6 = 10$
(c) $3[4(3 + 2) - 4] = 3(4 \times 5 - 4) = 3(20 - 4) = 3(16) = 48$
(d) $-3 - [2 + 4 \times (-6)] = -3 - [2 + (-24)] = -3 - (-22) = -3 + 22 = 19$

Notice that when we simplified (c) to $3(4 \times 5 - 4)$, we then applied Rule 1–7 and performed the multiplication, $(4 \times 5 = 20)$, first. The problem then simplified to $3(20 - 4)$. Then $3(20 - 4) = 3(16) = 48$. Had we *subtracted* first $(5 - 4 = 1)$, we would have had $3(4 \times 1) = 3(4) = 12$—which, of course, is a wrong answer.

PRACTICE PROBLEMS 1–10

Perform the indicated operations:

1. $-6 - (-3) + 3(3 - 4)$
2. $6(-3) - [6(-3)]$
3. $14 - (-4 + 18 \div 3)$
4. $(4 + 3)(6 - 3)$
5. $4 - [6(7 - 3)]$
6. $6[18 \div (-6 + 3)] + 7 \times 4$
7. $10 - (6 + 3) \times 4$
8. $18 - 3[(3 + 4)(8 - 12)]$
9. $4[7 - (-3) + (2 - 6)(3 - 7)]$
10. $4 - 3\{-2[6(2 + 3)] - 6\}$

SOLUTIONS

1. -6
2. 0
3. 12
4. 21
5. -20
6. -8
7. -26
8. 102
9. 104
10. 202

SELF-TEST 1–2

Perform the indicated operations:

1. $14 - (+4) + (-5)$
2. $-16 - (-4) - (+6)$
3. -8×6
4. $-9 \times (-4)$
5. $4 \times (-3)$
6. $51 \div (-3)$
7. $-64 \div (-4)$
8. $-36 \div 4$
9. $5 \times (-3) + (-6) \times 7$
10. $-9 \div 3 + 4 \times (-6)$
11. $16 - (-6) + 27 \div (-3)$
12. $4 - (-6) + 4(5 - 2)$
13. $7 \times (-2) - [4(-6)]$
14. $7[15 \div (-3 + 6)] + 8 \div 4$
15. $6 - 5\{-4[8(4 + 2)] - 4\}$

Answers to Self-test 1–2 are at the end of the chapter.

CHAPTER I AT A GLANCE

PAGE	KEY POINT OR RULE	EXAMPLE
9	*Rule 1–1:* To convert decimal fractions to decimal numbers, determine the value of the denominator. Place the rightmost digit of the numerator in this position.	$\frac{37}{1000} = 0.037$ The digit 7 goes in thousandths position, which puts the 3 in the hundredths position. Zero is a place holder and goes in the tenths position.
9	*Rule 1–2:* To convert decimal numbers to decimal fractions, determine the position of the rightmost digit. This tells us the value of the denominator. The numerator is the decimal number with the decimal point removed.	$0.0019 = \frac{19}{10,000}$ The 9 is in the ten-thousandths position, so the denominator is 10,000 and the numerator is 19.
12	*Key point:* Mixed numbers are numbers that are made up of a whole-number part and a fraction. When reading the number, the word *and* is used to connect the whole number and decimal number parts.	$6\frac{3}{10}$ 6 and 3 tenths $40\frac{11}{100}$ 40 and 11 hundredths
14	*Rule 1–3:* When rounding numbers, round *up* if the digit is 5, 6, 7, 8, 9. To round up, the digit is replaced with 0, and 1 is added to the digit to the left. Round *down* if the digit is 1, 2, 3, or 4. To round down, the digit is replaced with zero.	4747 rounded to the nearest ten is 4750. To the nearest hundred, it is 4700, and to the nearest 1000, it is 5000.
16	Apply Rule 1–3 to nonwhole numbers.	4.074 rounded to the nearest hundredth is 4.07. To the nearest tenth, it is 4.1, and to the nearest unit, it is 4.
16	*Key point:* Significant digits are the digits in a number (not counting leading zeroes) that are known to be accurate.	0.04
18	*Key point:* The most significant digit (MSD) is the leftmost nonzero digit. The least significant digit (LSD) is the rightmost digit.	In the number 2740, the MSD is 2 and the LSD is 0.

20	*Rule 1–4:* To add numbers with like signs, add the numbers and affix the sign to the answer. To add numbers with unlike signs, subtract the smaller from the larger and affix the sign of the larger to the answer.	$-4 + (-3) = -7$ $-4 + 8 = 4$
20	*Rule 1–5:* To subtract, change the sign of the subtrahend and then add, following the rule for addition.	$-4 - (-3) = -4 + 3 = -1$
21	*Rule 1–6:* When multiplying or dividing signed numbers, if all numbers are positive, the answer is positive. If there is an even number of negative signs, the answer is positive. If there is an odd number of negative signs, the answer is negative.	$(+) \cdot (+) = +$ $(-) \cdot (-) = +$ $(+) \cdot (-) = -$
23	*Key points:* A mathematical *term* is a number preceded by a $+$ or $-$ sign. A mathematical *expression* is made up of one or more terms.	-53 $4 + 7$
23	*Rule 1–7:* Unless directed by other mathematical operators, such as parentheses, a mathematical expression can be simplified by performing the multiplications and divisions first and then the additions and subtractions.	$(4 \times 3) - (9 \div 3) = 12 - 3 = 9$

When working *End of Chapter Problems,* if you need to go back to the section where the material was presented, just look at the problem set number. For example, if you are working problems in *End of Chapter Problems 1–3* and you need help, go back to section 1–3 and review the material and examples presented there.

END OF CHAPTER PROBLEMS 1–1

1. In the number 8042:
   - (a) What digit occupies the hundreds position?
   - (b) What digit occupies the units position?
   - (c) What weight does the 8 have?
   - (d) How many tens are there?

2. In the number 2074:
   - (a) What digit occupies the hundreds position?
   - (b) What digit occupies the units position?
   - (c) What weight does the 4 have?
   - (d) How many tens are there?

3. In the number 67,403:
   (a) What digit occupies the thousands position?
   (b) What digit occupies the tens position?
   (c) What weight does the 7 have?
   (d) How many hundreds are there?
5. In the number 894,307:
   (a) What digit occupies the hundreds position?
   (b) What digit occupies the ten-thousands position?
   (c) What weight does the 8 have?
   (d) How many thousands are there?

4. In the number 75,403:
   (a) What digit occupies the thousands position?
   (b) What digit occupies the tens position?
   (c) What weight does the 5 have?
   (d) How many hundreds are there?
6. In the number 190,348
   (a) What digit occupies the hundreds position?
   (b) What digit occupies the ten-thousands position?
   (c) What weight does the 4 have?
   (d) How many thousands are there?

END OF CHAPTER PROBLEMS 1–2

1. Convert the following fractions to decimal numbers:
   (a) $\dfrac{3}{10}$ (b) $\dfrac{16}{1000}$ (c) $\dfrac{278}{100,000}$
   (d) $\dfrac{1763}{10,000}$ (e) $\dfrac{435}{1000}$ (f) $\dfrac{2060}{10,000}$
3. Write the following numbers as decimal fractions:
   (a) 0.007   (b) 0.0432   (c) 0.174
   (d) 0.000065   (e) 0.00016
   (f) 0.01234
5. Write each of the following first as a decimal fraction and then as a decimal number:
   (a) 17 thousandths
   (b) 4 hundredths
   (c) 460 ten-thousandths
   (d) 27 millionths
   (e) 1780 hundred-thousandths
   (f) 65 thousandths
7. In the number 0.001642, in which place does the
   (a) one appear?   (b) two appear?
   (c) six appear?
9. In the number 0.508743, in which place does the
   (a) seven appear?   (b) three appear?
   (c) zero appear?

2. Convert the following fractions to decimal numbers:
   (a) $\dfrac{36}{1000}$ (b) $\dfrac{9}{10,000}$ (c) $\dfrac{289}{1000}$
   (d) $\dfrac{83}{100,000}$ (e) $\dfrac{27}{100}$ (f) $\dfrac{980}{10,000}$
4. Write the following numbers as decimal fractions:
   (a) 0.00273   (b) 0.0906
   (c) 0.00005   (d) 0.067
   (e) 0.417   (f) 0.000813
6. Write each of the following first as a decimal fraction and then as a decimal number:
   (a) 6 tenths
   (b) 83 hundred-thousandths
   (c) 48 thousandths
   (d) 230 millionths
   (e) 73 hundredths
   (f) 895 ten-thousandths
8. In the number 0.000867, in which place does the
   (a) seven appear?   (b) six appear?
   (c) eight appear?
10. In the number 0.40257, in which place does the
    (a) four appear?   (b) zero appear?
    (c) five appear?

11. Write the names of the following:
    (a) 0.006   (b) 0.147   (c) 0.00092
    (d) 0.000007   (e) 0.0413
    (f) 0.0101

12. Write the names of the following:
    (a) 0.175   (b) 0.00065
    (c) 0.00463   (d) 0.00007
    (e) 0.076   (f) 0.000047

END OF CHAPTER PROBLEMS 1–3

1. Convert each of the following decimal numbers to mixed numbers:
   (a) 7.14   (b) 50.02   (c) 710.143
   (d) 9.099   (e) 73.653   (f) 207.7834
   (g) 28.00736   (h) 8.0706

2. Convert each of the following decimal numbers to mixed numbers:
   (a) 3.07   (b) 38.4   (c) 17.002
   (d) 170.506   (e) 48.6043
   (f) 20.003   (g) 706.2007
   (h) 7.40300

3. Convert each of the mixed numbers to decimal numbers:
   (a) $5\dfrac{68}{100}$ (b) $25\dfrac{7}{1000}$

   (c) $7\dfrac{165}{10,000}$ (d) $70\dfrac{4}{10}$

   (e) $473\dfrac{25}{1000}$ (f) $80\dfrac{743}{100,000}$

   (g) $2475\dfrac{35}{1,000,000}$ (h) $307\dfrac{8}{100,000}$

4. Convert each of the following mixed numbers to decimal numbers:
   (a) $4\dfrac{27}{1000}$ (b) $275\dfrac{7}{100}$

   (c) $73\dfrac{4}{10}$ (d) $780\dfrac{400}{1000}$

   (e) $7\dfrac{73}{100,000}$ (f) $107\dfrac{703}{1,000,000}$

   (g) $76\dfrac{14}{100}$ (h) $45\dfrac{406}{10,000}$

5. Express the following numbers as decimal numbers, and mixed numbers:
   (a) Ninety-three and seven-tenths
   (b) Thirty and four-hundredths
   (c) Eleven and one ten-thousandth
   (d) Nine hundred five and fifty-two thousandths
   (e) Seventy-eight and thirty-four thousandths

6. Express the following numbers as decimal numbers, and mixed numbers:
   (a) Fourteen and seventeen hundredths
   (b) Seventy and twenty-five thousandths
   (c) Six and seven tenths
   (d) Forty-six and four thousandths
   (e) Four hundred eight and three hundredths

7. Express the following numbers as decimal numbers, and mixed numbers:
   (a) Two hundred seventy-three and twenty-five hundred-thousandths
   (b) Seven hundred four and seven hundred four millionths
   (c) Two thousand forty-four and five hundred four ten-thousandths
   (d) Ten thousand one hundred one and eighty-nine hundred-thousandths
   (e) Ninety and four hundred sixty-six ten-thousandths
   (f) Two hundred seven and one hundred ten-millionths

8. Express the following numbers as decimal numbers, and mixed numbers:
   (a) Three hundred seven and eight hundred four hundred-thousandths
   (b) Four and thirty ten-thousandths
   (c) Four hundred fifty-two and seventy-two millionths
   (d) Ninety-four and seventy-three ten-thousandths
   (e) Eight and three thousand four hundred nine millionths
   (f) Seven hundred two and fourteen ten-thousandths

END OF CHAPTER PROBLEMS 1-4

Round the following numbers to the nearest ten:

1. 17	2. 28
3. 45	4. 12
5. 64	6. 86
7. 127	8. 465
9. 874	10. 713

Round the following numbers to the nearest ten and hundred:

11. 273	12. 904
13. 356	14. 277
15. 1377	16. 7046
17. 1407	18. 1406
19. 8706	20. 7855

Round the following numbers to the nearest ten, hundred, and thousand:

21. 4817	22. 9616
23. 85,468	24. 44,066
25. 78,673	26. 28,445
27. 27,847	28. 30,073
29. 35,486	30. 98,789
31. 68,448	32. 26,755
33. 73,654	34. 18,506
35. Round 465,432 to the nearest ten-thousand.	36. Round 498,475 to the nearest ten-thousand.

END OF CHAPTER PROBLEMS 1-5

Round the following numbers to the nearest hundredth, tenth, unit, and ten:

1. 163.782	2. 243.647	3. 9.464
4. 8.746	5. 88.888	6. 44.545
7. 749.493	8. 915.605	9. 39.278
10. 83.655	11. 63.7478	12. 12.4507
13. 478.6706	14. 307.3525	15. 47.474
16. 27.047	17. 16.545	18. 33.554
19. 76.746	20. 47.456	

END OF CHAPTER PROBLEMS 1-6

1. How many significant digits are in the following numbers?
   (a) 70,940   (b) 0.04320
   (c) 300.0   (d) 9007
   (e) 0.000275   (f) 0.05090

2. How many significant digits are in the following numbers?
   (a) 0.05043   (b) 670.0
   (c) 0.000740   (d) 50,304
   (e) 7040   (f) 0.007040

3. In the number 8042:
   (a) What is the MSD?
   (b) What is the LSD?
5. In the number 67,403:
   (a) What is the MSD?
   (b) What is the LSD?
7. In the number 894,307:
   (a) What is the MSD?
   (b) What is the LSD?

4. In the number 2074:
   (a) What is the MSD?
   (b) What is the LSD?
6. In the number 75,403:
   (a) What is the MSD?
   (b) What is the LSD?
8. In the number 190,348:
   (a) What is the MSD?
   (b) What is the LSD?

END OF CHAPTER PROBLEMS 1–7

Perform the indicated operations:

1. $12 - (+4) + (-3)$
3. $16 + 4 + (-6)$
5. $-20 - (-4) - (-10)$
7. $30 + (-3) + 7$
9. $-10 - (+4) + (-14)$
11. $7 - (-7) - (+7)$
13. $-5 - (-3) - (+12)$
15. $14 + (-3) - (-7)$
17. $9 - (-4) - (-8)$
19. $-20 - (+6) + (-7)$

2. $-3 - (+6) - (-2)$
4. $-12 - (-6) - (+8)$
6. $-20 - (-8) - (+12)$
8. $40 + (-17) + 13$
10. $-16 - (+9) + (-18)$
12. $14 + (-5) - (+7)$
14. $-8 - (-10) - (+16)$
16. $30 + (-11) - (-8)$
18. $7 - (-14) - (-12)$
20. $-25 - (+10) + (-8)$

END OF CHAPTER PROBLEMS 1–8

Perform the indicated operations:

1. $14 \times (-3)$
3. $-9 \times (-6)$
5. -12×7
7. $-45 \times (-5)$
9. $-310 \times (-3)$
11. $28 \div (-4)$
13. $-144 \div 6$
15. $-72 \div (-6)$
17. $512 \div (-16)$
19. $-700 \div (-25)$

2. -15×6
4. $25 \times (-4)$
6. $-12 \times (-12)$
8. $180 \times (-4)$
10. -19×40
12. $135 \div (-5)$
14. $-512 \div 64$
16. $-441 \div (-21)$
18. $-136 \div (-8)$
20. $-900 \div 30$

END OF CHAPTER PROBLEMS 1–9

Perform the indicated operations:

1. $(-7) \times 3 + 4$
3. $6 \times (-3) - 6 \times (-3)$

2. $-6 - (-4) + 3 \times 4 - 4$
4. $14 - (-9) + 15 \div 3$

5. $6 + 3 \times 5 - 4$
7. $6 \times 4 - 6 \div 2$
9. $(-7) - (-4) \times 6 + 1$
11. $-2 - (-2) \times 4 - 5 \times (-4)$
13. $16 \div (-4) - 7 \times 3 + 4$

15. $-4 \times (-5) - 3 + 15 \div (-3)$
17. $(-35) \div 5 + 4 + 3 \times 6$
19. $7 \times (-3) - 4 \times (-5) + 6 \times (-3)$

6. $21 + 6 \times 3 - 21 \div 7$
8. $-5 + 8 \times (-4) - 3$
10. $-18 \div 3 + 6 \times 6$
12. $-4 - (-8) \times 3 + 10 \div (-5)$
14. $-10 \times (-3) - 6 \times (-10) \div$
    $(-5) - 5$
16. $17 - 5 \times 4 - 24 \div 6$
18. $48 \div (-6) - 6 \times (-7) - 3$
20. $-3 \times (-9) - 4 + 27 \div (-3)$

END OF CHAPTER PROBLEMS 1–10

Perform the indicated operations:

1. $-4 - (-2) + 6(6 - 3)$
3. $7(-3) - [4(-7)]$
5. $(5 + 4)(-6 + 2)$
7. $5 - 2 - 3[4(3 + 4)] - 2$
9. $4[24 \div (-8 + 4)] + 7 \times 4$
11. $15 - 4[(-3 + 6)(9 - 12)]$
13. $5[6 - (-4) + (3 - 7)(2 - 6)]$
15. $4[-6 - (-3)(4)] + 7 \times 3$
17. $[9 + (3 \times -7)][18 - (2 \times 3)]$
19. $14 - [(-3) \times 4] + [3(4 - 5)(7 - 3)$
    $+ 6]$

2. $-9 - (-4) + 6(7 - 9)$
4. $5(-6) + [-3(-5)]$
6. $(-8 + 12)(7 - 3)$
8. $-6 - 4 - [8(-3 + 1)] - 7$
10. $8[33 \div (-9 + 6)] + 6 \times 3$
12. $12 - 7[(4 - 7)(-9 - 3)]$
14. $3[8 - (-8) + (2 - 7)(4 - 8)]$
16. $(4 - 6)(-3 \times 4) - [-9(2 - 4)]$
18. $[24 - (6 \times 5)][14 - (8 \times 4)]$
20. $[-8 - (-4 \times 7)][-15 - (14 \div -2)]$

ANSWERS TO SELF-TESTS

1. (a) 2 (b) 8 (c) 0 (d) 7
   (e) 300 (f) 8

3. (a) $\dfrac{3}{10} = 0.3$ (b) $\dfrac{85}{100} = 0.85$

   (c) $\dfrac{18}{10,000} = 0.0018$

5. (a) $7\dfrac{46}{100}$ (b) $18\dfrac{6}{1000}$

7. (a) $3.07 = 3\dfrac{7}{100}$

   (b) $28.063 = 28\dfrac{63}{1000}$

2. (a) 0.41 (b) 0.009 (c) 0.01783

4. (a) Thirty-three hundredths
   (b) Four-thousandths

6. (a) 76.14 (b) 6.023

8.

	Ten	Hundred	Thousand
(a)	4770	4800	5000
(b)	9710	9700	10,000
(c)	15,790	15,800	16,000
(d)	37,050	37,000	37,000

9.		Tenth	Hundredth	Thousandth
	(a)	0.1	0.07	0.075
	(b)	0.5	0.46	0.461
	(c)	0.4	0.41	0.406
	(d)	0.4	0.37	0.375

10.		Ten	Unit	Tenth	Hundredth
	(a)	20	17	17.5	17.49
	(b)	20	23	23.5	23.46
	(c)	40	37	36.5	36.55
	(d)	20	21	20.7	20.71

SELF-TEST 1–2

1. 5
2. −18
3. −48
4. 36
5. −12
6. −17
7. 16
8. −9
9. −57
10. −27
11. 13
12. 22
13. 10
14. 37
15. 986

Powers of Ten

2

Introduction

In our study of computers and electronics, we often work with very large numbers or very small numbers. For example, computer data bases can store over 1,000,000,000,000 bits of information. At the other end of the quantity spectrum, some of the components in a computer have leakage currents of less than 0.0000000001 amperes. Mathematical tools have been developed to make it easier to work with these numbers. These tools are a math shorthand called *powers of ten* or *scientific notation.* The use of powers of ten makes working with very large or very small numbers more manageable. In this chapter we will develop an understanding of how to use these tools to help in solving electronic problems.

The concepts of *rounding* covered in Chapter 1 will be used with scientific notation to show how to estimate answers to complex problems. Learning how to estimate answers is important to students because it gives them quick ballpark answers to problems, helps them learn how to follow complex procedures, and helps them to think logically.

The student will also learn how to use a scientific calculator in this chapter. It is important that students learn both how to estimate answers and how to find exact answers to complex problems. Algorithms will be used where appropriate in the chapter to assist the student in the solution of more complex problems. Technologists who can follow complex procedures, think logically, and make reasonable estimates are usually good problem solvers, are always in demand, and command large salaries.

Chapter Objectives

In this chapter you will learn how to:

1. Convert numbers to powers of ten form.
2. Express numbers in scientific notation rounded to some given number of significant digits.
3. Add, subtract, multiply, and divide numbers in powers of ten form using a calculator.
4. Solve problems with complex denominators using a calculator.
5. Find the reciprocal of numbers using a calculator.
6. Find the square and square root of numbers using a calculator.

CONVERTING NUMBERS TO POWERS OF TEN FORM

To simplify working with large numbers, we use a math shorthand called *powers of ten* or *scientific notation*. Let's see what these terms mean.

In Table 2–1 we have converted numbers that are 10 or multiples of 10 to powers of ten form. We could continue the table indefinitely, but this table should be long enough to understand the concept.

In Table 2–1, notice the symmetry between rows. The number in any row is ten times larger than the number below it. Conversely, every number is ten times smaller than the number above it. In the rightmost column, the exponents decrease by one as we go down the rows. Also notice the row where $1 = 10^0$. Any number to the zero power is equal to one. This table should help you visualize why this is true.

What about numbers that are not integer multiples of ten? Later, in the chapters on logarithms, we will show how all numbers can be represented as exponents of ten.

Let's look at $10 \times 10 \times 10 = 1000 = 10^3$. In the expression 10^3, the number 10 is the *base* and 3 is the *exponent*. When a multiple of 10 is expressed in this way, it is said

Table 2–1   Conversions from Numbers That Are Multiples of Ten to Powers of Ten Form.

DECIMAL NUMBER	NUMBER IN POWERS OF TEN FORM
$1{,}000{,}000 = 10 \times 10 \times 10 \times 10 \times 10 \times 10$	10^6
$100{,}000 = 10 \times 10 \times 10 \times 10 \times 10$	10^5
$10{,}000 = 10 \times 10 \times 10 \times 10$	10^4
$1000 = 10 \times 10 \times 10$	10^3
$100 = 10 \times 10$	10^2
$10 = 10$	10^1
$1 = \dfrac{10}{10}$	10^0
$0.1 = \dfrac{1}{10}$	$\dfrac{1}{10^1} = 10^{-1}$
$0.01 = \dfrac{1}{100} = \dfrac{1}{10 \times 10}$	$\dfrac{1}{10^2} = 10^{-2}$
$0.001 = \dfrac{1}{1000} = \dfrac{1}{10 \times 10 \times 10}$	$\dfrac{1}{10^3} = 10^{-3}$
$0.0001 = \dfrac{1}{10{,}000} = \dfrac{1}{10 \times 10 \times 10 \times 10}$	$\dfrac{1}{10^4} = 10^{-4}$
$0.00001 = \dfrac{1}{100{,}000} = \dfrac{1}{10 \times 10 \times 10 \times 10 \times 10}$	$\dfrac{1}{10^5} = 10^{-5}$
$0.000001 = \dfrac{1}{1{,}000{,}000} = \dfrac{1}{10 \times 10 \times 10 \times 10 \times 10 \times 10}$	$\dfrac{1}{10^6} = 10^{-6}$

to be in the powers of ten form. 10^3 is read "ten to the third power." 10^3 occurs often in science and engineering and therefore has another name: ten cubed.

The exponent tells us how many times 10 is used as a factor. We may also think of the exponent as telling us how many zeros are included to the right of the digit 1. Thus in the number 1000, 10 is used as a factor 3 times ($10 \times 10 \times 10$) so the exponent is 3. Counting the number of zeros to the right of the 1 in 1000 we again get 3, which also tells us the exponent is 3. $1000 = 10^3$.

In the number 100, ten is used as a factor 2 times ($10 \times 10 = 100$) so the exponent is 2. Or, in the number 100 there are 2 zeros following the 1, so the exponent is 2. $100 = 10^2$. 10^2 occurs often in science and engineering and therefore has another name: ten squared. The rule is:

> ☞ **RULE 2–1** In converting some multiple of 10 to powers of ten form, count the number of zeros. This number is the exponent. The answer is 10 raised to that power.

For example, 10,000 has four zeros after the one, therefore $10,000 = 10^4$.

PRACTICE PROBLEMS 2–1

Convert the following numbers to powers of ten form:

1. 100	2. 100,000	3. 10
4. 10,000,000	5. 1000	6. 10,000
7. 1,000,000	8. 100,000,000	

SOLUTIONS

1. 10^2	2. 10^5	3. 10^1
4. 10^7	5. 10^3	6. 10^4
7. 10^6	8. 10^8	

The answer to problem 3 could be written simply as 10 since $10^1 = 10$. In the future, when the number is raised to the first power, we will not write the exponent.

Additional practice problems are at the end of the chapter.[1]

---

1 At the end of each chapter there are additional problems labeled **End of Chapter Problems.** These problems are additional practice problems. For easy reference these *End of Chapter Problems* are numbered by section just like the Practice Problems. Thus, Practice Problems 2–1 and *End of Chapter Problems 2–1* relate to the material covered in section 2–1. Practice Problems 2–4 and *End of Chapter Problems 2–4* relate to the material covered in section 2–4, and so on.

CONVERTING FROM POWERS OF TEN FORM TO NUMBERS

Now let's consider a problem in which we wish to convert from powers of ten form to a number. Remember that the exponent tells us how many times 10 is used as a factor. Or, said another way, the exponent tells us how many zeros follow the 1. For example, let's change 10^4 to a number. 10^4 is read "ten to the fourth power." 4 is the exponent. Therefore, 10 is multiplied by itself four times: $10 \times 10 \times 10 \times 10 = 10,000$. Again, this is the same as saying that the answer is 1 followed by four zeros. $10^4 = 10,000$. For positive exponents the rule is:

> ☞ **RULE 2–2** In converting from powers of ten form to a number, write the digit 1 and follow with as many zeros as indicated by the exponent.

For example: with 10^3 the exponent is three so we place three zeros after the one ($10^3 = 1000$). When the exponent is zero we do not place any zeros after the one ($10^0 = 1$).

PRACTICE PROBLEMS 2–2

Convert the following numbers from powers of ten form to regular numbers.

1. 10^2	2. 10^1	3. 10^6
4. 10^7	5. 10^9	

SOLUTIONS

1. 100	2. 10	3. 1,000,000
4. 10,000,000	5. 1,000,000,000	

Additional practice problems are at the end of the chapter.

CONVERTING DECIMAL FRACTIONS TO POWERS OF TEN FORM

When we dealt with numbers that were multiples of 10, we learned that the exponents were 1 or greater. If the number is 1, the exponent is zero. $10^0 = 1$. There are no zeros following the number 1; therefore, in powers of ten form, the exponent is zero.

Numbers whose values are between zero and one have negative exponents. Table 2–2, which is a portion of Table 2–1, shows some of these numbers as decimal

Table 2–2 Conversions of Numbers That Are between Zero and One to Powers of Ten Form.

DECIMAL NUMBER	DECIMAL FRACTIONS	POWERS OF TEN FORM
0.1	$\dfrac{1}{10}$	$\dfrac{1}{10^1} = 10^{-1}$
0.01	$\dfrac{1}{100} = \dfrac{1}{10 \times 10}$	$\dfrac{1}{10^2} = 10^{-2}$
0.001	$\dfrac{1}{1000} = \dfrac{1}{10 \times 10 \times 10}$	$\dfrac{1}{10^3} = 10^{-3}$

numbers, decimal fractions, and expressed in powers of ten form. This table could be extended to infinitely small numbers; as the numbers get smaller, the exponent becomes a larger and larger negative number.

From these relationships, we can develop the following rules:

☞ **RULE 2–3** In converting from powers of ten to decimal fractions, the negative exponent tells how many zeros follow the 1 in the denominator.

$$10^{-3} = \frac{1}{1000}$$

$$10^{-4} = \frac{1}{10,000}$$

☞ **RULE 2–4** In converting from powers of ten to decimal numbers, the negative exponent tells how far the 1 is located to the right of the decimal point.

$$10^{-3} = 0.001$$

$$10^{-4} = 0.0001$$

PRACTICE PROBLEMS 2–3

Convert the following powers of ten to (a) decimal numbers and (b) decimal fractions:

1. 10^{-4} 2. 10^{-2}
3. 10^{-1} 4. 10^{-8}
5. 10^{-3} 6. 10^{-7}
7. 10^{-6} 8. 10^{-5}

1. (a) 0.0001 (b) $\dfrac{1}{10,000}$ 2. (a) 0.01 (b) $\dfrac{1}{100}$

3. (a) 0.1 (b) $\dfrac{1}{10}$ 4. (a) 0.00000001 (b) $\dfrac{1}{100,000,000}$

5. (a) 0.001 (b) $\dfrac{1}{1000}$ 6. (a) 0.0000001 (b) $\dfrac{1}{10,000,000}$

7. (a) 0.000001 (b) $\dfrac{1}{1,000,000}$ 8. (a) 0.00001 (b) $\dfrac{1}{100,000}$

Additional practice problems are at the end of the chapter.

2-4 CONVERTING POWERS OF TEN TO DECIMAL FRACTIONS

Let's turn it around and consider problems in which we wish to convert decimal numbers and decimal fractions to powers of ten form.

$$0.01 = \frac{1}{100} = \frac{1}{10^2} = 10^{-2}$$

$$0.001 = \frac{1}{1000} = \frac{1}{10^3} = 10^{-3}$$

$$0.0001 = \frac{1}{10,000} = \frac{1}{10^4} = 10^{-4}$$

☞ **RULE 2-5** In converting from a decimal fraction to powers of ten form, count the number of zeros in the denominator. This number is the negative power to which 10 is raised.

$$\frac{1}{10,000} = 10^{-4}$$

$$\frac{1}{100} = 10^{-2}$$

☞ **RULE 2-6** In converting from a decimal number to powers of ten form, count how far the digit 1 is located to the right of the decimal point. This is the value of the negative exponent.

$$0.0001 = 10^{-4}$$

$$0.01 = 10^{-2}$$

Convert the following decimal numbers to powers of ten form:

1. 0.001	2. 0.1	3. 0.000001
4. 0.01	5. 0.00001	6. 0.0000001
7. 0.0001	8. 0.00000001	

Convert the following decimal fractions to powers of ten form:

9. $\dfrac{1}{1000}$ 10. $\dfrac{1}{100,000}$ 11. $\dfrac{1}{10}$

12. $\dfrac{1}{1,000,000}$ 13. $\dfrac{1}{100}$ 14. $\dfrac{1}{10,000}$

SOLUTIONS

1. 10^{-3}	2. 10^{-1}	3. 10^{-6}
4. 10^{-2}	5. 10^{-5}	6. 10^{-7}
7. 10^{-4}	8. 10^{-8}	9. 10^{-3}
10. 10^{-5}	11. 10^{-1}	12. 10^{-6}
13. 10^{-2}	14. 10^{-4}	

MULTIPLICATION IN POWERS OF TEN FORM   2–5

☞ **RULE 2–7**   To multiply numbers in powers of ten form, *add the exponents.*

EXAMPLE 2–1

Find $10^6 \times 10^{-2}$.

SOLUTION

$$10^6 \times 10^{-2} = 10^{6+(-2)} = 10^4$$

or
$$1,000,000 \times 0.01 = 10^4$$

or
$$1,000,000 \times \frac{1}{100} = 10,000 = 10^4$$

EXAMPLE 2–2

Find $10^{-4} \times 10^3$.

$$10^{-4} \times 10^{3} = 10^{-4+3} = 10^{-1}$$

or
$$0.0001 \times 1000 = 0.1 = 10^{-1}$$

or
$$\frac{1}{10,000} \times 1000 = \frac{1}{10} = 0.1 = 10^{-1}$$

EXAMPLE 2–3

Find $10^{-5} \times 10^{-2}$.

SOLUTION

$$10^{-5} \times 10^{-2} = 10^{-5+(-2)} = 10^{-7}$$

or
$$\frac{1}{100,000} \times \frac{1}{100} = \frac{1}{10,000,000} = 10^{-7}$$

or
$$0.00001 \times 0.01 = 0.0000001 = 10^{-7}$$

PRACTICE PROBLEMS 2–5

Multiply the following numbers. Express the products in powers of ten form.

1. $10^{2} \times 10^{3}$ 2. $10^{4} \times 10^{6}$ 3. $10^{-1} \times 10^{-3}$
4. $10^{-3} \times 10^{-5}$ 5. $10^{5} \times 10^{-2}$ 6. $10^{6} \times 10^{-4}$
7. $10^{2} \times 10^{-3}$ 8. $10^{3} \times 10^{-3}$ 9. $10^{-4} \times 10^{7}$
10. $10^{-2} \times 10^{6}$

SOLUTIONS

1. 10^{5} 2. 10^{10} 3. 10^{-4}
4. 10^{-8} 5. 10^{3} 6. 10^{2}
7. 10^{-1} 8. 10^{0} 9. 10^{3}
10. 10^{4}

Additional practice problems are at the end of the chapter.

| 2–6 | **DIVISION IN POWERS OF TEN FORM** |

☞ **RULE 2–8** To divide numbers in powers of ten form, *subtract the exponents.*

EXAMPLE 2–4

Find $\dfrac{10^5}{10^2}$.

SOLUTION

$$\frac{10^5}{10^2} = 10^{5-2} = 10^3$$

or

$$\frac{100,000}{100} = 1000 = 10^3$$

EXAMPLE 2–5

Find $\dfrac{10^{-5}}{10^2}$.

SOLUTION

$$\frac{10^{-5}}{10^2} = 10^{-5-2} = 10^{-7}$$

or

$$\frac{0.00001}{100} = 0.0000001 = 10^{-7}$$

EXAMPLE 2–6

Find $\dfrac{10^5}{10^{-2}}$.

SOLUTION:

$$\frac{10^5}{10^{-2}} = 10^{5-(-2)} = 10^{5+2} = 10^7$$

or

$$\frac{100,000}{0.01} = 10,000,000 = 10^7$$

EXAMPLE 2–7

Find $\dfrac{10^{-5}}{10^{-2}}$.

SOLUTION

$$\frac{10^{-5}}{10^{-2}} = 10^{-5-(-2)} = 10^{-5+2} = 10^{-3}$$

or

$$\frac{0.00001}{0.01} = 0.001 = 10^{-3}$$

Perform the indicated operations. Express your answers in powers of ten form.

1. $\dfrac{10^4}{10^2}$ 2. $\dfrac{10^6}{10^3}$ 3. $\dfrac{10^2}{10^5}$

4. $\dfrac{10^1}{10^6}$ 5. $\dfrac{10^2}{10^{-2}}$ 6. $\dfrac{10^3}{10^{-5}}$

7. $\dfrac{10^{-2}}{10^{-4}}$ 8. $\dfrac{10^{-3}}{10^1}$ 9. $\dfrac{10^{-4}}{10^{-6}}$

10. $\dfrac{10^{-1}}{10^{-4}}$ 11. $\dfrac{10^{-5}}{10^{-3}}$ 12. $\dfrac{10^{-9}}{10^{-6}}$

13. $\dfrac{1}{10^2}$ 14. $\dfrac{1}{10^{-4}}$ 15. $\dfrac{1}{10^6}$

SOLUTIONS

1. 10^2 2. 10^3 3. 10^{-3}
4. 10^{-5} 5. 10^4 6. 10^8
7. 10^2 8. 10^{-4} 9. 10^2
10. 10^3 11. 10^{-2} 12. 10^{-3}
13. 10^{-2} 14. 10^4 15. 10^{-6}

Additional practice problems are at the end of the chapter.

2-7 COMBINED MULTIPLICATION AND DIVISION IN POWERS OF TEN FORM

Multiplication and division may be performed in the same problem. We must take care to observe the signs of the exponents as we add or subtract.

$$\frac{10^4 \times 10^2}{10^3} = \frac{10^{4+2}}{10^3} = \frac{10^6}{10^3} = 10^{6-3} = 10^3$$

Notice that we performed the indicated multiplication by adding the exponents. Next we performed the indicated division by subtracting the exponent in the denominator from the exponent in the numerator.

$$\frac{10^{-5} \times 10^3}{10^{-2}} = \frac{10^{-5+3}}{10^{-2}} = \frac{10^{-2}}{10^{-2}} = 10^{-2+2} = 10^0 = 1$$

Let's try two more.

$$\frac{10^3 \times 10^4}{10^{-2} \times 10^6} = \frac{10^{3+4}}{10^{-2+6}} = \frac{10^7}{10^4} = 10^{7-4} = 10^3$$

$$\frac{1}{10^2 \times 10^4} = \frac{1}{10^{2+4}} = \frac{1}{10^6} = 10^{-6}$$

We arrive at the answer 10^{-6} because the numerator (1) can be written in powers of ten form as 10^0. Then we have

$$\frac{10^0}{10^6} = 10^{0-6} = 10^{-6}$$

A very useful rule to remember is:

☞ **RULE 2-9** Whenever a power of ten is in the denominator, it can be moved to the numerator by changing the sign of the exponent.

Consider the example above in which we had

$$\frac{10^3 \times 10^4}{10^{-2} \times 10^6}$$

We could have solved by this method:

$$\frac{10^3 \times 10^4}{10^{-2} \times 10^6} = 10^3 \times 10^4 \times 10^2 \times 10^{-6} = 10^3$$

Notice that 10^{-2} in the denominator became 10^2 when it was moved to the numerator. 10^6 in the denominator became 10^{-6} in the numerator. When all powers of ten are in the numerator, we simply add the exponents.

$$10^{3+4+2-6} = 10^3$$

PRACTICE PROBLEMS 2–7

Perform the indicated operations. Express your answers in powers of ten form.

1. $\dfrac{10^2 \times 10^{-3}}{10^4}$

2. $\dfrac{10^{-1} \times 10^{-6}}{10^3}$

3. $\dfrac{10^1 \times 10^0 \times 10^6}{10^{-3}}$

4. $\dfrac{10^{-3} \times 10^4 \times 10^{-6}}{10^{-3}}$

5. $\dfrac{10^{-2} \times 10^4}{10^3 \times 10^2}$

6. $\dfrac{10^2 \times 10^6}{10^3 \times 10^{-1}}$

7. $\dfrac{1}{10^2 \times 10^4 \times 10^1}$

8. $\dfrac{1}{10^{-2} \times 10^6 \times 10^{-8}}$

9. $\dfrac{10^4 \times 10^{-1} \times 10^6}{10^2 \times 10^{-4} \times 10^5}$

10. $\dfrac{10^{-6} \times 10^{-3} \times 10^3}{10^{-9} \times 10^6 \times 10^{-3}}$

SOLUTIONS

1. 10^{-5} 2. 10^{-10}
3. 10^{10} 4. 10^{-2}
5. 10^{-3} 6. 10^6
7. 10^{-7} 8. 10^4
9. 10^6 10. $10^0 = 1$

Additional practice problems are at the end of the chapter.

SELF-TEST 2–1

Convert the following numbers to powers of ten form:

1. 1000   2. 1,000,000   3. 0.001
4. 0.0000001   5. 0.0001   6. 0.00001

Determine the numbers represented by the following powers of ten:

7. 10^3 8. 10^5 9. 10^2
10. 10^0 11. 10^{-4} 12. 10^{-1}
13. 10^{-5} 14. 10^{-8}

Convert the following powers of ten to decimal fractions and to decimal numbers:

15. 10^{-2} 16. 10^{-3} 17. 10^{-5}
18. 10^{-1}

Perform the indicated operations. Express your answers in powers of ten form.

19. $10^2 \times 10^5$ 20. $10^3 \times 10^{-4}$
21. $10^6 \times 10^{-2}$ 22. $10^{-3} \times 10^{-6}$
23. $\dfrac{10^2}{10^5}$ 24. $\dfrac{10^4}{10^{-6}}$
25. $\dfrac{10^{-3}}{10^6}$ 26. $\dfrac{10^{-2}}{10^{-5}}$
27. $\dfrac{10^{-2} \times 10^4}{10^{-3}}$ 28. $\dfrac{10^1 \times 10^{-5}}{10^4 \times 10^{-9}}$
29. $\dfrac{10^{-4} \times 10^{-1} \times 10^6}{10^6 \times 10^{-3} \times 10^1}$ 30. $\dfrac{10^{-2} \times 10^{-3} \times 10^9}{10^2 \times 10^3 \times 10^4}$

31. $\dfrac{10^4 \times 10^{-4} \times 10^6}{10^3 \times 10^{-3} \times 10^{-2}}$

32. $\dfrac{10^1 \times 10^{-7} \times 10^{-2}}{10^{-3} \times 10^{-4} \times 10^2}$

33. $\dfrac{1}{10^2 \times 10^{-3} \times 10^4}$

Answers to Self-test 2–1 are at the end of the chapter.

CONVERTING BETWEEN REGULAR NUMBERS AND POWERS OF TEN NOTATION $\quad$ 2-8

Consider the number 470. Let's convert the number to powers of ten notation with positive exponents.

$$470 = 47.0 \times 10^1 = 4.70 \times 10^2 = 0.470 \times 10^3 = 0.0470 \times 10^4$$

and so on. This is the same as writing

$$470 = 47.0 \times 10 = 4.70 \times 100 = 0.470 \times 1000 = 0.0470 \times 10{,}000$$

and so on. As we move the decimal point *left,* the exponent becomes more positive.

$$470 = 47.0 \times 10^1$$
$$47.0 \times 10^1 = 4.70 \times 10^{1+1} = 4.70 \times 10^2$$
$$4.70 \times 10^2 = 0.470 \times 10^{2+1} = 0.470 \times 10^3$$
$$0.470 \times 10^3 = 0.0470 \times 10^{3+1} = 0.0470 \times 10^4$$

☞ **RULE 2-10** For each place we move the decimal point left in a number, the exponent increases by 1.

Said another way, as we make the exponent more positive, the decimal point moves left.

☞ **RULE 2-11** For each digit we make the exponent more positive, the decimal point moves 1 place left.

EXAMPLE 2-8

Change 275 to powers of ten form with an exponent of (a) 2, (b) 4.

SOLUTION

(a) $275 = 2.75 \times 10^2$. The exponent changes by 2 in a positive direction, so the decimal point moves 2 places *left*.

(b) $275 = 0.0275 \times 10^4$. The exponent changes by 4 in a positive direction, so the decimal point moves 4 places *left*.

Or, in part (a) we changed the number so that the exponent is 2. If we start from there and change the exponent to 4, we just move the decimal point 2 places farther left.

$$2.75 \times 10^2 = 0.0275 \times 10^4.$$

Now let's convert the same number to powers of ten form with negative exponents.

$$470 = 4700 \times 10^{-1} = 47,000 \times 10^{-2} = 470,000 \times 10^{-3}$$

and so on. As we move the decimal point *right*, the exponent becomes more *negative*. For each place we move the point *right*, the exponent decreases by 1 (becomes more negative).

$$470 = 4700 \times 10^{-1}$$

$$4700 \times 10^{-1} = 47,000 \times 10^{-1+(-1)} = 47,000 \times 10^{-2}$$

$$47,000 \times 10^{-2} = 470,000 \times 10^{-2+(-1)} = 470,000 \times 10^{-3}$$

☞ **RULE 2-12** For each place we move the decimal point *right* in a number, the exponent decreases by 1.

Said another way, as we make the exponent more negative, the decimal point moves *right*.

☞ **RULE 2-13** For each digit we make the exponent more negative, the decimal point moves 1 place *right*.

EXAMPLE 2-9

Change 275 to powers of ten form with an exponent of (a) -2, (b) -4.

SOLUTION

(a) $275 = 27,500 \times 10^{-2}$. The exponent changes by 2 in a negative direction, so the decimal point moves 2 places *right*.

(b) $275 = 2,750,000 \times 10^{-4}$. The exponent changes by 4 in a negative direction, so the decimal point moves 4 places *right*.

Or, in part (a) we changed the number so that the exponent is -2. To make the exponent -4, we just move the decimal point 2 places farther right.

$$27{,}500 \times 10^{-2} = 2{,}750{,}000 \times 10^{-4}$$

PRACTICE PROBLEMS 2–8

1. Change 38.6 to powers of ten form with an exponent of (a) 2, (b) 3, and (c) 4.

2. Change 5600 to powers of ten form with an exponent of (a) 2, (b) 3, and (c) 4.

3. Change 0.0000905 to powers of ten form with an exponent of (a) -2, (b) -6, and (c) -8.

4. Change 0.000000277 to powers of ten form with an exponent of (a) -5, (b) -7, and (c) -9.

5. Change 1.80 to powers of ten form with an exponent of (a) 1, (b) 3, (c) -1, and (d) -3.

6. Change 0.833 to powers of ten form with an exponent of (a) 6, (b) 3, (c) -3, and (d) -6.

SOLUTIONS

1. (a) 0.386×10^2,   (b) 0.0386×10^3,   (c) 0.00386×10^4
2. (a) 56.0×10^2,   (b) 5.60×10^3,   (c) 0.560×10^4
3. (a) 0.00905×10^{-2},   (b) 90.5×10^{-6},   (c) 9050×10^{-8}
4. (a) 0.0277×10^{-5},   (b) 2.77×10^{-7},   (c) 277×10^{-9}
5. (a) 0.180×10,   (b) 0.00180×10^3,   (c) 18.0×10^{-1}, (d) 1800×10^{-3}
6. (a) 0.000000833×10^6, (b) 0.000833×10^3, (c) 833×10^{-3},   (d) $833{,}000 \times 10^{-6}$

Additional practice problems are at the end of the chapter.

EXPRESSING NUMBERS GREATER THAN ONE IN SCIENTIFIC NOTATION   2–9

In science and engineering it is common practice to round answers to two, three, or four significant figures and then to change the number to a number between 1 and 10 times some power of ten. When a number is written in this manner, it is said to be expressed in *scientific notation*. The following examples are numbers expressed in scientific notation:

$$1.76 \times 10^4 \qquad 4.067 \times 10^{-2} \qquad 9.2 \times 10^6$$

As we mentioned in Chapter 1, great accuracy in problem solving is not realistic in electronics because we usually work with components whose values are given to two significant figures (color bands on resistors, for example), and they may be as much as $\pm 10\%$ different than that value. In addition, measuring instruments such as voltmeters and ohmmeters, if of the analog type like a multimeter, do not provide accuracy beyond two figures. Even digital meters often provide just 1% accuracies.

Therefore, throughout the text, with few exceptions, we will round to three significant figures, even though your calculator can provide eight or more.

Consider the number 764,832. Let's express this number in scientific notation rounded to three significant digits or simply "three places." Let's first round to three places. The number becomes 765,000. We rounded up because the fourth digit was 8. Now let's change the number to a number between 1 and 10. The answer is 7.65. To get 7.65, we had to move the decimal point five places *to the left*. To have an answer that equals 765,000, we must multiply 7.65 by 100,000 or 10^5. $765,000 = 7.65 \times 10^5$. For each place we move the decimal point left, we must multiply by 10. Since we moved the decimal point five places to the left, we multiplied by 10^5.

EXAMPLE 2–10

Express the number 4703 in scientific notation rounded to three places.

SOLUTION   We must move the decimal point three places *left* in order to have a number between 1 and 10. That makes the exponent 3.

$$4703 = 4.703 \times 10^3$$

$$4.703 \times 10^3 = 4.70 \times 10^3, \quad \text{rounded to three places}$$

☞ **RULE 2–14** To express numbers greater than 1 in scientific notation, count the number of places the decimal point must move left. This number is the exponent and is positive.

PRACTICE PROBLEMS 2–9

Round the following numbers to three significant figures and express your answers in scientific notation:

1. 1730	2. 26,745	3. 1,654,736
4. 42.69	5. 173,426	6. 47,394
7. 79.063	8. 678,091	9. 189.83
10. 6746		

SOLUTIONS

1. 1.73×10^3	2. 2.67×10^4	3. 1.65×10^6
4. 4.27×10^1	5. 1.73×10^5	6. 4.74×10^4
7. 7.91×10^1	8. 6.78×10^5	9. 1.90×10^2
10. 6.75×10^3		

Additional practice problems are at the end of the chapter.

EXPRESSING NUMBERS LESS THAN ONE IN SCIENTIFIC NOTATION   2–10

Consider the number 0.00715. To change to scientific notation, we must move the decimal point three places to the *right* so that we have a number between 1 and 10. The number is now 7.15. We must multiply this new number by 10^{-3} so that the value of the number in scientific notation equals the value of the original number. For each place we move the decimal point *right,* we must multiply by 0.1 or 10^{-1}. We moved the point *right* three places, so the exponent is −3.

EXAMPLE 2–11

Express the number 0.00008706 in scientific notation rounded to three places.

SOLUTION   We must move the decimal point five places right to have a number between 1 and 10. That makes the exponent −5.

$$0.00008706 = 8.706 \times 10^{-5}$$

$$8.706 \times 10^{-5} = 8.71 \times 10^{-5}, \quad \text{rounded to three places}$$

☞   RULE 2–15   To express numbers that are less than 1 in scientific notation, count the number of places the decimal point must move *right*. This number is the exponent and is negative.

PRACTICE PROBLEMS 2–10

Round the following numbers to three significant figures and express your answers in scientific notation:

1. 0.0167	2. 0.640732	3. 0.0003988
4. 0.00607	5. 0.1086	6. 0.00000706
7. 0.00001763	8. 0.009096	9. 0.07646
10. 0.000567		

SOLUTIONS

1. 1.67×10^{-2}	2. 6.41×10^{-1}	3. 3.99×10^{-4}
4. 6.07×10^{-3}	5. 1.09×10^{-1}	6. 7.06×10^{-6}
7. 1.76×10^{-5}	8. 9.10×10^{-3}	9. 7.65×10^{-2}
10. 5.67×10^{-4}		

Additional practice problems are at the end of the chapter.

In the following section we introduce the calculator as a tool used in problem solving. The calculator keystrokes shown are usually the same for two of the most popular, low-cost scientific calculators: the TI-36X and the Casio fx-115W.

The biggest difference between the two is the display. The Casio has a 3-line display: the top line shows the keystrokes you have entered, the middle line shows the results, and the bottom line shows information about the calculator mode. The TI has a two-line display: the top line shows information about the calculator mode and the second line shows the numbers entered or the result of the calculation.

With the Casio, it is easy to display the last 79 keystrokes to find and correct data entry errors or to modify equations; the TI does not have this feature. The TI's 40 keys are a bit easier to learn than the Casio's 54 keys. The TI does metric conversions; this model of the Casio does not. See the Appendix for more information about these calculators.

We will show the TI keystrokes in the examples in the remainder of the text and point out Casio keystrokes when they differ.

Let's work the following problems together.

EXAMPLE 2-12

Find the product of 76.3 and 446 using the calculator. Express the answer first in floating-point notation and then in scientific notation rounded to three places.

SOLUTION   Your calculator should be capable of displaying answers in various notations. The TI default notation (the notation it is in when first turned on or when cleared with the $\boxed{\text{AC}}$ key) usually displays answers as ten-digit numbers. This notation is called *floating-decimal notation* or *floating-point* notation. Let's solve the problem using this notation.

$\boxed{\text{AC}}$	All Clear key clears everything and puts us in the floating-decimal notation.

(For Casio press $\boxed{\text{MODE}}$ $\boxed{\text{MODE}}$ $\boxed{\text{MODE}}$ $\boxed{\text{MODE}}$ $\boxed{1}$ $\boxed{9}$. The calculator will stay in this display mode, even when it is turned OFF and ON, until changed by the user.)

76.3	Key in the first number.
$\boxed{\times}$	The function is *multiply*, so we press the $\boxed{\times}$ key.
446	Enter the second number.
$\boxed{=}$	Press the $\boxed{=}$ key to display the answer. If you didn't make any mistakes, the number 34029.8 is displayed. Because the last four digits in this answer are zero, only six digits are displayed.

With the TI calculator we can display answers in Scientific Notation and we can choose how many digits we want to display to the right of the decimal point. To change the TI calculator display mode to scientific notation with 2 digits to the right of the decimal point:

1. Press the light blue key $\left(\text{which is labeled } \boxed{\text{3rd}}\right)$ in the upper left corner of the keyboard.

2. Press the key labeled $\overset{\text{SCI}}{\boxed{6}}$, which is near the lower right corner of the keyboard.

3. Press the yellow key $\left(\text{which is labeled } \boxed{\text{2nd}}\right)$ and then the $\boxed{2}$ key to display two digits to the right of the decimal point.

For the Casio the set-up is different:

1. Press $\boxed{\text{MODE}} \boxed{\text{MODE}} \boxed{\text{MODE}} \boxed{\text{MODE}} \boxed{2}$ to set to scientific notation.

2. Press $\boxed{3}$ to make it round calculations to three digits.

Now, let's work the problem again. This time the answer will be displayed in scientific notation rounded to three places.

76.3	Enter the first number.
$\boxed{\times}$	The function is *multiply*.
446	Enter the second number.
$\boxed{=}$	Display the answer, which is 3.40×10^4.

If you make a mistake along the way, do not clear with the *all clear* key $\left(\text{usually labeled } \boxed{\text{AC}}\right)$. Clear with the *clear entry* key labeled $\boxed{\text{CE}}$ on the TI and $\boxed{\text{DEL}}$ on the Casio. Remember, on the TI the $\boxed{\text{AC}}$ key clears everything and puts us in floating-decimal notation. The $\boxed{\text{CE}}$ key clears the entry *only*. Use $\boxed{\text{AC}}$ only when you want to clear everything.

EXAMPLE 2–13

Perform the indicated operation. Express your answer in scientific notation rounded to three places (see Example 2–12 to set the display mode).

$$\frac{57{,}950 \times 10^3}{0.267}$$

57950	Enter the first number.
$\boxed{\text{EE}}$ 3	Enter the exponent. $\left(\text{On the Casio use } \boxed{\text{EXP}}.\right)$
$\boxed{\div}$	The function is *divide*.
.267	Enter the number.
$\boxed{=}$	The answer is 2.17×10^8.

Most problems containing three or more operations can be solved many different ways using the calculator. We will solve Example 2-14 three ways. These three ways will show the student some additional features of the calculator.

EXAMPLE 2–14

Perform the indicated operation. Express your answer in scientific notation rounded to three places (see Example 2–12 to set the display mode).

$$\frac{0.00743 \times 10^2 \times 14}{5384 \times 0.04}$$

SOLUTION

.00743	Enter the first number.
$\boxed{\text{EE}}$ 2	Enter the exponent. $\left(\text{On the Casio use } \boxed{\text{EXP}}.\right)$
$\boxed{\times}$	The function is *multiply*.
14	Enter the next number.
$\boxed{\div}$	The function is *divide*.
5384	Enter the next number.
$\boxed{\div}$	The function is *divide*. The answer to
.04	this point is being *divided* by 0.04.
$\boxed{=}$	The answer is 4.83×10^{-2}.

Let's leave the calculator in this notation mode and solve the problem again.

.00743	Enter the first number.
$\boxed{\text{EE}}$ 2	Enter the exponent. $\left(\text{On the Casio use } \boxed{\text{EXP}}.\right)$
$\boxed{\times}$	The function is *multiply*.
14	Enter the second number.

$\div$	The function is *divide*.
(	Start a group.
5384	Enter one number of the denominator.
$\times$	*Multiply*
.04	by the other number in the denominator.
)	End the group.
=	Again the answer is 4.83×10^{-2}.

The parenthesis is a sign of grouping. The calculator will perform all arithmetic operations within the parentheses and then, when the $=$ key is pressed, division by this quantity is performed.

Let's try a third way. This time we will use the memory feature of the calculator. We will use $\boxed{STO}$ to indicate memory storage and $\boxed{RCL}$ to indicate memory recall.

5384	Enter one number of the denominator.
$\times$	The function is *multiply*.
.04	Enter the other number in the denominator.
=	The product is 2.15×10^2.
$\boxed{STO}$ $\boxed{1}$	Store this number $\left(\text{for Casio } \boxed{STO}\ \boxed{(-)}\right)$ (the value of the denominator).
.00743	Enter one number in the numerator.
$\boxed{EE}$ 2	Enter the exponent $\left(\text{for Casio } \boxed{EXP}\right)$.
$\times$	The function is *multiply*.
14	Enter the second number in the numerator.
=	The value of the numerator is 1.04×10^1.
$\div$	The function is *divide*.
$\boxed{RCL}$ $\boxed{1}$	Recall the value previously stored in memory (the value of the denominator) $\left(\text{for Casio } \boxed{ALPHA}\ \boxed{(-)}\right)$.
=	Again, the answer is 4.83×10^{-2}.

Any of these solutions is valid. It is left up to the student to determine which method he or she prefers. Usually, one way is quicker (requires fewer steps) than another. Students should also refer to their calculator user's manual for additional help when needed.

Work the following problems by using your calculator. Because all the operations are multiplications and divisions, there is no preferred choice of the order in which they are performed.

PRACTICE PROBLEMS 2–11

Perform the indicated operations. Express your answers in scientific notation rounded to three places.

1. 47×106
2. 789×206
3. 24.6×0.00173
4. $473 \times 10^{-3} \times 26$
5. 2700×0.743
6. $33 \times 10^3 \times 65 \times 10^{-4}$
7. $17 \times 43 \times 0.003$
8. $7800 \times 10^{-3} \times 27 \times 10^{-2}$
9. $680 \times 10^3 \times 473 \times 10^{-7}$
10. $1.8 \times 10^4 \times 2.3 \times 10^{-6}$
11. $9 \div 0.000742$
12. $25 \div 5600$
13. $15 \div 24 \times 10^{-3}$
14. $12 \div 680 \times 10^2$
15. $\dfrac{20 \times 22 \times 10^3}{720}$
16. $\dfrac{40 \times 213 \times 10^{-6}}{27 \times 10^{-4}}$
17. $\dfrac{18 \times 10^3 \times 27 \times 10^3}{45 \times 10^3}$
18. $\dfrac{18 \times 0.00473}{673 \times 10^{-6}}$
19. $\dfrac{6700 \times 876}{914 \times 732}$
20. $\dfrac{743 \times 10^2 \times 0.00365}{65 \times 10^{-1} \times 0.403}$
21. $\dfrac{0.000467 \times 10^4 \times 7680}{2.6 \times 10^{-3} \times 526 \times 10^2}$
22. $\dfrac{14 \times 5600}{61.6 \times 10^4}$
23. $\dfrac{479 \times 10^4 \times 5600 \times 10^{-2}}{48.7 \times 10^{-3} \times 9800 \times 10^4}$
24. $\dfrac{0.0406 \times 10^{-4} \times 5600 \times 10^{-2}}{7960 \times 10^2 \times 0.00315 \times 10^{-5}}$
25. $\dfrac{37 \times 10^3 \times 4.7 \times 10^4}{0.000183 \times 10^2 \times 836 \times 10^{-3}}$

SOLUTIONS

1. 4.98×10^3
2. 1.63×10^5
3. 4.26×10^{-2}
4. 1.23×10
5. 2.01×10^3
6. 2.15×10^2
7. $2.19 \times 10^0 = 2.19$
8. $2.11 \times 10^0 = 2.11$
9. 3.22×10
10. 4.14×10^{-2}
11. 1.21×10^4
12. 4.46×10^{-3}
13. 6.25×10^2
14. 1.76×10^{-4}

15. 6.11×10^2

17. 1.08×10^4

19. $8.77 \times 10^0 = 8.77$

21. 2.62×10^2

23. 5.62×10

25. 1.14×10^{11}

16. $3.16 \times 10^0 = 3.16$

18. 1.27×10^2

20. 1.04×10^2

22. 1.27×10^{-1}

24. 9.07×10^{-3}

Usually an answer like the answer to problem 7 (2.19×10^0) is simply written 2.19, since $10^0 = 1$ ($2.19 \times 10^0 = 2.19 \times 1 = 2.19$).

Additional practice problems are at the end of the chapter.

ADDITION AND SUBTRACTION   2–12

We can add or subtract numbers in powers of ten form or when using scientific notation *provided the exponents are the same*. For example, let's perform the following addition: $3 \times 10^2 + 4 \times 10^3$. The decimal numbers are 300 and 4000. If we add 300 and 4000, we get 4300, or 4.3×10^3. To perform the addition using powers of ten, we must first move the decimal point of one number (it doesn't matter which) so that the exponents are the same. Let's change 3×10^2 so that the exponent is 3. We are making the exponent one digit more positive, so the decimal point must move one place *left* (Rule 2–11: For each digit we make the exponent more positive, the decimal point moves 1 place left.)

$$3 \times 10^2 = 0.3 \times 10^3$$

Now we can perform the addition:

$$
\begin{array}{r}
0.3 \times 10^3 \\
+4 \times 10^3 \\
\hline
4.3 \times 10^3
\end{array}
$$

Addition or subtraction is performed by adding the numbers together and multiplying by the indicated power of ten. The *exponents* are not added. The exponents merely tell us that we are adding 0.3 thousands to 4 thousands to get 4.3 *thousands*. We could have changed 4×10^3 to 40×10^2 and added.

$$
\begin{array}{r}
3 \times 10^2 \\
+40 \times 10^2 \\
\hline
43 \times 10^2
\end{array}
$$

$$43 \times 10^2 = 4.3 \times 10 \times 10^2 = 4.3 \times 10^3$$

The result is the same but more steps are involved.

EXAMPLE 2–15

Let's subtract 6×10^{-3} from 10^{-2}.

$$1 \times 10^{-2}$$
$$(-)6 \times 10^{-3}$$

10^{-2} is written in scientific notation form ($10^{-2} = 1 \times 10^{-2}$) because addition or subtraction of the decimal numbers is to be performed. In this problem we must again change one of the numbers so that both exponents are the same. Let's change 6×10^{-3} so that the exponent is -2. We are making the exponent one digit more *positive*, so the decimal point must move one place *left* (Rule 2–11).

$$6 \times 10^{-3} = 0.6 \times 10^{-2}$$

Now we can perform the subtraction:

$$1 \quad \times 10^{-2}$$
$$(-)0.6 \times 10^{-2}$$
$$0.4 \times 10^{-2}$$

$$0.4 \times 10^{-2} = 4 \times 10^{-1} \times 10^{-2} = 4 \times 10^{-3}$$

We could have changed 1×10^{-2} to a number whose exponent is -3. $1 \times 10^{-2} = 0.01 = 10 \times 10^{-3}$. Because we wanted the exponent to equal -3, we had to move the decimal point one place to the right.

$$10 \times 10^{-3}$$
$$(-)6 \times 10^{-3}$$
$$4 \times 10^{-3}$$

We can check our work by performing the subtraction in decimal.

$$1 \times 10^{-2} = 0.01 \qquad 0.01$$
$$6 \times 10^{-3} = 0.006 \qquad (-)0.006$$
$$0.004$$

$$0.004 = 4 \times 10^{-3}$$

☞ **RULE 2–16** To add or subtract numbers in powers of ten form, the exponents must be the same.

PRACTICE PROBLEMS 2–12

Perform the indicated operations. Express your answers in scientific notation rounded to three places.

1. $10^2 + 10^3$
2. $10^1 + 10^{-1}$
3. $10^3 - 10^2$
4. $2 \times 10^2 + 3 \times 10^3$
5. $47 \times 10^{-2} + 560 \times 10^{-3}$
6. $40 \times 10^3 - 300 \times 10^2$
7. $3 \times 10^{-2} - 45 \times 10^{-4}$
8. $270 \times 10^{-1} + 46.3 \times 10^1$
9. $6.8 \times 10^3 + 6.8 \times 10^4 + 56,000$
10. $213 \times 10^{-6} + 0.000043 + 51.3 \times 10^{-5}$

1. 1.10×10^3
2. 1.01×10
3. 9.00×10^2
4. 3.20×10^3
5. 1.03
6. 1.00×10^4
7. 2.55×10^{-2}
8. 4.90×10^2
9. 1.31×10^5
10. 7.69×10^{-4}

Additional practice problems are at the end of the chapter.

ESTIMATING 2–13

Students should include estimating as one of their problem solving tools. There are times when a ballpark answer is all that's required. Also, if you can estimate answers, you have the means to quickly check answers as found using the calculator.

Most electronic and computer related problems can be solved to a reasonable degree of accuracy (usually better than 10%) without the use of a calculator by utilizing scientific notation and rounding. Here are a couple of examples.

EXAMPLE 2–16

Multiply 3752 × 8439. First, estimate the answer and then solve using the calculator.

SOLUTION

When estimating, what we want to do is round to the point where we can perform the multiplication in our heads. In this problem we will round to one place. Using powers of ten and rounding to one place we get

$$3752 \approx 4 \times 10^3$$
$$8439 \approx 8 \times 10^3$$

The resulting multiplication expressed in scientific notation is

$$4 \times 10^3 \times 8 \times 10^3 = 4 \times 8 \times 10^6 = 32 \times 10^6 = 3.2 \times 10^7$$

Using the calculator: $3752 \times 8439 = 3.17 \times 10^7$ accurate to three places

The estimated answer is within about 1% of the answer to three places found using the calculator. Typically, we will express estimated answers in scientific notation accurate to two places.

EXAMPLE 2–17

Divide $57,950 \times 10^3$ by 0.267. First estimate the answer and then solve using the calculator.

Solution Round to the point where you can do the division in your head. In this problem we will round to one place. Using powers of ten and rounding to one place we get

$$\frac{57{,}950 \times 10^3}{0.267} = \frac{6 \times 10^7}{3 \times 10^{-1}} = 2 \times 10^{7+1} = 2 \times 10^8$$

Using the calculator:

$$\frac{57{,}950 \times 10^3}{0.267} = 2.17 \times 10^8 \quad \text{rounded to three places}$$

Sometimes you have to fudge a little bit in estimating. For example, let's estimate the answer to 3.5×3.5. Since both numbers are between 3 and 4, we know that the answer has to be between 9 and 16. $3 \times 3 = 9$ and $4 \times 4 = 16$. Since both numbers are half way between 3 and 4, we should expect the answer to be about half way between 9 and 16 or somewhere around 12. Lets round our numbers to one place, according to the rules of rounding and see what happens. According to the rules of rounding. 3.5 rounds up to 4. But if we round *both* numbers up to 4 our estimated answer $(4 \times 4 = 16)$ won't be very close. When the second numbers to be rounded are close to 5, lets change the rules of rounding so that our estimate will be more accurate. If we round one number up and the other number down we get, to one place, 3 and 4. $3 \times 4 = 12$, which is pretty close to the real answer which is 12.3 to 3 places. With practice, you get a feel for when to fudge a little bit with the rule for rounding when estimating answers.

When rounding, we use the following steps:

1. Convert all numbers to scientific notation rounded to one place. Two places is better (results in greater accuracy) if you can still readily find the answer without using the calculator.
2. Perform the indicated multiplication or division.
3. Express answer in scientific notation accurate to two places.

PRACTICE PROBLEMS 2-13

Estimate the answers to the following problems. Express answers in scientific notation accurate to two places. Then solve the problems using a calculator. Express calculator answers in scientific notation accurate to three places.

1. 587×43.8
2. 713×8.53
3. 832.51×47.956
4. 397.63×0.00052439
5. $694 \div 137.9$
6. $3870 \div 88.72$
7. $55.374 \div 0.0356$
8. $0.9327 \div 9832.7$

SOLUTIONS

1. $587 \approx 6 \times 10^2 \qquad 43.8 \approx 4 \times 10 \qquad 6 \times 10^2 \times 4 \times 10 = 24 \times 10^3 = 2.4 \times 10^4$
   $587 \times 43.8 = 2.57 \times 10^4$

2. $713 \approx 7 \times 10^2$ $8.53 \approx 9$ $7 \times 10^2 \times 9 = 63 \times 10^2 = 6.3 \times 10^3$
   $713 \times 8.53 = 6.08 \times 10^3$
3. $832.51 \approx 8 \times 10^2$ $47.956 \approx 5 \times 10$ $8 \times 10^2 \times 5 \times 10 = 40 \times 10^3 = 4.0 \times 10^4$
   $832.51 \times 47.956 = 3.99 \times 10^4$
4. $397.63 \approx 4 \times 10^2$ $0.00052439 \approx 5 \times 10^{-4}$ $4 \times 10^2 \times 5 \times 10^{-4} = 20 \times 10^{-2} =$
   2.0×10^{-1} $397.63 \times 0.00052439 = 2.09 \times 10^{-1}$
5. $694 \approx 7 \times 10^2$ $137.9 \approx 1 \times 10^2$ $7 \times 10^2 \div 1 \times 10^2 = 7.0$ $694 \div 137.9 = 5.03$
6. $3870 \approx 4 \times 10^3$ $88.72 \approx 9 \times 10$ $4 \times 10^3 \div 9 \times 10 \approx 0.5 \times 10^2 = 5.0 \times 10$
   $3870 \times 88.72 \approx 4.36 \times 10$
7. $55.374 \approx 6 \times 10$ $0.0356 \approx 4 \times 10^{-2}$ $6 \times 10 \div 4 \times 10^{-2} = 1.5 \times 10^3$
   $55.374 \div 0.0356 = 1.56 \times 10^3$
8. $0.9327 \approx 9 \times 10^{-1}$ $9832.7 \approx 1 \times 10^4$ $9 \times 10^{-1} \div 1 \times 10^4 = 9.0 \times 10^{-5}$
   $0.9327 \div 9832.7 = 9.49 \times 10^{-5}$

In the following sections of this chapter and in the following chapters, get into the habit of estimating your answers. Often, mistakes you have made using the calculator (such as keying in the wrong number or selecting the wrong function) can be detected by estimating.

PROBLEMS WITH COMPLEX DENOMINATORS   2-14

A typical electronics problem might look like this:

$$\frac{25 \times 3.3 \times 10^3}{3.3 \times 10^3 + 4.7 \times 10^4}$$

To solve this problem, we must first perform the indicated addition in the denominator.

$$\begin{aligned} 3.3 \times 10^3 &= 0.33 \times 10^4 \\ +4.7 \times 10^4 &= \underline{4.7 \ \ \times 10^4} \\ &\ \ \ 5.03 \times 10^4 \end{aligned}$$

Now

$$\frac{25 \times 3.3 \times 10^3}{5.03 \times 10^4} = 1.64$$

All this math may be done by using the calculator:

25 $\boxed{\times}$ 3.3 $\boxed{\text{EE}}$ 3 $\boxed{\div}$ $\boxed{(}$ 3.3 $\boxed{\text{EE}}$ 3 $\boxed{+}$ 4.7 $\boxed{\text{EE}}$ 4 $\boxed{)}$ $\boxed{=}$

(For Casio press $\boxed{\text{EXP}}$ instead of $\boxed{\text{EE}}$ to indicate a multiplication by a power of ten.)

Answer: 1.64

When using a calculator with the parentheses feature, enclosing the denominator is the easiest way to solve this kind of problem.

EXAMPLE 2–18

Perform the indicated operations. Express your answers in scientific notation rounded to three places.

(a) $\dfrac{9 \times 18 \times 10^3}{18 \times 10^3 + 56 \times 10^4}$

(b) $\dfrac{150 \times 6800}{6800 + 39 \times 10^3}$

(c) $\dfrac{20 \times 1.2 \times 10^6}{1.2 \times 10^6 + 8.2 \times 10^6}$

(d) $\dfrac{14 \times 56 \times 10^3}{1.5 \times 10^3 + 5.6 \times 10^4}$

SOLUTIONS

(a) 2.80×10^{-1}

(b) 2.23×10

(c) 2.55

(d) 1.36×10

PRACTICE PROBLEMS 2–14

Perform the indicated operations. Express your answers in scientific notation.

1. $\dfrac{5 \times 33 \times 10^3}{3.3 \times 10^3 + 33 \times 10^3}$

2. $\dfrac{9 \times 6.8 \times 10^3}{470 + 6.8 \times 10^3}$

3. $\dfrac{20 \times 220 \times 10^3}{75 \times 10^3 + 220 \times 10^3}$

4. $\dfrac{12 \times 15 \times 10^3}{4.7 \times 10^3 + 1.5 \times 10^4}$

5. $\dfrac{9 \times 5.6 \times 10^3}{820 + 5.6 \times 10^3}$

6. $\dfrac{5 \times 2700}{6.8 \times 10^3 + 2700}$

7. $\dfrac{25 \times 4700}{4700 + 47 \times 10^3}$

8. $\dfrac{20 \times 3.3 \times 10^3}{3.3 \times 10^3 + 3.3 \times 10^4}$

SOLUTIONS

1. 4.55

2. 8.42

3. 1.49×10

4. 9.14

5. 7.85

6. 1.42

7. 2.27

8. 1.82

Additional practice problems are at the end of the chapter.

2–15   RECIPROCALS

☞ **KEY POINT**   The *reciprocal* of a number is that number divided into 1.

$\frac{1}{3}$ is the reciprocal of 3, $\frac{1}{50}$ is the reciprocal of 50, and so on. In electronics it is often necessary to find the reciprocal of a number and then perform some other mathematical operation with that reciprocal.

EXAMPLE 2–19

Find the reciprocal of 27×10^3. Express the answer in scientific notation rounded to three places.

SOLUTION Using the calculator, key in 27×10^3 and then press the reciprocal key $\boxed{1/x}$ (for the Casio use $\boxed{x^{-1}}$). The answer is 3.70×10^{-5}.

If we estimate the answer we get

$$\frac{1}{27 \times 10^3} = \frac{1}{2.7 \times 10^4} \approx \frac{1}{3 \times 10^4} = 3.3 \times 10^{-5} \quad \text{carried out to two places}$$

EXAMPLE 2–20

Perform the indicated operation. Express your answer in scientific notation rounded to three places.

$$\frac{1}{22 \times 10^3} + \frac{1}{47 \times 10^3}$$

SOLUTION

$$22 \boxed{EE} 3 \boxed{1/x} + 47 \boxed{EE} 3 \boxed{1/x} =$$

Answer: 6.67 × 10^{-5}

If we estimate the answer we get

$$\frac{1}{22 \times 10^3} + \frac{1}{47 \times 10^3} \approx \frac{1}{2 \times 10^4} + \frac{1}{5 \times 10^4} =$$

$$0.5 \times 10^{-4} + 0.2 \times 10^{-4} = 7 \times 10^{-5}$$

EXAMPLE 2–21

Perform the indicated operation. Express your answer in scientific notation rounded to three places.

$$\frac{1}{\dfrac{1}{3.3 \times 10^3} + \dfrac{1}{6.8 \times 10^3}}$$

SOLUTION

$$3.3 \boxed{EE} 3 \boxed{1/x} + 6.8 \boxed{EE} 3 \boxed{1/x} = \boxed{1/x}$$

Answer: 2.22 × 10^3

We took the reciprocal of the first number in the denominator and then added the reciprocal of the second number. At the end of this step $\left(\boxed{=}\right)$, you should have displayed 4.50×10^{-4}. The reciprocal of this number is the answer, which is 2.22×10^{3}.

PRACTICE PROBLEMS 2–15

Perform the indicated operations. Express your answers in scientific notation rounded to three places.

1. $\dfrac{1}{470}$

2. $\dfrac{1}{33 \times 10^{4}}$

3. $\dfrac{1}{68 \times 10^{-9}}$

4. $\dfrac{1}{3.7 \times 10^{-3}}$

5. $\dfrac{1}{2760}$

6. $\dfrac{1}{5.6 \times 10^{3}}$

7. $\dfrac{1}{50 \times 10^{3}}$

8. $\dfrac{1}{370 \times 10^{-6}}$

9. $\dfrac{1}{4 \times 10^{-3}}$

10. $\dfrac{1}{10 \times 10^{3}}$

11. $\dfrac{1}{2700} + \dfrac{1}{1800}$

12. $\dfrac{1}{68 \times 10^{3}} + \dfrac{1}{47 \times 10^{3}}$

13. $\dfrac{1}{12 \times 10^{3}} + \dfrac{1}{7.5 \times 10^{3}}$

14. $\dfrac{1}{5 \times 10^{4}} + \dfrac{1}{2.5 \times 10^{4}}$

15. $\dfrac{1}{\dfrac{1}{27 \times 10^{3}} + \dfrac{1}{18 \times 10^{3}}}$

16. $\dfrac{1}{\dfrac{1}{6800} + \dfrac{1}{5600}}$

17. $\dfrac{1}{27.6 \times 10^{3}} - \dfrac{1}{47 \times 10^{3}}$

18. $\dfrac{1}{\dfrac{1}{270} + \dfrac{1}{470} + \dfrac{1}{560}}$

SOLUTIONS

1. 2.13×10^{-3}

2. 3.03×10^{-6}

3. 1.47×10^{7}

4. 2.70×10^{2}

5. 3.62×10^{-4}

6. 1.79×10^{-4}

7. 2.00×10^{-5}

8. 2.70×10^{3}

9. 2.50×10^{2}

10. 1.00×10^{-4}

11. 9.26×10^{-4}

12. 3.60×10^{-5}

13. 2.17×10^{-4}

14. 6.00×10^{-5}

15. 1.08×10^{4}

16. 3.07×10^{3}

17. 1.50×10^{-5}

18. 1.31×10^{2}

Additional practice problems are at the end of the chapter.

Perform the indicated operations. Express your answers in scientific notation rounded to three places.

1. 27×65
2. 0.00173×3900
3. $47,000 \times 0.0000423$
4. $1765 \times 24,673$
5. $33 \times 10^4 \times 27 \times 10^3$
6. $5.6 \times 10^3 + 4.7 \times 10^4$
7. $147 \times 10^{-7} + 0.833 \times 10^{-5} + 21.3 \times 10^{-6}$
8. $2 \times 3.14 \times 2500 \times 500 \times 10^{-9}$
9. $\dfrac{47.3 \times 10^2}{0.00846}$
10. $\dfrac{0.025 \times 10^{-6}}{483 \times 10^{-9}}$
11. $\dfrac{7800 \times 0.00273}{1.28 \times 10^{-4} \times 8000}$
12. $\dfrac{0.0000064 \times 84 \times 10^2}{6730 \times 0.045 \times 10^{-3}}$
13. $\dfrac{12 \times 3900}{56,000 + 3900}$
14. $\dfrac{1}{270} + \dfrac{1}{390}$
15. $\dfrac{25 \times 4700}{4700 + 47,000}$
16. $\dfrac{1}{3900} + \dfrac{1}{56 \times 10^3} + \dfrac{1}{20 \times 10^3}$
17. $\dfrac{1}{6.28 \times 12 \times 10^3 \times 470 \times 10^{-12}}$
18. $\dfrac{25 \times 2700}{2.7 \times 10^3 + 3.3 \times 10^4}$
19. $\dfrac{1}{\dfrac{1}{560 \times 10^2} + \dfrac{1}{47 \times 10^3}}$
20. $\dfrac{1}{\dfrac{1}{2700} + \dfrac{1}{3300} + \dfrac{1}{4700}}$

Answers to Self-test 2–2 are at the end of the chapter.

POWERS AND ROOTS IN BASE TEN   2–16

When we wrote 10^3, we learned that the exponent (3) told us to multiply 10 by itself three times.

$$10^3 = 10 \times 10 \times 10$$

If we wrote $(10^3)^2$, we would be raising 10^3 to the second power. That is,

$$(10^3)^2 = 10^3 \times 10^3 = 10^6$$

We need the parentheses to identify the quantity we are acting on. Consider the quantity $(10^{-2})^3$.

$$(10^{-2})^3 = 10^{-2} \times 10^{-2} \times 10^{-2} = 10^{-6}.$$

In each case the exponent can be thought of as a multiplier.

$$(10^3)^2 = 10^{3 \times 2} = 10^6 \qquad \text{and} \qquad (10^{-2})^3 = 10^{-2 \times 3} = 10^{-6}$$

EXAMPLE 2–22

Perform the indicated operations.

(a) $(10^4)^{-3}$

(b) $\left(\dfrac{1}{10^4}\right)^2$

SOLUTIONS

(a) $(10^4)^{-3} = 10^{4 \times (-3)} = 10^{-12}$

(b) $\left(\dfrac{1}{10^4}\right)^2 = \dfrac{1^2}{10^{4 \times 2}} = \dfrac{1}{10^8} = 10^{-8}$

When the exponent is in the denominator, we usually move it to the numerator.

☞ **RULE 2–17** When raising a power to a power, multiply the exponents.

PRACTICE PROBLEMS 2–16

Perform the indicated operations. Express your answers in powers of ten form.

1. $(10^2)^3$
2. $(10^4)^2$
3. $(10^3)^3$
4. $(10^{-1})^2$
5. $(10^{-4})^5$
6. $(10^{-6})^2$
7. $(10^{-2})^{-3}$
8. $(10^{-3})^{-1}$
9. $(10^{-1})^{-4}$
10. $(10^3)^{-5}$
11. $(10^2)^{-1}$
12. $(10^6)^{-2}$
13. $\left(\dfrac{1}{10^2}\right)^4$
14. $\left(\dfrac{1}{10^{-3}}\right)^2$
15. $\left(\dfrac{1}{10^3}\right)^{-3}$

SOLUTIONS

1. 10^6
2. 10^8
3. 10^9
4. 10^{-2}
5. 10^{-20}
6. 10^{-12}
7. 10^6
8. 10^3
9. 10^4
10. 10^{-15}
11. 10^{-2}
12. 10^{-12}
13. 10^{-8}
14. 10^6
15. 10^9

Additional practice problems are at the end of the chapter.

SQUARE AND CUBE ROOTS
2–17 OF POWERS OF TEN

To find the square root or cube root of powers of ten, we multiply the exponents. $(10^4)^{1/2} = 10^{4 \times 1/2} = 10^2$. $(10^{12})^{1/3} = 10^{12 \times 1/3} = 10^4$.

$$\left(\dfrac{1}{10^6}\right)^{1/2} = \dfrac{1}{10^{6 \times 1/2}} = \dfrac{1}{10^3} = 10^{-3}$$

PRACTICE PROBLEMS 2–17

Perform the indicated operations. Express your answers in powers of ten form.

1. $(10^6)^{1/2}$
2. $(10^9)^{1/3}$
3. $(10^{-6})^{1/2}$
4. $(10^{-12})^{1/3}$
5. $\left(\dfrac{1}{10^{-3}}\right)^{1/3}$
6. $(10^{-6})^{1/3}$
7. $(10^3)^{1/3}$
8. $\left(\dfrac{1}{10^4}\right)^{1/2}$
9. $(10^{-15})^{1/3}$
10. $(10^{10})^{1/2}$

SOLUTIONS

1. 10^3
2. 10^3
3. 10^{-3}
4. 10^{-4}
5. 10
6. 10^{-2}
7. 10
8. 10^{-2}
9. 10^{-5}
10. 10^5

Additional practice problems are at the end of the chapter.

SQUARING NUMBERS   2–18

Let's square some numbers other than powers of ten and express the answers in scientific notation rounded to three places.

EXAMPLE 2–23

Find 14.7^2.

SOLUTION   This problem requires the use of another function key on the calculator. This key is usually labeled $\boxed{x^2}$. Pressing this key will square the contents of the x register (the value displayed). Use whatever keystrokes are necessary to perform this function. Using the calculator, 14.7 $\boxed{x^2}$. The answer, 2.16×10^2, is displayed. On the Casio, you must press $\boxed{=}$ after $\boxed{x^2}$.

EXAMPLE 2–24

Find $(0.0236 \times 10^2)^2$.

SOLUTION   Using the calculator, 0.0236 $\boxed{EE}$ 2 $\boxed{x^2}$. *Answer:* 5.57

Estimating yields:

$$(0.0236 \times 10^2)^2 \approx (2 \times 10^0)^2 = 4$$

We expected the estimated answer to be low because we rounded both numbers *down*.

EXAMPLE 2–25

Find $\dfrac{278^2}{16}$.

SOLUTION　Using the calculator, 278 $\boxed{x^2}$ $\boxed{\div}$ 16 $\boxed{=}$. *Answer:* 4.83×10^3.
Notice that only the number 278 is squared. The number 16 is not squared.
　Estimating yields:

$$\frac{278^2}{16} = \frac{(2.78 \times 10^2)^2}{1.6 \times 10} \approx \frac{(3 \times 10^2)^2}{2 \times 10} = \frac{9 \times 10^4}{2 \times 10} = 4.5 \times 10^3$$

PRACTICE PROBLEMS 2–18

Perform the indicated operations. Express your answers in scientific notation rounded to three places.

1. 6.3^2
2. 43^2
3. 0.0372^2
4. 0.00047^2
5. $(6.73 \times 10^2)^2$
6. $(1.04 \times 10^{-3})^2$
7. $(796 \times 10^{-2})^2$
8. 8764^2
9. 0.973^2
10. $(0.00567 \times 10^{-4})^2$
11. $(0.0000873 \times 10^4)^2$
12. $(1260 \times 10^{-3})^2$
13. $0.0073^2 \times 4700$
14. $(27 \times 10^{-6})^2 \times 68{,}000$
15. $20^2 \div 3300$
16. $9^2 \div 270$

SOLUTIONS

1. 3.97×10
2. 1.85×10^3
3. 1.38×10^{-3}
4. 2.21×10^{-7}
5. 4.53×10^5
6. 1.08×10^{-6}
7. 6.34×10
8. 7.68×10^7
9. 9.47×10^{-1}
10. 3.21×10^{-13}
11. 7.62×10^{-1}
12. 1.59
13. 2.50×10^{-1}
14. 4.96×10^{-5}
15. 1.21×10^{-1}
16. 3.00×10^{-1}

Additional practice problems are at the end of the chapter.

The following problems require using the square root key $\boxed{\sqrt{x}}$. We often use the notation $x^{1/2}$ instead of $\sqrt{x}$. Raising a number to the $\frac{1}{2}$ power is just another way of saying that the operation is "square root."

EXAMPLE 2–26

Find the square root of 376.

SOLUTION   Using the calculator, $376 \boxed{\sqrt{x}}$. The answer is 1.94×10.

EXAMPLE 2–27

Find $\dfrac{1}{6.28(0.05 \times 5 \times 10^{-9})^{1/2}}$.

SOLUTION   Using the calculator:

We first find the square root of the quantity inside the parentheses.
TI: 0.05

$\boxed{\times}$

5

$\boxed{\text{EE}}$

$\boxed{+/-}$

9

$\boxed{=}$ $0.05 \times 5 \times 10^{-9} = 2.50 \times 10^{-10}$

$\boxed{\sqrt{x}}$ The square root of 2.50×10^{-10} is 1.58×10^{-5}.

$\boxed{\times}$

6.28   Multiply by 6.28.

$\boxed{=}$ 9.93×10^{-5}.

$\boxed{1/x}$ The reciprocal of 9.93×10^{-5} is 1.01×10^{4}.

CASIO: $1 \boxed{\div} \boxed{(} 6.28 \boxed{\sqrt{}} \boxed{(} 0.05 \boxed{\times} 5 \boxed{\text{EXP}}$

$\boxed{(-)} 9 \boxed{)} \boxed{)} \boxed{=}$ Answer: 1.01×10^{4}

PRACTICE PROBLEMS 2–19

Perform the indicated operations. Express your answer in scientific notation rounded to three places.

1. $36.5^{1/2}$
2. $0.925^{1/2}$
3. $(635 \times 10^{-2})^{1/2}$
4. $(1870 \times 10^{3})^{1/2}$
5. $(0.000675 \times 10^{2})^{1/2}$
6. $(27.3 \times 10^{2})^{1/2}$
7. $(5.46 \times 10^{-4})^{1/2}$
8. $\left(\dfrac{734 \times 10^{2}}{423}\right)^{1/2}$
9. $\left(\dfrac{0.00726 \times 10^{-3}}{77 \times 10^{-6}}\right)^{1/2}$
10. $\dfrac{1}{6.28(0.2 \times 3 \times 10^{-9})^{1/2}}$

SOLUTIONS

1. 6.04
2. 9.62×10^{-1}
3. 2.52
4. 1.37×10^{3}
5. 2.60×10^{-1}
6. 5.22×10
7. 2.34×10^{-2}
8. 1.32×10
9. 3.07×10^{-1}
10. 6.50×10^{3}

Additional practice problems are at the end of the chapter.

SELF-TEST 2–3

Perform the indicated operations. Express your answers in scientific notation rounded to three places.

1. $(10^{3})^{5}$
2. $(10^{2})^{-4}$
3. $\left(\dfrac{1}{10^{-2}}\right)^{3}$
4. $(10^{-6})^{1/2}$
5. $(10^{6})^{1/3}$
6. $(10^{-3})^{1/3}$
7. 78.3^{2}
8. 0.000293^{2}
9. $(8796 \times 10^{3})^{2}$
10. $0.0173^{2} \times 1800$
11. $(39 \times 10^{-4})^{2} \times 120$
12. $\dfrac{15^{2}}{1500}$
13. $273^{1/2}$
14. $(2.65 \times 10^{-4})^{1/2}$
15. $\left(\dfrac{0.0000467 \times 10^{3}}{63}\right)^{1/2}$
16. $\left(\dfrac{645 \times 10^{2}}{14}\right)^{1/2}$
17. $\dfrac{(24.5 \times 10^{-3})^{1/2}}{17}$
18. $(300 \times 10^{-3} \times 220)^{1/2}$
19. $\left(\dfrac{273 \times 10^{4}}{0.0012}\right)^{1/2}$
20. $\dfrac{1}{6.28(50 \times 10^{-3} \times 2 \times 10^{-6})^{1/2}}$

Answers to Self-test 2–3 are at the end of the chapter.

CHAPTER 2 AT A GLANCE

PAGE	KEY POINT OR RULE	EXAMPLE
37	*Rule 2–1:* In converting some multiple of 10 to powers of ten form, count the number of zeros. This number is the exponent. The answer is 10 raised to that power.	$1000 = 10^3$
38	*Rule 2–2:* In converting from powers of ten form to a number, write the digit 1 and follow with as many zeros as indicated by the exponent.	$10^3 = 1000$
39	*Rule 2–3:* In converting from powers of ten to decimal fractions, the negative exponent tells how many zeros follow the 1 in the denominator.	$10^{-2} = \dfrac{1}{100}$
39	*Rule 2–4:* In converting from powers of ten to decimal numbers, the negative exponent tells how far the 1 is located to the right of the decimal point.	$10^{-2} = 0.01$
40	*Rule 2–5:* In converting from a decimal fraction to powers of ten form, count the number of zeros in the denominator. This number is the negative power to which 10 is raised.	$\dfrac{1}{100} = 10^{-2}$
40	*Rule 2–6:* In converting from a decimal number to powers of ten form, count how far the digit 1 is located to the right of the decimal point. This is the value of the negative exponent.	$0.01 = 10^{-2}$
41	*Rule 2–7:* To multiply numbers in powers of ten form, add the exponents.	$10^2 \times 10^3 = 10^{2+3} = 10^5$ $10^2 \times 10^{-3} = 10^{2+(-3)}$ $\qquad\qquad = 10^{-1}$
42	*Rule 2–8:* To divide numbers in powers of ten form, subtract the exponents.	$\dfrac{10^4}{10^2} = 10^{4-2} = 10^2$
45	*Rule 2–9:* Whenever a power of ten is in the denominator, it can be moved to the numerator by changing the sign of the exponent.	$\dfrac{10^2}{10^{-3}} = 10^2 \times 10^3 = 10^{2+3} = 10^5$

47	*Rule 2–10:* For each place we move the decimal point *left,* the exponent increases by 1.	$\begin{aligned} 1750 &= 175 \times 10 \\ &= 17.5 \times 10^2 \\ &= 1.75 \times 10^3 \end{aligned}$
48	*Rule 2–12:* For each place we move the decimal point *right,* the exponent decreases by 1.	$\begin{aligned} 0.00425 &= 0.0425 \times 10^{-1} \\ &= 0.425 \times 10^{-2} \\ &= 4.25 \times 10^{-3} \end{aligned}$
50	*Rule 2–14:* To express numbers greater than 1 in scientific notation, count the number of places the decimal point must move left. This number is the exponent and is positive.	$\begin{aligned} 704 &= 7.04 \times 10^2 \\ 37{,}654 &= 3.77 \times 10^4 \end{aligned}$ (rounded to three places)
51	*Rule 2–15:* To express numbers that are less than 1 in scientific notation, count the number of places the decimal point must move right. This number is the exponent and is negative.	$\begin{aligned} 0.00156 &= 1.56 \times 10^{-3} \\ 0.00027643 &= 2.76 \times 10^{-4} \end{aligned}$ (rounded to three places)
58	*Rule 2–16:* To add or subtract numbers in powers of ten form, the exponents must be the same.	$\begin{aligned} 3 \times 10^4 &= 3 \times 10^4 \\ +3 \times 10^3 &= 0.3 \times 10^4 \\ \hline &= 3.3 \times 10^4 \end{aligned}$
62	*Key Point:* The reciprocal of a number is that number divided into 1.	Find the reciprocal of 150. $\dfrac{1}{150} = 6.67 \times 10^{-3}$
66	*Rule 2–17:* When raising a power to a power, multiply the exponents.	$\begin{aligned} (10^2)^3 &= 10^{2 \times 3} = 10^6 \\ (10^6)^{1/2} &= 10^{6 \times 1/2} = 10^3 \\ (3.6 \times 10^3)^2 &= 1.30 \times 10^7 \\ (746 \times 10^{-3})^{1/2} &= 8.64 \times 10^{-1} \end{aligned}$

When working **End of Chapter Problems,** if you need to go back to the section where the material was presented, just look at the problem set number. For example, if you are working problems in *End of Chapter Problems 2–13* and you need help, go back to section 2–13 and review the material and examples presented there.

END OF CHAPTER PROBLEMS 2–1

Convert the following numbers to powers of ten form.

1. 10
2. 100,000
3. 1000
4. 10,000,000
5. 100
6. 1,000,000

END OF CHAPTER PROBLEMS 2–2

Convert the following power of ten numbers to decimal numbers.

1. 10^3
2. 10^5
3. 10^0
4. 10^8
5. 10^6
6. 10^4

END OF CHAPTER PROBLEMS 2–3

Convert the following powers of ten numbers to (a) decimal fractions and (b) decimal numbers.

1. 10^{-2}
2. 10^{-5}
3. 10^{-3}
4. 10^{-6}
5. 10^{-1}
6. 10^{-4}

END OF CHAPTER PROBLEMS 2–4

Convert the following decimal numbers to powers of ten form.

1. 0.0001
2. 0.01
3. 0.000001
4. 0.001

Convert the following decimal fractions to powers of ten form.

5. $\dfrac{1}{10,000}$
6. $\dfrac{1}{10,000,000}$
7. $\dfrac{1}{1000}$
8. $\dfrac{1}{1,000,000}$

END OF CHAPTER PROBLEMS 2–5

Perform the indicated operations. Express your answers in powers of ten form.

1. $10^4 \times 10^6$
2. $10^3 \times 10^2$
3. $10^2 \times 10^4$
4. $10^3 \times 10^5$
5. $10^5 \times 10^{-1}$
6. $10^6 \times 10^{-3}$
7. $10^3 \times 10^{-7}$
8. $10^2 \times 10^{-6}$
9. $10^{-1} \times 10^4$
10. $10^{-2} \times 10^6$
11. $10^{-6} \times 10^{-4}$
12. $10^{-7} \times 10^{-3}$
13. $10^{-3} \times 10^{-9}$
14. $10^{-2} \times 10^{-5}$
15. $10^{-4} \times 10^{-5}$
16. $10^{-3} \times 10^{-6}$

END OF CHAPTER PROBLEMS 2–6

Perform the indicated operations. Express your answers in powers of ten form.

1. $\dfrac{10^7}{10^3}$
2. $\dfrac{10^9}{10^4}$
3. $\dfrac{10^3}{10^8}$
4. $\dfrac{10^2}{10^5}$
5. $\dfrac{10^4}{10^{-4}}$
6. $\dfrac{10^7}{10^{-3}}$

7. $\dfrac{10^3}{10^{-9}}$

8. $\dfrac{10^4}{10^{-6}}$

9. $\dfrac{10^{-2}}{10^{-9}}$

10. $\dfrac{10^{-1}}{10^{-7}}$

11. $\dfrac{10^{-8}}{10^{-5}}$

12. $\dfrac{10^{-6}}{10^{-5}}$

13. $\dfrac{10^{-3}}{10^3}$

14. $\dfrac{10^{-6}}{10^1}$

15. $\dfrac{10^{-2}}{10^5}$

16. $\dfrac{10^{-4}}{10^7}$

17. $\dfrac{1}{10^{-3}}$

18. $\dfrac{1}{10^{-7}}$

19. $\dfrac{1}{10^4}$

20. $\dfrac{1}{10^6}$

END OF CHAPTER PROBLEMS 2–7

Perform the indicated operations. Express your answers in powers of ten form.

1. $\dfrac{10^{-3} \times 10^{-6}}{10^{-2}}$

2. $\dfrac{10^2 \times 10^6}{10^4}$

3. $\dfrac{10^{-6} \times 10^4}{10^5}$

4. $\dfrac{10^{-4} \times 10^9}{10^3}$

5. $\dfrac{1}{10^2 \times 10^{-4} \times 10^{-9}}$

6. $\dfrac{1}{10^{-2} \times 10^4 \times 10^{-7}}$

7. $\dfrac{1}{10^{-3} \times 10^{-6} \times 10^{-1}}$

8. $\dfrac{1}{10^2 \times 10^6 \times 10^{-3}}$

9. $\dfrac{10^{-4}}{10^6 \times 10^{-2} \times 10^4}$

10. $\dfrac{10^3}{10^4 \times 10^6 \times 10^{-1}}$

11. $\dfrac{10^{-3} \times 10^2 \times 10^4}{10^2 \times 10^{-7}}$

12. $\dfrac{10^{-1} \times 10^2 \times 10^4}{10^4 \times 10^2}$

13. $\dfrac{10^4 \times 10^6 \times 10^3}{10^2 \times 10^{-7} \times 10^5}$

14. $\dfrac{10^{-1} \times 10^{-7} \times 10^6}{10^1 \times 10^0 \times 10^{-4}}$

15. $\dfrac{10^3 \times 10^{-5} \times 10^6}{10^4 \times 10^{-3} \times 10^7}$

16. $\dfrac{10^{-2} \times 10^{-6} \times 10^3}{10^3 \times 10^6 \times 10^{-4}}$

17. $\dfrac{10^{-3} \times 10^{-6} \times 10^{-9}}{10^3 \times 10^6 \times 10^9}$

18. $\dfrac{10^{-9} \times 10^3 \times 10^{-3}}{10^4 \times 10^{-1} \times 10^5}$

END OF CHAPTER PROBLEMS 2–8

Change the following numbers to powers of ten form with an exponent of (a) 2, (b) 4, and (c) 6.

1. 475
2. 1500
3. 65.9
4. 478
5. 93,400
6. 180,000
7. 4780
8. 4700
9. 41,800
10. 18,000

Change the following numbers to powers of ten form with an exponent of (a) -2, (b) -4, and (c) -6.

11. 0.00465	12. 0.000758	13. 0.00000555
14. 0.0000906	15. 0.000673	16. 0.00825
17. 0.0000000108	18. 0.00000745	19. 0.000000233
20. 0.0000000560		

Change the following numbers to powers of ten form with an exponent of (a) 3, and (b) -3.

21. 325	22. 470	23. 722
24. 0.0157	25. 2700	26. 4.39
27. 180	28. 8060	29. 0.00450
30. 0.000850		

END OF CHAPTER PROBLEMS 2–9

Round the following numbers to three significant figures and express your answers in scientific notation.

1. 27,640	2. 43,966	3. 47.8
4. 277.7	5. 1,765,400	6. 3,716,500
7. 273.46	8. 35.986	9. 173,460
10. 406,600	11. 78.88	12. 15.756
13. 1,670,000	14. 12,673	15. 57,945
16. 83,876	17. 80,975	18. 40,713
19. 470,560	20. 199,999	

END OF CHAPTER PROBLEMS 2–10

Round the following numbers to three significant figures and express your answers in scientific notation.

1. 0.004783	2. 0.017600	3. 0.7474
4. 0.002087	5. 0.01637	6. 0.001037
7. 0.0045446	8. 0.0078449	9. 0.000050085
10. 0.000030067	11. 0.25465	12. 0.056374
13. 0.040045	14. 0.0740347	15. 0.0008964
16. 0.0006794	17. 0.0170983	18. 0.07096
19. 0.000002778	20. 0.000005278	

END OF CHAPTER PROBLEMS 2–11

Perform the indicated operations. Express your answers in scientific notation rounded to three places.

1. 273×78	2. 793×647
3. 0.0000273×4700	4. 0.00675×0.0147

5. $\dfrac{50}{0.00475}$

6. $\dfrac{12}{0.000273}$

7. $\dfrac{468 \times 10^3}{76,300}$

8. $\dfrac{52,800 \times 10^{-2}}{3960}$

9. $\dfrac{176 \times 473}{27,800}$

10. $\dfrac{4630 \times 138}{4360}$

11. $\dfrac{0.00173 \times 287}{379}$

12. $\dfrac{0.000814 \times 706}{55.6}$

13. $\dfrac{76 \times 10^{-2} \times 906 \times 10^4}{0.000559}$

14. $\dfrac{283 \times 10^{-3} \times 990 \times 10^5}{0.00466}$

15. $\dfrac{47,600 \times 0.00316 \times 10^{-2}}{689 \times 10^3}$

16. $\dfrac{9240 \times 0.00643 \times 10^4}{0.00329 \times 10^3}$

17. $\dfrac{0.00790 \times 835 \times 10^2}{9840 \times 10^{-2}}$

18. $\dfrac{0.00238 \times 650 \times 10^3}{365 \times 10^{-3}}$

19. $\dfrac{651 \times 0.00179}{73.4 \times 0.874}$

20. $\dfrac{522 \times 0.000196}{0.673 \times 5800}$

21. $\dfrac{127 \times 10^2 \times 0.024 \times 10^{-4}}{79 \times 10^{-6} \times 410 \times 10^4}$

22. $\dfrac{470 \times 10^4 \times 0.00715 \times 10^{-2}}{27 \times 10^{-4} \times 330 \times 10^{-3}}$

23. $\dfrac{7630 \times 10^{-2} \times 875 \times 10^3}{43 \times 10^3 \times 600 \times 10^{-6}}$

24. $\dfrac{2740 \times 10^{-3} \times 475 \times 10^2}{68 \times 10^3 \times 400 \times 10^{-7}}$

END OF CHAPTER PROBLEMS 2–12

Perform the indicated operations. Express your answers in scientific notation rounded to three places.

1. $10^3 + 10^4$
2. $10^2 + 10$
3. $10^{-1} + 10^{-1}$
4. $10^{-1} + 10^{-2}$
5. $33 \times 10^2 + 56 \times 10$
6. $56 \times 10^3 + 33 \times 10^4$
7. $470 \times 10^3 + 1.2 \times 10^4$
8. $750 \times 10^3 + 1.5 \times 10^4$
9. $27 \times 10^{-2} + 180 \times 10^{-4}$
10. $56 \times 10^{-2} + 470 \times 10^{-4}$
11. $45 \times 10^{-3} - 63 \times 10^{-4}$
12. $75 \times 10^{-3} - 47 \times 10^{-4}$
13. $8.2 \times 10^3 + 8.2 \times 10^4 + 9100$
14. $2.7 \times 10^2 + 6.8 \times 10^3 + 1200$
15. $43 \times 10^4 - 3.9 \times 10^3$
16. $18 \times 10^4 - 8.2 \times 10^3$

END OF CHAPTER PROBLEMS 2–13

In the following problems, first estimate the answer by rounding each term to one-place accuracy and using scientific notation. Then calculate the answer using a calculator set to scientific notation with three-place accuracy.

1. 737×53.9
2. 3.15×0.324
3. 0.0059×0.005892
4. 0.0932×0.00337
5. 5957.32×0.01317
6. 384.7×0.846
7. $620 \times 270 \times 190$
8. $434 \times 597 \times 850$
9. $32572 \div 0.329$
10. $5227.8 \div 0.0322$
11. $0.0987 \div 0.00475$
12. $0.0529 \div 0.00289$

13. $392.73 \div 8388.2$ 14. $213.71 \div 7569.3$ 15. $283 \times 489 \div 511$

16. $849 \times 334 \div 731$

END OF CHAPTER PROBLEMS 2–14

Perform the indicated operations. Express your answers in scientific notation rounded to three places.

1. $\dfrac{12 \times 47 \times 10^3}{4.7 \times 10^3 + 4.7 \times 10^4}$

2. $\dfrac{12 \times 56 \times 10^3}{3.3 \times 10^3 + 5.6 \times 10^4}$

3. $\dfrac{5 \times 4700}{470 + 4700}$

4. $\dfrac{5 \times 5600}{680 + 5600}$

5. $\dfrac{15 \times 82 \times 10^4}{10 \times 10^3 + 82 \times 10^4}$

6. $\dfrac{15 \times 33 \times 10^4}{10 \times 10^3 + 33 \times 10^4}$

7. $\dfrac{25 \times 750 \times 10^3}{91 \times 10^3 + 750 \times 10^3}$

8. $\dfrac{25 \times 680}{82 \times 10 + 680}$

9. $\dfrac{18 \times 12 \times 10^6}{12 \times 10^6 + 8.2 \times 10^6}$

10. $\dfrac{18 \times 18 \times 10^6}{18 \times 10^6 + 4.7 \times 10^6}$

11. $\dfrac{25 \times 3300}{3300 + 56,000}$

12. $\dfrac{20 \times 4700}{39,000 + 4700}$

13. $\dfrac{12 \times 5600}{5600 + 33,000}$

14. $\dfrac{5 \times 680}{680 + 1500}$

15. $\dfrac{12 \times 4.7 \times 10^3}{4.7 \times 10^4 + 4.7 \times 10^3}$

16. $\dfrac{15 \times 1.8 \times 10^3}{4.7 \times 10^4 + 1.8 \times 10^3}$

17. $\dfrac{20 \times 6.8 \times 10^3}{6.8 \times 10^3 + 4.7 \times 10^4}$

18. $\dfrac{5 \times 1.2 \times 10^3}{1.2 \times 10^3 + 1.8 \times 10^4}$

19. $\dfrac{25 \times 10^{-3} \times 213 \times 10^{-6}}{213 \times 10^{-6} + 303 \times 10^{-6}}$

20. $\dfrac{130 \times 10^{-6} \times 2 \times 10^{-3}}{2 \times 10^{-3} + 4 \times 10^{-3}}$

21. $\dfrac{15 \times 10^{-3} \times 303 \times 10^{-6}}{303 \times 10^{-6} + 370 \times 10^{-6}}$

22. $\dfrac{12 \times 10^{-3} \times 17.9 \times 10^{-6}}{17.9 \times 10^{-6} + 15 \times 10^{-6}}$

END OF CHAPTER PROBLEMS 2–15

Perform the indicated operations. Express your answers in scientific notation rounded to three places.

1. $\dfrac{1}{6.28 \times 50 \times 10^3 \times 30 \times 10^{-9}}$

2. $\dfrac{1}{6.28 \times 600 \times 10 \times 10^{-6}}$

3. $\dfrac{1}{6.28 \times 25 \times 10^3 \times 50 \times 10^{-3}}$

4. $\dfrac{1}{6.28 \times 2 \times 10^3 \times 7 \times 10^{-6}}$

5. $\dfrac{1}{680} + \dfrac{1}{560}$

6. $\dfrac{1}{470} + \dfrac{1}{560}$

7. $\dfrac{1}{3300} + \dfrac{1}{2700} + \dfrac{1}{1000}$

8. $\dfrac{1}{6800} + \dfrac{1}{5600} + \dfrac{1}{9100}$

9. $\dfrac{1}{56 \times 10^3} + \dfrac{1}{47 \times 10^3}$

10. $\dfrac{1}{27 \times 10^5} + \dfrac{1}{10 \times 10^5}$

11. $\dfrac{1}{470 \times 10^3} + \dfrac{1}{330 \times 10^3}$

12. $\dfrac{1}{68 \times 10^3} + \dfrac{1}{75 \times 10^3}$

13. $\dfrac{1}{27 \times 10^3} + \dfrac{1}{56 \times 10^3} + \dfrac{1}{18 \times 10^3}$

14. $\dfrac{1}{820 \times 10^3} + \dfrac{1}{390 \times 10^3} + \dfrac{1}{270 \times 10^3}$

15. $\dfrac{1}{\dfrac{1}{4700} + \dfrac{1}{5600}}$

16. $\dfrac{1}{\dfrac{1}{270} + \dfrac{1}{560}}$

17. $\dfrac{1}{\dfrac{1}{270} + \dfrac{1}{180} + \dfrac{1}{390}}$

18. $\dfrac{1}{\dfrac{1}{7500} + \dfrac{1}{5600} + \dfrac{1}{3300}}$

19. $\dfrac{1}{\dfrac{1}{56 \times 10^3} + \dfrac{1}{68 \times 10^3}}$

20. $\dfrac{1}{\dfrac{1}{2.2 \times 10^3} + \dfrac{1}{810}}$

21. $\dfrac{1}{\dfrac{1}{9.1 \times 10^3} + \dfrac{1}{8.2 \times 10^3} + \dfrac{1}{5.6 \times 10^3}}$

22. $\dfrac{1}{\dfrac{1}{18 \times 10^3} + \dfrac{1}{39 \times 10^3} + \dfrac{1}{56 \times 10^3}}$

END OF CHAPTER PROBLEMS 2–16

Perform the indicated operations.

1. $(10^2)^3$
2. $(10^2)^5$
3. $(10^5)^2$
4. $(10^{-2})^{-3}$
5. $(10^3)^{-4}$
6. $(10^2)^{-7}$
7. $(10^{-4})^{-3}$
8. $\left(\dfrac{1}{10^3}\right)^3$
9. $\left(\dfrac{1}{10^{-4}}\right)^{-2}$
10. $\left(\dfrac{1}{10^{-4}}\right)^{-2}$

END OF CHAPTER PROBLEMS 2–17

Perform the indicated operations.

1. $(10^6)^{1/2}$
2. $(10^8)^{1/2}$
3. $(10^{-4})^{1/2}$
4. $(10^{-2})^{1/2}$
5. $\left(\dfrac{1}{10^4}\right)^{1/2}$
6. $\left(\dfrac{1}{10^8}\right)^{1/2}$
7. $\left(\dfrac{1}{10^{-10}}\right)^{1/2}$
8. $\left(\dfrac{1}{10^{-6}}\right)^{1/2}$
9. $\left(\dfrac{10^2}{10^{-4}}\right)^{1/2}$
10. $\left(\dfrac{10^5}{10^{-1}}\right)^{1/2}$

END OF CHAPTER PROBLEMS 2–18

Perform the indicated operations. Express your answers in scientific notation rounded to three places.

1. 79.3^2
2. 446^2
3. 1780^2
4. 5600^2
5. 0.04758^2
6. 0.0652^2
7. 0.00873^2
8. 0.00165^2
9. $(45 \times 10^3)^2$
10. $(264 \times 10^2)^2$
11. $(300 \times 10^3)^2$
12. $(400 \times 10^3)^2$
13. $(0.296 \times 10^{-3})^2$
14. $(525 \times 10^{-4})^2$
15. $(0.00903 \times 10^{-2})^2$
16. $(0.033 \times 10^{-2})^2$
17. $(50 \times 10^{-3})^2 \times 2700$
18. $(375 \times 10^{-9})^2 \times 120$
19. $(0.0765 \times 10^2)^2 \times 0.037$
20. $(0.000821 \times 10^3)^2 \times 0.976$

END OF CHAPTER PROBLEMS 2–19

Perform the indicated operations. Express your answers in scientific notation rounded to three places.

1. $93^{1/2}$
2. $783^{1/2}$
3. $237^{1/2}$
4. $5756^{1/2}$
5. $0.00705^{1/2}$
6. $0.0085^{1/2}$
7. $0.000105^{1/2}$
8. $0.000187^{1/2}$
9. $(270 \times 10^5)^{1/2}$
10. $(432 \times 10^4)^{1/2}$
11. $(4500 \times 10^4)^{1/2}$
12. $(74.5 \times 10^3)^{1/2}$
13. $(0.00923 \times 10^{-3})^{1/2}$
14. $(0.00176 \times 10^4)^{1/2}$
15. $(0.000346 \times 10^5)^{1/2}$
16. $(0.000906 \times 10^{-6})^{1/2}$
17. $\left(\dfrac{0.289 \times 10^{-1}}{254}\right)^{1/2}$
18. $\left(\dfrac{0.00573 \times 10^{-4}}{330}\right)^{1/2}$
19. $\dfrac{1}{6.28(20 \times 10^{-3} \times 20 \times 10^{-9})^{1/2}}$
20. $\dfrac{1}{6.28(0.0047 \times 560 \times 10^{-12})^{1/2}}$
21. $\dfrac{1}{6.28(0.150 \times 680 \times 10^{-12})^{1/2}}$
22. $\dfrac{1}{6.28(100 \times 10^{-3} \times 20 \times 10^{-9})^{1/2}}$

ANSWERS TO SELF-TESTS

SELF-TEST 2–1

1. 10^3
2. 10^6
3. 10^{-3}
4. 10^{-7}
5. 10^{-4}
6. 10^{-5}
7. 1000
8. $100,000$
9. 100
10. 1
11. 0.0001
12. 0.1
13. 0.00001
14. 0.00000001
15. $\dfrac{1}{100} = 0.01$
16. $\dfrac{1}{1000} = 0.001$
17. $\dfrac{1}{100,000} = 0.00001$
18. $\dfrac{1}{10} = 0.1$

19. 10^7
20. 10^{-1}
21. 10^4
22. 10^{-9}
23. 10^{-3}
24. 10^{10}
25. 10^{-9}
26. 10^3
27. 10^5
28. 10
29. 10^{-3}
30. 10^{-5}
31. 10^8
32. 10^{-3}
33. 10^{-3}

SELF-TEST 2–2

1. 1.76×10^3
2. $6.75 \times 10^0 = 6.75$
3. $1.99 \times 10^0 = 1.99$
4. 4.35×10^7
5. 8.91×10^9
6. 5.26×10^4
7. 4.43×10^{-5}
8. 7.85×10^{-3}
9. 5.59×10^5
10. 5.18×10^{-2}
11. 2.08×10
12. 1.78×10^{-1}
13. 7.81×10^{-1}
14. 6.27×10^{-3}
15. $2.27 \times 10^0 = 2.27$
16. 3.24×10^{-4}
17. 2.82×10^4
18. $1.89 \times 10^0 = 1.89$
19. 2.56×10^4
20. 1.13×10^3

SELF-TEST 2–3

1. 10^{15}
2. 10^{-8}
3. 10^6
4. 10^{-3}
5. 10^2
6. 10^{-1}
7. 6.13×10^3
8. 8.58×10^{-8}
9. 7.74×10^{13}
10. 5.39×10^{-1}
11. 1.83×10^{-3}
12. 1.50×10^{-1}
13. 1.65×10^1
14. 1.63×10^{-2}
15. 2.72×10^{-2}
16. 6.79×10
17. 9.21×10^{-3}
18. 8.12
19. 4.77×10^4
20. 5.04×10^2

Units and Prefixes

3

Introduction

In this chapter the use of prefixes is introduced, and we show how the use of prefixes makes reading large and small quantities easy. In Chapter 2, we learned how to express answers in scientific notation. In this chapter, we will learn how to express answers in *engineering* notation. Engineering notation makes converting to prefixes easier, since both prefixes and engineering notation move the decimal point in increments of three. In the study of computers and electronics, we will find that prefixes are used to express the values of memory size, current, resistance, capacitance, frequency, and many other quantities. Because prefixes are so widely used, it is essential that you understand how to use them.

The metric system is the international system of weights and measurements. Because of the need for a single internationally accepted system, the metric system is gradually replacing the English system in this country.

In electronics, we already use the international system (SI) of measurements for most quantities, such as voltage, current, capacitance, and conductance, to name a few; but some measurements, such as distance, velocity, and weight, may be specified in either the English system or the metric system. For this reason, we must be able to convert from one system to the other.

Chapter Objectives

In this chapter you will learn how to:

1. Convert numbers to engineering notation.
2. Apply the various prefixes used in the electronics field.
3. Convert from one prefix to another.
4. Affix the correct prefix to electrical quantities.
5. Use the calculator in the engineering notation mode.
6. Apply the various prefixes used in the metric system.
7. Convert between the English and metric systems of measurements.

Most of the units we deal with are either very large or very small. To deal with these quantities more easily, prefixes are used. The prefixes relate to powers of ten that are multiples of 3 or -3. *Engineering notation* is the process of moving the decimal point in a number so that the resulting power of ten is changed to a multiple of 3 or -3. Some examples are 10^3, 10^6, 10^{-3}, and 10^{-6}. Once we have changed a number into engineering notation, we can replace the power of ten with the prefix that represents that power of ten.

Before we begin to use engineering notation and these prefixes, let's review powers of ten by changing a number to powers of ten notation with positive exponents. The rule is:

☞ **RULE 3-1** If we make the exponent more positive, the decimal point moves to the left.

Consider the number 6700. Let's change the number using powers of ten notation with positive exponents.

$$6700 = 670 \times 10^1 = 67 \times 10^2 = 6.7 \times 10^3 = 0.67 \times 10^4$$

and so on. Notice that as we make the exponent more positive, the decimal point moves farther left.

Now let's convert some numbers to powers of ten form using engineering notation. We will use Rule 3–1 and work some example problems.

EXAMPLE 3–1

Change 4700 to a number times 10^3.

SOLUTION We are changing the number to powers of ten form and we want the exponent to be 3, so the decimal point moves *left* three places.

$$4700 = 4.7 \times 10^3$$

EXAMPLE 3–2

Change 670×10^2 to a number times 10^3.

SOLUTION We can solve this two ways. First, we could change 670×10^2 to a regular number. $670 \times 10^2 = 67,000$. Then, applying the same rule as before, $67,000 = 67 \times 10^3$. Another approach would be to note that the exponent is already 2. The difference between 10^2 and 10^3 is 10^1. Therefore, we need to move the decimal point one place to the left.

$$670 \times 10^2 = 67 \times 10^3$$

EXAMPLE 3–3

Change 330,000 to some number times 10^3 and then to some number times 10^6.

SOLUTION Again, we move the decimal point three places to the left because the exponent is 3. Then we move the decimal point six places to the left because the exponent is 6.

$$330,000 = 330 \times 10^3$$
$$330,000 = 0.330 \times 10^6$$

Once we have 330×10^3, we can do this:

$$330 \times 10^3 = 0.330 \times 10^6$$

In this case, the exponent increases from 3 to 6, a change of 3 in a positive direction. Therefore, the decimal point moves three places to the left.

Now let's change numbers to powers of ten form with negative exponents. The rule is:

☞ **RULE 3–2** If the exponent is made more negative the decimal point moves to the right.

Consider the number 0.00275. Let's change to powers of ten notation using negative exponents and following Rule 3–2.

$$0.00275 = 0.0275 \times 10^{-1} = 0.275 \times 10^{-2} = 2.75 \times 10^{-3}$$

and so on. Notice that as we make the exponent more negative, the decimal point moves farther right.

Now, let's convert some numbers to powers of ten form using engineering notation. We will use Rule 3–2 and work some example problems.

EXAMPLE 3–4

Change 1.73 to some number times 10^{-3}.

SOLUTION The exponent is -3. Therefore, we move the decimal point three places to the *right*.

$$1.73 = 1730 \times 10^{-3}$$

EXAMPLE 3–5

Change 27.7×10^{-4} to some number times 10^{-6}.

SOLUTION We can solve the problem by changing 27.7×10^{-4} to a regular number and then moving the decimal point six places to the right.

$$27.7 \times 10^{-4} = 0.00277 = 2770 \times 10^{-6}$$

We could also solve the problem by noting that the exponent is already −4. Increasing its value to −6 is a change of 2 in a negative direction. Therefore, the decimal point moves two places to the right. Remember, if we make the exponent more negative, the decimal point moves to the *right*.

$$27.7 \times 10^{-4} = 2770 \times 10^{-6}$$

Now let's consider numbers for which we make the exponent more positive and then more negative.

EXAMPLE 3–6

Change 0.05×10^{-7} to some number times 10^{-9} and then to some number times 10^{-6}.

SOLUTION When we change the power of ten from 10^{-7} to 10^{-9}, we are making the exponent more negative. The difference is −2. That is, $10^{-7-2} = 10^{-9}$. Therefore, the decimal point must move to the *right*.

$$0.05 \times 10^{-7} = 5 \times 10^{-9}$$

Next we need to change the exponent so that it is −6. We must add 1 to −7. $-7 + 1 = -6$. That is, $10^{-7+1} = 10^{-6}$. According to our first rule, if we *add* 1 to the exponent, we must move the decimal point to the *left*.

$$0.05 \times 10^{-7} = 0.005 \times 10^{-6}$$

EXAMPLE 3–7

Change 4.7 to some number times 10^{3} and then to some number times 10^{-3}.

SOLUTION

$$4.7 = 0.0047 \times 10^{3}$$

Since the exponent is 3, the decimal point moves to the *left*.

$$4.7 = 4700 \times 10^{-3}$$

Since the exponent is −3, the decimal point moves to the *right*.

☞ **KEY POINT** Expressing numbers as numbers between 1 and 10 times the proper power of ten is called *scientific* notation. Expressing numbers as numbers times powers of ten that are multiples of 3 is called *engineering* notation.

PRACTICE PROBLEMS 3–1

Change the following numbers to numbers times 10^3 and to numbers times 10^6:

1. 27,000
2. 330,000
3. 5600
4. 390×10^2
5. 68×10^4
6. 1,200,000
7. 180×10^4
8. 1500×10^5
9. 1800×10^2
10. 51×10^5

Change the following numbers to numbers times 10^{-3} and to numbers times 10^{-6}.

11. 0.000423
12. 0.00716
13. 0.000014
14. 28.3×10^{-4}
15. 173×10^{-2}
16. 8.3×10^{-1}
17. 173×10^{-5}
18. 1.73×10^{-1}
19. 17.3×10^{-4}
20. 0.0706

Change the following numbers to numbers times 10^{-9} and to numbers times 10^{-12}:

21. 0.0046×10^{-4}
22. 0.413×10^{-7}
23. 6.73×10^{-10}
24. 6.73×10^{-8}
25. 67.3×10^{-11}
26. 0.0173×10^{-6}
27. 0.0563×10^{-8}
28. $76,500 \times 10^{-15}$
29. 6.43×10^{-10}
30. 906×10^{-10}

Change the following numbers to regular numbers, to numbers times 10^{-3}, and to numbers times 10^3:

31. 17×10^{-1}
32. 0.637×10^2
33. 716×10^{-2}
34. 47×10
35. 27.3×10^1
36. 3.93×10^2
37. 3.93×10^{-2}
38. 0.043×10^4
39. 0.706×10^{-1}
40. 89.1×10^2

SOLUTIONS

1. $27,000 = 27 \times 10^3 = 0.027 \times 10^6$
2. $330,000 = 330 \times 10^3 = 0.330 \times 10^6$
3. $5600 = 5.60 \times 10^3 = 0.0056 \times 10^6$
4. $390 \times 10^2 = 39 \times 10^3 = 0.039 \times 10^6$
5. $68 \times 10^4 = 680 \times 10^3 = 0.680 \times 10^6$
6. $1,200,000 = 1200 \times 10^3 = 1.20 \times 10^6$
7. $180 \times 10^4 = 1800 \times 10^3 = 1.8 \times 10^6$
8. $1500 \times 10^5 = 150,000 \times 10^3 = 150 \times 10^6$
9. $1800 \times 10^2 = 180 \times 10^3 = 0.180 \times 10^6$
10. $51 \times 10^5 = 5100 \times 10^3 = 5.1 \times 10^6$
11. $0.000423 = 0.423 \times 10^{-3} = 423 \times 10^{-6}$
12. $0.00716 = 7.16 \times 10^{-3} = 7160 \times 10^{-6}$

13. $0.000014 = 0.014 \times 10^{-3} = 14 \times 10^{-6}$

14. $28.3 \times 10^{-4} = 2.83 \times 10^{-3} = 2830 \times 10^{-6}$

15. $173 \times 10^{-2} = 1730 \times 10^{-3} = 1{,}730{,}000 \times 10^{-6}$

16. $8.3 \times 10^{-1} = 830 \times 10^{-3} = 830{,}000 \times 10^{-6}$

17. $173 \times 10^{-5} = 1.73 \times 10^{-3} = 1730 \times 10^{-6}$

18. $1.73 \times 10^{-1} = 173 \times 10^{-3} = 173{,}000 \times 10^{-6}$

19. $17.3 \times 10^{-4} = 1.73 \times 10^{-3} = 1730 \times 10^{-6}$

20. $0.0706 = 70.6 \times 10^{-3} = 70{,}600 \times 10^{-6}$

21. $0.0046 \times 10^{-4} = 460 \times 10^{-9} = 460{,}000 \times 10^{-12}$

22. $0.413 \times 10^{-7} = 41.3 \times 10^{-9} = 41{,}300 \times 10^{-12}$

23. $6.73 \times 10^{-10} = 0.673 \times 10^{-9} = 673 \times 10^{-12}$

24. $6.73 \times 10^{-8} = 67.3 \times 10^{-9} = 67{,}300 \times 10^{-12}$

25. $67.3 \times 10^{-11} = 0.673 \times 10^{-9} = 673 \times 10^{-12}$

26. $0.0173 \times 10^{-6} = 17.3 \times 10^{-9} = 17{,}300 \times 10^{-12}$

27. $0.0563 \times 10^{-8} = 0.563 \times 10^{-9} = 563 \times 10^{-12}$

28. $76{,}500 \times 10^{-15} = 0.0765 \times 10^{-9} = 76.5 \times 10^{-12}$

29. $6.43 \times 10^{-10} = 0.643 \times 10^{-9} = 643 \times 10^{-12}$

30. $906 \times 10^{-10} = 90.6 \times 10^{-9} = 90{,}600 \times 10^{-12}$

31. $17 \times 10^{-1} = 1.7 = 1700 \times 10^{-3} = 0.0017 \times 10^{3}$

32. $0.637 \times 10^{2} = 63.7 = 63{,}700 \times 10^{-3} = 0.0637 \times 10^{3}$

33. $716 \times 10^{-2} = 7.16 = 7160 \times 10^{-3} = 0.00716 \times 10^{3}$

34. $47 \times 10^{1} = 470 = 470{,}000 \times 10^{-3} = 0.47 \times 10^{3}$

35. $27.3 \times 10^{1} = 273 = 273{,}000 \times 10^{-3} = 0.273 \times 10^{3}$

36. $3.93 \times 10^{2} = 393 = 393{,}000 \times 10^{-3} = 0.393 \times 10^{3}$

37. $3.93 \times 10^{-2} = 0.0393 = 39.3 \times 10^{-3} = 0.0000393 \times 10^{3}$

38. $0.043 \times 10^{4} = 430 = 430{,}000 \times 10^{-3} = 0.43 \times 10^{3}$

39. $0.706 \times 10^{-1} = 0.0706 = 70.6 \times 10^{-3} = 0.0000706 \times 10^{3}$

40. $89.1 \times 10^{2} = 8910 = 8{,}910{,}000 \times 10^{-3} = 8.91 \times 10^{3}$

Additional practice problems are at the end of the chapter.

3–2   PREFIXES

Table 3–1 is a partial list of the International System of Units (SI). Only those units of interest to electronics students are included. Prefixes used with these electrical units are listed in Table 3–2. Prefixes are used to make it easier for us to communicate with each other. For example, suppose the value of a capacitance was 0.000000003 F. This is a very clumsy number to work with. If we express it in scientific notation, we get 3×10^{-9} F. That's better, but still awkward. Looking at our table of prefixes, we see that 10^{-9} is a *nano* unit and the symbol is n (lowercase n). We would write the value as 3 nF. "Three nanofarads" is the way we would say it.

2,700,000 Ω would be simpler to express if we used the prefix equal to 10^{6}, which is *mega* and is symbolized as M.

$$2{,}700{,}000 \ \Omega = 2.7 \times 10^{6} \ \Omega = 2.7 \ \text{M}\Omega$$

Table 3–1   Partial List of the International System of Units (SI).

SI PHYSICAL CHARACTERISTIC	NAME OF SI UNIT	SYMBOL
Capacitance (C)	farad	F
Admittance (Y)	siemens	S
Conductance (G)	siemens	S
Current (I, i)	ampere	A
Energy (w)	joule	J
Frequency (f)	hertz	Hz
Impedance (Z)	ohm	Ω
Inductance (L)	henry	H
Power (P)	watt	W
Reactance (X)	ohm	Ω
Resistance (R)	ohm	Ω
Susceptance (B)	siemens	S
Time (t)	second	s
Wavelength (λ)	meter	m
Electromotive force (E)	volt	V
Difference in potential (V)	volt	V

Suppose we calculate a circuit current to be 5.6×10^{-4} A. We need to replace the power of ten with some prefix to make the quantity easier to read. Since our prefixes are multiples of 10^3 or 10^{-3}, we have to change 10^{-4} to either 10^{-3} or 10^{-6}. Let's do both. First change 5.6×10^{-4} to a number times 10^{-3}.

$$5.6 \times 10^{-4} \text{ A} = 0.56 \times 10^{-3} \text{ A}$$

Table 3–2   Prefixes Used with Electrical Units Listed in Table 3–1.

PREFIX	SYMBOL	POWER OF TEN
Exa	E	10^{18}
Peta	P	10^{15}
Tera	T	10^{12}
Giga	G	10^{9}
Mega	M	10^{6}
kilo	k	10^{3}
Units	none	10^{0}
milli	m	10^{-3}
micro	μ	10^{-6}
nano	n	10^{-9}
pico	p	10^{-12}
femto	f	10^{-15}
atto	a	10^{-18}

To change the exponent from -4 to -3, we must add one ($10^{-4+1} = 10^{-3}$). Because we made the exponent more positive, the decimal point moved to the *left* (Rule 3–1). The prefix for 10^{-3} is *milli* (m). Then

$$5.6 \times 10^{-4}\,\text{A} = 0.56 \times 10^{-3}\,\text{A} = 0.56\ \text{mA}$$

Now let's change 5.6×10^{-4} to a number times 10^{-6}.

$$5.6 \times 10^{-4}\,\text{A} = 560 \times 10^{-6}\,\text{A}$$

To change the exponent from -4 to -6, we had to add -2. $10^{-4+(-2)} = 10^{-6}$. Because we made the exponent more negative, the decimal point moved to the *right* (Rule 3–2). The prefix for 10^{-6} is *micro* (the Greek letter μ is the symbol).

$$5.6 \times 10^{-4}\,\text{A} = 560 \times 10^{-6}\,\text{A} = 560\ \mu\text{A}$$

PRACTICE PROBLEMS 3–2

Using the indicated prefixes, change the following quantities:

1. $2700\ \Omega =$ _____ $\text{k}\Omega =$ _____ $\text{M}\Omega$
2. $12{,}000\ \text{Hz} =$ _____ $\text{kHz} =$ _____ MHz
3. $0.00076\ \text{A} =$ _____ $\text{mA} =$ _____ μA
4. $0.000023\ \text{A} =$ _____ $\text{mA} =$ _____ μA
5. $0.0000002\ \text{F} =$ _____ $\mu\text{F} =$ _____ nF
6. $0.000004\ \text{F} =$ _____ $\mu\text{F} =$ _____ nF
7. $68{,}000\ \Omega =$ _____ $\text{k}\Omega =$ _____ $\text{M}\Omega$
8. $120{,}000\ \Omega =$ _____ $\text{k}\Omega =$ _____ $\text{M}\Omega$
9. $3{,}500{,}000\ \text{Hz} =$ _____ $\text{kHz} =$ _____ MHz
10. $0.00037\ \text{S} =$ _____ $\text{mS} =$ _____ μS
11. $5 \times 10^{-4}\ \text{S} =$ _____ $\text{mS} =$ _____ μS
12. $5.6 \times 10^{4}\ \Omega =$ _____ $\text{k}\Omega =$ _____ $\text{M}\Omega$
13. $21 \times 10^{-2}\ \text{A} =$ _____ $\text{mA} =$ _____ μA
14. $68 \times 10^{-10}\ \text{F} =$ _____ $\text{nF} =$ _____ pF
15. $45 \times 10^{-4}\ \text{H} =$ _____ $\text{mH} =$ _____ μH
16. $0.15 \times 10^{-2}\ \text{V} =$ _____ $\text{mV} =$ _____ μV

SOLUTIONS

1. $2700\ \Omega = 2.7\ \text{k}\Omega = 0.0027\ \text{M}\Omega$
2. $12{,}000\ \text{Hz} = 12\ \text{kHz} = 0.012\ \text{MHz}$
3. $0.00076\ \text{A} = 0.76\ \text{mA} = 760\ \mu\text{A}$
4. $0.000023\ \text{A} = 0.023\ \text{mA} = 23\ \mu\text{A}$
5. $0.0000002\ \text{F} = 0.2\ \mu\text{F} = 200\ \text{nF}$
6. $0.000004\ \text{F} = 4\ \mu\text{F} = 4000\ \text{nF}$
7. $68{,}000\ \Omega = 68\ \text{k}\Omega = 0.068\ \text{M}\Omega$
8. $120{,}000\ \Omega = 120\ \text{k}\Omega = 0.12\ \text{M}\Omega$
9. $3{,}500{,}000\ \text{Hz} = 3500\ \text{kHz} = 3.5$ MHz
10. $0.00037\ \text{S} = 0.37\ \text{mS} = 370\ \mu\text{S}$
11. $5 \times 10^{-4}\ \text{S} = 0.5\ \text{mS} = 500\ \mu\text{S}$
12. $5.6 \times 10^{4}\ \Omega = 56\ \text{k}\Omega = 0.056\ \text{M}\Omega$

13. 21×10^{-2} A $= 210$ mA $= 210,000$ μA

14. 68×10^{-10} F $= 6.8$ nF $= 6800$ pF

15. 45×10^{-4} H $= 4.5$ mH $= 4500$ μH

16. 0.15×10^{-2} V $= 1.5$ mV $= 1500$ μV

Additional practice problems are at the end of the chapter.

CONVERTING NUMBERS WITH PREFIXES TO BASIC UNITS | 3-3

Next, let's change quantities to basic units. The rule is:

☞ **RULE 3-3** To change numbers with prefixes to regular numbers, replace the prefix with the power of ten that it represents and then change to a regular number.

Some examples are:

$$2200 \ \mu S = 2200 \times 10^{-6} \ S = 0.0022 \ S$$

$$68 \ k\Omega = 68 \times 10^{3} \ \Omega = 68,000 \ \Omega$$

$$30 \ mA = 30 \times 10^{-3} \ A = 0.03 \ A$$

$$25 \ kHz = 25 \times 10^{3} \ Hz = 25,000 \ Hz$$

PRACTICE PROBLEMS 3-3

Perform the indicated operations:

1. 46.7 mA = _____ A
3. 68 kΩ = _____ Ω
5. 2.73 kHz = _____ Hz
7. 670 μS = _____ S
9. 55 mH = _____ H
11. 120 kΩ = _____ Ω
13. 150 mA = _____ A
15. 0.68 kΩ = _____ Ω

2. 407 mA = _____ A
4. 4.7 kΩ = _____ Ω
6. 300 kHz = _____ Hz
8. 37 mS = _____ S
10. 465 μH = _____ H
12. 2 MΩ = _____ Ω
14. 0.43 mA = _____ A

SOLUTIONS

1. 46.7 mA = 0.0467 A
3. 68 kΩ = 68,000 Ω
5. 2.73 kHz = 2730 Hz
7. 670 μS = 0.000670 S
9. 55 mH = 0.055 H
11. 120 kΩ = 120,000 Ω

2. 407 mA = 0.407 A
4. 4.7 kΩ = 4700 Ω
6. 300 kHz = 300,000 Hz
8. 37 mS = 0.037 S
10. 465 μH = 0.000465 H
12. 2 MΩ = 2,000,000 Ω

13. 150 mA = 0.150 A   14. 0.43 mA = 0.00043 A
15. 0.68 kΩ = 680 Ω

Additional practice problems are at the end of the chapter.

3-4   CHANGING PREFIXES

Now let's change from one prefix to another.

EXAMPLE 3-8

$$20 \text{ mA} = \underline{\hspace{3cm}} \mu A$$

SOLUTION   In changing from mA to μA, we are really changing the exponent from -3 to -6; so let's rewrite the problem substituting powers of ten for prefixes.

$$20 \times 10^{-3} = \underline{\hspace{3cm}} \times 10^{-6}$$

We are making the exponent more *negative*. Remember Rule 3–2: *If we make the exponent more negative, the decimal point moves to the right*. The change in the value of the exponent from 10^{-3} to 10^{-6} is a change of -3. Therefore, the decimal point moves *three places to the right*.

$$20 \text{ mA} = 20,000 \ \mu A$$

EXAMPLE 3-9

Change 450 μA to mA.

SOLUTION   In changing from μA to mA, we are really changing the exponent from -6 to -3; so let's rewrite the problem substituting powers of ten for prefixes.

$$450 \times 10^{-6} \text{ A} = \underline{\hspace{3cm}} \times 10^{-3} \text{ A}$$

We are making the exponent more *positive*. Remember Rule 3–1: *If we make the exponent more positive, the decimal point moves to the left*. The change in the value of the exponent from 10^{-6} to 10^{-3} is a change of $+3$. Therefore, the decimal point moves *three places to the left*.

$$450 \times 10^{-6} \text{ A} = 0.450 \times 10^{-3} \text{ A}$$

$$450 \ \mu A = 0.450 \text{ mA}$$

EXAMPLE 3-10

Change 680 kΩ to MΩ.

SOLUTION   Substituting powers of ten for prefixes, we get

$$680 \times 10^3 \ \Omega = \underline{\hspace{3cm}} \times 10^6 \ \Omega$$

CHAPTER 3

In going from kΩ to MΩ, 10^3 Ω to 10^6 Ω, we are making the exponent more positive. Therefore, the decimal point moves to the *left*. The change from 10^3 to 10^6 is a change in the exponent of $+3$. Using Rule 3–1, the decimal point moves three places to the left.

$$680 \text{ k}\Omega = 0.680 \text{ M}\Omega$$

EXAMPLE 3–11

Change 0.01 μF to pF.

SOLUTION   Substituting powers of ten for prefixes, we get

$$0.01 \times 10^{-6} \text{ F} = \underline{\hspace{2cm}} \times 10^{-12} \text{ F}$$

In going from μF to pF, 10^{-6} to 10^{-12}, we are making the exponent more negative. Therefore, the decimal point moves to the *right*. The change from 10^{-6} to 10^{-12} is a change in the exponent of -6. Using Rule 3–2, the decimal point moves six places to the right.

$$0.01 \ \mu\text{F} = 10,000 \text{ pF}$$

In the following practice problems you will either change prefixes or change to basic units.

PRACTICE PROBLEMS 3–4

Perform the indicated operations.

1. 20 mA = _____ μA = _____ A
2. 0.01 μF = _____ nF = _____ pF
3. 2 H = _____ mH = _____ μH
4. 2.5 kΩ = _____ Ω = _____ MΩ
5. 680 Ω = _____ kΩ = _____ MΩ
6. 1.8 V = _____ mV = _____ μV
7. 20 μA = _____ mA = _____ A
8. 1.5 MΩ = _____ kΩ = _____ Ω
9. 0.02 S = _____ mS = _____ μS
10. 50 nF = _____ μF = _____ pF
11. 4.7×10^4 Ω = _____ kΩ = _____ MΩ
12. 16×10^{-7} F = _____ μF = _____ nF
13. 2.4×10^{-2} A = _____ mA = _____ μA
14. 8.4×10^2 Ω = _____ kΩ = _____ MΩ
15. 56×10^5 Ω = _____ kΩ = _____ MΩ
16. 27.3 mS = _____ S = _____ μS
17. 5600 μS = _____ S = _____ mS
18. 170 mW = _____ W = _____ μW

19. 0.000642 V = _____ μV = _____ mV

20. 30 kHz = _____ MHz = _____ Hz

SOLUTIONS

1. 20 mA = 20,000 μA = 0.020 A
2. 0.01 μF = 10 nF = 10,000 pF
3. 2 H = 2000 mH = 2,000,000 μH
4. 2.5 kΩ = 2500 Ω = 0.0025 MΩ
5. 680 Ω = 0.680 kΩ = 0.00068 MΩ
6. 1.8 V = 1800 mV = 1,800,000 μV
7. 20 μA = 0.020 mA = 0.000020 A
8. 1.5 MΩ = 1500 kΩ = 1,500,000 Ω
9. 0.02 S = 20 mS = 20,000 μS
10. 50 nF = 0.050 μF = 50,000 pF
11. 4.7×10^4 Ω = 47 kΩ = 0.047 MΩ
12. 16×10^{-7} F = 1.6 μF = 1600 nF
13. 2.4×10^{-2} A = 24 mA = 24,000 μA
14. 8.4×10^2 Ω = 0.84 kΩ = 0.00084 MΩ
15. 56×10^5 Ω = 5600 kΩ = 5.6 MΩ
16. 27.3 mS = 0.0273 S = 27,300 μS
17. 5600 μS = 0.00560 S = 5.60 mS
18. 170 mW = 0.170 W = 170,000 μW
19. 0.000642 V = 642 μV = 0.642 mV
20. 30 kHz = 0.030 MHz = 30,000 Hz

Additional practice problems are at the end of the chapter.

SELF-TEST 3–1

Express your answers in scientific notation:

1. 4760 = _____ $\times 10^3$ = _____ $\times 10^6$
2. 32.4×10^4 = _____ $\times 10^3$ = _____ $\times 10^6$
3. 2.71×10^{-1} = _____ $\times 10^{-3}$ = _____ $\times 10^3$
4. 46.7×10^{-2} = _____ $\times 10^{-3}$ = _____ $\times 10^3$
5. 76.3 mA = _____ A = _____ μA
6. 0.0055 μF = _____ nF = _____ pF
7. 8340 μS = _____ mS = _____ S
8. 20 kΩ = _____ Ω = _____ MΩ
9. 75,000 Hz = _____ MHz = _____ kHz
10. 146 mS = _____ S = _____ μS

Answers to Self-test 3–1 are at the end of the chapter.

3–5 APPLICATIONS

In problem solving we perform many different operations and express answers in many different forms. In this section we will do some of these operations and express answers by using prefixes whenever practical.

For example, when resistors are connected in parallel, we usually convert from resistance to conductance for problem-solving purposes. (Refer to Table 3–1 for the symbols and units of resistance and conductance.)

The calculator instructions shown in this chapter are for the T1-36X Solar calculator. If you have a Casio calculator, then remember these differences:

To calculate reciprocals with the TI use the $\boxed{1/x}$ key, with the Casio use $\boxed{x^{-1}}$ $\boxed{=}$.

To set the exponent for powers of ten with the TI use the $\boxed{EE}$ key, with the Casio use the $\boxed{EXP}$ key.

To set up the display mode for either calculator, please refer to Example 2-12 in Chapter 2.

EXAMPLE 3–12

Two resistors are in parallel. $R_1 = 27 \text{ k}\Omega$ and $R_2 = 18 \text{ k}\Omega$. Find the total conductance (G_T) and the total resistance (R_T). Round answers to three places.

SOLUTION To find the total resistance, we use the equation

$$R_T = \frac{1}{G_T}$$

where

$$G_T = G_1 + G_2$$

$$G_1 = \frac{1}{R_1}$$

$$G_2 = \frac{1}{R_2}$$

Since

$$G_1 = \frac{1}{R_1} \quad \text{and} \quad G_2 = \frac{1}{R_2}$$

we can write the equation like this:

$$R_T = \frac{1}{G_1 + G_2} = \frac{1}{\dfrac{1}{R_1} + \dfrac{1}{R_2}}$$

$$= \frac{1}{\dfrac{1}{27 \text{ k}\Omega} + \dfrac{1}{18 \text{ k}\Omega}} = \frac{1}{37.0 \text{ } \mu\text{S} + 55.6 \text{ } \mu\text{S}}$$

$$= \frac{1}{92.6 \text{ } \mu\text{S}} = 10.8 \text{ k}\Omega$$

The total conductance is 92.6 μS. The total resistance is 10.8 kΩ.

Now work the problem:

$$27 \boxed{\text{EE}} 3 \boxed{1/x} \qquad \frac{1}{R_1} = 37.0 \times 10^{-6}\,\text{S} = 37.0\,\mu\text{S}$$

$$\boxed{+}$$

$$18 \boxed{\text{EE}} 3 \boxed{1/x} \qquad \frac{1}{R_2} = 55.6 \times 10^{-6}\,\text{S} = 55.6\,\mu\text{S}$$

$$\boxed{=}$$

Total conductance.
$$G_T = 92.6 \times 10^{-6} = 92.6\,\mu\text{S}$$

$$\boxed{1/x}$$

Total resistance.
$$R_T = 10.8 \times 10^{3}\,\Omega = 10.8\,\text{k}\Omega$$

Example 3–13 offers an alternate method of solution.

EXAMPLE 3–13

When two resistors are connected in parallel, their total resistance can be calculated with the equation:

$$R_T = \frac{R_1 R_2}{R_1 + R_2} \qquad (3\text{–}3)$$

If $R_1 = 10\,\text{k}\Omega$ and $R_2 = 5\,\text{k}\Omega$, what is R_T in ohms? in kohms?

SOLUTION

$$R_T = \frac{R_1 R_2}{R_1 + R_2} = \frac{10\text{k} \times 5\text{k}}{10\text{k} + 5\text{k}}\,\Omega = \frac{10 \times 10^{3} \times 5 \times 10^{3}}{10 \times 10^{3} + 5 \times 10^{3}}\,\Omega = \frac{50 \times 10^{6}}{15 \times 10^{3}}\,\Omega$$

$$= \frac{50}{15} \times 10^{3}\,\Omega = 3.33 \times 10^{3}\,\Omega = 3.33\,\text{k}\Omega$$

EXAMPLE 3–14

Three resistors are connected in parallel. $R_1 = 470\,\Omega$, $R_2 = 220\,\Omega$, and $R_3 = 680\,\Omega$. Find the total conductance and the total resistance. Round answers to three places.

SOLUTION

$$R_T = \frac{1}{G_T} = \frac{1}{G_1 + G_2 + G_3} = \frac{1}{\dfrac{1}{R_1} + \dfrac{1}{R_2} + \dfrac{1}{R_3}} \qquad (3\text{–}4)$$

Since $R_1 = 470 \ \Omega$, $R_2 = 220 \ \Omega$, and $R_3 = 680 \ \Omega$, then

$$R_T = \frac{1}{\dfrac{1}{470 \ \Omega} + \dfrac{1}{220 \ \Omega} + \dfrac{1}{680 \ \Omega}} = \frac{1}{2.13 \text{ mS} + 4.55 \text{ mS} + 1.47 \text{ mS}}$$

$$= \frac{1}{8.14 \text{ mS}} = 123 \ \Omega \quad \text{(rounded to three places)}$$

The total conductance is 8.14 mS. The total resistance is 123 Ω.

With most calculators, all operations can be performed without having to store partial answers. Only the final answer need be written down:

470 $\boxed{1/x}$ + 220 $\boxed{1/x}$ + 680 $\boxed{1/x}$ $\boxed{=}$ *Answer:* $G_T = 8.14$ mS

                                      $\boxed{1/x}$ *Answer:* $R_T = 123 \ \Omega$

To add units together when the prefixes are different, it is necessary to convert so that all prefixes are the same. For example, let's add 6300 μS and 15.3 mS. We cannot add milliunits and microunits together, just as we could not add 10^{-3} and 10^{-6} together. We must convert 6300 μS to milliunits or convert 15.3 mS to microunits.

$$6300 \ \mu\text{S} + 15.3 \text{ mS} =$$
$$6.30 \text{ mS} + 15.3 \text{ mS} = 21.6 \text{ mS}$$

or

$$6300 \ \mu\text{S} + 15.3 \text{ mS} =$$
$$6300 \ \mu\text{S} + 15,300 \ \mu\text{S} = 21,600 \ \mu\text{S}$$

In this case, converting to milliunits is better because it is easier to work with that number.

Example 3–15 offers an alternate solution.

EXAMPLE 3–15

When three resistors are connected in parallel, their total resistance can be calculated with the equation:

$$R_T = \frac{R_1 R_2 R_3}{R_1 R_2 + R_1 R_3 + R_2 R_3} \tag{3–5}$$

If $R_1 = 10 \text{ k}\Omega$, $R_2 = 5 \text{ k}\Omega$, and $R_3 = 10 \text{ k}\Omega$, what is R_T in ohms? in kohms?

$$R_T = \frac{R_1 R_2 R_3}{R_1 R_2 + R_1 R_3 + R_2 R_3} = \frac{10k \times 5k \times 10k}{10k \times 5k + 10k \times 10k + 5k \times 10k} \, \Omega$$

$$= \frac{10 \times 10^3 \times 5 \times 10^3 \times 10 \times 10^3}{10 \times 10^3 \times 5 \times 10^3 + 10 \times 10^3 \times 10 \times 10^3 + 5 \times 10^3 \times 10 \times 10^3} \, \Omega$$

$$= \frac{500 \times 10^9}{50 \times 10^6 + 100 \times 10^6 + 50 \times 10^6} \, \Omega = \frac{500 \times 10^9}{200 \times 10^6} \, \Omega$$

$$= 2.5 \times 10^3 \Omega = 2.5 \, k\Omega$$

PRACTICE PROBLEMS 3–5

Perform the indicated operations:

1. 27 mA + 0.037 A = _____ A = _____ mA

2. 370 μS + 0.060 mS = _____ mS = _____ μS

3. $\dfrac{1}{68 \text{ k}\Omega} + \dfrac{1}{47 \text{ k}\Omega}$ = _____ mS = _____ μS

4. $\dfrac{1}{\dfrac{1}{4.7 \text{ k}\Omega} + \dfrac{1}{1.8 \text{ k}\Omega}}$ = _____ Ω = _____ kΩ

5. $\dfrac{20 \text{ V}}{27 \text{ k}\Omega}$ = _____ A = _____ mA = _____ μA

6. $\dfrac{25 \text{ V}}{2.36 \text{ mA}}$ = _____ Ω = _____ kΩ = _____ MΩ

7. $\dfrac{1}{6.28 \times 12 \text{ kHz} \times 50 \text{ nF}}$ = _____ Ω = _____ kΩ

8. $\dfrac{1}{2.7 \text{ k}\Omega} + \dfrac{1}{3.3 \text{ k}\Omega} + \dfrac{1}{4.7 \text{ k}\Omega}$ = _____ mS = _____ μS = _____ S

9. $\dfrac{1}{\dfrac{1}{22 \text{ k}\Omega} + \dfrac{1}{33 \text{ k}\Omega} + \dfrac{1}{56 \text{ k}\Omega}}$ = _____ Ω = _____ kΩ = _____ MΩ

10. $\dfrac{1}{6340 \text{ }\mu\text{S}}$ = _____ Ω = _____ kΩ

11. 2.73 kΩ × 0.43 mA = _____ V = _____ mV

12. 1.27 V + 48 mV + 5630 μV = _____ V = _____ mV = _____ μV

When two resistors are connected in parallel, their total resistance can be calculated. Using equation (3–2) or (3–3):

13. If $R_1 = 2 \text{ M}\Omega$ and $R_2 = 2 \text{ M}\Omega$, what is R_T in kohms and in MΩ?

14. If $R_1 = 2 \text{ M}\Omega$ and $R_2 = 4 \text{ M}\Omega$, what is R_T in kohms and in MΩ?

When three resistors are connected in parallel, their total resistance can be calculated. Using equation (3–4) or (3–5):

15. If $R_1 = 10$ kΩ, $R_2 = 20$ kΩ, and $R_3 = 30$ kΩ, what is R_T in ohms and in kohms?
16. If $R_1 = 10$ kΩ, $R_2 = 20$ kΩ, and $R_3 = 30$ MΩ, what is R_T in kohms and in Mohms?

SOLUTIONS

1. 27 mA + 0.037 A = 0.064 A = 64 mA
2. 370 μS + 0.060 mS = 0.430 mS = 430 μS
3. $\dfrac{1}{68 \text{ kΩ}} + \dfrac{1}{47 \text{ kΩ}} = 0.036$ mS = 36 μS
4. $\dfrac{1}{\dfrac{1}{4.7 \text{ kΩ}} + \dfrac{1}{1.8 \text{ kΩ}}} = 1300$ Ω = 1.30 kΩ
5. $\dfrac{20 \text{ V}}{27 \text{ kΩ}} = 0.000741$ A = 0.741 mA = 741 μA
6. $\dfrac{25 \text{ V}}{2.36 \text{ mA}} = 10,600$ Ω = 10.6 kΩ = 0.0106 MΩ
7. $\dfrac{1}{6.28 \times 12 \text{ kHz} \times 50 \text{ nF}} = 265$ Ω = 0.265 kΩ
8. $\dfrac{1}{2.7 \text{ kΩ}} + \dfrac{1}{3.3 \text{ kΩ}} + \dfrac{1}{4.7 \text{ kΩ}} = 0.886$ mS = 886 μS = 0.000886 S
9. $\dfrac{1}{\dfrac{1}{22 \text{ kΩ}} + \dfrac{1}{33 \text{ kΩ}} + \dfrac{1}{56 \text{ kΩ}}} = 10,700$ Ω = 10.7 kΩ = 0.0107 MΩ
10. $\dfrac{1}{6340 \text{ μS}} = 158$ Ω = 0.158 kΩ
11. 2.73 kΩ × 0.43 mA = 1.17 V = 1170 mV
12. 1.27 V + 48 mV + 5630 μV = 1.32 V = 1320 mV = 1,320,000 μV
13. 1000 kohms = 1 MΩ
14. 1330 kohms = 1.33 MΩ
15. 5450 Ω = 5.45 kΩ
16. 6.67 kΩ = 0.00667 MΩ

Additional practice problems are at the end of the chapter.

SYSTEMS OF MEASUREMENT   3-6

The systems of measurement with which we must become familiar are the metric system and the English system. The metric system uses the meter for linear measurement and the gram for mass or weight. The English system uses the yard for linear measurement and the pound for weight.

The metric system is the international system of weights and measurements. Because of the need for a single, internationally accepted system, the metric system is gradually replacing the English system in this country.

In electronics, we already use the international system (SI) of measurements for most quantities, such as voltage, current, capacitance, and conductance, to name a few; but some measurements, such as distance, velocity, and weight, may be specified in either the English system or the metric system. For this reason, we must be able to convert from one system to the other.

The most commonly used units of linear measurement in metric are the millimeter (mm), centimeter (cm), meter (m), and kilometer (km). The relationships between these units are

$$1 \text{ millimeter} = 10^{-3} \text{ meter} = 0.001 \text{ meter}$$

$$1 \text{ centimeter} = 10^{-2} \text{ m} = 0.01 \text{ m}$$

$$1 \text{ kilometer} = 10^{3} \text{ m} = 1000 \text{ m}$$

In the English system, we have

$$12 \text{ inches (in)} = 1 \text{ foot (ft)}$$

$$3 \text{ ft} = 1 \text{ yard (yd)}$$

$$5280 \text{ ft} = 1760 \text{ yd} = 1 \text{ mile (mi)}$$

To convert from English to metric,

$$1 \text{ in} = 2.540 \text{ cm}$$

$$1 \text{ yd} = 0.9144 \text{ m}$$

$$1 \text{ mi} = 1.609 \text{ km}$$

To convert from metric to English,

$$1 \text{ cm} = 0.3937 \text{ in}$$

$$1 \text{ m} = 39.37 \text{ in}$$

$$1 \text{ m} = 1.094 \text{ yd}$$

$$1 \text{ km} = 0.6214 \text{ mi}$$

EXAMPLE 3–16

Convert the following: (a) 14 cm to inches; (b) 40 cm to inches; (c) 1.7 m to inches and yards; (d) 3 km to yards and miles. Round answers to four places.

SOLUTIONS

(a) $14 \text{ cm} \times 0.3937 \text{ in/cm} = 5.512 \text{ in}$
(b) $40 \text{ cm} \times 0.3937 \text{ in/cm} = 15.75 \text{ in}$
(c) $1.7 \text{ m} \times 39.37 \text{ in/m} = 66.93 \text{ in}$
     $1.7 \text{ m} \times 1.094 \text{ yd/m} = 1.860 \text{ yd}$

(d) 3000 m $\times$ 1.094 yd/m = 3282 yd

3000 m = 3 km; 3 km $\times$ 0.6214 mi/km = 1.864 mi

EXAMPLE 3–17

Convert the following: (a) 20 in to cm; (b) 60 in to cm and m; (c) 100 yd to m; (d) 2000 yd to m and km. Round answers to four places.

SOLUTIONS

(a) 20 in $\times$ 2.54 cm/in = 50.80 cm

(b) 60 in $\times$ 2.54 cm/in = 152.4 cm = 1.524 m

(c) 100 yd $\times$ 0.9144 m/yd = 91.44 m

(d) 2000 $\times$ 0.9144 m/yd = 1829 m = 1.829 km

The units of velocity are meters/second or kilometers/hour in metric and feet/second or miles/hour in English.

EXAMPLE 3–18

Convert the following: (a) 300 meters/second (m/s) to feet/second (ft/s); (b) 35 miles/hour (mi/h) to kilometers/hour (km/h); (c) 700 feet/second (ft/s) to meters/second (m/s); (d) 108 kilometers/hour (km/h) to miles/hour (mi/h). Round answers to four places.

SOLUTIONS

(a) 39.37 in = 1 m

39.37 in/m $\div$ 12 in/ft = 3.281 ft/m

3.281 ft/m $\times$ 300 m/s = 984.3 ft/s

(b) 35 mi/h $\times$ 1.609 km/mi = 56.32 km/h

(c) 0.3048 ft = 1 m

700 ft/s $\times$ 0.3048 m/ft = 213.4 m/s

(d) 108 km/h $\times$ 0.6214 mi/km = 67.11 mi/h

The most commonly used units of mass or weight in metric are the milligram, gram, and kilogram. In the English system we have the ounce, pound, and ton.

$$16 \text{ ounces (oz)} = 1 \text{ pound (lb)}$$

$$2000 \text{ pounds} = 1 \text{ ton (t)}$$

To convert from metric to English,

$$1 \text{ gram (g)} = 0.03527 \text{ ounces}$$

$$1 \text{ kilogram (kg)} = 35.27 \text{ ounces} = 2.205 \text{ pounds}$$

To convert from English to metric,

$$1 \text{ ounce} = 28.35 \text{ grams}$$

$$1 \text{ pound} = 453.6 \text{ grams} = 0.4536 \text{ kilogram}$$

EXAMPLE 3–19

Convert the following: (a) 30 g to ounces; (b) 700 g to ounces and pounds; (c) 2.35 kg to ounces and pounds; (d) 0.255 oz to grams; (e) 7.5 oz to kilograms; (f) 1.5 lb to kilograms. Round answers to four places.

SOLUTIONS

(a) 30 g × 0.03527 oz/g = 1.058 oz
(b) 700 g × 0.03527 oz/g = 24.69 oz
  24.69 oz ÷ 16 oz/lb = 1.543 lb
(c) 2.35 kg × 35.27 oz/kg = 82.88 oz
  2.35 kg × 2.204 lb/kg = 5.180 lb
(d) 0.255 oz × 28.35 g/oz = 7.229 g
(e) 7.5 oz × 0.02835 kg/oz = 0.2126 kg (1 oz = 28.35 g = 0.02835 kg)
(f) 1.5 lb × 0.4536 kg/lb = 0.6804 kg

PRACTICE PROBLEMS 3–6

Make the following conversions. Round answers to four places.

1. 27 centimeters to inches
2. 3.4 meters to inches and yards
3. 5000 meters to yards and miles
4. 27 inches to centimeters
5. 440 yards to meters and kilometers
6. 2 miles to meters and kilometers
7. 400 meters/second to feet/second
8. 200 kilometers/hour to miles/hour
9. 300 feet/second to meters/second
10. 65 miles/hour to kilometers/hour
11. 70 grams to ounces
12. 1000 grams to ounces and pounds
13. 3.2 kilograms to ounces and pounds
14. 2 pounds to kilograms
15. 2.45 ounces to grams

SOLUTIONS

1. 10.63 in
2. 133.9 in = 3.718 yd
3. 5470 yd = 3.107 mi
4. 68.58 cm
5. 402.3 m = 0.4023 km
6. 3.219 km = 3219 m
7. 1312 ft/s
8. 124.3 mi/h
9. 91.44 m/s
10. 104.6 km/h
11. 2.469 oz
12. 35.27 oz = 2.205 lb
13. 112.9 oz = 7.055 lb
14. 0.9072 kg
15. 69.46 g

Additional practice problems are at the end of the chapter.

CHAPTER 3 AT A GLANCE

PAGE	KEY POINT OR RULE	EXAMPLE
82	*Rule 3–1:* If we make the exponent more positive, the decimal point moves to the *left*.	$6700 = 670 \times 10$ $= 67.0 \times 10^2$ $= 6.70 \times 10^3$
83	*Rule 3–2:* If the exponent is made more negative, the decimal point moves to the *right*.	$0.00275 = 0.0275 \times 10^{-1}$ $= 0.275 \times 10^{-2}$ $= 2.75 \times 10^{-3}$
84	*Key Point:* Expressing numbers as numbers between 1 and 10 times the proper power of ten is called *scientific* notation. Expressing numbers as numbers times powers of ten that are multiples of 3 is called *engineering* notation.	$47,000 = 4.70 \times 10^4$ in scientific notation to three places. $47,000 = 47.0 \times 10^3$ in engineering notation to three places.
89	*Rule 3–3:* To change numbers with prefixes to regular numbers, replace the prefix with the power of ten that it represents and then change to a regular number.	$30 \, \text{mA} = 30 \times 10^{-3} \, \text{A}$ $= 0.030 \, \text{A}$ $33 \, \text{k}\Omega = 33 \times 10^3 \, \Omega$ $= 33,000 \, \Omega$
98	Some conversions of linear measure between English and metric are: 1 in = 2.54 cm 1 cm = 0.3937 in 1 yd = 0.9144 m 1 m = 1.094 yd 1 mi = 1.609 km 1 km = 0.6214 mi	20 cm = 7.874 in 12 in = 30.48 cm 100 m = 109.4 yd 5 mi = 8.045 km
99	The units of velocity are meters/second or kilometers/hour in metric and feet/second or miles/hour in English.	15 m/s = 49.22 ft/s
99	Common units of weight in metric are the milligram, gram, and kilogram. In the English system, we have the ounce, pound, and ton. 16 oz = 1 lb 1 kg = 35.27 oz 2000 lb = 1 ton 1 lb = 453.6 gm	100 lb = 45.36 kg 100 kg = 220.5 lb

END OF CHAPTER PROBLEMS 3–1

Change the following numbers to numbers times 10^3 and to numbers times 10^6:

1. 56,000	2. 15,000	3. 220,000
4. 270,000	5. 39×10^4	6. 730×10^4
7. 1.8×10^5	8. 6.8×10^4	9. 4300×10
10. 3300×10		

Change the following numbers to numbers times 10^{-3} and to numbers times 10^{-6}:

11. 0.00022	12. 0.00012	13. 0.00213
14. 0.0000179	15. 0.00556×10^{-2}	16. 0.0133×10^{-1}
17. 1.22×10^{-4}	18. 40×10^{-4}	19. 25.6×10^{-5}
20. 11×10^{-2}		

Change the following numbers to numbers times 10^{-9} and to numbers times 10^{-12}:

21. 0.00667×10^{-4}	22. 0.00233×10^{-4}	23. 0.179×10^{-7}
24. 0.00041×10^{-8}	25. 67.4×10^{-10}	26. 273×10^{-10}
27. 1.77×10^{-11}	28. 3.75×10^{-11}	29. 700×10^{-10}
30. 0.034×10^{-7}		

Change the following numbers to regular numbers, to numbers times 10^{-3}, and to numbers times 10^3:

31. 73×10^{-1}	32. 4.20×10^2	33. 0.567×10
34. 17.4×10^{-2}	35. 1.78×10^{-2}	36. 92.5×10^{-1}
37. 783×10^{-1}	38. 0.000945×10^2	39. 0.0845×10^4
40. 32.5×10^{-2}		

END OF CHAPTER PROBLEMS 3–2

Using the indicated prefixes, change the following quantities:

1. $0.00026 \text{ A} = $ _____ mA = _____ μA
2. $0.0736 \text{ A} = $ _____ mA = _____ μA
3. $0.000632 \text{ S} = $ _____ mS = _____ μS
4. $0.00024 \text{ S} = $ _____ mS = _____ μS
5. $7630 \text{ Ω} = $ _____ kΩ = _____ MΩ
6. $56,000 \text{ Ω} = $ _____ kΩ = _____ MΩ
7. $17.3 \times 10^{-3} \text{ A} = $ _____ mA = _____ μA
8. $64 \times 10^{-6} \text{ A} = $ _____ mA = _____ μA
9. $71.3 \times 10^4 \text{ Ω} = $ _____ kΩ = _____ MΩ
10. $6.8 \times 10^4 \text{ Ω} = $ _____ kΩ = _____ MΩ
11. $5.63 \times 10^6 \text{ Hz} = $ _____ kHz = _____ MHz
12. $48.7 \times 10^2 \text{ Hz} = $ _____ kHz = _____ MHz
13. $2,000,000 \text{ Ω} = $ _____ kΩ = _____ MΩ

14. 470,000 Ω = _____ kΩ = _____ MΩ
15. 23.7 × 10^{-5} S = _____ mS = _____ μS
16. 5.63 × 10^{-2} S = _____ mS = _____ μS
17. 30 × 10^{-8} F = _____ μF = _____ nF
18. 1.2 × 10^{-7} F = _____ μF = _____ nF
19. 0.000062 A = _____ mA = _____ μA
20. 0.0075 A = _____ mA = _____ μA

END OF CHAPTER PROBLEMS 3–3

Using the indicated prefixes, change the following quantities.

1. 800 mA = _____ A
3. 2.5 mS = _____ S
5. 33 kΩ = _____ Ω
7. 0.47 kΩ = _____ Ω
9. 12.5 kHz = _____ Hz
11. 100 μF = _____ F
13. 0.25 mA = _____ A
15. 900 μS = _____ S
17. 750 kΩ = _____ Ω
19. 30 mA = _____ A

2. 7.03 mA = _____ A
4. 400 μS = _____ S
6. 680 kΩ = _____ Ω
8. 0.56 MΩ = _____ Ω
10. 73 kHz = _____ Hz
12. 500 μF = _____ F
14. 50 mA = _____ A
16. 2.5 mS = _____ S
18. 12 kΩ = _____ Ω
20. 30.5 mA = _____ A

END OF CHAPTER PROBLEMS 3–4

Using the indicated prefixes, change the following quantities.

1. 0.00065 A = _____ mA = _____ μA
2. 0.0175 A = _____ mA = _____ μA
3. 0.00000805 S = _____ mS = _____ μS
4. 0.00417 A = _____ mA = _____ μA
5. 56.2 × 10^{-3} S = _____ mS = _____ μS
6. 245 × 10^{-3} A = _____ mA = _____ μA
7. 613 × 10^{-6} S = _____ mS = _____ μS
8. 2.27 × 10^{-6} A = _____ mA = _____ μA
9. 7500 Ω = _____ kΩ = _____ MΩ
10. 1250 Hz = _____ kHz = _____ MHz
11. 510,000 Ω = _____ kΩ = _____ MΩ
12. 1,800,000 Ω = _____ kΩ = _____ MΩ
13. 12.7 × 10^{4} Hz = _____ kHz = _____ MHz
14. 77.5 × 10^{6} Hz = _____ kHz = _____ MHz
15. 713 × 10^{3} Ω = _____ kΩ = _____ MΩ
16. 910 × 10^{5} Ω = _____ kΩ = _____ MΩ
17. 46.3 mA = _____ μA = _____ A
18. 415 mS = _____ S = _____ μS
19. 0.05 nF = _____ μF = _____ pF
20. 157 nF = _____ pF = _____ μF

21. $3.2 \text{ k}\Omega = $ _____ $\Omega = $ _____ $\text{M}\Omega$

22. $35.9 \text{ kHz} = $ _____ $\text{Hz} = $ _____ MHz

23. $270 \text{ k}\Omega = $ _____ $\text{M}\Omega = $ _____ Ω

24. $4.7 \text{ k}\Omega = $ _____ $\text{M}\Omega = $ _____ Ω

25. $403 \ \mu\text{S} = $ _____ $\text{S} = $ _____ mS

26. $704 \text{ mS} = $ _____ $\text{S} = $ _____ μS

27. $1.03 \text{ MHz} = $ _____ $\text{Hz} = $ _____ kHz

28. $42.3 \text{ kHz} = $ _____ $\text{MHz} = $ _____ Hz

29. $1.43 \text{ nF} = $ _____ $\text{pF} = $ _____ μF

30. $0.025 \ \mu\text{F} = $ _____ $\text{pF} = $ _____ nF

31. $55 \times 10^{-4} \text{ S} = $ _____ $\text{mS} = $ _____ μS

32. $4.67 \times 10^{-4} \text{ S} = $ _____ $\text{mS} = $ _____ μS

33. $106 \times 10^{-2} \text{ nF} = $ _____ $\mu\text{F} = $ _____ pF

34. $7.36 \times 10^{3} \text{ nF} = $ _____ $\mu\text{F} = $ _____ pF

35. $4.63 \times 10^{2} \text{ k}\Omega = $ _____ $\Omega = $ _____ $\text{M}\Omega$

36. $66.7 \times 10^{2} \text{ k}\Omega = $ _____ $\Omega = $ _____ $\text{M}\Omega$

37. $9.63 \times 10^{4} \text{ Hz} = $ _____ $\text{kHz} = $ _____ MHz

38. $82.3 \times 10^{2} \text{ Hz} = $ _____ $\text{kHz} = $ _____ MHz

39. $0.0078 \text{ A} = $ _____ $\text{mA} = $ _____ μA

40. $0.00104 \text{ A} = $ _____ $\text{mA} = $ _____ μA

41. $176 \text{ mV} = $ _____ $\text{V} = $ _____ μV

42. $0.000425 \text{ V} = $ _____ $\text{mV} = $ _____ μV

43. $1.73 \text{ W} = $ _____ $\text{mW} = $ _____ kW

44. $1.67 \text{ W} = $ _____ $\text{mW} = $ _____ kW

45. $0.025 \text{ H} = $ _____ $\text{mH} = $ _____ μH

46. $0.0005 \text{ H} = $ _____ $\text{mH} = $ _____ μH

47. $173 \text{ mS} = $ _____ $\text{S} = $ _____ μS

48. $146 \ \mu\text{S} = $ _____ $\text{mS} = $ _____ S

49. $25 \times 10^{-7} \text{ F} = $ _____ $\mu\text{F} = $ _____ nF

50. $250 \text{ mH} = $ _____ $\text{H} = $ _____ μH

END OF CHAPTER PROBLEMS 3–5

Perform the indicated operations. Round answers to three places.

1. $367 \ \mu\text{A} + 1.67 \text{ mA} = $ _____ $\text{mA} = $ _____ μA

2. $0.417 \text{ mA} + 630 \ \mu\text{A} = $ _____ $\text{mA} = $ _____ μA

3. $0.0173 \text{ mA} + 78.4 \ \mu\text{A} = $ _____ $\text{mA} = $ _____ μA

4. $650 \ \mu\text{A} + 0.235 \text{ mA} = $ _____ $\text{mA} = $ _____ μA

5. $\dfrac{1}{10 \text{ k}\Omega} + \dfrac{1}{15 \text{ k}\Omega} = $ _____ $\text{mS} = $ _____ μS

6. $\dfrac{1}{6.8 \text{ k}\Omega} + \dfrac{1}{4.7 \text{ k}\Omega} = $ _____ $\text{mS} = $ _____ μS

7. $\dfrac{1}{33 \text{ k}\Omega} + \dfrac{1}{56 \text{ k}\Omega} + \dfrac{1}{100 \text{ k}\Omega} = $ _____ $\text{mS} = $ _____ μS

8. $\dfrac{1}{33\ k\Omega} + \dfrac{1}{100\ k\Omega} + \dfrac{1}{68\ k\Omega} =$ _____ mS = _____ μS

9. $\dfrac{1}{1.2\ k\Omega} + \dfrac{1}{4.7\ k\Omega} + \dfrac{1}{2.7\ k\Omega} =$ _____ mS = _____ μS

10. $\dfrac{1}{270\ \Omega} + \dfrac{1}{470\ \Omega} + \dfrac{1}{680\ \Omega} =$ _____ mS = _____ μS

11. $\dfrac{1}{\dfrac{1}{2.7\ k\Omega} + \dfrac{1}{1.5\ k\Omega}} =$ _____ $\Omega =$ _____ kΩ

12. $\dfrac{1}{\dfrac{1}{8.2\ k\Omega} + \dfrac{1}{6.8\ k\Omega}} =$ _____ $\Omega =$ _____ kΩ

13. $\dfrac{1}{\dfrac{1}{100\ k\Omega} + \dfrac{1}{220\ k\Omega}} =$ _____ $\Omega =$ _____ k$\Omega =$ _____ MΩ

14. $\dfrac{1}{\dfrac{1}{470\ k\Omega} + \dfrac{1}{680\ k\Omega}} =$ _____ $\Omega =$ _____ k$\Omega =$ _____ MΩ

15. $\dfrac{1}{\dfrac{1}{39\ k\Omega} + \dfrac{1}{12\ k\Omega} + \dfrac{1}{27\ k\Omega}} =$ _____ $\Omega =$ _____ k$\Omega =$ _____ MΩ

16. $\dfrac{1}{\dfrac{1}{1.2\ M\Omega} + \dfrac{1}{2.7\ M\Omega} + \dfrac{1}{1\ M\Omega}} =$ _____ $\Omega =$ _____ k$\Omega =$ _____ MΩ

17. $\dfrac{1}{\dfrac{1}{180\ \Omega} + \dfrac{1}{470\ \Omega} + \dfrac{1}{1\ k\Omega}} =$ _____ $\Omega =$ _____ kΩ

18. $\dfrac{1}{\dfrac{1}{330\ \Omega} + \dfrac{1}{820\ \Omega} + \dfrac{1}{910\ \Omega}} =$ _____ $\Omega =$ _____ kΩ

19. $\dfrac{1}{213\ \mu S} =$ _____ $\Omega =$ _____ kΩ

20. $\dfrac{1}{2.75\ mS} =$ _____ $\Omega =$ _____ kΩ

21. 500 μS + 1.76 mS + 0.000043 S = _____ mS = _____ μS

22. 670 μS + 2 mS + 0.00002 S = _____ mS = _____ μS

23. $\dfrac{9\ V}{33\ k\Omega} =$ _____ mA = _____ μA

24. $\dfrac{12\ V}{2.7\ k\Omega} =$ _____ mA = _____ μA

25. $\dfrac{14 V}{200\ \mu A} =$ _____ $\Omega =$ _____ kΩ

26. $\dfrac{10\,\text{V}}{2.5\,\text{mA}} =$ _____ $\Omega =$ _____ kΩ

27. $\dfrac{50\,\text{V}}{1.2\,\text{M}\Omega} =$ _____ mA $=$ _____ μA

28. $\dfrac{40\,\text{V}}{15\,\mu\text{A}} =$ _____ k$\Omega =$ _____ MΩ

29. 68 k$\Omega \times$ 140 μA $=$ _____ V $=$ _____ mV

30. 270 k$\Omega \times$ 0.17 mA $=$ _____ V $=$ _____ mV

31. 1.2 k$\Omega \times$ 3.73 mA $=$ _____ V $=$ _____ mV

32. 4.7 k$\Omega \times$ 634 μA $=$ _____ V $=$ _____ mV

33. $\dfrac{1}{6.28 \times 75\,\text{kHz} \times 25\,\text{pF}} =$ _____ $\Omega =$ _____ kΩ

34. $\dfrac{1}{6.28 \times 2.5\,\text{kHz} \times 20\,\text{nF}} =$ _____ $\Omega =$ _____ kΩ

35. $\dfrac{1}{6.28 \times 10\,\text{kHz} \times 0.05\,\mu\text{F}} =$ _____ $\Omega =$ _____ kΩ

36. $\dfrac{1}{6.28 \times 37.2\,\text{kHz} \times 200\,\text{nF}} =$ _____ $\Omega =$ _____ kΩ

37. $\dfrac{1}{6.28 \times 12\,\text{kHz} \times 2.7\,\text{k}\Omega} =$ _____ μF $=$ _____ nF

38. $\dfrac{1}{6.28 \times 200\,\text{Hz} \times 600\,\Omega} =$ _____ μF $=$ _____ nF

39. $\dfrac{1}{6.28(150\,\text{mH} \times 200\,\text{nF})^{1/2}} =$ _____ Hz $=$ _____ kHz

40. $\dfrac{1}{6.28(0.75\,\text{mH} \times 0.5\,\text{nF})^{1/2}} =$ _____ Hz $=$ _____ kHz

When two resistors are connected in parallel, their total resistance can be calculated. Using equation (3–2) or (3–3):

41. If $R_1 = 10$ kΩ and $R_2 = 10$ kΩ, then R_T equals _____ $\Omega =$ _____ kΩ?

42. If $R_1 = 10$ kΩ and $R_2 = 9$ kΩ, then R_T equals _____ $\Omega =$ _____ kΩ?

43. If $R_1 = 10$ MΩ and $R_2 = 10$ kΩ, then R_T equals _____ $\Omega =$ _____ kΩ?

44. If $R_1 = 1$ MΩ and $R_2 = 5$ MΩ, then R_T equals _____ k$\Omega =$ _____ MΩ?

When three resistors are connected in parallel, their total resistance can be calculated. Using equation (3–4) or (3–5):

45. If $R_1 = 10$ kΩ, $R_2 = 10$ kΩ, and $R_3 = 5$ kΩ, then R_T equals _____ $\Omega =$ _____ kΩ?

46. If $R_1 = 500\,\Omega$, $R_2 = 1$ kΩ, and $R_3 = 2$ kΩ, then R_T equals _____ $\Omega =$ _____ kΩ?

47. If $R_1 = 1$ kΩ, $R_2 = 2$ kΩ, and $R_3 = 4$ kΩ, then R_T equals _____ Ω = _____ kΩ?

48. If $R_1 = 1$ MΩ, $R_2 = 5$ MΩ, and $R_3 = 25$ MΩ, then R_T equals _____ Ω = _____ kΩ?

END OF CHAPTER PROBLEMS 3–6

Make the following conversions. Round answers to four places.

1. 100 centimeters to inches
2. 9 centimeters to inches
3. 40 centimeters to inches
4. 130 centimeters to inches
5. 2.3 meters to inches and yards
6. 2.25 meters to inches and yards
7. 10 meters to yards
8. 27 meters to yards
9. 10,000 meters to yards and miles
10. 15,000 meters to yards and miles
11. 10 inches to centimeters
12. 15 inches to centimeters
13. 35.6 inches to centimeters
14. 50.3 inches to centimeters
15. 880 yards to meters and km
16. 1000 yards to meters and kilometers
17. 55 mi/h to km/h
18. 65 mi/h to km/h
19. 100 mi/h to km/h
20. 75 mi/h to km/h
21. 500 meters/s to feet/s
22. 200 meters/s to feet/s
23. 100 meters/s to feet/s
24. 400 meters/s to feet/s
25. 60 grams to ounces
26. 85 grams to ounces
27. 150 grams to ounces
28. 225 grams to ounces
29. 800 grams to ounces and pounds
30. 1400 grams to ounces and pounds
31. 5 kg to oz and lb
32. 7.5 kg to oz and lb
33. 1.35 lb to kg
34. 0.85 lb to kg
35. 4 oz to g
36. 55 oz to g
37. 0.275 oz to g
38. 0.45 oz to g

ANSWERS TO SELF-TESTS

SELF-TEST 3–1

1. $4760 = 4.76 \times 10^3 = 0.00476 \times 10^6$
2. $32.4 \times 10^4 = 324 \times 10^3 = 0.324 \times 10^6$
3. $2.71 \times 10^{-1} = 271 \times 10^{-3} = 0.000271 \times 10^3$
4. $46.7 \times 10^{-2} = 467 \times 10^{-3} = 0.000467 \times 10^3$
5. 76.3 mA = 0.0763 A = 76,300 μA
6. 0.0055 μF = 5.5 nF = 5500 pF
7. 8340 μS = 8.34 mS = 0.00834 S
8. 20 kΩ = 20,000 Ω = 0.020 MΩ
9. 75,000 Hz = 0.075 MHz = 75 kHz
10. 146 mS = 0.146 S = 146,000 μS

Fractions, Decimals, and Percents

4

Introduction

Any study of electronics involves the use of fractions, decimal numbers, and percents. In electronics, we constantly use equations that contain fractions or deal with numbers that are nonwhole numbers. In this chapter we will review how to add, subtract, multiply, and divide fractions. We will also learn how to convert between fractions, decimals, and percents.

Only fractions that are real numbers are considered in this chapter. Fractions containing literal numbers are discussed in Chapter 6.

Chapter Objectives

In this chapter you will learn how to:

1. Identify prime numbers.
2. Find the prime factors of real numbers.
3. Find the lowest common multiple of numbers.
4. Multiply and divide whole and fractional numbers.
5. Add and subtract fractions with common denominators and with unlike denominators.
6. Express numbers as either mixed numbers or as improper fractions.
7. Multiply, divide, add, and subtract improper fractions and mixed numbers.
8. Convert between fractions, decimals, and percents.

4–1 PRIME NUMBERS AND PRIME FACTORS

When two (or more) whole numbers are multiplied, the result is called the product of those numbers. The two (or more) original numbers are called the factors of the product. For example in the equation $3 \times 7 = 21$, the three and seven are the factors of twenty-one. Another example is in the equation $5 \times 6 = 30$; the five and six are factors of thirty. There are also other factors of 30:

2 and 15 are factors because $2 \times 15 = 30$
3 and 10 are factors because $3 \times 10 = 30$
2, 3 and 5 are factors because $2 \times 3 \times 5 = 30$
1 and 30 are factors because $1 \times 30 = 30$

A number that has no whole-number factors except 1 and itself is called a *prime number*. Some examples of prime numbers are 2, 3, 5, 7, and 11. A number that is not prime is divisible by more than one prime number. For example, 4, 6, 8, 9, and 10 are not prime numbers because they are all divisible by more than one prime number. 4 is divisible by 2, 6 is divisible by 2 or 3, and so on. The prime numbers that make up a nonprime number are called *prime factors*. In determining prime factors, we ignore 1 as a factor.

Let's look at some nonprime numbers and their prime factors.

EXAMPLE 4–1

Find the prime factors of 10, 15, 21, and 26.

SOLUTION

$$10 = 2 \cdot 5$$
$$15 = 3 \cdot 5$$
$$21 = 3 \cdot 7$$
$$26 = 2 \cdot 13$$

Often a prime factor appears more than once as a factor, as shown in the following examples.

EXAMPLE 4–2

Find the prime factors of 8, 9, 36, and 150.

SOLUTIONS

$$8 = 2 \cdot 2 \cdot 2 = 2^3$$
$$9 = 3 \cdot 3 = 3^2$$
$$36 = 2 \cdot 2 \cdot 3 \cdot 3 = 2^2 \cdot 3^2$$
$$150 = 2 \cdot 3 \cdot 5 \cdot 5 = 2 \cdot 3 \cdot 5^2$$

Notice that we expressed the prime factors of 8 as $2 \cdot 2 \cdot 2$ or 2^3 and the prime factors of 9 as $3 \cdot 3$ or 3^2. Either form is acceptable and both forms are used.

PRACTICE PROBLEMS 4–1

Find the prime factors of:

1. 12
2. 24
3. 39
4. 45
5. 70
6. 72
7. 100
8. 120
9. 210
10. 300

SOLUTIONS

1. $2 \cdot 2 \cdot 3$
2. $2 \cdot 2 \cdot 2 \cdot 3$
3. $3 \cdot 13$
4. $3 \cdot 3 \cdot 5$
5. $2 \cdot 5 \cdot 7$
6. $2^3 \cdot 3^2$
7. $2^2 \cdot 5^2$
8. $2^3 \cdot 3 \cdot 5$
9. $2 \cdot 3 \cdot 5 \cdot 7$
10. $2^2 \cdot 3 \cdot 5^2$

Additional practice problems are at the end of the chapter.

4–2   REDUCING FRACTIONS TO LOWEST TERMS

Fractions have been reduced to lowest terms when there are no common factors in the numerator and denominator. For example $\frac{5}{7}$ has been reduced to its lowest terms because it has no common factors in the numerator and denominator. To reduce fractions to lowest terms, we first find the factors that are in both numerator and denominator and then cancel them. It is often necessary to break up the numerators and denominators into their prime factors to see if there are any common factors. The product of the remaining factors is the fraction reduced to lowest terms. Let's look at some fractions and reduce them to lowest terms.

EXAMPLE 4–3

Examine the following fractions and reduce to lowest terms when necessary:

(a) $\dfrac{4}{5}$ (b) $\dfrac{11}{130}$ (c) $\dfrac{12}{21}$ (d) $\dfrac{39}{45}$ (e) $\dfrac{54}{96}$ (f) fifteen one hundred twenty-fifths

SOLUTIONS

(a) $\dfrac{4}{5} = \dfrac{2 \cdot 2}{5}$

There are no common factors, so the fraction is already in lowest terms.

(b) $\dfrac{11}{130} = \dfrac{11}{2 \cdot 5 \cdot 13}$

There are no common factors, so the fraction is already at lowest terms.

(c) $\dfrac{12}{21} = \dfrac{2 \cdot 2 \cdot 3}{3 \cdot 7}$

Three is a factor common to both numerator and denominator, so cancel these factors.

$$\frac{2 \cdot 2 \cdot \cancel{3}}{\cancel{3} \cdot 7} = \frac{4}{7}$$

The fraction is now reduced to lowest terms.

(d) $\dfrac{39}{45} = \dfrac{3 \cdot 13}{3 \cdot 3 \cdot 5}$

Cancel like terms:

$$\frac{\cancel{3} \cdot 13}{\cancel{3} \cdot 3 \cdot 5} = \frac{13}{15}$$

The fraction is now reduced to lowest terms.

(e) $\dfrac{54}{96} = \dfrac{2 \cdot 3 \cdot 3 \cdot 3}{2 \cdot 2 \cdot 2 \cdot 2 \cdot 2 \cdot 3}$

Cancel like terms:

$$\frac{\cancel{2} \cdot \cancel{3} \cdot 3 \cdot 3}{\cancel{2} \cdot 2 \cdot 2 \cdot 2 \cdot 2 \cdot \cancel{3}} = \frac{9}{16}$$

The fraction is now reduced to lowest terms.

(f) A fraction consists of two parts: the numerator and the denominator. The first part given is the numerator and the second is the denominator. The words "fifteen one hundred twenty-fifths" must be split into these two parts. Since "fifteen one" is not a number, we must conclude that the numerator is "fifteen" and the remaining words "one hundred twenty-fifths" are the denominator. Restated numerically:

$$\text{fifteen one hundred twenty-fifths equals } \frac{15}{125}.$$

Reduced to lowest terms:

$$\frac{15}{125} = \frac{3}{25} \text{ and is written three twenty-fifths.}$$

Often with experience we can quickly recognize common factors that can be cancelled. For example the fraction $\frac{140}{150}$ has 10 as a common factor because both numerator and denominator end in zero. After cancelling the 10 in both numerator and denominator we are left with $\frac{14}{15}$. Now break the numerator and denominator into its prime factors: $\frac{14}{15} = \frac{2 \cdot 7}{3 \cdot 5}$. We can see that there are no factors that are common to the numerator and denominator so $\frac{14}{15}$ has been reduced to its lowest terms.

If the fraction was $\frac{105}{135}$, we might recognize that since both the numerator and denominator end in 5, then 5 is a common factor and can be cancelled. Dividing both numerator and denominator by 5 leaves us with $\frac{21}{27}$. If we do not recognize any common factor then we must find their prime factors: $\frac{21}{27} = \frac{3 \cdot 7}{3 \cdot 3 \cdot 3}$. Now we can see that 3 is common to both numerator and denominator and can be cancelled. So reducing $\frac{105}{135}$ to its lowest terms results in $\frac{7}{9}$.

PRACTICE PROBLEMS 4–2

Examine the following fractions and reduce to lowest terms when necessary:

1. $\dfrac{18}{39}$

2. $\dfrac{3}{8}$

3. $\dfrac{20}{32}$

4. $\dfrac{15}{45}$

5. $\dfrac{9}{16}$

6. $\dfrac{12}{33}$

7. $\dfrac{15}{34}$

8. $\dfrac{7}{35}$

9. $\dfrac{18}{32}$

10. $\dfrac{24}{64}$

11. $\dfrac{48}{108}$

12. $\dfrac{125}{625}$

13. $\dfrac{156}{195}$

14. $\dfrac{495}{3300}$

15. Thirty-nine one hundred sixty-fifths

SOLUTIONS

1. $\dfrac{18}{39} = \dfrac{2 \cdot \cancel{3} \cdot 3}{\cancel{3} \cdot 13} = \dfrac{6}{13}$

2. $\dfrac{3}{8} = \dfrac{3}{2 \cdot 2 \cdot 2} = \dfrac{3}{8}$

Fraction is already in lowest terms.

3. $\dfrac{20}{32} = \dfrac{\cancel{2} \cdot \cancel{2} \cdot 5}{\cancel{2} \cdot \cancel{2} \cdot 2 \cdot 2 \cdot 2} = \dfrac{5}{8}$

4. $\dfrac{15}{45} = \dfrac{\cancel{3} \cdot \cancel{5}}{\cancel{3} \cdot 3 \cdot \cancel{5}} = \dfrac{1}{3}$

5. $\dfrac{9}{16} = \dfrac{3 \cdot 3}{2 \cdot 2 \cdot 2 \cdot 2} = \dfrac{9}{16}$

6. $\dfrac{12}{33} = \dfrac{2 \cdot 2 \cdot \cancel{3}}{\cancel{3} \cdot 11} = \dfrac{4}{11}$

Fraction is already in lowest terms.

7. $\dfrac{15}{34} = \dfrac{3 \cdot 5}{2 \cdot 17} = \dfrac{15}{34}$

8. $\dfrac{7}{35} = \dfrac{\cancel{7}}{5 \cdot \cancel{7}} = \dfrac{1}{5}$

Fraction is already in lowest terms.

9. $\dfrac{18}{32} = \dfrac{\cancel{2} \cdot 3 \cdot 3}{\cancel{2} \cdot 2 \cdot 2 \cdot 2 \cdot 2} = \dfrac{9}{16}$

10. $\dfrac{24}{64} = \dfrac{\cancel{2} \cdot \cancel{2} \cdot \cancel{2} \cdot 3}{\cancel{2} \cdot \cancel{2} \cdot \cancel{2} \cdot 2 \cdot 2 \cdot 2} = \dfrac{3}{8}$

11. $\dfrac{48}{108} = \dfrac{\cancel{2} \cdot \cancel{2} \cdot 2 \cdot 2 \cdot \cancel{3}}{\cancel{2} \cdot \cancel{2} \cdot \cancel{3} \cdot 3 \cdot 3} = \dfrac{4}{9}$

12. $\dfrac{125}{625} = \dfrac{\cancel{5} \cdot \cancel{5} \cdot \cancel{5}}{\cancel{5} \cdot \cancel{5} \cdot \cancel{5} \cdot 5} = \dfrac{1}{5}$

13. $\dfrac{156}{195} = \dfrac{2 \cdot 2 \cdot \cancel{3} \cdot \cancel{13}}{\cancel{3} \cdot 5 \cdot \cancel{13}} = \dfrac{4}{5}$

14. $\dfrac{495}{3300} = \dfrac{3 \cdot 3 \cdot \cancel{5} \cdot \cancel{11}}{2 \cdot 2 \cdot 3 \cdot \cancel{5} \cdot 5 \cdot \cancel{11}} = \dfrac{3}{20}$

15. $\dfrac{39}{165} = \dfrac{13}{55}$ or thirteen fifty-fifths

SELF-TEST 4–1

Find the prime factors of the following numbers:

1. 40   2. 48   3. 56
4. 84   5. 180   6. 220

Examine the following fractions and reduce to lowest terms when necessary:

7. $\dfrac{16}{25}$ 8. $\dfrac{14}{21}$ 9. $\dfrac{24}{56}$

10. $\dfrac{70}{150}$

Answers to Self-test 4–1 are at the end of the chapter.

MULTIPLICATION OF FRACTIONS   4–3

We learned how to multiply and divide fractions in arithmetic. At that time we learned the following rule:

☞ **RULE 4–1** To multiply fractions, we (1) cancel factors that are in both numerators and denominators, (2) multiply the remaining numerators (their product is the numerator of the answer), (3) multiply the remaining denominators (their product is the denominator of the answer).

Let's try a few together:

EXAMPLE 4–4

Find (a) $\dfrac{2}{3} \times \dfrac{4}{5}$ (b) $\dfrac{1}{4} \times \dfrac{3}{4} \times \dfrac{3}{8}$ (c) $\dfrac{3}{8} \times \dfrac{4}{5}$ (d) $\dfrac{3}{4} \times \dfrac{2}{3} \times \dfrac{5}{6}$

Express answers in lowest terms.

SOLUTIONS

(a) $\dfrac{2}{3} \times \dfrac{4}{5} = \dfrac{2 \times 4}{3 \times 5} = \dfrac{8}{15} = \dfrac{2 \cdot 2 \cdot 2}{3 \cdot 5}$

There are no like factors in the numerator and denominator, so $\dfrac{8}{15}$ is the answer reduced to lowest terms.

(b) $\dfrac{1}{4} \times \dfrac{3}{4} \times \dfrac{3}{8} = \dfrac{1 \times 3 \times 3}{4 \times 4 \times 8} = \dfrac{9}{128} = \dfrac{3 \cdot 3}{2^7}$

There are no like factors, so the answer is $\dfrac{9}{128}$.

(c) $\dfrac{3}{8} \times \dfrac{4}{5} = \dfrac{3 \times 4}{8 \times 5} = \dfrac{3 \times \cancel{4}}{2 \times \cancel{4} \times 5} = \dfrac{3}{2 \times 5} = \dfrac{3}{10}$

Like terms are canceled, and the result, $\dfrac{3}{10}$, is the answer reduced to lowest terms.

(d) $\dfrac{3}{4} \times \dfrac{2}{3} \times \dfrac{5}{6} = \dfrac{3 \times 2 \times 5}{4 \times 3 \times 6} = \dfrac{\cancel{2} \cdot \cancel{3} \cdot 5}{\cancel{2} \cdot 2 \cdot 2 \cdot \cancel{3} \cdot 3} = \dfrac{5}{12}$

Like terms are canceled, and the result, $\dfrac{5}{12}$, is the answer reduced to lowest terms.

PRACTICE PROBLEMS 4–3

Find the product of the following fractions and reduce to lowest terms:

1. $\dfrac{1}{8} \times \dfrac{1}{2}$ 2. $\dfrac{2}{5} \times \dfrac{1}{6}$

3. $\dfrac{2}{3} \times \dfrac{3}{10}$ 4. $\dfrac{3}{16} \times \dfrac{1}{3}$

114   CHAPTER 4

5. $\dfrac{4}{15} \times \dfrac{3}{5} \times \dfrac{1}{2}$

6. $\dfrac{4}{21} \times \dfrac{3}{5} \times \dfrac{5}{8}$

7. $\dfrac{1}{2} \times \dfrac{2}{3} \times \dfrac{3}{5}$

8. $\dfrac{7}{8} \times \dfrac{1}{3} \times \dfrac{1}{2}$

9. $\dfrac{5}{12} \times \dfrac{4}{25} \times \dfrac{9}{16}$

10. $\dfrac{7}{30} \times \dfrac{3}{14} \times \dfrac{10}{21}$

11. Three-fourths and two-fifths

SOLUTIONS

1. $\dfrac{1}{8} \times \dfrac{1}{2} = \dfrac{1}{16}$

Fraction is already in lowest terms.

2. $\dfrac{2}{5} \times \dfrac{1}{6} = \dfrac{\not{2}}{\not{2} \cdot 3 \cdot 5} = \dfrac{1}{15}$

3. $\dfrac{2}{3} \times \dfrac{3}{10} = \dfrac{\not{2} \cdot \not{3}}{\not{2} \cdot \not{3} \cdot 5} = \dfrac{1}{5}$

4. $\dfrac{3}{16} \times \dfrac{1}{3} = \dfrac{\not{3}}{2 \cdot 2 \cdot 2 \cdot 2 \cdot \not{3}} = \dfrac{1}{16}$

5. $\dfrac{4}{15} \times \dfrac{3}{5} \times \dfrac{1}{2} = \dfrac{2 \cdot 2 \cdot \not{3}}{\not{2} \cdot \not{3} \cdot 5 \cdot 5} = \dfrac{2}{25}$

6. $\dfrac{4}{21} \times \dfrac{3}{5} \times \dfrac{5}{8} = \dfrac{\not{2} \cdot \not{2} \cdot \not{3} \cdot \not{5}}{\not{2} \cdot \not{2} \cdot 2 \cdot \not{3} \cdot \not{5} \cdot 7} = \dfrac{1}{14}$

7. $\dfrac{1}{2} \times \dfrac{2}{3} \times \dfrac{3}{5} = \dfrac{\not{2} \cdot \not{3}}{\not{2} \cdot \not{3} \cdot 5} = \dfrac{1}{5}$

8. $\dfrac{7}{8} \times \dfrac{1}{3} \times \dfrac{1}{2} = \dfrac{7}{2 \cdot 2 \cdot 2 \cdot 2 \cdot 3} = \dfrac{7}{48}$

Fraction is already in lowest terms.

9. $\dfrac{5}{12} \times \dfrac{4}{25} \times \dfrac{9}{16} = \dfrac{\not{2} \cdot \not{2} \cdot \not{3} \cdot 3 \cdot \not{5}}{\not{2} \cdot \not{2} \cdot 2 \cdot 2 \cdot 2 \cdot 2 \cdot \not{3} \cdot \not{5} \cdot 5} = \dfrac{3}{80}$

10. $\dfrac{7}{30} \times \dfrac{3}{14} \times \dfrac{10}{21} = \dfrac{\not{2} \cdot \not{3} \cdot \not{5} \cdot \not{7}}{\not{2} \cdot 2 \cdot \not{3} \cdot 3 \cdot \not{5} \cdot \not{7} \cdot 7} = \dfrac{1}{42}$

11. $\dfrac{3}{4} \times \dfrac{2}{5} = \dfrac{6}{20} = \dfrac{3}{10}$ or three-tenths

Additional practice problems are at the end of the chapter.

DIVISION OF FRACTIONS　　4-4

In arithmetic we learned that when we divide fractions we simply invert the divisor and then use the rules for multiplying fractions. Here are some examples.

EXAMPLE 4–5

Perform the indicated divisions and reduce to lowest terms:

(a) $\dfrac{1}{5} \div \dfrac{1}{3}$ (b) $\dfrac{5}{8} \div 3$ (c) $\dfrac{3}{8} \div \dfrac{3}{4}$ (d) $\dfrac{7}{16} \div \dfrac{21}{40}$

SOLUTIONS

(a) $\dfrac{1}{5} \div \dfrac{1}{3} = \dfrac{1}{5} \times 3 = \dfrac{3}{5}$ (b) $\dfrac{5}{8} \div 3 = \dfrac{5}{8} \times \dfrac{1}{3} = \dfrac{5}{24}$

(c) $\dfrac{3}{8} \div \dfrac{3}{4} = \dfrac{3}{8} \times \dfrac{4}{3} = \dfrac{1}{2}$

(d) $\dfrac{7}{16} \div \dfrac{21}{40} = \dfrac{7}{16} \times \dfrac{40}{21} = \dfrac{7 \times 40}{16 \times 21} = \dfrac{7 \cdot 2 \cdot 2 \cdot 2 \cdot 5}{2 \cdot 2 \cdot 2 \cdot 2 \cdot 3 \cdot 7} = \dfrac{5}{6}$

Notice that in each example we inverted the divisor and then multiplied. Like factors in the numerators and denominators were cancelled and the remaining factors were multiplied.

PRACTICE PROBLEMS 4–4

Perform the indicated divisions. Reduce answers to lowest terms.

1. $\dfrac{1}{4} \div \dfrac{3}{5}$

2. $\dfrac{5}{7} \div \dfrac{5}{3}$

3. $\dfrac{2}{15} \div \dfrac{4}{5}$

4. $\dfrac{4}{9} \div \dfrac{2}{3}$

5. $\dfrac{7}{30} \div \dfrac{5}{6}$

6. $\dfrac{1}{3} \div \dfrac{2}{5}$

7. $\dfrac{2}{5} \div \dfrac{4}{5}$

8. $\dfrac{3}{7} \div 4$

9. $\dfrac{9}{16} \div \dfrac{189}{200}$

10. $\dfrac{56}{69} \div \dfrac{112}{99}$

11. Two-fifteenths by four-fifths

SOLUTIONS

1. $\dfrac{1}{4} \div \dfrac{3}{5} = \dfrac{1}{4} \times \dfrac{5}{3} = \dfrac{5}{12}$

2. $\dfrac{5}{7} \div \dfrac{5}{3} = \dfrac{5}{7} \times \dfrac{3}{5} = \dfrac{3}{7}$

3. $\dfrac{2}{15} \div \dfrac{4}{5} = \dfrac{2}{15} \times \dfrac{5}{4} = \dfrac{1}{6}$

4. $\dfrac{4}{9} \div \dfrac{2}{3} = \dfrac{4}{9} \times \dfrac{3}{2} = \dfrac{2}{3}$

5. $\dfrac{7}{30} \div \dfrac{5}{6} = \dfrac{7}{30} \times \dfrac{6}{5} = \dfrac{7}{25}$

6. $\dfrac{1}{3} \div \dfrac{2}{5} = \dfrac{1}{3} \times \dfrac{5}{2} = \dfrac{5}{6}$

7. $\dfrac{2}{5} \div \dfrac{4}{5} = \dfrac{2}{5} \times \dfrac{5}{4} = \dfrac{1}{2}$

8. $\dfrac{3}{7} \div 4 = \dfrac{3}{7} \times \dfrac{1}{4} = \dfrac{3}{28}$

9. $\dfrac{9}{16} \div \dfrac{189}{200} = \dfrac{9}{16} \times \dfrac{200}{189} = \dfrac{9 \times 200}{16 \times 189} = \dfrac{\not2 \cdot \not2 \cdot \not2 \cdot \not3 \cdot \not3 \cdot 5 \cdot 5}{\not2 \cdot \not2 \cdot \not2 \cdot 2 \cdot \not3 \cdot \not3 \cdot 3 \cdot 7} = \dfrac{25}{42}$

10. $\dfrac{56}{69} \div \dfrac{112}{99} = \dfrac{56}{69} \times \dfrac{99}{112} = \dfrac{56 \times 99}{69 \times 112} = \dfrac{\not2 \cdot \not2 \cdot \not2 \cdot \not7 \cdot \not3 \cdot 3 \cdot 11}{3 \cdot 23 \cdot \not2 \cdot \not2 \cdot \not2 \cdot 2 \cdot \not7} = \dfrac{33}{46}$

11. $\dfrac{2}{15} \div \dfrac{4}{5} = \dfrac{2}{15} \times \dfrac{5}{4} = \dfrac{1}{6}$ or one-sixth

Additional practice problems are at the end of the chapter.

SELF-TEST 4–2

Perform the indicated operations:

1. $\dfrac{2}{5} \times \dfrac{7}{8}$

2. $\dfrac{3}{8} \times \dfrac{2}{3}$

3. $\dfrac{2}{3} \times \dfrac{1}{5} \times \dfrac{1}{4}$

4. $\dfrac{3}{4} \times \dfrac{2}{5} \times \dfrac{4}{15}$

5. $\dfrac{1}{5} \div \dfrac{2}{3}$

6. $\dfrac{4}{15} \div \dfrac{2}{3}$

7. $\dfrac{3}{5} \div 3$

8. $\dfrac{11}{32} \div \dfrac{33}{64}$

9. $\dfrac{105}{242} \div \dfrac{220}{363}$

Answers to Self-test 4–2 are at the end of the chapter.

ADDITION AND SUBTRACTION OF FRACTIONS

4-5-1 Adding Fractions with Common Denominators

Two or more fractions can be combined into one when the denominators are identical. If fractions to be added have the same denominator, the denominator in the answer is that number. The numerator in the answer is found by adding all the numerators together.

EXAMPLE 4-6

Perform the indicated additions. Reduce answers to lowest terms:

(a) $\dfrac{1}{5} + \dfrac{2}{5}$ (b) $\dfrac{3}{8} + \dfrac{1}{8}$ (c) $\dfrac{4}{11} + \dfrac{3}{11} + \dfrac{2}{11}$ (d) $\dfrac{2}{15} + \dfrac{4}{15} + \dfrac{4}{15}$

SOLUTIONS

(a) $\dfrac{1}{5} + \dfrac{2}{5} = \dfrac{1+2}{5} = \dfrac{3}{5}$ (b) $\dfrac{3}{8} + \dfrac{1}{8} = \dfrac{3+1}{8} = \dfrac{4}{8} = \dfrac{1}{2}$

(c) $\dfrac{4}{11} + \dfrac{3}{11} + \dfrac{2}{11} = \dfrac{4+3+2}{11} = \dfrac{9}{11}$

(d) $\dfrac{2}{15} + \dfrac{4}{15} + \dfrac{4}{15} = \dfrac{2+4+4}{15} = \dfrac{10}{15} = \dfrac{2}{3}$

In each of the examples above, the denominators are the same. That is, each fraction had a denominator of 5 in (a), a denominator of 8 in (b), and so on. Thus, 5 is the *common denominator* in (a) and 8 is the common denominator in (b). Said another way:

> ☞ **KEY POINT** When the denominators of fractions are identical, that number is the *common denominator.*

When adding fractions, then, the first step is to determine if the denominators are identical. If they are, the next step is to add the numerators, as in the examples above, and reduce to lowest terms.

4-5-2 Subtracting Fractions with Common Denominators

Up to this point we have been working with positive fractions. Let's consider fractions where part or all of the fraction is negative.

There are three signs associated with any fraction: the sign of the numerator, the sign of the denominator, and the sign of the entire fraction. In the fraction $\frac{3}{8}$, we know the signs are positive because no sign is given. In the fraction $-\frac{3}{8}$, the sign of the fraction is negative but the numerator and denominator are positive. In the fraction $\frac{-3}{8}$, the numerator is negative and the denominator and the fraction are positive. In the fraction $\frac{3}{-8}$, only the denominator is negative.

We don't leave fractions with negative numerators or denominators if we can avoid it. We change the signs so that the fraction in the answer is either positive or negative and both numerator and denominator are positive. To change the signs, the rule is:

> ☞ **RULE 4–2** Any two signs of a fraction may be changed without changing the value of the fraction.

EXAMPLE 4–7

Apply rule 4–2 to the following fractions, so that all numerators and denominators are positive:

(a) $\dfrac{-3}{8}$ (b) $\dfrac{3}{-8}$ (c) $\dfrac{-3}{-8}$

(d) $-\dfrac{-3}{8}$ (e) $-\dfrac{3}{-8}$ (f) $-\dfrac{-3}{-8}$

SOLUTIONS

(a) $+\dfrac{-3}{+8} = -\dfrac{+3}{+8}$ (b) $+\dfrac{+3}{-8} = -\dfrac{+3}{+8}$ (c) $+\dfrac{-3}{-8} = +\dfrac{+3}{+8}$

(d) $-\dfrac{-3}{+8} = +\dfrac{+3}{+8}$ (e) $-\dfrac{+3}{-8} = +\dfrac{+3}{+8}$ (f) $-\dfrac{-3}{-8} = -\dfrac{+3}{+8}$

(The plus signs were included here to help us change signs.) Notice that in each case *two* signs were changed.

Suppose we want to subtract one fraction from another as in Example 4–8.

EXAMPLE 4–8

Perform the indicated operation:

$$\frac{7}{8} - \frac{3}{8}$$

SOLUTION The subtraction is accomplished by changing the sign of the fraction and adding. However, if we change the sign of the fraction, we must also change *one more sign*. Remember, we must change *two* signs. The other sign we will change is the sign of the numerator. We don't want to change the sign of the denominator because, to add the fractions, the denominators must be identical. Eight and -8 are not identical.

To solve the problem in Example 4–8 then, we must first change the sign of the second fraction and add (remembering to also change the sign of the numerator).

$$\frac{7}{8} - \frac{3}{8} = \frac{7}{8} + \frac{-3}{8} = \frac{7 - 3}{8} = \frac{4}{8} = \frac{1}{2}$$

We perform the addition as before. When we add the numerators, we are adding $+7$ and -3. The result is 4. The resultant fraction, $\frac{4}{8}$, is reduced to lowest terms, or $\frac{1}{2}$.

Here are some more examples:

EXAMPLE 4–9

Perform the indicated operations:

(a) $\dfrac{7}{9} - \dfrac{3}{9}$ (b) $\dfrac{4}{15} + \dfrac{3}{15} - \dfrac{2}{15}$ (c) $\dfrac{3}{8} - \dfrac{7}{8} + \dfrac{1}{8}$ (d) $\dfrac{11}{12} + \dfrac{5}{12} - \dfrac{7}{12}$

SOLUTIONS

(a) $\dfrac{7}{9} - \dfrac{3}{9} = \dfrac{7}{9} + \dfrac{-3}{9} = \dfrac{7 - 3}{9} = \dfrac{4}{9}$

(b) $\dfrac{4}{15} + \dfrac{3}{15} - \dfrac{2}{15} = \dfrac{4}{15} + \dfrac{3}{15} + \dfrac{-2}{15} = \dfrac{4 + 3 - 2}{15} = \dfrac{5}{15} = \dfrac{1}{3}$

Of course, we soon recognize that whenever a subtraction is indicated, we merely subtract the numerator instead of adding, so when we see a problem like (b) above, we can skip a step.

(b) $\dfrac{4}{15} + \dfrac{3}{15} - \dfrac{2}{15} = \dfrac{4 + 3 - 2}{15} = \dfrac{5}{15} = \dfrac{1}{3}$

(c) $\dfrac{3}{8} - \dfrac{7}{8} + \dfrac{1}{8} = \dfrac{3 - 7 + 1}{8} = \dfrac{-3}{8} = -\dfrac{3}{8}$

Instead of leaving the answer as a positive fraction with a negative numerator, it is standard practice to change the signs so that the numerator is positive and the fraction is negative.

(d) $\dfrac{11}{12} + \dfrac{5}{12} - \dfrac{7}{12} = \dfrac{11 + 5 - 7}{12} = \dfrac{9}{12} = \dfrac{3}{4}$

PRACTICE PROBLEMS 4–5

Perform the indicated operations.

1. $\dfrac{1}{3} + \dfrac{2}{3}$

2. $\dfrac{1}{5} + \dfrac{2}{5}$

3. $\dfrac{5}{7} - \dfrac{2}{7}$

4. $\dfrac{9}{16} - \dfrac{5}{16}$

5. $\dfrac{2}{24} + \dfrac{3}{24} - \dfrac{11}{24}$

6. $\dfrac{2}{15} + \dfrac{3}{15} - \dfrac{13}{15}$

7. $\dfrac{9}{32} - \dfrac{4}{32} + \dfrac{5}{32}$

8. $\dfrac{13}{16} - \dfrac{3}{16} - \dfrac{5}{16}$

9. $\dfrac{4}{5} - \dfrac{3}{5} - \dfrac{2}{5}$

10. $\dfrac{7}{32} + \dfrac{4}{32} - \dfrac{15}{32}$

11. Add one-eighth and five-eighths

12. Subtract three-eighths from seven-eighths

SOLUTIONS

1. $\dfrac{1+2}{3} = \dfrac{3}{3} = 1$

2. $\dfrac{1+2}{5} = \dfrac{3}{5}$

3. $\dfrac{5-2}{7} = \dfrac{3}{7}$

4. $\dfrac{9-5}{16} = \dfrac{4}{16} = \dfrac{1}{4}$

5. $\dfrac{2+3-11}{24} = -\dfrac{6}{24} = -\dfrac{1}{4}$

6. $\dfrac{2+3-13}{15} = -\dfrac{8}{15}$

7. $\dfrac{9-4+5}{32} = \dfrac{10}{32} = \dfrac{5}{16}$

8. $\dfrac{13-3-5}{16} = \dfrac{5}{16}$

9. $\dfrac{4-3-2}{5} = -\dfrac{1}{5}$

10. $\dfrac{7+4-15}{32} = -\dfrac{4}{32} = -\dfrac{1}{8}$

11. $\dfrac{1}{8} + \dfrac{5}{8} = \dfrac{6}{8} = \dfrac{3}{4}$ or three-fourths

12. $\dfrac{7}{8} - \dfrac{3}{8} = \dfrac{4}{8} = \dfrac{1}{2}$ or one-half

Additional practice problems are at the end of the chapter.

LOWEST COMMON MULTIPLE

When adding or subtracting fractions, it is often necessary to find the *lowest common multiple* of the numbers that make up the denominators. This lowest common multiple will be the common denominator of the fractions.

☞ **KEY POINT** The *lowest common multiple* of two or more numbers is the lowest multiple that is common to each of them.

For example, consider the numbers 6 and 8. Some multiples of 6 are 12, 18, 24, 36, 42, and 48. Some multiples of 8 are 16, 24, 32, 40, and 48. Of these groups, the only *common* multiples are 24 and 48. The *lowest* common multiple of 6 and 8 is 24.

☞ **RULE 4-3** To find the lowest common multiple (LCM) of two or more numbers:

1. Factor each term.
2. Determine the *maximum* number of times a prime factor appears in any one term.
3. Multiply the resulting prime factors found in step 2. This number is the LCM.

☞ **KEY POINT** The lowest common multiple is the *lowest* multiple of each number that is common to all numbers.

EXAMPLE 4-10

Find the LCM of 6 and 8.

SOLUTION Find the prime factors of each number:

$$6 = 2 \cdot 3$$
$$8 = 2 \cdot 2 \cdot 2$$

The maximum number of times 2 appears as a factor is three times; therefore, $2 \cdot 2 \cdot 2$ is part of the LCM. Three appears a maximum of one time; therefore, 3 is part of the LCM. This takes care of all the prime factors. Our next step is to multiply these factors together to get the LCM.

$$2 \cdot 2 \cdot 2 \cdot 3 = 24$$

24 is the lowest number that contains both 6 and 8 as factors. That is, 6 is a factor of 24 ($6 \times 4 = 24$), and 8 is a factor of 24 ($8 \times 3 = 24$).

There are other common multiples of 6 and 8: 48, 120, and 240 to name a few. But 24 is the *lowest*.

EXAMPLE 4–11

Find the LCM of 30 and 36.

SOLUTION

$$30 = 2 \cdot 3 \cdot 5$$
$$36 = 2 \cdot 2 \cdot 3 \cdot 3$$

Two appears as a factor a maximum of two times, 3 appears as a factor a maximum of two times, and 5 appears as a factor once; therefore,

$$\text{LCM} = 2 \cdot 2 \cdot 3 \cdot 3 \cdot 5 = 180$$

180 is the lowest number that has both 30 and 36 as factors.

EXAMPLE 4–12

Find the LCM of 50, 60, and 80.

SOLUTION

$$50 = 2 \cdot 5 \cdot 5$$
$$60 = 2 \cdot 2 \cdot 3 \cdot 5$$
$$80 = 2 \cdot 2 \cdot 2 \cdot 2 \cdot 5$$

Two appears as a factor a maximum of four times, 3 appears as a factor a maximum of one time, and 5 appears as a factor a maximum of two times. Therefore,

$$\text{LCM} = 2 \cdot 2 \cdot 2 \cdot 2 \cdot 3 \cdot 5 \cdot 5 = 1200$$

1200 is the lowest number that has 50, 60, and 80 as factors.

PRACTICE PROBLEMS 4–6

Find the lowest common multiple (LCM) of the following numbers:

1. 3, 8	2. 9, 15	3. 3, 4, 8
4. 4, 8, 12	5. 12, 15, 25	6. 11, 12, 33, 132
7. 10, 15, 18	8. 10, 26, 65	9. 14, 70, 210
10. 52, 65, 78		

SOLUTIONS

1. 24	2. 45	3. 24
4. 24	5. 300	6. 132
7. 90	8. 130	9. 210
10. 780		

Additional practice problems are at the end of the chapter.

ADDING AND SUBTRACTING FRACTIONS WITH UNLIKE DENOMINATORS

When the fractions we are adding or subtracting have unlike denominators, we cannot add them directly as before. We must first find a common denominator. We do this by finding the lowest common multiple (LCM) of all the denominators. Then we change each fraction so that each has this LCM as its denominator. *The LCM is the common denominator.* Once we have a common denominator (CD), we can add and subtract as before.

> ☞ **RULE 4-4** To add (or subtract) fractions with unlike denominators: (1) find the CD, (2) change the denominator of all fractions to this denominator, (3) divide the CD by the old denominator of each fraction, then multiply this number by the old numerator to find the new numerator of the fraction, and (4) add (or subtract) the fractions, combining all like terms in the numerator. Reduce the answer to lowest terms.

EXAMPLE 4-13

Perform the indicated operations:

$$\frac{1}{4} + \frac{1}{6}$$

SOLUTION   Since the denominators in this example are different, we find the CD:

$$4 = 2 \cdot 2$$
$$6 = 2 \cdot 3$$
$$CD = 2 \cdot 2 \cdot 3 = 12$$

Next we change each fraction so that its denominator is 12 (the CD).

$$\frac{1}{4} = \frac{?}{12}$$

If we change the denominator of a fraction, we must also change the numerator; otherwise, the fractions are not equal.

$$\frac{1}{4} \neq \frac{1}{12}$$

If we divide the new denominator (12) by the old denominator (4), we get 3. Multiplying this number by the old numerator (1) we get $1 \times 3 = 3$.

$$\frac{1}{4} = \frac{1 \times 3}{12} = \frac{3}{12}$$

If we reduce $\frac{3}{12}$, we get the original fraction $\frac{1}{4}$. The fractions are equal.
The second fraction is changed in the same manner.

$$\frac{1}{6} = \frac{?}{12}$$

$$\frac{1}{6} = \frac{1 \times 2}{12} = \frac{2}{12}$$

$$\frac{1}{4} + \frac{1}{6} = \frac{3}{12} + \frac{2}{12} = \frac{3+2}{12} = \frac{5}{12}$$

EXAMPLE 4–14

Perform the indicated operation:

$$\frac{1}{6} + \frac{3}{8}$$

SOLUTION

Step 1. Find the common denominator (remember, the LCM is the CD):

$$6 = 2 \cdot 3$$

$$8 = 2 \cdot 2 \cdot 2$$

$$CD = 2 \cdot 2 \cdot 2 \cdot 3 = 24$$

Step 2. Change each fraction so that its denominator is the CD.

$$\frac{1}{6} = \frac{1 \times 4}{24} = \frac{4}{24}$$

$$\frac{3}{8} = \frac{3 \times 3}{24} = \frac{9}{24}$$

Step 3. Add the new fractions together. Reduce to lowest terms where possible.

$$\frac{4}{24} + \frac{9}{24} = \frac{4+9}{24} = \frac{13}{24}$$

EXAMPLE 4–15

Perform the indicated operation:

$$\frac{1}{4} + \frac{6}{18} - \frac{9}{24}$$

Step 1. Find the common denominator:

$$4 = 2 \cdot 2$$
$$18 = 2 \cdot 3 \cdot 3$$
$$24 = 2 \cdot 2 \cdot 2 \cdot 3$$
$$CD = 2^3 \cdot 3^2 = 72$$

Step 2. Change each fraction so that its denominator is the CD.

$$\frac{1}{4} = \frac{1 \times 18}{72} = \frac{18}{72}$$

$$\frac{6}{18} = \frac{6 \times 4}{72} = \frac{24}{72}$$

$$\frac{9}{24} = \frac{9 \times 3}{72} = \frac{27}{72}$$

Step 3. Perform the indicated operation. Reduce to lowest terms where possible.

$$\frac{1}{4} + \frac{6}{18} - \frac{9}{24} = \frac{18 + 24 - 27}{72} = \frac{15}{72} = \frac{5}{24}$$

EXAMPLE 4–16

A circuit consists of two components. One dissipates one-half watt of power and the other dissipates one-quarter watt. What is the total power dissipated? (Total power dissipated is the sum of the individual powers dissipated.)

SOLUTION   The total power dissipated is the sum of all the individual component dissipations so we must add one-half watt and one-quarter watt.

$$\frac{1}{2} + \frac{1}{4} = \frac{2}{4} + \frac{1}{4} = \frac{3}{4}$$

The total power dissipated is three-quarters of a watt or 3/4 W.

PRACTICE PROBLEMS 4–7

Perform the indicated operations. Reduce to lowest terms.

1. $\dfrac{7}{20} + \dfrac{5}{24}$

2. $\dfrac{7}{12} - \dfrac{5}{28}$

3. $\dfrac{6}{35} - \dfrac{3}{14}$

4. $\dfrac{3}{16} + \dfrac{2}{5}$

5. $\dfrac{1}{3} + \dfrac{1}{4} + \dfrac{1}{5}$

6. $\dfrac{1}{6} + \dfrac{1}{9} + \dfrac{1}{10}$

7. $\dfrac{5}{12} + \dfrac{5}{8} - \dfrac{1}{3}$

8. $\dfrac{5}{16} + \dfrac{5}{12} - \dfrac{1}{6}$

9. $\dfrac{7}{12} + \dfrac{5}{8} - \dfrac{5}{6}$

10. $\dfrac{5}{12} + \dfrac{5}{18} - \dfrac{5}{36}$

11. $\dfrac{4}{15} - \dfrac{5}{6} - \dfrac{1}{4}$

12. $\dfrac{7}{20} - \dfrac{7}{30} - \dfrac{3}{10}$

13. Subtract eight-ninths from four-fifths.

SOLUTIONS

1. CD = 120. $\dfrac{42}{120} + \dfrac{25}{120} = \dfrac{42 + 25}{120} = \dfrac{67}{120}$

2. CD = 84. $\dfrac{49}{84} - \dfrac{15}{84} = \dfrac{49 - 15}{84} = \dfrac{34}{84} = \dfrac{17}{42}$

3. CD = 70. $\dfrac{12}{70} - \dfrac{15}{70} = \dfrac{12 - 15}{70} = -\dfrac{3}{70}$

4. CD = 80. $\dfrac{15}{80} + \dfrac{32}{80} = \dfrac{15 + 32}{80} = \dfrac{47}{80}$

5. CD = 60. $\dfrac{20}{60} + \dfrac{15}{60} + \dfrac{12}{60} = \dfrac{20 + 15 + 12}{60} = \dfrac{47}{60}$

6. CD = 90. $\dfrac{15}{90} + \dfrac{10}{90} + \dfrac{9}{90} = \dfrac{15 + 10 + 9}{90} = \dfrac{34}{90} = \dfrac{17}{45}$

7. CD = 24. $\dfrac{10}{24} + \dfrac{15}{24} - \dfrac{8}{24} = \dfrac{10 + 15 - 8}{24} = \dfrac{17}{24}$

8. CD = 48. $\dfrac{15}{48} + \dfrac{20}{48} - \dfrac{8}{48} = \dfrac{15 + 20 - 8}{48} = \dfrac{27}{48} = \dfrac{9}{16}$

9. CD = 24. $\dfrac{14}{24} + \dfrac{15}{24} - \dfrac{20}{24} = \dfrac{14 + 15 - 20}{24} = \dfrac{9}{24} = \dfrac{3}{8}$

10. CD = 36. $\dfrac{15}{36} + \dfrac{10}{36} - \dfrac{5}{36} = \dfrac{15 + 10 - 5}{36} = \dfrac{20}{36} = \dfrac{5}{9}$

11. CD = 60. $\dfrac{16}{60} - \dfrac{50}{60} - \dfrac{15}{60} = \dfrac{16 - 50 - 15}{60} = -\dfrac{49}{60}$

12. CD = 60. $\dfrac{21}{60} - \dfrac{14}{60} - \dfrac{18}{60} = \dfrac{21 - 14 - 18}{60} = -\dfrac{11}{60}$

13. CD = 45. $\dfrac{4}{5} - \dfrac{8}{9} = \dfrac{36 - 40}{45} = -\dfrac{4}{45}$

Additional practice problems are at the end of the chapter.

SELF-TEST 4–3

Find the LCM of the following numbers:

1. 12, 15
2. 20, 24
3. 35, 42
4. 45, 75
5. 10, 21, 30
6. 14, 35, 105
7. 40, 60, 80
8. 45, 65, 91

Perform the indicated operations. Reduce to lowest terms.

9. $\dfrac{1}{8} + \dfrac{3}{8}$

10. $\dfrac{5}{24} + \dfrac{7}{24} + \dfrac{1}{24}$

11. $\dfrac{5}{12} - \dfrac{7}{12}$

12. $\dfrac{4}{15} + \dfrac{7}{15} - \dfrac{2}{15}$

13. $\dfrac{5}{12} - \dfrac{7}{15}$

14. $\dfrac{3}{8} + \dfrac{1}{6}$

15. $\dfrac{3}{10} + \dfrac{5}{12} - \dfrac{13}{16}$

16. $\dfrac{5}{24} - \dfrac{11}{18} - \dfrac{11}{36}$

Answers to Self-test 4–3 are at the end of the chapter.

4–8   IMPROPER FRACTIONS AND MIXED NUMBERS

☞ **KEY POINT** When the numerator is greater than the denominator, the fraction is called an *improper* fraction.

Some examples are $\frac{4}{3}, \frac{6}{5},$ and $\frac{17}{12}$. Improper fractions always have a value greater than 1.

☞ **KEY POINT** Numbers that have a whole-number part and a fractional part are called *mixed numbers*.

Some examples of mixed numbers are $2\frac{2}{3}$ and $5\frac{1}{4}$. When we write the number $2\frac{2}{3}$, we are really saying that we have 2 plus $\frac{2}{3}$: $2 + \frac{2}{3}$. Even though we don't write the "+", we know it is implied.

We can express nonwhole numbers as either mixed numbers or as improper fractions. In problem solving, it is easier to work with improper fractions. Answers, however, are usually written as mixed numbers. The conversion from one form to the other is shown in the following examples.

EXAMPLE 4–17

Change $\dfrac{4}{3}$ to a mixed number.

SOLUTION

Step 1. Perform the indicated division. Write the remainder as a fraction.

$$
\begin{array}{r}
1 + \frac{1}{3} \\
3\overline{)4} \\
\frac{3}{1}
\end{array}
$$

Step 2. Write the answer as a mixed number.

$$1 + \frac{1}{3} = 1\frac{1}{3}$$

Remember, $1\frac{1}{3}$ means $1 + \frac{1}{3}$.

EXAMPLE 4–18

Change $\dfrac{17}{12}$ to a mixed number.

SOLUTION

$$
\begin{array}{r}
1 + \frac{5}{12} \\
12\overline{)17} \\
\frac{12}{5}
\end{array}
\qquad
1 + \frac{5}{12} = 1\frac{5}{12}
$$

EXAMPLE 4–19

Change $\dfrac{16}{3}$ to a mixed number.

$$3\overline{)16} \qquad 5 + \frac{1}{3} = 5\frac{1}{3}$$

$$5 + \frac{1}{3}$$

$$\begin{array}{r} 15 \\ \hline 1 \end{array}$$

EXAMPLE 4–20

Change $2\frac{2}{3}$ to an improper fraction.

SOLUTION

Step 1. Separate the whole-number and fractional parts.

$$2\frac{2}{3} = 2 + \frac{2}{3}$$

Step 2. Find the CD and add. The denominators are $1\left(2 = \frac{2}{1}\right)$ and 3.

$$2 = \frac{6}{3}$$

Therefore, $\dfrac{6}{3} + \dfrac{2}{3} = \dfrac{6 + 2}{3} = \dfrac{8}{3}$

$$2\frac{2}{3} = \frac{8}{3}$$

A shortcut method is: Multiply the whole number and the denominator of the fraction.

$$2 \times 3 = 6$$

Add the numerator of the fraction to this number.

$$6 + 2 = 8.$$

This number is the numerator in the answer. The denominator is unchanged.

$$2\frac{2}{3} = \frac{6 + 2}{3} = \frac{8}{3}$$

EXAMPLE 4–21

Using the shortcut method, change $5\frac{1}{4}$ to an improper fraction.

$$5\frac{1}{4} = \frac{5 \times 4 + 1}{4} = \frac{20 + 1}{4} = \frac{21}{4}$$

EXAMPLE 4–22

Change $-8\frac{7}{8}$ to an improper fraction.

SOLUTION

$$-8\frac{7}{8} = -\frac{8 \times 8 + 7}{8} = -\frac{64 + 7}{8} = -\frac{71}{8}$$

PRACTICE PROBLEMS 4–8

Change the following improper fractions to mixed numbers:

1. $\dfrac{27}{4}$

2. $\dfrac{23}{8}$

3. $-\dfrac{27}{5}$

4. $\dfrac{17}{3}$

5. $\dfrac{17}{6}$

6. $-\dfrac{21}{4}$

7. $\dfrac{28}{3}$

8. $\dfrac{13}{4}$

Change the following mixed numbers to improper fractions:

9. $3\dfrac{5}{8}$

10. $5\dfrac{3}{4}$

11. $-7\dfrac{1}{3}$

12. $2\dfrac{1}{2}$

13. $4\dfrac{1}{8}$

14. $-6\dfrac{1}{3}$

15. $3\dfrac{1}{12}$

16. $5\dfrac{3}{8}$

SOLUTIONS

1. $6\dfrac{3}{4}$

2. $2\dfrac{7}{8}$

3. $-5\dfrac{2}{5}$

4. $5\dfrac{2}{3}$

5. $2\dfrac{5}{6}$

6. $-5\dfrac{1}{4}$

7. $9\dfrac{1}{3}$

8. $3\dfrac{1}{4}$

9. $\dfrac{29}{8}$

10. $\dfrac{23}{4}$

11. $-\dfrac{22}{3}$

12. $\dfrac{5}{2}$

13. $\dfrac{33}{8}$

14. $-\dfrac{19}{3}$

15. $\dfrac{37}{12}$

16. $\dfrac{43}{8}$

Additional practice problems are at the end of the chapter.

MULTIPLICATION AND DIVISION OF IMPROPER FRACTIONS

Multiplying and dividing improper fractions is no different from multiplying and dividing proper fractions. When the answer is an improper fraction, we change it to a mixed number.

EXAMPLE 4-23

Perform the indicated operation. If the answer is an improper fraction, change it to a mixed number.

(a) $\dfrac{4}{3} \times \dfrac{8}{7} = \dfrac{32}{21} = 1\dfrac{11}{21}$

(b) $\dfrac{5}{3} \times 2 = \dfrac{10}{3} = 3\dfrac{1}{3}$

(c) $\dfrac{5}{2} \div \dfrac{6}{5} = \dfrac{5}{2} \times \dfrac{5}{6} = \dfrac{25}{12} = 2\dfrac{1}{12}$

(d) $\dfrac{8}{3} \div \dfrac{1}{4} = \dfrac{8}{3} \times 4 = \dfrac{32}{3} = 10\dfrac{2}{3}$

☞ **KEY POINT** When we multiply or divide mixed numbers, we change the mixed numbers to improper fractions and perform the multiplication or division as before.

EXAMPLE 4-24

Perform the indicated operations. If the answer is an improper fraction, change it to a mixed number.

(a) $2\dfrac{1}{3} \times 3\dfrac{7}{8} = \dfrac{7}{3} \times \dfrac{31}{8} = \dfrac{217}{24} = 9\dfrac{1}{24}$

(b) $3\frac{1}{4} \times 4\frac{1}{3} = \frac{13}{4} \times \frac{13}{3} = \frac{169}{12} = 14\frac{1}{12}$

(c) $5\frac{1}{5} \div 3\frac{2}{3} = \frac{26}{5} \div \frac{11}{3} = \frac{26}{5} \times \frac{3}{11} = \frac{78}{55} = 1\frac{23}{55}$

(d) $2\frac{1}{3} \div 6\frac{1}{4} = \frac{7}{3} \div \frac{25}{4} = \frac{7}{3} \times \frac{4}{25} = \frac{28}{75}$

No change is necessary in (d) since the answer is a proper fraction.

PRACTICE PROBLEMS 4–9

Perform the indicated operation. If the answer is an improper fraction, change it to a mixed number.

1. $1\frac{7}{8} \times 3\frac{3}{4}$

2. $\frac{7}{8} \times \frac{16}{9}$

3. $3 \times \frac{8}{3}$

4. $\frac{5}{3} \times 3\frac{1}{2}$

5. $\frac{9}{5} \times 2\frac{2}{5}$

6. $\frac{9}{7} \div \frac{4}{3}$

7. $2\frac{3}{4} \div \frac{4}{3}$

8. $4 \div 1\frac{2}{3}$

9. $5\frac{1}{6} \div 4\frac{5}{12}$

10. $5\frac{1}{2} \div \frac{3}{4}$

11. Multiply the numbers three and one-half and eight and three-sevenths.

12. Divide twenty one and seven-eighths by five and three-eighths.

SOLUTIONS

1. $7\frac{1}{32}$

2. $1\frac{5}{9}$

3. 8

4. $5\frac{5}{6}$

5. $4\frac{8}{25}$

6. $\frac{27}{28}$

7. $2\frac{1}{16}$

8. $2\frac{2}{5}$

9. $1\dfrac{9}{53}$
10. $7\dfrac{1}{3}$

11. $3\dfrac{1}{2} \times 8\dfrac{3}{7} = \dfrac{7}{2} \times \dfrac{59}{7} = \dfrac{59}{2} = 29\dfrac{1}{2}$

12. $21\dfrac{7}{8} \div 5\dfrac{3}{8} = \dfrac{175}{8} \div \dfrac{43}{8} = \dfrac{175}{8} \times \dfrac{8}{43} = \dfrac{175}{43} = 4\dfrac{3}{43}$

Additional practice problems are at the end of the chapter.

4–10 ADDITION AND SUBTRACTION OF IMPROPER FRACTIONS

☞ **KEY POINT** To add and subtract, we change all mixed numbers to improper fractions and perform the additions and subtractions as we did with proper fractions.

Let's work some together.

EXAMPLE 4–25

(a) $\dfrac{9}{8} + \dfrac{5}{4}$

Find the common denominator and add. The CD is 8.

$$\dfrac{9}{8} + \dfrac{5}{4} = \dfrac{9}{8} + \dfrac{10}{8} = \dfrac{9 + 10}{8} = \dfrac{19}{8} = 2\dfrac{3}{8}$$

(b) $\dfrac{7}{3} + 2\dfrac{5}{6}$

Change $2\frac{5}{6}$ to an improper fraction:

$$2\dfrac{5}{6} = \dfrac{17}{6}$$

Find the CD. The CD is 6.

$$\dfrac{7}{3} + \dfrac{17}{6} = \dfrac{14}{6} + \dfrac{17}{6} = \dfrac{14 + 17}{6} = \dfrac{31}{6} = 5\dfrac{1}{6}$$

(c) $3\dfrac{3}{4} - 1\dfrac{7}{16}$

Change to improper fractions:

$$3\frac{3}{4} = \frac{15}{4} \text{ and } -1\frac{7}{16} = -\frac{23}{16}$$

Find the CD. The CD is 16.

$$\frac{15}{4} - \frac{23}{16} = \frac{60}{16} - \frac{23}{16} = \frac{60 - 23}{16} = \frac{37}{16} = 2\frac{5}{16}$$

(d) $\dfrac{9}{4} - 7\dfrac{2}{3}$

Change $-7\frac{2}{3}$ to an improper fraction:

$$-7\frac{2}{3} = -\frac{23}{3}$$

Find the CD. The CD is 12.

$$\frac{9}{4} - \frac{23}{3} = \frac{27}{12} - \frac{92}{12} = \frac{27 - 92}{12} = \frac{-65}{12} = -\frac{65}{12} = -5\frac{5}{12}$$

Remember, when the numerator or denominator is negative, we change *two* signs. In this case we change the sign of the numerator and the sign of the fraction.

PRACTICE PROBLEMS 4–10

Perform the indicated operations. Express your answers as mixed numbers or as proper fractions.

1. $\dfrac{16}{9} + \dfrac{4}{3}$

2. $3\dfrac{1}{3} + \dfrac{8}{15}$

3. $4\dfrac{7}{8} + \dfrac{9}{4}$

4. $5\dfrac{1}{3} + 6\dfrac{5}{8}$

5. $\dfrac{9}{8} - \dfrac{1}{2}$

6. $2\dfrac{1}{7} - \dfrac{8}{3}$

7. $4\dfrac{2}{5} - 2\dfrac{1}{10}$

8. $2\dfrac{9}{10} - 4\dfrac{1}{15}$

9. $\dfrac{9}{7} + \dfrac{8}{3} - 1\dfrac{2}{3}$

10. $5\dfrac{5}{6} - 1\dfrac{2}{3} + 3\dfrac{3}{4}$

SOLUTIONS

1. $\dfrac{28}{9} = 3\dfrac{1}{9}$

2. $\dfrac{58}{15} = 3\dfrac{13}{15}$

3. $\dfrac{57}{8} = 7\dfrac{1}{8}$

4. $\dfrac{287}{24} = 11\dfrac{23}{24}$

5. $\dfrac{5}{8}$

6. $-\dfrac{11}{21}$

7. $\dfrac{23}{10} = 2\dfrac{3}{10}$

8. $-\dfrac{35}{30} = -\dfrac{7}{6} = -1\dfrac{1}{6}$

9. $\dfrac{48}{21} = 2\dfrac{6}{21} = 2\dfrac{2}{7}$

10. $\dfrac{190}{24} = \dfrac{95}{12} = 7\dfrac{11}{12}$

Additional practice problems are at the end of the chapter.

SELF-TEST 4–4

Change to mixed numbers:

1. $\dfrac{29}{8}$

2. $\dfrac{33}{5}$

Change to improper fractions:

3. $3\dfrac{3}{16}$

4. $-8\dfrac{2}{3}$

Perform the indicated operation (if the answer is an improper fraction, change to a mixed number):

5. $2\dfrac{5}{8} \times \dfrac{9}{4}$

6. $3\dfrac{1}{3} \times 4\dfrac{1}{2}$

7. $\dfrac{9}{8} \div 1\dfrac{7}{8}$

8. $5\dfrac{1}{3} \div 2\dfrac{5}{6}$

9. $\dfrac{8}{3} + 1\dfrac{1}{2}$

10. $2\dfrac{7}{8} + 5\dfrac{1}{4}$

11. $6\dfrac{1}{2} - 1\dfrac{3}{32}$

12. $8\dfrac{1}{3} - 4\dfrac{5}{12}$

13. $1\dfrac{1}{4} + \dfrac{9}{8} - \dfrac{10}{3}$

14. $6\dfrac{1}{2} + 4\dfrac{1}{3} - 5\dfrac{1}{5}$

Answers to Self-test 4–4 are at the end of the chapter.

When working with fractions, it is often desirable or necessary to express answers as decimal numbers rather than as mixed numbers. We worked with decimal numbers in Chapter 1. Let's perform all four functions on two fractions and give the answers as decimal numbers and then as proper fractions or as mixed numbers.

EXAMPLE 4–26

Add, subtract, multiply, and divide the following two fractions: $\frac{3}{8}$ and $\frac{2}{3}$. Express answers as decimal numbers rounded to three places and then as proper fractions or as mixed numbers.

SOLUTIONS This problem is easily solved expressing the answers as decimal numbers using your calculator, but don't forget the hierarchy: Multiply, divide, add, and subtract.

$$\frac{3}{8} + \frac{2}{3} \quad 3 \div 8 + 2 \div 3 = \qquad \textit{Answer: } 1.04$$

$$\frac{3}{8} - \frac{2}{3} \quad 3 \div 8 - 2 \div 3 = \qquad \textit{Answer: } -0.292$$

$$\frac{3}{8} \times \frac{2}{3} \quad 3 \div 8 \times 2 \div 3 = \qquad \textit{Answer: } 0.250$$

$$\frac{3}{8} \div \frac{2}{3} \quad 3 \div 8 \div (2 \div 3) = \qquad \textit{Answer: } 0.563$$

(We had to use parentheses here to show that $3 \div 8$ was to be divided by the fraction $\frac{2}{3}$.)

As proper fractions or mixed numbers,

$$\frac{3}{8} + \frac{2}{3} = \frac{9 + 16}{24} = \frac{25}{24} = 1\frac{1}{24}$$

$$\frac{3}{8} - \frac{2}{3} = \frac{9 - 16}{24} = \frac{-7}{24} = -\frac{7}{24}$$

$$\frac{3}{8} \times \frac{2}{3} = \frac{6}{24} = \frac{1}{4}$$

$$\frac{3}{8} \div \frac{2}{3} = \frac{3}{8} \times \frac{3}{2} = \frac{9}{16}$$

EXAMPLE 4–27

Add, subtract, multiply, and divide the following two mixed numbers: $3\frac{1}{4}$ and $2\frac{1}{6}$. Express answers first as decimal numbers rounded to three places and then as proper fractions or as mixed numbers.

SOLUTIONS As decimal numbers,

$$3\frac{1}{4} + 2\frac{1}{6} = 5.24$$

$$3 \boxed{+} 1 \boxed{\div} 4 \boxed{+} 2 \boxed{+} 1 \boxed{\div} 6 \boxed{=} \qquad \text{\textit{Answer:} 5.42}$$

When there are a number of functions to perform, the use of parentheses is often necessary to let the calculator know how to handle the numbers.

$$3\frac{1}{4} - 2\frac{1}{6} = 1.08$$

$$3 \boxed{+} 1 \boxed{\div} 4 \boxed{-} \boxed{(} 2 \boxed{+} 1 \boxed{\div} 6 \boxed{)} \boxed{=} \qquad \text{\textit{Answer:} 1.08}$$

The parentheses tell the calculator that some function is to be performed on a *group*. In this case the group is the mixed number $2\frac{1}{6}$. Without the parentheses, the calculator would follow the algebraic logic and subtract 2 and then *add* $\frac{1}{6}$. To subtract the entire number $2\frac{1}{6}$, we put the number inside the parentheses. Once we perform the operations within the parentheses $(2 + \frac{1}{6})$, then that answer is subtracted from $3\frac{1}{4}$. When in doubt as to whether or not you need the parentheses, it is a good idea to go ahead and use them.

$$3\frac{1}{4} \times 2\frac{1}{6} = 7.04$$

$$\boxed{(} 3 \boxed{+} 1 \boxed{\div} 4 \boxed{)} \boxed{\times} \boxed{(} 2 \boxed{+} 1 \boxed{\div} 6 \boxed{)} \boxed{=} \qquad \text{\textit{Answer:} 7.04}$$

and, finally,

$$3\frac{1}{4} \div 2\frac{1}{6} = 1.50$$

$$\boxed{(} 3 \boxed{+} 1 \boxed{\div} 4 \boxed{)} \boxed{\div} \boxed{(} 2 \boxed{+} 1 \boxed{\div} 6 \boxed{)} \boxed{=} \qquad \text{\textit{Answer:} 1.50}$$

As proper fractions or as mixed numbers,

$$3\frac{1}{4} + 2\frac{1}{6} = \frac{13}{4} + \frac{13}{6} = \frac{39 + 26}{12} = \frac{65}{12} = 5\frac{5}{12}$$

$$3\frac{1}{4} - 2\frac{1}{6} = \frac{13}{4} - \frac{13}{6} = \frac{39 - 26}{12} = \frac{13}{12} = 1\frac{1}{12}$$

$$3\frac{1}{4} \times 2\frac{1}{6} = \frac{13}{4} \times \frac{13}{6} = \frac{169}{24} = 7\frac{1}{24}$$

$$3\frac{1}{4} \div 2\frac{1}{6} = \frac{13}{4} \div \frac{13}{6} = \frac{13}{4} \times \frac{6}{13} = \frac{78}{52} = \frac{3}{2} = 1\frac{1}{2}$$

PRACTICE PROBLEMS 4–11

Add, subtract, multiply, and divide the following numbers. Express answers as decimal numbers rounded to three places, and then as proper fractions or as mixed numbers.

1. $\frac{1}{3}$ and $\frac{5}{8}$

2. $\frac{6}{7}$ and $\frac{3}{5}$

3. $\frac{9}{16}$ and $\frac{3}{32}$

4. $\frac{7}{16}$ and $\frac{2}{3}$

5. $1\frac{3}{5}$ and $7\frac{1}{8}$

6. $9\frac{3}{16}$ and $2\frac{3}{4}$

7. $3\frac{1}{3}$ and $1\frac{5}{32}$

8. $4\frac{2}{9}$ and $2\frac{1}{3}$

9. $2\frac{11}{16}$ and $8\frac{1}{8}$

10. $6\frac{1}{2}$ and $4\frac{7}{8}$

SOLUTIONS

1. $\frac{1}{3} + \frac{5}{8} = \frac{8+15}{24} = \frac{23}{24} = 0.958$

$\frac{1}{3} - \frac{5}{8} = \frac{8-15}{24} = -\frac{7}{24} = -0.292$

$\frac{1}{3} \times \frac{5}{8} = \frac{5}{24} = 0.208$

$\frac{1}{3} \div \frac{5}{8} = \frac{1}{3} \times \frac{8}{5} = \frac{8}{15} = 0.533$

2. $\frac{6}{7} + \frac{3}{5} = \frac{30+21}{35} = \frac{51}{35} = 1\frac{16}{35} = 1.46$

$\frac{6}{7} - \frac{3}{5} = \frac{30-21}{35} = \frac{9}{35} = 0.257$

$\frac{6}{7} \times \frac{3}{5} = \frac{18}{35} = 0.514$

$\frac{6}{7} \div \frac{3}{5} = \frac{6}{7} \times \frac{5}{3} = \frac{30}{21} = \frac{10}{7} = 1\frac{3}{7} = 1.43$

3. $\frac{9}{16} + \frac{3}{32} = \frac{18+3}{32} = \frac{21}{32} = 0.656$

$\frac{9}{16} - \frac{3}{32} = \frac{18-3}{32} = \frac{15}{32} = 0.469$

$\frac{9}{16} \times \frac{3}{32} = \frac{27}{512} = 0.0527$

$\frac{9}{16} \div \frac{3}{32} = \frac{9}{16} \times \frac{32}{3} = \frac{18}{3} = 6$

4. $\frac{7}{16} + \frac{2}{3} = \frac{21+32}{48} = \frac{53}{48} = 1\frac{5}{48} = 1.10$

$\frac{7}{16} - \frac{2}{3} = \frac{21-32}{48} = -\frac{11}{48} = -0.229$

$\frac{7}{16} \times \frac{2}{3} = \frac{14}{48} = \frac{7}{24} = 0.292$

$\frac{7}{16} \div \frac{2}{3} = \frac{7}{16} \times \frac{3}{2} = \frac{21}{32} = 0.656$

5. $1\frac{3}{5} + 7\frac{1}{8} = \frac{8}{5} + \frac{57}{8} = \frac{64 + 285}{40}$

$\qquad = \frac{349}{40} = 8\frac{29}{40} = 8.73$

$\frac{8}{5} - \frac{57}{8} = \frac{64 - 285}{40} = -\frac{221}{40} = -5\frac{21}{40}$

$\qquad\qquad\qquad\qquad\qquad = -5.53$

$\frac{8}{5} \times \frac{57}{8} = \frac{57}{5} = 11\frac{2}{5} = 11.4$

$\frac{8}{5} \div \frac{57}{8} = \frac{8}{5} \times \frac{8}{57} = \frac{64}{285} = 0.225$

6. $9\frac{3}{16} + 2\frac{3}{4} = \frac{147}{16} + \frac{11}{4} = \frac{147 + 44}{16}$

$\qquad = \frac{191}{16} = 11\frac{15}{16} = 11.9$

$\frac{147}{16} - \frac{11}{4} = \frac{147 - 44}{16} = \frac{103}{16} = 6\frac{7}{16}$

$\qquad\qquad\qquad\qquad\qquad = 6.44$

$\frac{147}{16} \times \frac{11}{4} = \frac{1617}{64} = 25\frac{17}{64} = 25.3$

$\frac{147}{16} \div \frac{11}{4} = \frac{147}{16} \times \frac{4}{11} = \frac{147}{44} = 3\frac{15}{44}$

$\qquad\qquad\qquad\qquad\qquad = 3.34$

7. $3\frac{1}{3} + 1\frac{5}{32} = \frac{10}{3} + \frac{37}{32} = \frac{320 + 111}{96}$

$\qquad = \frac{431}{96} = 4\frac{47}{96} = 4.49$

$\frac{10}{3} - \frac{37}{32} = \frac{320 - 111}{96} = \frac{209}{96} = 2\frac{17}{96}$

$\qquad\qquad\qquad\qquad\qquad = 2.18$

$\frac{10}{3} \times \frac{37}{32} = \frac{5}{3} \times \frac{37}{16} = \frac{185}{48} = 3\frac{41}{48}$

$\qquad\qquad\qquad\qquad\qquad = 3.85$

$\frac{10}{3} \div \frac{37}{32} = \frac{10}{3} \times \frac{32}{37} = \frac{320}{111} = 2\frac{98}{111}$

$\qquad\qquad\qquad\qquad\qquad = 2.88$

8. $4\frac{2}{9} + 2\frac{1}{3} = \frac{38}{9} + \frac{7}{3} = \frac{38 + 21}{9}$

$\qquad = \frac{59}{9} = 6\frac{5}{9} = 6.56$

$\frac{38}{9} - \frac{7}{3} = \frac{38 - 21}{9} = \frac{17}{9} = 1\frac{8}{9} = 1.89$

$\frac{38}{9} \times \frac{7}{3} = \frac{266}{27} = 9\frac{23}{27} = 9.85$

$\frac{38}{9} \div \frac{7}{3} = \frac{38}{9} \times \frac{3}{7} = \frac{38}{21} = 1\frac{17}{21} = 1.81$

9. $2\frac{11}{16} + 8\frac{1}{8} = \frac{43}{16} + \frac{65}{8} = \frac{43 + 130}{16}$

$\qquad = \frac{173}{16} = 10\frac{13}{16} = 10.8$

$\frac{43}{16} - \frac{65}{8} = \frac{43 - 130}{16} = -\frac{87}{16} = -5\frac{7}{16}$

$\qquad\qquad\qquad\qquad\qquad = -5.44$

$\frac{43}{16} \times \frac{65}{8} = \frac{2795}{128} = 21\frac{107}{128} = 21.8$

$\frac{43}{16} \div \frac{65}{8} = \frac{43}{16} \times \frac{8}{65} = \frac{43}{130} = 0.331$

10. $6\frac{1}{2} + 4\frac{7}{8} = \frac{13}{2} + \frac{39}{8} = \frac{52 + 39}{8}$

$\qquad = \frac{91}{8} = 11\frac{3}{8} = 11.4$

$\frac{13}{2} - \frac{39}{8} = \frac{52 - 39}{8} = \frac{13}{8} = 1\frac{5}{8} = 1.63$

$\frac{13}{2} \times \frac{39}{8} = \frac{507}{16} = 31\frac{11}{16} = 31.7$

$\frac{13}{2} \div \frac{39}{8} = \frac{13}{2} \times \frac{8}{39} = \frac{4}{3} = 1\frac{1}{3} = 1.33$

FRACTION TO DECIMAL TO PERCENTAGE
CONVERSIONS   4–12

Numbers are expressed as fractions, decimal numbers, or as percentages. We have worked with fractions and decimals and have converted between the two. In this section, we will convert both fractions and decimals to their percent equivalents.

To change a fraction to a percentage, we must first change the fraction to a decimal number, and then change the decimal number to a percentage.

> ☞ **RULE 4–5** To change a decimal number to a percent, multiply by 100 and add the % sign.

EXAMPLE 4–28

Change the following fractions to percentages. Round the decimal numbers to three places if the fourth place number is zero. Round to four places if the fourth place number is not zero.   (a) 3/4   (b) 7/25   (c) 11/12   (d) 47/100

SOLUTIONS   To change the decimal number to a percentage, multiply by 100 (this moves the decimal point two places *right*) and add the percent sign. (a) 3/4 = 0.750 = 75.0%; (b) 7/25 = 0.280 = 28.0%; (c) 11/12 = 0.9167 = 91.67%; (d) 47/100 = 0.470 = 47.0%

To change from a percent to a fraction we must first change from a percent to a decimal number and then convert the decimal number to a fraction.

> ☞ **RULE 4–6** To change from a percent to a decimal number, divide the number by 100 and remove the % sign.

EXAMPLE 4–29

Change the following percentages to fractions. (a) 30% (b) 6.25% (c) 52% (d) 162.5%

SOLUTIONS   First change from a percent to a decimal by dividing by 100 (move the decimal point *left*) and removing the percent sign. Then change the decimal number to a fraction.

(a) 30% = 0.30 = 30/100 = 3/10;   (b) 6.25% = 0.0625 = 625/10,000 = 1/16
(c) 52% = 0.52 = 52/100 = 13/25   (d) 162.5% = 1.625 = 1625/1000 = 1 5/8

Complete the table. Round the decimal number to four places if the fourth-place number is not zero.

	Fraction	Decimal	Percent
1.	5/8		
2.	1/6		
3.	5/12		
4.	1/32		
5.	3/4		
6.		0.1875	
7.		0.875	
8.		0.350	
9.		0.800	
10.		0.0375	
11.			12%
12.			9.375%
13.			42.5%
14.			30%
15.			26.5%

SOLUTIONS

	Fraction	Decimal	Percent
1.	5/8	0.625	62.5%
2.	1/6	0.1667	16.67%
3.	5/12	0.4167	41.67%
4.	1/32	0.03125	3.125%
5.	3/4	0.750	75.0%
6.	3/16	0.1875	18.75%
7.	7/8	0.875	87.5%
8.	7/20	0.350	35.0%
9.	4/5	0.800	80.0%
10.	3/80	0.0375	3.75%
11.	3/25	0.120	12.0%
12.	3/32	0.09375	9.375%
13.	17/40	0.425	42.5%
14.	3/10	0.300	30%
15.	53/200	0.265	26.5%

Additional practice problems are at the end of the chapter.

Add, subtract, multiply, and divide the following numbers. Express answers as decimal numbers rounded to three places and then as proper fractions or as mixed numbers.

1. $\dfrac{9}{16}$ and $\dfrac{3}{8}$

2. $\dfrac{2}{3}$ and $\dfrac{1}{6}$

3. $5\dfrac{7}{8}$ and $2\dfrac{1}{4}$

4. $3\dfrac{3}{16}$ and $6\dfrac{1}{3}$

Convert the following fraction to a decimal and a percent.

5. 13/16

Convert the following decimal to a fraction and a percent.

6. 0.550

Answers to Self-test 4–5 are at the end of the chapter.

CHAPTER 4 AT A GLANCE

PAGE	KEY POINT OR RULE	EXAMPLE
108	*Key Point:* When two (or more) whole numbers are multiplied, the result is called the *product* of those numbers. The two (or more) original numbers are called the *factors* of the product.	3 and 7 are factors of 21 because $3 \times 7 = 21$.
113	*Rule 4-1:* To multiply fractions, we (1) cancel factors that are in both numerators and denominators, (2) multiply the remaining numerators (their product is the numerator of the answer), (3) multiply the remaining denominators (their product is the denominator of the answer).	$\dfrac{5}{13} \times \dfrac{26}{50} = \dfrac{\cancel{5} \times \cancel{2} \times \cancel{13}}{\cancel{13} \times \cancel{5} \times 5 \times \cancel{2}} = \dfrac{1}{5}$
118	*Key Point:* When the denominators of fractions are identical, that number is the *common denominator* (CD).	In the fractions $\dfrac{1}{8}, \dfrac{3}{8},$ and $\dfrac{7}{8},$ 8 is the CD.
119	*Rule 4–2:* Any two signs of a fraction may be changed without changing the value of the fraction.	$-\dfrac{+3}{+8} = +\dfrac{-3}{+8} = -\dfrac{-3}{-8} = +\dfrac{+3}{-8}$

122	*Key Point:* The *lowest common multiple* (LCM) of two or more numbers is the lowest multiple that is common to each of them.	Some multiples of 4 and 6 are 12, 24, 36, and 48. The LCM is 12.

122   *Rule 4–3:* To find the LCM of two or more numbers: (1) Factor each term; (2) determine the *maximum* number of times a prime factor appears in any one term; (3) multiply the resulting prime factors found in step 2. This number is the LCM.

Find the LCM of 30 and 36.

$$30 = 2 \cdot 3 \cdot 5$$
$$36 = 2 \cdot 2 \cdot 3 \cdot 3$$
$$\text{LCM} = 2 \cdot 2 \cdot 3 \cdot 3 \cdot 5$$
$$= 180$$

124   *Rule 4–4:* To add or subtract fractions with unlike denominators: (1) find the CD (2) change the denominator of all fractions to this denominator, add the fractions, combining all like terms in the numerator. Reduce to lowest terms.

$$\frac{1}{3} + \frac{1}{4} = \frac{4}{12} + \frac{3}{12}$$
$$= \frac{4 + 3}{12} = \frac{7}{12}$$

128   *Key Point:* When the numerator is greater than the denominator, the fraction is called an *improper* fraction.

$\frac{9}{8}, \frac{25}{16},$ and $\frac{17}{3}$ are examples.

128   *Key Point:* Numbers that have a whole number part and a fractional part are called *mixed numbers*.

$4\frac{3}{4}$ and $6\frac{7}{8}$ are examples.

132   *Key Point:* When we multiply or divide mixed numbers, we change the mixed numbers to improper fractions and perform the function as before.

$$2\frac{2}{3} \times 3\frac{1}{4} = \frac{8}{3} \times \frac{13}{4}$$
$$= \frac{104}{12} = 8\frac{2}{3}$$

134   *Key Point:* To add and subtract, we change all mixed numbers to improper fractions and perform the function as with proper fractions.

$$2\frac{2}{3} + 3\frac{1}{4} = \frac{8}{3} + \frac{13}{4}$$
$$= \frac{32 + 39}{12} = \frac{71}{12} = 5\frac{11}{12}$$

141   *Rule 4–5:* To change a decimal number to a percent, multiply by 100 and add the % sign.

$0.175 = 17.5\%$

141   *Rule 4–6:* To change from a percent to a decimal number, divide the number by 100 and remove the % sign.

$62.5\% = 0.625$

END OF CHAPTER PROBLEMS 4–1

Find the prime factors of the following:

1. 18	2. 36	3. 44
4. 46	5. 63	6. 35
7. 92	8. 110	9. 231
10. 240	11. 56	12. 68
13. 147	14. 294	15. 84
16. 130	17. 210	18. 380
19. 455	20. 925	

END OF CHAPTER PROBLEMS 4–2

Examine the following fractions and reduce to lowest terms:

1. $\dfrac{3}{21}$	2. $\dfrac{5}{35}$	3. $\dfrac{5}{8}$
4. $\dfrac{12}{48}$	5. $\dfrac{22}{32}$	6. $\dfrac{18}{90}$
7. $\dfrac{18}{45}$	8. $\dfrac{24}{68}$	9. $\dfrac{21}{40}$
10. $\dfrac{45}{135}$	11. $\dfrac{66}{110}$	12. $\dfrac{195}{780}$
13. $\dfrac{182}{195}$	14. $\dfrac{330}{540}$	15. $\dfrac{273}{441}$
16. $\dfrac{220}{660}$	17. $\dfrac{294}{330}$	18. $\dfrac{390}{5005}$
19. $\dfrac{1190}{3570}$	20. $\dfrac{3675}{3850}$	

21. Reduce the fraction sixty-four one hundred twenty-eighths to its lowest terms.
22. Reduce the fraction twenty-eight one hundred twenty-eighths to its lowest terms.

END OF CHAPTER PROBLEMS 4–3

Find the product of the following fractions and reduce to lowest terms:

1. $\dfrac{1}{2} \times \dfrac{1}{3}$	2. $\dfrac{3}{4} \times \dfrac{1}{5}$	3. $\dfrac{3}{8} \times \dfrac{2}{9}$
4. $\dfrac{3}{4} \times \dfrac{1}{3}$	5. $\dfrac{9}{16} \times \dfrac{8}{15}$	6. $\dfrac{5}{8} \times \dfrac{5}{16}$
7. $\dfrac{22}{33} \times \dfrac{15}{28}$	8. $\dfrac{21}{40} \times \dfrac{20}{35}$	9. $\dfrac{3}{4} \times \dfrac{7}{8}$
10. $\dfrac{5}{8} \times \dfrac{2}{3}$	11. $\dfrac{3}{8} \times \dfrac{4}{9} \times \dfrac{2}{3}$	12. $\dfrac{3}{5} \times \dfrac{4}{9} \times \dfrac{5}{6}$

13. $\dfrac{1}{3} \times \dfrac{1}{4} \times \dfrac{2}{5}$

14. $\dfrac{5}{9} \times \dfrac{2}{3} \times \dfrac{1}{5}$

15. $\dfrac{1}{5} \times \dfrac{4}{7} \times \dfrac{1}{2}$

16. $\dfrac{2}{3} \times \dfrac{3}{4} \times \dfrac{2}{5}$

17. $\dfrac{6}{35} \times \dfrac{14}{15} \times \dfrac{5}{12}$

18. $\dfrac{10}{21} \times \dfrac{7}{12} \times \dfrac{18}{25}$

19. $\dfrac{8}{15} \times \dfrac{5}{16} \times \dfrac{22}{25}$

20. $\dfrac{15}{39} \times \dfrac{14}{30} \times \dfrac{3}{8}$

21. Multiply three-eighths and two-sevenths.
22. Multiply three-fourths and one-sixth.

END OF CHAPTER PROBLEMS 4–4

Perform the following divisions. Reduce your answers to lowest terms.

1. $\dfrac{9}{16} \div \dfrac{3}{4}$

2. $\dfrac{2}{5} \div \dfrac{2}{3}$

3. $\dfrac{3}{16} \div \dfrac{7}{8}$

4. $\dfrac{3}{20} \div \dfrac{5}{6}$

5. $\dfrac{5}{9} \div \dfrac{2}{3}$

6. $\dfrac{4}{15} \div \dfrac{2}{5}$

7. $\dfrac{9}{16} \div \dfrac{21}{32}$

8. $\dfrac{9}{20} \div \dfrac{7}{15}$

9. $\dfrac{8}{25} \div \dfrac{16}{35}$

10. $\dfrac{3}{10} \div \dfrac{15}{32}$

11. $\dfrac{5}{36} \div \dfrac{15}{54}$

12. $\dfrac{25}{48} \div \dfrac{25}{28}$

13. $\dfrac{7}{16} \div \dfrac{15}{32}$

14. $\dfrac{9}{32} \div \dfrac{33}{64}$

15. $\dfrac{35}{64} \div \dfrac{25}{32}$

16. $\dfrac{25}{64} \div \dfrac{25}{32}$

17. $\dfrac{23}{225} \div \dfrac{92}{175}$

18. $\dfrac{19}{325} \div \dfrac{76}{425}$

19. $\dfrac{105}{121} \div \dfrac{320}{363}$

20. $\dfrac{120}{338} \div \dfrac{756}{845}$

21. Divide three-fourths by six.
22. Divide two-ninths by four-ninths.

END OF CHAPTER PROBLEMS 4–5

Perform the indicated operations. Reduce to lowest terms where possible.

1. $\dfrac{1}{8} + \dfrac{3}{8}$

2. $\dfrac{5}{16} + \dfrac{9}{16}$

3. $\dfrac{2}{7} + \dfrac{3}{7}$

4. $\dfrac{9}{32} + \dfrac{15}{32}$

5. $\dfrac{11}{12} - \dfrac{5}{12}$

6. $\dfrac{5}{8} - \dfrac{1}{8}$

7. $\dfrac{9}{14} - \dfrac{5}{14}$

8. $\dfrac{5}{9} - \dfrac{3}{9}$

9. $\dfrac{5}{12} + \dfrac{5}{12} + \dfrac{1}{12}$

10. $\dfrac{3}{18} + \dfrac{5}{18} + \dfrac{1}{18}$

11. $\dfrac{3}{16} - \dfrac{5}{16} + \dfrac{7}{16}$

12. $\dfrac{1}{12} + \dfrac{7}{12} - \dfrac{5}{12}$

13. $\dfrac{7}{24} + \dfrac{11}{24} - \dfrac{5}{24}$

14. $\dfrac{23}{32} - \dfrac{17}{32} + \dfrac{9}{32}$

15. $\dfrac{7}{9} - \dfrac{5}{9} - \dfrac{4}{9}$

16. $\dfrac{5}{11} - \dfrac{9}{11} + \dfrac{3}{11}$

17. $\dfrac{11}{16} + \dfrac{5}{16} - \dfrac{7}{16}$

18. $\dfrac{3}{10} - \dfrac{5}{10} + \dfrac{9}{10}$

19. $\dfrac{17}{18} - \dfrac{11}{18} + \dfrac{5}{18}$

20. $\dfrac{7}{30} + \dfrac{11}{30} - \dfrac{23}{30}$

21. $\dfrac{7}{8} - \dfrac{3}{8} - \dfrac{5}{8}$

22. $\dfrac{1}{16} - \dfrac{11}{16} - \dfrac{3}{16}$

END OF CHAPTER PROBLEMS 4–6

Find the lowest common multiple (LCM) of the following numbers:

1. 6, 15
2. 20, 36
3. 55, 65
4. 45, 63
5. 84, 90
6. 45, 60
7. 22, 26
8. 34, 102
9. 8, 24, 36
10. 15, 18, 21
11. 63, 75, 105
12. 16, 24, 52
13. 18, 36, 72
14. 14, 56, 196
15. 20, 52, 65
16. 21, 70, 105
17. 50, 70, 175
18. 54, 132, 99
19. 45, 75, 225
20. 60, 51, 85

END OF CHAPTER PROBLEMS 4–7

Perform the indicated operations. Reduce to lowest terms.

1. $\dfrac{1}{4} + \dfrac{1}{6}$

2. $\dfrac{1}{3} + \dfrac{1}{5}$

3. $\dfrac{2}{5} + \dfrac{4}{15}$

4. $\dfrac{4}{7} + \dfrac{5}{21}$

5. $\dfrac{2}{3} - \dfrac{1}{6}$

6. $\dfrac{5}{12} - \dfrac{5}{8}$

7. $\dfrac{7}{8} - \dfrac{3}{16}$

8. $\dfrac{15}{16} - \dfrac{2}{3}$

9. $\dfrac{17}{24} - \dfrac{5}{6}$

10. $\dfrac{7}{18} - \dfrac{5}{6}$

11. $\dfrac{5}{12} - \dfrac{13}{16}$

12. $\dfrac{7}{30} - \dfrac{11}{15}$

13. $\dfrac{2}{3} + \dfrac{1}{6} + \dfrac{1}{12}$

14. $\dfrac{2}{5} + \dfrac{4}{15} + \dfrac{7}{30}$

15. $\dfrac{4}{15} + \dfrac{3}{25} + \dfrac{1}{3}$

16. $\dfrac{7}{30} + \dfrac{6}{35} + \dfrac{1}{5}$

17. $\dfrac{11}{42} + \dfrac{2}{3} - \dfrac{5}{28}$

18. $\dfrac{7}{36} + \dfrac{1}{2} - \dfrac{7}{48}$

19. $\dfrac{9}{56} - \dfrac{13}{42} - \dfrac{1}{21}$

20. $\dfrac{5}{33} - \dfrac{11}{66} - \dfrac{5}{8}$

21. $\dfrac{4}{55} - \dfrac{3}{25} + \dfrac{3}{5}$

22. $\dfrac{4}{39} + \dfrac{15}{78} - \dfrac{9}{26}$

23. $\dfrac{3}{65} + \dfrac{2}{15} - \dfrac{4}{39}$

24. $\dfrac{1}{3} + \dfrac{7}{16} + \dfrac{5}{24}$

25. Add two-fifths and one-third.
26. Add three-sevenths and seven-fourteenths.
27. Subtract three sixty-fourths from nine thirty-seconds.
28. Subtract five-eighteenths from seven-ninths.

END OF CHAPTER PROBLEMS 4–8

Change the following improper fractions to mixed numbers:

1. $\dfrac{17}{2}$ 2. $\dfrac{9}{4}$ 3. $\dfrac{37}{4}$

4. $\dfrac{26}{5}$ 5. $\dfrac{19}{5}$ 6. $\dfrac{17}{8}$

7. $\dfrac{43}{7}$ 8. $\dfrac{38}{8}$ 9. $\dfrac{27}{4}$

10. $\dfrac{37}{16}$ 11. $\dfrac{67}{7}$ 12. $\dfrac{57}{7}$

13. $\dfrac{87}{16}$ 14. $\dfrac{59}{16}$ 15. $\dfrac{89}{12}$

16. $\dfrac{107}{12}$ 17. $\dfrac{97}{32}$ 18. $\dfrac{113}{32}$

19. $\dfrac{37}{15}$ 20. $\dfrac{53}{15}$

Change the following mixed numbers to improper fractions:

21. $10\dfrac{1}{2}$ 22. $7\dfrac{3}{4}$ 23. $4\dfrac{3}{16}$

24. $3\dfrac{3}{5}$ 25. $3\dfrac{5}{8}$ 26. $4\dfrac{1}{7}$

27. $9\dfrac{1}{3}$ 28. $8\dfrac{3}{4}$ 29. $5\dfrac{1}{6}$

30. $6\dfrac{3}{8}$ 31. $8\dfrac{7}{8}$ 32. $5\dfrac{3}{16}$

33. $12\dfrac{1}{3}$ 34. $12\dfrac{5}{8}$ 35. $7\dfrac{1}{3}$

36. $8\dfrac{3}{32}$ 37. $10\dfrac{3}{5}$ 38. $10\dfrac{3}{5}$

39. $7\dfrac{5}{8}$ 40. $5\dfrac{11}{16}$

END OF CHAPTER PROBLEMS 4–9

Perform the indicated operations. If the answer is an improper fraction, change to a mixed number.

1. $\dfrac{3}{8} \times \dfrac{10}{3}$ 2. $\dfrac{8}{5} \times \dfrac{5}{2}$ 3. $\dfrac{3}{4} \times \dfrac{8}{5}$

4. $\dfrac{2}{3} \times \dfrac{16}{9}$ 5. $\dfrac{3}{8} \times \dfrac{5}{3}$ 6. $\dfrac{7}{16} \times \dfrac{32}{11}$

7. $\dfrac{9}{4} \times 2\dfrac{1}{5}$

8. $\dfrac{8}{5} \times 2\dfrac{7}{8}$

9. $3 \times 3\dfrac{1}{4}$

10. $2 \times 5\dfrac{5}{16}$

11. $2\dfrac{1}{3} \times 3\dfrac{1}{4}$

12. $2\dfrac{5}{6} \times 2\dfrac{7}{8}$

13. $3\dfrac{1}{3} \times 4\dfrac{1}{5}$

14. $2\dfrac{7}{8} \times 5\dfrac{3}{32}$

15. $3\dfrac{3}{8} \times 6\dfrac{1}{3}$

16. $4\dfrac{3}{8} \times 2\dfrac{3}{16}$

17. $2\dfrac{5}{6} \times 3\dfrac{2}{3}$

18. $4\dfrac{1}{3} \times 3\dfrac{3}{4}$

19. $5\dfrac{5}{6} \times 4\dfrac{2}{5}$

20. $6\dfrac{2}{3} \times 5\dfrac{5}{8}$

21. $\dfrac{7}{2} \div \dfrac{4}{3}$

22. $\dfrac{9}{5} \div \dfrac{7}{3}$

23. $\dfrac{7}{16} \div \dfrac{1}{4}$

24. $\dfrac{5}{9} \div \dfrac{2}{3}$

25. $3 \div \dfrac{7}{5}$

26. $4 \div \dfrac{11}{3}$

27. $6 \div \dfrac{5}{8}$

28. $5 \div \dfrac{1}{3}$

29. $\dfrac{9}{4} \div 1\dfrac{2}{3}$

30. $\dfrac{8}{7} \div 2\dfrac{3}{4}$

31. $\dfrac{7}{4} \div 2\dfrac{4}{5}$

32. $\dfrac{11}{9} \div 3\dfrac{2}{3}$

33. $3\dfrac{1}{2} \div 4\dfrac{1}{3}$

34. $5\dfrac{2}{3} \div 3\dfrac{1}{3}$

35. $2\dfrac{1}{3} \div 6\dfrac{1}{2}$

36. $1\dfrac{9}{16} \div 3\dfrac{3}{8}$

37. $5\dfrac{3}{5} \div 3\dfrac{1}{3}$

38. $3\dfrac{2}{3} \div 1\dfrac{13}{18}$

39. $4\dfrac{2}{5} \div 1\dfrac{8}{25}$

40. $6\dfrac{3}{8} \div 1\dfrac{3}{16}$

41. Multiply the numbers three and seven-eights and six and two-thirds.
42. Multiply the numbers one and four-fifths and two and six-sevenths.
43. Divide twenty-six and five-sixths by five and one-fourth.
44. Divide five and five-sixteenths by three and three-fourths.

END OF CHAPTER PROBLEMS 4–10

Perform the indicated operations. If the answer is an improper fraction, change to a mixed number.

1. $\dfrac{16}{5} + \dfrac{10}{3}$

2. $\dfrac{15}{2} + \dfrac{10}{7}$

3. $4\dfrac{5}{6} + \dfrac{5}{7}$

4. $6\dfrac{1}{10} + \dfrac{2}{5}$

5. $2\dfrac{1}{3} + \dfrac{9}{5}$

6. $3\dfrac{3}{8} + \dfrac{13}{9}$

7. $3\dfrac{7}{8} + 4\dfrac{1}{2}$

8. $4\dfrac{2}{5} + 3\dfrac{7}{10}$

9. $\dfrac{7}{8} - \dfrac{7}{32}$

10. $\dfrac{23}{32} - \dfrac{5}{16}$

11. $\dfrac{10}{3} - \dfrac{1}{4}$

12. $\dfrac{12}{5} - \dfrac{4}{3}$

13. $3\dfrac{5}{6} - \dfrac{9}{4}$

14. $7\dfrac{1}{2} - \dfrac{13}{8}$

15. $3\dfrac{1}{8} - 1\dfrac{1}{4}$

16. $3\dfrac{7}{12} - 5\dfrac{3}{8}$ 17. $4\dfrac{5}{6} - 2\dfrac{2}{3}$ 18. $6\dfrac{9}{16} - 3\dfrac{5}{32}$

19. $\dfrac{9}{4} + \dfrac{7}{2} - 2\dfrac{1}{3}$ 20. $\dfrac{8}{3} + \dfrac{9}{5} - 1\dfrac{1}{2}$ 21. $7\dfrac{1}{2} - 2\dfrac{1}{3} + 3\dfrac{1}{4}$

22. $6\dfrac{2}{3} + 3\dfrac{1}{5} - 2\dfrac{1}{8}$ 23. $2\dfrac{7}{8} + 4\dfrac{3}{4} - 3\dfrac{11}{16}$ 24. $7\dfrac{1}{3} - 4\dfrac{5}{6} + 4\dfrac{2}{3}$

25. $3\dfrac{7}{16} - \dfrac{15}{32} + 2\dfrac{3}{8}$ 26. $6\dfrac{7}{8} - 2\dfrac{3}{16} + 5\dfrac{7}{16}$ 27. $7\dfrac{2}{3} - 3\dfrac{2}{5} + 2\dfrac{4}{15}$

28. $2\dfrac{2}{7} + 3\dfrac{2}{3} - 5\dfrac{3}{7}$ 29. $8\dfrac{3}{8} + 4\dfrac{2}{3} - 3\dfrac{1}{12}$ 30. $6\dfrac{1}{6} + 4\dfrac{1}{3} - 2\dfrac{1}{9}$

31. Add one and five-eighths to three and one-half.
32. Add five-eighths to two and two-thirds.
33. Subtract two and three-fifths from nine-tenths.
34. Subtract one and five-eighths from three-fourths.
35. One part of a circuit dissipates three-eighths of a watt of power. The remaining parts of the circuit dissipate five-sixths of a watt. What is the total power dissipation? Express the answer as a proper fraction or as a mixed number.
36. One part of a circuit dissipates four-ninths of a watt of power. The remaining parts of the circuit dissipate nine-tenths of a watt. What is the total power dissipation in watts? Express your answer as a proper fraction or as a mixed number.

END OF CHAPTER PROBLEMS 4–11

Add, subtract, multiply, and divide the following numbers. Express answers as proper fractions or as mixed numbers, and then as decimal numbers rounded to three places.

1. $\dfrac{3}{8}$ and $\dfrac{3}{32}$ 2. $\dfrac{3}{5}$ and $\dfrac{2}{3}$ 3. $\dfrac{7}{16}$ and $\dfrac{2}{7}$

4. $\dfrac{13}{32}$ and $\dfrac{5}{16}$ 5. $\dfrac{5}{8}$ and $\dfrac{37}{64}$ 6. $\dfrac{5}{16}$ and $\dfrac{13}{64}$

7. $\dfrac{3}{5}$ and $\dfrac{3}{10}$ 8. $\dfrac{5}{12}$ and $\dfrac{3}{10}$ 9. $3\dfrac{3}{8}$ and $2\dfrac{1}{3}$

10. $7\dfrac{5}{32}$ and $4\dfrac{1}{16}$ 11. $4\dfrac{2}{5}$ and $2\dfrac{15}{64}$ 12. $9\dfrac{7}{16}$ and $4\dfrac{9}{32}$

13. $7\dfrac{1}{2}$ and $3\dfrac{5}{9}$ 14. $8\dfrac{9}{16}$ and $5\dfrac{23}{64}$ 15. $4\dfrac{7}{8}$ and $8\dfrac{7}{16}$

16. $5\dfrac{4}{7}$ and $9\dfrac{5}{12}$ 17. $3\dfrac{2}{3}$ and $6\dfrac{2}{3}$ 18. $6\dfrac{3}{64}$ and $9\dfrac{1}{8}$

19. $1\dfrac{7}{12}$ and $3\dfrac{2}{3}$ 20. $2\dfrac{9}{16}$ and $4\dfrac{13}{32}$

END OF CHAPTER PROBLEMS 4–12

Complete the table. Round the decimal number to four places if the fourth-place number is not zero.

	Fraction	Decimal	Percent
1.	3/8		
2.	7/25		
3.	19/400		
4.	9/400		
5.	7/16		
6.	9/20		
7.	1/15		
8.	4/15		
9.	1/12		
10.	7/12		
11.	1/3		
12.	2/3		
13.		0.700	
14.		0.400	
15.		0.0500	
16.		0.920	
17.		0.200	
18.		0.165	
19.		0.00400	
20.		0.900	
21.		0.175	
22.		0.0900	
23.		0.6875	
24.		0.9375	
25.		0.600	
26.		0.8125	
27.			7.5%
28.			45%
29.			15%
30.			55%
31.			0.5%
32.			900%
33.			250%
34.			0.45%
35.			56.25%
36.			12.5%
37.			31.25%
38.			67.5%

ANSWERS TO SELF-TESTS

SELF-TEST 4–1

1. $2 \cdot 2 \cdot 2 \cdot 5 = 2^3 \cdot 5$
2. $2 \cdot 2 \cdot 2 \cdot 2 \cdot 3 = 2^4 \cdot 3$
3. $2 \cdot 2 \cdot 2 \cdot 7 = 2^3 \cdot 7$
4. $2 \cdot 2 \cdot 3 \cdot 7 = 2^2 \cdot 3 \cdot 7$
5. $2 \cdot 2 \cdot 3 \cdot 3 \cdot 5 = 2^2 \cdot 3^2 \cdot 5$
6. $2 \cdot 2 \cdot 5 \cdot 11 = 2^2 \cdot 5 \cdot 11$
7. $\dfrac{16}{25}$
8. $\dfrac{2}{3}$
9. $\dfrac{3}{7}$
10. $\dfrac{7}{15}$

SELF-TEST 4–2

1. $\dfrac{7}{20}$
2. $\dfrac{1}{4}$
3. $\dfrac{1}{30}$
4. $\dfrac{2}{25}$
5. $\dfrac{3}{10}$
6. $\dfrac{2}{5}$
7. $\dfrac{1}{5}$
8. $\dfrac{2}{3}$
9. $\dfrac{63}{88}$

SELF-TEST 4–3

1. 60
2. 120
3. 210
4. 225
5. 210
6. 210
7. 240
8. 4095
9. $\dfrac{1}{2}$
10. $\dfrac{13}{24}$
11. $-\dfrac{1}{6}$
12. $\dfrac{3}{5}$
13. $-\dfrac{1}{20}$
14. $\dfrac{13}{24}$
15. $-\dfrac{23}{240}$
16. $-\dfrac{17}{24}$

SELF-TEST 4–4

1. $3\dfrac{5}{8}$
2. $6\dfrac{3}{5}$
3. $\dfrac{51}{16}$
4. $-\dfrac{26}{3}$
5. $5\dfrac{29}{32}$
6. 15
7. $\dfrac{3}{5}$
8. $1\dfrac{15}{17}$
9. $4\dfrac{1}{6}$
10. $8\dfrac{1}{8}$
11. $5\dfrac{13}{32}$
12. $3\dfrac{11}{12}$
13. $-\dfrac{23}{24}$
14. $5\dfrac{19}{30}$

1. $\dfrac{9}{16} + \dfrac{3}{8} = \dfrac{15}{16} = 0.938$

   $\dfrac{9}{16} - \dfrac{3}{8} = \dfrac{3}{16} = 0.188$

   $\dfrac{9}{16} \times \dfrac{3}{8} = \dfrac{27}{128} = 0.211$

   $\dfrac{9}{16} \div \dfrac{3}{8} = 1\dfrac{1}{2} = 1.50$

2. $\dfrac{2}{3} + \dfrac{1}{6} = \dfrac{5}{6} = 0.833$

   $\dfrac{2}{3} - \dfrac{1}{6} = \dfrac{1}{2} = 0.500$

   $\dfrac{2}{3} \times \dfrac{1}{6} = \dfrac{1}{9} = 0.111$

   $\dfrac{2}{3} \div \dfrac{1}{6} = 4$

3. $5\dfrac{7}{8} + 2\dfrac{1}{4} = 8\dfrac{1}{8} = 8.13$

   $5\dfrac{7}{8} - 2\dfrac{1}{4} = 3\dfrac{5}{8} = 3.63$

   $5\dfrac{7}{8} \times 2\dfrac{1}{4} = 13\dfrac{7}{32} = 13.2$

   $5\dfrac{7}{8} \div 2\dfrac{1}{4} = 2\dfrac{11}{18} = 2.61$

4. $3\dfrac{3}{16} + 6\dfrac{1}{3} = 9\dfrac{25}{48} = 9.52$

   $3\dfrac{3}{16} - 6\dfrac{1}{3} = -3\dfrac{7}{48} = -3.15$

   $3\dfrac{3}{16} \times 6\dfrac{1}{3} = 20\dfrac{3}{16} = 20.2$

   $3\dfrac{3}{16} \div 6\dfrac{1}{3} = \dfrac{153}{304} = 0.503$

5. $\dfrac{13}{16} = 0.8125 = 81.25\%$

6. $0.550 = \dfrac{11}{20} = 55.0\%$

Algebra Fundamentals

Algebra is the part of mathematics where letters of the alphabet are used to represent numerical quantities and procedures are developed for performing mathematical operations on these letters.

In Chapter 5, we define several algebraic words and show how algebraic procedures are used to solve some basic electronics problems. In Chapter 6, we describe procedures for multiplying and dividing algebraic fractions that are useful to solve electronics problems. In Chapter 7, we define linear and second degree equations and show how to solve these types of equations. In Chapter 8, we discuss the multiplication, division, and factoring of polynomials. In Chapter 9, we use the algebraic tools we learned in the previous chapters to solve practical electrical problems.

The algebraic procedures learned in this section must be mastered in order to advance to the next section that deals with analysis of basic electrical circuits.

Algebraic Terms: Roots and Powers

5

Introduction

When we move into areas new to us, whether that area is mathematics or electronics or some other technology, there are always new words to learn. To communicate with other people in the field, the same words in our vocabularies must have the same meaning if we are to understand each other. The words that we will define, such as *algebraic expression, term,* and *literal number,* are probably words you have heard and used before.

We will learn how *variables* are used, and we will assign values to variables and solve some simple electronics problems. We will also solve problems that include squares and square roots.

The skills we develop in this chapter and the concepts we learn will be used throughout the text.

Chapter Objectives

In this chapter, you will learn how to:

1. Identify algebraic expressions and terms.
2. Solve problems that contain both real and literal numbers.
3. Find the square and square root of both real and literal numbers.
4. Solve problems using various forms of Ohm's law.
5. Develop problem-solving techniques.

NUMBERS, EXPRESSIONS, AND TERMS 5-1

5-1-1 Literal and Real Numbers

We may divide numbers into two groups. The first group we can define as *real* numbers. These are the numbers we use all the time: 7, 468, 3.25, -6, and $2\frac{1}{2}$ are all examples of real numbers. Real numbers always have the same value. The number 6 always has a value of 6. It can never mean some other amount. Real numbers are also referred to as *constants*.

Often, in electronics, we have quantities that are not constant. For example, circuit current will depend on the values we assign to circuit voltage and resistance. Voltages, currents, resistances, and other circuit parameters are called *circuit variables*. Letters called *literal numbers* are used to represent these quantities. $E = IR$ is Ohm's law written as an equation, where E represents the voltage source, I represents circuit current, and R is the circuit resistance. These letters, E, I, and R, represent the variables voltage, current, and resistance. These literal numbers can represent different values. The letter E can stand for one value in one circuit and an entirely different value in another circuit. E might equal 10 volts in one circuit, 25 volts in a second circuit, and so on. Literal numbers can also be used to represent constants, but we usually use them to represent variables. Later in this chapter, and in other chapters, we will solve many equations that contain variables.

5-1-2 Algebraic Expressions and Terms

An *algebraic expression* is a group of numbers that contains a variable. The numbers can be real numbers, literal numbers, or both. $6A$, $3 + 2x$, $7(x - 2)$, $1.6 + 7x$, and $3ab + 4c - 5ac$ are all examples of algebraic expressions. If we knew the value of the literal numbers, then we could determine the numerical value of the expression.

When numbers are separated into groups with $+$ or $-$ signs, each group in the expression is called a *term*. The algebraic expression $3ab + 4c - 5ac$ contains three terms: $3ab$, $4c$, and $5ac$. $6A$ is one term. The expression $1.6 + 7x$ contains two terms: 1.6 and $7x$.

When an algebraic expression contains one term, it is called a *monomial*. When an expression contains more than one term, it is called a *polynomial*. Two-term expressions are also called *binomials*. Three-term expressions are also called *trinomials*. These words, binomial and trinomial, are more descriptive than polynomial.

So, an algebraic expression can be a monomial or a polynomial. Furthermore, monomials and polynomials can contain either real or literal numbers, or both.

5-1-3 Numerical Coefficients

The *numerical coefficient* of an algebraic term is the real-number part of the term. If no numerical value is given, it is understood to be 1. In the term $6xy$, 6 is the numerical coefficient. In the term ax, 1 is the coefficient.

PRACTICE PROBLEMS 5-1

List the number of terms in the following expressions. Also indicate the kind of expression (monomial, binomial, and so on).

1. $2a + 3y$
2. $2x + y + 1$
3. $4(x + 1) - 3$
4. $\dfrac{3a}{8} + 26$

5. $\dfrac{4x}{5} + 1$

6. $3(a - 1) + 2b$

7. $\dfrac{3a - 1}{4} + 6$

8. $\dfrac{4c + 3}{2d} - b$

9. $\dfrac{7y}{8} + 6 - 4(x - 3)$

10. $3a + 6b - 4c - 3$

11. $4c - 3a + 4$

12. $\dfrac{4x - 3y}{4} - 6x + 3$

In the following expressions, list the numerical coefficients and the literal numbers in each term.

13. $a + 2b + 4c$

14. $\dfrac{3x}{4} + y$

15. $-2a + 3b - \dfrac{c}{4}$

SOLUTIONS

1. 2, binomial
3. 2, binomial
5. 2, binomial
7. 2, binomial
9. 3, trinomial
11. 3, trinomial

2. 3, trinomial
4. 2, binomial
6. 2, binomial
8. 2, binomial
10. 4, polynomial
12. 3, trinomial

The binomials and trinomials may also be called polynomials.

13. 1, 2, and 4 are the numerical coefficients. a, b, and c are the literal numbers.
14. $\frac{3}{4}$ and 1 are the numerical coefficients. x and y are the literal numbers.
15. -2, 3, and $-\frac{1}{4}$ are the numerical coefficients. a, b and c are the literal numbers.

Additional practice problems are at the end of the chapter.

EXPONENTS 5-2

In Chapter 2, we discussed the exponent whose base was 10. The exponent told us how many times 10, the base, was multiplied by itself. This is the function of an exponent in any base. The base may be any real number. For example,

$$5^3 = 5 \cdot 5 \cdot 5 = 125$$

The base can also be a literal number:

$$x^3 = x \cdot x \cdot x = ?$$

We were able to find the value of 5^3 because 5 is a real number and we can deal with real numbers directly. We can find x^3 only if we are given the value for x. Suppose we assign a value of 4 to x.

Then

$$x^3 = x \cdot x \cdot x = 4 \cdot 4 \cdot 4 = 64$$

☞ **KEY POINT** We can solve problems that contain literal numbers only if we know their value.

PRACTICE PROBLEMS 5–2

Solve the following problems. Round your answers to three places.

1. 7^2 2. 6^3 3. 4.5^2
4. 8.73^2 5. 24^2

Solve the following problems where $x = 3$, $y = 4$:

6. x^3 7. y^2 8. y^3
9. x^2 10. x^4

SOLUTIONS

1. 49.0   2. 216   3. 20.3
4. 76.2   5. 576   6. 27.0
7. 16.0   8. 64.0   9. 9.00
10. 81.0

Additional practice problems are at the end of the chapter.

5–3   FINDING VALUES OF ALGEBRAIC EXPRESSIONS

When real numbers and literal numbers appear in the same term of an algebraic expression, we can determine the value of the term if we know the value of the literal numbers. For example, in the term $6x^2$, if $x = 3$, then $6x^2 = 6 \cdot x^2 = 6 \cdot 3^2 = 6 \cdot 9 = 54$. When one or more of the numbers in a term is raised to a power, only that number is affected.

When 3 is substituted for x in our example, we get $6 \cdot 3^2$. Our first step is to find 3^2. Then we complete the multiplication. We can find the value of any term by multiplying the real numbers times the known values of the literal numbers.

EXAMPLE 5–1

Find (a) 5^3; (b) a^2, where $a = 7.3$; (c) b^3, where $b = 14.5$. Round your answers to three places.

(a) $5^3 = 5 \times 5 \times 5 = 125$.

Using the calculator,

$$5 \boxed{\times} 5 \boxed{\times} 5 \boxed{=} \qquad \textit{Answer:} 125$$

or $\qquad 5 \boxed{y^x} 3 \boxed{=} \qquad \textit{Answer:} 125$

When using the $\boxed{y^x}$ key, the number is keyed in *first,* and then the exponent is keyed in. $\left(\text{Casio calculators use } \boxed{x^y} \text{ instead of } \boxed{y^x}.\right)$

(b) $a^2 = 7.3^2 = 7.3 \times 7.3 = 53.3$.

Using the calculator, $7.3 \boxed{x^2}$. The answer is displayed and is 53.3. Using the $\boxed{x^2}$ key saves keystrokes, which decreases the chance for errors.

(c) $b^3 = 14.5^3 = 14.5 \times 14.5 \times 14.5 = 3050$.

Using the calculator,

$$\text{TI:} \qquad 14.5 \boxed{y^x} 3 \boxed{=} \qquad \textit{Answer:} 3050$$

$$\text{Casio:} \quad 14.5 \boxed{x^y} 3 \boxed{=} \qquad \textit{Answer:} 3050$$

Notice in (b) and (c) that we had to be given a value for a and a value for b in order to solve the problems.

EXAMPLE 5–2

Let $x = 3$ and $y = 4$. Find (a) xy; (b) x^2y; (c) x^2y^2; (d) $4x^3y^2$. Round answers to 3 places.

(a) $xy = 3 \cdot 4 = 12$
(b) $x^2y = 3^2 \cdot 4 = 9 \cdot 4 = 36$
(c) $x^2y^2 = 3^2 \cdot 4^2 = 9 \cdot 16 = 144$
(d) $4x^3y^2 = 4 \cdot 3^3 \cdot 4^2 = 4 \cdot 27 \cdot 16 = 1730$

The algorithm for (d) looks like this:

$$4 \boxed{\times} 3 \boxed{y^x} 3 \boxed{\times} 4 \boxed{x^2} \boxed{=} \qquad \textit{Answer:} 1728 = 1730 \text{ to 3 places.}$$

☞ **KEY POINT** If a sign of grouping is used, then all numbers within that group are acted on by the exponent.

For example, in the term $(ab)^2$, both a and b are raised to the second power: $(ab)^2 = a^2 \cdot b^2$.

EXAMPLE 5–3

Find $(ab)^2$, where $a = 3$ and $b = 4$.

SOLUTION

$$(ab)^2 = a^2 \cdot b^2 = 3^2 \cdot 4^2 = 9 \cdot 16 = 144$$

It would also be correct to find the product of a and b and then raise that value to the second power.

$$(ab)^2 = (3 \cdot 4)^2 = 12^2 = 144$$

Either method is correct and either method may be used in problem solving.
Using the calculator,

TI: 3 $\boxed{\times}$ 4 $\boxed{=}$ $\boxed{x^2}$ *Answer: 144*

Casio: 3 $\boxed{\times}$ 4 $\boxed{=}$ $\boxed{x^2}$ $\boxed{=}$ *Answer: 144*

If there is more than one term in an algebraic expression, we find the value of each term and then perform the indicated addition or subtraction.

EXAMPLE 5–4

Find $3a^2b - ab^2$, where $a = 5$ and $b = 6$.

SOLUTION

$$3a^2b - ab^2 = 3 \cdot 5^2 \cdot 6 - 5 \cdot 6^2$$
$$= 3 \cdot 25 \cdot 6 - 5 \cdot 36$$
$$= 450 - 180 = 270$$

Using the calculator,

3 $\boxed{\times}$ 5 $\boxed{x^2}$ $\boxed{\times}$ 6 $\boxed{-}$ 5 $\boxed{\times}$ 6 $\boxed{x^2}$ $\boxed{=}$ *Answer: 270*

Find the values of the following algebraic expressions where $x = 3$, $y = 4$, and $z = 5$. Round answers to 3 places:

1. xy
2. x^2y
3. $(xy)^2$
4. $2y^2z$
5. $3(xz)^2$
6. $x^3y - z$
7. $y^3 - z^2$
8. $(x^2y - z^2)^2$
9. $(xy^2z^3 + 4x^2y^3z^2)^2$
10. $(2x^2yz^4 - 8xy^3z^2)^3$

SOLUTIONS

1. 12
2. 36
3. 144
4. 160
5. 675
6. 103
7. 39
8. 121
9. 4.04×10^9
10. 2.87×10^{11}

Additional practice problems are at the end of the chapter.

ROOTS | 5–4

The *square root* of a number is one of its two equal factors. $6 \cdot 6 = 36$. The two equal factors are 6; therefore, 6 is the square root of 36. Actually, there are *two* square roots of 36 because $(-6)(-6)$ also equals 36. All positive numbers have two square roots: one negative and one positive. Usually, in solving electrical problems, the negative root is not considered.

The symbol used to denote the square root of a number is $\sqrt{}$ and is called the *radical.* $\sqrt{36} = 6$. Raising a number to the one-half power is another way of denoting square root. $36^{1/2} = 6$. Both notations will be used in this text.

We can extract from memory the square root of numbers like 25, 36, 64, and 81. If we go much beyond $\sqrt{100}$, most of us have to use the calculator.

There is a significant difference between the TI and Casio calculators' usage of the square root key. The TI calculator uses the displayed number when the square root key is pressed. The Casio uses the number entered after the square root key is pressed.

When using either calculator, if the value for which the square root is sought contains more than one term, then use parentheses to enclose those terms.

After finding the square root of a number, we can check our answer by squaring it and comparing it to the original number.

EXAMPLE 5–5

Find $\sqrt{40}$.

$$\sqrt{40} = 6.32 \quad \text{(rounded to three places)}$$

TI: $\boxed{40}\ \boxed{\sqrt{x}}$ *Answer:* 6.32

Casio: $\boxed{\sqrt{\ }}\ \boxed{40}\ \boxed{=}$ *Answer:* 6.32

Check: $6.32 \cdot 6.32 \simeq 40$

Because we rounded to three places, we don't get exactly 40 when we check.

The cube root of a number is one of its three equal factors. The symbol is $\sqrt[3]{\ }$. Raising a number to the one-third power is another way of indicating cube root. $\sqrt[3]{x} = x^{1/3}$. Because we seldom work with cube roots or higher roots in electronics, we will limit problem solving in this chapter to square roots.

☞ **KEY POINT** When we extract the square root of an expression, we perform all mathematical functions under the radical *first*. Then we find the square root.

EXAMPLE 5–6

Find the value of the following algebraic expressions, where $x = 4$, $y = 5$, and $z = 6$. Express answers in scientific notation rounded to three places.

(a) $\sqrt{2x + 3y}$ (b) $\sqrt{3xy^2}$ (c) $\sqrt{x^2y^2 + 4}$
(d) $\sqrt{x^2y^3z^3 - x^3y^4z - 4y}$

SOLUTIONS

(a) $\sqrt{2x + 3y} = \sqrt{8 + 15} = \sqrt{23} = 4.80$

TI: $\boxed{2}\ \boxed{\times}\ \boxed{4}\ \boxed{+}\ \boxed{3}\ \boxed{\times}\ \boxed{5}\ \boxed{=}\ \boxed{\sqrt{x}}$ *Answer:* 4.80

Casio: $\boxed{\sqrt{\ }}\ \boxed{(}\ \boxed{2}\ \boxed{\times}\ \boxed{4}\ \boxed{+}\ \boxed{3}\ \boxed{\times}\ \boxed{5}\ \boxed{)}\ \boxed{=}$ *Answer:* 4.80

(b) $\sqrt{3xy^2} = \sqrt{3 \cdot 4 \cdot 5^2} = \sqrt{12 \cdot 25} = \sqrt{300} = 1.73 \times 10$

TI: $\boxed{3}\ \boxed{\times}\ \boxed{4}\ \boxed{\times}\ \boxed{5}\ \boxed{x^2}\ \boxed{=}\ \boxed{\sqrt{x}}$ *Answer:* 1.73 × 10

Casio: $\boxed{\sqrt{\ }}\ \boxed{(}\ \boxed{3}\ \boxed{\times}\ \boxed{4}\ \boxed{\times}\ \boxed{5}\ \boxed{x^2}\ \boxed{)}\ \boxed{=}$ *Answer:* 1.73 × 10

(c) $\sqrt{x^2y^2} + 4 = \sqrt{4^2 \cdot 5^2} + 4 = \sqrt{16 \cdot 25} + 4$

$$= \sqrt{400} + 4 = 20 + 4 = 2.40 \times 10$$

TI: $4\boxed{x^2}\boxed{\times}5\boxed{x^2}\boxed{=}\boxed{\sqrt{x}}\boxed{+}4\boxed{=}$ Answer: 2.40×10

Casio: $\boxed{\sqrt{}}\boxed{(}4\boxed{x^2}\boxed{\times}5\boxed{x^2}\boxed{)}\boxed{+}4\boxed{=}$ Answer: 2.40×10

(d) $\sqrt{x^2y^3z^3 - x^3y^4z} - 4y = \sqrt{4^2 \cdot 5^3 \cdot 6^3 - 4^3 \cdot 5^4 \cdot 6} - 4 \cdot 5$

$$= \sqrt{16 \cdot 125 \cdot 216 - 64 \cdot 625 \cdot 6} - 20 = \sqrt{4.32 \times 10^5 - 2.40 \times 10^5} - 20$$

$$= 438 - 20 = 4.18 \times 10^2$$

TI: $4\boxed{x^2}\boxed{\times}5\boxed{y^x}3\boxed{\times}6\boxed{y^x}3\boxed{-}\boxed{(}4\boxed{y^x}3\boxed{\times}$

$5\boxed{y^x}4\boxed{\times}6\boxed{)}\boxed{=}\boxed{\sqrt{x}}\boxed{-}4\boxed{\times}5\boxed{=}$
Answer: 4.18×10^2

Casio: $\boxed{\sqrt{}}\boxed{(}4\boxed{x^2}\boxed{\times}5\boxed{x^y}3\boxed{\times}6\boxed{x^y}3\boxed{-}4\boxed{x^y}3\boxed{\times}$

$5\boxed{x^y}4\boxed{\times}6\boxed{)}\boxed{-}4\boxed{\times}5\boxed{=}$ Answer: 4.18×10^2

PRACTICE PROBLEMS 5–4

Solve the following problems rounded to three places:

1. $5^{1/2}$

2. $7^{1/2}$

3. $27^{1/2}$

4. $256^{1/2}$

5. $512^{1/2}$

6. $1000^{1/2}$

Find the value of the following algebraic expressions, where $x = 3$, $y = 4$, and $z = 5$. Round answers to three places.

7. $\sqrt{xy}$

8. $\sqrt{z^2}$

9. $\sqrt{y^2z^2}$

10. $\sqrt{5x^2z}$

11. $\sqrt{x^2y^2}$

12. $\sqrt{4y^2 + z^2}$

13. $\sqrt{x^4 - 8y}$

14. $\sqrt{x^2yz^2 + 10}$

15. $\sqrt{3x^2y^3 - x^3z^2} + 3y$

16. $\sqrt{2x^4yz^2 - 3xy^3z} - 4xz$

1. 2.24   2. 2.65
3. 5.20   4. 16.0
5. 22.6   6. 31.6
7. 3.46   8. 5.00
9. 20.0   10. 15.0
11. 12.0   12. 9.43
13. 7.00   14. 30.2
15. 44.4   16. 55.4

Additional practice problems are at the end of the chapter.

SELF-TEST 5–1

Find the value of the following expressions, where $x = 3$, $y = 5$, and $z = 4$. Consider only the positive root. Round answers to three places.

1. x^2y 2. $\sqrt{xz}$

3. $(2xy)^2$ 4. $\sqrt{3xy^2}$

5. $xy + 2z$ 6. $3y^2 - x^2z$

7. $\sqrt{3x^3 + 2yz}$ 8. $\sqrt{3xy + yz^2}$

9. $\sqrt{2xz^2 + 3y^2z}$ 10. $\sqrt{4y^2z - 2xz^2}$

Answers to Self-test 5–1 are at the end of the chapter.

5–5   PRACTICAL APPLICATIONS—DC CIRCUITS

Literal numbers are used in equations to represent the relationship between variables in electronics circuits. In this section we will use some of these equations and solve for circuit variables. Where square roots are found, only the positive root is considered.

One form of Ohm's law is $E = IR$. E is the applied voltage and is measured in volts; I is the circuit current and is measured in amperes; R is the circuit resistance and is measured in ohms.

To solve the problems in the following examples with your calculator, set the calculator up in engineering notation if you have that feature; otherwise, set it up in scientific notation rounded to three significant figures.

EXAMPLE 5–7

Find E when $I = 5$ mA and $R = 2.7$ kΩ.

SOLUTION   Recall from Chapter 3 that the prefix milli (m) is 10^{-3} and kilo (k) is 10^3. Then

$$E = IR = 5 \text{ mA} \times 2.7 \text{ k}\Omega = 5 \times 10^{-3} \times 2.7 \times 10^3 = 13.5 \text{ V}$$

5 $\boxed{\text{EE}}$ $\boxed{+/-}$ 3 $\boxed{\times}$ 2.7 $\boxed{\text{EE}}$ 3 $\boxed{=}$ *Answer:* 13.5 V

Remember, on the Casio, to set the exponent for a power of ten, the keys are $\boxed{\text{EXP}}$ and $\boxed{(-)}$ rather than $\boxed{\text{EE}}$ and $\boxed{+/-}$.

The power dissipated in a circuit can be found by using the equation $P = I^2R$. Power is measured in watts.

EXAMPLE 5–8

Find P when $I = 50$ mA and $R = 220$ Ω.

SOLUTION

$$P = I^2R = (50 \text{ mA})^2 \times 220 \ \Omega = (50 \times 10^{-3})^2 \times 220 = 0.055 \text{ W}$$

50 $\boxed{\text{EE}}$ $\boxed{+/-}$ 3 $\boxed{x^2}$ $\boxed{\times}$ 220 $\boxed{=}$ *Answer:* 550 $\times$ 10^{-3} = 550 mW

When the power dissipated and the resistance are known, circuit current may be found by using the equation $I = \sqrt{\dfrac{P}{R}}$

EXAMPLE 5–9

Find I when $P = 700$ mW and $R = 3.3$ kΩ.

SOLUTION

$$I = \sqrt{\frac{P}{R}} = \sqrt{\frac{700 \text{ mW}}{3.3 \text{ k}\Omega}} = \sqrt{\frac{700 \times 10^{-3}}{3.3 \times 10^3}} = 1.46 \times 10^{-2} \text{ A}$$

$$= 14.6 \times 10^{-3} \text{ A} = 14.6 \text{ mA}$$

TI: 700 $\boxed{\text{EE}}$ $\boxed{+/-}$ 3 $\boxed{\div}$ 3.3 $\boxed{\text{EE}}$ 3 $\boxed{=}$ $\boxed{\sqrt{x}}$

Answer: 14.6 $\times$ 10^{-3} = 14.6 mA

Casio: $\boxed{\sqrt{}}$ $\boxed{(}$ 700 $\boxed{\text{EXP}}$ $\boxed{(-)}$ 3 $\boxed{\div}$ 3.3 $\boxed{\text{EXP}}$

3 $\boxed{)}$ $\boxed{=}$

Answer: 14.6 $\times$ 10^{-3} = 14.6 mA

☞ **KEY POINT** Prefixes are changed to their powers of ten values to solve problems, but it is standard practice to use prefixes instead of powers of ten in answers.

The applied voltage may be found when P and R are known by using the equation $E = \sqrt{PR}$.

EXAMPLE 5–10

Find E when $P = 100$ mW and $R = 10$ kΩ.

SOLUTION

$$E = \sqrt{PR} = \sqrt{100 \text{ mW} \times 10 \text{ k}\Omega}$$
$$= \sqrt{100 \times 10^{-3} \times 10 \times 10^{3}} = 31.6 \text{ V}$$

TI: 100 [EE] [+/−] 3 [×] 10 [EE] 3 [=] [√x̄]

Answer: 31.6 = 31.6 V

Casio: [√] [(] 100 [EXP] [(−)] 3 [×] 10 [EXP]

3 [)] [=]   *Answer:* 31.6 V

P can be found when E and R are known by using the equation $P = \dfrac{E^2}{R}$.

EXAMPLE 5–11

Find P when $E = 15$ V and $R = 2$ kΩ.

SOLUTION

$$P = \frac{E^2}{R} = \frac{15^2}{2 \text{ k}\Omega} = \frac{15^2}{2 \times 10^3} = 0.113 \text{ W} = 113 \text{ mW}$$

TI: 15 [x²] [÷] 2 [EE] 3 [=]

Answer: $1.13 \times 10^{-1} = 113$ mW

Casio: 15 [x²] [÷] 2 [EXP] 3 [=]

Answer: $1.13 \times 10^{-1} = 113$ mW

In the following problems, round answers to three places. Use prefixes where appropriate.

$E = IR$. Find E when:

1. $I = 2$ mA, $R = 2.7$ kΩ
3. $I = 400$ μA, $R = 22$ kΩ
5. $I = 5.6$ mA, $R = 4.7$ kΩ

2. $I = 4.3$ mA, $R = 3.3$ kΩ
4. $I = 40$ mA, $R = 680$ Ω

$P = I^2R$. Find P when:

6. $I = 40$ mA, $R = 1$ kΩ
8. $I = 2.3$ A, $R = 33$ Ω
10. $I = 400$ μA, $R = 56$ kΩ

7. $I = 170$ mA, $R = 470$ Ω
9. $I = 1.76$ mA, $R = 27$ kΩ

$I = \sqrt{\dfrac{P}{R}}$. Find I when:

11. $P = 250$ mW, $R = 120$ Ω
13. $P = 37.3$ mW, $R = 680$ Ω
15. $P = 680$ mW, $R = 1.8$ kΩ

12. $P = 1.2$ W, $R = 22$ kΩ
14. $P = 4.73$ mW, $R = 12$ kΩ

$E = \sqrt{PR}$. Find E when:

16. $P = 56$ mW, $R = 2$ kΩ
18. $P = 2.7$ W, $R = 150$ Ω
20. $P = 30$ mW, $R = 3.9$ kΩ

17. $P = 475$ mW, $R = 100$ Ω
19. $P = 780$ mW, $R = 2.7$ kΩ

$P = \dfrac{E^2}{R}$. Find P when:

21. $E = 12$ V, $R = 470$ Ω
23. $E = 20$ V, $R = 4.7$ kΩ
25. $E = 15$ V, $R = 12$ kΩ

22. $E = 9$ V, $R = 1.8$ kΩ
24. $E = 30$ V, $R = 120$ Ω

SOLUTIONS

1. 5.40 V
3. 8.80 V
5. 26.3 V
7. 13.6 W

2. 14.2 V
4. 27.2 V
6. 1.60 W
8. 175 W

9. 83.6 mW   10. 8.96 mW

11. 45.6 mA   12. 7.39 mA

13. 7.41 mA   14. 628 μA

15. 19.4 mA   16. 10.6 V

17. 6.89 V   18. 20.1 V

19. 45.9 V   20. 10.8 V

21. 306 mW   22. 45.0 mW

23. 85.1 mW   24. 7.50 W

25. 18.8 mW

Additional practice problems are at the end of the chapter.

5–6   PRACTICAL APPLICATIONS—AC CIRCUITS

Capacitive reactance in an ac circuit is measured in ohms. The equation is

$$X_C = \frac{1}{2\pi fC}$$

where π (the Greek letter pi) is a constant and equals 3.14 rounded to three places, f is the frequency in hertz (Hz), and C is the capacitance in farads (F).

EXAMPLE 5–12

Find the capacitive reactance (X_C) when f = 7.5 kHz and C = 0.5 μF.

SOLUTION

$$X_C = \frac{1}{2\pi fC} = \frac{1}{2 \times \pi \times 7.5 \text{ kHz} \times 0.5 \ \mu\text{F}} = 42.4 \ \Omega$$

TI: 2 ⨯ π ⨯ 7.5 EE 3 ⨯ .5 EE

+/− 6 = 1/x   *Answer:* 42.4 Ω

Casio: 1 ÷ (2 π ⨯ 7.5 EXP 3 ⨯

.5 EXP (−) 6) =   *Answer:* 42.4 Ω

An important parameter in ac circuits is the resonant frequency (f_r). This frequency is found by using the following equation:

$$f_r = \frac{1}{2\pi\sqrt{LC}}$$

where L is the circuit inductance in henries (H) and C is the circuit capacitance in farads (F).

EXAMPLE 5-13

Find the resonant frequency when $L = 200$ mH and $C = 300$ pF.

$$f_r = \frac{1}{2\pi\sqrt{LC}} = \frac{1}{2 \times \pi \times \sqrt{200 \text{ mH} \times 300 \text{ pF}}}$$

$$= \frac{1}{6.28 \times \sqrt{6 \times 10^{-11}}} = \frac{1}{6.28 \times 7.75 \times 10^{-6}}$$

$$= \frac{1}{4.87 \times 10^{-5}} = 2.05 \times 10^4 \text{ Hz} = 20.5 \text{ kHz}$$

TI: $2 \boxed{\times} \boxed{\pi} \boxed{\times} \boxed{(}~ 200 \boxed{\text{EE}} \boxed{+/-}~ 3 \boxed{\times}$

$300 \boxed{\text{EE}} \boxed{+/-}~ 12 \boxed{)} \boxed{\sqrt{x}} \boxed{=} \boxed{1/x}$

Answer: 20.5 × 10³ = 20.5 kHz

Casio: $1 \boxed{\div} \boxed{(}~ 2 \boxed{\pi} \boxed{\times} \boxed{\sqrt{}} \boxed{(}~ 200 \boxed{\text{EXP}}$

$\boxed{(-)}~ 3 \boxed{\times}~ 300 \boxed{\text{EXP}} \boxed{(-)}~ 12 \boxed{)} \boxed{)}$

$\boxed{=}$ *Answer:* 2.05 × 10⁴ = 20.5 kHz

If you were going to solve a number of problems where you had to find $2 \times \pi$, you might save time by storing the product:

TI: $2 \boxed{\times} \boxed{\pi} \boxed{=} \boxed{\text{STO}}~ 1$

Casio: $2 \boxed{\pi} \boxed{\text{STO}} \boxed{\text{M+}}$

Now, with 2π stored, we could do this:

TI: $200 \boxed{\text{EE}} \boxed{+/-}~ 3 \boxed{\times}~ 300 \boxed{\text{EE}} \boxed{+/-}$

$12 \boxed{=} \boxed{\sqrt{x}} \boxed{\times} \boxed{\text{RCL}}~ 1 \boxed{=} \boxed{1/x}$

Answer: 20.5 × 10³

Casio: $1 \boxed{\div} \boxed{(} \boxed{\text{RCL}} \boxed{\text{M+}} \boxed{\times} \boxed{\sqrt{}} \boxed{(}~ 200$

$\boxed{\text{EXP}} \boxed{(-)}~ 3 \boxed{\times}~ 300 \boxed{\text{EXP}} \boxed{(-)}~ 12 \boxed{)}$

$\boxed{)} \boxed{=}$ *Answer:* 20.5 × 10³ = 20.5 kHz

Again the answer is 20.5×10^3. Notice that if you key in *RCL* 1 *first* you will have to use parentheses, as we did in the original solution.

Always use prefixes instead of powers of ten form in your final answers.

Another important parameter in ac circuits is *impedance*. Impedance is measured in ohms. One method used to find impedance when both resistance and reactance are known is

$$Z = \sqrt{R^2 + X^2}$$

where R is the circuit resistance and X is the circuit reactance.

EXAMPLE 5–14

Find Z when $R = 5.6$ kΩ and $X = 4$ kΩ.

SOLUTION

$$Z = \sqrt{R^2 + X^2} = \sqrt{(5.6 \text{ k}\Omega)^2 + 4 \text{ (k}\Omega)^2} = \sqrt{(5.6 \times 10^3)^2 + (4 \times 10^3)^2}$$
$$= \sqrt{3.14 \times 10^7 + 1.6 \times 10^7} = \sqrt{4.74 \times 10^7}$$
$$= 6.88 \times 10^3 \ \Omega = 6.88 \text{ k}\Omega$$

TI: 5.6 $\boxed{\text{EE}}$ 3 $\boxed{x^2}$ $\boxed{+}$ 4 $\boxed{\text{EE}}$ 3 $\boxed{x^2}$ $\boxed{=}$ $\boxed{\sqrt{x}}$

Answer: **6.88** $\times$ **10³** = **6.88 kΩ**

Casio: $\boxed{\sqrt{}}$ $\boxed{(}$ 5.6 $\boxed{\text{EXP}}$ 3 $\boxed{x^2}$ $\boxed{+}$ 4 $\boxed{\text{EXP}}$ 3

$\boxed{x^2}$ $\boxed{)}$ $\boxed{=}$ *Answer:* **6.88** $\times$ **10³** = **6.88 kΩ**

Remember, when adding numbers, the exponents to the base 10 must be the same. (When using the calculator to perform the addition, no changes need to be made.)

PRACTICE PROBLEMS 5–6

In the following problems, round answers to three places. Use prefixes where appropriate.

$X_C = \dfrac{1}{2\pi f C}$. Find X_C when:

1. $f = 120$ Hz, $C = 2$ μF
3. $f = 420$ Hz, $C = 0.5$ μF
5. $f = 1.2$ kHz, $C = 50$ nF

2. $f = 2.5$ kHz, $C = 300$ nF
4. $f = 60$ Hz, $C = 10$ μF

$$f_r = \frac{1}{2\pi\sqrt{LC}}.$$ Find f_r when:

6. $L = 300$ mH, $C = 100$ nF
7. $L = 500$ mH, $C = 470$ pF
8. $L = 200$ mH, $C = 2$ μF
9. $L = 1.7$ H, $C = 20$ nF
10. $L = 37$ mH, $C = 75$ nF

$$Z = \sqrt{R^2 + X^2}.$$ Find Z when:

11. $R = 1.8$ kΩ, $X = 3$ kΩ
12. $R = 27$ kΩ, $X = 50$ kΩ
13. $R = 1$ kΩ, $X = 600$ Ω
14. $R = 4.7$ kΩ, $X = 3.1$ kΩ
15. $R = 12$ kΩ, $X = 16$ kΩ

SOLUTIONS

1. 663 Ω
2. 212 Ω
3. 758 Ω
4. 265 Ω
5. 2.65 kΩ
6. 919 Hz
7. 10.4 kHz
8. 252 Hz
9. 863 Hz
10. 3.02 kHz
11. 3.50 kΩ
12. 56.8 kΩ
13. 1.17 kΩ
14. 5.63 kΩ
15. 20.0 kΩ

SELF-TEST 5–2

1. $E = IR$. Find E if $I = 7.5$ mA and $R = 2.7$ kΩ.

2. $P = I^2R$. Find P if $I = 27$ mA and $R = 150$ Ω.

3. $I = \sqrt{\dfrac{P}{R}}$. Find I when $P = 750$ mW and $R = 3.3$ kΩ.

4. $E = \sqrt{PR}$. Find E when $P = 60$ mW and $R = 6.8$ kΩ.

5. $P = \dfrac{E^2}{R}$. Find P when $E = 15$ V and $R = 1.8$ kΩ.

6. $X_C = \dfrac{1}{2\pi fC}$. Find X_C when $f = 3.2$ kHz and $C = 150$ nF.

7. $f_r = \dfrac{1}{2\pi\sqrt{LC}}$. Find f_r when $L = 700$ mH and $C = 470$ pF.

8. $Z = \sqrt{R^2 + X^2}$. Find Z when $R = 5.6$ kΩ and $X = 8$ kΩ.

Answers to Self-test 5–2 are at the end of the chapter.

CHAPTER 5 AT A GLANCE

PAGE	KEY POINT OR RULE	EXAMPLE
160	*Key Point:* We can solve problems that contain literal numbers only if we know their value.	Find x^3 if $x = 5$. $$x^3 = 5^3 = 125$$
162	*Key Point:* If a sign of grouping is used, then all numbers within that group are acted on by the exponent.	Find $(ab)^2$ if $a = 3$ and $b = 4$. $$(3 \times 4)^2 = 12^2 = 144$$
164	*Key Point:* When we extract the square root of an expression, we perform all mathematical functions under the radical *first*.	Let $a = 4$, $b = 5$. $$\sqrt{a + b^2} = 5.39$$
168	*Key Point:* Prefixes are changed to their powers of ten values to solve problems, but use prefixes in answers.	Find I when $P = 700$ mW and $R = 3.3$ kΩ. $$I = \sqrt{\frac{P}{R}} = \sqrt{\frac{700 \text{ mW}}{3.3 \text{ k}\Omega}}$$ $$= \sqrt{\frac{700 \times 10^{-3}}{3.3 \times 10^3}}$$ $$= 1.46 \times 10^{-2} \text{ A}$$ $$= 14.6 \times 10^{-3} \text{ A}$$ $$= 14.6 \text{ mA}$$

END OF CHAPTER PROBLEMS 5–1

List the number of terms in the following expressions. Also indicate the kind of expression (monomial, binomial, and so on).

1. $\dfrac{a}{2} + 2b$

2. $4a - \dfrac{3b}{4}$

3. $4 + 6b + c$

4. $6 + 2a - 3b$

5. $\dfrac{6a}{7b} + \dfrac{4c}{2}$

6. $2x + \dfrac{7z}{3}$

7. $\dfrac{4(3 - x)}{a} + 7b$

8. $\dfrac{6(2 - y)}{z} + 4b$

9. $\dfrac{2a + 3b}{4} + 3 - \dfrac{c}{7}$

10. $\dfrac{3x - 4y}{5} + 6 + \dfrac{z}{2}$

In the following expressions, list the numerical coefficients and the literal numbers in each term:

11. $3a - 4b + 2c$

12. $5x + 2y - 7z$

13. $\dfrac{2x}{3} + 7y$

14. $\dfrac{2a}{3} - 3b$

15. $\dfrac{5}{2a} - \dfrac{4}{b}$

16. $\dfrac{3}{x} + \dfrac{7y}{8}$

END OF CHAPTER PROBLEMS 5–2

Solve the following problems. Round to three places.

1. 2.3^2

2. 5.6^2

3. 7.4^2

4. 7.6^2

5. 16^2

6. 21^2

7. 8^3

8. 7^3

9. 3.6^3

10. 6.6^3

11. 5.5^3

12. 10.7^3

Solve the following problems where $a = 4$ and $b = 9$:

13. a^2

14. b^2

15. a^3

16. b^3

17. a^4

18. b^4

END OF CHAPTER PROBLEMS 5–3

Solve the following problems, where $a = 5$, $b = 3$, and $c = 6$. Round answers to three places.

1. ac

2. bc^2

3. a^2c

4. $(ab)^2$

5. $(ac)^2$

6. $a^3 - b^2$

7. $2(ab)^2$

8. $4(bc)^2$

9. $a^2b^2 - 3c^2$

10. $abc^2 - a^2b^2$

11. $ac^2 - 2abc$

12. $b^2c + a^3$

13. $(a^2b - 2c^2)^2$

14. $(2a^2b - 4b^2)^2$

15. $3a^2b^3c - 2a^3b^2c^2$

16. $4a^3b^3c^4 - 6a^3b^4c^3$

17. $3ab^3c + 4a^2bc^3 - 2a^4bc^2$

18. $2a^3b^3c - 4a^2b^4c^2 + 5b^3c^3$

19. $(3a^2c^4 - 2a^3b^2c)^3$

20. $(2a^3b^4c^3 - 3ab^2c^3)^3$

END OF CHAPTER PROBLEMS 5–4

Solve the following problems. Round to three places.

1. $30^{1\backslash 2}$

2. $18^{1\backslash 2}$

3. $40^{1\backslash 2}$

4. $80^{1\backslash 2}$

5. $200^{1\backslash2}$

6. $676^{1\backslash2}$

7. $1040^{1\backslash2}$

8. $1500^{1\backslash2}$

9. $2000^{1\backslash2}$

10. $2700^{1\backslash2}$

Solve the following problems, where $x = 3$, $y = 4$, and $z = 5$. Round answers to three places.

11. $\sqrt{yz}$

12. $\sqrt{xz}$

13. $\sqrt{y^2}$

14. $\sqrt{3y^2}$

15. $\sqrt{6xz^2}$

16. $\sqrt{5y^2z^2}$

17. $\sqrt{x^2 + 3y^2}$

18. $\sqrt{6y^2 + z^2}$

19. $\sqrt{6.4y^3 - z^2}$

20. $\sqrt{3.7y^3 - x^2}$

21. $\sqrt{y^2z^2 + x}$

22. $\sqrt{x^2z^2 + y}$

23. $\sqrt{x^3 + 2z}$

24. $\sqrt{y^3 + 3z}$

25. $\sqrt{x^3z^2 + 3x^4y^2}$

26. $\sqrt{x^3y^3 + 4x^4z^3}$

27. $\sqrt{3y^3z^3 - 5x^3z^3}$

28. $\sqrt{8x^3z^3 - 3y^3z^3}$

29. $\sqrt{4x^3y^4z^2 - 2x^2y^3z - z}$

30. $\sqrt{3x^4y^4z^3 - 4x^3y^3z^2 - y^3}$

31. $\sqrt{5.6x^4y^3z^4 - 3x^2y^4z^3 - 4z^3}$

32. $\sqrt{2.7x^3y^3z^4 - 6x^3y^4z^2 - 8y^3}$

END OF CHAPTER PROBLEMS 5–5

In the following problems, round answers to three places. Use prefixes where appropriate.

$E = IR$. Find E when:

1. $I = 660\ \mu\text{A}, R = 6.8\ \text{k}\Omega$

2. $I = 1.73\ \text{mA}, R = 18\ \text{k}\Omega$

3. $I = 3.7\ \text{mA}, R = 18\ \text{k}\Omega$

4. $I = 3.7\ \text{mA}, R = 1.5\ \text{k}\Omega$

5. $I = 50\ \text{mA}, R = 2.7\ \text{k}\Omega$

6. $I = 125\ \mu\text{A}, R = 68\ \text{k}\Omega$

7. $I = 430\ \text{mA}, R = 560\ \Omega$

8. $I = 84.5\ \text{mA}, R = 82\ \Omega$

$P = I^2R$. Find P when:

9. $I = 7.5\ \text{mA}, R = 680\ \Omega$

10. $I = 760\ \mu\text{A}, R = 2.2\ \text{k}\Omega$

11. $I = 36.5\ \text{mA}, R = 3.3\ \text{k}\Omega$

12. $I = 2\ \text{mA}, R = 150\ \Omega$

13. $I = 1.2\ \text{A}, R = 5\ \Omega$

14. $I = 620\ \mu\text{A}, R = 1.2\ \text{k}\Omega$

15. $I = 23.5\ \text{mA}, R = 810\ \Omega$

16. $I = 65.5\ \text{mA}, R = 750\ \Omega$

$I = \sqrt{\dfrac{P}{R}}$. Find I when:

17. $P = 65.4\ \text{mW}, R = 5.6\ \text{k}\Omega$

18. $P = 100\ \text{mW}, R = 1.5\ \text{k}\Omega$

19. $P = 50\ \text{mW}, R = 18\ \text{k}\Omega$

20. $P = 500\ \text{mW}, R = 470\ \Omega$

21. $P = 4.6\ \text{W}, R = 10\ \Omega$

22. $P = 30\ \text{W}, R = 8\ \Omega$

23. $P = 650\ \mu\text{W}, R = 56\ \Omega$

24. $P = 375\ \mu\text{W}, R = 100\ \Omega$

$E = \sqrt{PR}$. Find E when:

25. $P = 8.3$ mW, $R = 27$ kΩ
27. $P = 130$ mW, $R = 750$ Ω
29. $P = 18$ mW, $R = 6.8$ kΩ
31. $P = 100$ mW, $R = 100$ Ω

26. $P = 78.3$ mW, $R = 4.7$ kΩ
28. $P = 500$ mW, $R = 2.5$ kΩ
30. $P = 1.2$ W, $R = 1800$ Ω
32. $P = 700$ μW, $R = 150$ kΩ

$P = \dfrac{E^2}{R}$. Find P when:

33. $E = 16$ V, $R = 6.8$ kΩ
35. $E = 13.3$ V, $R = 33$ kΩ
37. $E = 18.2$ V, $R = 10$ kΩ
39. $E = 150$ V, $R = 4.7$ kΩ

34. $E = 4.7$ V, $R = 680$ Ω
36. $E = 8.73$ V, $R = 33$ kΩ
38. $E = 25$ V, $R = 1.2$ kΩ
40. $E = 360$ mV, $R = 8.2$ kΩ

END OF CHAPTER PROBLEMS 5–6

In the following problems, round answers to three places. Use prefixes where appropriate.

$X_C = \dfrac{1}{2\pi f C}$. Find X_C when:

1. $f = 5$ kHz, $C = 100$ nF
3. $f = 200$ Hz, $C = 200$ nF
5. $f = 6.3$ kHz, $C = 2$ μF
7. $f = 65.4$ kHz, $C = 250$ pF
9. $f = 130$ kHz, $C = 750$ pF

2. $f = 10$ kHz, $C = 1$ μF
4. $f = 15$ kHz, $C = 400$ pF
6. $f = 500$ Hz, $C = 750$ nF
8. $f = 57.5$ kHz, $C = 470$ pF
10. $f = 250$ kHz, $C = 250$ pF

$f_r = \dfrac{1}{2\pi\sqrt{LC}}$. Find f_r when:

11. $L = 80$ mH, $C = 150$ nF
13. $L = 150$ μH, $C = 50$ nF
15. $L = 50$ mH, $C = 250$ nF
17. $L = 400$ mH, $C = 5$ nF
19. $L = 0.25$ H, $C = 25$ μF

12. $L = 1$ mH, $C = 10$ μF
14. $L = 6.5$ mH, $C = 500$ nF
16. $L = 300$ mH, $C = 60$ nF
18. $L = 550$ mH, $C = 10$ nF
20. $L = 0.75$ H, $C = 75$ μF

$Z = \sqrt{R^2 + X^2}$. Find Z when:

21. $R = 120$ kΩ, $X = 80$ kΩ
23. $R = 200$ Ω, $X = 100$ Ω
25. $R = 33$ kΩ, $X = 20$ kΩ
27. $R = 4.7$ kΩ, $X = 7.35$ kΩ
29. $R = 2.2$ MΩ, $X = 6.75$ MΩ

22. $R = 100$ Ω, $X = 100$ Ω
24. $R = 3$ kΩ, $X = 4$ kΩ
26. $R = 27$ kΩ, $X = 33$ kΩ
28. $R = 8.2$ kΩ, $X = 12$ kΩ
30. $R = 910$ kΩ, $X = 1.25$ MΩ

ANSWERS TO SELF-TESTS

SELF-TEST 5–1

1. 45.0	2. 3.46	3. 900
4. 15.0	5. 23.0	6. 39.0
7. 11.0	8. 11.2	9. 19.9
10. 17.4		

SELF-TEST 5–2

1. 20.3 V	2. 109 mW	3. 15.1 mA
4. 20.2 V	5. 125 mW	6. 332 Ω
7. 8.77 kHz	8. 9.77 kΩ	

Fractions and Literal Numbers

6

Introduction

In Chapter 4, we worked with fractions that contained only real numbers. In this chapter, we will continue our study of fractions and will include literal numbers. The study of fractions that contain literal numbers is extremely important to the electronics student because almost all fractions we encounter contain literal numbers. To deal with the various equations we work with in problem solving, a basic understanding of fractions and literal numbers is necessary.

Chapter Objectives

In this chapter you will learn how to:

1. Find the prime factors of literal numbers.
2. Find the lowest common multiple of literal numbers.
3. Multiply and divide whole and fractional numbers.
4. Add and subtract fractions with common denominators.
5. Add and subtract fractions with unlike denominators.

PRIME NUMBERS　6–1

In Chapter 4, we learned how to define and recognize prime numbers. Literal numbers are prime when they are raised to the first power. x is prime. x^2 is not prime because $x^2 = x \cdot x$. The prime factors of $x^3 y^2$ are $x \cdot x \cdot x \cdot y \cdot y$.

Real numbers and literal numbers often appear in the same term. Such terms can be factored by finding the prime factors of the real numbers as before, and then finding the prime factors of the literal numbers. Let's factor some of these numbers.

EXAMPLE 6–1

Factor $40a^2b^3$.

SOLUTION

$$40a^2b^3 = 2 \cdot 2 \cdot 2 \cdot 5 \cdot a \cdot a \cdot b \cdot b \cdot b$$

EXAMPLE 6–2

Factor $33x^2y^2z$.

SOLUTION

$$33x^2y^2z = 3 \cdot 11 \cdot x \cdot x \cdot y \cdot y \cdot z$$

PRACTICE PROBLEMS 6–1

Find the prime factors of the following:

1. $6ab^2$
2. $8a^2bc^2$
3. $22x^3$
4. $38x^2y^2$
5. $30x^2y^3$
6. $115z^2$
7. $136x^2z^3$
8. $168a^2c^2$
9. $175b^2c^3$
10. $200y^3$

SOLUTIONS

1. $2 \cdot 3 \cdot a \cdot b \cdot b$
2. $2 \cdot 2 \cdot 2 \cdot a \cdot a \cdot b \cdot c \cdot c$
3. $2 \cdot 11 \cdot x \cdot x \cdot x$
4. $2 \cdot 19 \cdot x \cdot x \cdot y \cdot y$
5. $2 \cdot 3 \cdot 5 \cdot x \cdot x \cdot y \cdot y \cdot y$
6. $5 \cdot 23 \cdot z \cdot z$
7. $2 \cdot 2 \cdot 2 \cdot 17 \cdot x \cdot x \cdot z \cdot z \cdot z$
8. $2 \cdot 2 \cdot 2 \cdot 3 \cdot 7 \cdot a \cdot a \cdot c \cdot c$
9. $5 \cdot 5 \cdot 7 \cdot b \cdot b \cdot c \cdot c \cdot c$
10. $2 \cdot 2 \cdot 2 \cdot 5 \cdot 5 \cdot y \cdot y \cdot y$

Additional practice problems are at the end of the chapter.

6–2   LOWEST COMMON MULTIPLE

Finding the lowest common multiple of numbers containing literal numbers is similar to finding the LCM of real numbers (see Rule 4–3). The rule is:

☞ **RULE 6–1** To find the lowest common multiple of numbers containing literal numbers:

1. Factor each term.
2. Determine the *maximum number* of times a prime factor appears in any one term.
3. Multiply the resulting prime factors found in step 2. This number is the LCM.

The lowest common multiple is the *lowest* multiple of each term that is common to all terms.

EXAMPLE 6–3

Find the LCM of $6a$ and $18a^2$.

SOLUTION   Find the prime factors of each term:

$$6a = 2 \cdot 3 \cdot a$$

$$18a^2 = 2 \cdot 3 \cdot 3 \cdot a \cdot a$$

The maximum number of times 2 appears as a factor is once; therefore, 2 is part of the LCM. Three appears a maximum of two times; therefore, $3 \cdot 3$ or 3^2 is part of the LCM. Since a appears a maximum of two times, $a \cdot a$ or a^2 is part of the LCM. This takes care of all the prime factors. Our next step is to multiply these factors together to get the LCM.

$$2 \cdot 3 \cdot 3 \cdot a \cdot a = 18a^2$$

$18a^2$ is the lowest number that contains both $6a$ and $18a^2$ as factors. That is, $6a$ is a factor of $18a^2$ ($6a \cdot 3a = 18a^2$) and $18a^2$ is a factor ($18a^2 \cdot 1 = 18a^2$). There are other common multiples: $36a^2$, $36a^3$, and $54a^3$ to name a few. But $18a^2$ is the *lowest*.

EXAMPLE 6–4

Find the LCM of $10xy^2$ and $15x^3y$.

SOLUTION

$$10xy^2 = 2 \cdot 5 \cdot x \cdot y \cdot y$$

$$15x^3y = 3 \cdot 5 \cdot x \cdot x \cdot x \cdot y$$

Two, three, and five each appear as factors once. Therefore, 2 and 3 and 5 are part of the LCM. Since x appears a maximum of three times, x^3 is part of the LCM. Since y appears a maximum of two times, y^2 is part of the LCM.

$$LCM = 2 \cdot 3 \cdot 5 \cdot x^3 \cdot y^2$$
$$= 30x^3y^2$$

$30x^3y^2$ is the lowest number that has both $10xy^2$ and $15x^3y$ as factors.

EXAMPLE 6–5

Find the LCM of $14a^2b$, $42abc^2$, and $12b$.

$$14a^2b = 2 \cdot 7 \cdot a \cdot a \cdot b$$

$$42abc^2 = 2 \cdot 3 \cdot 7 \cdot a \cdot b \cdot c \cdot c$$

$$12b = 2 \cdot 2 \cdot 3 \cdot b$$

$$\text{LCM} = 2 \cdot 2 \cdot 3 \cdot 7 \cdot a \cdot a \cdot b \cdot c \cdot c = 84a^2bc^2$$

$84a^2bc^2$ is the smallest number that has $14a^2b$, $42abc^2$, and $12b$ as factors. That is, $84a^2bc^2$ is the smallest number that is divisible by all three factors. Just as with real numbers, literal numbers do not always appear as factors in each term. Remember the rule: *Determine the maximum number of times a prime factor appears in **any** one term.*

PRACTICE PROBLEMS 6–2

Find the LCM of the following numbers:

1. a^2b^3, a^3b^2
2. a^2b^4, a^2b^3
3. a^2b^2, a^3b, ab
4. a^3b^2, a^2b^3, a^3b^3
5. $a^3b^3c^3$, abc, $a^2b^3c^2$
6. $12a^2b$, $18ab^2$
7. $3ab^2$, $8a^2b^2$
8. $12xyz^2$, $30x^2yz^2$, $45xyz$
9. $11xz^2$, $44x^2y$, y^2z
10. $108xy^4$, $72y^4$, x^2z
11. $81a^3b^2$, $99a^3b$, $77ab$
12. $64ab^2$, $96a^2b^2$, $160a^3b$

SOLUTIONS

1. a^3b^3
2. a^2b^4
3. a^3b^2
4. a^3b^3
5. $a^3b^3c^3$
6. $36a^2b^2$
7. $24a^2b^2$
8. $180x^2yz^2$
9. $44x^2y^2z^2$
10. $216x^2y^4z$
11. $6237a^3b^2$
12. $960a^3b^2$

Additional practice problems are at the end of the chapter.

SELF-TEST 6–1

Find the prime factors of the following numbers:

1. $28xy^2$
2. $54a^3b^2$
3. $76xy^3$
4. $120x^2z^2$
5. $130a^3b^2$
6. $315a^2b^2$

MULTIPLICATION OF MONOMIALS 6-3

To multiply monomials, we multiply the real numbers and then multiply the literal numbers by using the laws of exponents learned in previous chapters. Recall from arithmetic that each part of a multiplication problem has a name.

$$\begin{array}{ll} 6a & \text{(multiplicand)} \\ \underline{\times 7a} & \text{(multiplier)} \\ 42a^2 & \text{(product)} \end{array}$$

The product cannot be simplified further because we have not been given a value for a. Remember that we can indicate the operation to be performed, in this case multiplication, in a number of ways. We indicate multiplication with $\times$ or $\cdot$ or ( ) or other signs of grouping.

In the problem $4a^2b \cdot 3ab^3$, we multiply 4×3 and get 12. $a^2 \cdot a = a^{2+1} = a^3$ and $b \times b^3 = b^{1+3} = b^4$. The answer is $12a^3b^4$. With practice, we can do the addition of exponents of like bases in our heads.

☞ **KEY POINT** In multiplication, all the literal numbers contained in the multiplicand and the multiplier must appear in the product unless the exponent is zero.

EXAMPLE 6–6

Perform the indicated operations.

1. $3x^3y^{-2}z \times 5x^2y^2z^2$
2. $4x^{-3}yz^{-2} \times 2xyz^2 \times 3x^{-1}y^{-1}z^2$

SOLUTIONS

1. If we put like bases together, we can more easily see the result:

$$3 \cdot 5 \cdot x^3 \cdot x^2 \cdot y^{-2} \cdot y^2 \cdot z \cdot z^2 = 15 \cdot x^{3+2} \cdot y^{-2+2} \cdot z^{1+2}$$
$$= 15x^5y^0z^3$$
$$= 15x^5z^3$$

Remember, any number raised to the zero power equals 1.

2. Again, putting like bases together:

$$4 \cdot 2 \cdot 3 \cdot x^{-3} \cdot x \cdot x^{-1} \cdot y \cdot y \cdot y^{-1} \cdot z^{-2} \cdot z^2 \cdot z^2$$
$$= 24 \cdot x^{-3+1-1} y^{1+1-1} \cdot z^{-2+2+2}$$
$$= 24 x^{-3} y z^2$$

PRACTICE PROBLEMS 6–3

Perform the indicated operations:

1. $2ab \cdot 3a^2 b$
2. $4a^3 b \cdot 3ab^2$
3. $3a^4 b \cdot 3a^{-3} b^{-4}$
4. $6x^{-2} y^4 \cdot 8x^4 y^{-1}$
5. $2x^3 y^2 \cdot 3x^{-2} y^{-1} \cdot 3x^4 y^3$
6. $5x^6 y^2 z^2 \cdot x^{-2} y^{-4} z^{-3} \cdot x^{-1} y^3$
7. $3x^3 y^{-3} z^{-2} \cdot 3x^{-3} y^3 \cdot 3xyz^4$
8. $7x^5 y^{-2} z^2 \cdot 3x^{-1} y^5 z^4 \cdot 2x^{-1} y^{-1} z^{-1}$

SOLUTIONS

1. $6a^3 b^2$
2. $12a^4 b^3$
3. $9ab^{-3}$
4. $48x^2 y^3$
5. $18x^5 y^4$
6. $5x^3 yz^{-1}$
7. $27xyz^2$
8. $42x^3 y^2 z^5$

Additional practice problems are at the end of the chapter.

6–4   DIVISION OF MONOMIALS

Recall from arithmetic that the parts of a division problem are as follows:

(a) $\dfrac{18a}{6a} = 3$ ⟵ dividend / quotient / divisor

(b) $6a\overline{)18a}$ — 3 ⟵ quotient, 18a ⟵ dividend, 6a ⟵ divisor

When written as a fraction as in (a), the dividend is the numerator and the divisor is the denominator.

In arithmetic we learned how to do "long division." Here, our long division is done by means of a calculator, but we must do the division of literal numbers ourselves because calculators can deal only with real numbers.

In dividing monomials, we find the prime factors of the dividend and the divisor. Then we cancel each factor that appears in both the dividend and the divisor. The resultant number is the quotient.

EXAMPLE 6–7

Perform the indicated operations.

1. $\dfrac{10a^3}{2a^2}$ 2. $\dfrac{6a^2b^3}{18a^4b}$ 3. $\dfrac{40x^2y^{-1}}{10x^{-1}y^{-3}}$

SOLUTIONS   1. Find the prime factors and cancel those appearing in both numerator and denominator.

$$\frac{\cancel{2} \cdot 5 \cdot \cancel{a} \cdot \cancel{a} \cdot a}{\cancel{2} \cdot \cancel{a} \cdot \cancel{a}} = \frac{5a}{1} = 5a$$

Another way to solve the problem would be to move all literal numbers into the numerator and multiply. Recall that we can move numbers from the numerator to the denominator or from the denominator to the numerator by changing the sign of the exponent. Using this method, we get

$$\frac{10a^3}{2a^2} = \frac{10a^3 \cdot a^{-2}}{2} = 5a^{3-2} = 5a$$

2.
$$\frac{\cancel{2} \cdot \cancel{3} \cdot \cancel{a} \cdot \cancel{a} \cdot \cancel{b} \cdot b \cdot b}{\cancel{2} \cdot \cancel{3} \cdot 3 \cdot \cancel{a} \cdot \cancel{a} \cdot a \cdot a \cdot \cancel{b}} = \frac{b^2}{3a^2} = \frac{a^{-2}b^2}{3}$$

or
$$\frac{6a^2a^{-4}b^3b^{-1}}{18} = \frac{6a^{-2}b^2}{18} = \frac{a^{-2}b^2}{3}$$

3.
$$\frac{40x \cdot x \cdot \cancel{y}}{10x^{-1} \cdot \cancel{y} \cdot y^{-1} \cdot y^{-1}} = \frac{4x^2}{x^{-1}y^{-2}}$$

$$\frac{4x^2 \cdot x}{y^{-2}} = \frac{4x^3}{y^{-2}} = 4x^3y^2$$

and also
$$\frac{40x^2y^{-1}}{10x^{-1}y^{-3}} = \frac{40x^2 \cdot x \cdot y^{-1} \cdot y^3}{10} = 4x^3y^2$$

The second method is usually quicker. Both methods of solution are important in that they help develop important math concepts and ideas. The first method helps us develop an understanding of factoring so that we may deal more easily with fractions. The second method helps us strengthen our understanding of exponents and bases.

PRACTICE PROBLEMS 6–4

Perform the following divisions by (a) canceling like factors and (b) moving all literal numbers to the numerator and multiplying:

1. $\dfrac{15a^4}{5a}$

2. $\dfrac{36a^2}{6a^3}$

3. $\dfrac{42a^3b^2}{7ab}$

4. $\dfrac{56ab^3}{7a^3b}$

5. $\dfrac{20a^4b^2c}{3a^2b^2c^2}$

6. $\dfrac{39a^4bc^4}{3a^2b^3c}$

7. $\dfrac{7x^{-2}y}{63xy^2}$

8. $\dfrac{72xy^2}{9x^{-3}y^{-4}}$

9. $\dfrac{84x^{-1}y^2z^{-3}}{6x^3y^{-1}z^{-2}}$

10. $\dfrac{46x^{-1}y^2z^{-3}}{23xy^{-1}z^2}$

SOLUTIONS

1. (a) $\dfrac{3 \cdot \cancel{5} \cdot \cancel{a} \cdot a \cdot a \cdot a}{\cancel{5}\cancel{a}} = 3a^3$

   (b) $\dfrac{15a^4 \cdot a^{-1}}{5} = 3a^3$

2. (a) $\dfrac{\cancel{2} \cdot 2 \cdot \cancel{3} \cdot 3 \cdot \cancel{a} \cdot \cancel{a}}{\cancel{2} \cdot \cancel{3} \cdot \cancel{a} \cdot \cancel{a} \cdot a} = \dfrac{6}{a} = 6a^{-1}$

   (b) $\dfrac{36a^2 \cdot a^{-3}}{6} = 6a^{-1}$

3. (a) $\dfrac{2 \cdot 3 \cdot \cancel{7} \cdot \cancel{a} \cdot a \cdot a \cdot \cancel{b} \cdot b}{\cancel{7} \cdot \cancel{a} \cdot \cancel{b}} = 6a^2b$

   (b) $\dfrac{42a^3 \cdot a^{-1} \cdot b^2 \cdot b^{-1}}{7} = 6a^2b$

4. (a) $\dfrac{2 \cdot 2 \cdot 2 \cdot \cancel{7} \cdot \cancel{a} \cdot \cancel{b} \cdot b \cdot b}{\cancel{7} \cdot \cancel{a} \cdot a \cdot a \cdot \cancel{b}} = \dfrac{8b^2}{a^2} = 8a^{-2}b^2$

   (b) $\dfrac{56a \cdot a^{-3} \cdot b^3 \cdot b^{-1}}{7} = 8a^{-2}b^2$

5. (a) $\dfrac{2 \cdot 2 \cdot 5 \cdot \cancel{a} \cdot \cancel{a} \cdot a \cdot a \cdot \cancel{b} \cdot \cancel{b} \cdot \cancel{c}}{3 \cdot \cancel{a} \cdot \cancel{a} \cdot \cancel{b} \cdot \cancel{b} \cdot \cancel{c} \cdot c} = \dfrac{20a^2}{3c} = \dfrac{20a^2c^{-1}}{3}$

   (b) $\dfrac{20a^4 \cdot a^{-2} \cdot b^2 \cdot b^{-2} \cdot c \cdot c^{-2}}{3} = \dfrac{20a^2b^0c^{-1}}{3} = \dfrac{20a^2c^{-1}}{3}$

6. (a) $\dfrac{\cancel{3} \cdot 13 \cdot \cancel{a} \cdot \cancel{a} \cdot a \cdot a \cdot \cancel{b} \cdot \cancel{c} \cdot c \cdot c \cdot c}{\cancel{3} \cdot \cancel{a} \cdot \cancel{a} \cdot \cancel{b} \cdot b \cdot b \cdot \cancel{c}} = \dfrac{13a^2c^3}{b^2} = 13a^2b^{-2}c^3$

   (b) $\dfrac{39a^4 \cdot a^{-2} \cdot b \cdot b^{-3} \cdot c^4 \cdot c^{-1}}{3} = 13a^2b^{-2}c^3$

7. (a) $\dfrac{\cancel{7} \cdot x^{-1} \cdot x^{-1} \cdot \cancel{y}}{3 \cdot 3 \cdot \cancel{7} \cdot x \cdot \cancel{y} \cdot y} = \dfrac{x^{-2}}{9xy} = \dfrac{x^{-2} \cdot x^{-1} \cdot y^{-1}}{9} = \dfrac{x^{-3}y^{-1}}{9}$

   (b) $\dfrac{7x^{-2} \cdot x^{-1} \cdot y \cdot y^{-2}}{63} = \dfrac{x^{-3}y^{-1}}{9}$

8. (a) $\dfrac{2 \cdot 2 \cdot 2 \cdot \cancel{3} \cdot \cancel{3} \cdot x \cdot y \cdot y}{\cancel{3} \cdot \cancel{3} \cdot x^{-1} \cdot x^{-1} \cdot x^{-1} \cdot y^{-1} \cdot y^{-1} \cdot y^{-1} \cdot y^{-1}} = \dfrac{8xy^2}{x^{-3} \cdot y^{-4}} = 8x \cdot x^3 \cdot y^2 \cdot y^4 = 8x^4 y^6$

   (b) $\dfrac{72x \cdot x^3 \cdot y^2 \cdot y^4}{9} = 8x^4 y^6$

9. (a) $\dfrac{\cancel{2} \cdot 2 \cdot \cancel{3} \cdot 7 \cdot x^{-1} \cdot y \cdot y \cdot \cancel{z^1} \cdot \cancel{z^1} \cdot z^{-1}}{\cancel{2} \cdot \cancel{3} \cdot x \cdot x \cdot x \cdot y^{-1} \cdot \cancel{z^1} \cdot \cancel{z^1}} = \dfrac{14x^{-1}y^2 z^{-1}}{x^3 y^{-1}} = 14x^{-1}x^{-3}y^2 y^1 z^{-1} = 14x^{-4}y^3 z^{-1}$

   (b) $\dfrac{84x^{-1} \cdot x^{-3} \cdot y^2 \cdot y \cdot z^{-3} \cdot z^2}{6} = 14x^{-4}y^3 z^{-1}$

10. (a) $\dfrac{2 \cdot \cancel{23} \cdot x^{-1} \cdot y \cdot y \cdot z^{-1} \cdot z^{-1} \cdot z^{-1}}{\cancel{23}x \cdot y^{-1} \cdot z \cdot z} = 2x^{-1} \cdot x^{-1} \cdot y^2 \cdot y \cdot z^{-3} \cdot z^{-2} = 2x^{-2}y^3 z^{-5}$

   (b) $\dfrac{46x^{-1} \cdot x^{-1} \cdot y^2 \cdot y \cdot z^{-3} \cdot z^{-2}}{23} = 2x^{-2}y^3 z^{-5}$

Additional practice problems are at the end of the chapter.

SELF-TEST 6–2

Perform the indicated operations:

1. $3x \cdot 5y$
2. $3x^2 y \cdot 4x^2 y^3$
3. $5a^{-1}b^2 \cdot 6a^3 b^{-1}$
4. $3a^2 b^{-1} \cdot 5ab^{-2} \cdot 2a^{-4}b$
5. $6ab^{-3}c^2 \cdot 7a^{-4}b^2 c \cdot a^{-1}$
6. $\dfrac{48x^2 y}{6y^2}$
7. $\dfrac{24x^{-2}y}{3xy^2}$
8. $\dfrac{56a^2 bc^{-3}}{8b}$
9. $\dfrac{63a^{-1}b^2 c^{-4}}{7a^3 c^{-3}}$
10. $\dfrac{48xyz^{-3}}{3x^{-1}y^3 z}$

Answers to Self-test 6–2 are at the end of the chapter.

MULTIPLICATION OF FRACTIONS   6–5

☞ **KEY POINT** To multiply fractions, we (1) cancel factors that are common to both numerators and denominators, (2) multiply the remaining numerators (their product is the numerator of the answer), (3) multiply the remaining denominators (their product is the denominator of the answer).

EXAMPLE 6–8

Multiply the following fractions:

(a) $\dfrac{3a}{4} \times \dfrac{6a^2}{10}$ (b) $\dfrac{7x^2y}{2x} \times \dfrac{3x^3y^3}{14x^2}$ (c) $\dfrac{3a^{-1}b^2}{4a^2b^3} \times \dfrac{4a^3b^{-3}}{9a^{-1}b^{-1}} \times \dfrac{2a^2b^4}{a^3b^2}$

SOLUTIONS

(a) $\dfrac{3a}{4} \times \dfrac{6a^2}{10} = \dfrac{3a \cdot 6a^2}{4 \cdot 10} = \dfrac{2 \cdot 3 \cdot 3 \cdot a^3}{2 \cdot 2 \cdot 2 \cdot 5} = \dfrac{9a^3}{20}$

(b) $\dfrac{7x^2y}{2x} \times \dfrac{3x^3y^3}{14x^2} = \dfrac{7x^2y \cdot 3x^3y^3}{2x \cdot 14x^2} = \dfrac{3 \cdot 7 \cdot x^5 \cdot y^4}{2 \cdot 2 \cdot 7 \cdot x^3}$

$= \dfrac{3 \times 7 \cdot x^5 \cdot x^{-3} \cdot y^4}{2 \cdot 2 \cdot 7} = \dfrac{3x^2y^4}{4}$

(c) $\dfrac{3a^{-1}b^2}{4a^2b^3} \times \dfrac{4a^3b^{-3}}{9a^{-1}b^{-1}} \times \dfrac{2a^2b^4}{a^3b^2} = \dfrac{3a^{-1}b^2 \cdot 4a^3b^{-3} \cdot 2a^2b^4}{4a^2b^3 \cdot 9a^{-1}b^{-1} \cdot a^3b^2}$

At this point we can save some time by examining both numerator and denominator for like factors. The factors don't have to be prime; they only have to be *like* factors. 4, a^2, a^3, b^2, a^{-1} are found in both numerator and denominator and can be canceled at this time. We are now left with

$$\dfrac{3a^{-1}b^2 \cdot 4a^3b^{-3} \cdot 2a^2b^4}{4a^2b^3 \cdot 9a^{-1}b^{-1} \cdot a^3b^2} = \dfrac{3 \cdot b^{-3} \cdot 2b^4}{b^3 \cdot 9b^{-1}} = \dfrac{6b}{9b^2}$$

$$= \dfrac{2 \cdot 3 \cdot b}{3 \cdot 3 \cdot b \cdot b} = \dfrac{2}{3b} = \dfrac{2b^{-1}}{3}$$

In each case, we reduce to lowest terms by canceling like terms. In part (c), the literal number, b, was moved to the numerator in the final answer. Although it is mathematically correct to leave literal numbers in the denominator, we normally give answers with all literal numbers in the numerator.

PRACTICE PROBLEMS 6–5

Find the product of the following numbers. Reduce answers to lowest terms.

1. $\dfrac{3x}{5} \times \dfrac{2x}{4}$

2. $\dfrac{4x^2}{7} \times \dfrac{3x}{4}$

3. $\dfrac{5x^2}{8y} \times \dfrac{6xy}{5}$

4. $\dfrac{2x^3}{3y^2} \times \dfrac{3y^3}{4x}$

5. $\dfrac{3xy^{-1}}{5z} \times \dfrac{5y^3z^2}{9x^{-2}}$

6. $\dfrac{4a^{-1}b^2}{5c^2} \times \dfrac{b^{-3}c}{3a^2} \times \dfrac{3a}{4b^{-1}c^3}$

7. $\dfrac{2x^2y^{-1}}{5z^2} \times \dfrac{20y^3z^3}{30xz^{-1}} \times \dfrac{xz^2}{y^3}$

SOLUTIONS

1. $\dfrac{6x^2}{20} = \dfrac{3x^2}{10}$

2. $\dfrac{12x^3}{28} = \dfrac{3x^3}{7}$

3. $\dfrac{30x^3y}{40y} = \dfrac{3x^3}{4}$

4. $\dfrac{6x^3y^3}{12xy^2} = \dfrac{x^2y}{2}$

5. $\dfrac{15xy^2z^2}{45x^{-2}z} = \dfrac{x^3y^2z}{3}$

6. $\dfrac{12b^{-1}c}{60a^2b^{-1}c^5} = \dfrac{a^{-2}c^{-4}}{5}$

7. $\dfrac{40x^3y^2z^5}{150xy^3z} = \dfrac{4x^2y^{-1}z^4}{15}$

Additional practice problems are at the end of the chapter.

DIVISION OF FRACTIONS 6-6

As with fractions that include only real numbers, division is performed by inverting the divisor and then multiplying.

EXAMPLE 6–9

Divide the following fractions:

(a) $\dfrac{3a}{4} \div \dfrac{4a}{5}$ (b) $\dfrac{15x^2y}{8z^2} \div \dfrac{12x^2z}{21y^3}$ (c) $\dfrac{28a^2c}{42b^3} \div \dfrac{12a^3b^2}{35c^2}$

SOLUTIONS

(a) $\dfrac{3a}{4} \div \dfrac{4a}{5} = \dfrac{3a}{4} \times \dfrac{5}{4a} = \dfrac{15a}{16a} = \dfrac{15}{16}$

(b) $\dfrac{15x^2y}{8z^2} \div \dfrac{12x^2z}{21y^3} = \dfrac{15x^2y}{8z^2} \times \dfrac{21y^3}{12x^2z}$

$= \dfrac{\cancel{3} \cdot 3 \cdot 5 \cdot 7\cancel{x^2}y^4}{\cancel{3} \cdot 2 \cdot 2 \cdot 2 \cdot 2 \cdot \cancel{x^2} \cdot z^3} = \dfrac{105y^4z^{-3}}{32}$

(c) $\dfrac{28a^2c}{42b^3} \div \dfrac{12a^3b^2}{35c^2} = \dfrac{28a^2c}{42b^3} \times \dfrac{35c^2}{12a^3b^2} = \dfrac{\cancel{2} \cdot 2 \cdot \cancel{7} \cdot 5 \cdot 7 \cdot a^2 \cdot c^3}{\cancel{2} \cdot 3 \cdot \cancel{7} \cdot \cancel{2} \cdot 2 \cdot 3 \cdot a^3 \cdot b^5}$

$= \dfrac{35a^{-1}b^{-5}c^3}{18}$

Perform the following divisions. Reduce answers to lowest terms.

1. $\dfrac{4a}{5} \div \dfrac{6a^2}{7}$

2. $\dfrac{5a^2}{6} \div \dfrac{15a}{18}$

3. $\dfrac{7xy^2}{9z} \div \dfrac{14x^2y}{3z^2}$

4. $\dfrac{42x^{-2}y}{15y^{-1}z^2} \div \dfrac{7xz}{3y}$

5. $\dfrac{20a^2b^{-1}c}{9} \div \dfrac{5ab^3c^{-2}}{27}$

6. $\dfrac{10a^{-2}b^3}{7c^2} \div \dfrac{5b^{-1}}{28ac^4}$

SOLUTIONS

1. $\dfrac{28a}{30a^2} = \dfrac{14a^{-1}}{15}$

2. $\dfrac{90a^2}{90a} = a$

3. $\dfrac{21xy^2z^2}{126x^2yz} = \dfrac{x^{-1}yz}{6}$

4. $\dfrac{126x^{-2}y^2}{105xy^{-1}z^3} = \dfrac{6x^{-3}y^3z^{-3}}{5}$

5. $\dfrac{540a^2b^{-1}c}{45ab^3c^{-2}} = 12ab^{-4}c^3$

6. $\dfrac{280a^{-1}b^3c^4}{35b^{-1}c^2} = 8a^{-1}b^4c^2$

Additional practice problems are at the end of the chapter.

SELF-TEST 6–3

Perform the indicated operations:

1. $\dfrac{3x^2}{7} \times \dfrac{2x^2}{5}$

2. $\dfrac{3ab^2}{5c} \times \dfrac{2bc^{-1}}{3a^2}$

3. $\dfrac{2a^2bc^{-1}}{3b^{-1}} \times \dfrac{4b^3}{2ac^2} \times \dfrac{7a^{-3}}{8}$

4. $\dfrac{5a}{6} \div \dfrac{3a^2}{4}$

5. $\dfrac{36a^2b}{42c^{-3}} \div \dfrac{9b^3}{21a^3c}$

6. $\dfrac{12b}{21a^2c} \div \dfrac{3b^2}{7a^{-1}c^2}$

7. $\dfrac{15x^{-1}y^{-1}}{27z^3} \div \dfrac{5x^{-1}y}{9z}$

Answers to Self-test 6–3 are at the end of the chapter.

ADDING AND SUBTRACTING FRACTIONS WITH COMMON DENOMINATORS

Recall that when adding or subtracting fractions we simply add or subtract the numerators if the denominators are identical. The denominator in the answer is the common denominator.

EXAMPLE 6–10

Perform the indicated operations. Reduce answers to lowest terms:

(a) $\dfrac{2a}{7} + \dfrac{3a}{7}$ (b) $\dfrac{4}{5a} + \dfrac{3}{5a}$ (c) $\dfrac{a}{12} + \dfrac{5a}{12} + \dfrac{7a}{12}$

(d) $\dfrac{7x}{9} - \dfrac{3x}{9}$ (e) $\dfrac{11a}{12} + \dfrac{5a}{12} - \dfrac{7a}{12}$

SOLUTIONS

(a) $\dfrac{2a}{7} + \dfrac{3a}{7} = \dfrac{2a + 3a}{7} = \dfrac{5a}{7}$ (b) $\dfrac{4}{5a} + \dfrac{3}{5a} = \dfrac{4 + 3}{5a} = \dfrac{7}{5a}$

(c) $\dfrac{a}{12} + \dfrac{5a}{12} + \dfrac{7a}{12} = \dfrac{a + 5a + 7a}{12} = \dfrac{13a}{12}$

(d) $\dfrac{7x}{9} - \dfrac{3x}{9} = \dfrac{7x}{9} + \dfrac{-3x}{9} = \dfrac{7x - 3x}{9} = \dfrac{4x}{9}$

(e) $\dfrac{11a}{12} + \dfrac{5a}{12} - \dfrac{7a}{12} = \dfrac{11a + 5a - 7a}{12} = \dfrac{16a - 7a}{12} = \dfrac{9a}{12} = \dfrac{3a}{4}$

In each of these examples, the denominators were identical (common denominators), so we added or subtracted the numerators to find the answer and then reduced to lowest terms.

When the numerators contain unlike terms, they cannot be added. In these fractions, all the unlike terms are written over the common denominator. In the following examples, the numerators contain unlike terms.

EXAMPLE 6–11

Perform the indicated operations. Reduce to lowest terms:

(a) $\dfrac{2a}{5} + \dfrac{3b}{5}$ (b) $\dfrac{3xy}{7} + \dfrac{2x}{7} + \dfrac{xy}{7}$ (c) $\dfrac{3}{8x} + \dfrac{5a}{8x} + \dfrac{2}{8x}$

(d) $\dfrac{7a}{12} - \dfrac{4b}{12}$ (e) $\dfrac{6ab}{3} + \dfrac{4b}{3} - \dfrac{3ac}{3}$ (f) $\dfrac{4a^2c}{7} + \dfrac{3ac}{7} - \dfrac{2a^2c}{7}$

(a) $\dfrac{2a}{5} + \dfrac{3b}{5} = \dfrac{2a + 3b}{5}$

The answer cannot be simplified further because $2a$ and $3b$ are unlike terms.

(b) $\dfrac{3xy}{7} + \dfrac{2x}{7} + \dfrac{xy}{7} = \dfrac{3xy + 2x + xy}{7} = \dfrac{4xy + 2x}{7}$

We can combine the like terms. $3xy + xy = 4xy$.

(c) $\dfrac{3}{8x} + \dfrac{5a}{8x} + \dfrac{2}{8x} = \dfrac{3 + 5a + 2}{8x} = \dfrac{5 + 5a}{8x}$

We can combine the like terms. $3 + 2 = 5$.

(d) $\dfrac{7a}{12} - \dfrac{4b}{12} = \dfrac{7a - 4b}{12}$

(e) $\dfrac{6ab}{3} + \dfrac{4b}{3} - \dfrac{3ac}{3} = \dfrac{6ab + 4b - 3ac}{3}$

(f) $\dfrac{4a^2c}{7} + \dfrac{3ac}{7} - \dfrac{2a^2c}{7} = \dfrac{4a^2c + 3ac - 2a^2c}{7} = \dfrac{2a^2c + 3ac}{7}$

We can combine the like terms. $4a^2c - 2a^2c = 2a^2c$.

If the denominators are common, we add or subtract the numerators and combine like terms. Remember, we can only combine *like* terms. In Example 6–11(f), we were able to combine $4a^2c$ and $-2a^2c$ because they were *like* terms. That is, the literal numbers were the same (a^2c).

PRACTICE PROBLEMS 6–7

Perform the indicated operations. Reduce to lowest terms.

1. $\dfrac{3}{5a} + \dfrac{1}{5a}$

2. $\dfrac{3}{8ab} + \dfrac{1}{8ab} + \dfrac{5}{8ab}$

3. $\dfrac{2}{9x} + \dfrac{5}{9x} - \dfrac{2}{9x}$

4. $\dfrac{5x}{11} + \dfrac{2x}{11}$

5. $\dfrac{7x}{20} - \dfrac{3x}{20}$

6. $\dfrac{3xy}{16} + \dfrac{xy}{16} + \dfrac{5xy}{16}$

7. $\dfrac{9ab}{20} - \dfrac{5ab}{20} - \dfrac{3ab}{20}$

8. $\dfrac{3a}{8b} + \dfrac{5}{8b} + \dfrac{2a}{8b}$

9. $\dfrac{2xy}{7} - \dfrac{3y}{7} + \dfrac{2y}{7}$

10. $\dfrac{3}{4y} - \dfrac{5}{4y}$

11. $\dfrac{8a}{21} - \dfrac{14a}{21} + \dfrac{2a}{21}$

SOLUTIONS

1. $\dfrac{3+1}{5a} = \dfrac{4}{5a} = \dfrac{4a^{-1}}{5}$

2. $\dfrac{3+1+5}{8ab} = \dfrac{9}{8ab} = \dfrac{9a^{-1}b^{-1}}{8}$

3. $\dfrac{2+5-2}{9x} = \dfrac{5}{9x} = \dfrac{5x^{-1}}{9}$

4. $\dfrac{5x+2x}{11} = \dfrac{7x}{11}$

5. $\dfrac{7x-3x}{20} = \dfrac{4x}{20} = \dfrac{x}{5}$

6. $\dfrac{3xy+xy+5xy}{16} = \dfrac{9xy}{16}$

7. $\dfrac{9ab-5ab-3ab}{20} = \dfrac{ab}{20}$

8. $\dfrac{3a+5+2a}{8b} = \dfrac{5a+5}{8b}$

9. $\dfrac{2xy-3y+2y}{7} = \dfrac{2xy-y}{7}$

10. $\dfrac{3-5}{4y} = \dfrac{-2}{4y} = \dfrac{-1}{2y} = -\dfrac{1}{2y} = -\dfrac{y^{-1}}{2}$

11. $\dfrac{8a-14a+2a}{21} = \dfrac{10a-14a}{21} = \dfrac{-4a}{21} = -\dfrac{4a}{21}$

Additional practice problems are at the end of the chapter.

ADDING AND SUBTRACTING FRACTIONS WITH UNLIKE DENOMINATORS 6–8

As with real numbers, when working with literal numbers and unlike denominators, we must first find the common denominator (the LCM) and then change each fraction so that all fractions have this common denominator. If we change the denominator, we must also change the numerator. We find the new numerator by using the rules given in Chapter 4 (Rule 4–4).

> ☞ **RULE 6-2** To add (or subtract) fractions with un-
> like denominators: (1) find the CD; (2) change the de-
> nominator of all fractions to this denominator; (3)
> divide the CD by the old denominator of each fraction
> and then multiply this number by the old numerator to
> find the new numerator of the fraction; (4) add (or sub-
> tract) the fractions, combining all like terms in the nu-
> merator. Reduce the answer to the lowest terms.

The following examples illustrate the process.

EXAMPLE 6–12

Perform the indicated operations. Reduce to lowest terms.

$$\frac{3a}{20} + \frac{2a}{15} + \frac{3a}{10}$$

SOLUTION

Step 1. Find the common denominator:

$$20 = 2 \cdot 2 \cdot 5$$
$$15 = 3 \cdot 5$$
$$10 = 2 \cdot 5$$
$$CD = 2 \cdot 2 \cdot 3 \cdot 5 = 60$$

Step 2. Change each fraction so that its denominator is the CD.

$$\frac{3a}{20} = \frac{3a \times 3}{60} = \frac{9a}{60}$$

$$\frac{2a}{15} = \frac{2a \times 4}{60} = \frac{8a}{60}$$

$$\frac{3a}{10} = \frac{3a \times 6}{60} = \frac{18a}{60}$$

Step 3. Add the new fractions together. Reduce to lowest terms where possible.

$$\frac{9a}{60} + \frac{8a}{60} + \frac{18a}{60} = \frac{9a + 8a + 18a}{60} = \frac{35a}{60} = \frac{7a}{12}$$

EXAMPLE 6–13

Perform the indicated operations. Reduce to lowest terms.

$$\frac{3}{10a^2} + \frac{4}{15a} + \frac{b}{6a^2}$$

SOLUTION

$$10a^2 = 2 \cdot 5 \cdot a \cdot a$$
$$15a = 3 \cdot 5 \cdot a$$
$$6a^2 = 2 \cdot 3 \cdot a \cdot a$$

$$CD = 2 \cdot 3 \cdot 5 \cdot a \cdot a = 30a^2$$

$$\frac{3}{10a^2} = \frac{3 \times 3}{30a^2} = \frac{9}{30a^2}$$

$$\frac{4}{15a} = \frac{4 \times 2a}{30a^2} = \frac{8a}{30a^2}$$

$$\frac{b}{6a^2} = \frac{b \times 5}{30a^2} = \frac{5b}{30a^2}$$

$$\frac{9}{30a^2} + \frac{8a}{30a^2} + \frac{5b}{30a^2} = \frac{9 + 8a + 5b}{30a^2}$$

No reduction of the numerator is possible here. We have no like terms.

PRACTICE PROBLEMS 6–8

Perform the indicated operations. Reduce to lowest terms where possible.

1. $\dfrac{3a}{4} + \dfrac{2b}{3}$

2. $\dfrac{2a}{3} + \dfrac{3a}{2}$

3. $\dfrac{b}{3a} + \dfrac{2b}{5}$

4. $\dfrac{x}{5y} + \dfrac{3x}{4}$

5. $\dfrac{x}{2y} + \dfrac{y}{5x}$

6. $\dfrac{3x}{y} + \dfrac{2y}{x}$

7. $\dfrac{2xy}{15z} + \dfrac{3xz}{5y}$

8. $\dfrac{5}{6a} + \dfrac{4}{15a^3} + \dfrac{1}{4a^2}$

9. $\dfrac{7a}{8} - \dfrac{2a}{5} + \dfrac{3a}{20}$

10. $\dfrac{3bc}{16a} - \dfrac{5ab}{8c}$

11. $\dfrac{11a}{15} - \dfrac{3a}{5} + \dfrac{2a}{3}$

12. $\dfrac{2a^2c}{3b} - \dfrac{4a^2c}{15b} + \dfrac{2a^2c}{5b}$

SOLUTIONS

1. $\dfrac{9a}{12} + \dfrac{8b}{12} = \dfrac{9a + 8b}{12}$

2. $\dfrac{4a}{6} + \dfrac{9a}{6} = \dfrac{4a + 9a}{6} = \dfrac{13a}{6}$

3. $\dfrac{5b}{15a} + \dfrac{6ab}{15a} = \dfrac{5b + 6ab}{15a}$

4. $\dfrac{4x}{20y} + \dfrac{15xy}{20y} = \dfrac{4x + 15xy}{20y}$

5. $\dfrac{5x^2}{10xy} + \dfrac{2y^2}{10xy} = \dfrac{5x^2 + 2y^2}{10xy}$

6. $\dfrac{3x^2}{xy} + \dfrac{2y^2}{xy} = \dfrac{3x^2 + 2y^2}{xy}$

7. $\dfrac{2xy^2}{15yz} + \dfrac{9xz^2}{15yz} = \dfrac{2xy^2 + 9xz^2}{15yz}$

8. $\dfrac{50a^2}{60a^3} + \dfrac{16}{60a^3} + \dfrac{15a}{60a^3} = \dfrac{50a^2 + 16 + 15a}{60a^3}$

9. $\dfrac{35a}{40} + \dfrac{-16a}{40} + \dfrac{6a}{40} = \dfrac{35a - 16a + 6a}{40} = \dfrac{25a}{40} = \dfrac{5a}{8}$

10. $\dfrac{3bc^2}{16ac} + \dfrac{-10a^2b}{16ac} = \dfrac{3bc^2 - 10a^2b}{16ac}$

11. $\dfrac{11a}{15} + \dfrac{-9a}{15} + \dfrac{10a}{15} = \dfrac{11a - 9a + 10a}{15} = \dfrac{12a}{15} = \dfrac{4a}{5}$

12. $\dfrac{10a^2c}{15b} + \dfrac{-4a^2c}{15b} + \dfrac{6a^2c}{15b} = \dfrac{10a^2c - 4a^2c + 6a^2c}{15b} = \dfrac{12a^2c}{15b} = \dfrac{4a^2c}{5b} = \dfrac{4a^2b^{-1}c}{5}$

Additional practice problems are at the end of the chapter.

SELF-TEST 6–4

Perform the indicated operations:

1. $\dfrac{8}{9x} - \dfrac{2}{9x}$

2. $\dfrac{2}{7a} + \dfrac{3}{7a}$

3. $\dfrac{2a}{9c} + \dfrac{4a}{9c} - \dfrac{a}{9c}$

4. $\dfrac{15a}{12} - \dfrac{a}{12}$

5. $\dfrac{3}{5x} + \dfrac{2}{7x^2} + \dfrac{3}{10x^3}$

6. $\dfrac{4a}{5b} + \dfrac{2}{3}$

7. $\dfrac{4}{15b} - \dfrac{5b}{18a^2c} + \dfrac{3}{10}$

Answers to Self-test 6–4 are at the end of the chapter.

CHAPTER 6 AT A GLANCE

PAGE	KEY POINT OR RULE	EXAMPLE
180	*Rule 6–1:* To find the lowest common multiple of numbers containing literal numbers: (1) factor each term; (2) determine the maximum number of times a prime factor appears in any one term; (3) multiply the resulting prime factors found in step 2. This number is the LCM.	Find the LCM of $10xy^2$ and $15x^3y$. $$10xy^2 = 2 \cdot 5 \cdot x \cdot y \cdot y$$ $$15x^3y = 3 \cdot 5 \cdot x \cdot x \cdot x \cdot y$$ $$\text{LCM} = 2 \cdot 3 \cdot 5 \cdot x^3 \cdot y^2 = 30x^3y^2$$
183	*Key Point:* In multiplication, all the literal numbers contained in the multiplicand and the multiplier must appear in the product unless the exponent is zero.	$3x^3y^{-2}z \times 5x^2y^2z^2$ $= 15x^5y^0z^3 = 15x^5z^3$

| 187 | *Key Point:* To multiply fractions, we (1) cancel factors that are common to both numerators and denominators, (2) multiply the remaining numerators (their product is the numerator of the answer), (3) multiply the remaining denominators (their product is the denominator of the answer). | $$\frac{3a}{4b} \times \frac{6a^2b}{10} = \frac{3a \cdot 6a^2b}{4b \cdot 10}$$ $$= \frac{\cancel{2} \cdot 3 \cdot 3 \cdot a^3 \cdot \cancel{b}}{\cancel{2} \cdot 2 \cdot 2 \cdot 5 \cdot \cancel{b}} = \frac{9a^3}{20}$$ |

| 193 | *Rule 6–2:* To add (or subtract) fractions with unlike denominators: (1) find the CD; (2) change the denominator of all fractions to this denominator; (3) divide the CD by the old denominator of each fraction and then multiply this number by the old numerator to find the new numerator of the fraction; (4) add (or subtract) the fractions, combining all like terms in the numerator. Reduce the answer to the lowest terms. | Add $\dfrac{3a}{2b}$ and $\dfrac{2b}{3a}$. Change to fractions with a CD; then add. $$\frac{3a}{2b} = \frac{9a^2}{6ab}$$ $$\frac{2b}{3a} = \frac{4b^2}{6ab}$$ $$\frac{3a}{2b} + \frac{2b}{3a} = \frac{9a^2 + 4b^2}{6ab}$$ |

END OF CHAPTER PROBLEMS 6–1

Find the prime factors of the following:

1. $8a^2b^2$
3. $24x^2y$
5. $42c^3$
7. $88y^2z$
9. $135a^2c^2$

2. $18x^2$
4. $21ab^2$
6. $90yz^3$
8. $154a^3b$
10. $210ab^2$

END OF CHAPTER PROBLEMS 6–2

Find the LCM of the following numbers:

1. ab^3, a^3b
3. a^3bc^3, ab^3c
5. $x^2y^3z^4, xy^2z^2$
7. ab^2, a^3b^3, a^3b^2
9. $a^2bc^3, a^4b^2c^2$
11. a^4b^2c, a^4b^3c, abc^2
13. $26xy^3, 39x^2y^2$
15. $16x^2y^2z, 32xy, 8x^3z^2$
17. $30a^2b, 75ab^2, 50abc$
19. $18x^2z^3, 27y^4z, 30x^2y^2z$

2. a^3b^3, a^4b^2
4. $a^3b^2c, a^2b^2c^3$
6. $x^2y^2z^3, xy^4z^2$
8. a^4b^2, a^4bc^3, a^2b^3c
10. $ab^2c^2, a^3bc^2, a^2b^2c^3$
12. $15x^2y^2, 35x^3y^3$
14. $8x^4y^2z, 18x^2y$
16. $64x^2y, 32x^2y^4, 128xy^2$
18. $15xz^2, 20x^2y^2, 28x^2yz^3$
20. $20x^3z^2, 30y^2z^3, 35x^2yz^2$

END OF CHAPTER PROBLEMS 6–3

Perform the indicated operations:

1. $3a^2b \cdot 7a^2b^2$
2. $4a^3b \cdot 4a^2b$
3. $5a^{-1}b^{-3} \cdot 4a^{-3}b^{-2}$
4. $6a^{-2}b^{-2} \cdot 5a^{-3}b^{-3}$
5. $10x^2yz \cdot x^3y^2z \cdot 2xz^2$
6. $8x^3y^2z \cdot x^2z^2 \cdot 3x^2yz$
7. $a^{-1}bc^2 \cdot 3a^2b^{-2}c^{-4} \cdot 2a^2b^{-1}c$
8. $x^{-2}y^{-1}z^2 \cdot 4x^2y^{-1}z \cdot 5x^3y^{-3}z^2$
9. $3x^2y^{-1}z \cdot 4x^4y^{-2}z \cdot 5x^{-1}y^{-1}z^2$
10. $4x^{-3}yz^2 \cdot 5x^{-3}y^{-3}z \cdot 2x^3y^2z^{-4}$

END OF CHAPTER PROBLEMS 6–4

Reduce to lowest terms:

1. $\dfrac{18a^3}{2a}$
2. $\dfrac{27a^6}{3a^2}$
3. $\dfrac{48a^2b^3}{6ab}$
4. $\dfrac{63a^5b^4}{9a^3b^2}$
5. $\dfrac{25a^4b^2c}{5abc^2}$
6. $\dfrac{8ab^3c^2}{32a^2bc^4}$
7. $\dfrac{9x^{-2}y^2}{45x^{-3}y}$
8. $\dfrac{49x^4y^{-3}}{7x^2y^{-4}}$
9. $\dfrac{44x^{-3}yz^{-2}}{4x^2y^2z^2}$
10. $\dfrac{36x^3y^{-4}z^2}{12x^{-3}y^2z^{-2}}$

END OF CHAPTER PROBLEMS 6–5

Perform the indicated multiplications. Reduce to lowest terms.

1. $\dfrac{4b}{15} \times \dfrac{3b}{8}$
2. $\dfrac{6b}{7} \times \dfrac{21b}{3}$
3. $\dfrac{3}{2b} \times \dfrac{5}{7b}$
4. $\dfrac{5}{6b} \times \dfrac{7}{9b}$
5. $\dfrac{5a}{9} \times \dfrac{6a^2}{7}$
6. $\dfrac{7a^2}{12} \times \dfrac{6a^3}{7}$
7. $\dfrac{3x^2}{10y} \times \dfrac{5x}{8y}$
8. $\dfrac{20a^2}{21b^2} \times \dfrac{3a^2}{5b^2}$
9. $\dfrac{3}{4y^2} \times \dfrac{8xy}{9}$
10. $\dfrac{10x^2y}{12} \times \dfrac{5y}{x}$
11. $\dfrac{8x^{-1}y}{5} \times \dfrac{10x^2y^{-3}}{3z^{-1}}$
12. $\dfrac{6xy^{-2}}{5} \times \dfrac{15x^{-4}y^2}{6z^{-2}}$
13. $\dfrac{11a^2}{32b^3} \times \dfrac{4a^2b}{33c}$
14. $\dfrac{7x^2y^{-2}}{8z^2} \times \dfrac{3z^3}{14x^3y^{-2}}$

15. $\dfrac{7x^2}{24y^2} \times \dfrac{4xy^{-2}}{21x}$

16. $\dfrac{9a^3}{16b^2} \times \dfrac{8ab^4}{18c^2}$

17. $\dfrac{4a^3c^{-2}}{5b^2} \times \dfrac{7b^{-1}c^3}{16a^2} \times \dfrac{20a^3b^{-1}}{21c}$

18. $\dfrac{3x^4}{5y^2} \times \dfrac{2y^3z^3}{9x^2} \times \dfrac{30y^2z^{-1}}{3x^{-1}}$

19. $\dfrac{6xy^{-2}}{15z^2} \times \dfrac{25x^{-3}z^2}{18y} \times \dfrac{12y^2}{5x^{-1}z^{-2}}$

20. $\dfrac{5x^{-3}z^2}{12y^2} \times \dfrac{24yz^3}{35x^2} \times \dfrac{14y}{6x^{-2}z^{-2}}$

END OF CHAPTER PROBLEMS 6–6

Perform the indicated divisions. Reduce to lowest terms.

1. $\dfrac{3x}{8} \div \dfrac{7x^3}{3}$

2. $\dfrac{7x^3}{9} \div \dfrac{42x}{21}$

3. $\dfrac{7x}{16} \div \dfrac{21x^2}{32}$

4. $\dfrac{5a^2}{16} \div \dfrac{35a^3}{24}$

5. $\dfrac{16b}{3} \div \dfrac{8b^4}{27}$

6. $\dfrac{27a^4}{5} \div \dfrac{9a}{45}$

7. $\dfrac{14ab^2}{9c^3} \div \dfrac{35b}{c^2}$

8. $\dfrac{35a^3b^{-1}}{24c^2} \div \dfrac{7ab^2}{36c^4}$

9. $\dfrac{3xy^3}{10x^{-1}} \div \dfrac{3xy^3}{10x^{-1}}$

10. $\dfrac{10a}{21b^2c^3} \div \dfrac{15a^{-3}b}{28b^{-1}c^2}$

11. $\dfrac{32a}{45b^2} \div \dfrac{16a^2}{25bc}$

12. $\dfrac{12xy^3}{42z^2} \div \dfrac{32x^3y}{35z^4}$

13. $\dfrac{15a^4b}{64c^3} \div \dfrac{75ab}{32c^2}$

14. $\dfrac{16a^5c^2}{35b^3} \div \dfrac{32a^3bc^2}{25}$

15. $\dfrac{20x^3y^2}{21z^2} \div \dfrac{30xy}{7z^2}$

16. $\dfrac{18x^2y^3}{35z^3} \div \dfrac{24xy^2}{14z^2}$

17. $\dfrac{21a^3b^{-2}}{4c^2} \div \dfrac{14a^{-1}b^2}{3c^{-4}}$

18. $\dfrac{16a^4b^{-1}}{15c^3} \div \dfrac{24a^{-2}b^2}{18c^2}$

19. $\dfrac{42yz^{-2}}{55x} \div \dfrac{21x^2y^2}{22z}$

20. $\dfrac{56y^2z^{-3}}{25x^2} \div \dfrac{21x^3y^2}{35z}$

END OF CHAPTER PROBLEMS 6–7

Perform the following operations. Reduce to lowest terms.

1. $\dfrac{2}{7x} + \dfrac{3}{7x}$

2. $\dfrac{2}{9a} + \dfrac{5}{9a}$

3. $\dfrac{6}{7x} - \dfrac{2}{7x}$

4. $\dfrac{5}{3x} - \dfrac{3}{3x}$

5. $\dfrac{1}{12ab} + \dfrac{5}{12ab} + \dfrac{3}{12ab}$

6. $\dfrac{1}{14ab} + \dfrac{3}{14ab} + \dfrac{6}{14ab}$

7. $\dfrac{3}{14a} + \dfrac{9}{14a} - \dfrac{5}{14a}$

8. $\dfrac{9}{16x} - \dfrac{4}{16x} + \dfrac{3}{16x}$

9. $\dfrac{5x}{24} + \dfrac{7x}{24}$

10. $\dfrac{3a}{20} + \dfrac{7a}{20}$

11. $\dfrac{8x}{15} - \dfrac{2x}{15}$

12. $\dfrac{7ab}{9} - \dfrac{2ab}{9}$

13. $\dfrac{7ab}{25} + \dfrac{6ab}{25} + \dfrac{2ab}{25}$

14. $\dfrac{2xy}{15} + \dfrac{4xy}{15} + \dfrac{8xy}{15}$

15. $\dfrac{9ab}{25} - \dfrac{2ab}{25} + \dfrac{8ab}{25}$

16. $\dfrac{3xy}{20} + \dfrac{7xy}{20} - \dfrac{5xy}{20}$

17. $\dfrac{2x}{9} + \dfrac{4x}{9} + \dfrac{4x}{9}$

18. $\dfrac{3ab}{14} + \dfrac{3ab}{14} + \dfrac{7ab}{14}$

19. $\dfrac{3x}{7} - \dfrac{5x}{7}$

20. $\dfrac{5}{11a} - \dfrac{9}{11a}$

21. $\dfrac{9x}{24} - \dfrac{15x}{24} + \dfrac{x}{24}$

22. $\dfrac{7a}{15} - \dfrac{7a}{15} - \dfrac{2a}{15}$

23. $\dfrac{4ab}{8} + \dfrac{3b}{8} - \dfrac{2ab}{8}$

24. $\dfrac{6ab}{4} + \dfrac{2a}{4} - \dfrac{5ab}{4}$

25. $\dfrac{6abc}{4d} - \dfrac{3ac}{4d} + \dfrac{abc}{4d}$

26. $\dfrac{14xyz}{6b} - \dfrac{6xz}{6b} + \dfrac{3xyz}{6b}$

27. $\dfrac{3abc}{4x} - \dfrac{5abc}{4x} - \dfrac{3ab}{4x}$

28. $\dfrac{4abc}{9} - \dfrac{7abc}{9} - \dfrac{3ab}{9}$

29. $\dfrac{6xy}{5} - \dfrac{4xz}{5} + \dfrac{3yz}{5}$

30. $\dfrac{6xz}{7} - \dfrac{5xy}{7} - \dfrac{6yz}{7}$

END OF CHAPTER PROBLEMS 6–8

Perform the indicated operations. Reduce to lowest terms where possible.

1. $\dfrac{3x}{5} + \dfrac{2x}{15}$

2. $\dfrac{3x}{8} + \dfrac{3x}{32}$

3. $\dfrac{2a}{3} + \dfrac{3a}{2}$

4. $\dfrac{b}{3} + \dfrac{2b}{7}$

5. $\dfrac{x}{2} - \dfrac{2x}{3}$

6. $\dfrac{2x}{3} - \dfrac{3x}{5}$

7. $\dfrac{4x}{5} - \dfrac{3x}{10}$

8. $\dfrac{5a}{12} - \dfrac{7a}{9}$

9. $\dfrac{y}{2x} + \dfrac{5y}{8}$

10. $\dfrac{2b}{3a} + \dfrac{3a}{2b}$

11. $\dfrac{3x}{4y} + \dfrac{2y}{3x}$

12. $\dfrac{11ab}{16c} - \dfrac{ac}{4b}$

13. $\dfrac{2b}{5x} + \dfrac{3}{10y}$

14. $\dfrac{3a}{4b} + \dfrac{3b}{8c}$

15. $\dfrac{5ab^2}{16c} - \dfrac{15ab}{32c^2}$

16. $\dfrac{5x^2z}{8y^2} - \dfrac{5z^3}{4y}$

17. $\dfrac{3b}{4} + \dfrac{3b}{8} - \dfrac{15b}{16}$

18. $\dfrac{3a}{5} - \dfrac{3a}{8} - \dfrac{3a}{4}$

19. $\dfrac{2a}{7} + \dfrac{7a}{15} - \dfrac{a}{3}$

20. $\dfrac{7x}{12} - \dfrac{3x}{8} + \dfrac{5x}{16}$

21. $\dfrac{x}{10y^2} - \dfrac{2x}{5y^2} + \dfrac{x}{3y^2}$

22. $\dfrac{3x}{5y^2} - \dfrac{3x}{10y^2} + \dfrac{7x}{20y^2}$

23. $\dfrac{4x}{3yz} + \dfrac{2y}{5x} - \dfrac{z}{6y}$

24. $\dfrac{7y}{9xz} - \dfrac{5x}{8y} + \dfrac{7z}{36x}$

25. $\dfrac{4a}{33b} + \dfrac{5b}{55c} - \dfrac{3c}{11a}$

26. $\dfrac{3b}{8c} - \dfrac{7b}{15c} + \dfrac{5b}{24a}$

27. $\dfrac{5y^2}{6x^2y} + \dfrac{4xy}{15y^2} - \dfrac{2x^2}{9y}$

28. $\dfrac{5}{12x^2y} - \dfrac{5z}{6xy^2} + \dfrac{2y}{15x}$

29. $\dfrac{3b}{20a^2} + \dfrac{7c}{10ab} - \dfrac{5a}{12b^3}$

30. $\dfrac{4a}{7b^2} + \dfrac{2c}{9ab^2} - \dfrac{4ab^2}{21c^2}$

ANSWERS TO SELF-TESTS

1. $\dfrac{2}{3x}$

2. $\dfrac{5}{7a}$

3. $\dfrac{5a}{9c}$

4. $\dfrac{7a}{6}$

5. $\dfrac{42x^2 + 20x + 21}{70x^3}$

6. $\dfrac{12a + 10b}{15b}$

7. $\dfrac{24a^2c - 25b^2 + 27a^2bc}{90a^2bc}$

Linear Equations

7

Introduction

When algebraic terms or expressions are separated by an equals sign, the resulting algebraic statement is called an *equation*. In the simplest equations, the literal numbers are raised to the first power. These equations are called *linear* equations. If the literal numbers are raised to the second power, we have *second-degree* equations. We will be solving both linear and second-degree equations in this chapter.

All linear equations can be solved provided there is only one unknown—one literal number. Ohm's law equations may be either linear or second-degree equations, depending on the variables used. $E = IR$ is a linear equation. If two of the variables are known, we can solve for the third. $P = I^2R$ is a second-degree equation because I is raised to the second power. These equations can be solved, again provided that there is only one unknown variable.

The use of equations such as these is part of any study of electronics. The solution of any electronics problem will involve the use of some sort of equation. It may be a simple linear equation or a second-degree equation, like the ones we will study in this chapter, or it might be a more complex equation of the type we will study in future chapters. Any course of study in problem solving must include a thorough understanding of how to solve linear equations and second-degree equations. In this chapter we will also show how the calculator is used in solving equations.

Chapter Objectives

In this chapter, you will learn how to:

1. Recognize and identify the parts of an equation.
2. Solve various kinds of linear equations containing one unknown variable.
3. Solve various second-degree equations containing one unknown variable.
4. Use the calculator to solve various kinds of electrical problems that involve the use of linear or second-degree equations.
5. Apply the rules that have been developed to help in solving equations.
6. Verify the solutions of equations that have been solved.

7-1-1 Identifying Equations

Algebraic expressions separated by an equals sign (=) are called equations. The equals sign implies that the expressions are equal. Let's call the expression to the left of the equals sign the *left side* and the expression to the right of the equals sign the *right side*. All the following algebraic statements are equations.

> **EXAMPLE 7-1**

Solve for the unknown in the following equations.

(a) $16 + 4 = 20$ (b) $16 + x = 20$ (c) $x + 4 = 20$

(d) $x - 4 = 20$ (e) $4x = 20$ (f) $\dfrac{x}{4} = 20$

(g) $x^2 + x = 20$ (h) $x^2 + 4 = 20$

In Example 7–1(a) all numbers are real numbers. Both sides have a value of 20. In all the other examples, a part of the left side is unknown. The value of the unknown (x) must be of such as to make the left side equal to 20. Sometimes the value of x can be found by inspection. This should be the case in Example 7–1(b) and (c). In (b), $x = 4$:

$$16 + x = 20$$
$$16 + 4 = 20$$
$$20 = 20$$

and in (c), x must equal 16.

$$x + 4 = 20$$
$$16 + 4 = 20$$
$$20 = 20$$

In other examples we must work a little harder. Notice in Examples 7–1(g) and (h) that we have x^2 terms. In all the other equations, x is raised to the first power. Equations in which the unknown is raised to the first power are called *linear* equations. Equations containing x^2 terms are called *second-degree* equations. If equations contained x^3 terms, they would be called *third-degree* equations.

7-1-2 Solving Linear Equations

Here is the basic rule to use when solving equations.

> ☞   **RULE 7-1**   We can perform any mathematical operation on one side of an equation, *provided we perform the same operation on the other side.*

This rule means that we can make any change we want in the value of one side as long as we make the same change in the other side. We can add, subtract, multiply, divide, square, and so on. We can perform any of these operations as long as we maintain the equality of sides.

In solving equations, we typically solve for some unknown value. In simple linear equations, there is only one unknown and only one value for that unknown.

In this and following chapters, we will use a five-step method of solution for both linear and second-degree equations.

☞ **RULE 7–2** To solve both linear and second-degree equations:

Step 1. Clear all fractions (if any).
Step 2. Move all terms containing the unknown to one side and combine them into one term. It is usually preferable to move unknowns to the left side.
Step 3. Move all constants to the other side.
Step 4. Perform the operation or operations necessary to make the coefficient of the unknown equal to 1 (don't forget to perform this operation on *both* sides).
Step 5. Substitute the value of the unknown back into the original equation to verify solution.

These five steps do not have to be done in the order stated, except that step 5 would always be the last step. Also, not all equations would require all five steps.

Consider Example 7–1(c). We know from inspection that $x = 16$, but let's work it mathematically. In solving equations we must somehow get the unknown on one side by itself. That is, we have to move all terms that are real numbers to one side and leave the term containing the unknown on the other side. It is standard practice to put the term containing the unknown on the left side. We would solve the equation in the following manner.

EXAMPLE 7–I(c)

$x + 4 = 20$. Solve for x.

SOLUTION   We want the unknown variable, x, on the left side by itself. This can be done only by subtracting 4. $x + 4 - 4 = x$. But if we subtract 4 from the left side, we must subtract 4 from the right side to maintain the equality of sides.

$$x + 4 - 4 = 20 - 4$$

$$x = 16$$

Then we replace the unknown with its value and check for equality:

$$x + 4 = 20$$

$$16 + 4 = 20$$

$$20 = 20$$

When we want to move a term from one side to the other, we do it by adding or subtracting. In the preceding example, we subtracted. In Example 7–1(d) we will add.

EXAMPLE 7–1(d)

$x - 4 = 20$. Solve for x.

SOLUTION

$$x - 4 + 4 = 20 + 4 \qquad \text{(add 4 to both sides)}$$

$$x = 24$$

CHECK

$$24 - 4 = 20$$

$$20 = 20$$

From these examples we can make up the following rule:

☞ **RULE 7-3** Terms may be moved from one side of an equation to the other side by changing their signs.

Notice in Example 7–1(c) that 4 was positive when it was on the left side, and it became negative when it was moved to the right side. In Example 7–1(d), 4 was negative when it was on the left side, and it became positive when it was moved to the right side.

$$x + 4 = 20$$

$$x = 20 - 4 \qquad \text{(move 4 to right side and change its sign)}$$

$$x = 16$$

$$x - 4 = 20$$

$$x = 20 + 4 \qquad \text{(move } -4 \text{ to right side and change its sign)}$$

$$x = 24$$

EXAMPLE 7–1(e)

$4x = 20$. Solve for x.

SOLUTION

$$4x = 20$$

$$\frac{4x}{4} = \frac{20}{4} \qquad \text{(divide } both\ sides \text{ by 4)}$$

$$x = 5$$

$$4(5) = 20$$

$$20 = 20$$

In this example, the left-side term is $4x$. Since we have to find the value of x, we must somehow get rid of the 4. That is, we want the left side to equal x. We can make the left side equal to x by dividing by 4. But if we divide the left side by 4 we must also divide the right side by 4 (Rule 7–1).

EXAMPLE 7–I(f)

$\dfrac{x}{4} = 20$. Solve for x.

SOLUTION   How can we make the left side equal to x? Multiplication by 4 is the only way it can be done.

$$\frac{x}{4} \times 4 = \frac{4x}{4} = x$$

Solving the equation for x, we get

$$\frac{x}{4} = 20$$

$$\frac{x}{4}(4) = 20(4)$$

$$x = 80$$

$$\frac{80}{4} = 20$$

$$20 = 20$$

These two examples (e and f) show how we can move parts of terms from one side to another by multiplying or dividing.

A shortcut method of multiplying or dividing terms in order to simplify the terms is called *cross-multiplication*. When we cross-multiply, we multiply the numerator of one term and the denominator of the other term. Consider Example 7–1(f) again. (We have shown the 1 in the denominator of the right side just to help you see the process.)

$$\frac{x}{4} = \frac{20}{1} \qquad \text{If we cross-multiply, we get}$$

$$x \cdot 1 = 20 \cdot 4$$

$$x = 80$$

Cross-multiplication in this example is really the multiplication of both sides by 4 and then by 1. Often the process can be done by inspection. It's a quick way of getting rid of a fraction when it is part of the equation.

EXAMPLE 7–2

$\dfrac{20}{x} = 5$. Find x.

SOLUTION   We must get x out of the denominator. There are two ways we can do this. The first way is to take the reciprocal of both sides:

$$\frac{20}{x} = 5$$

$$\frac{x}{20} = \frac{1}{5}$$

Next we multiply both sides by 20. This is the only way we can move 20 to the right side.

$$\frac{x}{20}(20) = \frac{1}{5}(20)$$

$$\frac{20x}{20} = \frac{20}{5}$$

$$x = 4$$

CHECK

$$\frac{20}{4} = 5$$

$$5 = 5$$

The other way to solve the problem is to multiply both sides by x.

$$\frac{20}{x}(x) = 5x$$

$$\frac{20x}{x} = 5x$$

$$20 = 5x$$

Next we divide both sides by 5.

$$\frac{20}{5} = \frac{5x}{5}$$

$$4 = x$$

or $$x = 4$$

Of course, we could have cross-multiplied here also, but we wanted to go through the actual mathematical operations.

EXAMPLE 7–3

$\dfrac{x}{3} = -5$. Find x.

SOLUTION

$$\frac{x}{3}(3) = -5(3) \qquad \text{(multiply both sides by 3)}$$

$$x = -15$$

CHECK

$$\frac{-15}{3} = -5$$

$$-5 = -5$$

EXAMPLE 7–4

$\dfrac{24}{-x} = -2$. Find x.

SOLUTION

$$\frac{-x}{24} = -\frac{1}{2} \qquad \text{(take the reciprocal of both sides)}$$

$$\frac{-x}{24}(24) = -\frac{1}{2}(24) \qquad \text{(multiply both sides by 24)}$$

$$-x = -12$$

$$x = 12 \qquad \text{(change the sign of both sides)}$$

CHECK

$$\frac{24}{-12} = -2$$

$$-2 = -2$$

EXAMPLE 7–5

$\dfrac{2x}{5} = \dfrac{3}{4}$. Find x.

$$2x \cdot 4 = 3 \cdot 5 \qquad \text{(cross-multiply)}$$

$$8x = 15$$

$$x = \frac{15}{8} = 1\frac{7}{8} \qquad \text{(divide both sides by 8 and reduce to lowest terms)}$$

EXAMPLE 7–6

$15 = 3xy$. Solve for each of the variables.

SOLUTION   Find x

$$\frac{15}{3y} = \frac{3xy}{3y} \qquad \text{(divide both sides by } 3y\text{)}$$

$$\frac{5}{y} = x$$

$$x = \frac{5}{y}$$

Find y

$$15 = 3xy$$

$$\frac{15}{3x} = \frac{3xy}{3x} \qquad \text{(divide both sides by } 3x\text{)}$$

$$\frac{5}{x} = y$$

$$y = \frac{5}{x}$$

CHECK

$$15 = 3\left(\frac{5}{y}\right)y$$

$$15 = \frac{3 \cdot 5 \cdot \cancel{y}}{\cancel{y}}$$

$$15 = 15$$

PRACTICE PROBLEMS 7–1

Solve for the unknown in the following equations. Check your answers by plugging the answer back into the original problem.

1. $x - 3 = 14$

2. $a + 5 = 6$

3. $4 - a = 6$

4. $5 + y = 6$

5. $3x = 12$

6. $5a = 30$

7. $6a = -42$

8. $-3a = 18$

9. $\dfrac{x}{3} = 4$

10. $\dfrac{a}{7} = 3$

11. $\dfrac{a}{5} = -3$

12. $\dfrac{-x}{6} = 7$

13. $\dfrac{24}{x} = 6$

14. $\dfrac{28}{a} = 7$

15. $\dfrac{-32}{b} = 4$

16. $\dfrac{-48}{y} = -6$

17. $\dfrac{2x}{3} = 4$

18. $\dfrac{4}{3x} = 8$

19. $5 + 6x = 23$

20. $3a + 4 = 25$

21. $\dfrac{7x}{4} = \dfrac{5}{8}$

22. $\dfrac{3}{7x} = \dfrac{3}{11}$

23. $6 = 3ab$

24. $2a = 3b$

25. $4\,k\Omega = R_1 + R_2$

26. $X_L = 2\pi fL$ (Do not solve for π. π is a constant, not a variable. π always equals 3.14, to three places.)

SOLUTIONS

1. $x = 17$

2. $a = 1$

3. $a = -2$

4. $y = 1$

5. $x = 4$

6. $a = 6$

7. $a = -7$

8. $a = -6$

9. $x = 12$

10. $a = 21$

11. $a = -15$

12. $x = -42$

13. $x = 4$

14. $a = 4$

15. $b = -8$

16. $y = 8$

17. $x = 6$

18. $x = \dfrac{1}{6}$

19. $x = 3$

20. $a = 7$

21. $x = \dfrac{5}{14}$

22. $x = \dfrac{11}{7} = 1\dfrac{4}{7}$

23. $a = \dfrac{2}{b}, b = \dfrac{2}{a}$

24. $a = \dfrac{3b}{2}, b = \dfrac{2a}{3}$

25. $R_1 = 4\,k\Omega - R_2, R_2 = 4\,k\Omega - R_1$

26. $f = \dfrac{X_L}{2\pi L}, L = \dfrac{X_L}{2\pi f}$

Additional practice problems are at the end of the chapter.

Up to this point we have solved equations by means of addition, subtraction, multiplication, or division. Some equations we use in electronics require additional operations. Let's consider some general case equations.

EXAMPLE 7–7

$x^2 = 81$. Find x.

SOLUTION   Whenever a number is squared, we can find that number by finding its square root. $\sqrt{6^2} = 6$, $\sqrt{7^2} = 7$, and $\sqrt{x^2} = x$. If we take the square root of the left side, we must also take the square root of the right side.

$$x^2 = 81$$
$$\sqrt{x^2} = \sqrt{81}$$
$$x = 9$$

As in previous chapters, we are not considering the negative root (-9 in this example).

The following mathematical manipulations are very useful in problem solving:

☞ **KEY POINT**   The square root of the square of a number is that number.

We can turn this rule around and say:

☞ **KEY POINT**   The square of the square root of a number is that number.

An example of how to apply this rule follows.

EXAMPLE 7–8

$\sqrt{x} = 5$. Find x.

SOLUTION   We need x where we have $\sqrt{x}$. We must square the term to get rid of the square root.

$$(\sqrt{x})^2 = x \quad \text{or} \quad (x^{1/2})^2 = x^1 = x$$

We must also square the right side.

$$\sqrt{x} = 5$$
$$(\sqrt{x})^2 = 5^2$$
$$x = 25$$

CHECK

$$\sqrt{25} = 5, \qquad 5 = 5$$

Let's look at some more examples of problems involving squares and square roots.

EXAMPLE 7–9

$\dfrac{x^2}{4} = 6$. Find x.

SOLUTION

$$\frac{x^2}{4}(4) = 6(4) \qquad \text{(multiply both sides by 4)}$$

$$x^2 = 24$$

$$\sqrt{x^2} = \sqrt{24} \qquad \text{(take the square root of both sides)}$$

$$x = 4.90$$

As we get into more complex problems, we need to use the calculator when we verify our answers.

CHECK

$$\frac{4.90^2}{4} = 6$$

4.90 $\boxed{x^2}$ $\boxed{\div}$ 4 $\boxed{=}$. The answer is 6.0025. The answer is not *exactly* 6 because x did not equal exactly 4.90. We rounded to 4.90. However, if you had stored $\sqrt{24}$ when you found it.

$$\sqrt{24} = 24 \;\boxed{\sqrt{x}}\;\boxed{\text{STO}}$$

and if you recall the stored value when you check it,

$$\boxed{\text{RCL}}\;\boxed{x^2}\;\boxed{\div}\;4\;\boxed{=}$$

you will get *exactly* 6.0000. So, get into the habit of storing your answers to these equations to make it easier and quicker to verify your answers.

EXAMPLE 7–10

$\dfrac{\sqrt{y}}{2} = 4$. Find y.

SOLUTION

$$\frac{\sqrt{y}}{2}(2) = 4(2) \qquad \text{(multiply both sides by 2)}$$

$$\sqrt{y} = 8$$

$$y = 64 \qquad \text{(square both sides)}$$

$$\frac{\sqrt{64}}{2} = 4$$

$$\frac{8}{2} = 4$$

$$4 = 4$$

EXAMPLE 7–11

$4 = \dfrac{1}{3a\sqrt{b}}$. Find a and then b.

SOLUTION

$$4(a) = \frac{1}{3a\sqrt{b}}(a) \qquad \text{(multiply by } a)$$

$$4a = \frac{1}{3\sqrt{b}}$$

$$\frac{4a}{4} = \frac{1}{3\sqrt{b}}\left(\frac{1}{4}\right) \qquad \text{(divide by 4)}$$

$$a = \frac{1}{12\sqrt{b}}$$

CHECK

$$4 = \frac{1}{3\left(\dfrac{1}{12\sqrt{b}}\right)(\sqrt{b})} = \frac{1}{3\left(\dfrac{1}{12}\right)} = \frac{1}{\dfrac{3}{12}} = \frac{12}{3} = 4$$

$4 = \dfrac{1}{3a\sqrt{b}}$. Find b.

$$4(\sqrt{b}) = \frac{1}{3a\sqrt{b}}(\sqrt{b}) \qquad \text{(multiply by } \sqrt{b})$$

$$4\sqrt{b} = \frac{1}{3a}$$

$$\frac{4\sqrt{b}}{4} = \frac{1}{3a}\left(\frac{1}{4}\right) \qquad \text{(divide by 4)}$$

$$\sqrt{b} = \frac{1}{12a}$$

$$(\sqrt{b})^2 = \left(\frac{1}{12a}\right)^2 \quad \text{(square both sides)}$$

$$b = \frac{1}{144a^2}$$

or

$$b = \frac{a^{-2}}{144}$$

CHECK

$$4 = \frac{1}{3a \sqrt{\dfrac{1}{144a^2}}}.$$

$$4 = \frac{1}{3a\left(\dfrac{1}{12a}\right)} = \frac{1}{\dfrac{3a}{12a}} = \frac{12}{3} = 4$$

We can't cover all the kinds of equations that might be encountered, but the preceding examples should provide enough problem-solving situations so that any similar equations can be solved.

PRACTICE PROBLEMS 7–2

Solve for the unknown in the following equations. Round real-number answers to three places:

1. $x^2 = 36$

2. $\dfrac{x^2}{5} = 7$

3. $25 = 5x^2$

4. $3x^2 = 48$

5. $\dfrac{a^2}{3} = 15$

6. $\dfrac{3x^2}{4} = 60$

7. $\dfrac{14}{5b^2} = 6$

8. $\dfrac{21}{5} = 4x^2$

9. $\sqrt{x} = 6$

10. $\dfrac{\sqrt{b}}{6} = 2$

11. $4 = \dfrac{1}{3\sqrt{x}}$

12. $\sqrt{a} = 9$

13. $\dfrac{\sqrt{b}}{5} = 5$

14. $4 = \dfrac{1}{2\sqrt{x}}$

15. $5 = \dfrac{5}{2\sqrt{x}}$

16. $3\sqrt{b} = \dfrac{3}{5}$

17. $I^2 = \dfrac{P}{R}$

18. $f_r = \dfrac{1}{2\pi\sqrt{LC}}$

19. $4 = \dfrac{1}{3a\sqrt{b}}$

20. $6 = \dfrac{2}{3\sqrt{ab}}$

21. $X_C = \dfrac{1}{2\pi fC}$

SOLUTIONS

1. $x = 6.00$ 2. $x = 5.92$

3. $x = 2.24$ 4. $x = 4.00$

5. $a = 6.71$ 6. $x = 8.94$

7. $b = 0.683$ 8. $x = 1.02$

9. $x = 36.0$ 10. $b = 144$

11. $x = 0.00694$ 12. $a = 81.0$

13. $b = 625$ 14. $x = 0.0156$

15. $x = 0.250$ 16. $b = 0.0400$

17. $I = \sqrt{\dfrac{P}{R}},\; R = \dfrac{P}{I^2},\; P = I^2R$ 18. $L = \dfrac{1}{(2\pi f_r)^2 C},\; C = \dfrac{1}{(2\pi f_r)^2 L}$

19. $a = \dfrac{1}{12\sqrt{b}},\; b = \dfrac{1}{144a^2}$ 20. $a = \dfrac{1}{81b},\; b = \dfrac{1}{81a}$

21. $f = \dfrac{1}{2\pi CX_C},\; C = \dfrac{1}{2\pi fX_C}$

SELF-TEST 7–1

Solve for the unknown in the following equations.

1. $y - 6 = 4$ 2. $4x = 20$

3. $\dfrac{-a}{4} = 2$ 4. $\dfrac{5}{2y} = 2$

5. $a^2 = 49$ 6. $\dfrac{a^2}{3} = 3$

7. $\sqrt{a} = 5$ 8. $\dfrac{\sqrt{y}}{5} = 6$

9. $2 = \dfrac{1}{4\sqrt{x}}$

Solve for each literal number in the following equations:

10. $3xy = 7$ 11. $R = \dfrac{E^2}{P}$

12. $3 = \dfrac{1}{4x\sqrt{y}}$

Answers to Self-test 7–1 are at the end of the chapter.

Often, in electronics problem solving, we are solving an equation—solving for the un-known. In this section we will change equations so that the unknown is on one side and all known values are on the other. The following are examples of problems that have to be solved when working with electrical circuits.

EXAMPLE 7–12

$E = IR$. Find I when $E = 12$ V and $R = 12$ kΩ.

SOLUTION First solve the equation for I.

$$E = IR$$

$$\frac{E}{R} = \frac{IR}{R} \qquad \text{(divide by } R\text{)}$$

$$\frac{E}{R} = I$$

$$I = \frac{E}{R} \qquad \text{(change sides so that the unknown is on the left side)}$$

Now key in the values of E and R and solve.

$$I = \frac{12 \text{ V}}{12 \text{ k}\Omega} = 1 \text{ mA}$$

EXAMPLE 7–13

$P = \dfrac{E^2}{R}$. Find E when $P = 300$ mW and $R = 1.5$ kΩ.

SOLUTION Solve for E.

$$PR = \frac{E^2}{R}(R) \qquad \text{(multiply by } R\text{)}$$

$$PR = E^2$$

$$\sqrt{PR} = \sqrt{E^2} \qquad \text{(take the square root of both sides)}$$

$$\sqrt{PR} = E$$

$$E = \sqrt{PR}$$

Key in the known values.

$$E = \sqrt{300 \text{ mW} \times 1.5 \text{ k}\Omega} = \sqrt{450}$$

$$E = 21.2 \text{ V}$$

EXAMPLE 7-14

$\frac{N_P}{N_S} = \sqrt{\frac{Z_P}{Z_S}}$. Find Z_P when $Z_S = 8\ \Omega$, $N_P = 500$, and $N_S = 25$.

SOLUTION Solve for Z_P.

$$\left(\frac{N_P}{N_S}\right)^2 = \left[\sqrt{\left(\frac{Z_P}{Z_S}\right)}\right]^2 \qquad \text{(square both sides)}$$

$$\left(\frac{N_P}{N_S}\right)^2 = \frac{Z_P}{Z_S}$$

$$\left(\frac{N_P}{N_S}\right)^2 \times Z_S = \frac{Z_P}{Z_S}(Z_S) \qquad \text{(multiply by } Z_S\text{)}$$

$$\left(\frac{N_P}{N_S}\right)^2 Z_S = Z_P$$

$$Z_P = \left(\frac{N_P}{N_S}\right)^2 Z_S$$

Key in the known values.

$$Z_P = \left(\frac{500}{25}\right)^2 \times 8\ \Omega$$

$$= 3200\ \Omega = 3.20\ k\Omega$$

Often the student has a choice when attempting a solution using the calculator. This is a good example of two different ways to go:

$$\boxed{(}\ \boxed{500}\ \boxed{\div}\ \boxed{25}\ \boxed{)}\ \boxed{x^2}\ \boxed{\times}\ \boxed{8}\ \boxed{=}$$

yields 3.20 kΩ, as does

$$\boxed{500}\ \boxed{\div}\ \boxed{25}\ \boxed{=}\ \boxed{x^2}\ \boxed{\times}\ \boxed{8}\ \boxed{=}$$

Casio users should use the first method to take advantage of Casio's playback feature. The Casio can store in memory an equation containing up to 72 keystrokes. The user can see 12 characters at a time on the top line of the display and can scroll forward or backward through all 72 keystrokes and edit any of them. This feature can be used to correct a data entry error or to modify the equation. After editing, the equals key can be pressed to get a new answer.

EXAMPLE 7-15

$f_r = \dfrac{1}{2\pi\sqrt{LC}}$. Find L if $C = 400$ pF and $f_r = 15$ kHz.

Solve for L.

$$f_r\sqrt{LC} = \frac{1}{2\pi} \quad \text{(multiply by } \sqrt{LC})$$

$$\sqrt{LC} = \frac{1}{2\pi f_r} \quad \text{(divide by } f_r)$$

$$LC = \left(\frac{1}{2\pi f_r}\right)^2 \quad \text{(square both sides)}$$

$$LC = \frac{1}{(2\pi f_r)^2} \quad (1^2 = 1)$$

$$L = \frac{1}{(2\pi f_r)^2 C} \quad \text{(divide by } C)$$

Put in the known values.

$$L = \frac{1}{(2\pi \times 15 \text{ kHz})^2 \times 400 \text{ pF}} = \frac{1}{3.55}$$

$$= 2.81 \times 10^{-1} \text{ H} = 281 \text{ mH}$$

(2 × π × 15 EE 3) x² × 400 EE +/− 12 = 1/x

Answer: 2.81×10^{-1}

Casio users should remember the differences from the TI calculator.

To set the power-of-ten exponent use key EXP; the TI used EE.

To make the power-of-ten exponent a negative number use keys EXP (−); the TI uses EE +/−.

To find a reciprocal of the number entered use x^{-1}; the TI uses 1/x.

EXAMPLE 7–16

A computer's hard drive requires 12 volts and 1.5 amps. Using the equation $E = IR_{eq}$ find the equivalent resistance (R_{eq}) of the hard drive.

SOLUTION

1. Write the applicable equation: $\quad E = I \times R_{eq}$
2. Algebraically solve for the unknown: $\quad R_{eq} = E/I$
3. Substitute all the known values: $\quad R_{eq} = \dfrac{12 \text{ V}}{1.5 \text{ A}}$
4. Calculate the answer: $\quad R_{eq} = 0.800 \ \Omega$

EXAMPLE 7–17

A computer's microprocessor dissipates 3 watts of power and has an equivalent resistance (R_{eq}) of 3 ohms. Using the equation $P = E^2/R_{eq}$, calculate the voltage.

SOLUTION

1. Write the applicable equation:

$$P = E^2/R_{eq}$$

2. Algebraically solve for the unknown:

$$E^2 = P \times R_{eq}$$
$$E = (P \times R_{eq})^{1/2}$$

3. Substitute all the known values:

$$E = (3 \text{ W} \times 3 \text{ }\Omega)^{1/2}$$

4. Calculate the answer:

$$E = 3 \text{ V}$$

EXAMPLE 7–18

Computers that do not run on batteries usually have a voltage transformer in their power supply. The relationship between the voltages and impedances of the primary and secondary windings of a transformer are given by the equation: $V_P/V_S = (Z_P/Z_S)^{1/2}$. When V_P is 120 volts, V_S is 28 volts, and Z_P is 100 ohms, what is Z_S?

SOLUTION

1. Write the applicable equation:

$$V_P/V_S = (Z_P/Z_S)^{1/2}$$

2. Algebraically solve for the unknown:

$$(V_P/V_S)^2 = (Z_P/Z_S)$$
$$Z_S = (Z_P(V_S)^2)/(V_P)^2$$

3. Substitute all the known values:

$$Z_S = \frac{100 \text{ }\Omega \times (28 \text{ V})^2}{(120 \text{ V})^2}$$

4. Calculate the answer:

$$Z_S = 5.44 \text{ }\Omega$$

EXAMPLE 7–19

A circuit that includes an inductor and a capacitor has a resonant frequency (f_r) of 10 kilohertz. Using the equation $f_r = \dfrac{1}{2\pi\sqrt{LC}}$, find the value of L when C is 1,000 picofarads.

SOLUTION

1. Write the applicable equation:

$$f_r = \frac{1}{2\pi\sqrt{LC}}$$

2. Algebraically solve for the unknown:

$$(LC)^{1/2} = 1/(2\pi f_r)$$
$$LC = (1/(2\pi f_r))^2$$
$$L = (1/(2\pi f_r))^2/C$$

3. Substitute all the known values:

$$L = (1/(2\pi \times 10 \text{ kHz}))^2/(10^3 \text{ pF})$$

4. Calculate the answer:

$$L = 253 \text{ mH}$$

PRACTICE PROBLEMS 7–3

Perform the indicated operations. Round answers to three places.

1. $R_T = R_1 + R_2$. Find R_2 when
   $R_T = 678\ \Omega$ and $R_1 = 238\ \Omega$.

2. $G_T = G_1 + G_2$. Find G_2 when
   $G_T = 600\ \mu S$ and $G_1 = 170\ \mu S$.

3. $A_V = \dfrac{V_o}{V_{in}}$. Find V_o if $A_V = 48$ and
   $V_{in} = 25\ mV$.

4. $A_V = \dfrac{V_o}{V_{in}}$. Find V_{in} if $A_V = 37$ and
   $V_o = 3.83\ V$.

5. $P = \dfrac{E^2}{R}$. Find R if $E = 30\ V$ and
   $P = 780\ mW$.

6. $P = \dfrac{E^2}{R}$. Find E if $P = 720\ mW$ and
   $R = 1.2\ k\Omega$.

7. $P = I^2R$. Find I if $P = 170\ mW$ and
   $R = 1.2\ k\Omega$.

8. $P = I^2R$. Find R if $P = 1.2\ W$ and
   $I = 500\ mA$.

9. $\dfrac{N_P^2}{N_S^2} = \dfrac{Z_P}{Z_S}$. Find N_S if $N_P = 200$,
   $Z_P = 2\ k\Omega$, and $Z_S = 16\ \Omega$.

10. $\dfrac{N_P^2}{N_S^2} = \dfrac{Z_P}{Z_S}$. Find N_P if $N_S = 50$,
    $Z_P = 800\ k\Omega$, and $Z_S = 25\ k\Omega$.

11. $X_C = \dfrac{1}{2\pi fC}$. Find C if $f = 75\ Hz$ and
    $X_C = 13.7\ k\Omega$.

12. $X_C = \dfrac{1}{2\pi fC}$. Find f if $C = 20\ nF$ and
    $X_C = 3.7\ k\Omega$.

13. $X_L = 2\pi fL$. Find L if $f = 10\ kHz$ and
    $X_L = 2.73\ k\Omega$.

14. $X_L = 2\pi fL$. Find f if $L = 2.3\ H$ and
    $X_L = 4.55\ k\Omega$.

15. $B_C = 2\pi fC$. Find C if $f = 2.7\ kHz$ and
    $B_C = 200\ \mu S$.

16. $B_C = 2\pi fC$. Find f if $C = 400\ nF$ and
    $B_C = 4.35\ mS$.

17. $f_r = \dfrac{1}{2\pi\sqrt{LC}}$. Find L if $f_r = 7.85\ kHz$
    and $C = 20\ nF$.

18. $f_r = \dfrac{1}{2\pi\sqrt{LC}}$. Find C if $f_r = 12\ kHz$
    and $L = 400\ mH$.

19. A computer's microprocessor requires
    5 volts and 1.7 amps. Using the
    equation $E = I \times R_{eq}$, calculate the
    equivalent resistance (R_{eq}) of the
    microprocessor.

20. A computer's modem requires 12
    volts and 0.5 amps. Using the
    equation $E = I \times R_{eq}$, calculate the
    equivalent resistance (R_{eq}) of the
    modem.

21. A computer's microprocessor
    dissipates 10 watts of power and has
    an equivalent resistance (R_{eq}) of 2.5
    ohms. Using the equation $P = E^2/R_{eq}$,
    calculate the voltage.

22. A computer's microprocessor requires
    a 5 volt supply and dissipates 7 watts
    of power. Using the equation
    $P = E^2/R_{eq}$, calculate the equivalent
    resistance (R_{eq}).

SOLUTIONS

1. $R_2 = 440\ \Omega$
3. $V_o = 1.2\ V$
5. $R = 1.15\ k\Omega$
7. $I = 11.9\ mA$
9. $N_S = 17.9$
11. $C = 155\ nF$

2. $G_2 = 430\ \mu S$
4. $V_{in} = 0.104\ V$
6. $E = 29.4\ V$
8. $R = 4.80\ \Omega$
10. $N_P = 283$
12. $f = 2.15\ kHz$

13. $L = 43.4$ mH

14. $f = 315$ Hz

15. $C = 11.8$ nF

16. $f = 1.73$ kHz

17. $L = 20.6$ mH

18. $C = 440$ pF

19. $R_{eq} = 2.94\ \Omega$

20. $R_{eq} = 24.0\ \Omega$

21. $E = 5$ V

22. $R_{eq} = 3.57\ \Omega$

SELF-TEST 7–2

Solve for the unknown in the following problems. Round answers to three places.

1. $P = \dfrac{E^2}{R}$. Find E if $P = 12.6$ W and $R = 470\ \Omega$.

2. $P = I^2R$. Find I if $P = 780$ mW and $R = 18$ kΩ.

3. $\dfrac{N_P^2}{N_S^2} = \dfrac{Z_P}{Z_S}$. Find N_P if $N_S = 2000$, $Z_P = 50\ \Omega$, and $Z_S = 4$ kΩ.

4. $X_L = 2\pi fL$. Find f if $L = 600$ mH and $X_L = 5.35$ kΩ.

5. $f_r = \dfrac{1}{2\pi\sqrt{LC}}$. Find L if $f_r = 10$ kHz and $C = 200$ pF.

Answers to Self-test 7–2 are at the end of the chapter.

CHAPTER 7 AT A GLANCE

PAGE	KEY POINT OR RULE	EXAMPLE
204	*Rule 7–1:* We can perform any mathematical operation on one side of an equation, *provided we perform the same operation on the other side.*	If $x = 5$, then $\dfrac{x}{2} = \dfrac{5}{2}$ $x + 3 = 5 + 3$ $x^2 = 5^2$
205	*Rule 7–2:* To solve both linear and second-degree equations: (1) Clear all fractions; (2) move all terms containing the unknown to one side; (3) move all constants to the other side; (4) perform the operations necessary to make the coefficient of the unknown equal to 1; (5) verify the solution.	$2x + 3x = 20$ $5x = 20$ $x = 4$ Verify: $2(4) + 3(4) = 20$ $8 + 12 = 20$

206	*Rule 7–3:* Terms may be moved from one side of an equation to the other side by changing their signs.	$x + 4 = 20$ $x = 20 - 4$
212	*Key Point:* The square root of the square of a number is that number.	$\sqrt{a^2} = a^{2 \cdot 1/2} = a$
212	*Key Point:* The square of the square root of a number is that number.	$\left(\sqrt{a}\right)^2 = a^{1/2 \cdot 2} = a$

END OF CHAPTER PROBLEMS 7–1

Express answers that are nonwhole numbers as fractions or mixed numbers.

1. $a - 7 = 3$
2. $a - 5 = 6$
3. $5 - x = 3$
4. $6 - x = -4$
5. $8 - x = 5$
6. $b + 3 = 12$
7. $7 - x = -2$
8. $a - 4 = 3$
9. $b + 2 = 6$
10. $2 - x = 5$
11. $3 - x = 7$
12. $4 = x - 9$
13. $5y = 20$
14. $6y = 30$

15. $2b = -18$
16. $6a = -18$
17. $\dfrac{a}{4} = 3$

18. $\dfrac{x}{5} = 4$
19. $-6x = 42$
20. $7b = -35$

21. $\dfrac{c}{3} = 4$
22. $-3x = 27$
23. $\dfrac{a}{5} = -2$

24. $\dfrac{c}{6} = 4$
25. $\dfrac{x}{-2} = -8$
26. $\dfrac{a}{4} = -5$

27. $\dfrac{x}{-4} = -3$
28. $\dfrac{4a}{8} = 4$
29. $\dfrac{3}{5y} = 6$

30. $\dfrac{5}{3y} = 1$
31. $\dfrac{3x}{4} = 6$
32. $\dfrac{3x}{5} = 4$

33. $\dfrac{7}{2y} = 14$
34. $\dfrac{5}{3a} = 6$
35. $\dfrac{5y}{-3} = 10$

36. $\dfrac{6y}{-5} = 7$
37. $\dfrac{-6}{7y} = 3$
38. $\dfrac{-3}{2a} = 12$

39. $\dfrac{-8}{3b} = 2$
40. $\dfrac{-12}{5b} = 2$
41. $\dfrac{4}{x} = \dfrac{2}{3}$

42. $\dfrac{5}{x} = \dfrac{3}{8}$
43. $\dfrac{2a}{5} = \dfrac{3}{7}$
44. $\dfrac{5a}{9} = \dfrac{2}{3}$

45. $\dfrac{14x}{5} = \dfrac{3}{10}$
46. $\dfrac{9x}{4} = \dfrac{9}{2}$
47. $\dfrac{7}{2a} = \dfrac{21}{4}$

48. $\dfrac{6}{5a} = \dfrac{5}{7}$
49. $\dfrac{4}{9} = \dfrac{3}{5x}$
50. $\dfrac{3}{16} = \dfrac{5}{4x}$

END OF CHAPTER PROBLEMS 7–2

Solve for the unknown in the following equations. Round answers to three places.

1. $x^2 = 25$

2. $x^2 = 36$

3. $\dfrac{a^2}{3} = 18$

4. $\dfrac{x^2}{6} = 7$

5. $\dfrac{a^2}{4} = 14$

6. $\dfrac{x^2}{5} = 6$

7. $\dfrac{6}{b^2} = 4$

8. $\dfrac{3}{a^2} = 8$

9. $12 = \dfrac{72}{a^2}$

10. $3 = \dfrac{48}{x^2}$

11. $\sqrt{a} = 7$

12. $\sqrt{a} = 5$

13. $\dfrac{\sqrt{x}}{3} = 2$

14. $\dfrac{\sqrt{y}}{4} = 6$

15. $\dfrac{\sqrt{a}}{6} = 2.3$

16. $\dfrac{\sqrt{b}}{1.7} = 8.3$

17. $2 = \dfrac{1}{2\sqrt{x}}$

18. $4 = \dfrac{1}{3\sqrt{x}}$

19. $3 = \dfrac{1}{4\sqrt{x}}$

20. $5 = \dfrac{1}{3\sqrt{a}}$

21. $\dfrac{3}{4\sqrt{a}} = 6$

22. $\dfrac{2}{3\sqrt{a}} = 6$

23. $\dfrac{2\sqrt{x}}{3} = 6$

24. $\dfrac{3.7\sqrt{x}}{4.2} = 9.2$

Solve for each literal number in the following equations:

25. $2ax = 12$

26. $6bc = 30$

27. $8bc = 14$

28. $12xy = 16$

29. $P = \dfrac{E^2}{R}$

30. $B_L = \dfrac{1}{2\pi fL}$

31. $B_C = 2\pi fC$

32. $6 = \dfrac{1}{5a\sqrt{x}}$

33. $3 = \dfrac{1}{2a\sqrt{b}}$

34. $3 = \dfrac{7}{\sqrt{bc}}$

35. $5 = \dfrac{6}{\sqrt{xy}}$

36. $P = I^2 R$

37. $X_L = 2\pi fL$

38. $X_C = \dfrac{1}{2\pi fC}$

39. $f_r = \dfrac{1}{2\pi\sqrt{LC}}$

40. $Z_P = \left(\dfrac{N_P}{N_S}\right)^2 Z_S$

END OF CHAPTER PROBLEMS 7–3

Perform the indicated operations. Round answers to three places.

1. $R_T = R_1 + R_2$. Find R_1 if $R_T = 12.4 \text{ k}\Omega$ and $R_2 = 5.6 \text{ k}\Omega$.

2. $R_T = R_1 + R_2$. Find R_2 if $R_T = 22 \text{ k}\Omega$ and $R_1 = 12 \text{ k}\Omega$.

3. $G_T = G_1 + G_2$. Find G_1 if $G_T = 54.9 \ \mu\text{S}$ and $G_2 = 37 \ \mu\text{S}$.

4. $G_T = G_1 + G_2$. Find G_2 if $G_T = 400 \ \mu\text{S}$ and $G_1 = 303 \ \mu\text{S}$.

5. $A_V = \dfrac{V_o}{V_{in}}$. Find V_o if $A_V = 150$ and $V_{in} = 55$ mV.

6. $A_V = \dfrac{V_o}{V_{in}}$. Find V_{in} if $A_V = 50$ and $V_o = 3.15$ mV.

For problems 7-18, use $P = \dfrac{E^2}{R}$.

7. Find R if $P = 2$ W and $E = 15$ V.

8. Find R if $P = 17.3$ W and $E = 30$ V.

9. Find R if $P = 150$ mW and $E = 450$ mV.

10. Find R if $P = 27.3$ W and $E = 5.5$ V.

11. Find R if $P = 30$ mW and $E = 32$ V.

12. Find R if $P = 132$ mW and $E = 12$ V.

13. Find E if $P = 6.73$ W and $R = 27$ Ω.

14. Find E if $P = 680$ mW and $R = 56$ kΩ.

15. Find E if $P = 10.3$ W and $R = 1.8$ kΩ.

16. Find E if $P = 300$ mW and $R = 330$ Ω.

17. Find E if $P = 1.73$ W and $R = 47$ kΩ.

18. Find E if $P = 635$ mW and $R = 4.7$ kΩ.

For problems 19-30, use the equation $P = I^2 R$.

19. Find I if $P = 600$ mW and $R = 680$ Ω.

20. Find I if $P = 3.35$ W and $R = 820$ Ω.

21. Find I if $P = 500$ mW and $R = 2.7$ kΩ.

22. Find I if $P = 3.83$ W and $R = 910$ Ω.

23. Find I if $P = 14.7$ W and $R = 100$ Ω.

24. Find I if $P = 100$ W and $R = 150$ Ω.

25. Find R if $P = 500$ mW and $I = 10$ mA.

26. Find R if $P = 2$ W and $I = 550$ mA.

27. Find R if $P = 600$ mW and $I = 47.3$ mA.

28. Find R if $P = 420$ mW and $I = 720$ μA.

29. Find R if $P = 104$ W and $I = 3.77$ A.

30. Find R if $P = 40$ W and $I = 350$ mA.

For problems 31-50, use the equation $\dfrac{N_P{}^2}{N_S{}^2} = \dfrac{Z_P}{Z_S}$.

31. Find N_P if $N_S = 100$, $Z_P = 2.5$ kΩ, and $Z_S = 8$ Ω.

32. Find N_P if $N_S = 200$, $Z_P = 400$ Ω, and $Z_S = 10$ Ω.

33. Find N_P if $N_S = 50$, $Z_P = 40$ Ω, and $Z_S = 500$ Ω.

34. Find N_P if $N_S = 100$, $Z_P = 75$ Ω, and $Z_S = 2.35$ kΩ.

35. Find N_P if $N_S = 25$, $Z_P = 750$ Ω, and $Z_S = 75$ Ω.

36. Find N_P if $N_S = 1000$, $Z_P = 45$ Ω, and $Z_S = 1.2$ kΩ.

37. Find Z_S if $N_P = 1200$, $N_S = 100$, and $Z_P = 1$ kΩ.

38. Find Z_S if $N_P = 2500$, $N_S = 50$, and $Z_P = 500$ Ω.

39. Find Z_S if $N_P = 100, N_S = 1800$, and $Z_P = 2.3\ \Omega$.

40. Find Z_S if $N_P = 250, N_S = 2500$, and $Z_P = 5\ \Omega$.

41. Find Z_S if $N_P = 800, N_S = 60$, and $Z_P = 325\ \Omega$.

42. Find Z_S if $N_P = 3000, N_S = 300$, and $Z_P = 1.5\ k\Omega$.

43. Find N_S if $N_P = 3000, Z_S = 12\ \Omega$, and $Z_P = 1.5\ k\Omega$.

44. Find N_S if $N_P = 250, Z_S = 120\ k\Omega$, and $Z_P = 1.25\ k\Omega$.

45. Find N_S if $N_P = 2000, Z_S = 4\ \Omega$, and $Z_P = 800\ \Omega$.

46. Find N_S if $N_P = 200, Z_S = 800\ \Omega$, and $Z_P = 2.4\ \Omega$.

47. Find Z_P if $N_P = 100, N_S = 2500$, and $Z_S = 2\ k\Omega$.

48. Find Z_P if $N_P = 250, N_S = 10{,}000$, and $Z_S = 5.5\ k\Omega$.

49. Find Z_P if $N_P = 5000, N_S = 1000$, and $Z_S = 40\ \Omega$.

50. Find Z_P if $N_P = 7500, N_S = 500$, and $Z_S = 15\ \Omega$.

For problems 51-54, use the equation $X_C = \dfrac{1}{2\pi f C}$.

51. Find C if $f = 2.7\ kHz$ and $X_C = 700\ \Omega$.

52. Find C if $f = 10.7\ kHz$ and $X_C = 500\ \Omega$.

53. Find f if $C = 10\ nF$ and $X_C = 5\ k\Omega$.

54. Find f if $C = 400\ nF$ and $X_C = 3\ k\Omega$.

For problems 55-58, use the equation $X_L = 2\pi f L$.

55. Find L if $f = 4.8\ kHz$ and $X_L = 3.85\ k\Omega$.

56. Find L if $f = 1.2\ kHz$ and $X_L = 30\ k\Omega$.

57. Find f if $L = 54.1\ mH$ and $X_L = 170\ \Omega$.

58. Find f if $L = 150\ mH$ and $X_L = 6.4\ k\Omega$.

For problems 59-62, use the equation $f_r = \dfrac{1}{2\pi\sqrt{LC}}$.

59. Find L if $f_r = 30\ kHz$ and $C = 100\ pF$.

60. Find L if $f_r = 3.5\ kHz$ and $C = 400\ nF$.

61. Find C if $f_r = 100\ Hz$ and $L = 100\ mH$.

62. Find C if $f_r = 20\ kHz$ and $L = 200\ mH$.

63. A computer's microprocessor requires 5 volts and 0.8 amps. Using the equation $E = I \times R_{eq}$, calculate the equivalent resistance (R_{eq}) of the microprocessor.

64. A computer's microprocessor requires 5 volts and 0.6 amps. Using the equation $E = I \times R_{eq}$, calculate the equivalent resistance of the microprocessor.

65. A computer's modem requires 12 volts and 0.3 amps. Using the equation $E = I \times R_{eq}$, calculate the equivalent resistance of the modem.

66. A computer's modem requires 12 volts and 0.4 amps. Using the equation $E = I \times R_{eq}$, calculate the equivalent resistance of the modem.

67. A computer's microprocessor dissipates 5 watts of power and has an equivalent resistance of 5 ohms. Using the equation $P = E^2/R_{eq}$, calculate the voltage.

68. A computer's microprocessor dissipates 1 watt of power and has an equivalent resistance of 25 ohms. Using the equation $P = E^2/R_{eq}$, calculate the voltage.

69. A computer's microprocessor requires a 5 volt supply and dissipates 3 watts of power. Using the equation $P = E^2/R_{eq}$, calculate the equivalent resistance.

70. A computer's microprocessor requires a 5 volt supply and dissipates 0.9 watts of power. Using the equation $P = E^2/R_{eq}$, calculate the equivalent resistance.

ANSWERS TO SELF-TESTS

SELF-TEST 7–1

1. $y = 10$

2. $x = 5$

3. $a = -8$

4. $y = \dfrac{5}{4}$

5. $a = 7$

6. $a = 3$

7. $a = 25$

8. $y = 900$

9. $x = \dfrac{1}{64}$

10. $x = \dfrac{7}{3y}, y = \dfrac{7}{3x}$

11. $P = \dfrac{E^2}{R}, E = \sqrt{PR}$

12. $x = \dfrac{1}{12\sqrt{y}}, y = \dfrac{1}{144x^2}$

SELF-TEST 7–2

1. $E = 77.0$ V

2. $I = 6.58$ mA

3. $N_P = 224$

4. $f = 1.42$ kHz

5. $L = 1.27$ H

Factoring Algebraic Expressions

8

Introduction

In previous chapters we multiplied or divided monomials. Many of the algebraic expressions we work with are polynomials, that is, algebraic expressions that contain more than one term.

As we encounter more complex equations, we must develop ways to simplify these equations so that we can solve them. Many times, equations can be simplified by factoring. Factoring is the process of finding the common factors of polynomials and then extracting these factors from each term. In this chapter you will learn how to factor polynomials.

Chapter Objectives

In this chapter you will learn how to:

1. Multiply polynomials by monomials.
2. Multiply binomials by binomials.
3. Divide polynomials by monomials and by binomials.
4. Determine common factors of polynomials.
5. Determine the factors of trinomials.

8-1   MULTIPLICATION OF POLYNOMIALS BY MONOMIALS—WHOLE NUMBERS

Let's multiply the polynomial $(4x + 2)$ by the monomial $3x$. There are two ways to multiply a polynomial by a monomial. We will show both ways in the following examples.

EXAMPLE 8–1

$$3x(4x + 2) = 3x \cdot 4x + 3x \cdot 2 = 12x^2 + 6x$$

The sign of grouping (the parentheses) tells us that each term within the group must be multiplied by $3x$. Since there are two terms, and each term is multiplied by $3x$, there will be two terms in the answer.

EXAMPLE 8–2

$$4x + 2$$
$$\underline{\times \qquad 3x}$$
$$12x^2 + 6x$$

In this example, we set up the problem the way we set up a multiplication problem in arithmetic. Each term in the multiplicand $(4x + 2)$ is multiplied by the multiplier $(3x)$.

Most problems are given in the form shown in Example 8–1. Most students prefer this method of solution, so this is the method we will use in the rest of the examples. Always keep in mind the following rule when working problems of this kind.

> ☞ **RULE 8–1** When multiplying a polynomial by a monomial, there must be as many terms in the answer as there are in the polynomial.

EXAMPLE 8–3

Multiply $4(2x^2 + 3x - 4)$.

SOLUTION   Multiply each term in the group by 4. This yields

$$4 \cdot 2x^2 + 4 \cdot 3x - 4 \cdot 4 = 8x^2 + 12x - 16$$

Notice that we put a minus sign in front of $4 \cdot 4$ because the sign of the term was negative and the sign of the multiplier was positive. Remember $(+) \cdot (-) = (-)$.

EXAMPLE 8–4

Perform the indicated multiplication and combine like terms.

$$4a(a + 2b) - 3a(a + 3b)$$

SOLUTION   We have two different multiplications to perform:

$$4a(a + 2b) \quad \text{and} \quad -3a(a + 3b)$$
$$4a(a + 2b) = 4a^2 + 8ab$$
$$-3a(a + 3b) = -3a^2 - 9ab$$

Putting them together, we get

$$4a(a + 2b) - 3a(a + 3b) = 4a^2 + 8ab - 3a^2 - 9ab$$

When we combine like terms, we get

$$4a^2 - 3a^2 + 8ab - 9ab = a^2 - ab$$

EXAMPLE 8–5

Perform the indicated multiplication and combine like terms:

$$3x^2 + 6x(2x - 3y - 2) - 2x(3x - 4y - 3)$$

SOLUTION

$$6x(2x - 3y - 2) = 12x^2 - 18xy - 12x$$

$$-2x(3x - 4y - 3) = -6x^2 + 8xy + 6x$$

Notice that in each part of the problem where there are three terms in the polynomial there are three terms in that part of the answer (Rule 8–1).

Putting it all together and combining like terms, we get

$$3x^2 + 6x(2x - 3y - 2) - 2x(3x - 4y - 3)$$
$$= 3x^2 + 12x^2 - 18xy - 12x - 6x^2 + 8xy + 6x$$
$$= 9x^2 - 10xy - 6x$$

PRACTICE PROBLEMS 8–1

Perform the indicated operations. Combine like terms.

1. $4(x + 2)$
2. $3(a - 4)$
3. $4a - 3(4a - 3)$
4. $5x + 2(3x - 4)$
5. $2x(3 + 4y)$
6. $3a(2 - 4b)$
7. $2x(3x + 2y)$
8. $3a(2a - 6b)$
9. $3x(4x - 2y + 4)$
10. $4(2x^2 + 3x - 4)$
11. $5(3a^2 - a - 3)$
12. $3a(a + b) - 2a(a + 4b)$
13. $2x(x - 2y) - 3x(x + 4y)$
14. $x(xy + 4y) - 2(x^2y - 3xy)$
15. $x(3xy - 3y) + 4(x^2y - 4xy)$
16. $2a(a^2 + 9a + 20) + 3a(a^2 - a - 12)$
17. $3x(2x^2 - 7x + 1) - 2x(2x^2 - 3x + 4)$
18. $4x^2 + 2x(x + y - 3) - 2x(2x - y)$
19. $4a(a - 2b + 3) - 2a^2 + 3a(2a - 2b - 2)$
20. $3a(a^2 - 2ab + 2) - 6(a^2 - a^2b + 2)$

SOLUTIONS

1. $4x + 8$
2. $3a - 12$
3. $-8a + 9$
4. $11x - 8$
5. $6x + 8xy$
6. $6a - 12ab$
7. $6x^2 + 4xy$
8. $6a^2 - 18ab$
9. $12x^2 - 6xy + 12x$
10. $8x^2 + 12x - 16$
11. $15a^2 - 5a - 15$
12. $a^2 - 5ab$
13. $-x^2 - 16xy$
14. $-x^2y + 10xy$
15. $7x^2y - 19xy$
16. $5a^3 + 15a^2 + 4a$

17. $2x^3 - 15x^2 - 5x$
19. $8a^2 - 14ab + 6a$

18. $2x^2 + 4xy - 6x$
20. $3a^3 - 6a^2 + 6a - 12$

Additional practice problems are at the end of the chapter.

MULTIPLICATION OF POLYNOMIALS BY MONOMIALS—FRACTIONAL NUMBERS

The following three examples show how we can multiply polynomials that contain fractions. Again, each term in the polynomial is multiplied by the monomial.

EXAMPLE 8–6

$$3a\left(2a + 3 - \frac{4a^2}{3a}\right) = 6a^2 + 9a - \frac{12a^3}{3a}$$

$$= 6a^2 + 9a - 4a^2 = 2a^2 + 9a$$

The third term, $-\frac{12a^3}{3a}$, reduces to $-4a^2$.

EXAMPLE 8–7

$$4x\left(\frac{1}{x} + \frac{3x}{2} + 4\right) = \frac{4x}{x} + \frac{12x^2}{2} + 16x = 4 + 6x^2 + 16x$$

$$= 6x^2 + 16x + 4 \qquad \text{(in descending order)}$$

EXAMPLE 8–8

$$3a\left(\frac{1}{3} + \frac{2a^2}{4a} - \frac{6a}{5}\right) = \frac{3a}{3} + \frac{6a^3}{4a} - \frac{18a^2}{5} = a + \frac{3a^2}{2} - \frac{18a^2}{5}$$

$$= \frac{10a + 15a^2 - 36a^2}{10} = \frac{10a - 21a^2}{10}$$

$$= -\frac{21a^2 - 10a}{10} \qquad \text{(in descending order)}$$

In Example 8–8, after reducing each term, we found the common denominator which is 10. Placing each term over the common denominator allows us to combine like terms ($15a^2$ and $-36a^2$).

Perform the indicated operations:

1. $5a\left(4a - \dfrac{3a^2}{5a}\right)$

2. $3a\left(2a + \dfrac{5a^2}{3a}\right)$

3. $5x\left(2x + 2 - \dfrac{3x^2}{5x}\right)$

4. $3x\left(\dfrac{1}{x} + \dfrac{2x}{3} + 1\right)$

5. $2y\left(\dfrac{2}{y} + \dfrac{4y}{2} - 3\right)$

6. $3y\left(\dfrac{2}{3} - \dfrac{3y^2}{2y} + \dfrac{6y}{4}\right)$

7. $2b\left(\dfrac{3}{5} + \dfrac{4b^2}{3b} - \dfrac{5b}{4}\right)$

8. $5b\left(\dfrac{3}{10} - \dfrac{5b^2}{15b} + \dfrac{2b}{5}\right)$

SOLUTIONS

1. $17a^2$

2. $11a^2$

3. $7x^2 + 10x$

4. $2x^2 + 3x + 3$

5. $4y^2 - 6y + 4$

6. $2y$

7. $\dfrac{b^2}{6} + \dfrac{6b}{5} = \dfrac{5b^2 + 36b}{30}$

8. $\dfrac{b^2}{3} + \dfrac{3b}{2} = \dfrac{2b^2 + 9b}{6}$

In problems 7 and 8, either answer is acceptable. Additional practice problems are at the end of the chapter.

8–3 MULTIPLICATION OF BINOMIALS BY BINOMIALS

The procedure for multiplying a binomial by a binomial is similar to the procedure for multiplying a two-digit number by a two-digit number.

👉 **KEY POINT** When multiplying a binomial by a binomial, multiply each term in the multiplicand by each term in the multiplier.

EXAMPLE 8–9

Perform the indicated operation.

$$\begin{array}{r} a + 4 \\ (\times)\, a - 2 \\ \hline \end{array}$$

SOLUTION We must multiply each term in the multiplicand by each term in the multiplier.

$$a + 4$$
$$a - 2$$
$$\overline{}$$
$$a^2 + 4a \qquad \text{result of } a(a + 4)$$
$$\underline{\quad - 2a - 8} \qquad \text{result of } -2(a + 4)$$
$$a^2 + 2a - 8$$

Notice that like terms are placed in the same column. This lets us combine like terms when the two rows are added together. No matter how many terms are in the multiplicand and multiplier, this method can be used.

Let's look at it again.

EXAMPLE 8–10

Perform the indicated operation.

$$(a + 4)(a - 2)$$

SOLUTION   This time let's do it mentally; the procedure is to multiply each term in the second binomial by each term in the first binomial and combine like terms:

$$a \cdot a = a^2$$

$$a \cdot -2 = -2a$$

$$(a + 4)(a - 2)$$

$$4 \cdot a = 4a$$

$$4 \cdot -2 = -8$$

The four resulting terms are $a^2 - 2a + 4a - 8$. Combining like terms, we get $a^2 + 2a - 8$.

In problems such as this, we may say that the middle term equals *the product of the means plus the product of the extremes*. The *means* refers to the product of the two inner terms (4 and a in this problem). The *extremes* refers to the product of the two outer terms (a and -2 in this problem). The result, $-2a + 4a$ yields the middle term, $2a$. This is often a quick way to find the value of the middle term.

Let's try two more.

EXAMPLE 8–11

Perform the indicated operation.

$$(x - 3y)(x + 7y)$$

Multiplying each term in the second binomial by each term in the first yields

$$x \cdot x = x^2$$

$$x \cdot 7y = 7xy$$

$$-3y \cdot x = -3xy$$

$$-3y \cdot 7y = -21y^2$$

$$(x - 3y)(x + 7y) = x^2 + 7xy - 3xy - 21y^2 = x^2 + 4xy - 21y^2$$

EXAMPLE 8–12

Perform the indicated operation.

$$(3a - 2b)(4a - 4b)$$

SOLUTION Multiplying each term in the second binomial by each term in the first binomial yields

$$3a \cdot 4a = 12a^2$$

$$3a \cdot -4b = -12ab$$

$$-2b \cdot 4a = -8ab$$

$$-2b \cdot -4b = 8b^2$$

$$(3a - 2b)(4a - 4b) = 12a^2 - 12ab - 8ab + 8b^2$$

$$= 12a^2 - 20ab + 8b^2$$

EXAMPLE 8–13

Perform the indicated operation.

$$(a + 3)^2$$

SOLUTION Identical binomials are often expressed in this manner. Remember that the exponent 2 means that the binomial is multiplied by itself two times.

$$(a + 3)^2 = (a + 3)(a + 3)$$

Now we can solve the problem just as we did in the previous examples by multiplying each term in the second binomial by each term in the first:

$$a \cdot a = a^2$$

$$3 \cdot a = 3a$$

$$a \cdot 3 = 3a$$

$$3 \cdot 3 = 9$$

$$(a + 3)^2 = a^2 + 3a + 3a + 9 = a^2 + 6a + 9$$

The middle terms will always be the same when multiplying identical binomials, so we can shorten the process by simply finding the product of one middle term and then multiplying by 2:

$$(a + 3)^2 = a^2 + 2(3a) + 9 = a^2 + 6a + 9$$

PRACTICE PROBLEMS 8–3

Perform the indicated operations:

1. $(a + 2)^2$
2. $(a - 4)^2$
3. $(2a - 3)^2$
4. $(3a - 2)^2$
5. $(x + 2y)^2$
6. $(2x - 3y)^2$
7. $(a + 6)(a + 2)$
8. $(a + 2)(a - 5)$
9. $(a - 3)(a + 4)$
10. $(a - 4)(a - 5)$
11. $(x + 3)(x - 3)$
12. $(x - 5)(x + 5)$
13. $(a + b)(a + 2b)$
14. $(a + 3b)(a - 4b)$
15. $(2a + 3b)^2$
16. $(4x - 3y)^2$

SOLUTIONS

1. $a^2 + 4a + 4$
2. $a^2 - 8a + 16$
3. $4a^2 - 12a + 9$
4. $9a^2 - 12a + 4$
5. $x^2 + 4xy + 4y^2$
6. $4x^2 - 12xy + 9y^2$
7. $a^2 + 8a + 12$
8. $a^2 - 3a - 10$
9. $a^2 + a - 12$
10. $a^2 - 9a + 20$
11. $x^2 - 9$
12. $x^2 - 25$
13. $a^2 + 3ab + 2b^2$
14. $a^2 - ab - 12b^2$
15. $4a^2 + 12ab + 9b^2$
16. $16x^2 - 24xy + 9y^2$

Additional practice problems are at the end of the chapter.

DIVISION OF POLYNOMIALS   8–4

The procedure for dividing a polynomial by a monomial is similar to the procedure for arithmetic long division.

> ☞ **KEY POINT** When dividing a polynomial by a monomial, divide each term in the polynomial by the monomial.

The following examples illustrate the process.

EXAMPLE 8–14

Perform the indicated operation.

$$(6x^3 + 12x^2 - 3x) \div 3x$$

SOLUTION   Each term in the dividend is divided by $3x$, the divisor. We could set up the problem as we would a long-division problem and solve like this:

$$
\begin{array}{r}
2x^2 + 4x - 1 \\
3x\overline{)6x^3 + 12x^2 - 3x} \\
6x^3 \qquad\qquad\qquad\quad 3x \cdot 2x^2 = 6x^3 \\
\hline
0 + 12x^2 \\
12x^2 \qquad\qquad\quad 3x \cdot 4x = 12x^2 \\
\hline
0 - 3x \\
- 3x \qquad\quad 3x \cdot -1 = -3x \\
\hline
0 \qquad\quad \text{no remainder}
\end{array}
$$

$3x$ is divided into the first term. The result of this division is $2x^2 \cdot 3x \cdot 2x^2 = 6x^3$. The process continues until all terms have been divided by $3x$. Usually, a problem like this can be done by inspection. If there is a remainder, it is written as a fraction. The following example illustrates this.

EXAMPLE 8–15

Perform the indicated operation.

$$(7x^2 + 4x) \div 4x$$

SOLUTION

$$
\begin{array}{r}
x + 1 + \dfrac{3x^2}{4x} \\
4x\overline{)7x^2 + 4x} \\
4x^2 \qquad\qquad\quad \\
\hline
3x^2 + 4x \\
4x \\
\hline
3x^2 + 0
\end{array}
$$

$3x^2$ is left over; therefore, the remainder is $\dfrac{3x^2}{4x}$, which reduces to $\dfrac{3x}{4}$.

Check the answer by multiplying the divisor and the quotient. The product should equal the dividend:

$$4x\left(x + 1 + \frac{3x}{4}\right) = 4x^2 + 4x + \frac{12x^2}{4}$$

$$= 4x^2 + 4x + 3x^2 = 7x^2 + 4x$$

Let's try dividing a polynomial by a binomial.

> ☞ **KEY POINT**   When dividing a polynomial by a binomial, set it up as a long-division problem with the literal numbers in descending order; then perform the division.

The following examples illustrate the process.

EXAMPLE 8–16

Perform the indicated operation.

$$(a^2 + 3a - 54) \div (a + 9)$$

SOLUTION   First, we set it up as a long-division problem, and then we make sure that the literal numbers in both the dividend and divisor are in descending order, as shown:

$$a + 9\overline{)a^2 + 3a - 54}$$

Then we determine how many times the first term in the divisor goes into the first term of the dividend. a goes into a^2 a times. Multiplying this a times each term in the divisor results in $a^2 + 9a$. Subtracting this result from the dividend leaves a remainder of $-6a$. After we bring the next term down, we have the partial solution shown:

$$
\begin{array}{r}
a \phantom{{}+ 3a - 54} \\
a + 9\overline{)a^2 + 3a - 54} \\
\underline{a^2 + 9a \phantom{{}- 54}} \\
0 - 6a - 54
\end{array}
$$

Next we determine how many times a goes into $-6a$. The answer is -6. Now we multiply -6 by $(a + 9)$. There is no remainder and the solution is complete.

$$
\begin{array}{r}
a - 6 \phantom{{}a - 54} \\
a + 9\overline{)a^2 + 3a - 54} \\
\underline{a^2 + 9a \phantom{{}- 54}} \\
-6a - 54 \\
\underline{-6a - 54} \\
0
\end{array}
$$

We can check our answer:

$$(a + 9)(a - 6) = a^2 + 3a - 54$$

Let's try another one.

EXAMPLE 8–17

Perform the indicated operation.

$$(6a^2 - 23ab + 20b^2) \div (3a - 4b)$$

SOLUTION

$$
\begin{array}{r}
2a - 5b \phantom{{}ab + 20b^2} \\
3a - 4b\overline{)6a^2 - 23ab + 20b^2} \\
\underline{6a^2 - 8ab \phantom{{}+ 20b^2}} \\
-15ab + 20b^2 \\
\underline{-15ab + 20b^2} \\
0
\end{array}
$$

$$(3a - 4b)(2a - 5b) = 6a^2 - 23ab + 20b^2$$

Let's do two more.

EXAMPLE 8–18

Perform the indicated operation.

$$(x^2 - 49) \div (x + 7)$$

SOLUTION

$$
\begin{array}{r}
x \qquad\quad - 7 \\
x + 7 \overline{)x^2 \qquad\quad - 49} \\
\underline{x^2 + 7x \qquad} \\
-7x - 49 \\
\underline{-7x - 49} \\
0
\end{array}
$$

CHECK

$$(x + 7)(x - 7) = x^2 - 49$$

EXAMPLE 8–19

Perform the indicated operation.
   This one has a remainder.

$$(x^2 + 4x + 7) \div (x - 3)$$

SOLUTION

$$
\begin{array}{r}
x \qquad\quad + 7 \\
x - 3 \overline{)x^2 + 4x + 7} \\
\underline{x^2 - 3x \qquad} \\
7x + 7 \\
\underline{7x - 21} \\
28
\end{array}
$$

The remainder is $\dfrac{28}{x - 3}$, so the answer is

$$x + 7 + \frac{28}{x - 3}$$

Check the answer:

$$(x - 3)\left(x + 7 + \frac{28}{x - 3}\right) = (x - 3)\left(\frac{x^2 - 3x + 7x - 21 + 28}{x - 3}\right)$$

$$= x^2 - 3x + 7x - 21 + 28 = x^2 + 4x + 7$$

PRACTICE PROBLEMS 8–4

Perform the indicated operations:

1. $(a^3 + 2a^2 - a) \div a$
2. $(4a^4 - 6a^3 + 8a^2) \div 2a$
3. $(9x^2 + 5x) \div 3x$
4. $(14x^2 - 3x) \div 3x$
5. $(x^2 + 13x + 42) \div (x + 6)$
6. $(x^2 - 36) \div (x + 6)$
7. $(9x^2 - 12x + 4) \div (3x - 2)$
8. $(4a^2 + 12ab + 9b^2) \div (2a + 3b)$
9. $(a^2 - ab - 12b^2) \div (a - 4b)$
10. $(x^2 - 9x + 20) \div (x - 5)$
11. $(3ab + a^2 + 2b^2) \div (b + a)$
12. $(9 + 4x^2 + 12x) \div (3 + 2x)$
13. $(2x^2 + 3x + 14) \div (x + 1)$
14. $(6x^2 + 2x - 7) \div (2x + 4)$
15. $(x^2 + 2xy - y^2) \div (x + y)$
16. $(2x^2 - 4xy + 2y^2) \div (2x + 2y)$

SOLUTIONS

1. $a^2 + 2a - 1$
2. $2a^3 - 3a^2 + 4a$
3. $3x + 1 + \dfrac{2x}{3x} = 3x + 1 + \dfrac{2}{3} = 3x + \dfrac{5}{3}$
4. $4x - 1 + \dfrac{2x^2}{3x} = 4x - 1 + \dfrac{2x}{3}$
5. $x + 7$
6. $x - 6$
7. $3x - 2$
8. $2a + 3b$
9. $a + 3b$
10. $x - 4$
11. $a + 2b$
12. $2x + 3$
13. $2x + 1 + \dfrac{13}{x + 1}$
14. $3x - 5 + \dfrac{13}{2x + 4}$
15. $x + y - \dfrac{2y^2}{x + y}$
16. $x - 3y + \dfrac{8y^2}{2x + 2y}$

Additional practice problems are at the end of the chapter.

SELF-TEST 8–1

Perform the indicated operations:

1. $3a(a - 6)$
2. $2xy(3x + 2y)$
3. $(a - 9)^2$
4. $(4a + 2)^2$
5. $(a + 6)(a - 1)$
6. $(x + 2y)(x - 6y)$
7. $(2x + 3y)(4x + 4y)$
8. $3(a - 2b)(2a - 2b)$
9. $(6a^3 - 12a^2) \div 3a$
10. $(x^2 + 2x - 48) \div (x + 8)$
11. $(x^2 + 13x + 36) \div (x + 9)$
12. $(a^2 + 12ab + 20b^2) \div (a + 2b)$
13. $(13x + 30 + x^2) \div (x + 9)$
14. $(10ab + a^2 + 20b^2) \div (a + 2b)$

Answers to Self-test 8–1 are at the end of the chapter.

Earlier in this chapter we multiplied monomials times polynomials. Here we are going to reverse the process. We are going to determine what the common factors are of various algebraic expressions. This is simply a process of determining what factors are common to each term in the expression.

☞ **RULE 8–2** To factor a polynomial:

1. Find the prime factors of each term.
2. Determine the factors common to all terms.
3. Factor the polynomial by dividing the polynomial by the common factors.

Consider the expression $6a^2 + 12a - 18a^3$. If we factor each term, we get

$$6a^2 = 2 \cdot 3 \cdot a^2$$

$$12a = 2 \cdot 2 \cdot 3 \cdot a$$

$$18a^3 = 2 \cdot 3 \cdot 3 \cdot a^3$$

The common factors are the factors that appear in all terms. These common factors are 2, 3, and a or $6a$. Since $6a$ is common to each term, we can divide the expression by $6a$. This yields $a + 2 - 3a^2$.

$$\frac{6a^2 + 12a - 18a^3}{6a} = a + 2 - 3a^2$$

The factors of $6a^2 + 12a - 18a^3$ are $6a$ and $a + 2 - 3a^2$.

$$6a^2 + 12a - 18a^3 = 6a(a + 2 - 3a^2)$$

Since we normally write algebraic expressions either in descending order or in ascending order, let's rearrange the answer so that it looks like this:

$$6a(-3a^2 + a + 2)$$

EXAMPLE 8–20

Factor $28a^3 - 35a^2$.

SOLUTION The prime factors are

$$28a^3 = 2 \cdot 2 \cdot 7 \cdot a^3$$

$$35a^2 = 5 \cdot 7 \cdot a^2$$

The common factors are 7 and a^2. Therefore, we divide each term by $7a^2$ to find the factors of the original problem.

$$\frac{28a^3 - 35a^2}{7a^2} = 4a - 5$$

The factors of $28a^3 - 35a^2$ are $7a^2$ and $4a - 5$.

$$28a^3 - 35a^2 = 7a^2(4a - 5)$$

EXAMPLE 8–21

Factor $36x^3y^2z^3 + 24x^2yz^2 - 18x^4y^4z^3$

SOLUTION   The prime factors are

$$36x^3y^2z^3 = 2 \cdot 2 \cdot 3 \cdot 3 \cdot x^3 \cdot y^2 \cdot z^3$$
$$24x^2yz^2 = 2 \cdot 2 \cdot 2 \cdot 3 \cdot x^2 \cdot y \cdot z^2$$
$$18x^4y^4z^3 = 2 \cdot 3 \cdot 3 \cdot x^4 \cdot y^2 \cdot z^3$$

The common factors are 2, 3, x^2, y, and z^2. $2 \cdot 3 \cdot x^2 \cdot y \cdot z^2 = 6x^2yz^2$. Dividing each term by the common factors, we get

$$\frac{36x^3y^2z^3 + 24x^2yz^2 - 18x^4y^4z^3}{6x^2yz^2} = 6xyz + 4 - 3x^2y^3z$$
$$36x^3y^2z^3 + 24x^2yz^2 - 18x^4y^4z^3 = 6x^2yz^2(6xyz + 4 - 3x^2y^3z)$$

EXAMPLE 8–22

Factor $\dfrac{20a^4b^2c^3}{27} + \dfrac{30a^3b^3c^2}{9}$

SOLUTION   Find the prime factors of both numerator and denominator.

$$20a^4b^2c^3 = 2 \cdot 2 \cdot 5 \cdot a^4 \cdot b^2 \cdot c^3$$
$$30a^3b^3c^2 = 2 \cdot 3 \cdot 5 \cdot a^3b^3c^2$$
$$27 = 3 \cdot 3 \cdot 3$$
$$9 = 3 \cdot 3$$

The common factors in the numerators are 2, 5, a^3, b^2, and c^2. The product of these common factors is $10a^3b^2c^2$. The common factors in the denominators are $3 \cdot 3$, which is equal to 9. Therefore the factor common to both terms is $\dfrac{10a^3b^2c^2}{9}$. Dividing each term by the common factor, we get:

$$\frac{20a^4b^2c^3}{27} \div \frac{10a^3b^2c^2}{9} = \frac{20a^4b^2c^3}{27} \times \frac{9}{10a^3b^2c^2} = \frac{2ac}{3}$$
$$\frac{30a^3b^3c^2}{9} \div \frac{10a^3b^2c^2}{9} = \frac{30a^3b^3c^2}{9} \times \frac{9}{10a^3b^2c^2} = 3b$$

Combining the above, we get:

$$\frac{20a^4b^2c^3}{27} + \frac{30a^3b^3c^2}{9} = \frac{10a^3b^2c^2}{9}\left(\frac{2ac}{3} + 3b\right)$$

We can always check the result of factoring by performing the multiplication and comparing the product with the original algebraic expression.

PRACTICE PROBLEMS 8–5

Factor the following problems:

1. $4a + 2$
2. $6a + 8$
3. $16 - 2y$
4. $21 - 3y$
5. $x^3 + 2x^2$
6. $4x^4 + 2x^2$
7. $18a^2 - 9a$
8. $15a^2 - 3a$
9. $14x^2y - 14x^2z + 20xyz$
10. $15a^2b - 21a^3b^2 + 9a^2b^2$
11. $\dfrac{28x^2y^3z^4}{15} - \dfrac{56x^3y^2z^3}{45}$
12. $\dfrac{42ab^3c^2}{25} + \dfrac{63a^2b^4c^3}{50} - \dfrac{147a^3b^2c}{100}$

SOLUTIONS

1. $2(2a + 1)$
2. $2(3a + 4)$
3. $2(8 - y)$
4. $3(7 - y)$
5. $x^2(x + 2)$
6. $2x^2(2x^2 + 1)$
7. $9a(2a - 1)$
8. $3a(5a - 1)$
9. $2x(7xy - 7xz + 10yz)$
10. $3a^2b(5 - 7ab + 3b)$
11. $\dfrac{28x^2y^2z^3}{15}\left(yz - \dfrac{2x}{3}\right)$
12. $\dfrac{21ab^2c}{25}\left(2bc + \dfrac{3ab^2c^2}{2} - \dfrac{7a^2}{4}\right)$

Additional practice problems are at the end of the chapter.

8-6   FACTORS OF TRINOMIALS

☞ **KEY POINT** A trinomial of the form $x^2 + bx + c$ can be factored if we can find two numbers whose sum is the coefficient of the second term and whose product is the third term.

For example, consider the expression $x^2 + 13x + 42$. From the work we did earlier in the chapter, we know that the answer will be in the form $(x + m)(x + n)$. In this expression the coefficient of the second term is 13, and the third term is 42. We must find two numbers whose sum is 13 and whose product is 42. We find these numbers by trial and error. That is, we consider all the combinations of numbers whose sum is 13 and select the one combination whose product is 42. In this example, the combinations are

$$\begin{array}{ll} 12 + 1 & 12 \times 1 = 12 \\ 11 + 2 & 11 \times 2 = 22 \\ 10 + 3 & 10 \times 3 = 30 \\ 9 + 4 & 9 \times 4 = 36 \\ 8 + 5 & 8 \times 5 = 40 \\ 7 + 6 & 7 \times 6 = 42 \end{array}$$

The two numbers must be 7 and 6:

$$x^2 + 13x + 42 = (x + 7)(x + 6)$$

It would also be correct to write the answer as $(x + 6)(x + 7)$. With practice, we can usually determine the correct combination by inspection.

EXAMPLE 8–23

Factor $x^2 - 6x - 27$.

SOLUTION   The coefficient of the middle term is -6 and the product is -27. Any time the third term is negative, the factors must be of this form: $(x + m)(x - n)$. The unknown number m must be less than n because the middle term is negative. There are so many combinations of numbers that equal -6 that it is simpler in this case to consider numbers whose product is -27.

$$\begin{array}{ll} 1 \cdot -27 = -27 & 1 + (-27) = -26 \\ -1 \cdot 27 = -27 & -1 + 27 = 26 \\ 3 \cdot -9 = -27 & 3 + (-9) = -6 \\ -3 \cdot 9 = -27 & -3 + 9 = 6 \end{array}$$

Of the four possibilities, only 3 and -9 equal -6 when added together:

$$x^2 - 6x - 27 = (x + 3)(x - 9)$$

EXAMPLE 8–24

Factor $x^2 + 2x - 8$.

SOLUTION   The third term is again negative; therefore, the answer is again of the form $(x + m)(x - n)$. Let's again consider all the possible combinations that yield a product of -8:

$$\begin{array}{ll} 2 \cdot -4 = -8 & 2 + (-4) = -2 \\ -2 \cdot 4 = -8 & -2 + 4 = 2 \\ 1 \cdot -8 = -8 & 1 + (-8) = -7 \\ -1 \cdot 8 = -8 & -1 + 8 = 7 \end{array}$$

Of the four possibilities, only -2 and 4 equal 2 when added together:

$$x^2 + 2x - 8 = (x + 4)(x - 2)$$

EXAMPLE 8–25

Factor $x^2 + 12x + 36$.

SOLUTION   Inspection shows that the factors are $(x + 6)$ and $(x + 6)$ or $(x + 6)^2$:

$$x^2 + 12x + 36 = (x + 6)^2$$

EXAMPLE 8–26

Factor $x^2 - 36$.

SOLUTION   Note that there is no middle term and that the last term is negative. The only way the middle term can drop out is for the two numbers to be equal. The only two equal numbers whose product is -36 are 6 and -6. Then

$$x^2 - 36 = (x + 6)(x - 6)$$

EXAMPLE 8–27

Factor $x^2 - 12x + 36$.

SOLUTION   The middle term is negative and the last term is positive. This combination can result only if both numbers are negative. The numbers must be -6 because $-6 + (-6) = -12$ and $-6 \cdot -6 = 36$.

$$x^2 - 12x + 36 = (x - 6)^2$$

In Example 8–28 we have added another squared term, y^2. When trinomials are of the form $ax^2 + bxy + cy^2$, the factors must be $(px + my)(qx + ny)$. Otherwise, the method of solution is the same.

EXAMPLE 8–28

Factor $x^2 + 2xy + y^2$.

SOLUTION   By inspection we see that the factors must be $(x + y)$ and $(x + y)$ or $(x + y)^2$.

EXAMPLE 8–29

Factor $x^2 - 4xy - 32y^2$.

SOLUTION   The coefficient of $y^2 = -32$ and the coefficient of the middle term is -4. The possible solutions are

$$-32 \cdot 1 \qquad -32 + 1 = -31$$
$$-16 \cdot 2 \qquad -16 + 2 = -14$$
$$-8 \cdot 4 \qquad -8 + 4 = -4$$

The correct combination is -8 and $+4$. Because the middle term is negative, the larger of the two digits must be the negative digit. Therefore, the three combinations in which the smaller of the two digits is negative were not considered.

$$x^2 - 4xy - 32y^2 = (x + 4y)(x - 8y)$$

EXAMPLE 8–30

Factor $4x^2 + 13xy + 10y^2$.

SOLUTION   This example shows a coefficient of x^2 that is some number other than 1, in this case 4. The coefficient of y^2 is 10. We are looking for two numbers whose product is 4 (because the first term is $4x^2$) and two numbers whose product is 10 (the last term is $10y^2$). The product of the means plus the product of the extremes must equal 13, the coefficient of the middle term. The possible combinations are:

$$(2 + 2)(2 + 5) \qquad 4 + 10 = 14$$
$$(2 + 5)(2 + 2) \qquad 10 + 4 = 14$$
$$(1 + 5)(4 + 2) \qquad 20 + 2 = 22$$
$$(1 + 2)(4 + 5) \qquad 8 + 5 = 13$$

In the right column, we added the product of the means and the product of the extremes. We tried different combinations until we finally got a sum equal to 13, the coefficient of the middle term. Of course, there are still other combinations we didn't consider, but we got the right answer on the fourth try, so there was no need to try any more.

$$4x^2 + 13xy + 10y^2 = (x + 2y)(4x + 5y)$$

In this chapter we have presented trinomials that were factorable. In reality, not all trinomials are factorable using the methods we have developed in this chapter. In such cases, we must resort to other methods of solution; the quadratic is the most common. The use of the quadratic in finding the factors of trinomials is covered in Chapter 9.

PRACTICE PROBLEMS 8–6

Factor the following:

1. $x^2 + 3x + 2$
2. $x^2 + 7x + 12$
3. $a^2 + a - 6$
4. $a^2 - a - 12$
5. $a^2 + 15a + 56$
6. $a^2 + 2a - 15$
7. $x^2 + 12x + 36$
8. $x^2 - 64$
9. $x^2 + 12xy + 35y^2$
10. $x^2 + 16xy + 63y^2$
11. $2a^2 - 12ab - 32b^2$
12. $2a^2 + 4ab - 48b^2$

1. $(x + 2)(x + 1)$ 2. $(x + 3)(x + 4)$ 3. $(a + 3)(a - 2)$
4. $(a - 4)(a + 3)$ 5. $(a + 7)(a + 8)$ 6. $(a + 5)(a - 3)$
7. $(x + 6)^2$ 8. $(x + 8)(x - 8)$ 9. $(x + 5y)(x + 7y)$
10. $(x + 7y)(x + 9y)$ 11. $2(a + 2b)(a - 8b)$ 12. $2(a + 6b)(a - 4b)$

Additional practice problems are at the end of the chapter.

SELF-TEST 8–2

Factor the following:

1. $9x - 36$ 2. $56y^2 + 8y$
3. $a^2 + 15a + 36$ 4. $a^2 - 7a - 30$
5. $x^2 + 2xy - 35y^2$ 6. $x^2 + 12xy + 36y^2$
7. $2x^2 + 16x + 32$ 8. $3a^2 - 3b^2$
9. $a^2 - 3ab - 54b^2$ 10. $a^2 - 6ab + 9b^2$

Answers to Self-test 8–2 are at the end of the chapter.

CHAPTER 8 AT A GLANCE

PAGE	KEY POINT OR RULE	EXAMPLE
229	*Rule 8–1:* When multiplying a polynomial by a monomial, there must be as many terms in the answer as there are in the polynomial.	$3x(4x - 2y + 4)$ $= 12x^2 - 6xy + 12x$
232	*Key Point:* When multiplying a binomial by a binomial, multiply each term in the multiplicand by each term in the multiplier.	$(a + 4)(a - 2)$ $= a^2 + 4a - 2a - 8$ $= a^2 + 2a - 8$
235	*Key Point:* When dividing a polynomial by a monomial, divide each term in the polynomial by the monomial.	$(6x^3 + 12x^2 - 3x) \div 3x$ $= 2x^2 + 4x - 1$
236	*Key Point:* When dividing a polynomial by a binomial, set it up as a long-division problem with the literal numbers in descending order; then perform the division.	$(a^2 + 3a - 54) \div (a + 9)$ $$\begin{array}{r} a - 6 \\ a + 9 \overline{)a^2 + 3a - 54} \\ \underline{a^2 + 9a} \\ -6a - 54 \\ \underline{-6a - 54} \\ 0 \end{array}$$

240   *Rule 8–2:* To factor a polynomial:

1. Find the prime factors of each term.
2. Determine the factors common to all terms.
3. Factor the polynomial by dividing the polynomial by the common factors.

Factor: $6a^2 + 12a - 18a^3$

$$6a^2 = 2 \cdot 3 \cdot a^2$$
$$12a = 2 \cdot 2 \cdot 3 \cdot a$$
$$18a^3 = 2 \cdot 3 \cdot 3 \cdot a^3$$

The common factors are
$$2 \cdot 3 \cdot a = 6a$$
$$(6a^2 + 12a - 18a^3) \div 6a$$
$$= a + 2 - 3a^2$$
$$6a^2 + 12a - 18a^3$$
$$= 6a(a + 2 - 3a^2)$$
$$= 6a(-3a^2 + a + 2)$$

242   *Key Point:* Trinomials can be factored if we can find two numbers whose sum is the coefficient of the second term and whose product is the third term.

Factor $x^2 + 13x + 42$.
$(x + 7)(x + 6)$

END OF CHAPTER PROBLEMS 8–1

Perform the indicated operations. Combine like terms.

1. $3(a + 3)$
2. $6(x + 2)$
3. $4(a - 1)$
4. $5(a - 3)$
5. $7(4 - b)$
6. $6(3 - b)$
7. $3a - 2(3a - 1)$
8. $4a - 3(2a - 4)$
9. $6a - 3(4a + 2)$
10. $5a - 5(2a + 3)$
11. $4(2x - 3) - 2x$
12. $5(3x + 5) - 8x$
13. $3x(2 + 2y)$
14. $8x(3 - 2y)$
15. $3a(2a + 3b)$
16. $4a(3a + 5b)$
17. $3(3x^2 + 6x - 12)$
18. $4(2x^2 + 5x - 6)$
19. $2(4a^2 - 3a + 2)$
20. $4(3a^2 + 6a - 3)$
21. $4a(a + b) - 3a(a + 5b)$
22. $5a(a + b) - 7a(a + 2b)$
23. $a(2ab + 2b) - 3(a^2b - 2ab)$
24. $2a(ab + 2b) - 6(2a^2b - ab)$
25. $2x(4xy - 3y) - 3(x^2y + xy)$
26. $3x(2xy + 4y) - 6(2x^2y - 2xy)$
27. $2x(3xy - 4y) - 4(x^2y + 3xy)$
28. $3x(xy - 2y) - 2(x^2y + xy)$
29. $x(x^2 + 8x + 15) + 2x(x^2 - x - 6)$
30. $2x(x^2 + 7x + 12) + x(x^2 - 4x + 9)$
31. $4a(a^2 - 3a + 6) - 3a(a^2 + 5a - 6)$
32. $5a(a^2 - 2a - 3) + 2a(a^2 - 5a + 4)$
33. $3a^2 + 2a(a - b + 2) - 2a(2a + b)$
34. $2x^2 + 3x(x - 2y + 3) - 3x(2x + 4y)$
35. $4x(2x - 3y + 2) - 2x^2 - 3x(x + 2y - 3)$
36. $3a(4a - 2 + 3b) + 4a^2 - 7a(2a - 3 + b)$

END OF CHAPTER PROBLEMS 8–2

Perform the indicated operations. Combine like terms.

1. $4x\left(3x + \dfrac{2x^2}{3x}\right)$

2. $5x\left(3x + \dfrac{2x^2}{10x}\right)$

3. $2y\left(3y - \dfrac{3y^2}{6y}\right)$

4. $6y\left(4y - \dfrac{6y^2}{3y}\right)$

5. $3a\left(2a + 3 + \dfrac{4a^2}{6a}\right)$

6. $4a\left(3a + 4 + \dfrac{4a^2}{8a}\right)$

7. $2b\left(4b - 3 - \dfrac{6b^2}{4b}\right)$

8. $3b\left(4b - 3 - \dfrac{4b^2}{6b}\right)$

9. $4a\left(\dfrac{2}{a} + \dfrac{3a}{2} - 4\right)$

10. $2a\left(\dfrac{6}{a} + \dfrac{6a}{4} - 3\right)$

11. $3a\left(\dfrac{3}{2a} + \dfrac{2a}{3} - 2\right)$

12. $5a\left(\dfrac{4}{3a} + \dfrac{3a}{5} - 3\right)$

13. $3x\left(\dfrac{1}{3} - \dfrac{2x^2}{3x} + 2\right)$

14. $2x\left(\dfrac{1}{8} - \dfrac{3x^2}{2x} + 5\right)$

15. $6x\left(\dfrac{1}{2} + \dfrac{4x^2}{3x} - 3\right)$

16. $10x\left(\dfrac{1}{5} + \dfrac{3x^2}{3x} - 2\right)$

17. $8b\left(\dfrac{1}{4} + \dfrac{5b^2}{8b} - 1\right)$

18. $9b\left(\dfrac{3}{9} + \dfrac{4b^2}{9b} - 2\right)$

19. $3b\left(\dfrac{3}{4} + \dfrac{2x^2}{5b} + 1\right)$

20. $5b\left(\dfrac{5}{6} + \dfrac{3b^2}{4b} + 3\right)$

END OF CHAPTER PROBLEMS 8–3

Perform the indicated operations. Combine like terms.

1. $(a + 3)^2$

2. $(a + 6)^2$

3. $(a - 6)^2$

4. $(a - 3)^2$

5. $(2x - 4)^2$

6. $(3x - 4)^2$

7. $(4x + 3)^2$

8. $(4x + 3)^2$

9. $(2a + 3b)^2$

10. $(a^4 + 3b)^2$

11. $(x^3 - 4y)^2$

12. $(x^4 - 5y)^2$

13. $(2x - 3y)^2$

14. $(5x - 4y)^2$

15. $(a + 1)(a + 3)$

16. $(y + 3)(y + 7)$

17. $(y + 3)(y - 6)$

18. $(a + 5)(a - 2)$

19. $(a - 2)(a + 6)$

20. $(a - 7)(a + 3)$

21. $(x - 4)(x - 1)$

22. $(x - 6)(x - 4)$

23. $(a + 4)(a - 4)$

24. $(a + 7)(a - 7)$

25. $(2x + 3)(2x - 3)$

26. $(3x - 4)(3x + 4)$

27. $(x + y)(x + 3y)$

28. $(x + 2y)(x + 2y)$

29. $(x - 5y)(x + 4y)$

30. $(x - 6y)(x + 5y)$

31. $(a + 2b)(4a - 3b)$
33. $(a - 4b)(2a - 2b)$
35. $(2a + 3b)(3a - 4b)$
37. $(3x + 4y)(2x + 5y)$
39. $(2x - 5y)(2x - 3y)$

32. $(a + 4b)(3a - 3b)$
34. $(a - 5b)(a - 6b)$
36. $(4a + 2b)(3a - 4b)$
38. $(2x + 3y)(3x + 5y)$
40. $(3x - 2y)(7x - 5y)$

END OF CHAPTER PROBLEMS 8–4

Perform the indicated operations:

1. $(x^3 + 3x^2 + 2x) \div x$
3. $(9x^2 + 5x) \div 3x$
5. $(9x^2 - 5x) \div 5x$
7. $(15x^2 + 6x) \div 5x$
9. $(x^2 + 4x - 21) \div (x + 7)$
11. $(x^2 - 49) \div (x + 7)$
13. $(9x^2 - 12x + 4) \div (3x - 2)$
15. $(a^2 - 25) \div (a + 5)$
17. $(6x^2 + xy - 12y^2) \div (3x - 4y)$
19. $(9a^2 - 24ab + 16b^2) \div (3a - 4b)$
21. $(x^2 - 6xy - 27y^2) \div (x + 3y)$
23. $(16 - 12x^2 - 16x) \div (2 - 3x)$
25. $(12x^2 - 4x - 5) \div (3x + 2)$
27. $(8x^2 + 20xy + 15y^2) \div (4x + 5y)$
29. $(4a^2 + 2a + 2) \div (a + 1)$
31. $(6a^2 - 3a - 2) \div (a - 1)$
33. $(a^2 + 4ab + b^2) \div (a + b)$
35. $(6x^3 + 3x^2 + 4x + 4) \div (x + 1)$

2. $(x^4 + 3x^3 + x^2) \div x^2$
4. $(14x^2 - 3x) \div 3x$
6. $(5x^2 + 8x) \div 2x$
8. $(8x^2 + 3x) \div 3x$
10. $(x^2 + 3x - 10) \div (x - 2)$
12. $(x^2 - 16) \div (x - 4)$
14. $(6x^2 + 15x + 9) \div (3x + 3)$
16. $(b^2 - 13b + 42) \div (b - 6)$
18. $(8x^2 + 20xy + 8y^2) \div (4x + 2y)$
20. $(15 + 35a + 20a^2) \div (3 + 4a)$
22. $(12b^2 + 15a^2 - 36ab) \div (5a - 2b)$
24. $(20 - 6x - 8x^2) \div (4 + 2x)$
26. $(24a^2 - 50a + 20) \div (3a - 4)$
28. $(10a^2 - 17ab + 4b^2) \div (5a - 4b)$
30. $(a^2 + 3a - 4) \div (a - 2)$
32. $(4a^2 - 4a + 4) \div (2a + 1)$
34. $(a^2 - 3ab - 2b^2) \div (a - b)$
36. $(4a^3 - 6a^2 + 4a - 3) \div (2a - 1)$

END OF CHAPTER PROBLEMS 8–5

Factor the following:

1. $3a + 9$
3. $7b - 21$
5. $4x^2y + 6xy$
7. $16a^2 + 2a$
9. $24a^2bc^3 - 16ab^2c^2$
11. $12a^3b^2 + 9a^2b^4 - 6a^4b^4$
13. $10x^3y^2 - 20x^4y^4 + 15xy^3$
15. $15a^3b^2c^4 - 3a^2b^3c^4 - 12b^2c$
17. $35x^4y^2z^3 + 25x^2y^4z^4 - 30x^2y^2z^2$
19. $24x^3y^2z^4 - 42x^2y^3z^2 - 72x^2yz^2$
21. $\dfrac{7x^2z^3}{12} - \dfrac{14x^3y^2z^2}{9}$

2. $9b - 18$
4. $6x + 18$
6. $6x^2y - 9xy$
8. $18a^3 + 3a$
10. $140a^3bc^2 - 35a^2b^2c^4$
12. $21a^4b^3 + 7ab^3 - 14a^2b$
14. $18xy^2 - 9x^3y^2 + 27x^2y$
16. $16a^2bc^3 - 8a^3b^2c^4 - 4a^3b^2c^3$
18. $42x^2y^3z^4 - 63x^4y^2z^4 + 21x^4y^3z^3$
20. $36x^2y^2z + 72x^3y^3z^2 - 144x^3y^2z^2$
22. $\dfrac{15x^3z^2}{42} - \dfrac{45x^2y^3z^3}{14}$

23. $\dfrac{12x^4y^2z}{35} + \dfrac{24x^2y^3z^3}{55} - \dfrac{18x^3y^3z^4}{25}$

24. $\dfrac{21x^2y^3z^4}{16} - \dfrac{63x^3y^3z^3}{32} + \dfrac{35x^4y^2z^4}{64}$

25. $\dfrac{21a^3b^2c^2}{22} + \dfrac{28a^2b^3c^2}{33} - \dfrac{42a^3b^2c^3}{44}$

26. $\dfrac{52a^3b^2c}{165} - \dfrac{26a^2b^3c^2}{100} + \dfrac{104a^2b^2c}{55}$

END OF CHAPTER PROBLEMS 8–6

Factor the following:

1. $x^2 + 5x + 6$
3. $x^2 + 4x - 32$
5. $a^2 + 6a + 9$
7. $a^2 - 16$
9. $x^2 - 49$
11. $x^2 - 8x + 15$
13. $a^2 - 12a + 35$
15. $a^2 - 11a + 28$
17. $y^2 + 5y - 36$
19. $x^2 - 2x - 35$
21. $x^2 - 2x - 24$
23. $6x^2 + 5x - 6$
25. $8a^2 - 20a + 8$
27. $9x^2 + 6x - 8$
29. $6b^2 - 18b + 12$
31. $10a^2 + 31a + 15$
33. $2x^2 + 7xy + 6y^2$
35. $20a^2 - 7ab - 6b^2$
37. $8x^2 + 20xy - 12y^2$
39. $12x^2 + 18xy + 6y^2$

2. $x^2 + 7x + 10$
4. $a^2 - a - 42$
6. $a^2 + 16a + 64$
8. $x^2 - 81$
10. $x^2 - 49$
12. $y^2 - 14y + 48$
14. $a^2 - 11a + 24$
16. $a^2 - 14a + 45$
18. $y^2 + 4y - 45$
20. $x^2 + 4x - 12$
22. $x^2 - 7x - 18$
24. $3x^2 + 13x + 12$
26. $4a^2 - 27a + 18$
28. $12x^2 - 7x - 12$
30. $18b^2 - 33b + 12$
32. $24a^2 + 4a - 4$
34. $20a^2 - 20b^2$
36. $12x^2 - 32xy - 12y^2$
38. $28x^2 + 14xy - 14y^2$
40. $12x^2 + 20xy + 8y^2$

ANSWERS TO SELF-TESTS

SELF-TEST 8–1

1. $3a^2 - 18a$
4. $16a^2 + 16a + 4$
7. $8x^2 + 20xy + 12y^2$
10. $x - 6$
13. $x + 4 - \dfrac{6}{x + 9}$

2. $6x^2y + 4xy^2$
5. $a^2 + 5a - 6$
8. $6a^2 - 18ab + 12b^2$
11. $x + 4$
14. $a + 10b - \dfrac{2ab}{a + 2b}$

3. $a^2 - 18a + 81$
6. $x^2 - 4xy - 12y^2$
9. $2a^2 - 4a$
12. $a + 10b$

SELF-TEST 8–2

1. $9(x - 4)$
4. $(a + 3)(a - 10)$
7. $2(x + 4)^2$
10. $(a - 3b)^2$

2. $8y(7y + 1)$
5. $(x + 7y)(x - 5y)$
8. $3(a + b)(a - b)$

3. $(a + 12)(a + 3)$
6. $(x + 6y)^2$
9. $(a + 6b)(a - 9b)$

Fractional Equations

Introduction

Many of the equations we work with in electronics are more complex than the ones we worked with in Chapter 6. Sometimes the unknown is part of a polynomial that can be in either the numerator or the denominator of a term.

In Chapters 7 and 8 we laid the groundwork for the solution of most complex equations. In this chapter we will be solving problems similar to most of the problem types you will encounter in your electronics studies. You will be given practical problems to solve and will be given various methods to solve them.

In solving second-degree equations, there are always two possible solutions or *roots*. Such equations are called *quadratic equations*. The roots may or may not be equal and may or may not have the same sign. We will learn how to use the quadratic equation to help in solving problems that we cannot factor.

Chapter Objectives

In this chapter you will learn how to:

1. Solve fractional linear equations.
2. Solve linear equations where the unknown is part of a polynomial.
3. Solve electrical problems where one of the quantities is unknown.
4. Use the quadratic equation to solve second-degree equations.

GENERAL EQUATIONS 9-1

When we have an equation to solve that involves fractions or mixed numbers, we usually have to get rid of the fraction in order to simplify the solution.

An easy rule to remember is:

> ☞ **RULE 9-1** To solve an equation that includes fractions or mixed numbers: (1) get rid of the fraction (or

mixed number); (2) solve for the unknown; (3) verify your answer.

Consider the equation in Example 9–1.

EXAMPLE 9–1

Solve for a.

$$\frac{1}{3} + \frac{a}{4} = 2$$

SOLUTION   The first step in the solution is to get rid of the fraction. We get rid of the fraction by first finding the CD (common denominator) and then multiplying each term by it. In this example, the CD is 12. Now we multiply each term by 12.

$$\frac{1}{3}(12) + \frac{a}{4}(12) = 2(12)$$

$$4 + 3a = 24$$

Remember, whatever we do to one term, we must do to *all* terms on both sides of the equation. The last step is to solve for a.

$$4 + 3a = 24$$

$$3a = 20$$

$$a = \frac{20}{3} = 6\frac{2}{3}$$

(When we have an improper fraction, we change it to a mixed number.)

CHECK

$$\frac{1}{3} + \frac{\frac{20}{3}}{4} = 2$$

$$\frac{1}{3} + \frac{5}{3} = 2$$

$$2 = 2$$

Note:

$$\frac{\frac{20}{3}}{4} = \frac{20}{3} \div 4 = \frac{20}{3} \times \frac{1}{4} = \frac{20}{12} = \frac{5}{3}$$

Let's try one in which the unknown is in the denominator.

EXAMPLE 9-2

Solve for x.

$$\frac{5}{x-1} - \frac{1}{2} = 2$$

SOLUTION   In this example the denominators are 2 and $x - 1$. This makes the CD $2(x - 1)$. To get rid of the fractions, we multiply both sides by this CD. Remember, we must multiply *each* term in each member by this CD.

$$\frac{5 \cdot 2(x - 1)}{x - 1} - \frac{2(x - 1)}{2} = 2 \cdot 2(x - 1)$$

Cancel like terms:

$$\frac{5 \cdot 2(x - 1)}{x - 1} - \frac{2(x - 1)}{2} = 2 \cdot 2(x - 1)$$

$$10 - (x - 1) = 4(x - 1)$$

$$10 - x + 1 = 4x - 4$$

$$11 - x = 4x - 4$$

$$11 = 5x - 4 \qquad \text{(move } -x \text{ to the right side and it becomes } +x; \text{ that is, } 4x + x = 5x)$$

$$15 = 5x \qquad \text{(move } -4 \text{ to the left side and it becomes } +4, \text{ that is, } 11 + 4 = 15)$$

$$3 = x$$

or

$$x = 3$$

CHECK

$$\frac{5}{3 - 1} - \frac{1}{2} = 2$$

$$\frac{5}{2} - \frac{1}{2} = 2$$

$$2 = 2$$

Let's consider an equation that includes more than one variable and solve for each of them.

EXAMPLE 9-3

$\dfrac{2}{x} - \dfrac{3}{y} = 4$. Solve for x and y.

SOLUTION   In this example the CD is xy. Multiplying each term by xy, we get

$$\frac{2xy}{x} - \frac{3xy}{y} = 4xy$$

(9–1)

$$2y - 3x = 4xy$$

This cleared the fractions and we now have a simple equation. Let's first solve for x. To do so, we first put all terms containing x on one side of the equation.

$$2y = 4xy + 3x \qquad \text{(move } -3x \text{ to the right side)}$$

or

$$4xy + 3x = 2y$$

Now we factor out the x.

$$x(4y + 3) = 2y$$

Next we move the $4y + 3$ to the right side.

$$\frac{x(4y + 3)}{4y + 3} = \frac{2y}{4y + 3} \qquad \text{(divide both sides by } 4y + 3)$$

$$\frac{x(\cancel{4y + 3})}{\cancel{4y + 3}} = \frac{2y}{4y + 3} \qquad \text{(cancel like terms)}$$

$$x = \frac{2y}{4y + 3}$$

Now let's solve for y. After multiplying all terms of the original equation by xy, we had

(9–1)

$$2y - 3x = 4xy$$

Now we put all terms containing y on one side (because we are solving for y) and move all other terms to the other side.

$$2y - 3x - 4xy = 0 \qquad \text{(move } 4xy \text{ to the left side)}$$

$$2y - 4xy = 3x \qquad \text{(move } -3x \text{ to the right side)}$$

Next we factor out y.

$$y(2 - 4x) = 3x$$

Now we move $2 - 4x$ to the right side.

$$\frac{y(2 - 4x)}{2 - 4x} = \frac{3x}{2 - 4x} \qquad \text{(divide each side by } 2 - 4x)$$

$$y = \frac{3x}{2 - 4x}$$

EXAMPLE 9–4

$\dfrac{2}{x + 2} + \dfrac{3}{x - 3} = \dfrac{4}{x + 2}$. Solve for x.

SOLUTION   In this example, the CD is $(x + 2)(x - 3)$. These terms, $(x + 2)$ and $(x - 3)$, are prime factors, so we leave them as they are. That is, we don't find their product. Multiply each term by the CD.

$$\frac{2(x + 2)(x - 3)}{x + 2} + \frac{3(x + 2)(x - 3)}{x - 3} = \frac{4(x + 2)(x - 3)}{x + 2}$$

Cancel like terms in numerators and denominators.

$$\frac{2(\cancel{x + 2})(x - 3)}{\cancel{x + 2}} + \frac{3(x + 2)(\cancel{x - 3})}{\cancel{x - 3}} = \frac{4(\cancel{x + 2})(x - 3)}{\cancel{x + 2}}$$

$$2(x - 3) + 3(x + 2) = 4(x - 3)$$

Combine like terms and solve for x.

$$2x - 6 + 3x + 6 = 4x - 12$$

$$2x + 3x - 4x = 6 - 6 - 12$$

$$x = -12$$

CHECK

$$\frac{2}{-12 + 2} + \frac{3}{-12 - 3} = \frac{4}{-12 + 2}$$

$$\frac{2}{-10} + \frac{3}{-15} = \frac{4}{-10}$$

$$-\frac{1}{5} - \frac{1}{5} = -\frac{2}{5}$$

$$-\frac{2}{5} = -\frac{2}{5}$$

PRACTICE PROBLEMS 9–1

Solve for the unknown in the following problems. Express your answers as fractions or mixed numbers.

1. $\dfrac{8x}{2} - 4 = 10$

2. $\dfrac{R}{2} + \dfrac{R}{4} = 3$

3. $\dfrac{Z + 1}{3} - Z = 4 - 2Z$

4. $\dfrac{4}{I + 2} + 2 = 5$

5. $\dfrac{3I}{2} - 2 = \dfrac{I}{2} + 4$

6. $\dfrac{4x}{2} - \dfrac{3+x}{5} = \dfrac{-2x+3}{4}$

7. $\dfrac{6-x}{x} - \dfrac{4}{x} = \dfrac{3}{x}$

8. $\dfrac{4}{x+3} - \dfrac{1}{3} = 4$

9. $\dfrac{x}{a} + \dfrac{x}{b} = 2$. Solve for a, b, and x.

10. $3(2x+1) = 2y + x - 3$.
    Solve for x and y.

11. $\dfrac{1}{a+4} - \dfrac{2}{a-2} = \dfrac{3}{a+4}$

12. $\dfrac{2}{3a+2} - \dfrac{3}{2a+4} = \dfrac{5}{3a+2}$

SOLUTIONS

1. $x = 3\dfrac{1}{2}$

2. $R = 4$

3. $Z = 2\dfrac{3}{4}$

4. $I = -\dfrac{2}{3}$

5. $I = 6$

6. $x = \dfrac{27}{46}$

7. $x = -1$

8. $x = -2\dfrac{1}{13}$

9. $a = \dfrac{bx}{2b-x}, b = \dfrac{ax}{2a-x}, x = \dfrac{2ab}{a+b}$

10. $x = \dfrac{2y-6}{5}, y = \dfrac{5x+6}{2}$

11. $a = -1$

12. $a = -1\dfrac{1}{5}$

Additional practice problems are at the end of the chapter.

9–2   SOME REAL EQUATIONS

In this section we will deal with equations that are used in circuit analysis. We have already used some of the equations in previous chapters. For instance, we have solved this equation for V_1, where we were given the values for the other variables.

(9–2)
$$V_1 = \dfrac{ER_1}{R_1 + R_2}$$

☞ **KEY POINT** When unknown variables appear in more than one term, we must factor out the unknown to arrive at a solution.

In the following example such a case is presented.

EXAMPLE 9–5

Given the equation $V_1 = \dfrac{ER_1}{R_1 + R_2}$, solve for R_1.

SOLUTION   Notice that R_1 appears in both the numerator and denominator of the right side. The R_1s can't cancel because there are two terms in the denominator, and R_1 appears in only one of them. R_1 would have to be common to *all* terms in the fraction to be canceled. Our first step will be to get rid of the fraction. We can do that by multiplying each term by $R_1 + R_2$.

$$V_1(R_1 + R_2) = \frac{ER_1(R_1 + R_2)}{R_1 + R_2} \qquad \text{(multiplying both sides by } R_1 + R_2 \\ \text{gets rid of the fraction)}$$

$$V_1(R_1 + R_2) = ER_1$$

Next we need to remove the parentheses on the left side because the variable we are solving for, R_1, is inside. We could remove the need for the parentheses by dividing each side by V_1, but we would still have R_1 in the right side as part of a fraction. A better step is to perform the indicated multiplication.

$$V_1R_1 + V_1R_2 = ER_1$$

Next we put all terms containing R_1 on the left side and move all other terms to the right side.

$$V_1R_1 + V_1R_2 - ER_1 = 0 \qquad \text{(move } ER_1 \text{ to the left side)}$$

$$V_1R_1 - ER_1 = -V_1R_2 \qquad \text{(move } V_1R_2 \text{ to the right side)}$$

We're solving for R_1, so we factor out R_1:

$$R_1(V_1 - E) = -V_1R_2$$

And finally we divide by $V_1 - E$.

$$\frac{R_1(V_1 - E)}{V_1 - E} = \frac{-V_1R_2}{V_1 - E}$$

$$R_1 = \frac{-V_1R_2}{V_1 - E} \qquad \text{(divide both sides by } V_1 - E)$$

We can get rid of the minus sign in the numerator by changing that sign and the sign of the denominator or the sign of the fraction. Remember, there are three signs associated with a fraction: the sign of the numerator, the sign of the denominator, and the sign of the fraction. We can change any two signs without changing the value of the fraction. Don't forget, we must change the sign of *each* term when changing the sign of the numerator or denominator.

We could have avoided the problem with the signs if we had moved V_1R_2 to the right side in our second step. Then

$$V_1R_1 + V_1R_2 = ER_1$$

$$V_1R_2 = ER_1 - V_1R_1$$

$$V_1R_2 = R_1(E - V_1)$$

$$\frac{V_1R_2}{E - V_1} = R_1$$

(9–3) or $$R_1 = \frac{V_1R_2}{E - V_1}$$

EXAMPLE 9–6

Given the equation $r_i = \beta\left(r_e + \dfrac{r_b}{\beta}\right)$. Solve for r_e.

SOLUTION   In this example, as in many others, we find that there is more than one approach to the solution. The trick is to find the easiest one—the one that requires the fewest steps. In this case, let's divide both sides by β. This step isolates r_e in one term.

$$\frac{r_i}{\beta} = r_e + \frac{r_b}{\beta} \qquad \text{(divide both sides by } \beta)$$

$$r_e = \frac{r_i}{\beta} - \frac{r_b}{\beta} \qquad \left(\text{subtract } \frac{r_b}{\beta} \text{ from both sides}\right)$$

$$= \frac{r_i - r_b}{\beta}$$

PRACTICE PROBLEMS 9–2

1. $I_2 = \dfrac{I_TG_2}{G_1 + G_2}$. Solve for I_T, G_1, and G_2.

2. $V_1 = \dfrac{ER_1}{R_1 + R_2}$. Solve for E, R_1, and R_2.

3. $F = \dfrac{9}{5}C + 32$. Solve for C.

4. $A_i = \dfrac{h_{fe}}{h_{oe}R_L + 1}$. Solve for R_L.

5. $S = \dfrac{R_E + R_B}{R_E + R_B(1 - \alpha)}$. Solve for R_B.

6. $R_S = \dfrac{R_iR_o}{R_i + R_o}$. Solve for R_o.

7. $r_i = \beta\left(r_e + \dfrac{r_b}{\beta}\right)$. Solve for r_b.

8. $A_V = \dfrac{r_e}{r_e + r_e'}$. Solve for r_e.

SOLUTIONS

1. $I_T = \dfrac{I_2(G_1 + G_2)}{G_2}$

$G_1 = \dfrac{I_TG_2 - I_2G_2}{I_2}$

$G_2 = \dfrac{I_2G_1}{I_T - I_2}$

2. $E = \dfrac{V_1(R_1 + R_2)}{R_1}$

$R_1 = \dfrac{V_1R_2}{E - V_1}$

$R_2 = \dfrac{ER_1 - V_1R_1}{V_1}$

3. $C = \dfrac{5F - 160}{9}$

4. $R_L = \dfrac{h_{fe} - A_i}{A_i h_{oe}}$

5. $R_B = \dfrac{R_E - SR_E}{S - S\alpha - 1}$

6. $R_o = \dfrac{R_S R_i}{R_i - R_S}$

7. $r_b = r_i - \beta r_e$

8. $r_e = \dfrac{A_V r'_e}{1 - A_V}$

Additional practice problems are at the end of the chapter.

APPLICATIONS   9–3

Now let's carry the work in the previous section one step further. Let's determine actual values of unknowns just as we would do in circuit analysis.

EXAMPLE 9–7

Given the equation $I_1 = \dfrac{I_T G_1}{G_1 + G_2}$. $\hfill$ (9–4)

Let $I_1 = 2.73$ mA, $I_T = 6.75$ mA, and $G_1 = 370\ \mu S$. Find G_2. Round your answer to three places.

SOLUTION   Let's first multiply both sides by $G_1 + G_2$ to get rid of the fraction. This results in

$$I_1(G_1 + G_2) = \frac{I_T G_1(G_1 + G_2)}{G_1 + G_2}$$

$$I_1(G_1 + G_2) = I_T G_1 \hspace{3cm} (9\text{–}5)$$

Now we have two possible solutions. Let's examine both.

First solution:

$$G_1 + G_2 = \frac{I_T G_1}{I_1} \hspace{1cm} \text{(divide both sides by } I_1\text{)}$$

$$G_2 = \frac{I_T G_1}{I_1} - G_1 \hspace{1cm} \text{(move } G_1 \text{ to the right side)}$$

Then we plug in known values.

$$G_2 = \frac{6.75\ \text{mA} \times 370\ \mu S}{2.73\ \text{mA}} - 370\ \mu S$$

Using the calculator,

$$6.75 \boxed{EE} \boxed{+/-} 3 \boxed{\times} 370 \boxed{EE} \boxed{+/-} 6 \boxed{\div}$$

$$2.73 \boxed{EE} \boxed{+/-} 3 \boxed{-} 370 \boxed{EE} \boxed{+/-} 6 \boxed{=}$$

Casio users must remember to use $\boxed{EXP}$ instead of $\boxed{EE}$ and $\boxed{(-)}$ instead of $\boxed{+/-}$.

Answer: 545×10^{-6}

$G_2 = 545 \ \mu S$

Second solution:

$$I_1(G_1 + G_2) = I_T G_1 \qquad\qquad (9\text{--}5)$$

$$I_1 G_1 + I_1 G_2 = I_T G_1 \qquad\qquad \text{(remove parentheses in left side)}$$

$$I_1 G_2 = I_T G_1 - I_1 G_1 \qquad\qquad \text{(move } I_1 G_1 \text{ term to right side)}$$

$$G_2 = \frac{I_T G_1 - I_1 G_1}{I_1} \qquad\qquad \text{(divide by } I_1\text{)}$$

Then we plug in known values.

(9–6)
$$G_2 = \frac{6.75 \text{ mA} \times 370 \ \mu S - 2.73 \text{ mA} \times 370 \ \mu S}{2.73 \text{ mA}}$$

Using the calculator,

$$6.75 \boxed{EE} \boxed{+/-} 3 \boxed{\times} 370 \boxed{EE} \boxed{+/-} 6 \boxed{-}$$

$$2.73 \boxed{EE} \boxed{+/-} 3 \boxed{\times} 370 \boxed{EE} \boxed{+/-} 6 \boxed{=} \boxed{\div}$$

$$2.73 \boxed{EE} \boxed{+/-} 3 \boxed{=}$$

Answer: 545×10^{-6}

$G_2 = 545 \ \mu S$

The first solution appears to be easier, but both resulted in the same answer. To check our answer we can use the estimating techniques discussed in Chapter 2. We can round each given value to one significant digit and then perform the mathematical operations with the intent of getting a ballpark estimate. Rounding yields:

$$\frac{7 \text{ mA} \times 4 \times 10^2 \mu S - 3 \text{ mA} \times 4 \times 10^2 \mu S}{3 \text{ mA}} = \frac{28 \times 10^2 \mu S - 12 \times 10^2 \mu S}{3} = 533 \mu S$$

An easier way to check our answer is to make an estimate by combining the tools we have learned about estimating and factoring. The steps are:

(1) Factor out G_1: $G_2 = \dfrac{I_T G_1}{I_1} - G_1 = G_1\left(\dfrac{I_T}{I_1} - 1\right)$

(2) Substitute approximate values: $G_2 \approx 400 \, \mu S \left(\dfrac{7mA}{3mA} - 1 \right)$

(3) Cancel common factors in the numerator and denominator:

$$G_2 \approx 400 \, \mu S \left(\dfrac{7}{3} - 1 \right)$$

(4) Do arithmetic: $G_2 \approx 400 \, \mu S \left(\dfrac{4}{3} \right) = \dfrac{1600 \mu S}{3} = 533 \mu S$

This approach maintained the μS units, cancelled out the mA units, avoided using powers-of-ten, minimized the chances of error, and produced an estimate that has an error of less than 3%.

EXAMPLE 9–8

Given the equation $R_{TH} = \dfrac{V_{OC} - V_L}{I_L}$. (9–7)

Solve for V_{OC} when $R_{TH} = 1.5 \, k\Omega$, $V_L = 3.75 \, V$, and $I_L = 5 \, mA$. Round your answer to three places.

SOLUTION We need to rearrange the equation and solve for V_{OC}; therefore, let's do it this way:

$$R_{TH}I_L = V_{OC} - V_L \qquad \text{(multiply by } I_L \text{)}$$

$$R_{TH}I_L + V_L = V_{OC} \qquad \text{(move } V_L \text{ to the left side)}$$

We then plug in known values.

$$V_{OC} = 1.5 \, k\Omega \times 5 \, mA + 3.75 \, V$$

$$= 7.5 \, V + 3.75 \, V = 11.3 \, V$$

PRACTICE PROBLEMS 9–3

Perform the indicated operations. Round answers to three places.

1. $V_1 = \dfrac{ER_1}{R_1 + R_2}$
   (a) Solve for E when $V_1 = 34 \, V$,
       $R_1 = 270 \, \Omega$, and $R_2 = 330 \, \Omega$.
   (b) Solve for R_1 when $E = 25 \, V$,
       $V_1 = 14.7 \, V$, and $R_2 = 3.3 \, k\Omega$.

2. $I_L = \dfrac{I_{SC}G_L}{G_L + G_N}$
   (a) Solve for G_L when $I_{SC} = 10 \, mA$,
       $I_L = 3 \, mA$, and $G_N = 300 \, \mu S$.
   (b) Solve for G_N when $I_{SC} = 600 \, \mu A$,
       $I_L = 200 \, \mu A$, and $G_L = 2.7 \, mS$.

3. $A_i = \dfrac{h_{fe}}{1 + h_{oe}R_L}$

Solve for h_{fe} when $A_i = 65$, $h_{oe} = 10\ \mu S$, and $R_L = 5\ k\Omega$.

5. $R_{TH} = \dfrac{V_{OC} - V_L}{I_L}$

Solve for V_L when $R_{TH} = 1.7\ k\Omega$, $V_{OC} = 12\ V$, and $I_L = 750\ \mu A$.

4. $\beta = \dfrac{\alpha}{1 - \alpha}$

Solve for α when $\beta = 90$.

SOLUTIONS

1. (a) $E = 75.6\ V$; (b) $R_1 = 4.71\ k\Omega$
2. (a) $G_L = 129\ \mu S$; (b) $G_N = 5.40\ mS$
3. $h_{fe} = 68.3$
4. 0.989
5. 10.7 V

Additional practice problems are at the end of the chapter.

SELF-TEST 9–1

1. $\dfrac{R + 2}{3} - R = 2 - R$

Solve for R.

2. $\dfrac{2}{I - 2} + 3 = 5$

Solve for I.

3. $\dfrac{a}{2} + \dfrac{3}{b} = \dfrac{2}{x}$

Solve for a, b, and x.

4. $I_L = \dfrac{I_{SC}G_L}{G_L + G_N}$

Solve for G_L when $I_{SC} = 750\ \mu A$, $I_L = 200\ \mu A$, and $G_N = 300\ \mu S$.

5. $R_T = \dfrac{R_1 R_2}{R_1 + R_2}$

Solve for R_2 when $R_T = 38.7\ k\Omega$ and $R_1 = 68\ k\Omega$.

Answers to Self-test 9–1 are at the end of the chapter.

9–4   QUADRATIC EQUATIONS

☞ **KEY POINT** Quadratic equations are second-degree equations and have two solutions or roots.

In other words, the unknown is raised to the second power. Some examples of these quadratic equations are $x^2 = 25$, $2x^2 + x + 2 = 0$, and $x^2 + x = 4$. A quadratic equation in standard form looks like this:

(9–8)
$$ax^2 + bx + c = 0$$

where a and b are the coefficients of x^2 and x and where c is a constant. If the equation is solved for x, the result is

$$x = \frac{-b \pm \sqrt{b^2 - 4ac}}{2a}$$ (9–9)

When we solve for the unknown in a second-degree equation, there will always be two possible solutions called *roots*. Sometimes we can factor the quadratic to find the two roots. If the quadratic is not factorable, then we must resort to the quadratic equation. In the following examples we will first solve for the unknown by factoring, and then we will solve by using the quadratic equation.

EXAMPLE 9–9

Solve the equation $x^2 - 25 = 0$.

SOLUTION   Quadratic equations that have no x term are called *pure quadratics*. Such equations are readily solved by moving the constant to the right member and finding the square root of both members.

$$x^2 = 25$$
$$x = \sqrt{25} = \pm 5$$

The equation is satisfied when $x = 5$ or -5.

$$5^2 = 25 \qquad (-5)^2 = 25$$

9-4-1 Solving by Factoring

In Chapter 8 we determined that *trinomials of the form $x^2 + bx + c = 0$ can be factored if we can find two numbers whose sum is the coefficient of the second term and whose product is the third term.* In the following examples, we show how to solve for the unknown when the equation is factorable.

EXAMPLE 9–10

Find x in the equation $x^2 + 7x + 12 = 0$.

SOLUTION

$$x^2 + 7x + 12 = 0$$
$$(x + 3)(x + 4) = 0 \qquad \text{(factor the left side)}$$

Let each factor equal zero in turn, and solve for x.

$$x + 3 = 0$$
$$x = -3$$
$$x + 4 = 0$$
$$x = -4$$

The equation is satisfied when $x = -3$ or $x = -4$.

EXAMPLE 9–11

$x^2 - 3x - 6 = 4$. Solve for x.

SOLUTION   Put the equation in standard form and then factor the left member.

$$x^2 - 3x - 10 = 0$$
$$(x + 2)(x - 5) = 0$$

Equating each factor to zero, we get

$$x + 2 = 0$$
$$x = -2$$
$$x - 5 = 0$$
$$x = 5$$

The equation is satisfied when $x = -2$ or $x = 5$.

9-4-2 Solving by the Quadratic Equation

When we have a second-degree equation to solve, we first put it into the form $ax^2 + bx + c = 0$. If the equation is factorable, we merely factor as before. If the equation is not factorable, then we resort to using the quadratic equation.

EXAMPLE 9–12

Given the equation $x^2 + 4x + 5 = 13$, solve for x.

SOLUTION   First, we put the equation in standard form:

$$x^2 + 4x - 8 = 0$$

Next we determine if the equation can be factored. Since this one does not factor, we will solve it by using the quadratic equation:

(9–9)
$$x = \frac{-b \pm \sqrt{b^2 - 4ac}}{2a}$$

An examination of the equation shows that a (the coefficient of x^2) = 1; b (the coefficient of x) = 4; and c (the constant) = -8. Substituting these values in our equation results in the following:

$$x = \frac{-4 \pm \sqrt{4^2 - (4)(1)(-8)}}{2(1)} = \frac{-4 \pm \sqrt{16 + 32}}{2}$$

$$= \frac{-4 \pm \sqrt{48}}{2} = \frac{-4 \pm 6.93}{2}$$

At this point we see that there are two solutions:

1. $x = \dfrac{-4 + 6.93}{2} = \dfrac{2.93}{2} = 1.46$

2. $x = \dfrac{-4 - 6.93}{2} = \dfrac{-10.93}{2} = -5.46$

EXAMPLE 9–13

Given the equation $4x^2 + 3x - 5 = 0$, solve for x.

SOLUTION

(1) Write the algebraic solution for the quadratic equation:

$$x = \frac{-b \pm \sqrt{b^2 - 4ac}}{2a}$$

(2) Substitute with numeric values: $x = \dfrac{-3 \pm \sqrt{3^2 - 4 \times 4 \times (-5)}}{2 \times 4}$

(3) Calculate the answer:

$$x = \frac{-3 \pm \sqrt{9 + 80}}{8} = \frac{-3 \pm \sqrt{89}}{8} = \frac{-3 \pm 9.43}{8} = \frac{-12.43}{8} \text{ or } \frac{6.43}{8}$$

$$= -1.55 \text{ or } 0.804$$

(4) Check your answers by substituting each answer in the original equation.

EXAMPLE 9–14

Given the equation $2x^2 + 8x - 7 = 13$, solve for x.

SOLUTION

(1) Convert the equation to the format $x^2 + bx + c = 0$ to see if it can be factored.
(2) Make the right side zero by subtracting 13 from each side of the equation: $2x^2 + 8x - 20 = 0$
(3) Divide each side of the equation by two: $x^2 + 4x - 10 = 0$
(4) The factors of 10 are: 1 and 10 or 2 and 5. The equation is therefore not factorable and we must use the quadratic equation to find the roots.

(5) Write the algebraic solution for the quadratic equation:

$$x = \frac{-b \pm \sqrt{b^2 - 4ac}}{2a}$$

(6) Substitute with numeric values: $x = \dfrac{-4 \pm \sqrt{4^2 - 4 \times (-10)}}{2}$

(7) Calculate the answer:

$$x = \frac{-4 \pm \sqrt{16 + 40}}{2} = \frac{-4 \pm \sqrt{56}}{2} = \frac{-4 \pm 7.48}{2} = \frac{-11.48}{2} \text{ or } \frac{+3.48}{2}$$

$$= -5.74 \text{ or } 1.74$$

(8) Check your answers by substituting each answer in the original equation.

PRACTICE PROBLEMS 9–4

Solve the following equations for x by factoring.

1. $x^2 = 49$
3. $x^2 + 5x + 6 = 0$
5. $x^2 + 10x + 12 = -12$

2. $x^2 - x - 30 = 0$
4. $x^2 - 4x - 10 = 11$

Solve the following equations for x using the quadratic equation.

6. $2x^2 + 3x - 9 = 0$
8. $3x^2 - 6x = 7$

7. $x^2 - 5x + 2 = 10$
9. $x^2 - 3x - 12 = -5$

SOLUTIONS

1. $x = 7, x = -7$
3. $x = -2, x = -3$
5. $x = -4, x = -6$
7. $x = 6.27, x = -1.27$
9. $x = -1.54, x = 4.54$

2. $x = -5, x = 6$
4. $x = 7, x = -3$
6. $x = 1.50, x = -3.00$
8. $x = 2.83, x = -0.826$

Additional practice problems are at the end of the chapter.

SELF-TEST 9–2

Solve for x in the following problems:

1. $x^2 - 121 = 0$
3. $x^2 - 10x + 30 = 3x - 12$
5. $2x^2 - 5x - 3 = 0$

2. $x^2 - 6x - 27 = 0$
4. $2x^2 + 4x - 5 = 0$

Answers to Self-test 9–2 are at the end of the chapter.

CHAPTER 9 AT A GLANCE

PAGE	KEY POINT OR RULE	EXAMPLE
251	*Rule 9–1:* To solve an equation that includes fractions or mixed numbers: (1) get rid of the fraction (or mixed number); (2) solve for the unknown; (3) verify your answer.	$\dfrac{1}{3} + \dfrac{a}{4} = 2$ $\dfrac{1}{3}(12) + \dfrac{a}{4}(12) = 2(12)$ $4 + 3a = 24$ $a = \dfrac{20}{3} = 6\dfrac{2}{3}$
256	*Key Point:* When unknown variables appear in more than one term, we must factor out the unknown to arrive at a solution.	$ER_1 = V_1R_1 + V_2R_2$ Solve for R_1. $ER_1 - V_1R_1 = V_2R_2$ $R_1(E - V_1) = V_2R_2$ $R_1 = \dfrac{V_2R_2}{E - V_1}$
262	*Key Point:* Quadratic equations are second-degree equations and have two solutions or *roots*.	$x^2 + 7x + 12 = 0$ $(x + 3)(x + 4) = 0$ $x = -3, x = -4$

END OF CHAPTER PROBLEMS 9–1

Solve for the unknown in the following problems. Express your answers as proper fractions or as mixed numbers.

1. $\dfrac{5x}{2} + 3 = 6$

2. $\dfrac{6x}{4} - 2 = 6$

3. $\dfrac{4a}{3} - 5 = 2$

4. $\dfrac{2b}{5} - 6 = 12$

5. $\dfrac{2I}{4} + \dfrac{I}{6} = 2$

6. $\dfrac{3R}{2} + \dfrac{2R}{3} = 8$

7. $\dfrac{2x}{3} - \dfrac{3x}{4} = 1$

8. $\dfrac{3x}{4} - \dfrac{7x}{2} = 5$

9. $\dfrac{4R}{2} = 3 + \dfrac{R}{3}$

10. $\dfrac{3R}{5} - 2 = \dfrac{2R}{3}$

11. $\dfrac{R + 3}{4} - 2 = \dfrac{3R + 1}{2} + 4$

12. $\dfrac{2R - 3}{3} - 5 = \dfrac{2R + 3}{5} - 3$

13. $\dfrac{R - 2}{5} + 3 = \dfrac{2R + 2}{3} + 4$

14. $\dfrac{a + 3}{2} + 4 = \dfrac{3a - 2}{5} + 1$

15. $\dfrac{c - 3}{4} - \dfrac{c + 1}{3} = 2 + 3c$

16. $\dfrac{R + 2}{3} - \dfrac{3R - 1}{2} = 3 - 2R$

17. $\dfrac{3}{I-3} + 3 = 4$

18. $\dfrac{4}{I-4} + 2 = 5$

19. $\dfrac{4}{R-2} + 2 = 3$

20. $\dfrac{5}{R+3} - 4 = 3$

21. $\dfrac{5x}{2} + \dfrac{x-2}{6} = \dfrac{3x-6}{4}$

22. $\dfrac{2x}{4} - \dfrac{3-x}{2} = \dfrac{3x-1}{6}$

23. $\dfrac{4a}{3} + \dfrac{a+2}{5} = \dfrac{2a-3}{5}$

24. $\dfrac{3c}{6} - \dfrac{2c-3}{3} = \dfrac{3c-2}{2}$

25. $\dfrac{4-x}{2x} - \dfrac{3}{x} = \dfrac{1}{x}$

26. $\dfrac{3x-1}{2x} + \dfrac{1}{x} = \dfrac{3}{x}$

27. $\dfrac{3-y}{2y} - \dfrac{2}{y} = \dfrac{3}{y}$

28. $\dfrac{4+y}{3y} + \dfrac{3}{y} = \dfrac{1}{2y}$

29. $\dfrac{3}{G-3} - \dfrac{1}{4} = 5$

30. $\dfrac{2}{R+3} - \dfrac{1}{2} = 1$

31. $\dfrac{2}{R+3} - \dfrac{1}{3} = 4$

32. $\dfrac{3}{R-5} - \dfrac{2}{3} = 5$

33. $\dfrac{1}{a} + \dfrac{1}{b} = 2.$ Solve for a and b.

34. $\dfrac{2}{a} + \dfrac{4}{b} = 4.$ Solve for a and b.

35. $\dfrac{2}{a} - 3 = \dfrac{1}{b}.$ Solve for a and b.

36. $\dfrac{3}{a} - 5 = \dfrac{2}{b}.$ Solve for a and b.

37. $\dfrac{x}{2a} + \dfrac{x}{3b} = 2.$ Solve for x, a, and b.

38. $\dfrac{2x}{3a} - \dfrac{x}{2b} = 3.$ Solve for x, a, and b.

39. $\dfrac{3x}{2a} - \dfrac{x}{3b} = 4.$ Solve for x, a, and b.

40. $\dfrac{2x}{6a} - \dfrac{4x}{5b} = 1.$ Solve for x, a, and b.

41. $\dfrac{3}{a-2} + \dfrac{2}{a+2} = \dfrac{b}{a+2}$
Solve for a and b.

42. $\dfrac{1}{x+4} - \dfrac{2}{x-4} = \dfrac{2y}{x-4}$
Solve for x and y.

43. $\dfrac{2}{2a-1} + \dfrac{3}{3a+2} = \dfrac{2b}{2a-1}$
Solve for a and b.

44. $\dfrac{3}{3a-3} + \dfrac{2}{2a+3} = \dfrac{5b}{2a+3}$
Solve for a and b.

45. $\dfrac{5}{3b-2} - \dfrac{4}{3b-1} = \dfrac{3}{3b-1}$
Solve for b.

46. $\dfrac{3}{2R+3} + \dfrac{5}{3R-3} = \dfrac{2}{3R-3}$
Solve for R.

END OF CHAPTER PROBLEMS 9–2

Perform the indicated operations.

1. $V_L = \dfrac{V_{OC}R_L}{R_{TH} + R_L}.$ Solve for V_{OC}, R_L, and R_{TH}.

2. $V_2 = \dfrac{ER_2}{R_1 + R_2}.$ Solve for E, R_1, and R_2.

3. $I_L = \dfrac{I_{SC}G_L}{G_L + G_N}.$ Solve for I_{SC}, G_L, and G_N.

4. $I_1 = \dfrac{I_T G_1}{G_1 + G_2}.$ Solve for I_T, G_1, and G_2.

5. $R_S = \dfrac{R_i R_o}{R_i + R_o}$. Solve for R_i and R_o.

6. $S = \dfrac{R_E + (R/\beta)}{R_E + I(1 - \alpha)}$. Solve for R and R_E.

7. $r_i = \beta\left(r_e + \dfrac{r_b}{\beta}\right)$. Solve for r_e and r_b.

8. $A_V = \dfrac{r_e}{r_e + r_i}$. Solve for r_e and r_i.

9. $R_T = \dfrac{R_1 R_2}{R_1 + R_2}$. Solve for R_1 and R_2.

10. $Z_T = \dfrac{Z_1 Z_2}{Z_1 + Z_2}$. Solve for Z_1 and Z_2.

11. $A = \dfrac{B}{1 + (1/A)}$. Solve for B.

12. $B = \dfrac{A}{2 + (1/B)}$. Solve for A.

13. $f = \dfrac{1}{2\pi C(R_1 + R_2)}$. Solve for R_1.

14. $I_C = \dfrac{V_{CC}}{R_E + (R_B/\beta)}$. Solve for β, R_E, and R_B.

15. $I_B = \dfrac{V_{CC} - V_B}{R_1 + R_2}$. Solve for V_B and R_1.

16. $K = \dfrac{1}{1 + (\beta R_E/R_B)}$. Solve for R_B and R_E.

END OF CHAPTER PROBLEMS 9–3

Perform the indicated operations. Round answers to three places.

1. $V_2 = \dfrac{ER_2}{R_1 + R_2}$
   (a) Solve for R_2 when $E = 35$ V, $V_2 = 10$ V, and $R_1 = 12$ kΩ.
   (b) Solve for R_1 when $E = 15$ V, $V_2 = 6.3$ V, and $R_2 = 560$ Ω.

2. $V_2 = \dfrac{ER_2}{R_1 + R_2}$
   (a) Solve for R_2 when $E = 15$ V, $V_2 = 11.1$ V, and $R_1 = 12$ kΩ.
   (b) Solve for R_1 when $E = 20$ V, $V_2 = 4.13$ V, and $R_2 = 330$ Ω.

3. $I_1 = \dfrac{I_T G_1}{G_1 + G_2}$
   (a) Solve for I_T when $I_1 = 4.95$ mA, $G_1 = 21.3$ μS, and $G_2 = 30.3$ μS.
   (b) Solve for G_2 when $I_T = 100$ μA, $I_1 = 45.1$ μA, and $G_1 = 1.47$ mS.

4. $I_1 = \dfrac{I_T G_1}{G_1 + G_2}$
   (a) Solve for I_T when $I_1 = 8.35$ mA, $G_1 = 3$ mS, and $G_2 = 5.83$ mS.
   (b) Solve for G_2 when $I_T = 250$ μA, $I_1 = 120$ μA, and $G_1 = 50$ μS.

5. $A_i = \dfrac{h_{fe}}{1 + h_{oe}R_L}$
   (a) Solve for h_{oe} when $A_i = 37$, $h_{fe} = 70$, and $R_L = 12$ kΩ (h_{oe} is measured in siemens).
   (b) Solve for R_L when $A_i = 80$, $h_{fe} = 100$, and $h_{oe} = 12$ μS.

6. $a = \dfrac{\beta}{1 + \beta}$
   Solve for β when $a = 0.996$.

7. $R_T = \dfrac{R_1 R_2}{R_1 + R_2}$
   Solve for R_1 when $R_T = 1.83$ kΩ and $R_2 = 4.7$ kΩ.

8. $R_T = \dfrac{R_1 R_2}{R_1 + R_2}$
   Solve for R_1 when $R_T = 11.5$ kΩ and $R_2 = 27$ kΩ.

9. $Z_T = \dfrac{Z_1 Z_2}{Z_1 + Z_2}$

Solve for Z_2 when $Z_T = 64.4$ kΩ and $Z_1 = 91$ kΩ.

10. $Z_T = \dfrac{Z_1 Z_2}{Z_1 + Z_2}$

Solve for Z_2 when $Z_T = 43$ kΩ and $Z_1 = 56$ kΩ.

11. $I_B = \dfrac{V_{CC} - V_B}{R_1 + R_2}$

Solve for R_2 when $V_{CC} = 20$ V, $V_B = 1.3$ V, $I_B = 312$ μA, and $R_1 = 56$ kΩ.

12. $I_B = \dfrac{V_{CC} - V_B}{R_1 + R_2}$

Solve for R_1 when $V_{CC} = 12$ V, $V_B = 1.54$ V, $I_B = 187$ μA, and $R_2 = 33$ kΩ.

13. $I_C = \dfrac{V_{CC}}{R_E + (R_B/\beta)}$

Find R_B when $V_{CC} = 15$ V, $I_C = 1.2$ mA, $R_E = 1.5$ kΩ and $\beta = 50$.

14. $I_C = \dfrac{V_{CC}}{R_E + (R_B/\beta)}$

Find R_E when $V_{CC} = 10$ V, $I_C = 700$ μA, $R_B = 560$ kΩ and $\beta = 70$.

15. $f = \dfrac{1}{2\pi C(R_1 + R_2)}$

Solve for R_1 when $f = 8.76$ kHz, $C = 500$ pF, and $R_2 = 27$ kΩ.

16. $f = \dfrac{1}{2\pi C(R_1 + R_2)}$

Solve for R_1 when $f = 3.18$ kHz, $C = 20$ nF, and $R_2 = 1$ kΩ.

END OF CHAPTER PROBLEMS 9–4

Solve the following quadratics by factoring:

1. $x^2 - 16 = 0$
3. $x^2 + 6x + 8 = 0$
5. $x^2 + 3x - 18 = 0$
7. $x^2 - 6x + 8 = 0$
9. $x^2 - x - 56 = 0$
11. $x^2 - 15 = 3x + 25$
13. $x^2 - 3x + 16 = 10x - 20$
15. $x^2 + 2x = 5x + 18$
17. $x^2 - 2x + 25 = 4 - 12x$
19. $x^2 = 3x - 2$

2. $x^2 = 64$
4. $x^2 + 12x + 32 = 0$
6. $x^2 - x - 30 = 0$
8. $x^2 - 19x + 48 = 0$
10. $x^2 - 4x - 45 = 0$
12. $x^2 - 2 = 10 - x$
14. $x^2 - 3x = 4x - 10$
16. $x^2 - 5x + 10 = 7x - 10$
18. $x^2 + 40 = 10 - 13x$
20. $x^2 + 10x - 8 = 16 + 8x$

Solve for x in the following problems:

21. $x^2 + 8x + 10 = 0$
23. $x^2 + 6x + 6 = 0$
25. $x^2 + 2x - 14 = 0$
27. $x^2 - 15x - 15 = 0$
29. $x^2 - 3x = 4x + 4$
31. $3x^2 + 2x - 1 = 0$
33. $2x^2 - 5x - 5 = 0$
35. $4x^2 + 12x + 8 = 0$
37. $3x^2 + 4x = -3x - 3$
39. $3x^2 + 6x = 20x + 20$

22. $x^2 + 7x + 4 = 0$
24. $x^2 + 16x + 9 = 0$
26. $x^2 + 7x - 8 = 0$
28. $x^2 - 20x - 28 = 0$
30. $3x^2 - 8x + 3 = 0$
32. $x^2 - 6x = 2x + 8$
34. $2x^2 - 9x + 5 = 0$
36. $3x^2 - 9x = 10$
38. $2x^2 + 7x = 25 - 4x$
40. $5x^2 - 7x + 5 = 15x - 13$

ANSWERS TO SELF-TESTS

SELF-TEST 9–1

1. $R = 4$

2. $I = 3$

3. $x = \dfrac{4b}{ab + 6}$, $a = \dfrac{4b - 6x}{bx}$,

   $b = \dfrac{6x}{4 - ax}$

4. $G_L = 109 \ \mu S$

5. $R_2 = 89.8 \ k\Omega$

SELF-TEST 9–2

1. $x = \pm 11$

2. $x = 9, x = -3$

3. $x = 6, x = 7$

4. $x = 0.871, x = -2.87$

5. $x = 3.00, x = -0.500$

Math for DC Electronics

The arithmetic and algebraic concepts we learned in Sections 1 and 2 will be used in this section to solve circuit equations related to simple DC circuits.

DC stands for *direct current*. A simple DC circuit is a group of electrical components connected in a closed loop in which electrical current flows in one direction. You can think of a DC circuit as a *one* **Direct**ion **Current** circuit. In the next section (Math for AC Electronics), you will find that in AC circuits the direction of the current alternates in two directions—therefore it is called **A**lternating **C**urrent. Electrical current can be visualized as water flowing through a recirculating filter system in a fish tank. A water pump that keeps the water flowing powers the recirculating system. The pump is analogous to the electrical power supply or battery in a DC circuit: they both provide the force to cause the current flow. The water flow is analogous to the electrical charges moving through a circuit. The filter screen is analogous to electrical resistance, since the filter can limit the rate of water flow like resistance limits current.

The water pressure, supplied by the recirculating pump, is analogous to the voltage in a circuit, which pushes the electrical charges around the circuit. This electrical pressure is referred to as an "electrical energy source", "electromotive force or EMF", "difference in potential", or "voltage". An important similarity in the water pump/DC circuit analogy is that both systems require a closed loop. The water pump provides the water pressure that causes water to flow from the tank through tubing, filters, the pump itself, more tubing and then back to the fish tank. The battery (or DC power supply) provides electrical pressure that causes charges to flow from one side of the battery, through the wires and resistors and back through the battery. In both cases, if we open the loop by disconnecting a tube or a wire in the circuit, the flow stops.

In Chapters 10, 11, and 12, we will use algebraic equations to analyze the currents and voltages in DC circuits. We will learn how to calculate any one of the circuit variables when the other values are known. We will learn how to describe circuit behavior by using algebraic equations based on scientific laws and theorems.

In Chapter 13, we will learn graphing, which will help us describe circuit behavior by plotting points on a graph. Graphing helps us visualize how currents and voltages can change in a circuit when component values change. It will also help us in Chapter 14 when we use a graphical technique to solve simultaneous linear equations, which are algebraic equations used to solve complex DC circuit problems.

The math skills learned in this section are the foundation upon which are built the math skills in the next section needed to solve alternating current problems.

dc Circuit Analysis: Kirchhoff's Laws

10

Introduction

Three basic laws deal with circuit analysis. These are Kirchhoff's current and voltage laws and Ohm's law. Theorems have been developed from these basic laws to help in solving the more complex problems. These theorems will be discussed in Chapter 12. Since Kirchhoff's laws and Ohm's law are basic to the understanding of all electrical and electronics circuits, these laws will be discussed in detail in this and the following chapter.

All electrical circuits consist of three basic parts: an *electrical energy source,* a *transfer network,* and a *load.* Some typical energy loads are lamps, motors, and radios. If a load is an energy user, it is said to have the electrical property of resistance. In this kind of load, electrical energy is converted to some other form, usually heat and light. If the load is an energy storer, it has the property of capacitance or inductance. In this chapter we will limit ourselves to loads that are energy users, that is, those loads that are resistive.

The transfer network in an electronic circuit can be as simple as a copper conductor, which provides a path for electrical current to flow between the voltage source and the load. An example of such a simple circuit is shown in Figure 10–1. In circuits that are more complex, the transfer network can contain many components and branches. In order to analyze the behavior of these complex circuits we can apply Kirchhoff's and Ohm's laws.

Chapter Objectives

In this chapter you will learn how to:

1. Use Kirchhoff's current law to find the currents in series, parallel, and series–parallel circuits.
2. Use Kirchhoff's voltage law to find the voltage drops in series, parallel, and series–parallel circuits.
3. Find the polarity of voltage drops in dc circuits.

FIGURE 10–1
Simple electrical
circuit.

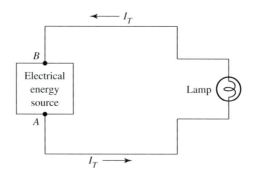

10–1   KIRCHHOFF'S CURRENT LAW

Kirchhoff's current law describes the behavior of electrical charges in a circuit. Current can be visualized by using the following analogy. Consider water flowing in a pipe. If we were to choose a point anywhere along the length of the pipe, the rate of water flow (say gallons per minute) *into* that point would have to equal the rate of water flow *leaving* it, as illustrated in Figure 10–2.

Charges moving in a given direction in an electrical conductor are analogous to water flowing in a pipe. When a charge in motion approaches a point already containing a charge, there is a repelling force between them. If there is a continuous supply of charges, incoming charges cause other charges to move because like charges repel each other. If charges, then, are removed from a point, other charges take their place. This flow of charges in an electrical circuit is called *current*.

Kirchhoff's current law essentially expresses the law of conservation of charges and further defines their behavior at a point in an electrical circuit. It can be stated as follows:

☞ **KEY POINT** The number of charges per second (current) flowing *to* a point in a conductor must equal the number of charges flowing *from* that point.

Stated another way:

☞ **KEY POINT** The algebraic sum of currents *into* and *out* of a point must equal zero if currents toward the point are given an arbitrary plus sign (+) and if currents leaving the point are given an arbitrary negative sign (−) (Figure 10–3).

FIGURE 10–2
Water flow is
analogous to
moving
electrical
charges.

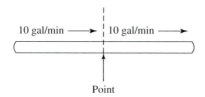

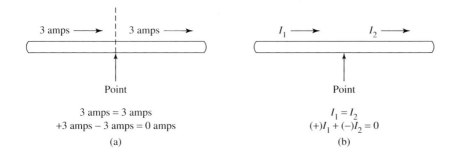

FIGURE 10–3
Current into
and out of a
point in an
electrical
conductor.

3 amps = 3 amps
+3 amps − 3 amps = 0 amps

(a)

$I_1 = I_2$
$(+)I_1 + (-)I_2 = 0$

(b)

10-1-1 Current in Series Circuits

Practical applications of Kirchhoff's current law may be further visualized by the following: Note that the current (I in Figure 10–1) has but one path. Therefore, according to Kirchhoff's current law, the magnitude of the current must be the same at any point in the closed circuit. If, for example, 1 amp leaves point A in the diagram, 1 amp must return to point B at that same instant. The current through the energy source, the lamp, and the conductor must be the same. Such an electrical circuit is called a *series circuit*. A series circuit is a circuit in which there is only one path for current.

The diagram in Figure 10–4 consists of two lamps tied end to end. Again, note that the current has but one path and its magnitude must be the same at any point in that path.

10-1-2 Current in Parallel Circuits

Let us now go back to our water pipe analogy. This time we will add a T to our pipe. At a given instant, if 10 gal/min enter the T, then 10 gal/min must leave at that same instant, as shown in Figure 10–5. The rate of water flowing *into* the T is equal to the rate of water flowing *out* of the T.

The electrical analogy of the T is called a *junction*. A junction is a point in an elec-

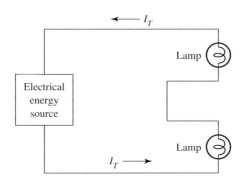

FIGURE 10–4
Two lamps
connected in
series.

FIGURE 10–5
Rate of water
into a junction
must equal rate
of water flowing
out of the
junction.

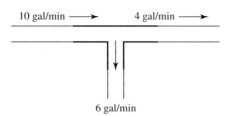

10 gal/min ⟶   4 gal/min ⟶

6 gal/min

trical circuit where three or more conductors meet. Kirchhoff's current law for a junction can be stated as follows:

> ☞ **KEY POINT** The sum of currents toward a junction must equal the sum of currents leaving that junction (as illustrated in Figure 10–6).

Stated another way:

> ☞ **KEY POINT** The algebraic sum of currents into and out of a junction must equal zero if currents toward the junction are given an arbitrary (+) sign and if currents leaving the point are given an arbitrary (−) sign.

In Figure 10–7 the current into the junction is

$$I_1 + I_2 = 2\,A + 3\,A = 5\,A$$

Then, according to Kirchhoff's current law, the current out of the junction (I_3) must equal 5 A. Algebraically,

$$I_1 + I_2 = I_3$$

or
$$I_1 + I_2 - I_3 = 0$$

FIGURE 10–6
Sum of currents
into a junction
must equal the
sum of currents
leaving the
junction.

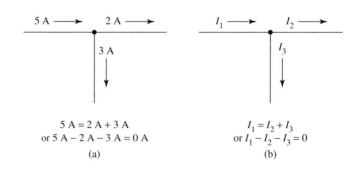

5 A ⟶   2 A ⟶

3 A

$$5\,A = 2\,A + 3\,A$$
or $5\,A - 2\,A - 3\,A = 0\,A$

(a)

I_1 ⟶ I_2 ⟶

I_3

$$I_1 = I_2 + I_3$$
or $I_1 - I_2 - I_3 = 0$

(b)

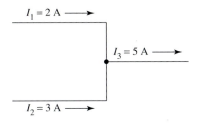

FIGURE 10–7
Current into
and out of the
junction equals
5 Amperes.

In Figure 10–8 the current into the junction, I_1, equals 8 A. Therefore, the current out of the junction must equal 8 A, or

$$I_2 + I_3 + I_4 = 8A$$

$$1\,A + 3\,A + I_4 = 8\,A$$

Therefore,

$$I_4 = 4\,A$$

Algebraically,

$$I_1 = I_2 + I_3 + I_4$$

or $\qquad I_1 - I_2 - I_3 - I_4 = 0$

Solve for I_1 and I_3 in Figure 10–9.

$$I_1 + I_2 + I_3 = I_4 + I_5$$

Then

$$I_1 = I_4 + I_5 - I_3 - I_2$$

and $\qquad I_3 = I_4 + I_5 - I_2 - I_1$

The diagram in Figure 10–10 consists of two lamps connected *across* an energy source. Note that the current, I_T, into junction A must equal the currents out of that junction ($I_1 + I_2$). Therefore, from Kirchhoff's current law,

$$I_T = I_1 + I_2$$

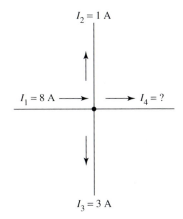

FIGURE 10–8
I_4 must equal 4
Amperes.

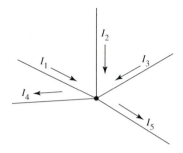

If the current, I_T, in Figure 10–10 is 5 A and I_1 is measured to be 2 A, then I_2 must equal 3 A.

$$I_T = I_1 + I_2$$

$$5 \text{ A} = 2 \text{ A} + I_2$$

$$5 \text{ A} - 2 \text{ A} = I_2$$

$$3 \text{ A} = I_2$$

The branch currents into and out of junction B must also be equal to 5 A. Circuits that have multiple current paths are called *parallel circuits.*

10-1-3 Series-Parallel Circuits

An electrical circuit that has both series- and parallel-circuit characteristics is called a *series–parallel circuit.* Any number of energy sources and loads may make up a series–parallel circuit. For example, Figure 10–11 shows three lamps and an energy source connected in a series–parallel arrangement. The part of the transfer network between points A and B carries the total current. At point B the current divides (parallel circuit): One part flows through lamp L_2 and the remainder flows through lamp L_3. The current recombines at point C, and total current flows in the transfer network between point C and lamp L_1 and on to point D (series circuit). So we have one circuit component (L_1) in which all the current flows (series portion) and two circuit components in which part of the total current flows (parallel portion).

FIGURE 10–10
Two lamps
connected in
parallel.

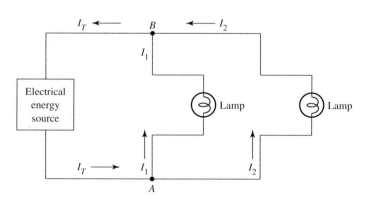

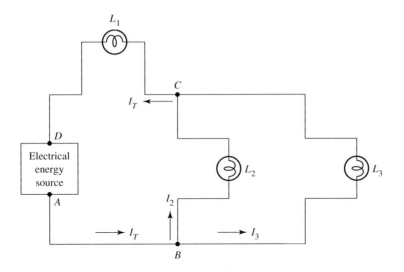

FIGURE 10–11
Series-parallel
circuit.

Figure 10–12 shows another circuit configuration. It is common practice to identify each resistor in a circuit by labeling with numeric subscripts. Figure 10–12 shows four resistors labeled R_1, R_2, R_3, and R_4. It is also common practice to label the current through each resistor with corresponding subscripts. For example: the current through R_1 is I_1, the current through R_2 is I_2 etc. We also identify the voltage drop across each resistor with corresponding subscripts. For example: V_1 is the voltage drop across R_1, V_2 is the voltage drop across R_2, etc. These conventions are used throughout this text-book.

Application of Kirchhoff's current law yields these current relationships:

$$I_T = I_1$$
$$I_T = I_4 + I_3$$
$$I_T = I_4 + I_2$$
$$I_2 = I_3$$
$$I_1 = I_2 + I_4$$

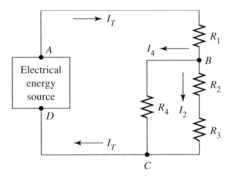

FIGURE 10–12

Inspection of the circuit shows that R_2 and R_3 are in series. Together they are in parallel with R_4. Finally, this whole combination is in series with R_1. In the circuit, then, total current flows through R_1, and a division takes place at point B, resulting in two currents, I_4 and $I_2 (I_T = I_2 + I_4)$. Since R_2 and R_3 are in series, the current through them is the same $(I_2 = I_3)$. A recombination of currents takes place at the junction of R_3 and R_4 (point C). Then total current flows from point C to the electrical energy source (point D).

EXAMPLE 10–1

In Figure 10–13, if $I_T = 1$ A, $I_3 = 0.4$ A, and $I_4 = 0.5$ A, find I_1 and I_2.

FIGURE 10–13

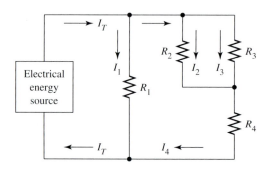

SOLUTION   Using Kirchhoff's current law, we can write equations for various currents:

$$I_T = I_1 + I_4$$

$$I_4 = I_2 + I_3$$

$$I_T = I_1 + I_2 + I_3$$

Choosing an equation containing only one unknown, we get

$$I_T = I_1 + I_4$$

or $$I_1 = I_T - I_4 = 1 \text{ A} - 0.5 \text{ A} = 0.5 \text{ A}$$

and $$I_4 = I_2 + I_3$$

or $$I_2 = I_4 - I_3 = 0.5 \text{ A} - 0.4 \text{ A} = 0.1 \text{ A}$$

PRACTICE PROBLEMS 10–1

Refer to Figure 10–14.

1. Find I_T if $I_1 = 1$ A, $I_2 = 2$ A, and $I_3 = 3$ A.
2. Find I_3 if $I_T = 12$ mA, $I_1 = 1$ mA, and $I_2 = 7$ mA.
3. Find I_2 if $I_T = 10$ μA, $I_1 = 4$ μA, and $I_3 = 2$ μA.
4. Find I_1 if $I_T = 2$ A, $I_2 = 1.2$ A, and $I_3 = 300$ mA.

FIGURE 10–14

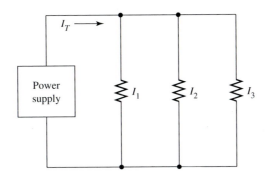

FIGURE 10–15

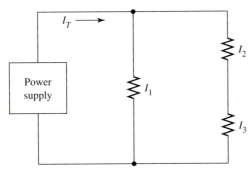

Refer to Figure 10–15.

5. $I_1 = 4$ mA and $I_2 = 3$ mA. Find I_T and I_3.

7. $I_T = 400$ mA and $I_1 = 300$ mA. Find I_2 and I_3.

6. $I_T = 2.3$ mA and $I_3 = 600$ μA. Find I_1 and I_2.

Refer to Figure 10–12.

8. $I_1 = 400$ mA, $I_2 = 150$ mA. Find I_T, I_3, and I_4.

9. $I_3 = 600$ μA, $I_4 = 800$ μA. Find I_T, I_1, and I_2.

Refer to Figure 10–13.

10. $I_T = 17$ mA, $I_2 = 6$ mA, $I_4 = 13$ mA. Find I_1 and I_3.

11. $I_1 = 500$ μA, $I_2 = 1.3$ mA, $I_3 = 850$ μA. Find I_T and I_4.

Refer to Figure 10–16.

12. In Figure 10–16: I_T is 11 A, I_1 is 6 A, and I_3 is 5 A. Does current flow through R_5 from A to B or from B to A? Find I_5, I_2, and I_4.

13. In Figure 10–16: I_T is 100 mA, I_1 is 45 mA, and I_3 is 55 mA. Does current flow through R_5 from A to B or from B to A? Find I_5, I_2 and I_4.

FIGURE 10–16

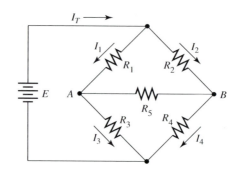

14. In Figure 10–16: I_3 is 200 mA, I_4 is 200 mA, and I_5 is 10 mA. The current flows through R_5 from A to B. Find I_T, I_1, and I_2.

SOLUTIONS

1. 6 A
3. 4 μA
5. $I_T = 7$ mA, $I_3 = 3$ mA
7. $I_2 = 100$ mA, $I_3 = 100$ mA

9. $I_T = 1.4$ mA, $I_1 = 1.4$ mA, $I_2 = 600$ μA
11. $I_T = 2.65$ mA, $I_4 = 2.15$ mA

13. From B to A. $I_5 = I_1 - I_3 = -10$ mA. $I_2 = I_T - I_1 = 55$ mA. $I_4 = I_2 + I_5 = 45$ mA.

2. 4 mA
4. 500 mA
6. $I_1 = 1.7$ mA, $I_2 = 600$ μA
8. $I_T = 400$ mA, $I_3 = 150$ mA, $I_4 = 250$ mA
10. $I_1 = 4$ mA, $I_3 = 7$ mA

12. From A to B. $I_5 = I_1 - I_3 = 1$ A. $I_2 = I_T - I_1 = 5$ A. $I_4 = I_2 + I_5 = 6$ A.
14. $I_T = I_3 + I_4 = 400$ mA. $I_1 = I_3 + I_5 = 210$ mA. $I_2 = I_4 - I_5 = 190$ mA.

Additional practice problems are at the end of the chapter.

10–2   KIRCHHOFF'S VOLTAGE LAW

If a charge travels around any closed path in an electrical network and returns to its starting point, its net potential energy change must be zero. That is, it must go through equal potential energy gains and potential energy losses.

Kirchhoff's voltage law expresses the law of conservation of energy and can be stated as follows:

☞ **KEY POINT** Around any closed path the sum of potential rises must equal the sum of potential drops ($\Sigma E = \Sigma V$).

Potential *rises* are the electrical energy sources and could be the 110 volts supplied by the local electric company, a battery, or some other source. A potential *rise* is also called an electromotive force (emf), source voltage, or applied voltage. The symbol is E and the unit is the volt.

Potential *drops* refer to the energy used in causing current through the energy users, like the lamps in Figure 10–17. These potential drops are also called *voltage drops*. The symbol is V, and the unit is the volt.

Stated another way:

> ☞ **KEY POINT** The algebraic sum of emf's (E) and voltage drops (V) around a closed path is equal to zero ($\Sigma E - \Sigma V = 0$).

In the circuit of Figure 10–17, both L_1 and L_2 are connected *across* the electrical energy source. Kirchhoff's voltage law tells us that in any closed path the emf's and voltage drops must be equal. For loop 1, $E = V_1$ (the voltage drop across L_1), and for loop 2, $E = V_2$ (the voltage drop across L_2). If $E = V_1 = V_2$, then $V_1 = V_2$.

There are two ways of referring to the voltages in a circuit. We can refer to the voltage drop across each component or we can refer to the voltage value at each place in a circuit relative to some reference point (see Fig. 10–17). Both designations are used in electronics. The reference point is often called "ground," "earth," or "common."

Consider a single charge at the reference point in the circuit. The electrical energy source raises the potential energy of the charge to 10 V at point A. For the charge to return to the reference point, it must travel through lamp 1 or lamp 2. No matter which path it takes, it must give up the potential energy that the electrical energy source provided it. Therefore, in parallel circuits the voltage drop is the same for each branch and equals the emf.

From this and previous discussions we may conclude the following:

> ☞ **KEY POINT** In a parallel circuit, (1) the voltage drop is the same across each branch; $V_1 = V_2 = V_N$; (2) the total current into and out of the branches equals the sum of the individual branch currents: $I_T = I_1 + I_2 + \cdots + I_N$.

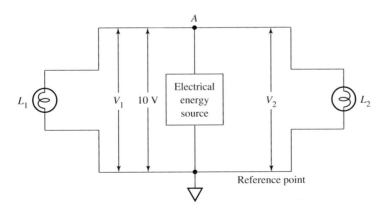

FIGURE 10–17
Both lamps are connected across the energy source.

In Figure 10–18 we have an electrical energy source connected in series with two lamps. Once again, let's consider a single charge at the reference point. The electrical energy source raises the potential energy of the charge to 10 V (point A). In this circuit there is only one path the charge may take. It must pass through lamp 1 and then through lamp 2. When it has passed through lamp 1, it has converted some, but not all, of the energy provided by the electrical energy source. The remainder of its energy is converted when it travels through lamp 2. The charge drops across a potential difference of 6 V for lamp 1. Since it drops across a total potential difference of 10 V, there remains 4 V to drop across lamp 2. When the charge finally passes through lamp 2, all the potential energy provided to it by the electrical energy source is converted, and the charge has made a complete trip around the circuit and returned to the reference point.

From this and previous discussions we may conclude the following:

☞ **KEY POINT** In a series circuit, (1) the current is the same throughout the circuit: $I_T = I_1 = I_2 = I_N$; (2) the sum of the voltage drops must equal the emf: $E = V_1 + V_2 + \cdots + V_N$.

According to Kirchhoff's voltage law, the total rise in potential (E) in Figure 10–18 must equal the total drop in potential:

$$E = V_1 + V_2$$

If, for the above example, the potential rise is equal to 10 V, then the sum of potential drops ($V_1 + V_2$) must also equal 10 V.

FIGURE 10–18
The sum of the voltage drops ($V_1 + V_2$) equals the source voltage (E).

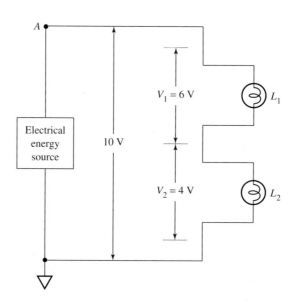

CHAPTER 10

EXAMPLE 10–2

In Figure 10–18, find the magnitude of V_2 if $E = 30$ V and V_1 equals 10 V.

SOLUTION

$$E = V_1 + V_2$$

Then
$$V_2 = E - V_1$$
$$V_2 = 30 \text{ V} - 10 \text{ V} = 20 \text{ V}$$

EXAMPLE 10–3

In Figure 10–19, determine V_2.

FIGURE 10–19

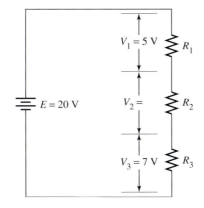

SOLUTION

$$E = V_1 + V_2 + V_3$$

Then
$$V_2 = E - V_1 - V_3 = 20 \text{ V} - 5 \text{ V} - 7 \text{ V} = 8 \text{ V}$$

EXAMPLE 10–4

In Figure 10–20 on the next page, determine V_4.

SOLUTION

$$E = V_1 + V_2 + V_3 + V_4$$

Then

$$V_4 = E - V_1 - V_2 - V_3 = 25 \text{ V} - 12 \text{ V} - 8 \text{ V} - 1 \text{ V} = 4 \text{ V}$$

In Figure 10–21, Kirchhoff's voltage law is satisfied because part of the source voltage is dropped across the circuit between points A and B (across R_2 and R_3) and part is dropped across R_1.

FIGURE 10–20

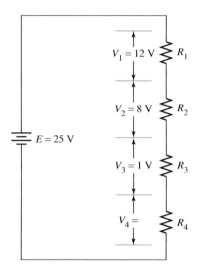

$$E = V_1 + V_2$$

and
$$E = V_1 + V_3$$

Since there are two paths, either may be chosen in writing the equation. In Figure 10–22, application of Kirchhoff's voltage law yields

$$E = V_1 + V_2 + V_3$$

or
$$E = V_1 + V_4$$

We can write it either way since R_4 is in parallel with the series circuit consisting of R_2 and R_3. Because of this relationship between R_2, R_3 and R_4, we can also say that

$$V_4 = V_2 + V_3$$

Recalling that *the algebraic sum of emf's and voltage drops around a closed path is equal to zero,* we could also write the equations like this:

$$E - V_1 - V_2 - V_3 = 0$$

and
$$E - V_1 - V_4 = 0$$

also
$$V_4 - V_2 - V_3 = 0$$

FIGURE 10–21

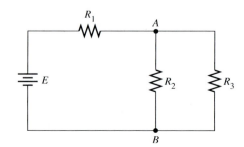

FIGURE 10–22

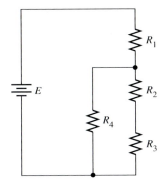

That is, we may choose *any* closed path (or loop), and the sum of the drops and emf's in that loop must equal zero.

EXAMPLE 10–5

In Figure 10–22, let $E = 10$ V, $V_2 = 5$ V and $V_3 = 2$ V. Find V_1 and V_4.

SOLUTION

$$V_4 = V_2 + V_3$$
$$V_4 = 5 \text{ V} + 2 \text{ V} = 7 \text{ V}$$
$$E = V_1 + V_2 + V_3$$
$$V_1 = E - V_2 - V_3 = 10 \text{ V} - 5 \text{ V} - 2 \text{ V} = 3 \text{ V}$$

PRACTICE PROBLEMS 10–2

1. In Figure 10–23, if $E = 25$ V and $V_1 = 10$ V, find V_2.
3. In Figure 10–19, if $V_1 = 3$ V, $V_2 = 30$ V, and $V_3 = 10$ V, find E.

2. In Figure 10–19, if $E = 20$ V, $V_1 = 5$ V, and $V_2 = 7$ V, find V_3.

FIGURE 10–23

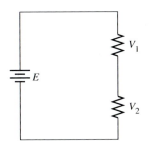

FIGURE 10–24

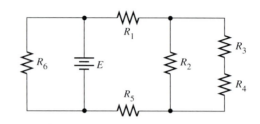

Refer to Figure 10–21.

4. $E = 50$ V, $V_3 = 20$ V. Find V_1 and V_2.
6. $V_1 = 7$ V, $V_3 = 14$ V. Find E and V_2.

5. $E = 50$ V, $V_1 = 15$ V. Find V_2 and V_3.

Refer to Figure 10–22.

7. $E = 10$ V, $V_2 = 5$ V, $V_3 = 2$ V. Find V_1 and V_4.
9. $V_1 = 7$ V, $V_2 = 5$ V, $V_4 = 8$ V. Find E and V_3.

8. $E = 150$ mV, $V_2 = 30$ mV, $V_4 = 80$ mV. Find V_1 and V_3.

Refer to Figure 10–24.

10. $E = 35$ V, $V_1 = 12$ V, $V_3 = 10$ V, $V_5 = 5$ V. Find V_2, V_4 and V_6.

SOLUTIONS

1. 15 V
3. 43 V
5. $V_2 = 35$ V, $V_3 = 35$ V
7. $V_1 = 3$ V, $V_4 = 7$ V
9. $E = 15$ V, $V_3 = 3$ V

2. 8 V
4. $V_1 = 30$ V, $V_2 = 20$ V
6. $E = 21$ V, $V_2 = 14$ V
8. $V_1 = 70$ mV, $V_3 = 50$ mV
10. $V_2 = 18$ V, $V_4 = 8$ V, $V_6 = 35$ V

10–3   POLARITY

☞ **KEY POINT** In an electrical circuit, the polarity of the voltage drops across the individual resistances depends on the direction of current through them.

If we use electron current, as in Figure 10–25, the polarities are as shown. The end of the resistor where current enters is assigned a negative (−) sign. This makes the other end positive (+). If we use conventional current, as in Figure 10–26, the end where current enters is assigned a positive (+) sign. This makes the other end negative (−). We see in Figures 10–25 and 10–26 that the resultant polarities are the same regardless of whether we use electron current or conventional current. In our circuits polarities are assigned, but usually we choose the direction of current with which we are most comfortable.

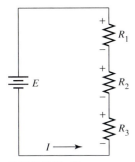

FIGURE 10–25
Series circuit showing direction of electron current.

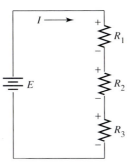

FIGURE 10–26
Series circuit showing direction of conventional current.

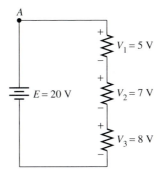

FIGURE 10–27   In a series circuit the algebraic sum of the emf's and voltage drops must equal zero.

In Figure 10–27, polarities are shown as dictated by the direction of current. Suppose that $V_1 = 5$ V, $V_2 = 7$ V, and $V_3 = 8$ V. The general equation states that

$$E + V_1 + V_2 + V_3 = 0$$

Observing polarity, let's start at point A and go around the loop in a counterclockwise direction. Using the sign we encounter first as we go through each source and resistance, we get

$$+20\text{ V} - 8\text{ V} - 7\text{ V} - 5\text{ V} = 0$$

$$20\text{ V} - 20\text{ V} = 0$$

In Figure 10–28, R_2 and R_3 are in parallel and they are in series with R_1. Because R_2 and R_3 are in parallel, $V_2 = V_3$. One loop includes E, R_1, and R_2. Therefore, $E = V_1 + V_2$. A second loop includes E, R_1, and R_3. Therefore, $E = V_1 + V_3$. In each case, 25 V = 15 V + 10 V.

Let's find V_2 and V_4 in Figure 10–29. One loop includes E, R_1, R_2, and R_4. Since we have two unknown voltage drops in this loop, we can't use it. Another loop includes E, R_3, and R_4. In this loop the only unknown is V_4.

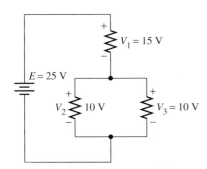

FIGURE 10–28   Series–parallel circuit with polarities shown.

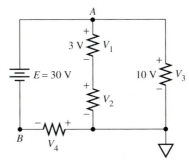

FIGURE 10–29   The loops used need not include an energy source.

$$E = V_3 + V_4$$

$$V_4 = E - V_3 = 30 \text{ V} - 10 \text{ V} = 20 \text{ V}$$

Now that we know V_4, we can use the loop that includes E, R_1, and R_4 to find V_2.

$$E = V_1 + V_2 + V_4$$

$$V_2 = E - V_1 - V_4$$

$$= 30 \text{ V} - 3 \text{ V} - 20 \text{ V} = 30 \text{ V} - 23 \text{ V} = 7 \text{ V}$$

We could also use the loop consisting of R_1, R_2, and R_3 to find V_2. Observing polarity and moving around the loop clockwise from R_2, we get

$$-V_2 - V_1 + V_3 = 0$$

$$V_2 = V_3 - V_1$$

$$= 10 \text{ V} - 3 \text{ V} = 7 \text{ V}$$

The loop we use may or may not include a source.

Suppose we were interested in the potential at one point in a circuit with reference to another point. For example, in Figure 10–29, we might ask the question, "What is the potential at point A with respect to common?" "With respect to" means "with reference to." Said another way, "If common is our reference point, what is the potential at point A?" The answer is 10 V. Point A is 10 V more positive than common. This can be determined three different ways:

1. The circuit from point A to common through R_3 shows that point A is the positive side of R_3 and common is the negative side. The drop across R_3 is 10 V; so point A is 10 V positive with respect to common.
2. Starting at point A and going to common through the path consisting of R_1 and R_2 yields $+3 \text{ V} + 7 \text{ V} = 10 \text{ V}$.
3. Starting at point A and going to common through the path consisting of E and R_4, we get $+30 \text{ V} - 20 \text{ V} = 10 \text{ V}$.

In each case we observed polarity and added (algebraically) the emf's and voltage drops.

What is the potential at point B with respect to common? The answer is -20 V. The easiest way to determine this is to observe the polarity of the drop across R_4. We could arrive at the same answer by using the path consisting of E and R_3 or by using the path consisting of E, R_1, and R_2. Using E and R_3, we get $-30 \text{ V} + 10 \text{ V} = -20 \text{ V}$. Using E, R_1, and R_2, we get $-30 \text{ V} + 3 \text{ V} + 7 \text{ V} = -20 \text{ V}$.

EXAMPLE 10–6

In Figure 10–30,

(a) What is the drop across R_2 and what is its polarity with respect to common?
(b) What is the drop across R_3 and what is its polarity with respect to common?
(c) What is the potential at point A with respect to common?

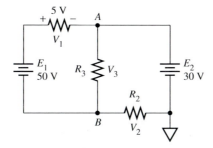

FIGURE 10–30

SOLUTIONS

(a) The drop across $R_2(V_2)$ can be found by going around the outer loop. The other choice is the loop consisting of E_2, R_2, and R_3, but we don't know V_3; therefore, we can't use that loop. The outer loop consists of E_1, E_2, R_1, and R_2.

   Let's start at the left side of R_2, which is point B, and go around the loop in a clockwise direction. The general equation for the loop is

$$E_1 + V_1 + E_2 + V_2 = 0$$

Observing polarity, we get

$$-50 \text{ V} + 5 \text{ V} + 30 \text{ V} + V_2 = 0$$

Notice that a positive polarity was assigned to V_2. Because we don't know the polarity of V_2 yet, we give it a + sign until we solve the equation. Solving, we get

$$-15 \text{ V} + V_2 = 0$$

$$V_2 = 15 \text{ V}$$

Notice that V_2 is positive. This tells us that the first sign we encounter when we get to R_2 is a + sign. We went in a clockwise direction, so the + sign must be assigned to the right side of R_2 (which makes the left side negative), as shown in Figure 10–31. If we had gone in a counterclockwise direction from point B, we would have this equation:

$$V_2 + E_2 + V_1 + E_1 = 0$$

FIGURE 10–31

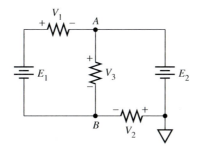

Assigning polarities and values, we get

$$V_2 - 30 \text{ V} - 5 \text{ V} + 50 \text{ V} = 0$$
$$V_2 + 15 \text{ V} = 0$$
$$V_2 = -15 \text{ V}$$

Going counterclockwise we see that V_2 is *negative* 15 V. That is, the left side of R_2 is negative (which makes the right side positive) and agrees with our first method.

(b) Knowing both V_1 and V_2 allows us to choose either the loop that includes R_3, E_1, and R_1 or the loop that includes R_3, R_2, and E_2. Let's use the loop that includes R_3, E_1, and R_1. Starting at point A and moving clockwise (remember we can go in either direction), we get

$$V_3 + E_1 + V_1 = 0 \qquad \text{(general equation)}$$
$$V_3 - 50 \text{ V} + 5 \text{ V} = 0 \qquad \text{(plug in known values and observe polarity)}$$
$$V_3 - 45 \text{ V} = 0$$
$$V_3 = 45 \text{ V}$$

We started at point A. Since the first sign we encountered was a + sign ($V_3 = 45$ V), the polarity is as shown in Figure 10–31.

(c) Notice that there are three paths from point A to common. One is through E_2. Another is through R_2 and R_3. A third is through R_1, E_1, and R_2. No matter which is chosen, the result should show that point A is 30 V positive with respect to common.

$$P_1A \text{ to common} = E_2 = 30 \text{ V}$$
$$= V_3 + V_2 = 45 \text{ V} - 15 \text{ V} = 30 \text{ V}$$
$$= V_1 + E_1 + V_2 = -5 \text{ V} + 50 \text{ V} - 15 \text{ V}$$
$$= 30 \text{ V}$$

PRACTICE PROBLEMS 10–3

1. Refer to Figure 10–29. Let $V_1 = 7$ V and $V_3 = 15$ V. Find (a) V_2, (b) V_4, (c) potential at point A with respect to common, (d) potential at point A with respect to point B, and (e) potential at point B with respect to common.

2. Refer to Figure 10–32. Find (a) V_1, (b) V_3, (c) potential at point A with respect to common, (d) potential at point B with respect to common, and (e) potential at point A with respect to point B.

3. Refer to Figure 10–33. Find (a) V_3, (b) V_4, (c) potential at point A with respect to common, and (d) potential at point B with respect to point A.

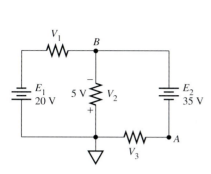

FIGURE 10–32

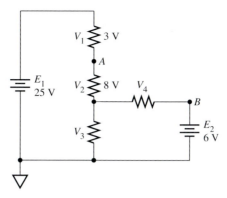

FIGURE 10–33

SOLUTIONS

1. (a) 8 V   (b) 15 V   (c) 15 V   (d) 30 V   (e) −15 V
2. (a) 15 V   (b) 40 V   (c) −40 V   (d) −5 V   (e) −35 V
3. (a) 14 V   (b) 8 V   (c) 22 V   (d) −16 V

Additional practice problems are at the end of the chapter.

SELF-TEST 10–1

1. Refer to Figure 10–14. Find I_3 if $I_T = 16$ mA, $I_1 = 4$ mA, and $I_2 = 2$ mA.
2. Refer to Figure 10–15. $I_1 = 40$ μA and $I_2 = 60$ μA. Find I_T and I_3.
3. Refer to Figure 10–15. $I_T = 3$ mA and $I_2 = 2.6$ mA. Find I_1 and I_3.
4. Refer to Figure 10–12. $I_T = 800$ μA, $I_2 = 308$ μA. Find I_1, I_3, and I_4.
5. Refer to Figure 10–13. $I_1 = 15$ mA, $I_2 = 5$ mA, $I_3 = 7$ mA. Find I_T and I_4.
6. Refer to Figure 10–19. $E = 30$ V, $V_1 = 7$ V, $V_2 = 10$ V. Find V_3.
7. Refer to Figure 10–19. $V_1 = 200$ mV, $V_2 = 1.3$ V, $V_3 = 400$ mV. Find E.
8. Refer to Figure 10–21. $E = 20$ V, $V_1 = 7$ V. Find V_2 and V_3.
9. Refer to Figure 10–22. $V_1 = 3$ V, $V_2 = 10$ V, $V_4 = 12$ V. Find E and V_3.
10. Refer to Figure 10–34. Find V_2 and V_3.
11. Refer to Figure 10–29. Let $V_1 = 13$ V and $V_3 = 21$ V. Find (a) V_2, (b) V_4, (c) potential at point B with respect to point A, and (d) potential at point A with respect to common.
12. Refer to Figure 10–33. Let $V_1 = 4$ V and $V_2 = 12$ V. Find (a) V_3, (b) V_4, (c) potential at common with respect to point A, and (d) potential at point B with respect to point A.

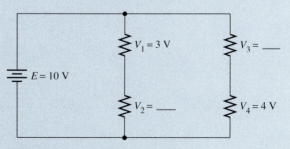

FIGURE 10–34

Answers to Self-test 10–1 are at the end of the chapter.

CHAPTER 10 AT A GLANCE

PAGE	KEY POINT	EXAMPLE
276	*Key Point:* The number of charges per second (current) flowing *to* a point in a conductor must equal the number of charges flowing *from* that point. (See Figure 10–35.)	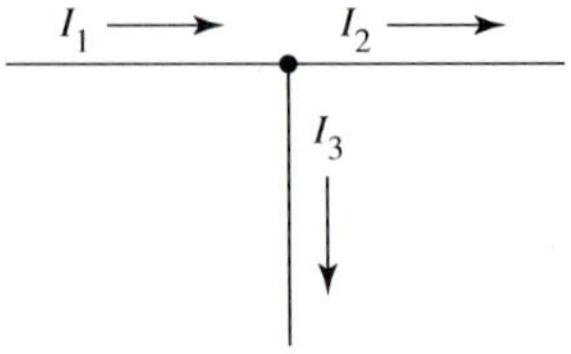 $I_1 = I_2 + I_3$ or $I_1 - I_2 - I_3 = 0$ FIGURE 10–35
284	*Key Point:* Around any closed path, the sum of potential rises must equal the sum of potential drops. $\Sigma E = \Sigma V$.	$E = V_1 + V_2$
285	*Key Point:* In a parallel circuit: (1) the voltage drop is the same across each branch; (2) the total current into and out of the branches equals the sum of the individual branch currents.	$V_1 = V_2 = V_N$ $I_T = I_1 + I_2 + \cdots + I_N$
286	*Key Point:* In a series circuit: (1) the current is the same throughout the circuit; (2) the sum of the voltage drops must equal the emf.	$I_T = I_1 = I_2 = I_N$ $E = V_1 + V_2 + \cdots + V_N$
290	*Key Point:* In an electrical circuit the polarity of the voltage drops across the individual resistances depends on the direction of current through them. (See Figure 10–36.)	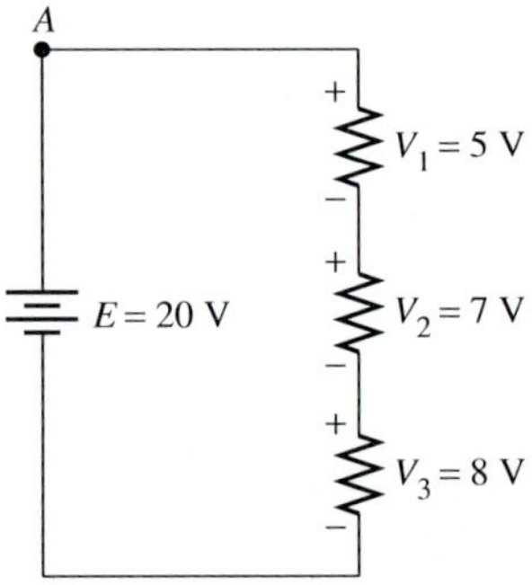 $E = 20$ V, $V_1 = 5$ V, $V_2 = 7$ V, $V_3 = 8$ V FIGURE 10–36

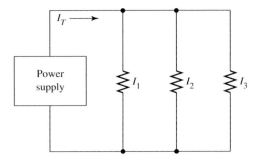

FIGURE 10–37

END OF CHAPTER PROBLEMS 10–1

Refer to Figure 10–37.

1. Find I_T if $I_1 = 3$ A, $I_2 = 2$ A, and $I_3 = 4$ A.
2. Find I_T if $I_1 = 10$ mA, $I_2 = 15$ mA, and $I_3 = 5$ mA.
3. Find I_3 if $I_T = 20$ mA, $I_1 = 6$ mA, and $I_2 = 11$ mA.
4. Find I_3 if $I_T = 450$ mA, $I_1 = 100$ mA, and $I_2 = 150$ mA.
5. Find I_2 if $I_T = 65$ μA, $I_1 = 10$ μA, and $I_3 = 10$ μA.
6. Find I_2 if $I_T = 7$ mA, $I_1 = 500$ μA, and $I_3 = 3.8$ mA.
7. Find I_1 if $I_T = 1.3$ A, $I_2 = 300$ mA, and $I_3 = 40$ mA.
8. Find I_1 if $I_T = 400$ mA, $I_2 = 120$ mA, and $I_3 = 180$ mA.

Refer to Figure 10–38.

9. $I_1 = 10$ mA and $I_2 = 15$ mA. Find I_T and I_3.
10. $I_1 = 600$ μA and $I_2 = 1.5$ mA. Find I_T and I_3.
11. $I_T = 1.6$ A and $I_3 = 800$ mA. Find I_1 and I_2.
12. $I_T = 400$ mA and $I_3 = 150$ mA. Find I_1 and I_2.
13. $I_T = 600$ μA and $I_1 = 400$ μA. Find I_2 and I_3.
14. $I_T = 2.1$ mA and $I_1 = 700$ μA. Find I_2 and I_3.

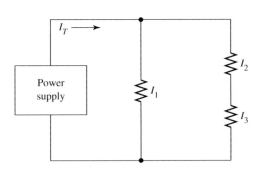

FIGURE 10–38

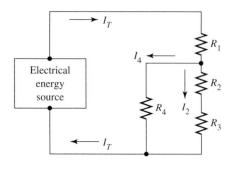

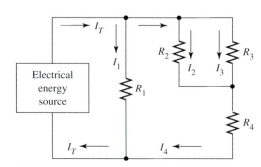

FIGURE 10–39   FIGURE 10–40

Refer to Figure 10–39.

15. $I_1 = 1.4$ A, $I_2 = 800$ mA. Find I_T, I_3, and I_4.

16. $I_1 = 750$ mA, $I_2 = 350$ mA. Find I_T, I_3, and I_4.

17. $I_3 = 500$ μA, $I_4 = 2.7$ mA. Find I_T, I_1, and I_2.

18. $I_3 = 3$ mA, $I_4 = 4.2$ mA. Find I_T, I_1, and I_2.

Refer to Figure 10–40.

19. $I_T = 6.7$ mA, $I_2 = 1.2$ mA, $I_4 = 5$ mA. Find I_1 and I_3.

20. $I_T = 224$ mA, $I_2 = 65$ mA, $I_4 = 160$ mA. Find I_1 and I_3.

21. $I_1 = 6$ mA, $I_3 = 3.4$ mA, $I_4 = 8.2$ mA. Find I_T and I_2.

22. $I_1 = 760$ mA, $I_3 = 340$ mA, $I_4 = 465$ mA. Find I_T and I_2.

23. Refer to Figure 10–24. $I_T = 100$ mA, $I_2 = 20$ mA, and $I_5 = 30$ mA. Find I_1, I_3, I_4, and I_6.

END OF CHAPTER PROBLEMS 10–2

Refer to Figure 10–41.

1. $E = 30$ V and $V_1 = 12$ V. Find V_2.
3. $V_1 = 7$ V and $V_2 = 6$ V. Find E.

2. $E = 12$ V and $V_1 = 4$ V. Find V_2.
4. $V_1 = 750$ mV and $V_2 = 1.2$ V. Find E.

FIGURE 10–41

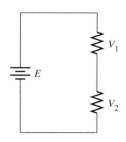

Refer to Figure 10–42.

5. $E = 70$ V, $V_1 = 22$ V, $V_2 = 37$ V. Find V_3.
7. $V_1 = 3$ V, $V_2 = 8$ V, $V_3 = 7$ V. Find E.

9. $E = 37$ V, $V_2 = 20$ V, $V_3 = 3$ V. Find V_1.

6. $E = 15$ V, $V_1 = 2$ V, $V_2 = 6$ V. Find V_3.
8. $V_1 = 40$ V, $V_2 = 22$ V, $V_3 = 30$ V. Find E.
10. $E = 55$ V, $V_2 = 15$ V, $V_3 = 20$ V. Find V_1.

Refer to Figure 10–43.

11. $E = 20$ V, $V_3 = 8$ V. Find V_1 and V_2.

13. $E = 18$ V, $V_1 = 5$ V. Find V_2 and V_3.
15. $V_1 = 12$ V, $V_3 = 5$ V. Find E and V_2.

12. $E = 8.5$ V, $V_3 = 3.6$ V. Find V_1 and V_2.
14. $E = 40$ V, $V_1 = 27$ V. Find V_2 and V_3.
16. $V_1 = 450$ mV, $V_3 = 1.1$ V. Find E and V_2.

Refer to Figure 10–44.

17. $E = 12$ V, $V_2 = 3$ V, $V_3 = 6$ V. Find V_1 and V_4.
19. $E = 3$ V, $V_2 = 500$ mV, $V_4 = 1.3$ V. Find V_1 and V_3.
21. $V_1 = 20$ V, $V_2 = 30$ V, $V_4 = 40$ V. Find E and V_3.

18. $E = 30$ V, $V_2 = 8$ V, $V_3 = 7$ V. Find V_1 and V_4.
20. $E = 35$ V, $V_2 = 10$ V, $V_4 = 16$ V. Find V_1 and V_3.
22. $V_1 = 3.8$ V, $V_2 = 1.6$ V, $V_4 = 4.6$ V. Find E and V_3.

Refer to Figure 10–45.

23. $E = 50$ V, $V_1 = 10$ V, $V_3 = 8$ V, $V_5 = 15$ V. Find V_2, V_4, and V_6.
25. $V_1 = 9$ V, $V_4 = 6.5$ V, $V_5 = 4$ V, $V_6 = 25$ V. Find E, V_2, and V_3.

24. $E = 25$ V, $V_1 = 6.7$ V, $V_3 = 3.2$ V, $V_5 = 4.7$ V. Find V_2, V_4, and V_6.
26. $V_1 = 12$ V, $V_4 = 12$ V, $V_5 = 12$ V, $V_6 = 50$ V. Find E, V_2, and V_3.

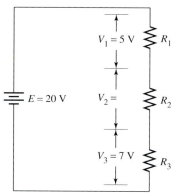

FIGURE 10–42

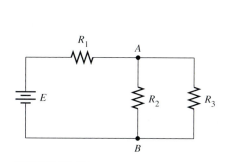

FIGURE 10–43

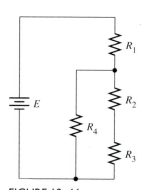

FIGURE 10–44

FIGURE 10–45

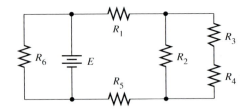

END OF CHAPTER PROBLEMS 10–3

1. Refer to Figure 10–46. Let $V_1 = 15$ V and $V_3 = 25$ V. Find (a) V_2, (b) V_4, (c) potential at point A with respect to common, (d) potential at point A with respect to point B, and (e) potential at point B with respect to common.

2. Refer to Figure 10–46. Let $V_1 = 6$ V and $V_3 = 18$ V. Find (a) V_2, (b) V_4, (c) potential at point A with respect to common, (d) potential at point A with respect to point B, and (e) potential at point B with respect to common.

3. Refer to Figure 10–47. Let $V_2 = 8$ V. Find (a) V_1, (b) V_3, (c) potential at point A with respect to common, (d) potential at point A with respect to point B, and (e) potential at point B with respect to common.

4. Refer to Figure 10–47. Let $V_2 = 10$ V. Find (a) V_1, (b) V_3, (c) potential at point A with respect to common, (d) potential at point A with respect to point B, and (e) potential at point B with respect to common.

5. Refer to Figure 10–48. Let $V_1 = 6$ V and $V_2 = 8$ V. Find (a) V_3, (b) V_4, (c) potential at point A with respect to common, and (d) potential at point B with respect to point A.

6. Refer to Figure 10–48. Let $V_1 = 3$ V and $V_2 = 5$ V. Find (a) V_3, (b) V_4, (c) potential at point A with respect to common, and (d) potential at point B with respect to point A.

7. Refer to Figure 10–49. Let $E_1 = 40$ V, $E_2 = 10$ V, and $V_3 = 6$ V with polarity as shown. Find (a) V_1, (b) V_2, (c) potential at point A with respect to common, and (d) potential at point C with respect to point B.

8. Refer to Figure 10–49. Let $E_1 = 30$ V, $E_2 = 5$ V, and $V_3 = 7$ V with polarity as shown. Find (a) V_1, (b) V_2, (c) potential at point A with respect to common, and (d) potential at point C with respect to point B.

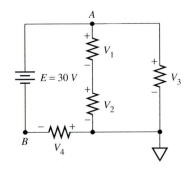

FIGURE 10–46

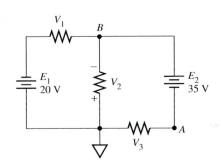

FIGURE 10–47

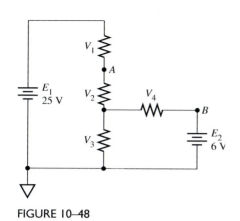

FIGURE 10–48

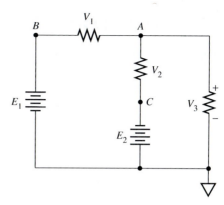

FIGURE 10–49

ANSWERS TO SELF-TEST

SELF-TEST 10–1

1. 10 mA

2. $I_T = 100\ \mu A,$
   $I_3 = 60\ \mu A$

3. $I_1 = 400\ \mu A,$
   $I_3 = 2.6$ mA

4. $I_1 = 800\ \mu A,$
   $I_3 = 308\ \mu A,$
   $I_4 = 492\ \mu A$

5. $I_T = 27$ mA,
   $I_4 = 12$ mA

6. 13 V

7. 1.9 V

8. $V_2 = 13$ V, $V_3 = 13$ V

9. $E = 15$ V, $V_3 = 2$ V

10. $V_2 = 7$ V, $V_3 = 6$ V

11. (a) 8 V   (b) 9 V
    (c) −30 V   (d) 21 V

12. (a) 9 V   (b) 3 V
    (c) −21 V
    (d) −15 V

dc Circuit Analysis: Ohm's Law

Introduction

In Chapter 10 we learned about the relationship between voltage drops and the source voltage in series, parallel, and series–parallel circuits. These relationships were developed using Kirchhoff's voltage law. Kirchhoff's current law explained to us the relationship between the source current and the individual branch currents in these same circuits. Now that we understand these relationships, we are ready to learn how to compute these voltage drops and currents and how voltage, current, and resistance relate one to the other.

Ohm's law expresses the relationship between voltage, current, and resistance in an electrical circuit. Ohm's law states that circuit current is directly proportional to voltage and inversely proportional to resistance. Expressed as an equation, Ohm's law is

$$I = \frac{E}{R}$$

where E is the source voltage, I is the circuit current, and R is the circuit resistance. If we know the source voltage and the circuit resistance, we can calculate the circuit current. If E and I are known, we can rearrange the equation to solve for R.

$$R = \frac{E}{I}$$

If I and R are known, we can solve for E.

$$E = IR$$

We must always know two of the variables in order to solve for the third.

These are the three basic forms of Ohm's law. In this chapter we will learn how to apply these equations in the solution of various kinds of electrical circuit problems. Ohm's law and Kirchhoff's laws are the basic laws that express the relationship between voltage, current, and resistance in electrical circuits. A thorough understanding of these laws and how to use them is essential to the study of electronics.

Chapter Objectives

In this chapter you will learn how to:

1. Compute total resistance in a series circuit.
2. Compute total conductance in a parallel circuit.
3. Relate between conductance and resistance.
4. Compute total resistance in a series–parallel circuit.
5. Use Ohm's law and Kirchhoff's laws to solve series, parallel, and series–parallel circuits.
6. Use the calculator in solving circuit problems.

CIRCUIT RESISTANCE IN SERIES AND PARALLEL CIRCUITS | 11-1

Before we get into solving circuit problems in terms of voltage drops and currents, let's learn how to calculate circuit resistance. If there is only one resistance in an electrical circuit, the total circuit resistance is equal to that resistance. Such a circuit is shown in Figure 11–1(a). $R_T = R$. In Figure 11–1(b), there are two resistances in series. To identify which resistance we are talking about, they are labeled R_1 and R_2.

☞ **KEY POINT** In a series circuit the total resistance equals the sum of the individual resistances.

$$R_T = R_1 + R_2 + R_3 + \cdots + R_N$$

Therefore, in Figure 11–1(b), $R_T = R_1 + R_2$.

EXAMPLE 11–1

In Figure 11–1(c), let $R_1 = 2.7 \text{ k}\Omega$ and $R_2 = 4.7 \text{ k}\Omega$. Find R_T.

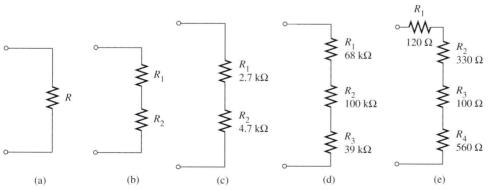

FIGURE 11–1 Resistors connected in series. The total resistance equals the sum of the individual resistances.

(a) (b) (c) (d) (e)

SOLUTION

$$R_T = R_1 + R_2$$
$$= 2.7 \text{ k}\Omega + 4.7 \text{ k}\Omega = 7.4 \text{ k}\Omega$$

EXAMPLE 11–2

In Figure 11–1(d), let $R_1 = 68 \text{ k}\Omega$, $R_2 = 100 \text{ k}\Omega$, and $R_3 = 39 \text{ k}\Omega$. Find R_T.

SOLUTION

$$R_T = R_1 + R_2 + R_3$$
$$= 68 \text{ k}\Omega + 100 \text{ k}\Omega + 39 \text{ k}\Omega = 207 \text{ k}\Omega$$

EXAMPLE 11–3

In Figure 11–1(e), let $R_1 = 120 \text{ }\Omega$, $R_2 = 330 \text{ }\Omega$, $R_3 = 100 \text{ }\Omega$, and $R_4 = 560 \text{ }\Omega$. Find R_T.

SOLUTION

$$R_T = R_1 + R_2 + R_3 + R_4$$
$$= 120 \text{ }\Omega + 330 \text{ }\Omega + 100 \text{ }\Omega + 560 \text{ }\Omega = 1.11 \text{ k}\Omega$$

In Figure 11–2(a), R_1 and R_2 are connected in parallel. Connecting resistances in parallel *reduces* total circuit resistance. This is so because as parallel paths are added, more circuit current flows. If total circuit current is increased, then total circuit resistance must decrease.

FIGURE 11–2
Resistors connected in parallel. The total conductance equals the sum of the individual branch conductances.

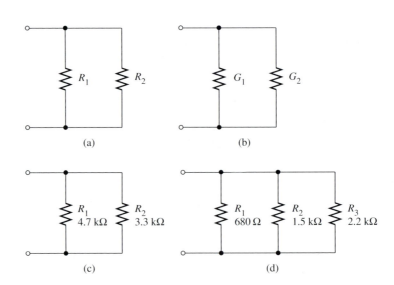

(a) (b)

(c) (d)

It is more logical to think of the parallel circuit in terms of circuit conductances. Conductance and resistance are reciprocals.

$$G = \frac{1}{R} \qquad R = \frac{1}{G}$$

Figure 11–2(b) is the same as Figure 11–2(a) except that we are using the symbol for conductance (G) to show that we will work with that unit in the parallel circuit.

☞ **KEY POINT** In a parallel circuit the total conductance equals the sum of the individual branch conductances.

$$G_T = G_1 + G_2 + G_3 + \cdots + G_N$$

EXAMPLE 11–4

In Figure 11–2(c), let $R_1 = 4.7\ k\Omega$ and $R_2 = 3.3\ k\Omega$. Find R_T.

SOLUTION Because we are dealing with a parallel circuit, we must first find G_1 and G_2.

$$G_1 = \frac{1}{R_1} = \frac{1}{4.7\ k\Omega} = 213\ \mu S$$

$$G_2 = \frac{1}{R_2} = \frac{1}{3.3\ k\Omega} = 303\ \mu S$$

Now we can find G_T:

$$G_T = G_1 + G_2 = 213\ \mu S + 303\ \mu S = 516\ \mu S$$

R_T is the reciprocal of G_T.

$$R_T = \frac{1}{G_T} = \frac{1}{516\ \mu S} = 1.94\ k\Omega$$

We can put the whole solution into one equation:

$$R_T = \frac{1}{\dfrac{1}{R_1} + \dfrac{1}{R_2}}$$

or

$$R_T = \frac{1}{G_1 + G_2}$$

Using the calculator,

TI: 4.7 $\boxed{\text{EE}}$ 3 $\boxed{1/x}$ $\boxed{+}$ 3.3 $\boxed{\text{EE}}$ 3 $\boxed{1/x}$ $\boxed{=}$ $\boxed{1/x}$

Answer: 1.94×10^3

Notice that when the $\boxed{=}$ sign was pressed G_T (516 μS) was displayed.

Casio: I $\div$ $($ 4.7 $\boxed{\text{EXP}}$ 3 $\boxed{x^{-1}}$ $+$ 3.3 $\boxed{\text{EXP}}$ 3 $\boxed{x^{-1}}$

$)$ $=$

EXAMPLE 11–5

In Figure 11–2(d), let $R_1 = 680\ \Omega$, $R_2 = 1.5\ \text{k}\Omega$, and $R_3 = 2.2\ \text{k}\Omega$. Find R_T.

SOLUTION

$$G_T = G_1 + G_2 + G_3$$

$$= 1.47\ \text{mS} + 667\ \mu\text{S} + 445\ \mu\text{S} = 2.59\ \text{mS}$$

$$R_T = \frac{1}{G_T} = \frac{1}{2.59\ \text{mS}} = 386\ \Omega$$

☞ **KEY POINT** It is always true that *in a parallel circuit, R_T is always less than the smallest branch resistance.*

Using the calculator:

TI: 680 $\boxed{1/x}$ $+$ 1.5 $\boxed{\text{EE}}$ 3 $\boxed{1/x}$ $+$ 2.2 $\boxed{\text{EE}}$ 3 $\boxed{1/x}$

$=$ $\boxed{1/x}$

Answer: **386**

$R_T = 386\ \Omega$

Casio: $($ 680 $\boxed{x^{-1}}$ $+$ 1.5 $\boxed{\text{EXP}}$ 3 $\boxed{x^{-1}}$ $+$ 2.2

$\boxed{\text{EXP}}$ 3 $\boxed{x^{-1}}$ $)$ $\boxed{x^{-1}}$ $=$

PRACTICE PROBLEMS 11–1

1. Refer to Figure 11–1(c). Find R_T when
   (a) $R_1 = 82\ \text{k}\Omega$, $R_2 = 100\ \text{k}\Omega$, and
   (b) $R_1 = 33\ \text{k}\Omega$, $R_2 = 15\ \text{k}\Omega$.

2. Refer to Figure 11–1(d). Find R_T when
   (a) $R_1 = 120\ \Omega$, $R_2 = 220\ \Omega$,
   $R_3 = 1\ \text{k}\Omega$, and (b) $R_1 = 68\ \text{k}\Omega$,
   $R_2 = 120\ \text{k}\Omega$, $R_3 = 39\ \text{k}\Omega$.

3. Refer to Figure 11–2(c). Find G_T and
   R_T when (a) $R_1 = 6.8\ \text{k}\Omega$, $R_2 = 2.7$
   $\text{k}\Omega$, (b) $R_1 = 7.5\ \text{k}\Omega$, $R_2 = 15\ \text{k}\Omega$,
   (c) $R_1 = 47\ \text{k}\Omega$, $R_2 = 120\ \text{k}\Omega$, and
   (d) $R_1 = 68\ \Omega$, $R_2 = 220\ \Omega$.

4. Refer to Figure 11–2(d). Find G_T and
   R_T when (a) $R_1 = 1.5\ \text{k}\Omega$, $R_2 = 4.7$
   $\text{k}\Omega$, $R_3 = 3.3\ \text{k}\Omega$, (b) $R_1 = 680\ \Omega$,
   $R_2 = 1.8\ \text{k}\Omega$, $R_3 = 470\ \Omega$,
   (c) $R_1 = 120\ \text{k}\Omega$, $R_2 = 120\ \text{k}\Omega$,
   $R_3 = 180\ \text{k}\Omega$, and (d) $R_1 = 22\ \text{k}\Omega$,
   $R_2 = 100\ \text{k}\Omega$, $R_3 = 10\ \text{k}\Omega$.

SOLUTIONS

1. (a) $R_T = R_1 + R_2 = 82\ k\Omega + 100\ k\Omega = 182\ k\Omega$
   (b) $R_T = R_1 + R_2 = 33\ k\Omega + 15\ k\Omega = 48\ k\Omega$
2. (a) $R_T = R_1 + R_2 + R_3 = 120\ \Omega + 220\ \Omega + 1\ k\Omega = 1.34\ k\Omega$
   (b) $R_T = R_1 + R_2 + R_3 = 68\ k\Omega + 120\ k\Omega + 39\ k\Omega = 227\ k\Omega$
3. (a) $G_T = G_1 + G_2 = 147\ \mu S + 370\ \mu S = 517\ \mu S$

   $$R_T = \frac{1}{G_T} = \frac{1}{517\ \mu S} = 1.93\ k\Omega$$

   (b) $G_T = G_1 + G_2 = 133\ \mu S + 66.7\ \mu S = 200\ \mu S$

   $$R_T = \frac{1}{G_T} = \frac{1}{200\ \mu S} = 5\ k\Omega$$

   (c) $G_T = G_1 + G_2 = 21.3\ \mu S + 8.33\ \mu S = 29.6\ \mu S$

   $$R_T = \frac{1}{G_T} = \frac{1}{29.6\ \mu S} = 33.8\ k\Omega$$

   (d) $G_T = G_1 + G_2 = 14.7\ mS + 4.55\ mS = 19.3\ mS$

   $$R_T = \frac{1}{G_T} = \frac{1}{19.3\ mS} = 51.9\ \Omega$$

4. (a) $G_T = G_1 + G_2 + G_3 = 667\ \mu S + 213\ \mu S + 303\ \mu S = 1.18\ mS$

   $$R_T = \frac{1}{G_T} = \frac{1}{1.18\ mS} = 846\ \Omega$$

   (b) $G_T = G_1 + G_2 + G_3 = 1.47\ mS + 556\ \mu S + 2.13\ mS = 4.15\ mS$

   $$R_T = \frac{1}{G_T} = \frac{1}{4.15\ mS} = 241\ \Omega$$

   (c) $G_T = G_1 + G_2 + G_3 = 8.33\ \mu S + 8.33\ \mu S + 5.56\ \mu S = 22.2\ \mu S$

   $$R_T = \frac{1}{G_T} = \frac{1}{22.2\ \mu S} = 45.0\ k\Omega$$

   (d) $G_T = G_1 + G_2 + G_3 = 45.5\ \mu S + 10\ \mu S + 100\ \mu S = 155\ \mu S$

   $$R_T = \frac{1}{G_T} = \frac{1}{155\ \mu S} = 6.43\ k\Omega$$

Additional practice problems are at the end of the chapter.

RESISTANCE IN SERIES–PARALLEL CIRCUITS 11-2

Now let's look at some series–parallel circuits.

> ☞ **RULE 11-1** To find the total resistance in a series–parallel circuit: (1) Determine which resistors are in parallel and reduce that part of the circuit to an equivalent resistance; (2) add series resistances when possible; (3) continue steps 1 and 2 until the circuit is reduced to one resistance.

Let's find the total resistance of the circuit in Figure 11–3(a), where $R_1 = 4.7$ kΩ, $R_2 = 6.8$ kΩ, and $R_3 = 12$ kΩ. Notice that R_2 and R_3 are in parallel and they are in series with R_1. The equation is

$$R_T = R_1 + R_2 \parallel R_3$$

(The symbol $\parallel$ means "in parallel with.") Let's first reduce R_2 and R_3 to an equivalent series resistance, which we will call R_X. We will solve for R_X the same way we solved for R_T in Figure 11–2(c).

$$G_X = G_2 + G_3 = 147 \ \mu S + 83.3 \ \mu S = 230 \ \mu S$$

$$R_X = \frac{1}{G_X} = \frac{1}{230 \ \mu S} = 4.34 \ k\Omega$$

We now have a circuit that looks like Figure 11–3(b). Then

$$R_T = R_1 + R_X = 4.7 \ k\Omega + 4.34 \ k\Omega = 9.04 \ k\Omega$$

We have reduced a complex circuit to a single resistance, as shown in Figure 11–3(c).

$$R_T = R_1 + \cfrac{1}{\cfrac{1}{R_2} + \cfrac{1}{R_3}}$$

Using the calculator,

TI: 4.7 $\boxed{EE}$ 3 $\boxed{+}$ $\boxed{(}$ 6.8 $\boxed{EE}$ 3 $\boxed{1/x}$ $\boxed{+}$ 12 $\boxed{EE}$ 3

$\boxed{1/x}$ $\boxed{)}$ $\boxed{1/x}$ $\boxed{=}$

Answer: **9.04** $\times$ **10^3**

$$R_T = 9.04 \ k\Omega$$

Casio: The TI key sequence will work on the Casio, but

use $\boxed{EXP}$ instead of $\boxed{EE}$ and $\boxed{x^{-1}}$ instead of $\boxed{1/x}$.

FIGURE 11–3
Reducing a
series–parallel
circuit to a
single equivalent
resistance.

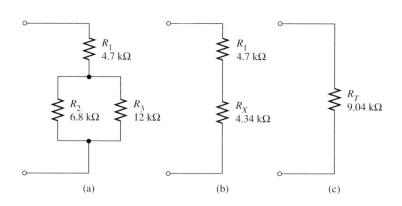

(a)   (b)   (c)

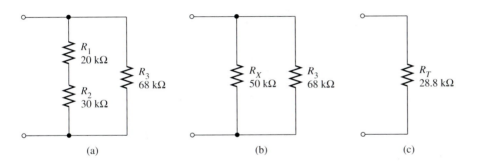

FIGURE 11–4
Reducing a
series-parallel
circuit to a
single equivalent
resistance.

(a)　　　　　　(b)　　　　　　(c)

Let's try another one. In Figure 11–4(a), let $R_1 = 20$ kΩ, $R_2 = 30$ kΩ, and $R_3 = 68$ kΩ. Notice that R_1 and R_2 are in series. The equivalent resistance (R_X) is equal to $R_1 + R_2$.

$$R_X = R_1 + R_2 = 20 \text{ k}\Omega + 30 \text{ k}\Omega = 50 \text{ k}\Omega$$

The circuit now looks like the one in Figure 11–4(b). The final step is to reduce the circuit to a single resistance, R_T.

$$G_T = G_X + G_3 = 20 \ \mu\text{S} + 14.7 \ \mu\text{S} = 34.7 \ \mu\text{S}$$

$$R_T = \frac{1}{G_T} = \frac{1}{34.7 \ \mu\text{S}} = 28.8 \text{ k}\Omega$$

Let's try one more. In Figure 11–5(a), let $R_1 = 1.2$ kΩ, $R_2 = 6.8$ kΩ, $R_3 = 10$ kΩ, and $R_4 = 3.3$ kΩ. Notice that R_2, R_3, and R_4 are connected in parallel. To find R_X, we must first find G_X.

$$G_X = G_2 + G_3 + G_4 = 147 \ \mu\text{S} + 100 \ \mu\text{S} + 303 \ \mu\text{S} = 550 \ \mu\text{S}$$

$$R_X = \frac{1}{G_X} = \frac{1}{550 \ \mu\text{S}} = 1.82 \text{ k}\Omega$$

This equivalent resistance, R_X, is in series with R_1, as shown in Figure 11–5(b). R_T then is equal to R_1 and R_X in series.

$$R_T = R_X + R_1 = 1.82 \text{ k}\Omega + 1.2 \text{ k}\Omega = 3.02 \text{ k}\Omega$$

The circuit reduces to the equivalent resistance shown in Figure 11–5(c).

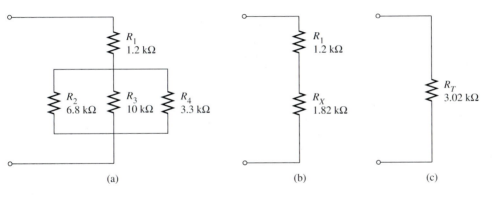

FIGURE 11–5
Reducing a
series-parallel
circuit to a
single equivalent
resistance.

(a)　　　　　　(b)　　　　　　(c)

PRACTICE PROBLEMS 11–2

1. Refer to Figure 11–3(a) and find R_T
   when (a) $R_1 = 330\ \Omega$, $R_2 = 680\ \Omega$,
   $R_3 = 910\ \Omega$, and (b) $R_1 = 82\ \mathrm{k}\Omega$,
   $R_2 = 270\ \mathrm{k}\Omega$, $R_3 = 180\ \mathrm{k}\Omega$.

2. Refer to Figure 11–4(a) and find R_T
   when (a) $R_1 = 150\ \Omega$, $R_2 = 560\ \Omega$,
   $R_3 = 1.5\ \mathrm{k}\Omega$, and (b) $R_1 = 22\ \mathrm{k}\Omega$,
   $R_2 = 56\ \mathrm{k}\Omega$, $R_3 = 39\ \mathrm{k}\Omega$.

3. Refer to Figure 11–5(a) and find R_T
   when (a) $R_1 = 750\ \mathrm{k}\Omega$, $R_2 = 1.2\ \mathrm{M}\Omega$,
   $R_3 = 2.2\ \mathrm{M}\Omega$, $R_4 = 910\ \mathrm{k}\Omega$, and
   (b) $R_1 = 2.7\ \mathrm{k}\Omega$, $R_2 = 1\ \mathrm{k}\Omega$,
   $R_3 = 10\ \mathrm{k}\Omega$, $R_4 = 8.2\ \mathrm{k}\Omega$.

4. Refer to Figure 11–5(a) and find R_T
   when (a) $R_1 = 27\ \mathrm{k}\Omega$, $R_2 = 47\ \mathrm{k}\Omega$,
   $R_3 = 56\ \mathrm{k}\Omega$, $R_4 = 100\ \mathrm{k}\Omega$, and
   (b) $R_1 = 120\ \mathrm{k}\Omega$, $R_2 = 330\ \mathrm{k}\Omega$,
   $R_3 = 680\ \mathrm{k}\Omega$, $R_4 = 820\ \mathrm{k}\Omega$.

SOLUTIONS

1. (a) $R_T = R_1 + R_2 \| R_3 = 330\ \Omega +$
   $389\ \Omega = 719\ \Omega$ (b) $R_T = R_1 + R_2 \|$
   $R_3 = 82\ \mathrm{k}\Omega + 108\ \mathrm{k}\Omega = 190\ \mathrm{k}\Omega$

2. (a) $R_T = (R_1 + R_2) \| R_3 = 710\ \Omega \|$
   $1.5\ \mathrm{k}\Omega = 482\ \Omega$
   (b) $R_T = (R_1 + R_2) \| R_3 = 78\ \mathrm{k}\Omega \|$
   $39\ \mathrm{k}\Omega = 26.0\ \mathrm{k}\Omega$

3. (a) $R_T = R_1 + R_2 \| R_3 \| R_4 =$
   $750\ \mathrm{k}\Omega + 419\ \mathrm{k}\Omega = 1.17\ \mathrm{M}\Omega$
   (b) $R_T = R_1 + R_2 \| R_3 \| R_4 =$
   $2.7\ \mathrm{k}\Omega + 818\ \Omega = 3.52\ \mathrm{k}\Omega$

4. (a) $R_T = R_1 + R_2 \| R_3 \| R_4 =$
   $27\ \mathrm{k}\Omega + 20.4\ \mathrm{k}\Omega = 47.4\ \mathrm{k}\Omega$
   (b) $R_T = R_1 + R_2 \| R_3 \| R_4 =$
   $120\ \mathrm{k}\Omega + 175\ \mathrm{k}\Omega = 295\ \mathrm{k}\Omega$

Additional practice problems are at the end of the chapter.

There are many kinds of series–parallel circuits. We have examined only a few. In all problem solving we must recognize which resistors are in parallel and reduce that part of the circuit to an equivalent resistance. We then add series resistances and continue both operations until we have reduced the circuit to one resistance (R_T).

SELF-TEST 11–1

1. Refer to Figure 11–1(d). Find R_T when
   $R_1 = 39\ \mathrm{k}\Omega$, $R_2 = 56\ \mathrm{k}\Omega$, and
   $R_3 = 27\ \mathrm{k}\Omega$.

2. Refer to Figure 11–2(c). Find R_T when
   $R_1 = 820\ \Omega$ and $R_2 = 1.2\ \mathrm{k}\Omega$.

3. Refer to Figure 11–2(d). Find R_T when
   $R_1 = 33\ \mathrm{k}\Omega$, $R_2 = 100\ \mathrm{k}\Omega$, and
   $R_3 = 82\ \mathrm{k}\Omega$.

4. Refer to Figure 11–3(a). Find R_T when
   $R_1 = 12\ \mathrm{k}\Omega$, $R_2 = 27\ \mathrm{k}\Omega$, and
   $R_3 = 47\ \mathrm{k}\Omega$.

5. Refer to Figure 11–4(a). Find R_T when
   $R_1 = 330\ \Omega$, $R_2 = 1.2\ \mathrm{k}\Omega$, and
   $R_3 = 910\ \Omega$.

6. Refer to Figure 11–5(a). Find R_T when
   $R_1 = 270\ \mathrm{k}\Omega$, $R_2 = 1.2\ \mathrm{M}\Omega$,
   $R_3 = 750\ \mathrm{k}\Omega$, and $R_4 = 470\ \mathrm{k}\Omega$.

Answers to Self-test 11–1 are at the end of the chapter.

Consider the circuit in Figure 11–6. Let's find the current, I, and the voltage drops across R_1 and R_2. The general Ohm's law equation is

$$I = \frac{E}{R}$$

When dealing with a specific situation, we must be more specific in labeling the variables in our equation. For example, in a series circuit we know from Kirchhoff's law that current is constant. $I_T = I_1 = I_2$. Therefore, we can label the current I and not use a subscript. There are three resistances though, R_1, R_2, and R_T, and there are three voltages, E, V_1, and V_2. If we use R_T in the equation, we must use E. If we use R_1, we must use V_1, and so on. To find I, we could then write the equation three ways, depending on known values:

$$I = \frac{E}{R_T}$$

$$I = \frac{V_1}{R_1}$$

$$I = \frac{V_2}{R_2}$$

In our circuit we know E and can find R_T:

$$R_T = R_1 + R_2 = 5 \text{ k}\Omega$$

Then

$$I = \frac{E}{R_T} = \frac{10 \text{ V}}{5 \text{ k}\Omega} = 2 \text{ mA}$$

Knowing I, we can solve for V_1 and V_2 by rearranging the equation and solving for the unknown:

$$V_1 = IR_1 = 2 \text{ mA} \times 2 \text{ k}\Omega = 4 \text{ V}$$

$$V_2 = IR_2 = 2 \text{ mA} \times 3 \text{ k}\Omega = 6 \text{ V}$$

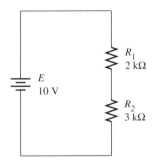

FIGURE 11–6

Two resistors connected in series.

Using Kirchhoff's voltage law to check our answers, we get

$$E = V_1 + V_2$$

$$10\,V = 4\,V + 6\,V$$

Let's look at the problem from another angle. Because the current is constant and $IR = V$, by inspection we can see that the greater voltage drops across R_2.

> ☞ **KEY POINT** In any series circuit the greater volt-
> age drops across the greater resistance.

Furthermore, the voltage drops are always in proportion to the resistances. In this problem the resistances are in the ratio of 2 to 3. Expressed as a proportion,

(11–1)
$$\frac{R_1}{R_2} = \frac{V_1}{V_2}$$

We could also say that

(11–2)
$$\frac{R_1}{R_T} = \frac{V_1}{E}$$

(11–3)
$$\frac{R_2}{R_T} = \frac{V_2}{E}$$

R_1 is 0.4 or 40% of total resistance. Therefore, 40% of the total voltage

$$\frac{R_1}{R_T} = \frac{2\,k\Omega}{5\,k\Omega} = 0.4$$

must drop across it. We can rearrange our proportion (Equation 11–2):

(11–2)
$$\frac{R_1}{R_T} = \frac{V_1}{E}$$

$$V_1 = E\left(\frac{R_1}{R_T}\right)$$

(11–4)   or
$$V_1 = \frac{ER_1}{R_T}$$

Then it follows that

(11–5)
$$V_2 = \frac{ER_2}{R_T}$$

We will use these equations often in problem solving. Looking at Equation 11–4 again, we see

$$V_1 = \left(\frac{E}{R_T}\right)R_1$$

or
$$V_1 = IR_1 \qquad \left(I = \frac{E}{R_T}\right)$$

EXAMPLE 11–6

In Figure 11–6, let $E = 12$ V, $R_2 = 1.5$ kΩ, and $V_1 = 4$ V. Find R_1, V_2, I, and R_T.

SOLUTION In this problem, V_2 is the only unknown variable that can be found from the information given. We can't find R_1 because we don't know I. We can't find R_T because we don't know I or R_1. We can find V_2 by using Kirchhoff's voltage law.

$$E = V_1 + V_2$$
$$12 \text{ V} = 4 \text{ V} + V_2$$
$$8 \text{ V} = V_2$$

Once we have determined the value of V_2 we can find I.

$$I = \frac{V_2}{R_2} = \frac{8 \text{ V}}{1.5 \text{ k}\Omega} = 5.33 \text{ mA}$$

Now that we know the value of I we can find R_1 and R_T.

$$R_1 = \frac{V_1}{I} = \frac{4 \text{ V}}{5.33 \text{ mA}} = 750 \ \Omega$$

$$R_T = R_1 + R_2 = 750 \ \Omega + 1.5 \text{ k}\Omega = 2.25 \text{ k}\Omega$$

Also,

$$R_T = \frac{E}{I} = \frac{12 \text{ V}}{5.33 \text{ mA}} = 2.25 \text{ k}\Omega$$

EXAMPLE 11–7

In Figure 11–7, let $E = 12$ V, $V_1 = 2.73$ V, $I = 38.7$ mA, and $R_2 = 100 \ \Omega$. Find V_2, V_3, R_1, R_3, and R_T.

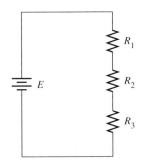

FIGURE 11–7
Three resistors
connected in
series.

SOLUTION   A check of given values shows that with the information given we can find R_T because we know E and I.

$$R_T = \frac{E}{I} = \frac{12 \text{ V}}{38.7 \text{ mA}} = 310 \text{ }\Omega$$

We can also find R_1 because we know I and V_1.

$$R_1 = \frac{V_1}{I} = \frac{2.73 \text{ V}}{38.7 \text{ mA}} = 70.5 \text{ }\Omega$$

We can find V_2 because we know I and R_2.

$$V_2 = IR_2 = 38.7 \text{ mA} \times 100 \text{ }\Omega = 3.87 \text{ V}$$

The given values plus the unknowns we have found provide us with enough information to solve for the rest of the unknowns, which are V_3 and R_3. Although other solutions may be possible, we will solve for these variables as follows:

$$V_3 = E - V_1 - V_2 = 12 \text{ V} - 2.73 \text{ V} - 3.87 \text{ V} = 5.40 \text{ V}$$

Then

$$R_3 = \frac{V_3}{I} = \frac{5.40 \text{ V}}{38.7 \text{ mA}} = 140 \text{ }\Omega$$

Check R_T:

$$R_T = R_1 + R_2 + R_3 = 70.5 \text{ }\Omega + 100 \text{ }\Omega + 140 \text{ }\Omega = 311 \text{ }\Omega$$

(rounding causes a difference of 1 in the LSD position)

Check E:

$$E = V_1 + V_2 + V_3 = 2.73 \text{ V} + 3.87 \text{ V} + 5.40 \text{ V} = 12 \text{ V}$$

EXAMPLE 11–8

Two resistors are connected in series across a 3-volt battery. One of the resistors has a value of 1 k-ohm and the other has a value of 2 k-ohms. What is the current through and the voltage drop across each resistor?

SOLUTION   First, let's draw the circuit schematic and write in all known values. The circuit schematic you draw should look like Figure 11–6 with the values changed to the values given in this example.

Now, apply Ohm's law and/or Kirchhoff's laws, as before, to determine the equations for the unknowns, then calculate the answers.

$$I = \frac{E}{R_T} = \frac{E}{R_1 + R_2} = \frac{3 \text{ V}}{1 \text{ k}\Omega + 2 \text{ k}\Omega} = \frac{3 \text{ V}}{3 \text{ k}\Omega} = 1 \text{ mA}$$

$$V_1 = IR_1 = 1 \text{ mA} \times 1 \text{ k}\Omega = 1 \text{ V} \text{ and } V_2 = IR_2 = 1 \text{ mA} \times 2 \text{ k}\Omega = 2 \text{ V}$$

PRACTICE PROBLEMS 11-3

1. Refer to Figure 11-6. Find I, R_T, V_1, and V_2 when (a) $E = 25$ V, $R_1 = 2.7$ kΩ, and $R_2 = 4.7$ kΩ, (b) $E = 30$ V, $R_1 = 10$ kΩ, and $R_2 = 15$ kΩ, and (c) $E = 12$ V, $R_1 = 470$ Ω, and $R_2 = 1.2$ kΩ.

2. Refer to Figure 11-6. Find R_T, R_1, V_1, and V_2 when $E = 10$ V, $I = 150$ μA, and $R_2 = 20$ kΩ.

3. Refer to Figure 11-6. Find E, R_T, V_2, and R_1 when $I = 12.6$ mA, $V_1 = 6.7$ V, and $R_2 = 810$ Ω.

4. Refer to Figure 11-7. Find I, R_T, V_1, V_2, and V_3 when (a) $E = 10$ V, $R_1 = 3.3$ kΩ, $R_2 = 6.8$ kΩ, and $R_3 = 2.2$ kΩ, (b) $E = 20$ V, $R_1 = 4.7$ kΩ, $R_2 = 10$ kΩ, and $R_3 = 12$ kΩ, and (c) $E = 15$ V, $R_1 = 12$ kΩ, $R_2 = 33$ kΩ, and $R_3 = 18$ kΩ.

5. Refer to Figure 11-7. Find V_2, V_3, R_1, R_3, and R_T when $E = 18$ V, $V_1 = 8.2$ V, $I = 40$ μA, and $R_2 = 180$ kΩ.

6. Refer to Figure 11-7. Find E, R_2, R_3, V_1, and V_3 when $I = 37.6$ mA, $R_1 = 47$ Ω, $V_2 = 1.92$ V, and $R_T = 137$ Ω.

7. Two resistors are connected in series across a nine-volt battery. One of the resistors has a value of three k-ohms. The other has a value of 6 k-ohms. What is the current through and the voltage drop across each resistor?

8. Three resistors are connected in series across a twelve-volt battery. The battery voltage is divided equally across each resistor. The battery current is one milliamp. What is the current through and the voltage drop across each resistor? What is the value of each resistor?

SOLUTIONS

1. (a) $R_T = R_1 + R_2 = 2.7$ kΩ + 4.7 kΩ = 7.4 kΩ

$$I = \frac{E}{R_T} = \frac{25 \text{ V}}{7.4 \text{ kΩ}} = 3.38 \text{ mA}$$

$V_1 = IR_1 = 3.38$ mA × 2.7 kΩ = 9.12 V

$V_2 = IR_2 = 3.38$ mA × 4.7 kΩ = 15.9 V

(b) $R_T = 25.0$ kΩ
$I = 1.20$ mA
$V_1 = 12.0$ V
$V_2 = 18.0$ V

(c) $R_T = 1.67$ kΩ
$I = 7.19$ mA
$V_1 = 3.38$ V
$V_2 = 8.62$ V

2. $R_T = 66.7$ kΩ, $R_1 = 46.7$ kΩ, $V_1 = 7$ V, $V_2 = 3$ V

3. $E = 16.9$ V, $V_2 = 10.2$ V, $R_1 = 532\ \Omega$, $R_T = 1.34\ k\Omega$

4. (a) $R_T = R_1 + R_2 + R_3 = 3.3\ k\Omega + 6.8\ k\Omega + 2.2\ k\Omega = 12.3\ k\Omega$

$$I = \frac{E}{R_T} = \frac{10\ V}{12.3\ k\Omega} = 813\ \mu A$$

$V_1 = IR_1 = 813\ \mu A \times 3.3\ k\Omega = 2.68$ V

$V_2 = IR_2 = 813\ \mu A \times 6.8\ k\Omega = 5.53$ V

$V_3 = IR_3 = 813\ \mu A \times 2.2\ k\Omega = 1.79$ V

(b) $R_T = 26.7\ k\Omega$ (c) $R_T = 63\ k\Omega$

    $I = 749\ \mu A$ $I = 238\ \mu A$

    $V_1 = 3.52$ V $V_1 = 2.86$ V

    $V_2 = 7.49$ V $V_2 = 7.86$ V

    $V_3 = 8.99$ V $V_3 = 4.29$ V

5. $R_T = 450\ k\Omega$, $V_2 = 7.20$ V, $V_3 = 2.60$ V, $R_3 = 65.0\ k\Omega$, $R_1 = 205\ k\Omega$

6. $E = 5.15$ V, $V_1 = 1.77$ V, $V_3 = 1.46$ V, $R_2 = 51.1\ \Omega$, $R_3 = 38.9\ \Omega$

7. $I = \dfrac{E}{R_T} = \dfrac{9\ V}{3\ k\Omega + 6\ k\Omega} = 1$ mA;

$V_1 = 1$ mA $\times 3\ k\Omega = 3$ V;

$V_2 = 1$ mA $\times 6\ k\Omega = 6$ V

8. $V_1 = V_2 = V_3 = \dfrac{E}{3} = \dfrac{12\ V}{3} = 4$ V;

$R_1 = R_2 = R_3 = \dfrac{4\ V}{1\ mA} = 4\ k\Omega$

Additional practice problems are at the end of the chapter.

11–4   OHM'S LAW: PARALLEL CIRCUITS

Consider the circuit in Figure 11–8.
Let's find V, G_T, R_T, I_1, and I_2. The general equation is

$$V = IR = I\left(\frac{1}{G}\right) = \frac{I}{G}$$

Specifically,

$$V = \frac{I_T}{G_T} = \frac{I_1}{G_1} = \frac{I_2}{G_2}$$

FIGURE 11–8
Two resistors connected in parallel.

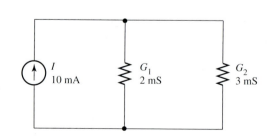

Since we are given I_T, we will find G_T and V by using

$$G_T = G_1 + G_2 \quad \text{and} \quad V = \frac{I_T}{G_T}$$

$$= \frac{10 \text{ mA}}{2 \text{ mS} + 3 \text{ mS}} = \frac{10 \text{ mA}}{5 \text{ mS}} = 2 \text{ V}$$

Now we can find I_1 and I_2:

$$I = \frac{V}{R} = VG$$

$$I_1 = VG_1 = 2 \text{ V} \times 2 \text{ mS} = 4 \text{ mA}$$

$$I_2 = VG_2 = 2 \text{ V} \times 3 \text{ mS} = 6 \text{ mA}$$

We must keep in mind that V is the constant in a parallel circuit. Since we know G_T,

$$R_T = \frac{1}{G_T} = \frac{1}{5 \text{ mS}} = 200 \ \Omega$$

We see that I_2 was the greater current. That was because G_2 was the greater conductance.

> ☞ **KEY POINT** In a parallel circuit the greater current flows through the greater conductance.

The currents are always in proportion to the conductance. In this problem the conductances are in the ratio of 2 to 3. Expressed as a proportion,

$$\frac{G_1}{G_2} = \frac{I_1}{I_2} \tag{11–6}$$

We could also say that

$$\frac{G_1}{G_T} = \frac{I_1}{I_T} \tag{11–7}$$

and

$$\frac{G_2}{G_T} = \frac{I_2}{I_T} \tag{11–8}$$

G_1 is 0.4 or 40% of the total conductance. Therefore, 40% of the total current flows through it. We can rearrange our proportion (Equation 11–7):

$$\frac{G_1}{G_T} = \frac{I_1}{I_T} \tag{11–7}$$

$$I_1 = I_T \left(\frac{G_1}{G_T} \right)$$

or

$$I_1 = \frac{I_T G_1}{G_T} \tag{11–9}$$

Then it follows that

(11–10)
$$I_2 = \frac{I_T G_2}{G_T}$$

We will use these equations very often in problem solving.

Notice the duality or similarity between Equations 11–4 and 11–9. The relationship between voltage drops and resistance in a series circuit is the same as the relationship between current and conductance in a parallel circuit.

EXAMPLE 11–9

In Figure 11–9, let $I_T = 400\ \mu A$, $R_1 = 8.1\ k\Omega$, and $I_2 = 150\ \mu A$. Solve for V, I_1, R_2, G_T, and R_T.

FIGURE 11–9
Two resistors connected in parallel.

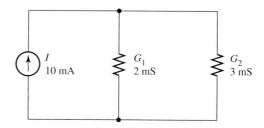

SOLUTION An examination of known variables shows that we have one unknown current, I_1. We can find this current by applying Kirchhoff's current law. We don't have enough information to find any of the other unknowns at this time.

$$I_T = I_1 + I_2$$

$$400\ \mu A = I_1 + 150\ \mu A$$

$$250\ \mu A = I_1$$

Now that we know I_1 we can find V.

$$V = I_1 R_1 = 250\ \mu A \times 8.1\ k\Omega = 2.03\ V$$

With V we can find R_T, G_T, and R_2.

$$R_2 = \frac{V}{I_2} = \frac{2.03\ V}{150\ \mu A} = 13.5\ k\Omega$$

$$G_T = \frac{I_T}{V} = \frac{400\ \mu A}{2.03\ V} = 198\ \mu S$$

$$R_T = \frac{1}{G_T} = \frac{1}{198\ \mu S} = 5.06\ k\Omega$$

There may be alternate solutions that result in the same answers. Typically, more than one valid approach can be used to solve problems like these.

EXAMPLE 11–10

Refer to Figure 11–10. Let $V = 50$ V, $G_1 = 90$ μS, $I_2 = 400$ μA, and $G_T = 250$ μS. Find R_1, R_2, R_3, R_T, I_1, I_3, and I_T.

SOLUTION From the given data we can find I_T.

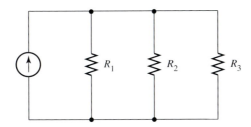

FIGURE 11–10
Three resistors
connected in
parallel.

$$I_T = VG_T = 50 \text{ V} \times 250 \text{ }\mu\text{S} = 12.5 \text{ mA}$$

We can also find R_2:

$$R_2 = \frac{V}{I_2} = \frac{50 \text{ V}}{400 \text{ }\mu\text{A}} = 125 \text{ k}\Omega$$

Since we know G_1, we can find R_1:

$$R_1 = \frac{1}{G_1} = \frac{1}{90 \text{ }\mu\text{S}} = 11.1 \text{ k}\Omega$$

Since we know G_1 and V, we can find I_1:

$$I_1 = VG_1 = 50 \text{ V} \times 90 \text{ }\mu\text{S} = 4.50 \text{ mA}$$

Since we know G_T, we can find R_T:

$$R_T = \frac{1}{G_T} = \frac{1}{250 \text{ }\mu\text{S}} = 4.00 \text{ k}\Omega$$

The rest of the unknowns can now be found:

$$I_3 = I_T - I_1 - I_2 = 12.5 \text{ mA} - 4.5 \text{ mA} - 400 \text{ }\mu\text{A} = 7.60 \text{ mA}$$

$$R_3 = \frac{V}{I_3} = \frac{50 \text{ V}}{7.6 \text{ mA}} = 6.58 \text{ k}\Omega$$

EXAMPLE 11–11

Two resistors in parallel divide a battery-supplied current so that there is seven hundred fifty microamps through a 12 k-ohm resistor. The current through the other resistor is seven milliamps. Determine the source voltage, the value of the other resistor, total current, and total resistance.

SOLUTION First, let's draw the circuit schematic and write in all known values. The circuit schematic should look like Figure 11–8 using the known values given in the example. Next, apply Ohm's law and Kirchhoff's laws, as before, to determine the equations for the unknowns, then calculate the answers.

We know two of the variables for one branch so we can use Ohm's law to determine the third.

$$V_1 = I_1 \times R_1 = 750 \ \mu A \times 12 \ k\Omega = 9.00 \ V$$
$$E = V_1 = V_2 \ (\text{Kirchhoff's voltage law}). \ \text{Then}$$
$$R_2 = \frac{V_2}{I_2} = \frac{9.00 \ V}{7 \ mA} = 1.29 \ k\Omega$$
$$I_T = I_1 + I_2 = 750 \ \mu A + 7 \ mA = 7.75 \ mA \ (\text{Kirchhoff's current law})$$

PRACTICE PROBLEMS 11–4

1. Refer to Figure 11–9. Find G_T, R_T, V, I_1, and I_2 when (a) $I_T = 20$ mA, $R_1 = 2.2$ kΩ, and $R_2 = 4.7$ kΩ, and (b) $I_T = 200 \ \mu A$, $R_1 = 15$ kΩ, and $R_2 = 27$ kΩ.

2. Refer to Figure 11–9. Find I_1, V, R_2, G_T, and R_T when $I_T = 1.5$ mA, $R_1 = 10$ kΩ and $I_2 = 800 \ \mu A$.

3. Refer to Figure 11–9. Find I_1, I_2, R_2, G_T, and R_T when $I_T = 1$ mA, $V = 2.7$ V, and $R_1 = 5.6$ kΩ.

4. Refer to Figure 11–10. Find G_T, R_T, V, I_1, I_2, and I_3 when (a) $I_T = 10$ mA, $R_1 = 330 \ \Omega$, $R_2 = 470 \ \Omega$, and $R_3 = 680 \ \Omega$, and (b) $I_T = 250 \ \mu A$, $R_1 = 5.6$ kΩ, $R_2 = 2.2$ kΩ, and $R_3 = 3.9$ kΩ.

5. Refer to Figure 11–10. Given $I_T = 50$ mA, $I_1 = 20$ mA, $V = 9$ V, and $R_2 = 1$ kΩ, find R_1, R_3, I_2, I_3, R_T, and G_T.

6. Refer to Figure 11–10. Given $V = 12$ V, $G_1 = 17.9 \ \mu S$, $I_2 = 255 \ \mu A$, and $G_T = 69.4 \ \mu S$, find R_1, R_2, R_3, I_1, I_3, and I_T.

7. Two resistors are connected in parallel across a battery. One resistor has a resistance of three k-ohms and the current through it is two milliamps. The current through the other resistor is six milliamps. What is the battery voltage, the total current, and the value of the second resistor?

8. Three resistors are connected in parallel across an energy source. The total current from the source is one hundred twenty milliamps. Ten percent of the current flows through the first branch, twenty percent through the second branch, and the remainder through the third branch. The resistance in the first branch is one k-ohm. Calculate the voltage drops, currents, and resistance of each branch. Calculate the total resistance.

SOLUTIONS

1. (a) $G_T = G_1 + G_2 = 455\ \mu S + 213\ \mu S = 667\ \mu S$

   $R_T = \dfrac{1}{G_T} = \dfrac{1}{667\ \mu S} = 1.50\ k\Omega$

   $V = \dfrac{I_T}{G_T} = I_T R_T = 30\ V$ (either expression yields 30 V)

   $I_1 = VG_1 = 30\ V \times 455\ \mu S = 13.6\ mA$ $\left(\text{we could have used } \dfrac{V}{R_1}\right)$

   $I_2 = VG_2 = 30\ V \times 213\ \mu S = 6.38\ mA$

   (b) $G_T = 104\ \mu S$
   $R_T = 9.64\ k\Omega$
   $V = 1.93\ V$
   $I_1 = 129\ \mu A$
   $I_2 = 71.4\ \mu A$

2. $I_1 = 700\ \mu A$, $V = 7.00\ V$, $R_2 = 8.75\ k\Omega$, $R_T = 4.67\ k\Omega$, $G_T = 214\ \mu S$
3. $I_1 = 482\ \mu A$, $I_2 = 518\ \mu A$, $R_2 = 5.21\ k\Omega$, $G_T = 370\ \mu S$, $R_T = 2.70\ k\Omega$
4. (a) $G_T = G_1 + G_2 + G_3 = 3.03\ mS + 2.13\ mS + 1.47\ mS = 6.63\ mS$

   $R_T = \dfrac{1}{G_T} = \dfrac{1}{6.63\ mS} = 151\ \Omega$

   $V = I_T R_T = 10\ mA \times 151\ \Omega = 1.51\ V$
   $I_1 = VG_1 = 1.51\ V \times 3.03\ mS = 4.57\ mA$
   $I_2 = VG_2 = 1.51\ V \times 2.13\ mS = 3.21\ mA$
   $I_3 = VG_3 = 1.51\ V \times 1.47\ mS = 2.22\ mA$

   (b) $G_T = 890\ \mu S$
   $R_T = 1.12\ k\Omega$
   $V = 281\ mV$
   $I_1 = 50.2\ \mu A$
   $I_2 = 128\ \mu A$
   $I_3 = 72.1\ \mu A$

5. $I_2 = 9\ mA$, $I_3 = 21\ mA$, $R_1 = 450\ \Omega$, $R_3 = 429\ \Omega$, $R_T = 180\ \Omega$, $G_T = 5.56\ mS$
6. $R_T = 14.4\ k\Omega$, $R_1 = 55.9\ k\Omega$, $R_2 = 47.1\ k\Omega$, $R_3 = 33.1\ k\Omega$, $I_1 = 215\ \mu A$, $I_3 = 363\ \mu A$, $I_T = 833\ \mu A$
7. $V_1 = I_1 \times R_1 = 2\ mA \times 3\ k\Omega = 6\ V$; $E = V_1 = V_2 = 6\ V$; $I_T = I_1 + I_2 = 2\ mA + 6\ mA = $

   $8\ mA$; $R_2 = \dfrac{V_2}{I_2} = \dfrac{6\ V}{6\ mA} = 1\ k\Omega$

8. $I_1 = 10\% \times 120\ mA = 12\ mA$; $I_2 = 20\% \times 120\ mA = 24\ mA$; $I_3 = I_T - (I_1 + I_2) = 84\ mA$;

   $V_1 = I_1 R_1 = 12\ mA \times 1\ k\Omega = 12\ V$; $E = V_1 = V_2 = V_3 = 12\ V$; $R_2 = \dfrac{V_2}{I_2} = \dfrac{12\ V}{24\ mA} = $

   $500\ \Omega$; $R_3 = \dfrac{V_3}{I_3} = \dfrac{12\ V}{84\ mA} = 143\ \Omega$; $R_T = \dfrac{E}{I_T} = \dfrac{12\ V}{120\ mA} = 100\ \Omega$

Additional practice problems are at the end of the chapter.

SELF-TEST 11–2

1. Refer to Figure 11–6. Let $E = 25\ V$, $R_1 = 18\ k\Omega$, and $V_2 = 6.5\ V$. Find I, R_T, R_2, and V_1.
2. Refer to Figure 11–7. Let $I = 800\ \mu A$, $R_1 = 68\ k\Omega$, $V_2 = 50\ V$, and $R_T = 200$ $k\Omega$. Find E, R_2, R_3, V_1, and V_3.
3. Refer to Figure 11–9. Let $I_T = 30\ mA$, $V = 1\ V$, and $R_1 = 100\ \Omega$. Find I_1, I_2, R_2, G_T, and R_T.
4. Refer to Figure 11–10. Let $I_T = 1\ mA$, $I_1 = 200\ \mu A$, $V = 10\ V$, and $R_2 = 27\ k\Omega$. Find R_1, R_3, I_2, I_3, G_T, and R_T.

Answers to Self-test 11–2 are at the end of the chapter.

Consider the circuit in Figure 11–11. Let's find the unknown variables, which are I_T, I_1, I_2, I_3, R_T, V_1, V_2, and V_3. Since we know all values of R and we know E, we will first find R_T. Then we can find I_T.

$$R_T = R_1 + R_2 \| R_3$$

Let

$$R_X = R_2 \| R_3$$

Then

$$G_X = G_2 + G_3 = 333 \ \mu S + 179 \ \mu S = 512 \ \mu S$$

$$R_X = \frac{1}{G_X} = \frac{1}{512 \ \mu S} = 1.95 \ k\Omega$$

$$R_T = R_1 + R_X = 2 \ k\Omega + 1.95 \ k\Omega = 3.95 \ k\Omega$$

$$I_T = \frac{E}{R_T} = \frac{20 \ V}{3.95 \ k\Omega} = 5.06 \ mA$$

Since R_1 is a series resistor, $I_T = I_1$.

$$I_1 = I_T = 5.06 \ mA$$

Now we can find V_1.

$$V_1 = I_1 R_1 = 5.06 \ mA \times 2 \ k\Omega = 10.1 \ V$$

Applying Kirchhoff's voltage law, we get

$$V_2 = V_3 = E - V_1 = 20 \ V - 10.1 \ V = 9.88 \ V$$

We could also find V_2 and V_3 by using

$$V_2 = V_3 = \frac{E R_X}{R_T} = \frac{20 \ V \times 1.95 \ k\Omega}{3.95 \ k\Omega} = 9.88 \ V$$

FIGURE 11–11
Resistors
connected to
form a series–
parallel circuit.

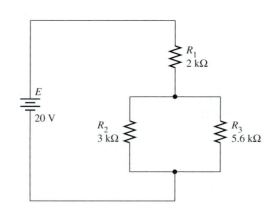

$$I_2 = \frac{V_2}{R_2} = \frac{9.88 \text{ V}}{3 \text{ k}\Omega} = 3.29 \text{ mA}$$

$$I_3 = \frac{V_3}{R_3} = \frac{9.88 \text{ V}}{5.6 \text{ k}\Omega} = 1.76 \text{ mA}$$

EXAMPLE 11–12

Solve for R_T, I_T, I_3, I_1, V_1, and V_2 in Figure 11–12.

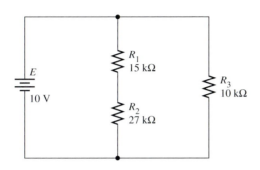

FIGURE 11–12
Resistors connected in a parallel–series circuit.

SOLUTION R_1 and R_2 are in series. This makes the branch resistance equal to $R_1 + R_2$. This branch is in parallel with R_3.

$$R_T = (R_1 + R_2) \| R_3$$

Let

$$R_X = R_1 + R_2$$

Then

$$R_X = 15 \text{ k}\Omega + 27 \text{ k}\Omega = 42 \text{ k}\Omega$$

$$G_T = G_X + G_3 = 23.8 \ \mu\text{S} + 100 \ \mu\text{S} = 124 \ \mu\text{S}$$

$$R_T = \frac{1}{G_T} = \frac{1}{124 \ \mu\text{S}} = 8.08 \text{ k}\Omega$$

Using the calculator,

TI: 15 [EE] 3 [+] 27 [EE] 3 [=] [1/x] [+] 10

[EE] 3 [1/x] [=] [1/x]

Answer: 8.08×10^3

$$R_T = 8.08 \text{ k}\Omega$$

Casio: The TI key sequence will work on the Casio, but

use [EXP] instead of [EE] and [x^{-1}] instead of [1/x].

Now we can find I_T:

$$I_T = \frac{E}{R_T} = \frac{10 \text{ V}}{8.08 \text{ k}\Omega} = 1.24 \text{ mA}$$

Knowing I_T, we can find I_3.

$$I_3 = \frac{I_T G_3}{G_T} = \frac{1.24 \text{ mA} \times 100 \text{ } \mu\text{S}}{124 \text{ } \mu\text{S}} = 1 \text{ mA}$$

Using the calculator,

1.24 $\boxed{\text{EE}}$ $\boxed{+/-}$ 3 $\boxed{\times}$ 100 $\boxed{\text{EE}}$ $\boxed{+/-}$ 6 $\boxed{\div}$

124 $\boxed{\text{EE}}$ $\boxed{+/-}$ 6 $\boxed{=}$

Answer: 1×10^{-3}

$$I_3 = 1 \text{ mA}$$

or

$$I_3 = \frac{V_3}{R_3} = \frac{10 \text{ V}}{10 \text{ k}\Omega} = 1 \text{ mA}$$

(We knew $V_3 = 10$ V because $V_3 = E$. The second method was easier. We must be aware that very often there is more than one path to take in finding the unknowns in complex circuits.) Since $I_3 = 1$ mA,

$$I_1 = I_2 = I_T - I_3 = 1.24 \text{ mA} - 1 \text{ mA} = 238 \text{ } \mu\text{A}$$

$$V_1 = \frac{ER_1}{R_X} = \frac{10 \text{ V} \times 15 \text{ k}\Omega}{42 \text{ k}\Omega} = 3.57 \text{ V}$$

$$V_2 = E - V_1 = 10 \text{ V} - 3.57 \text{ V} = 6.43 \text{ V}$$

EXAMPLE 11–13

Consider the circuit in Figure 11–13. Let's find all the currents and voltage drops and the total resistance.

FIGURE 11–13
Resistors connected in a series–parallel circuit.

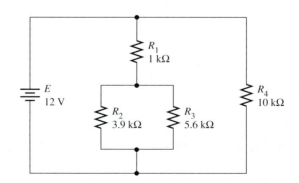

SOLUTION Since we are given all the circuit resistances, we can find R_T. A look at the circuit shows that R_2 and R_3 are in parallel. Let

$$R_A = R_2 \parallel R_3$$

$$R_A = \frac{1}{G_2 + G_3} = \frac{1}{435\ \mu S} = 2.3\ k\Omega$$

This resistance, R_A, is in series with R_1. The total branch resistance is $R_A + R_1$. Let's call this resistance R_B.

$$R_B = R_1 + R_A = 1\ k\Omega + 2.3\ k\Omega = 3.3\ k\Omega$$

R_B is in parallel with R_4.

$$R_T = R_B \parallel R_4 = \frac{1}{G_B + G_4} = \frac{1}{403\ \mu S} = 2.48\ k\Omega$$

Now we can find I_T.

$$I_T = \frac{E}{R_T} = \frac{12\ V}{2.48\ k\Omega} = 4.84\ mA$$

We can see that R_4 is connected across the source. Therefore, $V_4 = E$. Solving for I_4, we get

$$I_4 = \frac{V_4}{R_4} = \frac{12\ V}{10\ k\Omega} = 1.20\ mA$$

Applying Kirchhoff's current law, we get

$$I_T = I_1 + I_4$$

$$I_1 = I_T - I_4 = 4.84\ mA - 1.20\ mA = 3.64\ mA$$

Knowing I_1, we can find I_2 and I_3.

$$I_2 = \frac{I_1 G_2}{G_2 + G_3} = 2.14\ mA$$

$$I_3 = I_T - I_2 = 1.49\ mA$$

We can find the remaining voltage drops now.

$$V_1 = I_1 R_1 = 3.64\ mA \times 1\ k\Omega = 3.64\ V$$

$$V_2 = I_2 R_2 = 2.14\ mA \times 3.9\ k\Omega = 8.36\ V$$

$$V_3 = V_2 = 8.36\ V$$

EXAMPLE 11–14

Three resistors are connected in a series–parallel arrangement across a twenty-five-volt source. The resistor values are eighteen k-ohms, twenty two k-ohms, and thirty three k-ohms. One end of the eighteen k-ohm resistor is connected to the positive terminal of the battery. One end of the other two resistors are connected to the negative terminal of the battery. To complete the circuit, the other end of all three resistors are connected together. Find the voltage drop across and the current through each resistor, the total resistance, and the total current.

SOLUTION Let's draw the circuit schematic and write in all known values. The circuit schematic should look like Figure 11–11 using the known values given in the example. We will call the eighteen k-ohm resistor R_1, the twenty-two k-ohm resistor R_2, and the thirty three k-ohm resistor R_3. Next, apply Ohm's and Kirchhoff's laws, as before, to solve for the unknowns.

$$R_T = R_1 + R_2 \| R_3 = R_1 + \frac{1}{G_2 + G_3} = 18 \text{ k}\Omega + 13.2 \text{ k}\Omega = 31.2 \text{ k}\Omega$$

$$I_T = \frac{E}{R_T} = \frac{25 \text{ V}}{31.2 \text{ k}\Omega} = 801 \ \mu\text{A} \quad V_1 = I_1 R_1 = 801 \ \mu\text{A} \times 18 \text{ k}\Omega = 14.4 \text{ V}$$

$$V_2 = V_3 = E - V_1 = 25 \text{ V} - 14.4 \text{ V} = 10.6 \text{ V}; \quad I_2 = \frac{V_2}{R_2} = \frac{10.6 \text{ V}}{22 \text{ k}\Omega} = 481 \ \mu\text{A};$$

$$I_3 = \frac{V_3}{R_3} = \frac{10.6 \text{ V}}{33 \text{ k}\Omega} = 321 \ \mu\text{A}$$

PRACTICE PROBLEMS 11–5

1. Refer to Figure 11–11. Let $E = 25$ V, $R_1 = 2.7$ kΩ, $R_2 = 6.8$ kΩ, and $R_3 = 4.7$ kΩ. Find V_1, V_2, V_3, I_1, I_2, I_3, I_T, and R_T.

2. Refer to Figure 11–12. Let $E = 40$ V, $R_1 = 680$ Ω, $R_2 = 1.2$ kΩ, and $R_3 = 2.7$ kΩ. Find V_1, V_2, V_3, I_1, I_2, I_3, I_T, and R_T.

3. Refer to Figure 11–11. Let $E = 20$ V, $I_T = 5$ mA, $R_1 = 1.5$ kΩ, and $R_2 = 6.8$ kΩ. Find R_3, R_T, V_1, V_2, V_3, I_2, and I_3.

4. Refer to Figure 11–13. Let $E = 9$ V, $R_1 = 9.1$ kΩ, $R_2 = 27$ kΩ, $R_3 = 56$ kΩ, and $R_4 = 120$ kΩ. Find V_1, V_2, V_3, V_4, I_T, I_1, I_2, I_3, I_4, and R_T.

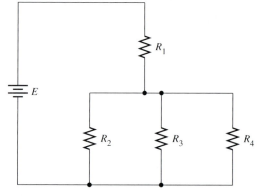

FIGURE 11–14   Series–parallel circuit.

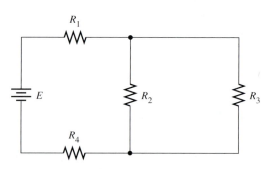

FIGURE 11–15   Series–parallel circuit.

5. Refer to Figure 11–14. Let $E = 10$ V, $I_T = 4$ mA, $V_2 = 3$ V, $R_3 = 1.5$ kΩ, and $R_4 = 6$ kΩ. Find I_2, I_3, I_4, R_1, R_2, and R_T.

7. Three resistors are connected in a series-parallel arrangement across a twelve-volt battery. The resistor values are one k-ohm, five k-ohms, and ten k-ohms. One end of the one k-ohm resistor is connected to the positive terminal of the battery. One end of the other two resistors are connected to the negative terminal of the battery. To complete the circuit, the other end of all three resistors are connected together. Find the voltage drop across and the current through each resistor, the total resistance, and the total current.

6. Refer to Figure 11–15. Let $E = 9$ V, $R_1 = 330$ Ω, $R_2 = 810$ Ω, $R_3 = 1$ kΩ, and $R_4 = 560$ Ω. Find V_1, V_2, V_3, V_4, I_1, I_2, I_3, I_4, I_T, and R_T.

8. Two branches are connected across a three-volt battery. The first branch consists of two three k-ohm resistors connected in series. The second branch consists of a one k-ohm resistor connected in series with a two k-ohm resistor. Calculate the voltage drop across and the current through each resistor. Calculate the total current and total resistance.

SOLUTIONS

1. $R_T = 5.48$ kΩ, $I_T = 4.56$ mA, $I_1 = 4.56$ mA, $I_2 = 1.86$ mA, $I_3 = 2.70$ mA, $V_1 = 12.3$ V, $V_2 = 12.7$ V, $V_3 = 12.7$ V

3. $R_T = 4$ kΩ, $V_1 = 7.5$ V, $V_2 = 12.5$ V, $V_3 = 12.5$ V, $I_2 = 1.84$ mA, $I_3 = 3.16$ mA, $R_3 = 3.95$ kΩ

5. $R_T = 2.5$ kΩ, $R_1 = 1.75$ kΩ, $R_2 = 2$ kΩ, $I_2 = 1.5$ mA, $I_3 = 2$ mA, $I_4 = 500$ μA

7. $R_T = R_1 + R_2 \| R_3 = 4.33$ kΩ; $I_T = 2.77$ mA; $I_1 = I_T = 2.77$ mA; $V_1 = 12.7$ V $V_2 = V_3 = E - V_1 = 12$ V $- 2.77$ V $= 9.23$ V; $I_2 = 1.85$ mA; $I_3 = 0.923$ mA

2. $R_T = 1.11$ kΩ, $I_T = 36.1$ mA, $I_1 = 21.3$ mA, $I_2 = 21.3$ mA, $I_3 = 14.8$ mA, $V_1 = 14.5$ V, $V_2 = 25.5$ V, $V_3 = 40$ V

4. $R_T = 22.3$ kΩ, $I_T = 404$ μA, $I_1 = 329$ μA, $I_2 = 222$ μA, $I_3 = 107$ μA, $I_4 = 7500$ μA, $V_1 = 3.00$ V, $V_2 = 6.00$ V, $V_3 = 6.00$ V, $V_4 = 9$ V.

6. $R_T = 1.34$ kΩ, $I_T = I_1 = I_4 = 6.73$ mA, $I_2 = 3.72$ mA, $I_3 = 3.01$ mA, $V_1 = 2.22$ V, $V_2 = V_3 = 3.01$ V, $V_4 = 3.77$ V

8. Let's identify the first branch resistors as R_1 and R_2. Then the other branch resistors would be R_3 and R_4.

$$I_1 = I_2 = \frac{E}{R_1 + R_2} = 0.500 \text{ mA};$$

$$I_3 = I_4 = \frac{E}{R_3 + R_4} = 1 \text{ mA};$$

$V_1 = I_1R_1 = 1.5$ V; $V_2 = I_2 \times R_2 = 1.5$ V; $V_3 = I_3R_3 = 1$ V; $V_4 = I_4R_4 = 2$ V; $I_T = I_1 + I_2 = 1.50$ mA;

$$R_T = \frac{E}{I_T} = 2 \text{ kΩ}$$

1. Refer to Figure 11–11. Let $E = 3$ V, $I_T = 70$ μA, $R_2 = 33$ kΩ, and $R_3 = 47$ kΩ. Find V_1, V_2, V_3, I_2, I_3, R_T, and R_1.

2. Refer to Figure 11–12. Let $E = 20$ V, $R_1 = 20$ kΩ, $R_3 = 10$ kΩ, $V_2 = 6.57$ V. Find I_T, I_1, I_2, I_3, V_1, V_3, R_T, and R_2.

3. Refer to Figure 11–14. Let $E = 15$ V, $R_1 = 12$ kΩ, $R_2 = 100$ kΩ, $R_3 = 22$ kΩ, and $R_4 = 68$ kΩ. Find I_T, I_1, I_2, I_3, I_4, V_1, V_2, V_3, V_4, and R_T.

4. Refer to Figure 11–15. Let $E = 25$ V, $R_1 = 2.7$ kΩ, $R_2 = 3.9$ kΩ, $R_3 = 6.8$ kΩ, and $R_4 = 2.2$ kΩ. Find V_1, V_2, V_3, V_4, I_T, I_1, I_2, I_3, I_4, and R_T.

Answers to Self-test 11–3 are at the end of the chapter.

CHAPTER 11 AT A GLANCE

PAGE	KEY POINT	EXAMPLE
303	*Key Point:* In a series circuit the total resistance equals the sum of the individual resistances.	$R_T = R_1 + R_2 + \cdots + R_N$
305	*Key Point:* In a parallel circuit the total conductance equals the sum of the individual branch conductances.	$G_T = G_1 + G_2 + \cdots + G_N$
312	*Key Point:* In a series circuit the greater voltage drops across the greater resistance.	If $R_1 > R_2$, then $V_1 > V_2$.
317	*Key Point:* In a parallel circuit the greater current flows through the greater conductance.	If $G_1 > G_2$, then $I_1 > I_2$.

END OF CHAPTER PROBLEMS 11–1

Refer to Figure 11–16(a).

1. Find R_T when (a) $R_1 = 30$ kΩ, $R_2 = 15$ kΩ, (b) $R_1 = 750$ Ω, $R_2 = 1.2$ kΩ, (c) $R_1 = 6.8$ kΩ, $R_2 = 10$ kΩ, and (d) $R_1 = 120$ kΩ, $R_2 = 180$ kΩ.

2. Find R_T when (a) $R_1 = 27$ kΩ, $R_2 = 43$ kΩ, (b) $R_1 = 100$ Ω, $R_2 = 270$ Ω, (c) $R_1 = 560$ kΩ, $R_2 = 470$ kΩ, and (d) $R_1 = 75$ kΩ, $R_2 = 91$ kΩ.

3. Find R_2 when (a) $R_1 = 1.2$ kΩ, $R_T = 3$ kΩ, and (b) $R_1 = 27$ kΩ, $R_T = 66$ kΩ.

4. Find R_2 when (a) $R_1 = 680$ Ω, $R_T = 1.43$ kΩ, and (b) $R_1 = 220$ kΩ, $R_T = 550$ kΩ.

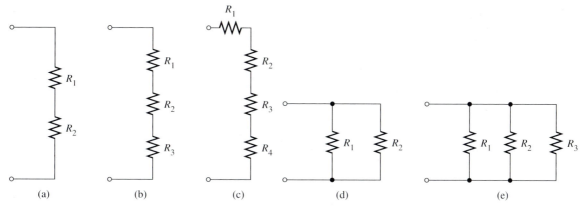

FIGURE 11–16

Refer to Figure 11–16(b).

5. Find R_T when (a) $R_1 = 470 \ \Omega$, $R_2 = 820 \ \Omega$, $R_3 = 1 \ k\Omega$, (b) $R_1 = 15 \ k\Omega$, $R_2 = 47 \ k\Omega$, $R_3 = 68 \ k\Omega$, (c) $R_1 = 47 \ \Omega$, $R_2 = 100 \ \Omega$, $R_3 = 150 \ \Omega$, and (d) $R_1 = 330 \ k\Omega$, $R_2 = 470 \ k\Omega$, $R_3 = 560 \ k\Omega$.

6. Find R_T when (a) $R_1 = 1.2 \ k\Omega$, $R_2 = 5.6 \ k\Omega$, $R_3 = 2.2 \ k\Omega$, (b) $R_1 = 82 \ k\Omega$, $R_2 = 75 \ k\Omega$, $R_3 = 100 \ k\Omega$, (c) $R_1 = 47 \ \Omega$, $R_2 = 39 \ \Omega$, $R_3 = 120 \ \Omega$, and (d) $R_1 = 330 \ \Omega$, $R_2 = 250 \ \Omega$, $R_3 = 510 \ \Omega$.

7. Find R_3 when (a) $R_1 = 3.3 \ k\Omega$, $R_2 = 4.7 \ k\Omega$, $R_T = 13.6 \ k\Omega$, and (b) $R_1 = 68 \ k\Omega$, $R_2 = 18 \ k\Omega$, $R_T = 168 \ k\Omega$.

8. Find R_3 when (a) $R_1 = 750 \ \Omega$, $R_2 = 910 \ \Omega$, $R_T = 2.48 \ k\Omega$, and (b) $R_1 = 1.5 \ k\Omega$, $R_2 = 2.7 \ k\Omega$, $R_T = 7.50 \ k\Omega$.

Refer to Figure 11–16(d).

9. Find G_T and R_T when (a) $R_1 = 10 \ k\Omega$, $R_2 = 20 \ k\Omega$, (b) $R_1 = 1.2 \ k\Omega$, $R_2 = 2.7 \ k\Omega$, (c) $R_1 = 47 \ k\Omega$, $R_2 = 18 \ k\Omega$, and (d) $R_1 = 75 \ k\Omega$, $R_2 = 150 \ k\Omega$.

10. Find G_T and R_T when (a) $R_1 = 75 \ \Omega$, $R_2 = 120 \ \Omega$, (b) $R_1 = 560 \ \Omega$, $R_2 = 1.8 \ k\Omega$, (c) $R_1 = 180 \ k\Omega$, $R_2 = 200 \ k\Omega$, and (d) $R_1 = 82 \ k\Omega$, $R_2 = 100 \ k\Omega$.

Refer to Figure 11–16(e).

11. Find G_T and R_T when (a) $R_1 = 2.2 \ k\Omega$, $R_2 = 6.8 \ k\Omega$, $R_3 = 12 \ k\Omega$, (b) $R_1 = 68 \ k\Omega$, $R_2 = 200 \ k\Omega$, $R_3 = 82 \ k\Omega$, (c) $R_1 = 910 \ \Omega$, $R_2 = 2.2 \ k\Omega$, $R_3 = 1.2 \ k\Omega$, and (d) $R_1 = 120 \ k\Omega$, $R_2 = 68 \ k\Omega$, $R_3 = 270 \ k\Omega$.

12. Find G_T and R_T when (a) $R_1 = 18 \ k\Omega$, $R_2 = 18 \ k\Omega$, $R_3 = 18 \ k\Omega$, (b) $R_1 = 470 \ \Omega$, $R_2 = 820 \ \Omega$, $R_3 = 910 \ \Omega$, (c) $R_1 = 33 \ k\Omega$, $R_2 = 22 \ k\Omega$, $R_3 = 8.2 \ k\Omega$, and (d) $R_1 = 220 \ k\Omega$, $R_2 = 1.2 \ M\Omega$, $R_3 = 68 \ k\Omega$.

13. Find R_2 when (a) $R_1 = 56$ kΩ, $R_3 = 33$ kΩ, $R_T = 3.83$ kΩ, and (b) $R_1 = 100$ Ω, $R_3 = 300$ Ω, $R_T = 54.5$ Ω.

14. Find R_3 when (a) $R_1 = 120$ kΩ, $R_2 = 470$ kΩ, $R_T = 78.2$ kΩ, and (b) $R_1 = 27$ kΩ, $R_2 = 39$ kΩ, $R_T = 12.9$ kΩ.

END OF CHAPTER PROBLEMS 11–2

1. Refer to Figure 11–17. Find R_T when (a) $R_1 = 560$ Ω, $R_2 = 1$ kΩ, $R_3 = 820$ Ω, (b) $R_1 = 7.5$ kΩ, $R_2 = 18$ kΩ, $R_3 = 27$ kΩ, (c) $R_1 = 27$ kΩ, $R_2 = 12$ kΩ, $R_3 = 47$ kΩ, and (d) $R_1 = 330$ kΩ, $R_2 = 680$ kΩ, $R_3 = 1$ MΩ.

2. Refer to Figure 11–17. Find R_T when (a) $R_1 = 150$ kΩ, $R_2 = 750$ kΩ, $R_3 = 1$ MΩ, (b) $R_1 = 150$ Ω, $R_2 = 470$ Ω, $R_3 = 270$ Ω, (c) $R_1 = 4.3$ kΩ, $R_2 = 12$ kΩ, $R_3 = 1.2$ kΩ, and (d) $R_1 = 68$ kΩ, $R_2 = 120$ kΩ, $R_3 = 75$ kΩ.

3. Refer to Figure 11–18. Find R_T when (a) $R_1 = 4.7$ kΩ, $R_2 = 6.8$ kΩ, $R_3 = 9.1$ kΩ, (b) $R_1 = 1.8$ kΩ, $R_2 = 1.5$ kΩ, $R_3 = 820$ Ω, (c) $R_1 = 27$ kΩ, $R_2 = 18$ kΩ, $R_3 = 56$ kΩ, and (d) $R_1 = 680$ Ω, $R_2 = 470$ Ω, $R_3 = 2.2$ kΩ.

4. Refer to Figure 11–18. Find R_T when (a) $R_1 = 430$ kΩ, $R_2 = 470$ kΩ, $R_3 = 330$ kΩ, (b) $R_1 = 2.5$ kΩ, $R_2 = 8.2$ kΩ, $R_3 = 15$ kΩ, (c) $R_1 = 27$ kΩ, $R_2 = 2$ kΩ, $R_3 = 20$ kΩ, and (d) $R_1 = 27$ Ω, $R_2 = 39$ Ω, $R_3 = 100$ Ω.

5. Refer to Figure 11–19. Find R_T when (a) $R_1 = 12$ kΩ, $R_2 = 47$ kΩ, $R_3 = 33$ kΩ, $R_4 = 27$ kΩ, (b) $R_1 = 120$ Ω, $R_2 = 470$ Ω, $R_3 = 1$ kΩ, $R_4 = 560$ Ω, (c) $R_1 = 390$ kΩ, $R_2 = 1.2$ MΩ, $R_3 = 820$ kΩ, $R_4 = 750$ kΩ, and (d) $R_1 = 5.1$ kΩ, $R_2 = 6.8$ kΩ, $R_3 = 10$ kΩ, $R_4 = 10$ kΩ.

6. Refer to Figure 11–19. Find R_T when (a) $R_1 = 910$ Ω, $R_2 = 2.7$ kΩ, $R_3 = 10$ kΩ, $R_4 = 4.7$ kΩ, (b) $R_1 = 100$ kΩ, $R_2 = 330$ kΩ, $R_3 = 820$ kΩ, $R_4 = 180$ kΩ, (c) $R_1 = 47$ Ω, $R_2 = 120$ Ω, $R_3 = 910$ Ω, $R_4 = 220$ Ω, and (d) $R_1 = 100$ kΩ, $R_2 = 33$ kΩ, $R_3 = 33$ kΩ, $R_4 = 47$ kΩ.

7. Refer to Figure 11–20. Find R_T when $R_1 = 1.2$ kΩ, $R_2 = 10$ kΩ, $R_3 = 15$ kΩ, $R_4 = 6.8$ kΩ, $R_5 = 4.7$ kΩ, and $R_6 = 5.6$ kΩ.

8. Refer to Figure 11–20. Find R_T when $R_1 = 680$ Ω, $R_2 = 2.2$ kΩ, $R_3 = 7.5$ kΩ, $R_4 = 330$ Ω, $R_5 = 820$ Ω, $R_6 = 750$ Ω.

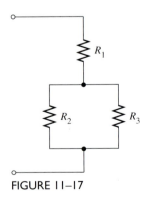

FIGURE 11–17

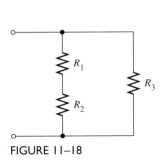

FIGURE 11–18

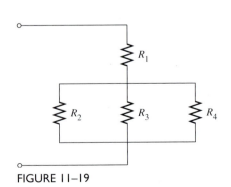

FIGURE 11–19

9. Refer to Figure 11–21. Find R_T when
$R_1 = 120 \text{ k}\Omega$, $R_2 = 270 \text{ k}\Omega$,
$R_3 = 91 \text{ k}\Omega$, $R_4 = 330 \text{ k}\Omega$,
$R_5 = 470 \text{ k}\Omega$, $R_6 = 560 \text{ k}\Omega$,
$R_7 = 1 \text{ M}\Omega$, $R_8 = 680 \text{ k}\Omega$, and
$R_9 = 820 \text{ k}\Omega$.

10. Refer to Figure 11–21. Find R_T when
$R_1 = 47 \text{ k}\Omega$, $R_2 = 100 \text{ k}\Omega$,
$R_3 = 39 \text{ k}\Omega$, $R_4 = 330 \text{ k}\Omega$,
$R_5 = 220 \text{ k}\Omega$, $R_6 = 180 \text{ k}\Omega$,
$R_7 = 250 \text{ k}\Omega$, $R_8 = 82 \text{ k}\Omega$, and
$R_9 = 75 \text{ k}\Omega$.

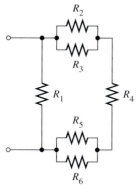

FIGURE 11–20

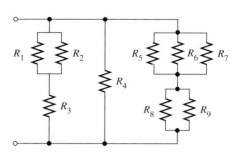

FIGURE 11–21

END OF CHAPTER PROBLEMS 11–3

Refer to Figure 11–22.

1. Find I, R_T, V_1, and V_2 when
(a) $E = 9 \text{ V}$, $R_1 = 270 \ \Omega$,
$R_2 = 680 \ \Omega$, (b) $E = 12 \text{ V}$,
$R_1 = 1.2 \text{ k}\Omega$, $R_2 = 3.3 \text{ k}\Omega$,
(c) $E = 40 \text{ V}$, $R_1 = 120 \text{ k}\Omega$,
$R_2 = 220 \text{ k}\Omega$, and (d) $E = 10 \text{ V}$,
$R_1 = 10 \text{ k}\Omega$, $R_2 = 3.3 \text{ k}\Omega$.

2. Find I, R_T, V_1, and V_2 when
(a) $E = 30 \text{ V}$, $R_1 = 4.7 \text{ k}\Omega$,
$R_2 = 22 \text{ k}\Omega$, (b) $E = 5 \text{ V}$,
$R_1 = 47 \text{ k}\Omega$, $R_2 = 33 \text{ k}\Omega$,
(c) $E = 18 \text{ V}$, $R_1 = 180 \ \Omega$,
$R_2 = 1.5 \text{ k}\Omega$, and (d) $E = 12 \text{ V}$,
$R_1 = 750 \ \Omega$, $R_2 = 680 \ \Omega$.

3. Find I, R_T, V_1, and R_2 when
(a) $E = 10 \text{ V}$, $R_1 = 1 \text{ k}\Omega$, $V_2 = 3 \text{ V}$,
(b) $E = 40 \text{ V}$, $R_1 = 6.8 \text{ k}\Omega$,
$V_2 = 23.5 \text{ V}$, (c) $E = 15 \text{ V}$,
$R_1 = 8.2 \text{ k}\Omega$, $V_2 = 12 \text{ V}$, and
(d) $E = 100 \text{ V}$, $R_1 = 220 \text{ k}\Omega$,
$V_2 = 29.3 \text{ V}$.

4. Find I, R_T, V_1, and R_2 when
(a) $E = 12 \text{ V}$, $R_1 = 470 \ \Omega$,
$V_2 = 3.71 \text{ V}$, (b) $E = 40 \text{ V}$,
$R_1 = 68 \text{ k}\Omega$, $V_2 = 22.3 \text{ V}$,
(c) $E = 60 \text{ V}$, $R_1 = 2.7 \text{ k}\Omega$,
$V_2 = 20 \text{ V}$, and (d) $E = 10 \text{ V}$,
$R_1 = 120 \ \Omega$, $V_2 = 3.13 \text{ V}$.

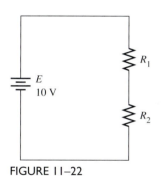

FIGURE 11–22

5. Find R_1, R_T, V_1, and V_2 when
   (a) $E = 20$ V, $I = 213$ μA,
   $R_2 = 33$ kΩ, (b) $E = 25$ V, $I = 1$ mA,
   $R_2 = 10$ kΩ, (c) $E = 12$ V,
   $I = 7.69$ mA, $R_2 = 1$ kΩ, and
   (d) $E = 9$ V, $I = 20$ μA,
   $R_2 = 180$ kΩ.
7. Find E, R_1, R_T, and V_2 when
   (a) $I = 75$ μA, $V_1 = 6.2$ mV,
   $R_2 = 3$ kΩ, (b) $I = 2$ mA,
   $V_1 = 14.3$ V, $R_2 = 12$ kΩ,
   (c) $I = 167$ μA, $V_1 = 4.50$ V,
   $R_2 = 33$ kΩ, and (d) $I = 14.9$ mA,
   $V_1 = 1.78$ V, $R_2 = 82$ Ω.

6. Find R_1, R_T, V_1, and V_2 when
   (a) $E = 100$ V, $I = 2.50$ mA,
   $R_2 = 25$ kΩ, (b) $E = 60$ V,
   $I = 6.67$ mA, $R_2 = 4.3$ kΩ,
   (c) $E = 5$ V, $I = 20$ mA,
   $R_2 = 120$ Ω, and (d) $E = 20$ V,
   $I = 364$ μA, $R_2 = 22$ kΩ.
8. Find E, R_1, R_T, and V_2 when
   (a) $I = 192$ μA, $V_1 = 15.7$ V,
   $R_2 = 56$ kΩ, (b) $I = 1.32$ mA,
   $V_1 = 6.44$ V, $R_2 = 3.9$ kΩ,
   (c) $I = 16.1$ mA, $V_1 = 19.3$ V,
   $R_2 = 1.5$ kΩ, and (d) $I = 14.5$ mA,
   $V_1 = 682$ mV, $R_2 = 91$ Ω.

Refer to Figure 11–23.

9. Find I, R_T, V_1, V_2, and V_3 when
   (a) $E = 9$ V, $R_1 = 33$ kΩ,
   $R_2 = 10$ kΩ, $R_3 = 18$ kΩ,
   (b) $E = 40$ V, $R_1 = 220$ kΩ,
   $R_2 = 100$ kΩ, $R_3 = 68$ kΩ,
   (c) $E = 25$ V, $R_1 = 820$ Ω,
   $R_2 = 1.2$ kΩ, $R_3 = 560$ Ω, and
   (d) $E = 15$ V, $R_1 = 1.8$ kΩ,
   $R_2 = 4.3$ kΩ, $R_3 = 2.5$ kΩ.
11. Find V_2, V_3, R_1, R_3, and R_T when
   (a) $E = 25$ V, $V_1 = 5$ V, $I = 1$ mA,
   $R_2 = 10$ kΩ, (b) $E = 20$ V,
   $V_1 = 6.7$ V, $I = 150$ μA, $R_2 = 47$ kΩ,
   (c) $E = 80$ V, $V_1 = 29.9$ V,
   $I = 1.66$ mA, $R_2 = 8.2$ kΩ, and
   (d) $E = 3$ V, $V_1 = 433$ mV,
   $I = 15.5$ mA, $R_2 = 91$ Ω.

10. Find I, R_T, V_1, V_2, and V_3 when
   (a) $E = 80$ V, $R_1 = 100$ kΩ,
   $R_2 = 20$ kΩ, $R_3 = 30$ kΩ,
   (b) $E = 100$ V, $R_1 = 22$ kΩ,
   $R_2 = 10$ kΩ, $R_3 = 5.1$ kΩ,
   (c) $E = 9$ V, $R_1 = 2$ kΩ,
   $R_2 = 10$ kΩ, $R_3 = 15$ kΩ, and
   (d) $E = 5$ V, $R_1 = 300$ Ω,
   $R_2 = 100$ Ω, $R_3 = 200$ Ω.
12. Find V_2, V_3, R_1, R_3, and R_T when
   (a) $E = 9$ V, $V_1 = 2.30$ V,
   $I = 19.2$ μA, $R_2 = 250$ kΩ,
   (b) $E = 12$ V, $V_1 = 5.30$ V,
   $I = 77.9$ μA, $R_2 = 47$ kΩ,
   (c) $E = 60$ V, $V_1 = 17.6$ V,
   $I = 58.8$ mA, $R_2 = 470$ Ω, and
   (d) $E = 10$ V, $V_1 = 2.50$ V,
   $I = 500$ μA, $R_2 = 7.5$ kΩ.

FIGURE 11–23

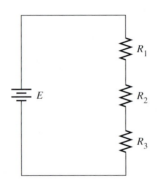

CHAPTER 11

13. Find E, R_2, R_3, V_1, and V_3 when
    (a) $I = 3$ mA, $R_1 = 4.7$ kΩ,
    $V_2 = 7.3$ V, $R_T = 12$ kΩ,
    (b) $I = 270$ μA, $R_1 = 15$ kΩ,
    $V_2 = 3.7$ V, $R_T = 43.2$ kΩ,
    (c) $I = 66.7$ μA, $R_1 = 200$ kΩ,
    $V_2 = 6.67$ V, $R_T = 450$ kΩ, and
    (d) $I = 2.30$ mA, $R_1 = 470$ Ω,
    $V_2 = 690$ mV, $R_T = 870$ Ω.

14. Find E, R_2, R_3, V_1, and V_3 when
    (a) $I = 250$ μA, $R_1 = 20$ kΩ,
    $V_2 = 4.40$ V, $R_T = 91$ kΩ,
    (b) $I = 1.58$ mA, $R_1 = 2.7$ kΩ,
    $V_2 = 15.8$ V, $R_T = 15.8$ kΩ,
    (c) $I = 61.7$ mA, $R_1 = 12$ Ω,
    $V_2 = 1.36$ V, $R_T = 90$ Ω, and
    (d) $I = 30.6$ μA, $R_1 = 1$ MΩ,
    $V_2 = 17.1$ V, $R_T = 2.29$ MΩ.

15. Two resistors are connected in series across a three-volt battery. One of the resistors has a value of five k-ohms. The other has a value of one k-ohms. What is the current through and the voltage drop across each resistor?

16. Two resistors are connected in series across a six-volt battery. One of the resistors has a value of two k-ohms. The other has a value of three k-ohms. What is the current through and the voltage drop across each resistor?

17. Three resistors are connected in series across a nine-volt battery. The battery voltage is divided equally across each resistor. The current is three milliamps. What is the current through and the voltage drop across each resistor? What is the value of each resistor?

18. Three resistors are connected in series across a six-volt battery. The battery voltage is divided equally across each resistor. The current is three hundred milliamps. What is the current through and the voltage drop across each resistor? What is the value of each resistor?

END OF CHAPTER PROBLEMS 11-4

Refer to Figure 11-24.

1. Find G_T, R_T, V, I_1, and I_2 when
   (a) $I_T = 3$ mA, $R_1 = 270$ Ω,
   $R_2 = 560$ Ω, (b) $I_T = 700$ μA,
   $R_1 = 68$ kΩ, $R_2 = 39$ kΩ,
   (c) $I_T = 37.6$ mA, $R_1 = 220$ Ω,
   $R_2 = 270$ Ω, and (d) $I_T = 100$ μA,
   $R_1 = 56$ kΩ, $R_2 = 33$ kΩ.

2. Find G_T, R_T, V, I_1, and I_2 when
   (a) $I_T = 3.55$ mA, $R_1 = 68$ kΩ,
   $R_2 = 47$ kΩ, (b) $I_T = 3$ mA,
   $R_1 = 47$ kΩ, $R_2 = 22$ kΩ,
   (c) $I_T = 22$ mA, $R_1 = 2.7$ kΩ,
   $R_2 = 3.9$ kΩ, and (d) $I_T = 460$ μA,
   $R_1 = 2.7$ kΩ, $R_2 = 3.3$ kΩ.

FIGURE 11-24

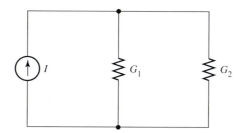

3. Find V, R_2, I_1, G_T, and R_T when
   (a) $I_T = 25$ mA, $R_1 = 680$ Ω,
   $I_2 = 15$ mA, (b) $I_T = 5$ mA,
   $R_1 = 1.8$ kΩ, $I_2 = 1$ mA,
   (c) $I_T = 2.3$ mA, $R_1 = 4.7$ kΩ,
   $I_2 = 1.05$ mA, and (d) $I_T = 30$ mA,
   $R_1 = 4.7$ kΩ, $I_2 = 1.23$ mA.

4. Find V, R_2, I_1, G_T, and R_T when
   (a) $I_T = 400$ μA, $R_1 = 390$ Ω,
   $I_2 = 174$ μA, (b) $I_T = 1$ mA,
   $R_1 = 12$ kΩ, $I_2 = 333$ μA,
   (c) $I_T = 10$ mA, $R_1 = 2.2$ kΩ,
   $I_2 = 4$ mA, and (d) $I_T = 200$ μA,
   $R_1 = 18$ kΩ, $I_2 = 80$ μA.

5. Find I_1, I_2, R_2, G_T, and R_T when
   (a) $I_T = 10$ mA, $V = 20$ V,
   $R_1 = 4.7$ kΩ, (b) $I_T = 350$ μA,
   $V = 25$ V, $R_1 = 180$ kΩ,
   (c) $I_T = 20$ mA, $R_1 = 820$ Ω,
   $V = 9.74$ V, and (d) $I_T = 2$ mA,
   $R_1 = 43$ kΩ, $V = 48.6$ V.

6. Find I_1, I_2, G_T, and R_T when
   (a) $I_T = 5$ mA, $R_1 = 33$ kΩ,
   $V = 85.7$ V, (b) $I_T = 100$ μA,
   $R_1 = 250$ kΩ, $V = 6.88$ V,
   (c) $I_T = 200$ μA, $R_1 = 560$ kΩ,
   $V = 81$ V, and (d) $I_T = 35$ mA,
   $R_1 = 120$ Ω, $V = 2.33$ V.

7. Find I_2, I_T, G_T, R_T, and R_1 when
   (a) $V = 12$ V, $I_1 = 530$ μA,
   $R_2 = 20$ kΩ, (b) $V = 5$ V,
   $I_1 = 1$ mA, $R_2 = 15$ kΩ,
   (c) $V = 30.9$ V, $I_1 = 1.71$ mA,
   $R_2 = 10$ kΩ, and (d) $V = 2.16$ V,
   $I_1 = 263$ μA, $R_2 = 91$ kΩ

8. Find I_2, I, G_T, R_T, and R_1 when
   (a) $V = 99.5$ V, $I_1 = 23.1$ mA,
   $R_2 = 3.3$ kΩ, (b) $V = 39.5$ V,
   $I_1 = 91.9$ mA, $R_2 = 470$ Ω,
   (c) $V = 245$ V, $I_1 = 164$ mA,
   $R_2 = 2.2$ kΩ, and (d) $V = 11.1$ V,
   $I_1 = 503$ μA, $R_2 = 68$ kΩ

Refer to Figure 11–25.

9. Find G_T, R_T, V, I_1, I_2, and I_3 when
   (a) $I_T = 1.73$ mA, $R_1 = 2.7$ kΩ,
   $R_2 = 1.2$ kΩ, $R_3 = 1$ kΩ,
   (b) $I_T = 500$ μA, $R_1 = 8.1$ kΩ,
   $R_2 = 7.5$ kΩ, $R_3 = 4.7$ kΩ,
   (c) $I_T = 500$ μA, $R_1 = 22$ kΩ,
   $R_2 = 39$ kΩ, $R_3 = 56$ kΩ, and
   (d) $I_T = 2$ mA, $R_1 = 2.7$ kΩ,
   $R_2 = 3.9$ kΩ, $R_3 = 1.8$ kΩ.

10. Find G_T, R_T, V, I_1, I_2, and I_3 when
    (a) $I_T = 1$ mA, $R_1 = 18$ kΩ,
    $R_2 = 8.2$ kΩ, $R_3 = 33$ kΩ,
    (b) $I_T = 100$ μA, $R_1 = 9.1$ kΩ,
    $R_2 = 6.8$ kΩ, $R_3 = 5.6$ kΩ,
    (c) $I_T = 5$ mA, $R_1 = 820$ Ω,
    $R_2 = 2$ kΩ, $R_3 = 1.2$ kΩ, and
    (d) $I_T = 10$ mA, $R_1 = 820$ Ω,
    $R_2 = 680$ Ω, $R_3 = 430$ Ω.

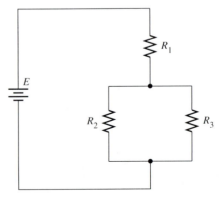

FIGURE 11–26

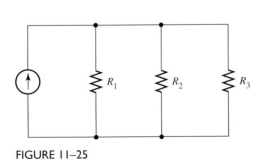

FIGURE 11–25

11. Find R_1, R_3, R_T, G_T, I_2, and I_3 when
    (a) $I_T = 4$ mA, $I_1 = 700$ μA,
    $V = 12.6$ V, $R_2 = 12$ kΩ,
    (b) $I_T = 300$ μA, $I_1 = 75$ μA,
    $V = 20$ V, $R_2 = 270$ kΩ,
    (c) $I_T = 300$ μA, $I_1 = 34.5$ μA,
    $V = 23.4$ V, $R_2 = 750$ kΩ, and
    (d) $I_T = 60$ mA, $I_1 = 15.1$ mA,
    $V = 13.8$ V, $R_2 = 680$ Ω.

12. Find R_1, R_3, R_T, G_T, I_2, and I_3 when
    (a) $I_T = 200$ μA, $I_1 = 36.4$ μA,
    $V = 36.4$ V, $R_2 = 680$ kΩ,
    (b) $I_T = 80$ mA, $I_1 = 19$ mA,
    $V = 88.9$ V, $R_2 = 2.2$ kΩ,
    (c) $I_T = 10$ mA, $I_1 = 4.16$ mA,
    $V = 162$ V, $R_2 = 47$ kΩ, and
    (d) $I_T = 2.5$ mA, $I_1 = 1.39$ mA,
    $V = 3.48$ V, $R_2 = 8.2$ kΩ.

13. Find I_1, I_3, I_T, R_1, R_2, R_3, and
    R_T when (a) $V = 15$ V, $G_1 = 213$ μS,
    $I_2 = 3$ mA, $G_T = 1$ mS, (b) $V = 30$ V,
    $G_1 = 2$ mS, $I_2 = 60$ mA, $G_T = 8$ mS,
    (c) $V = 6.09$ V, $I_2 = 10.9$ mA,
    $G_1 = 1$ mS, $G_T = 3.29$ mS, and
    (d) $V = 5.29$ V, $G_1 = 13.3$ μS,
    $I_2 = 53$ μA, $G_T = 28.3$ μS.

14. Find I_1, I_3, I_T, R_1, R_2, R_3, and R_T when
    (a) $V = 12$ V, $G_1 = 4.55$ μS,
    $I_2 = 31.4$ μA, $G_T = 34.1$ μS,
    (b) $V = 80$ V, $G_1 = 21.3$ μS,
    $I_2 = 1.34$ mA, $G_T = 64.9$ μS,
    (c) $V = 9$ V, $I_2 = 123$ μA,
    $G_1 = 100$ μS, $G_T = 146$ μS, and
    (d) $V = 25$ V, $G_1 = 455$ μS,
    $I_2 = 5.14$ mA, $G_T = 858$ μS.

15. Two resistors are connected in parallel across a battery. One resistor has a resistance of twelve k-ohms and the current through it is two milliamps. The current through the other resistor is six milliamps. What is the battery voltage, the total current, and the value of the second resistor?

16. Two resistors are connected in parallel across a battery. One resistor has a resistance of three k-ohms and the current through it is four milliamps. The current through the other resistor is eight hundred microamps. What is the battery voltage, the total current, and the value of the second resistor?

17. Three resistors are connected in parallel across an energy source. The total current from the source is one hundred eighty milliamps. Fifty percent of the current flows through the first branch, thirty percent through the second branch, and the remainder through the third branch. The resistance in the first branch is one k-ohm. Calculate the voltage drops, currents, and resistance of each branch. Calculate the total resistance.

18. Three resistors are connected in parallel across an energy source. The total current from the source is three milliamps. Ten percent of the current flows through the first branch, thirty percent through the second branch, and the remainder through the third branch. The resistance in the first branch is ten thousand ohms. Calculate the voltage drops, currents, and resistance of each branch. Calculate the total resistance.

END OF CHAPTER PROBLEMS 11–5

Refer to Figure 11–26 on the previous page for Problems 1 through 4.

1. Find V_1, V_2, V_3, I_1, I_2, I_3, I_T, and
   R_T when (a) $E = 15$ V, $R_1 = 10$ kΩ,
   $R_2 = 33$ kΩ, $R_3 = 47$ kΩ, and
   (b) $E = 9$ V, $R_1 = 180$ Ω, $R_2 = $
   560 Ω, $R_3 = 680$ Ω.

2. Find V_1, V_2, V_3, I_1, I_2, I_3, I_T, and
   R_T when (a) $E = 50$ V, $R_1 = 8.2$ kΩ,
   $R_2 = 3.3$ kΩ, $R_3 = 4.7$ kΩ, and
   (b) $E = 5$ V, $R_1 = 1.2$ kΩ, $R_2 = 3$ kΩ,
   and $R_3 = 2.2$ kΩ.

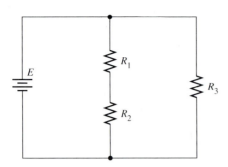

FIGURE 11–27

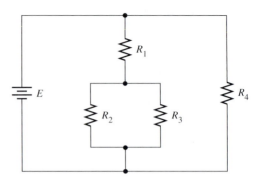

FIGURE 11–28

3. (a) Let $E = 10$ V, $I_T = 500$ μA, $R_1 = 12$ kΩ, and $R_2 = 47$ kΩ. Find R_3, R_T, V_1, V_2, V_3, I_2, and I_3.
   (b) Let $E = 12$ V, $I_T = 600$ μA, $R_2 = 39$ kΩ, and $R_3 = 27$ kΩ. Find R_1, R_T, V_1, V_2, V_3, I_2, and I_3.

5. Refer to Figure 11–27. Find V_1, V_2, V_3, I_1, I_2, I_3, I_T, and R_T when
   (a) $E = 60$ V, $R_1 = 12$ kΩ, $R_2 = 22$ kΩ, $R_3 = 18$ kΩ, and
   (b) $E = 3$ V, $R_1 = 7.5$ kΩ, $R_2 = 27$ kΩ, $R_3 = 47$ kΩ.

7. Refer to Figure 11–28. (a) Let $E = 40$ V, $R_1 = 680$ Ω, $R_2 = 1.5$ kΩ, $R_3 = 2.7$ kΩ, and $R_4 = 4.7$ kΩ. Find V_1, V_2, V_3, V_4, I_T, I_1, I_2, I_3, I_4, and R_T. (b) Let $E = 30$ V, $R_1 = 150$ kΩ, $R_2 = 470$ kΩ, $R_3 = 680$ kΩ, and $R_4 = 680$ kΩ. Find V_1, V_2, V_3, V_4, I_T, I_1, I_2, I_3, I_4, and R_T.

9. Refer to Figure 11–29. (a) Let $E = 25$ V, $R_1 = 3$ kΩ, $R_2 = 15$ kΩ, $R_3 = 22$ kΩ, and $R_4 = 18$ kΩ. Find I_T, I_1, I_2, I_3, I_4, R_T, V_1, V_2, V_3, and V_4.
   (b) Let $E = 25$ V, $I_T = 250$ μA, $V_2 = 10$ V, $R_3 = 150$ kΩ, and $R_4 = 200$ kΩ. Find I_1, I_2, I_3, I_4, R_1, R_2, and R_T.

4. (a) Let $E = 20$ V, $I_T = 1.34$ mA, $R_1 = 12$ kΩ, and $R_2 = 10$ kΩ. Find R_3, R_T, V_1, V_2, V_3, I_2, and I_3.
   (b) Let $E = 25$ V, $I_T = 48.3$ μA, $R_2 = 470$ kΩ, and $R_3 = 330$ kΩ. Find R_1, R_T, V_1, V_2, V_3, I_2, and I_3.

6. Refer to Figure 11–27. Find V_1, V_2, V_3, I_1, I_2, I_3, I_T, and R_T when
   (a) $E = 25$ V, $R_1 = 15$ kΩ, $R_2 = 10$ kΩ, $R_3 = 20$ kΩ, and
   (b) $E = 100$ V, $R_1 = 2.5$ kΩ, $R_2 = 4.3$ kΩ, $R_3 = 5.6$ kΩ.

8. Refer to Figure 11–28. (a) Let $E = 70$ V, $R_1 = 1$ MΩ, $R_2 = 470$ kΩ, $R_3 = 750$ kΩ, and $R_4 = 680$ kΩ. Find V_1, V_2, V_3, V_4, I_T, I_1, I_2, I_3, I_4, and R_T. (b) Let $E = 45$ V, $R_1 = 910$ Ω, $R_2 = 1.5$ kΩ, $R_3 = 470$ Ω, and $R_4 = 1.2$ kΩ. Find V_1, V_2, V_3, V_4, I_T, I_1, I_2, I_3, I_4, and R_T.

10. Refer to Figure 11–29. (a) Let $E = 20$ V, $R_1 = 6.8$ kΩ, $R_2 = 10$ kΩ, $R_3 = 15$ kΩ, and $R_4 = 8.2$ kΩ. Find I_T, I_1, I_2, I_3, I_4, R_T, V_1, V_2, V_3, and V_4.
    (b) Let $E = 9$ V, $I_T = 156$ μA, $V_2 = 4.79$ V, $R_3 = 75$ kΩ, and $R_4 = 68$ kΩ. Find I_1, I_2, I_3, I_4, R_1, R_2, and R_T.

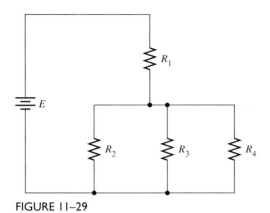

FIGURE 11–29

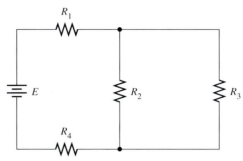

FIGURE 11–30

11. Refer to Figure 11–30. (a) Let $E =$ 40 V, $R_1 = 10$ kΩ, $R_2 = 27$ kΩ, $R_3 =$ 56 kΩ, and $R_4 = 33$ kΩ. Find V_1, V_2, V_3, V_4, I_1, I_2, I_3, I_4, I_T, and R_T. (b) Let $E = 15$ V, $R_1 = 150$ kΩ, $R_2 = 470$ kΩ, $R_3 = 680$ kΩ, and $R_4 = 150$ kΩ. Find V_1, V_2, V_3, V_4, I_1, I_2, I_3, I_4, I_T, and R_T.

13. Three resistors are connected in a series–parallel arrangement across a twelve-volt battery. The resistor values are one hundred ohms, four hundred seventy ohms, and eight hundred twenty ohms. One end of the one hundred ohm resistor is connected to the positive terminal of the battery. One end of the other two resistors are connected to the negative terminal of the battery. To complete the circuit, the other end of all three resistors are connected together. Find the voltage drop across and the current through each resistor, the total resistance, and the total current.

15. Two branches are connected across a forty-five-volt source. The first branch consists of two three-k-ohm resistors connected in series. The second branch consists of a one k-ohm resistor connected in series with a two k-ohm resistor. Calculate the voltage drop across and the current through each resistor. Calculate the total current and total resistance.

12. Refer to Figure 11–30. (a) Let $E = 15$ V, $R_1 = 1$ kΩ, $R_2 = 12$ kΩ, $R_3 = 18$ kΩ, and $R_4 = 3$ kΩ. Find V_1, V_2, V_3, V_4, I_1, I_2, I_3, I_4, I_T, and R_T. (b) Let $E = 50$ V, $R_1 = 56$ kΩ, $R_2 = 120$ kΩ, $R_3 = 1$ MΩ, and $R_4 = 270$ kΩ. Find V_1, V_2, V_3, V_4, I_1, I_2, I_3, I_4, I_T, and R_T.

14. Three resistors are connected in a series–parallel arrangement across a twelve-volt battery. The resistor values are fifteen hundred ohms, twenty-five hundred ohms, and fifty-six hundred ohms. One end of the fifteen hundred ohm resistor is connected to the positive terminal of the battery. One end of the other two resistors are connected to the negative terminal of the battery. To complete the circuit, the other end of all three resistors are connected together. Find the voltage drop across and the current through each resistor, the total resistance, and the total current.

16. Two branches are connected across a thirty-six-volt source. The first branch consists of two fifteen-k-ohm resistors connected in series. The second branch consists of a ten k-ohm resistor connected in series with a twenty k-ohm resistor. Calculate the voltage drop across and the current through each resistor. Calculate the total current and total resistance.

ANSWERS TO SELF-TESTS

1. 122 kΩ
2. 487 Ω
3. 19.0 kΩ
4. 29.1 kΩ
5. 571 Ω
6. 503 kΩ

1. $R_T = 24.3$ kΩ,
   $I = 1.03$ mA,
   $R_2 = 6.32$ kΩ,
   $V_1 = 18.5$ V

2. $E = 160$ V,
   $V_1 = 54.4$ V,
   $V_3 = 55.6$ V,
   $R_3 = 69.5$ kΩ,
   $R_2 = 62.5$ kΩ

3. $R_T = 33.3$ Ω,
   $G_T = 30$ mS,
   $I_1 = 10$ mA,
   $I_2 = 20$ mA,
   $R_2 = 50$ Ω

4. $G_T = 100$ μS,
   $R_T = 10$ kΩ,
   $I_2 = 370$ μA,
   $I_3 = 430$ μA,
   $R_1 = 50$ kΩ,
   $R_3 = 23.3$ kΩ

1. $R_T = 42.9$ kΩ,
   $V_1 = 1.64$ V,
   $V_2 = V_3 = 1.36$ V,
   $I_2 = 41.1$ μA,
   $I_3 = 28.9$ μA,
   $R_1 = 23.5$ kΩ

2. $V_1 = 13.4$ V,
   $V_3 = 20$ V,
   $I_1 = I_2 = 672$ μA,
   $I_3 = 2$ mA,
   $I_T = 2.67$ mA,
   $R_T = 7.49$ kΩ,
   $R_2 = 9.78$ kΩ

3. $I_T = I_1 = 571$ μA,
   $I_2 = 81.4$ μA,
   $I_3 = 370$ μA,
   $I_4 = 120$ μA,
   $V_1 = 6.86$ V,
   $V_2 = V_3 = V_4 = 8.14$ V,
   $R_T = 26.3$ kΩ

4. $V_1 = 9.15$ V,
   $V_2 = V_3 = 8.40$ V,
   $V_4 = 7.45$ V,
   $I_T = 3.39$ mA,
   $I_1 = I_4 = 3.39$ mA,
   $I_2 = 2.15$ mA,
   $I_3 = 1.23$ mA,
   $R_T = 7.38$ kΩ

dc Circuit Analysis: Circuit Theorems

12

Introduction

It is often necessary to determine the internal resistance (or conductance) of an electrical energy source, amplifier, or system. We are also frequently asked to determine current and voltage distribution in electrical circuits for which Ohm's law and Kirchhoff's laws are not adequate. A variety of theorems have been developed over the years to assist the technician and engineer in solving these complex circuit problems. Thévenin's theorem, Norton's theorem, and the superposition theorem are some of the more widely used theorems.

Chapter Objectives

In this chapter you will learn how to:

1. Use the superposition theorem to solve complex circuit problems.
2. Use Thévenin's theorem to solve complex circuit problems.
3. Use Norton's theorem to solve complex circuit problems.

SUPERPOSITION THEOREM   12–1

The superposition theorem is one of the theorems we can use to simplify circuits that contain more than one source. Let's assume that R_2 is the load in Figure 12–1 and find the load current using the superposition theorem. The steps are (1) *short-circuit one supply and calculate the current through the load as a result of the remaining source;* (2) *place the short across the other source and again calculate the current through the load.* The algebraic sum of the two currents is the load current. The polarity of the drop across the load is determined by the direction of the greater current, if they are opposing. (3) *The load current found in step 2 is the load current in the original circuit.* Using this known current, find the other currents and voltage drops using Ohm's law and Kirchhoff's laws.

FIGURE 12–1
Series–parallel
circuit with two
sources.
Solution by
using the
superposition
theorem.

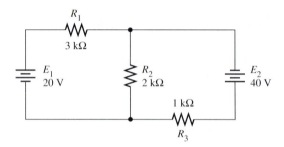

Using this method, we will find the current through and the voltage drop across R_2 in Figure 12–1. Then we will find the other currents and voltage drops. First, replace one source with a short circuit and determine the magnitude and direction of I_2. If we replace E_2 with a short circuit, the circuit shown in Figure 12–2 results.

$$I_T = \frac{E_1}{R_1 + R_X}$$

where $R_X = R_2 \| R_3$.

$$I_T = \frac{20 \text{ V}}{3 \text{ k}\Omega + 667 \text{ }\Omega} = 5.45 \text{ mA}$$

$$I_2 = \frac{I_T R_X}{R_2} = \frac{5.45 \text{ mA} \times 667 \text{ }\Omega}{2 \text{ k}\Omega} = 1.82 \text{ mA}$$

The current is 1.82 mA and has the direction shown. Now let's replace E_1 with a short circuit and compute the current due to E_2. This is done in Figure 12–3.

$$I_T = \frac{E_2}{R_3 + R_X}$$

where $R_X = R_1 \| R_2$.

$$I_T = \frac{40 \text{ V}}{1 \text{ k}\Omega + 1.2 \text{ k}\Omega} = 18.2 \text{ mA}$$

$$I_2 = \frac{I_T R_X}{R_2} = \frac{18.2 \text{ mA} \times 1.2 \text{ k}\Omega}{2 \text{ k}\Omega} = 10.9 \text{ mA}$$

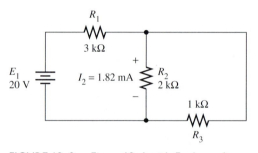

FIGURE 12–2   Figure 12–1 with E_2 shorted.

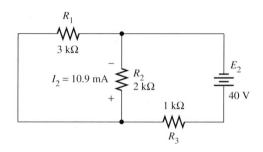

FIGURE 12–3   Figure 12–1 with E_1 shorted.

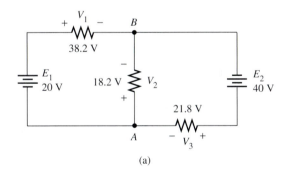

FIGURE 12–4
Figure 12–1
with the drop
across R_2
shown.

The current due to E_2 alone is 10.9 mA, and the polarity is as shown in Figure 12–3. Now, if we go back to the original circuit and superimpose the currents through R_2 as in Figure 12–4, we get one current whose value is 1.82 mA, causing a polarity at point A which is negative with respect to point B, and we get a second current of 10.9 mA, causing a polarity at point A which is positive with respect to point B. The currents are opposing. If we add the currents algebraically, we get a resulting current that is the difference between the two: 10.9 mA − 1.82 mA = 9.09 mA. That is, the actual current through R_2 is 9.09 mA, and point A is positive with respect to point B. The voltage drop across R_2 is 9.09 mA × 2 kΩ, which equals 18.2 V.

Knowing V_2, we can determine V_1 and V_3 by using Kirchhoff's voltage law. Using the loop containing E_1, R_1, and R_2 in Figure 12–5(a), we get the general equation $E_1 + V_1 + V_2 = 0$. Starting at point A and moving clockwise, we get

$$-E_1 + V_1 - V_2 = 0$$

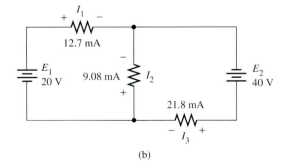

FIGURE 12–5
Figure 12–1
with all voltage
drops and
currents shown.

At this point we don't know the polarity of the drop across V_1; therefore, we assigned positive on the left side. If our assumption is wrong, when we solve for V_1, its magnitude will be correct but its value will be negative. If that turns out to be the case, we simply reverse the assigned polarity. Plugging in known values,

$$-20 \text{ V} + V_1 - 18.2 \text{ V} = 0$$

$$V_1 - 38.2 \text{ V} = 0$$

$$V_1 = 38.2 \text{ V}$$

Because the computed value for V_1 is positive, we have assigned the correct polarity and no change is necessary. Then

$$I_1 = \frac{V_1}{R_1} = \frac{38.2 \text{ V}}{3 \text{ k}\Omega} = 12.7 \text{ mA}$$

Now let's look at the loop containing E_2, R_2, and R_3. Referring again to Figure 12–5(a), the general equation is $E_2 + V_2 + V_3 = 0$. Starting at point B and moving counterclockwise, we get $-V_2 + V_3 + E_2 = 0$. Again, the polarity of the drop across V_3 is not known, and we have assumed it to be positive. Plugging in known values and solving for V_3, we get

$$-18.2 \text{ V} + V_3 + 40 \text{ V} = 0$$

$$V_3 + 21.8 \text{ V} = 0$$

$$V_3 = -21.8 \text{ V}$$

V_3 equals 21.8 V, but the polarity assigned must be reversed since the computed value was negative. Solving for I_3, we get

$$I_3 = \frac{V_3}{R_3} = \frac{21.8 \text{ V}}{1 \text{ k}\Omega} = 21.8 \text{ mA}$$

The complete circuit showing currents is in Figure 12–5(b).

Let's try another one. Find the currents and voltage drops in the circuit shown in Figure 12–6.

FIGURE 12–6
Series–parallel circuit with two sources. Again solution is by using the superposition theorem.

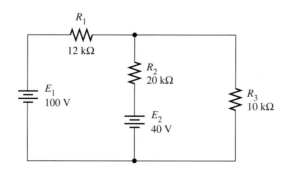

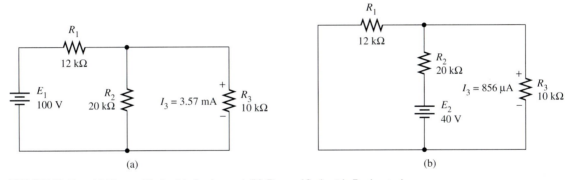

FIGURE 12–7   (a) Figure 12–6 with E_2 shorted. (b) Figure 12–6 with E_1 shorted.

In Figure 12–7(a), the circuit is redrawn with E_2 replaced by a short circuit. The current through R_3 is calculated:

$$I_T = \frac{E_1}{R_1 + R_X}$$

where $R_X = R_2 \parallel R_3$.

$$I_T = \frac{100 \text{ V}}{12 \text{ k}\Omega + 6.67 \text{ k}\Omega} = 5.36 \text{ mA}$$

$$I_3 = \frac{I_T R_X}{R_3} = \frac{5.36 \text{ mA} \times 6.67 \text{ k}\Omega}{10 \text{ k}\Omega} = 3.57 \text{ mA}$$

In Figure 12–7(b) the circuit is again redrawn, but this time E_1 is replaced with a short circuit. I_3 in this circuit is calculated:

$$I_T = \frac{E_2}{R_2 + R_X}$$

where $R_X = R_1 \parallel R_3$.

$$I_T = \frac{40 \text{ V}}{20 \text{ k}\Omega + 5.45 \text{ k}\Omega} = 1.57 \text{ mA}$$

$$I_3 = \frac{I_T R_X}{R_3} = \frac{1.57 \text{ mA} \times 5.45 \text{ k}\Omega}{10 \text{ k}\Omega} = 857 \text{ } \mu\text{A}$$

The current through R_3 in the original circuit (Figure 12–6) is the algebraic sum of these two currents. Since the two currents are in the same direction, the resulting current is the sum of the two individual currents: $I_3 = 3.57 \text{ mA} + 857 \text{ } \mu\text{A} = 4.43 \text{ mA}$. The voltage drop across R_3 can now be determined:

$$V_3 = I_3 R_3 = 4.43 \text{ mA} \times 10 \text{ k}\Omega = 44.3 \text{ V}$$

FIGURE 12–8
Figure 12–6
with polarities
assigned.

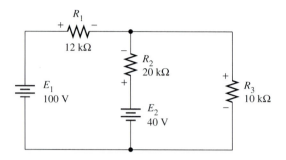

In Figure 12–8, polarities have been assigned. The polarity of the voltage drop across R_3 had to be as shown. This was dictated by the resulting direction of I_3. The drops across R_1 and R_2 were chosen arbitrarily. Remember, if we assign the wrong polarity to V_1 and V_2, the calculated values will be correct but they will be negative. Using Kirchhoff's voltage law, we get

$$E_1 + V_1 + V_3 = 0$$
$$100\text{ V} - V_1 - 44.3\text{ V} = 0$$
$$-V_1 + 55.7\text{ V} = 0$$
$$V_1 = 55.7\text{ V}$$
$$E_2 - V_3 - V_2 = 0$$
$$40\text{ V} - 44.3\text{ V} - V_2 = 0$$
$$V_2 = -4.29\text{ V}$$

V_2 is negative, which tells us that the assigned polarity is wrong. The polarity is corrected in Figure 12–9. Now we can find I_1 and I_2 by using Ohm's law:

$$I_1 = \frac{V_1}{R_1} = \frac{55.7\text{ V}}{12\text{ k}\Omega} = 4.64\text{ mA}$$

$$I_2 = \frac{V_2}{R_2} = \frac{4.29\text{ V}}{20\text{ k}\Omega} = 214\ \mu\text{A}$$

FIGURE 12–9
Figure 12–6
showing all
voltage drops.

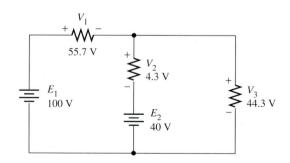

Kirchhoff's laws are now used to check our work. I_1 is total current. Therefore,

$$I_1 = I_2 + I_3$$

$$4.64 \text{ mA} = 214 \text{ } \mu\text{A} + 4.43 \text{ mA}$$

Also,

$$E_1 - V_3 - V_1 = 0$$

$$100 \text{ V} - 44.3 \text{ V} - 55.7 \text{ V} = 0$$

And

$$E_1 - E_2 - V_2 - V_1 = 0$$

$$100 \text{ V} - 40 \text{ V} - 4.29 \text{ V} - 55.7 \text{ V} = 0$$

PRACTICE PROBLEMS 12–1

Find the various voltage drops and currents in the circuits of Figure 12–10.

SOLUTIONS

(a) In Figure 12–10(a) the current through R_2 is 45.3 mA. Replacing E_2 with a short circuit, we get

$$R_T = R_1 + R_2 \| R_3 = 100 \text{ } \Omega + 132 \text{ } \Omega$$

$$= 232 \text{ } \Omega$$

$$I_T = \frac{E_1}{R_T} = \frac{40 \text{ V}}{232 \text{ } \Omega} = 172 \text{ mA}$$

$$I_2 = \frac{I_T R_X}{R_2} = \frac{172 \text{ mA} \times 132 \text{ } \Omega}{330 \text{ } \Omega}$$

$$= 69.0 \text{ mA}$$

I_2 due to E_1 is 69.0 mA.

Replacing E_1 with a short circuit, we get

$$R_T = R_3 + R_1 \| R_2 = 220 \text{ } \Omega$$

$$+ 76.7 \text{ } \Omega = 297 \text{ } \Omega$$

$$I_T = \frac{E_2}{R_T} = \frac{30 \text{ V}}{297 \text{ } \Omega} = 101 \text{ mA}$$

$$I_2 = \frac{I_T R_X}{R_2} = \frac{101 \text{ mA} \times 76.7 \text{ } \Omega}{330 \text{ } \Omega}$$

$$= 23.5 \text{ mA}$$

I_2 due to E_2 is 23.5 mA.

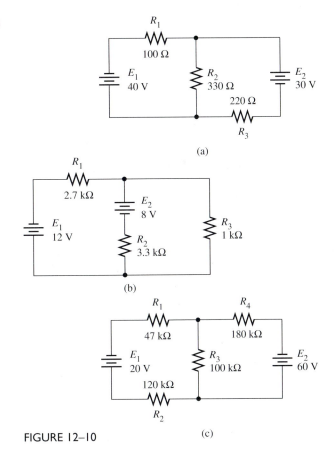

(a)

(b)

(c)

FIGURE 12–10

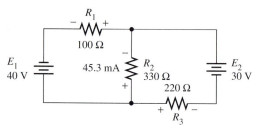

FIGURE 12–11

The currents are opposing. Therefore, the resulting current is the difference between the two currents, or 45.3 mA. The polarities are shown in Figure 12–11. Using Ohm's law and Kirchhoff's laws, we can find the rest of the currents and voltage drops.

$$V_2 = I_2R_2 = 45.3 \text{ mA} \times 330 \text{ }\Omega$$

$$= 15.0 \text{ V}$$

$$V_1 = E_1 - V_2 = 40 \text{ V} - 15.0 \text{ V}$$

$$= 25.0 \text{ V}$$

$$V_3 = E_2 + V_2 = 30 \text{ V} + 15.0 \text{ V}$$

$$= 45.0 \text{ V}$$

$$I_1 = \frac{V_1}{R_1} = \frac{25.1 \text{ V}}{100 \text{ }\Omega} = 250 \text{ mA}$$

$$I_3 = \frac{V_3}{R_3} = \frac{44.9 \text{ V}}{220 \text{ }\Omega} = 205 \text{ mA}$$

(b) In Figure 12–10(b) the current through R_3 is 1.2 mA. Replacing E_1 with a short circuit, we get

$$R_T = R_2 + R_1 \| R_3 = 3.3 \text{ k}\Omega$$

$$+ 730 \text{ }\Omega = 4.03 \text{ k}\Omega$$

$$I_T = \frac{E_2}{R_T} = \frac{8 \text{ V}}{4.03 \text{ k}\Omega} = 1.99 \text{ mA}$$

$$I_3 = \frac{I_TR_X}{R_3} = \frac{1.99 \text{ mA} \times 730 \text{ }\Omega}{1 \text{ k}\Omega}$$

$$= 1.45 \text{ mA}$$

I_3 due to E_2 is 1.45 mA.

Replacing E_2 with a short circuit, we get

$$R_T = R_1 + R_2 \| R_3$$

$$= 2.7 \text{ k}\Omega + 767 \text{ }\Omega$$

$$= 3.47 \text{ k}\Omega$$

$$I_T = \frac{E_1}{R_T} = \frac{12 \text{ V}}{3.47 \text{ k}\Omega} = 3.46 \text{ mA}$$

$$I_3 = \frac{I_TR_X}{R_3} = \frac{3.46 \text{ mA} \times 767 \text{ }\Omega}{1 \text{ k}\Omega}$$

$$= 2.66 \text{ mA}$$

I_3 due to E_1 is 2.66 mA.

The two currents are opposing and the resulting current through R_3 is 1.2 mA. The polarities are shown in Figure 12–12. Using Ohm's law and Kirchhoff's laws, we can find the rest of the currents and voltage drops.

$$V_3 = I_3R_3 = 1.2 \text{ mA} \times 1 \text{ k}\Omega$$

$$= 1.21 \text{ V}$$

$$V_2 = E_2 + V_3 = 8 \text{ V} + 1.21 \text{ V}$$

$$= 9.21 \text{ V}$$

$$V_1 = E_1 - V_3 = 12 \text{ V} - 1.21 \text{ V}$$

$$= 10.8 \text{ V}$$

$$I_2 = \frac{V_2}{R_2} = \frac{9.21 \text{ V}}{3.3 \text{ k}\Omega} = 2.79 \text{ mA}$$

$$I_1 = \frac{V_1}{R_1} = \frac{10.8 \text{ V}}{2.7 \text{ k}\Omega} = 4.00 \text{ mA}$$

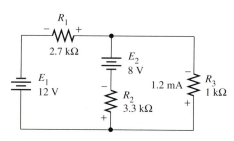

FIGURE 12–12

(c) In Figure 12–10(c) the current through R_3 is 211 μA. Replacing E_1 with a short circuit, we get

$$R_T = R_4 + R_3 \| (R_1 + R_2)$$

$$= 180 \text{ k}\Omega + 62.5 \text{ k}\Omega = 243 \text{ k}\Omega$$

$$I_T = \frac{E_2}{R_T} = \frac{60 \text{ V}}{243 \text{ k}\Omega} = 247 \text{ }\mu\text{A}$$

$$I_3 = \frac{I_T R_X}{R_3} = \frac{247 \text{ }\mu\text{A} \times 62.5 \text{ k}\Omega}{100 \text{ k}\Omega}$$

$$= 155 \text{ }\mu\text{A}$$

I_3 due to E_2 is 155 μA.
   Replacing E_2 with a short circuit,

$$R_T = R_1 + R_2 + R_3 \| R_4$$

$$= 47 \text{ k}\Omega + 120 \text{ k}\Omega + 64.3 \text{ k}\Omega$$

$$= 231 \text{ k}\Omega$$

$$I_T = \frac{E_1}{R_T} = \frac{20 \text{ V}}{231 \text{ k}\Omega} = 86.5 \text{ }\mu\text{A}$$

$$I_3 = \frac{I_T R_X}{R_3} = \frac{86.5 \text{ }\mu\text{A} \times 64.3 \text{ k}\Omega}{100 \text{ k}\Omega}$$

$$= 55.6 \text{ }\mu\text{A}$$

I_3 due to E_1 is 55.6 μA.
   The two currents are aiding in making the resulting current through R_3 the sum of the two currents, or 210 μA. The polarities are shown in Figure 12–13. Using Ohm's law and Kirchhoff's laws, the rest of the currents and voltage drops may be found.

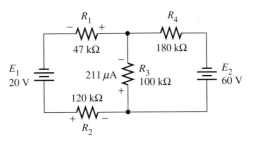

FIGURE 12–13

$$V_3 = I_3 R_3 = 210 \text{ }\mu\text{A} \times 100 \text{ k}\Omega$$

$$= 21.0 \text{ V}$$

$$V_4 = E_2 - V_3 = 60 \text{ V} - 21.0 \text{ V}$$

$$= 39.0 \text{ V}$$

$$I_4 = \frac{V_4}{R_4} = \frac{39.0 \text{ V}}{180 \text{ k}\Omega} = 216 \text{ }\mu\text{A}$$

$$V_1 + V_2 = E_2 - V_4 - E_1$$

$$= 60 \text{ V} - 39.0 \text{ V} - 20 \text{ V}$$

$$= 1.03 \text{ V}$$

$$I_1 = I_2 = \frac{V_1 + V_2}{R_1 + R_2} = \frac{1.03 \text{ V}}{167 \text{ k}\Omega}$$

$$= 6.18 \text{ }\mu\text{A}$$

$$V_1 = I_1 R_1 = 6.18 \text{ }\mu\text{A} \times 47 \text{ k}\Omega$$

$$= 0.290 \text{ V}$$

$$V_2 = I_2 R_2 = 6.18 \text{ }\mu\text{A} \times 120 \text{ k}\Omega$$

$$= 0.741 \text{ V}$$

Additional practice problems are at the end of the chapter.

SELF-TEST 12–1

Use the superposition theorem to find the various currents and voltage drops in the circuits in Figure 12–14.

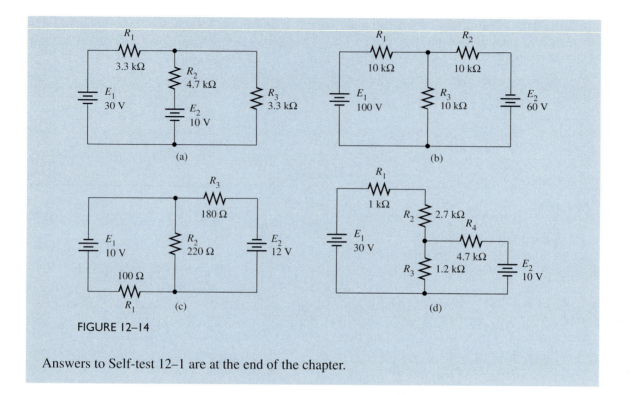

FIGURE 12-14

Answers to Self-test 12–1 are at the end of the chapter.

12–2   THÉVENIN'S THEOREM

Thévenin's theorem states that any network, no matter how complex, can be reduced to an equivalent voltage source and series resistance. The voltage source is labeled V_{OC} and the series resistance is R_{TH}, as illustrated in Figure 12–15. Consider the circuit in Figure 12–16. Because this is a simple series–parallel circuit, the various currents and voltage drops can be found easily by using Ohm's law and Kirchhoff's laws.

$$R_T = R_1 + R_2 \parallel R_3 = 1 \text{ k}\Omega + 500 \ \Omega = 1.5 \text{ k}\Omega$$

$$I_T = \frac{E}{R_T} = \frac{10 \text{ V}}{1.5 \text{ k}\Omega} = 6.67 \text{ mA}$$

$$I_1 = I_T = 6.67 \text{ mA}$$

$$V_1 = I_1 R_1 = 6.67 \text{ mA} \times 1 \text{ k}\Omega = 6.67 \text{ V}$$

$$V_2 = V_3 = E - V_1 = 10 \text{ V} - 6.67 \text{ V} = 3.33 \text{ V}$$

$$I_2 = \frac{V_2}{R_2} = \frac{3.33 \text{ V}}{1 \text{ k}\Omega} = 3.33 \text{ mA}$$

$$I_3 = \frac{V_3}{R_3} = \frac{3.33 \text{ V}}{1 \text{ k}\Omega} = 3.33 \text{ mA}$$

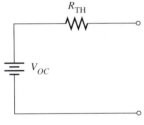

FIGURE 12–15   Thévenin's
equivalent circuit.

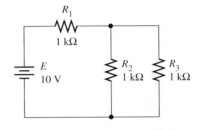

FIGURE 12–16   Series–parallel
circuit to be Thévenized.

In circuits such as this, circuit theorems are not needed. Let's go ahead and Thévenize this circuit anyway just to develop a basic understanding of Thévenin's theorem. Remember that we said Thévenin's circuit is an *equivalent* circuit. This means that the Thévenin equivalent circuit and the circuit it replaces furnish the same energy to the load. Suppose in Figure 12–16 we consider that R_3 is the load. We may consider that E, R_1, and R_2 are the circuit that supplies energy to R_3, just like V_{OC} and R_{TH}, as indicated in Figure 12–17.

In Thévenizing, we first determine from which points we wish to examine the circuit. In Figure 12–17 we are considering that R_3 is the load. Therefore, the circuit to the left of terminals x–y will be Thévenized.

To find Thévenin's equivalent circuit, we use the following procedure:

1. *Remove the load and calculate the difference in potential between the open-circuit terminals (x–y). This is the voltage source V_{OC} in our equivalent circuit.* In Figure 12–18 the load (R_3) has been removed. Looking back from the open-circuit terminals, we see that the circuit has been reduced to a simple series circuit and $V_{OC} = V_2$. Applying Ohm's law, we get $V_{OC} = 5$ V.
2. *With the load removed as in step 1, replace the source with a short circuit and determine the resistance at the open-circuit terminals.*[1] This resistance is R_{TH} in

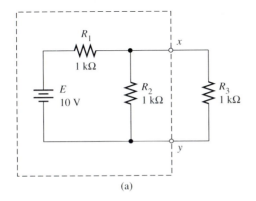

(a)

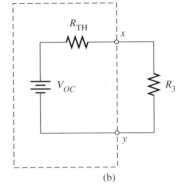

(b)

FIGURE 12–17
Circuit in Figure
12–16 (a) using
R_3 as the load;
(b) Equivalent
circuit.

---

[1]We have assumed that the source is an ideal voltage source and $R_{int} = 0 \ \Omega$. If the source resistance were some finite value, then the source would be replaced with that resistance instead of a short circuit.

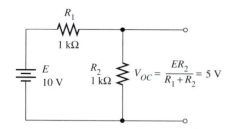

FIGURE 12–18   Circuit used to find V_{OC} for the circuit in Figure 12–16.

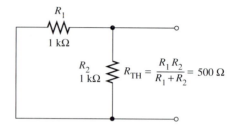

FIGURE 12–19   Circuit used to find R_{TH} for the circuit in Figure 12–16.

our equivalent circuit. Replacing the source with a short circuit as in Figure 12–19 results in a circuit with R_1 and R_2 in parallel. This parallel circuit is seen looking back into the circuit from the open-circuit terminals. The equivalent resistance (R_{TH}) is 500 Ω. V_{OC} and R_{TH} are connected in series to form Thévenin's equivalent circuit in Figure 12–20.

3. *Connect the load across the output terminals of the Thévenin equivalent circuit. Calculate V_L and I_L, the voltage drop across and the current through the load. Using Ohm's law in Figure 12–21, we get*

$$I_L = \frac{V_{OC}}{R_{TH} + R_L} \tag{12–1}$$

$$I_L = \frac{5 \text{ V}}{1.5 \text{ k}\Omega} = 3.33 \text{ mA}$$

$$V_L = \frac{V_{OC}R_L}{R_{TH} + R_L} \tag{12–2}$$

$$V_L = \frac{5 \text{ V} \times 1 \text{ k}\Omega}{1.5 \text{ k}\Omega} = 3.33 \text{ V}$$

Thévenin's equivalent circuit causes the same current through R_3 and the same voltage drop across it as does the original circuit. Furthermore, the polarity associated with

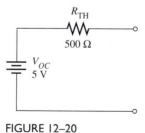

FIGURE 12–20
Thévenin's equivalent circuit.

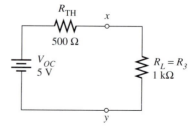

FIGURE 12–21   Equivalent circuit with the load connected.

V_{OC} in Figure 12–18 is the polarity of the voltage drop across R_3 in the original circuit. (In Figure 12–17 it was obvious that point y was negative with respect to point x, but in some complex circuits it is not so obvious.)

Knowing V_3 and I_3 (remember we called R_3 the load), we could find the rest of the voltage drops and currents just as we did in Section 12–1 (superposition theorem).

Let's Thévenize some rather simple series–parallel circuits. We will find only V_L and I_L.

PRACTICE PROBLEMS 12–2

For the problems in Figure 12–22:

1. Develop Thévenin's equivalent circuit.
2. Find V_L and I_L.

SOLUTIONS

(a) In Figure 12–22(a), if we open circuit the load, $V_{OC} = V_2$.
Thévenin's equivalent circuit with the load connected is shown in Figure 12–23. Applying Ohm's law, we get

FIGURE 12–22

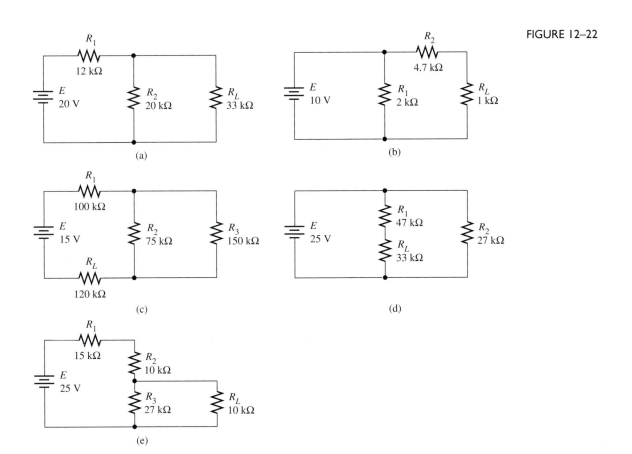

(a)

(b)

(c)

(d)

(e)

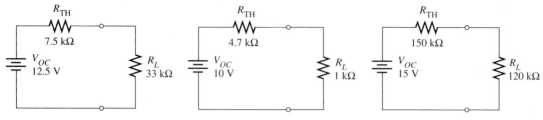

FIGURE 12–23   FIGURE 12–24   FIGURE 12–25

$$V_{OC} = V_2 = \frac{ER_2}{R_1 + R_2} = \frac{20 \text{ V} \times 20 \text{ k}\Omega}{32 \text{ k}\Omega}$$

$$= 12.5 \text{ V}$$

$$R_{TH} = R_1 \parallel R_2 = 7.5 \text{ k}\Omega$$

$$I_L = \frac{V_{OC}}{R_{TH} + R_L} = \frac{12.5 \text{ V}}{40.5 \text{ k}\Omega} = 309 \text{ } \mu A$$

$$V_L = \frac{V_{OC}R_L}{R_{TH} + R_L} = \frac{12.5 \text{ V} \times 33 \text{ k}\Omega}{40.5 \text{ k}\Omega}$$

$$= 10.2 \text{ V}$$

or $V_L = I_L R_L = 309 \text{ } \mu A \times 33 \text{ k}\Omega$

$$= 10.2 \text{ V}$$

(b) In Figure 12–22(b), if we open circuit the load, $V_{OC} = V_1 = E$. Shorting the source places a short circuit across R_1 so that $R_{TH} = R_2 = 4.7 \text{ k}\Omega$. Thévenin's equivalent circuit with the load connected is shown in Figure 12–24. Applying Ohm's law, we get

$$I_L = \frac{V_{OC}}{R_{TH} + R_L} = \frac{10 \text{ V}}{5.7 \text{ k}\Omega} = 1.75 \text{ mA}$$

$$V_L = I_L R_L = 1.75 \text{ mA} \times 1 \text{ k}\Omega$$

$$= 1.75 \text{ V}$$

(c) In Figure 12–22(c), if we open circuit the load, there is no circuit current. Therefore, $V_{OC} = E = 15 \text{ V}$.

$$R_{TH} = R_1 + R_2 \parallel R_3 = 100 \text{ k}\Omega$$

$$+ 50 \text{ k}\Omega = 150 \text{ k}\Omega$$

Thévenin's equivalent circuit with the load connected is shown in Figure 12–25.

$$I_L = \frac{V_{OC}}{R_{TH} + R_L} = \frac{15 \text{ V}}{270 \text{ k}\Omega} = 55.6 \text{ } \mu A$$

$$V_L = I_L R_L = 55.6 \text{ } \mu A \times 120 \text{ k}\Omega$$

$$= 6.67 \text{ V}$$

(d) In Figure 12–22(d), if we open circuit the load, V_{OC} again equals E since V_1 would equal 0 V. Shorting the source to find R_{TH} causes a short circuit across R_2. Therefore, $R_{TH} = R_1 = 47 \text{ k}\Omega$. Thévenin's equivalent circuit with the load connected is shown in Figure 12–26.

$$I_L = \frac{V_{OC}}{R_{TH} + R_L} = \frac{25 \text{ V}}{80 \text{ k}\Omega} = 313 \text{ } \mu A$$

$$V_L = I_L R_L = 313 \text{ } \mu A \times 33 \text{ k}\Omega$$

$$= 10.3 \text{ V}$$

(e) In Figure 12–22(e), if we open circuit the load, $V_{OC} = V_3$.

$$V_{OC} = V_3$$

$$= \frac{ER_3}{R_1 + R_2 + R_3} = \frac{25 \text{ V} \times 27 \text{ k}\Omega}{52 \text{ k}\Omega}$$

$$= 13 \text{ V}$$

$$R_{TH} = R_3 \parallel (R_1 + R_2) = 13 \text{ k}\Omega$$

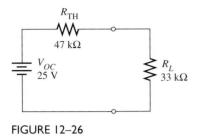

FIGURE 12–26

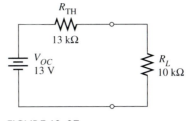

FIGURE 12–27

Thévenin's equivalent circuit with the load connected is shown in Figure 12–27.

$$I_L = \frac{V_{OC}}{R_{TH} + R_L} = \frac{13\ \text{V}}{23\ \text{k}\Omega} = 565\ \mu\text{A}$$

$$V_L = I_L R_L = 565\ \mu\text{A} \times 10\ \text{k}\Omega$$

$$= 5.65\ \text{V}$$

Additional practice problems are at the end of the chapter.

THÉVENIN'S THEOREM AND COMPLEX CIRCUITS   12–3

Now let's Thévenize some circuits that are more complex. Consider the circuit in Figure 12–28. We have already found the various currents and voltage drops by using the superposition theorem. We will consider that R_2 is the load and find I_2 and V_2 by using Thévenin's theorem.

First, let's open circuit the load, as in Figure 12–29(a), and calculate V_{OC}. Because E_1 and E_2 are connected series aiding, $E_T = E_1 + E_2 = 60$ V. Therefore, we may consider that there is an emf of 60 V in series with R_1 and R_3.

$$V_1 = \frac{E_T R_1}{R_1 + R_3} = \frac{60\ \text{V} \times 3\ \text{k}\Omega}{4\ \text{k}\Omega} = 45\ \text{V}$$

$$V_3 = E_T - V_1 = 60\ \text{V} - 45\ \text{V} = 15\ \text{V}$$

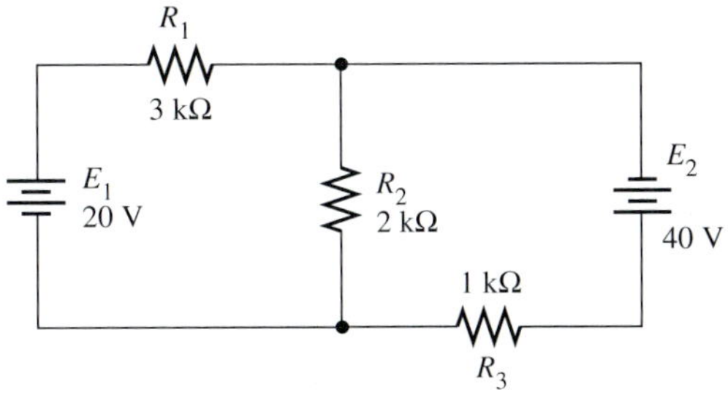

FIGURE 12–28
Complex circuit to be solved using Thévenin's theorem.

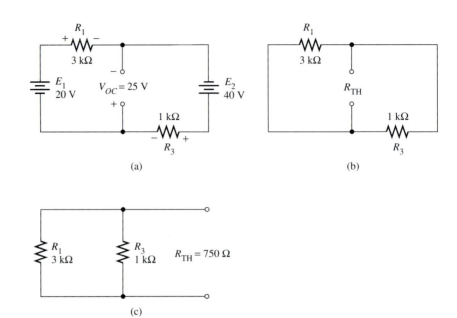

FIGURE 12–29
(a) Remove the
load and find
V_{OC}; (b) replace
voltage sources
with short
circuits and find
R_{TH}; (c) Figure
12–29(b)
redrawn.

Using Kirchhoff's voltage law, we get $V_{OC} = E_2 - V_3 = 25$ V (or $V_{OC} = V_1 - E_1 = 25$ V) and the polarity is as shown.

Next we replace E_1 and E_2 with short circuits as in Figure 12–29(b) and calculate R_{TH}. In Figure 12–29(c) the circuit has been redrawn so that we can see that R_1 and R_3 are in parallel.

$$R_{TH} = R_1 \parallel R_3 = 750 \ \Omega$$

Now let's draw Thévenin's equivalent circuit and connect the load across it as in Figure 12–30.

$$I_L = \frac{V_{OC}}{R_{TH} + R_L} = \frac{25 \text{ V}}{2.75 \text{ k}\Omega} = 9.09 \text{ mA}$$

$$V_L = I_L R_L = 9.09 \text{ mA} \times 2 \text{ k}\Omega = 18.2 \text{ V}$$

Knowing V_L and I_L (V_2 and I_2), we could find the rest of the voltage drops and currents as before. Notice that when we found V_{OC} we also found the polarity of the voltage drop across the load in the original circuit.

Let's find I_3 and V_3 in Figure 12–31 by using Thévenin's theorem. In Figure 12–32 the load has been removed, resulting in a series circuit. The sources are connected series opposing, so $E_T = E_1 - E_2 = 60$ V

$$V_2 = \frac{E_T R_2}{R_1 + R_2} = \frac{60 \text{ V} \times 20 \text{ k}\Omega}{32 \text{ k}\Omega} = 37.5 \text{ V}$$

Using Kirchhoff's voltage law, we get $V_{OC} = E_2 + V_2 = 77.5$ V.

CHAPTER 12

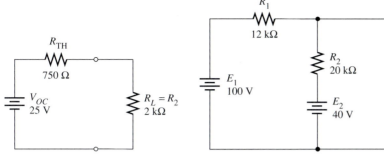

FIGURE 12–30   Equivalent circuit
with the load connected.

FIGURE 12–31   Find values of V_3 and I_3
using Thévenin's theorem.

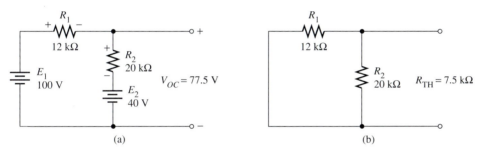

(a)

(b)

FIGURE 12–32   (a) The load (R_3) is removed to find V_{OC}; (b) the sources are shorted to find R_{TH}.

In Figure 12–32(b), E_1 and E_2 are replaced with short circuits and R_{TH} is calculated.

$$R_{TH} = R_1 \parallel R_2 = 7.50 \text{ k}\Omega$$

R_L is connected across Thévenin's equivalent circuit in Figure 12–33.

$$I_L = \frac{V_{OC}}{R_{TH} + R_L} = \frac{77.5 \text{ V}}{17.5 \text{ k}\Omega} = 4.43 \text{ mA}$$

$$V_L = I_L R_L = 4.43 \text{ mA} \times 10 \text{ k}\Omega = 44.3 \text{ V}$$

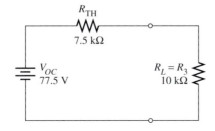

FIGURE 12–33
Equivalent
circuit with the
load corrected.

FIGURE 12–34
Bridge circuit.

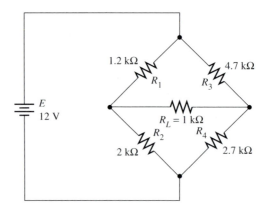

Consider the bridge circuit in Figure 12–34. Let's remove the load and redraw it slightly to help us see the circuit. With the load removed (Figure 12–35), we can treat the circuit as a series–parallel circuit, where $E = V_1 + V_2 = V_3 + V_4$.

$$V_1 = \frac{ER_1}{R_1 + R_2} = 4.5 \text{ V}$$

$$V_2 = E - V_1 = 7.5 \text{ V}$$

$$V_3 = \frac{ER_3}{R_3 + R_4} = 7.62 \text{ V}$$

$$V_4 = E - V_3 = 4.38 \text{ V}$$

Applying Kirchhoff's voltage law starting at point x and moving clockwise, we get

$$V_{OC} + V_4 - V_2 = 0$$

$$V_{OC} + 4.38 \text{ V} - 7.5 \text{ V} = 0$$

$$V_{OC} = 3.12 \text{ V}$$

We assumed that point x was positive with respect to point y. The fact that V_{OC} is positive in our calculation proves that we were right. If V_{OC} had been -3.12 V in our equation, it would simply have meant that point x was *negative* with respect to point y.

FIGURE 12–35
Load removed
to find V_{OC}.

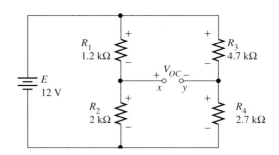

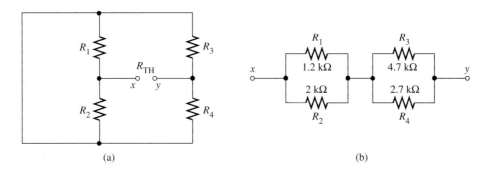

FIGURE 12–36
Source shorted
to find R_{TH}.

(a)   (b)

Now let's short E and calculate R_{TH}. Figure 12–36(a) shows the circuit with E shorted. Since it may be difficult to see the series–parallel circuit, let's redraw the circuit as in Figure 12–36(b).

$$R_{TH} = R_1 \parallel R_2 + R_3 \parallel R_4 = 750 \ \Omega + 1.71 \ \text{k}\Omega = 2.46 \ \text{k}\Omega$$

Thévenin's equivalent circuit with the load connected is shown in Figure 12–37.

$$I_L = \frac{V_{OC}}{R_{TH} + R_L} = \frac{3.12 \ \text{V}}{3.46 \ \text{k}\Omega} = 0.901 \ \text{mA}$$

$$V_L = I_L R_L = 0.902 \ \text{mA} \times 1 \ \text{k}\Omega = 0.901 \ \text{V}$$

Let's look at a typical bias arrangement for a bipolar transistor. Such a circuit is drawn in Figure 12–38. Let's find V_C and I_C. If we assume that $I_B = 0 \ \mu\text{A}$, then $I_C = I_E$.

$$I_E = \frac{\dfrac{V_{CC} R_2}{R_1 + R_2} - V_{BE}}{R_4} = \frac{\dfrac{25 \ \text{V} \times 3.3 \ \text{k}\Omega}{47 \ \text{k}\Omega + 3.3 \ \text{k}\Omega} - 0.6 \ \text{V}}{1 \ \text{k}\Omega} = 1.04 \ \text{mA}$$

$$I_C = I_E = 1.04 \ \text{mA}$$

$$V_C = V_{CC} - I_C R_3 = 25 \ \text{V} - 7.8 \ \text{V} = 17.2 \ \text{V}$$

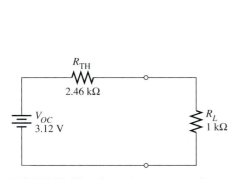

FIGURE 12–37   Equivalent circuit with the load connected.

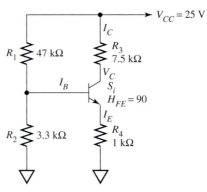

FIGURE 12–38   Transistor amplifier circuit.

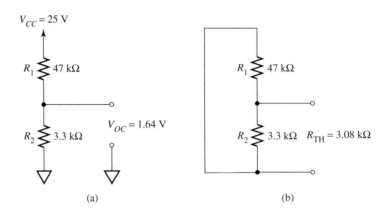

(a)   (b)

However, $I_B = 0\ \mu A$ only if H_{FE} is infinitely large. Since H_{FE} is 90, I_B is some finite value and $I_C \neq I_E$. Using Thévenin's theorem, we can find the exact value of I_C for this transistor. Figure 12–39 shows the circuit with the base lead disconnected. That is, the circuit looking into the base is the load.

$$V_{OC} = \frac{V_{CC}R_2}{R_1 + R_2} = \frac{25\ V \times 3.3\ k\Omega}{47\ k\Omega + 3.3\ k\Omega} = 1.64\ V$$

In Figure 12–39(b) the source is shorted and

$$R_{TH} = R_1 \parallel R_2 = 3.08\ k\Omega$$

Next we will connect the load to the equivalent circuit as in Figure 12–40. We needn't consider the collector since the collector–base junction is reverse biased. Therefore, the load consists of the emitter–base junction in series with some resistance. This resistance, labeled R_L, is the resistance we see looking across the junction from base to emitter.

$$R_L = R_4(H_{FE} + 1) = 91\ k\Omega$$

The base current is the current in this circuit. Therefore,

$$I_B = \frac{V_{OC} - V_{BE}}{R_{TH} + R_L} = \frac{1.64\ V - 0.6\ V}{3.08\ k\Omega + 91\ k\Omega} = 11.1\ \mu A$$

$$I_C = H_{FE}I_B = 90 \times 11.1\ \mu A = 995\ \mu A$$

$$V_C = V_{CC} - I_C R_3 = 25\ V - 7.46\ V = 17.5\ V$$

FIGURE 12–40
The equivalent circuit with the load connected.

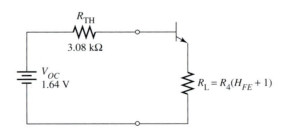

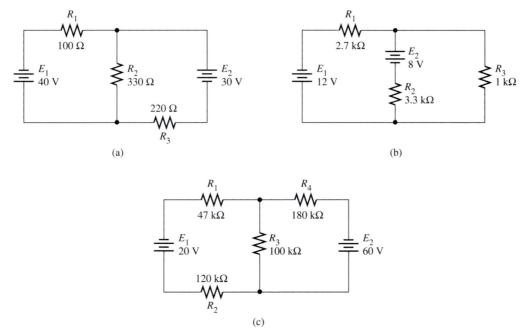

FIGURE 12–41

PRACTICE PROBLEMS 12–3

Develop Thévenin's equivalent circuit for the following problems and find V_L and I_L:

1. Figure 12–41(a). Assume that $R_2 = R_L$.
2. Figure 12–41(b). Assume that $R_3 = R_L$.
3. Figure 12–41(c). Assume that $R_3 = R_L$.
4. Figure 12–42.
5. Figure 12–43. Find I_B, I_C, and V_C.

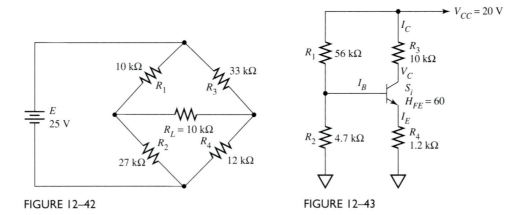

FIGURE 12–42   FIGURE 12–43

SOLUTIONS

1. With R_2 removed, as in Figure 12–44(a), a series circuit results. E_1 and E_2 are connected series aiding. Therefore, 70 V will drop across R_1 and R_3.

$$V_1 = \frac{70 \text{ V} \times R_1}{R_1 + R_3} = \frac{70 \text{ V} \times 100 \text{ }\Omega}{320 \text{ }\Omega}$$

$$= 21.9 \text{ V}$$

Applying Kirchhoff's voltage law to find V_{OC}, we see that the loop equation

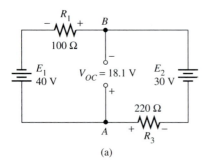

(a)

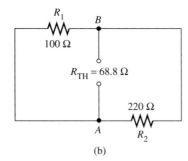

(b)

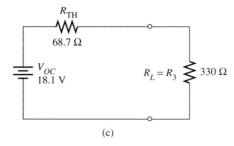

(c)

FIGURE 12–44

is $V_{OC} + V_1 + E = 0$. Moving counterclockwise from point A, we get

$$V_{OC} + 21.9 \text{ V} - 40 \text{ V} = 0$$

$$V_{OC} = 18.1 \text{ V}$$

We assumed that the potential at point A was positive with respect to point B. The fact that V_{OC} is positive indicates that we chose the correct polarity. In Figure 12–44(b) the sources have been shorted. Looking into the circuit from the open-circuit terminals, we see that R_1 and R_3 are in parallel and the equivalent resistance (R_{TH}) is 68.8 Ω. The Thévenin's equivalent circuit is shown in series with the load in Figure 12–44(c).

$$I_L = \frac{V_{OC}}{R_{TH} + R_L} = \frac{18.1 \text{ V}}{399 \text{ }\Omega} = 45.5 \text{ mA}$$

$$V_L = I_L R_L = 45.5 \text{ mA} \times 330 \text{ }\Omega$$

$$= 15.0 \text{ V}$$

2. With R_3 removed, a series circuit results. E_1 and E_2 are connected series aiding, and 20 V will drop across R_1 and R_2.

$$V_2 = \frac{E_T R_2}{R_1 + R_2} = \frac{20 \text{ V} \times 3.3 \text{ k}\Omega}{6 \text{ k}\Omega} = 11 \text{ V}$$

Applying Kirchhoff's voltage law to find V_{OC} in Figure 12–45(a), we see that the loop equation is $V_{OC} + E_2 + V_2 = 0$. Moving counterclockwise from point A, we get:

$$8 \text{ V} - 11 \text{ V} + V_{OC} = 0$$

$$V_{OC} = 3 \text{ V}$$

Shorting E_1 and E_2 results in R_1 in parallel with R_2 looking back from the open-circuit terminals as shown in Fig-

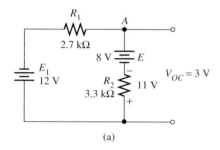

(a)

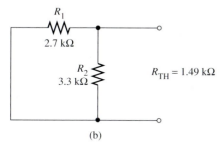

(b)

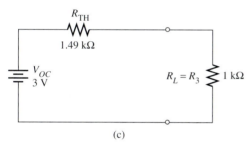

(c)

FIGURE 12–45

ure 12–45(b). This results in an R_{TH} of 1.49 kΩ. Thévenin's equivalent circuit is shown in series with the load in Figure 12–45(c).

$$I_L = \frac{V_{OC}}{R_{TH} + R_L} = \frac{3 \text{ V}}{2.49 \text{ k}\Omega} = 1.21 \text{ mA}$$

$$V_L = I_L R_L = 1.21 \text{ mA} \times 1 \text{ k}\Omega = 1.21 \text{ V}$$

3. E_1 and E_2 are connected series opposing in Figure 12–41(c). Therefore, with the load (R_3) removed as in Figure 12–46(a), 40 V will drop across R_1, R_2, and R_4 in series. Let's work with the loop containing V_{OC}, R_4, and E_2.

$$V_4 = \frac{40 \text{ V} \times R_4}{R_T} = \frac{40 \text{ V} \times 180 \text{ k}\Omega}{347 \text{ k}\Omega}$$

$$= 20.7 \text{ V}$$

$$V_{OC} = E_2 - V_4 = 60 \text{ V} - 20.7 \text{ V}$$

$$= 39.3 \text{ V}$$

The source is replaced with a short circuit, and the resulting circuit is shown in Figure 12–46(b).

$$R_{TH} = R_4 \| (R_1 + R_2) = 86.6 \text{ k}\Omega$$

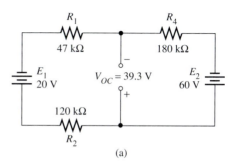

(a)

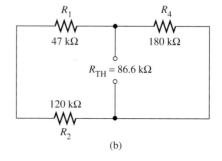

(b)

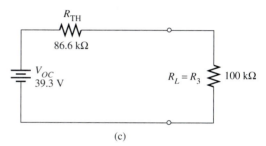

(c)

FIGURE 12–46

Connecting the load resistor to Thévenin's equivalent circuit in Figure 12–46(c), we get:

$$I_L = \frac{V_{OC}}{R_{TH} + R_L} = \frac{39.3\text{ V}}{187\text{ k}\Omega} = 210\ \mu A$$

$$V_L = I_L R_L = 210\ \mu A \times 100\text{ k}\Omega = 21.0\text{ V}$$

4. In Figure 12–47(a) the load has been re-moved and V_2 and V_4 are calculated.

$$V_2 = \frac{ER_2}{R_1 + R_2} = \frac{25\text{ V} \times 27\text{ k}\Omega}{37\text{ k}\Omega}$$

$$= 18.2\text{ V}$$

$$V_4 = \frac{ER_4}{R_3 + R_4} = \frac{25\text{ V} \times 12\text{ k}\Omega}{45\text{ k}\Omega}$$

$$= 6.67\text{ V}$$

Using the circuit consisting of V_2, V_4, and V_{OC} in Figure 12–47(b), we see that $V_{OC} = 11.6$ V and the polarity is

as shown. The circuit, with the source shorted, is redrawn in Figure 12–47(c) and R_{TH} is calculated.

$$R_{TH} = R_1 \parallel R_2 + R_3 \parallel R_4$$

$$= 7.3\text{ k}\Omega + 8.8\text{ k}\Omega = 16.1\text{ k}\Omega$$

Thévenin's equivalent circuit with the load connected is shown in Figure 12–47(d).

$$I_L = \frac{V_{OC}}{R_{TH} + R_L} = \frac{11.6\text{ V}}{26.1\text{ k}\Omega} = 444\ \mu A$$

$$V_L = I_L R_L = 444\ \mu A \times 10\text{ k}\Omega$$

$$= 4.44\text{ V}$$

5. In Figure 12–48(a), with the load re-moved, $V_{OC} = V_2$.

$$V_2 = \frac{V_{CC}R_2}{R_1 + R_2} = \frac{20\text{ V} \times 4.7\text{ k}\Omega}{56\text{ k}\Omega + 4.7\text{ k}\Omega}$$

$$= 1.55\text{ V}$$

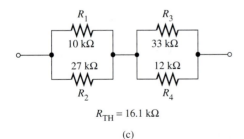

(a)

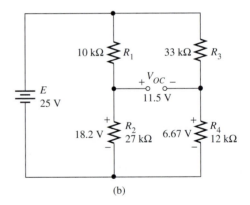

(b)

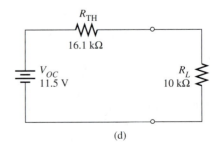

(c)

(d)

FIGURE 12–47

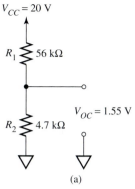

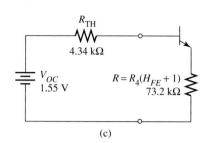

(a)   (b)   (c)

FIGURE 12–48

With the source shorted as in Figure 12–48(b), R_{TH} = 4.34 kΩ.

$$R_{TH} = R_1 \| R_2 = 4.34 \text{ k}\Omega$$

Thévenin's equivalent circuit with the load connected is shown in Figure 12–48(c).

$$I_L = I_B = \frac{V_{OC} - V_{BE}}{R_{TH} + R_L}$$

$$= \frac{1.55 \text{ V} - 0.6 \text{ V}}{77.5 \text{ k}\Omega} = 12.2 \text{ } \mu\text{A}$$

$$I_C = H_{FE}I_B = 60 \times 12.2 \text{ } \mu\text{A}$$

$$= 734 \text{ } \mu\text{A}$$

$$V_C = V_{CC} - V_3 = 20 \text{ V}$$

$$- (734 \text{ } \mu\text{A} \times 10 \text{ k}\Omega)$$

$$= 20 \text{ V} - 7.35 \text{ V} = 12.7 \text{ V}$$

Additional practice problems are at the end of the chapter.

SELF-TEST 12–2

1. In Figure 12–49, assume that R_4 is the load. Thévenize the circuit and find V_L and I_L.

2. In Figure 12–50, assume that R_3 is the load. Thévenize the circuit and find V_L and I_L.

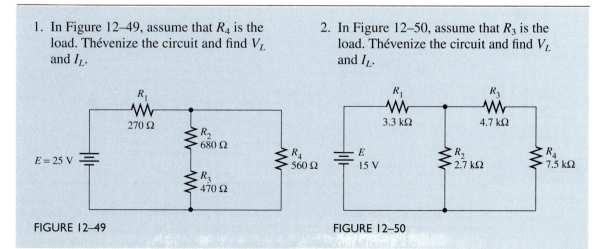

FIGURE 12–49   FIGURE 12–50

3. In Figure 12–51, assume that R_3 is the load. Thévenize the circuit and find V_L and I_L.

4. In Figure 12–52, assume that R_3 is the load. Thévenize the circuit and find V_L and I_L.

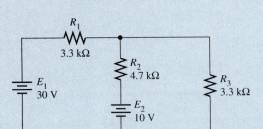

FIGURE 12–51

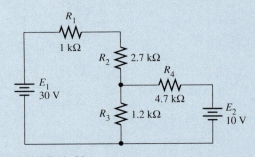

FIGURE 12–52

5. In Figure 12–53, Thévenize the circuit and find V_L and I_L.

6. In Figure 12–54, Thévenize the circuit at the base and find I_B, I_C, and V_C.

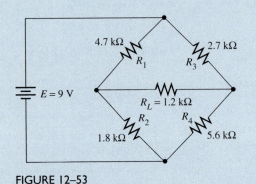

FIGURE 12–53

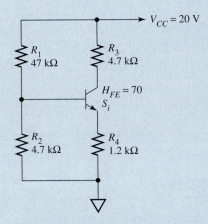

FIGURE 12–54

Answers to Self-test 12–2 are at the end of the chapter.

12–4   NORTON'S THEOREM

Norton's theorem provides the means by which we can reduce a complex circuit to an equivalent current source in parallel with some conductance. The current source is labeled I_{SC} and the parallel conductance is labeled G_N, as illustrated in Figure 12–55.

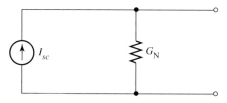

FIGURE 12–55   Norton's equivalent circuit.

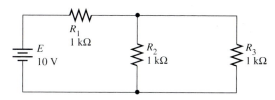

FIGURE 12–56   Circuit to be solved using Norton's theorem.

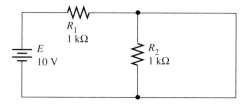

FIGURE 12–57   Circuit in Figure 12–56 with the load shorted.

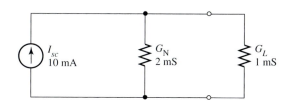

FIGURE 12–58   Equivalent circuit with the load connected.

Consider the circuit in Figure 12–56. This circuit was shown in Figure 12–16 and was Thévenized in Figure 12–17. Assuming that R_3 is the load, Norton's equivalent circuit is found by using the following procedure:

1. Replace the load with a short circuit and calculate the current through the short circuit. This is the current source in our equivalent circuit. In Figure 12–57 the load (R_3) has been replaced with a short circuit. This reduces the circuit resistance to R_1 and the current is 10 mA.

$$ I = \frac{V_1}{R_1} = \frac{10\ \text{V}}{1\ \text{k}\Omega} = 10\ \text{mA} $$

2. Open circuit the load. Replace the source with a short circuit and determine the conductance at the open-circuit terminals. This conductance is the reciprocal of R_{TH}. In some circuits it will be easier to find R_{TH} and then G_N. $G_N = 1/R_{\text{TH}}$. The conductance, G_N, is 2 mS.

3. Connect the load across the output terminals of the Norton's equivalent circuit. Calculate V_L and I_L. Norton's equivalent circuit of Figure 12–56 with the load connected is shown in Figure 12–58.

Using Ohm's law, we get

$$ V_L = \frac{I_{SC}}{G_N + G_L} $$

$$ = \frac{10\ \text{mA}}{3\ \text{mS}} = 3.33\ \text{V} $$

$$ I_L = \frac{I_{SC}G_L}{G_N + G_L} $$

$$= \frac{10 \text{ mA} \times 1 \text{ mS}}{3 \text{ mS}} = 3.33 \text{ mA}$$

or

$$= V_L G_L = 3.33 \text{ V} \times 1 \text{ mS} = 3.33 \text{ mA}$$

Norton's equivalent circuit causes the same current through R_3 and the same voltage drop across it as does the original circuit. Furthermore, the direction of current through the short circuit in Figure 12–57 is the same as in the original circuit.

Knowing V_3 and I_3 (remember we called R_3 the load), we can find the rest of the voltage drops and currents by using Ohm's law and Kirchhoff's laws.

Let's Nortonize some rather simple series–parallel circuits. We will find only V_L and I_L.

PRACTICE PROBLEMS 12–4

For the problems in Figure 12–59:

1. Develop Norton's equivalent circuit.

2. Find V_L and I_L.

FIGURE 12–59

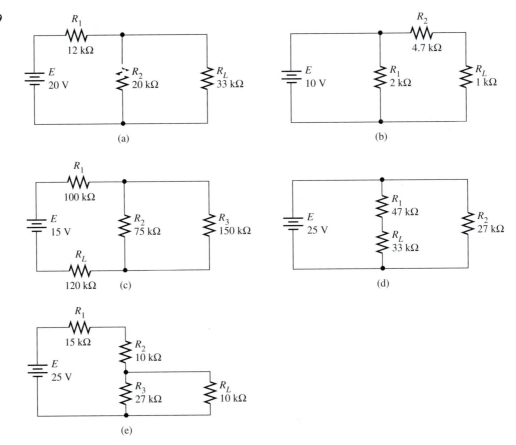

CHAPTER 12

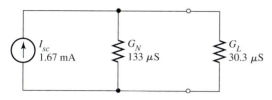

FIGURE 12–60

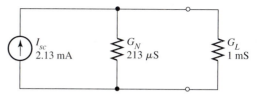

FIGURE 12–61

SOLUTIONS

(a) In Figure 12–59(a), if we short circuit the load, we put a short circuit across R_2 and reduce the circuit resistance to R_1. All the circuit current flows through the short circuit.

$$I_{SC} = \frac{V_1}{R_1} = \frac{20 \text{ V}}{12 \text{ k}\Omega} = 1.67 \text{ mA}$$

$$G_N = G_1 + G_2 = 83.3 \text{ } \mu\text{S} + 50 \text{ } \mu\text{S}$$
$$= 133 \text{ } \mu\text{S}$$

Norton's equivalent circuit with the load connected is shown in Figure 12–60. Applying Ohm's law, we get

$$V_L = \frac{I_{SC}}{G_N + G_L} = \frac{1.67 \text{ mA}}{164 \text{ } \mu\text{S}} = 10.2 \text{ V}$$

$$I_L = V_L G_L = 10.2 \text{ V} \times 30.3 \text{ } \mu\text{A}$$
$$= 309 \text{ } \mu\text{A}$$

(b) In Figure 12–59(b), if we short circuit the load, the short circuit current is the current through R_2.

$$I_{SC} = I_2 = \frac{V_2}{R_2} = \frac{10 \text{ V}}{4.7 \text{ k}\Omega} = 2.13 \text{ mA}$$

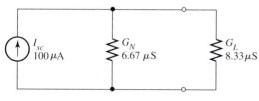

FIGURE 12–62

If we open circuit the load and short circuit the source,

$$R_{TH} = R_2 = 4.7 \text{ k}\Omega$$

$$G_N = \frac{1}{R_{TH}} = 213 \text{ } \mu\text{S}$$

Norton's equivalent circuit with the load connected is shown in Figure 12–61. Applying Ohm's law, we get

$$V_L = \frac{I_{SC}}{G_N + G_L} = \frac{2.13 \text{ mA}}{1.21 \text{ mS}} = 1.75 \text{ V}$$

$$I_L = V_L G_L = 1.75 \text{ mA}$$

(c) In Figure 12.59(c), if we short circuit the load, $R_T = R_1 + R_2 \| R_3$ or

$$R_T = 100 \text{ k}\Omega + 50 \text{ k}\Omega = 150 \text{ k}\Omega$$

$$I_{SC} = I_T = \frac{E}{R_T} = \frac{15 \text{ V}}{150 \text{ k}\Omega} = 100 \text{ } \mu\text{A}$$

$$R_{TH} = R_1 + R_2 \| R_3 = 150 \text{ k}\Omega$$

$$G_N = \frac{1}{R_{TH}} = \frac{1}{150 \text{ k}\Omega} = 6.67 \text{ } \mu\text{S}$$

Norton's equivalent circuit with the load connected is shown in Figure 12–62.

$$V_L = \frac{I_{SC}}{G_N + G_L} = \frac{100 \text{ } \mu\text{A}}{15 \text{ } \mu\text{S}} = 6.67 \text{ V}$$

$$I_L = V_L G_L = 6.67 \text{ V} \times 8.33 \text{ } \mu\text{S}$$
$$= 55.6 \text{ } \mu\text{A}$$

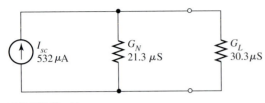

FIGURE 12–63

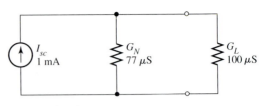

FIGURE 12–64

(d) If we short circuit the load in Figure 12–59(d), the current through R_1 is the short-circuit current.

$$I_{SC} = I_1 = \frac{V_1}{R_1} = \frac{25 \text{ V}}{47 \text{ k}\Omega} = 532 \text{ }\mu\text{A}$$

$$G_N = \frac{1}{R_1} = 21.3 \text{ }\mu\text{S}$$

Norton's equivalent circuit with the load connected is shown in Figure 12–63.

$$V_L = \frac{I_{SC}}{G_N + G_L} = \frac{532 \text{ }\mu\text{A}}{51.6 \text{ }\mu\text{S}} = 10.3 \text{ V}$$

$$I_L = V_L G_L = 10.3 \text{ V} \times 30.3 \text{ }\mu\text{S}$$
$$= 313 \text{ }\mu\text{A}$$

(e) If we short circuit the load in Figure 12–59(e), R_3 is also shorted. Then $R_T = R_1 + R_2$. I_T in the resulting circuit is I_{SC}.

$$I_{SC} = \frac{E}{R_T} = \frac{25 \text{ V}}{25 \text{ k}\Omega} = 1 \text{ mA}$$

$$G_N = \frac{1}{R_3} + \frac{1}{R_1 + R_2} = 37 \text{ }\mu\text{S}$$
$$+ 40 \text{ }\mu\text{S} = 77 \text{ }\mu\text{S}$$

Norton's equivalent circuit with the load connected is shown in Figure 12–64.

$$V_L = \frac{I_{SC}}{G_N + G_L} = \frac{1 \text{ mA}}{177 \text{ }\mu\text{S}} = 5.65 \text{ V}$$

$$I_L = V_L G_L = 5.56 \text{ V} \times 100 \text{ }\mu\text{S}$$
$$= 565 \text{ }\mu\text{A}$$

Additional problems are at the end of the chapter.

SELF-TEST 12–3

1. Refer to Figure 12–59(a). Let $E = 15$ V, $R_1 = 68$ kΩ, $R_2 = 100$ kΩ, and $R_L = 75$ kΩ. Nortonize the circuit and find V_L and I_L.

2. Refer to Figure 12–59(d). Let $E = 20$ V, $R_1 = 680$ Ω, $R_2 = 1.2$ kΩ, and $R_L = 560$ Ω. Nortonize the circuit and find V_L and I_L.

Answers to Self-test 12–3 are at the end of the chapter.

CHAPTER 12 AT A GLANCE

PAGE	THEOREM	EXAMPLE

339   *Superposition Theorem:* To find currents and voltage drops using the superposition theorem:

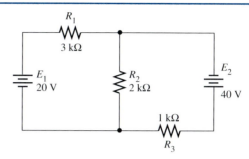

FIGURE 12–65

(1) short-circuit one supply and calculate the current through the load as a result of the remaining source;

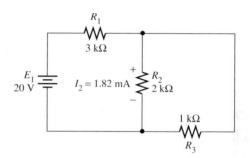

FIGURE 12–66

(2) place the short across the other source and again calculate the current through the load;

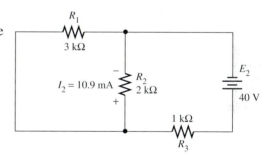

FIGURE 12–67

(3) the algebraic sum of load currents in (2) is the load current in the original circuit. Using this current, calculate other currents and voltage drops.

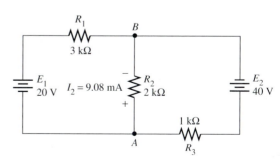

FIGURE 12–68

349   *Thévenin's Theorem:* To find currents and voltage drops using Thévenin's theorem:

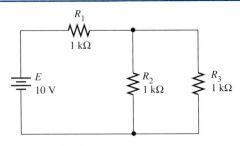

FIGURE 12–69

(1) remove the load and calculate the difference in potential between the open-circuit terminals;

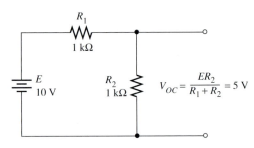

FIGURE 12–70

(2) with the load removed as in step 1, replace the source with a short circuit and determine the resistance at the open-circuit terminals;

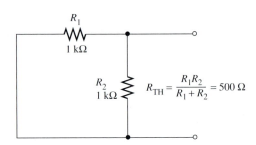

FIGURE 12–71

(3) connect the load across the output terminals of the Thévenin's equivalent circuit. Calculate V_L and I_L.

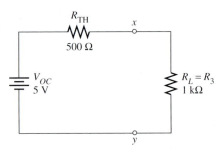

FIGURE 12–72

364   *Norton's Theorem:* To find currents and voltage drops using Norton's theorem:

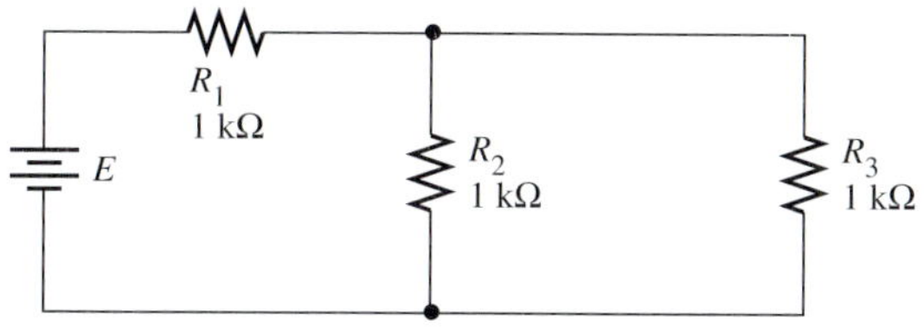

FIGURE 12–73

(1) replace the load with a short circuit and calculate the current through the short circuit;

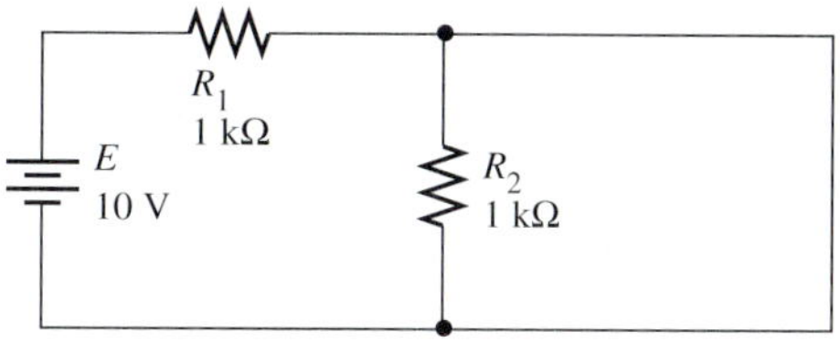

FIGURE 12–74

(2) open circuit the load and replace the source with a short circuit and determine the conductance at the open-circuit terminals;

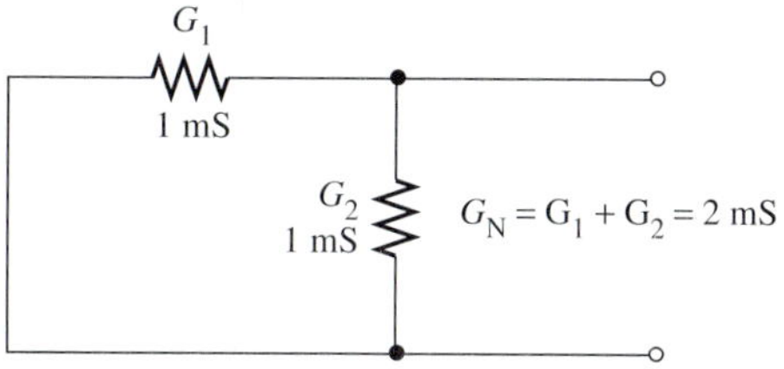

$G_N = G_1 + G_2 = 2$ mS

FIGURE 12–75

(3) connect the load across the output terminals of the equivalent circuit and calculate V_L and I_L.

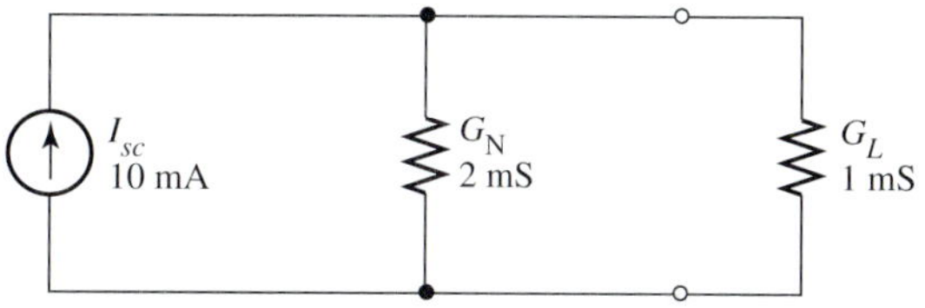

FIGURE 12–76

END OF CHAPTER PROBLEMS 12–1

Use the superposition theorem to find the various currents and voltage drops in the following problems:

1. Refer to Figure 12–77(a). Let
   $E_1 = 25$ V, $E_2 = 50$ V, $R_1 = 2.2$ kΩ,
   $R_2 = 1.8$ kΩ, and $R_3 = 1.2$ kΩ.
3. Refer to Figure 12–77(c). Let
   $E_1 = 100$ V, $E_2 = 40$ V, $R_1 = 2.7$ kΩ,
   $R_2 = 3.3$ kΩ, $R_3 = 1$ kΩ, and
   $R_4 = 4.7$ kΩ

2. Refer to Figure 12–77(b). Let
   $E_1 = 10$ V, $E_2 = 20$ V, $R_1 = 5.6$ kΩ,
   $R_2 = 1$ kΩ, and $R_3 = 4.7$ kΩ.
4. Refer to Figure 12–77(b). Let
   $E_1 = 30$ V, $E_2 = 20$ V, $R_1 = 15$ kΩ,
   $R_2 = 12$ kΩ, and $R_3 = 10$ kΩ.

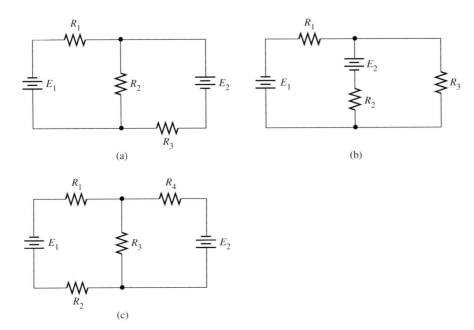

(a)

(b)

(c)

FIGURE 12–77

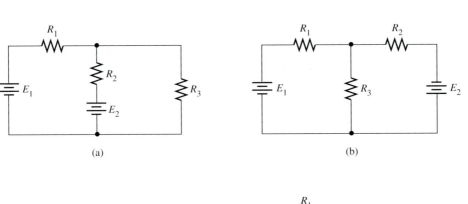

(a)

(b)

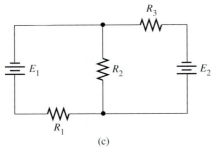

(c)

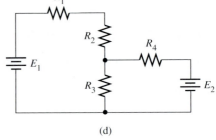

(d)

FIGURE 12–78

5. Refer to Figure 12–78(a) on the previous page. Let $E_1 = 9$ V, $E_2 = 18$ V, $R_1 = 3.3$ kΩ, $R_2 = 4.7$ kΩ, and $R_3 = 3.3$ kΩ.

6. Refer to Figure 12–78(b). Let $E_1 = 40$ V, $E_2 = 30$ V, $R_1 = 4.7$ kΩ, $R_2 = 7.5$ kΩ, and $R_3 = 10$ kΩ.

7. Refer to Figure 12–78(c). Let $E_1 = 25$ V, $E_2 = 10$ V, $R_1 = 680$ Ω, $R_2 = 470$ Ω, and $R_3 = 1$ kΩ.

8. Refer to Figure 12–78(c). Let $E_1 = 12$ V, $E_2 = 30$ V, $R_1 = 18$ kΩ, $R_2 = 22$ kΩ, and $R_3 = 47$ kΩ.

9. Refer to Figure 12–78(d). Let $E_1 = 40$ V, $E_2 = 15$ V, $R_1 = 2.2$ kΩ, $R_2 = 1.2$ kΩ, $R_3 = 3.3$ kΩ, and $R_4 = 3.3$ kΩ.

10. Refer to Figure 12–78(d). Let $E_1 = 50$ V, $E_2 = 20$ V, $R_1 = 10$ kΩ, $R_2 = 20$ kΩ, $R_3 = 30$ kΩ, and $R_4 = 27$ kΩ.

END OF CHAPTER PROBLEMS 12–2

Thévenize each circuit and find V_L and I_L.

1. Refer to Figure 12–79(a), where $E = 30$ V, $R_1 = 22$ kΩ, $R_2 = 47$ kΩ, and $R_L = 10$ kΩ.

2. Refer to Figure 12–79(b), where $E = 20$ V, $R_1 = 20$ kΩ, $R_2 = 47$ kΩ, and $R_L = 33$ kΩ.

FIGURE 12–79

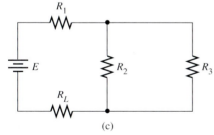

(a)

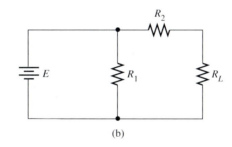

(b)

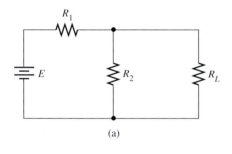

(c)

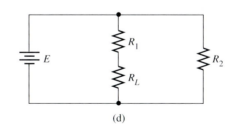

(d)

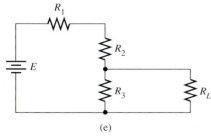

(e)

3. Refer to Figure 12–79(c), where
   $E = 12$ V, $R_1 = 680$ Ω, $R_2 = 470$ Ω,
   $R_3 = 1$ kΩ, and $R_L = 750$ Ω.
5. Refer to Figure 12–79(e), where
   $E = 50$ V, $R_1 = 22$ kΩ, $R_2 = 10$ kΩ,
   $R_3 = 33$ kΩ, and $R_L = 15$ kΩ.

4. Refer to Figure 12–79(d), where
   $E = 40$ V, $R_1 = 5.6$ kΩ, $R_2 = 3.9$ kΩ,
   and $R_L = 2.2$ kΩ.

END OF CHAPTER PROBLEMS 12–3

In the following problems, Thévenize each circuit and find V_L and I_L.

1. Refer to Figure 12–77(a), where
   $E_1 = 25$ V, $E_2 = 50$ V, $R_1 = 2.2$ kΩ,
   $R_2 = 2.5$ kΩ, and $R_3 = 1.2$ kΩ.
   Assume that R_2 is the load.
3. Refer to Figure 12–77(c), where
   $E_1 = 100$ V, $E_2 = 40$ V, $R_1 = 2.7$ kΩ,
   $R_2 = 3.3$ kΩ, $R_3 = 2$ kΩ, and
   $R_4 = 4.7$ kΩ. Assume that R_3 is the
   load.
5. Refer to Figure 12–80, where
   $E = 20$ V, $R_1 = 4.7$ kΩ, $R_2 = 6.8$ kΩ,
   $R_3 = 1.5$ kΩ, $R_4 = 3$ kΩ, and
   $R_L = 1$ kΩ.
7. Refer to Figure 12–81, where
   $V_{CC} = 20$ V, $R_1 = 47$ kΩ,
   $R_2 = 6.8$ kΩ, $R_3 = 2$ kΩ,
   $R_4 = 470$ Ω, and $H_{FE} = 80$. Thévenize
   and find I_B, I_C, and V_C.

2. Refer to Figure 12–77(b), where
   $E_1 = 10$ V, $E_2 = 25$ V, $R_1 = 5.6$ kΩ,
   $R_2 = 1$ kΩ, and $R_3 = 4.7$ kΩ. Assume
   that R_2 is the load.
4. Refer to Figure 12–77(c), where
   $E_1 = 12$ V, $E_2 = 9$ V, $R_1 = 680$ Ω,
   $R_2 = 560$ Ω, $R_3 = 1$ kΩ, and
   $R_4 = 820$ Ω. Assume that R_3 is the
   load.
6. Refer to Figure 12–80, where
   $E = 25$ V, $R_1 = 680$ Ω, $R_2 = 1$ kΩ,
   $R_3 = 2$ kΩ, $R_4 = 1.8$ kΩ, and
   $R_L = 1.2$ kΩ.
8. Refer to Figure 12–81, where
   $V_{CC} = 25$ V, $R_1 = 56$ kΩ,
   $R_2 = 10$ kΩ, $R_3 = 8.2$ kΩ,
   $R_4 = 2$ kΩ, and $H_{FE} = 60$. Thévenize
   and find I_B, I_C, and V_C.

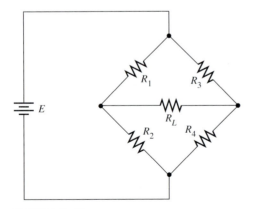

FIGURE 12–80

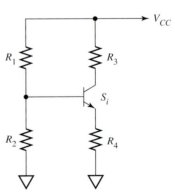

FIGURE 12–81

END OF CHAPTER PROBLEMS 12–4

1. Refer to Figure 12–79(a). Nortonize the circuit and find V_L and I_L where (a) $E = 30$ V, $R_1 = 22$ kΩ, $R_2 = 47$ kΩ, and $R_L = 10$ kΩ, and (b) $E = 50$ V, $R_1 = 18$ kΩ, $R_2 = 33$ kΩ, and $R_L = 39$ kΩ.

2. Refer to Figure 12–79(b). Nortonize the circuit and find V_L and I_L where (a) $E = 20$ V, $R_1 = 27$ kΩ, $R_2 = 56$ kΩ, and $R_L = 33$ kΩ, and (b) $E = 25$ V, $R_1 = 330$ Ω, $R_2 = 270$ Ω, and $R_L = 560$ Ω.

3. Refer to Figure 12–79(c). Nortonize the circuit and find V_L and I_L, where (a) $E = 12$ V, $R_1 = 680$ Ω, $R_2 = 470$ Ω, $R_3 = 1$ kΩ, and $R_L = 750$ Ω, and (b) $E = 10$ V, $R_1 = 12$ kΩ, $R_2 = 15$ kΩ, $R_3 = 18$ kΩ, and $R_L = 10$ kΩ.

4. Refer to Figure 12–79(d). Nortonize the circuit and find V_L and I_L, where (a) $E = 40$ V, $R_1 = 2.2$ kΩ, $R_2 = 3.9$ kΩ, and $R_L = 2.7$ kΩ, and (b) $E = 9$ V, $R_1 = 330$ Ω, $R_2 = 270$ Ω, and $R_L = 390$ Ω.

5. Refer to Figure 12–79(e). Nortonize the circuit and find V_L and I_L where (a) $E = 50$ V, $R_1 = 22$ kΩ, $R_2 = 10$ kΩ, $R_3 = 33$ kΩ, $R_L = 15$ kΩ, and (b) $E = 10$ V, $R_1 = 3.3$ kΩ, $R_2 = 3.3$ kΩ, $R_3 = 6.8$ kΩ, and $R_L = 2$ kΩ.

ANSWERS TO SELF-TESTS

SELF-TEST 12–1

(a) Short E_1:

$$R_T = 4.7 \text{ k}\Omega + 1.65 \text{ k}\Omega = 6.35 \text{ k}\Omega$$

$$I_T = \frac{10 \text{ V}}{6.35 \text{ k}\Omega} = 1.57 \text{ mA}$$

$$I_3 = \frac{1.57 \text{ mA} \times 1.65 \text{ k}\Omega}{3.3 \text{ k}\Omega} = 787 \text{ }\mu\text{A}$$

Short E_2:

$$R_T = 3.3 \text{ k}\Omega + 1.94 \text{ k}\Omega = 5.24 \text{ k}\Omega$$

$$I_T = \frac{30 \text{ V}}{5.24 \text{ k}\Omega} = 5.73 \text{ mA}$$

$$I_3 = \frac{5.73 \text{ mA} \times 1.94 \text{ k}\Omega}{3.3 \text{ k}\Omega}$$

$$= 3.36 \text{ mA}$$

In the original circuit:

$$I_3 = 787 \text{ }\mu\text{A} + 3.37 \text{ mA} = 4.15 \text{ mA}$$

$V_3 = 13.7$ V

$V_2 = 3.70$ V

$I_2 = 787$ μA

$V_1 = 16.3$ V

$I_1 = 4.94$ mA

(b) Short E_1:

$$R_T = 10 \text{ k}\Omega + 5 \text{ k}\Omega = 15 \text{ k}\Omega$$

$$I_T = \frac{60 \text{ V}}{15 \text{ k}\Omega} = 4 \text{ mA}$$

$$I_3 = 2 \text{ mA}$$

Short E_2:

$$R_T = 15 \text{ k}\Omega$$

$$I_T = \frac{100 \text{ V}}{15 \text{ k}\Omega} = 6.67 \text{ mA}$$

$$I_3 = 3.33 \text{ mA}$$

In the original circuit:

$$I_3 = 3.33 \text{ mA} - 2 \text{ mA}$$
$$= 1.33 \text{ mA}$$
$$V_3 = 13.3 \text{ V}$$
$$V_2 = 73.3 \text{ V}$$
$$I_2 = 7.33 \text{ mA}$$
$$V_1 = 86.7 \text{ V}$$
$$I_1 = 8.67 \text{ mA}$$

(c) Short E_1:

$$R_T = 180 \ \Omega + 68.8 \ \Omega = 249 \ \Omega$$

$$I_T = \frac{12 \text{ V}}{249 \ \Omega} = 48.2 \text{ mA}$$

$$I_2 = \frac{48.2 \text{ mA} \times 68.7 \ \Omega}{220 \ \Omega}$$
$$= 15.1 \text{ mA}$$

Short E_2:

$$R_T = 100 \ \Omega + 99 \ \Omega = 199 \ \Omega$$

$$I_T = \frac{10 \text{ V}}{199 \ \Omega} = 50.3 \text{ mA}$$

$$I_2 = \frac{50.3 \text{ mA} \times 99 \ \Omega}{220 \ \Omega}$$
$$= 22.6 \text{ mA}$$

In the original circuit:

$$I_2 = 15.1 \text{ mA} + 22.6 \text{ mA}$$
$$= 37.7 \text{ mA}$$
$$V_2 = 8.29 \text{ V}$$
$$V_3 = 3.71 \text{ V}$$
$$I_3 = 20.6 \text{ mA}$$
$$V_1 = 1.71 \text{ V}$$
$$I_1 = 17.1 \text{ mA}$$

(d) Short E_2:

$$R_T = R_1 + R_2 + R_3 \parallel R_4$$
$$= 1 \text{ k}\Omega + 2.7 \text{ k}\Omega + 956 \ \Omega$$
$$= 4.66 \text{ k}\Omega$$

$$I_T = \frac{E_1}{R_T} = \frac{30 \text{ V}}{4.66 \text{ k}\Omega} = 6.44 \text{ mA}$$

$$I_3 = \frac{I_T R_x}{R_3} = \frac{6.44 \text{ mA} \times 956 \ \Omega}{1.2 \text{ k}\Omega}$$
$$= 5.13 \text{ mA}$$

Short E_1:

$$R_T = R_4 + R_3 \parallel (R_1 + R_2)$$
$$= 4.7 \text{ k}\Omega + 906 \ \Omega = 5.61 \text{ k}\Omega$$

$$I_T = \frac{E_2}{R_T} = \frac{10 \text{ V}}{5.61 \text{ k}\Omega} = 1.78 \text{ mA}$$

$$I_3 = \frac{I_T R_X}{R_3} = \frac{1.78 \text{ mA} \times 906 \ \Omega}{1.2 \text{ k}\Omega}$$
$$= 1.35 \text{ mA}$$

In the original circuit:

$$I_3 = 1.35 \text{ mA} + 5.13 \text{ mA} = 6.48 \text{ mA}$$
$$V_3 = 6.48 \text{ mA} \times 1.2 \text{ k}\Omega = 7.78 \text{ V}$$
$$V_4 = E_2 - V_3 = 10 \text{ V} - 7.78 \text{ V}$$
$$= 2.22 \text{ V}$$

$$I_4 = \frac{2.22 \text{ V}}{4.7 \text{ k}\Omega} = 473 \ \mu\text{A}$$

$$I_1 = I_2 = I_3 - I_4 = 6.01 \text{ mA}$$
$$V_1 = 6.01 \text{ mA} \times 1 \text{ k}\Omega = 6.01 \text{ V}$$
$$V_2 = 6.01 \text{ mA} \times 2.7 \text{ k}\Omega = 16.2 \text{ V}$$

1. The Thévenized circuit with the load connected is shown in Figure 12–82.

$$V_{OC} = V_2 + V_3 = \frac{E(R_2 + R_3)}{R_1 + R_2 + R_3}$$

$$= 20.2 \text{ V}$$

$$R_{TH} = R_1 \| (R_2 + R_3) = 219 \ \Omega$$

$$V_L = \frac{V_{OC}R_L}{R_{TH} + R_L} = \frac{20.2 \text{ V} \times 560 \ \Omega}{779 \ \Omega}$$

$$= 14.6 \text{ V}$$

$$I_L = \frac{V_{OC}}{R_{TH} + R_L} = \frac{20.2 \text{ V}}{779 \ \Omega} = 26.0 \text{ mA}$$

2. The Thévenized circuit with the load connected is shown in Figure 12–83.

$$V_{OC} = V_2 = \frac{ER_2}{R_1 + R_2}$$

$$= \frac{15 \text{ V} \times 2.7 \text{ k}\Omega}{6 \text{ k}\Omega} = 6.75 \text{ V}$$

$$R_{TH} = R_4 + R_1 \| R_2 = 7.5 \text{ k}\Omega$$

$$+ 1.49 \text{ k}\Omega = 8.99 \text{ k}\Omega$$

$$V_L = \frac{V_{OC}R_L}{R_{TH} + R_L} = \frac{6.75 \text{ V} \times 4.7 \text{ k}\Omega}{13.7 \text{ k}\Omega}$$

$$= 2.32 \text{ V}$$

$$I_L = \frac{V_{OC}}{R_{TH} + R_L} = \frac{6.75 \text{ V}}{13.7 \text{ k}\Omega}$$

$$= 493 \ \mu\text{A}$$

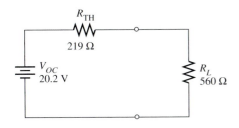

FIGURE 12–82

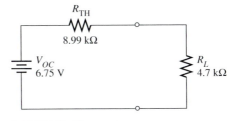

FIGURE 12–83

FIGURE 12–84

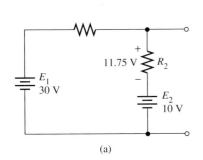

(a)

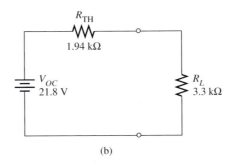

(b)

3. The Thévenized circuit with the load connected is shown in Figure 12–84(b) on the previous page. The circuit used to find V_{OC} is shown in Figure 12–84(a).

$$E_T = E_1 - E_2 = 20 \text{ V}$$

$$V_2 = \frac{20 \text{ V} \times 4.7 \text{ k}\Omega}{8 \text{ k}\Omega} = 11.8 \text{ V}$$

$$V_{OC} = E_2 + V_2 = 21.8 \text{ V}$$

$$R_{TH} = R_1 \| R_2 = 1.94 \text{ k}\Omega$$

$$V_L = \frac{V_{OC}R_L}{R_{TH} + R_L} = \frac{21.8 \text{ V} \times 3.3 \text{ k}\Omega}{5.24 \text{ k}\Omega}$$

$$= 13.7 \text{ V}$$

$$I_L = \frac{V_{OC}}{R_{TH} + R_L} = \frac{21.8 \text{ V}}{5.24 \text{ k}\Omega}$$

$$= 4.15 \text{ mA}$$

4. The Thévenized circuit with the load connected is shown in Figure 12–85(b). The circuit used to find V_{OC} is shown in Figure 12–85(a).

$$E_T = E_1 - E_2 = 20 \text{ V}$$

$$V_4 = \frac{E_T R_4}{R_1 + R_2 + R_4}$$

$$= \frac{20 \text{ V} \times 4.7 \text{ k}\Omega}{8.4 \text{ k}\Omega}$$

$$= 11.2 \text{ V}$$

$$V_{OC} = V_4 + E_2 = 21.2 \text{ V}$$

$$R_{TH} = R_4 \| (R_1 + R_2)$$

$$= 4.7 \text{ k}\Omega \| 3.7 \text{ k}\Omega$$

$$= 2.07 \text{ k}\Omega$$

$$V_L = \frac{V_{OC}R_L}{R_{TH} + R_L} = \frac{21.2 \text{ V} \times 1.2 \text{ k}\Omega}{3.27 \text{ k}\Omega}$$

$$= 7.78 \text{ V}$$

$$I_L = \frac{V_{OC}}{R_{TH} + R_L} = \frac{21.2 \text{ V}}{3.27 \text{ k}\Omega}$$

$$= 6.48 \text{ mA}$$

FIGURE 12–85

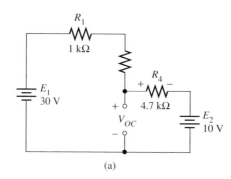

(a)

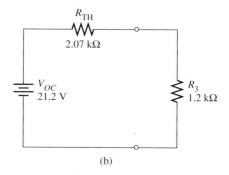

(b)

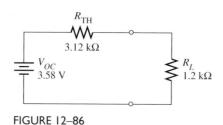

FIGURE 12–86

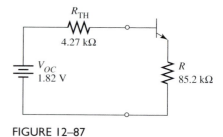

FIGURE 12–87

5. The Thévenized circuit with the load connected is shown in Figure 12–86 on the previous page. Various solutions are possible. We will use a circuit that includes E, R_4, R_L, and R_2 with the load removed:

$$V_2 = \frac{ER_2}{R_1 + R_2} = \frac{9 \text{ V} \times 1.8 \text{ k}\Omega}{6.5 \text{ k}\Omega}$$

$$= 2.49 \text{ V}$$

$$V_4 = \frac{ER_4}{R_3 + R_4} = \frac{9 \text{ V} \times 5.6 \text{ k}\Omega}{8.3 \text{ k}\Omega}$$

$$= 6.07 \text{ V}$$

$$V_{OC} = V_4 - V_2 = 6.07 \text{ V} - 2.49 \text{ V}$$

$$= 3.58 \text{ V}$$

$$R_{TH} = R_1 \parallel R_2 + R_3 \parallel R_4$$

$$= 1.3 \text{ } k\Omega + 1.82 \text{ k}\Omega$$

$$= 3.12 \text{ k}\Omega$$

$$I_L = \frac{V_{OC}}{R_{TH} + R_L} = \frac{3.58 \text{ V}}{3.12 \text{ k}\Omega + 1.2 \text{ k}\Omega}$$

$$= 828 \text{ } \mu\text{A}$$

$$V_L = I_L R_L = 828 \text{ } \mu\text{A} \times 1.2 \text{ k}\Omega$$

$$= 994 \text{ mV}$$

6. The Thévenized circuit with the load connected is shown in Figure 12–87 on the previous page.

$$V_{OC} = \frac{V_{CC}R_2}{R_1 + R_2} = \frac{20 \text{ V} \times 4.7 \text{ k}\Omega}{47 \text{ k}\Omega + 4.7 \text{ k}\Omega}$$

$$= 1.82 \text{ V}$$

$$R_{TH} = R_1 \parallel R_2 = 4.27 \text{ k}\Omega$$

$$R_L = R_4(H_{FE} + 1) = 85.2 \text{ k}\Omega$$

$$I_B = \frac{V_{OC} - V_{BE}}{R_{TH} + R} = \frac{1.22 \text{ V}}{89.5 \text{ k}\Omega}$$

$$= 13.6 \text{ } \mu\text{A}$$

$$I_C = H_{FE}I_B = 953 \text{ } \mu\text{A}$$

$$V_C = V_{CC} - V_3 = 20 \text{ V} - 4.49 \text{ V}$$

$$= 15.5 \text{ V}$$

SELF-TEST 12–3

1. $I_{SC} = 221 \text{ } \mu\text{A}$
   $G_N = 24.7 \text{ } \mu\text{S}$
   $V_L = 5.80 \text{ V}$
   $I_L = 77.3 \text{ } \mu\text{A}$

2. $I_{SC} = 29.4 \text{ mA}$
   $G_N = 1.47 \text{ mS}$
   $V_L = 9.03 \text{ V}$
   $I_L = 16.1 \text{ mA}$

Graphing

13

Introduction

In electronics, we are often trying to find how one quantity will change as a direct result of changing some other quantity. The most common example is to analyze a circuit to see how the circuit current will change when we change the voltage. To do this we have the option of doing a theoretical analysis or a practical analysis. A theoretical analysis might involve using a mathematical model of a circuit and solving circuit equations to determine values of currents and voltages. A practical analysis might involve constructing a circuit and making measurements of currents and voltages. In either case, we can record the results of our analysis in a table whose values represent the electrical characteristics of the circuit. We can take this analysis one step farther by plotting the table values on a graph so that we can visualize the behavior of the circuit.

Suppose we wanted to learn about the electrical characteristics of a silicone semiconductor diode, which is the first step in learning about transistors. In order to test a diode to find how the current through the diode changes when the diode voltage changes, we could construct the DC circuit shown in Figure 13–1. We would then adjust the power supply voltage so that the voltage across the diode was 0.1 volts and then measure and record the current. Next, we would adjust the supply so the voltage across the diode was 0.2 volts and then measure and record that current. We could repeat these measurements every 0.1 volts up to 0.8 volts and record the diode current for each voltage. Figure 13–2 shows a table with all the measurements entered.

Our next step would be to plot each of these values on a graph. Figure 13–3(a) shows a graph with the voltage scale along the horizontal axis and the current scale along the vertical axis. Both axes have a linear scale, which means that the difference between each grid line is a constant value. In Figure 13–3(a), the difference between each gridline crossing the horizontal axis is 0.1 volts; the difference between each gridline crossing the vertical axis is 2 milliamps. An X is placed on the graph for each

FIGURE 13–1
Diagram of the circuit used to test a silicone diode.

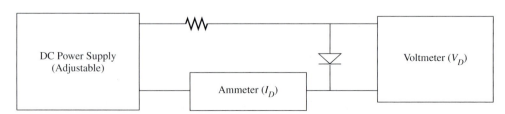

DC Power Supply (Adjustable)

Ammeter (I_D)

Voltmeter (V_D)

V_D	I_D
0.1 V	0.06 μA
0.2 V	0.16 μA
0.3 V	1.0 μA
0.4 V	9 μA
0.5 V	70 μA
0.6 V	0.6 mA
0.7 V	4.1 mA
0.8 V	18 mA

FIGURE 13–2
Table of measured diode currents (I_D) at specified diode voltages (V_D) from the circuit in Figure 13–1.

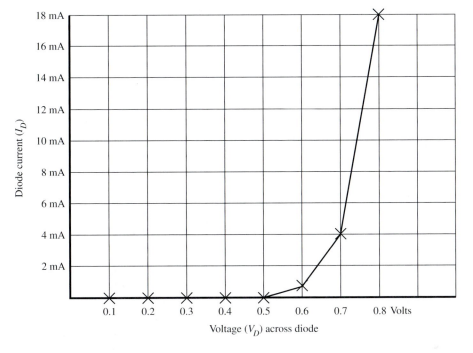

FIGURE 13–3(a)
The linear graph of the data in Figure 13–2.

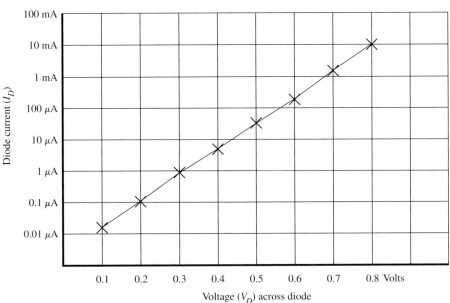

FIGURE 13–3(b)
The semi-log graph of the data in Figure 13–2.

measurement in the table in Figure 13–2. The Xs are then connected, creating a curve that approximates the voltage-current characteristic of the diode tested. This curve is typical of all silicone diodes.

The first five current measurements are so small that they seem to fall on the horizontal axis in Figure 13–3(a). To create a graph that shows how the current is changing for the lower voltages, we have constructed the semi-log graph in Figure 13–3(b). It is called "semi-log" because the vertical axis is logarithmic and the horizontal is linear. Each gridline that crosses the vertical axis represents a ten-fold change in diode current. The same measurement points from Figure 13–2 are plotted on the semi-log graph. Now more details about the diode characteristics at the low voltages are discernable. Whether you find Figure 13–3(a) or (b) more useful would depend on how the diode is used in a circuit.

All silicone diodes have a similar current versus voltage curve. We could improve the accuracy of this curve by making measurements every 10 mV instead of every 100 mV, but this picture is sufficiently accurate to give us a good idea of the electrical characteristics of a silicone diode.

These graphs, as is true of most graphs, are not exactly correct. For example, we had to estimate where on the graphs to plot the points when the current did not fall on one of the horizontal grid lines. Furthermore, we cannot tell exactly what the current will be for any voltages that fall between our measurement points. The important thing that these graphs accomplish is to provide us with a clear picture of how the current changes in a silicone diode when the voltage across it changes.

Technical data is often presented in graphical form. The ability to construct and read graphs is an important part of our study of electronics.

In this chapter you will learn how to create a graph from collected data. You will also learn how to graph an equation. Finally, you will learn how to interpret a graph that has already been created.

Chapter Objectives

In this chapter you will learn how to:

1. Plot points on a system of rectangular coordinates.
2. Graph linear equations.
3. Put equations in the slope–intercept form.
4. Interpret graphs.
5. Plot curves from given data.

13–1 PLOTTING POINTS ON A SYSTEM OF RECTANGULAR COORDINATES

Consider the graph in Figure 13–4. This is called a system of *rectangular coordinates*. This kind of graph will be used many times throughout the text. The point where the *x*-axis (the horizontal axis) and the *y*-axis (the vertical axis) intersect (cross) is called

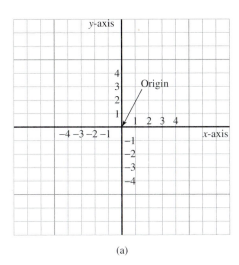

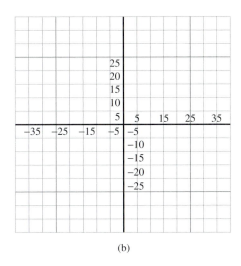

FIGURE 13–4
Graphs divided
into equal
divisions.

(a)   (b)

the *origin*. Notice that the origin is 0 for both vertical and horizontal axes. Along the horizontal or *x*-axis, positive values are plotted to the right of the origin and negative values are plotted to the left of the origin. Along the vertical, or *y*-axis, positive values are plotted from the origin upward and negative values are plotted from the origin downward. The graph is divided into equal parts or divisions. The value of each division depends on the requirements of the problem. In Figure 13–4(a), the value of each division both horizontally and vertically is 1. In Figure 13–4(b) the value of each division is 5.

Let's identify some points using the graph in Figure 13–5. We identify points by giving the *x* and *y* values or coordinates. Suppose we were to identify a point whose coordinates are (5, −2). The first number identifies the horizontal or *x* value and the second number identifies the vertical or *y* value. The point is located by moving 5 units to the right along the *x*-axis because the number is positive. We move 2 units down along the *y*-axis because the number is negative. The point is shown as point *A*. Let's find the point whose coordinates are (−4, 6). This point is identified as point *B* in

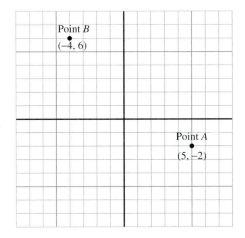

FIGURE 13–5
Identifying
points on a
graph.

Figure 13–5. The point was found by moving 4 units *left* from the origin along the *x*-axis. We moved left because $x = -4$. Then we moved upward 6 units. We moved upward because $y = 6$.

PRACTICE PROBLEMS 13–1

Identify the following points on the graph of Figure 13–4(a).

1. (2,6)
3. (0,−8)
5. (−1,2)
7. (−3,−4)

2. (−1,7)
4. (−6,0)
6. (6,−3)

Identify the following points on the graph of Figure 13–4(b).

8. (−15,0)
10. (25,25)
12. (−10,20)

9. (5,−10)
11. (−15,−20)

SOLUTIONS

See Figure 13–6(a) for problems 1 through 7; see Figure 13–6(b) for problems 8 through 12.

Additional practice problems are at the end of the chapter.

FIGURE 13–6

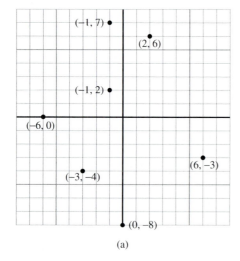

(a)

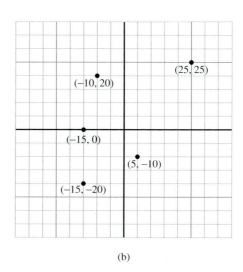

(b)

CHAPTER 13

Recall from previous chapters that linear equations are equations in which a term contains only *one* variable. Furthermore, the variable exponent is 1. For example, $2x + 1 = 9$, $y - 6 = 0$, and $5 + 6a = 23$ are all linear equations. $6x + 7y = 14$ is also a linear equation. Even though the variables x and y are different, they appear in different terms. $x + 6 + 7y + 14z = 10$ would be a linear equation because no term contains more than one variable. $6xy + 7 = 0$ and $4x^2 + 2y = 14$ are *not* linear equations. In the first equation one term contains two variables (x and y). In the second equation one of the variables is squared ($4x^2$).

☞ **KEY POINT** *The graph of any linear equation is a straight line.*

Consider the equation $2x - y = 6$. Usually, the graph of a linear equation intersects (crosses) both the horizontal and vertical axes. Therefore, we may assume that one point on the graph will be where $y = 0$ and x equals a value to be determined by solving the resulting equation. That is,

$$2x - y = 6$$

$$\text{If } y = 0$$

$$\text{then } 2x = 6$$

$$\text{and } x = 3$$

One point on the graph then is where $y = 0$ and $x = 3$. On the graph the point would be identified as $(3, 0)$. (The x value is always given first.) Another point would be where $x = 0$.

$$2x - y = 6$$

$$\text{If } x = 0$$

$$\text{then } -y = 6$$

$$y = -6$$

A second point on the graph is where $x = 0$ and $y = -6$. On the graph the point would be identified as $(0, -6)$. The graph is shown in Figure 13–7. Point A is called the *x-intercept* because that is where the line crosses the *x*-axis. Its coordinates are $(3, 0)$. The *x*-coordinate is called the *abscissa*. The *y*-coordinate is called the *ordinate*.

Point B is called the *y-intercept*. Its coordinates are $(0, -6)$. Many points along the graph may be plotted or identified. These points may be found by letting x equal some

FIGURE 13–7
Plot of the
equation
$2x - y = 6$.

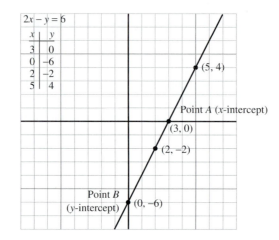

value and then solving for y. Or we could let y equal some value and solve for x. For example, using the same equation, we get

$$2x - y = 6$$

$$\text{If } x = 2$$

$$\text{then } 2(2) - y = 6$$

$$4 - y = 6$$

$$-y = 2$$

$$y = -2$$

The coordinates are $(2, -2)$.

$$\text{If } y = 4$$

$$\text{then } 2x - (4) = 6$$

$$2x = 10$$

$$x = 5$$

The coordinates are $(5, 4)$. These two points are also plotted in Figure 13–7.

A table showing values of x and corresponding values of y is also included in Figure 13–7. Such a table should be drawn whenever points are to be plotted because it helps identify coordinates.

EXAMPLE 13–1

Given the equation $y = 2x + 2$, draw the graph. Find the x- and y-intercepts. Identify four points on the graph.

SOLUTION First, let's rearrange the terms so that both variables are in the left-hand member. This will make it easier for us to solve for unknowns as we go along.

$$y = 2x + 2$$

$$y - 2x = 2$$

Next let's find the *x*-intercept. This is the point where the line passes through the *x*-axis.

$$y - 2x = 2$$

$$\text{Let } y = 0$$

$$\text{then } -2x = 2$$

$$x = -1$$

The *x*-intercept is at coordinates $(-1, 0)$. Now find the *y*-intercept.

$$y - 2x = 2$$

$$\text{Let } x = 0$$

$$\text{then } y = 2$$

The *y*-intercept is at coordinates $(0, 2)$. Use a straightedge and draw a line extending through the two points as shown in Figure 13–8. A third point is found by letting *x* = some number. Upon examination of the straight line, we see that the point $(2, 6)$ should fall on the line. Let's see if it does.

$$y - 2x = 2$$

Let *x* = 2. Then

$$y - 2(2) = 2$$

$$y - 4 = 2$$

$$y = 6$$

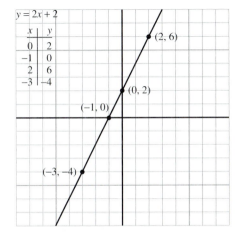

FIGURE 13–8
Plot of the equation
$y = 2x + 2$.

The fact that the point $(2,6)$ falls on the line tells us that our line is in the correct location. Let $x = -3$. Then

$$y - 2(-3) = 2$$
$$y + 6 = 2$$
$$y = -4$$

When we plot $(-3,-4)$, we see that this point falls on our line. The point $(2,6)$ or the point $(-3,-4)$ would not be proof enough that our line is in the correct location. *Three* points are needed to verify our line. If the line passes through all three points, we can assume it to be correct.

PRACTICE PROBLEMS 13–2

Plot the following equations on linear graph paper. Identify the x- and y-intercepts. Plot at least three points on the graph:

1. $x - 4y = 8$
2. $3x + 6y = 12$
3. $y = 4x + 4$
4. $x = 3y - 6$
5. $2y + 4 = -x$
6. $5y + 28 = -7x$

SOLUTIONS

1. The graph is shown in Figure 13–9(a). The x-intercept $= (8,0)$. The y-intercept $= (0,-2)$.

2. The graph is shown in Figure 13–9(b). The x-intercept $= (4,0)$. The y-intercept $= (0,2)$.

FIGURE 13–9

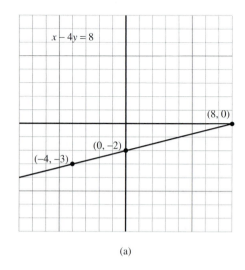

(a)

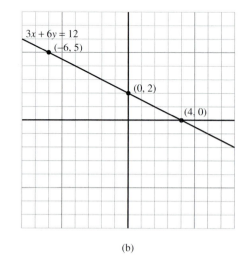

(b)

FIGURE 13–9
(cont.)

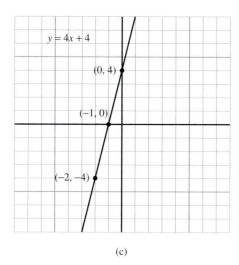

(c)

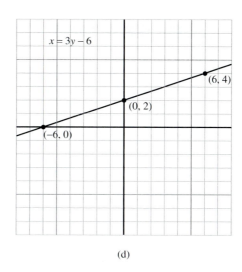

(d)

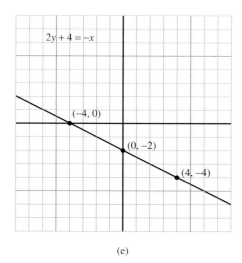

(e)

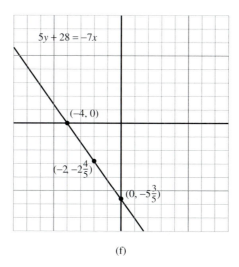

(f)

3. The graph is shown in Figure
    13–9(c). The x-intercept $= (-1,0)$.
    The y-intercept $= (0,4)$.
4. The graph is shown in Figure
    13–9(d). The x-intercept $= (-6,0)$.
    The y-intercept $= (0,2)$.
5. The graph is shown in Figure
    13–9(e). The x-intercept $= (-4,0)$.
    The y-intercept $= (0,-2)$.
6. The graph is shown in Figure
    13–9(f). The x-intercept $= (-4,0)$.
    The y-intercept $= (0,-5\frac{3}{5})$.

Additional practice problems are at the end of the chapter.

SLOPE OF A LINE

The linear equation in Figure 13–7 is redrawn in Figure 13–10. Consider the point (3,0). If x is changed from 3 to 5, y changes from 0 to 4. The change in x is 2. This would be written $\Delta x = 2$ (*delta x* = 2). The change in y is 4. This would be written $\Delta y = 4$ (*delta y* = 4). Upon inspection of the curve, we find that for any increase of 2 units horizontally ($\Delta x = 2$) y increases 4 units ($\Delta y = 4$). For example, if x increases from -2 units to 0, y increases from -10 units to -6 units. If x increases from 1 unit to 3 units, y increases from -4 units to 0. The slope, then, is:

$$\frac{\Delta y}{\Delta x} = \frac{2}{1} = 2$$

That is, any increase of 1 unit in the value of x causes a 2 unit *increase* in the value of y.

Consider the straight line in Figure 13–11, which results from the equation $x + 2y = 4$. If x increases from 0 to 4 ($\Delta x = 4$), y *decreases* from 2 to 0 ($\Delta y = -2$). Any increase in the value of x causes a corresponding decrease in the value of y in the ratio of 2 to -1. Any 4-unit increase in x causes a 2-unit decrease in y. Any 8-unit increase in x causes a 4-unit decrease in y and so on.

☞ **KEY POINT**   The *slope* of a line is defined as the change in *y* that results from an *increase* in *x*.

$$\text{Slope} = \frac{\Delta y}{\Delta x} = \frac{\text{change in } y}{\text{increase in } x}$$

When we read a graph, we always make Δx positive. The resulting Δy will be either positive or negative depending on whether y increased or decreased as x increased.

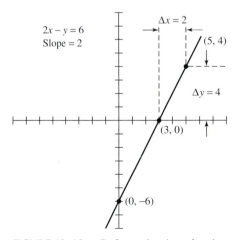

FIGURE 13–10   Defining the slope for the equation $2x - y = 6$.

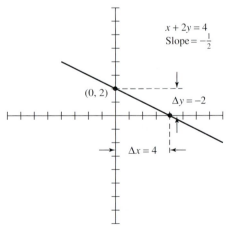

FIGURE 13–11   Defining the slope for the equation $x + 2y = 4$.

The slope is positive if Δy is positive and it is negative if Δy is negative. In Figure 13–11, the slope equals

$$\frac{\Delta y}{\Delta x} = \frac{-2}{4} = -\frac{1}{2}$$

When the slope is positive, as in Figure 13–10, the line *rises* from left to right. When the slope is negative, as in Figure 13–11, the line *falls* from left to right.

EXAMPLE 13–2

Find the slopes of the lines in Practice Problems 13–2.

SOLUTION   Using the x- and y-intercepts to determine the slope:

$$\text{In problem 1 Slope} = \frac{\Delta y}{\Delta x} = \frac{1}{4}$$

The slope is positive. As x is increased 4 units, y increases 1 unit.

$$\text{In problem 2 Slope} = \frac{\Delta y}{\Delta x} = -\frac{1}{2}$$

The slope is negative. As x is increased 2 units, y decreases 1 unit.

$$\text{In problem 3 Slope} = \frac{\Delta y}{\Delta x} = \frac{4}{1} = 4$$

The slope is positive. As x is increased 1 unit, y increases 4 units.

$$\text{In problem 4 Slope} = \frac{\Delta y}{\Delta x} = \frac{1}{3}$$

The slope is positive. As x is increased 3 units, y increases 1 unit.

$$\text{In problem 5 Slope} = \frac{\Delta y}{\Delta x} = -\frac{1}{2}$$

The slope is negative. As x is increased 2 units, y decreases 1 unit.

$$\text{In problem 6 Slope} = \frac{\Delta y}{\Delta x} = -\frac{5\frac{3}{5}}{4} = -\frac{28}{20} = -\frac{7}{5}$$

The slope is negative. As x is increased 4 units, y decreases $5\frac{3}{5}$ units.

PRACTICE PROBLEMS 13–3

Plot the following equations on linear graph paper. Identify the x- and y-intercepts. Determine the slope.

1. $x + y = 4$
2. $x - 3y = -6$
3. $4x + y = -8$
4. $8x - 4y = 16$
5. $6 - 4x = 12y$
6. $\dfrac{2x}{3} + 4 = 2y$

SOLUTIONS

1. Figure 13–12(a) shows the graph of $x + y = 4$. The x-intercept is at coordinates $(4,0)$. The y-intercept is at coordinates $(0,4)$.

$$\text{Slope} = \frac{\Delta y}{\Delta x} = \frac{-4}{4} = -1$$

2. Figure 13–12(b) shows the graph of $x - 3y = -6$. The x-intercept is at coordinates $(-6,0)$. The y-intercept is at coordinates $(0,2)$.

$$\text{Slope} = \frac{\Delta y}{\Delta x} = \frac{2}{6} = \frac{1}{3}$$

3. Figure 13–12(c) shows the graph of $4x + y = -8$. The x-intercept is at coordinates $(-2,0)$. The y-intercept is at coordinates $(0,-8)$.

$$\text{Slope} = \frac{\Delta y}{\Delta x} = \frac{-8}{2} = -4$$

4. Figure 13–12(d) shows the graph of $8x - 4y = 16$. The x-intercept is $(2,0)$. The y-intercept is $(0,-4)$.

$$\text{Slope} = \frac{\Delta y}{\Delta x} = \frac{4}{2} = 2$$

5. Figure 13–12(e) shows the graph of $6 - 4x = 12y$. Rearranging the equation, we get: $4x + 12y = 6$. We can simplify the equation by dividing both sides by 2 to get: $2x + 6y = 3$. (Both equations yield the same curve. Usually, simplifying the equation when possible will result in an equation that is easier to work with.) The x-intercept is at coordinates $(\frac{3}{2},0)$. The y-intercept is at coordinates $(0,\frac{1}{2})$.

$$\text{Slope} = \frac{\Delta y}{\Delta x} = \frac{-\dfrac{1}{2}}{\dfrac{3}{2}} = -\frac{1}{3}$$

6. Figure 13–12(f) shows the graph of $\dfrac{2x}{3} + 4 = 2y$. Rearranging the equation and simplifying, we get: $x - 3y = -6$. The x-intercept is at coordinates $(-6,0)$. The y-intercept is at coordinates $(0,2)$.

$$\text{Slope} = \frac{\Delta y}{\Delta x} = \frac{2}{6} = \frac{1}{3}$$

Additional practice problems are at the end of the chapter.

FIGURE 13–12

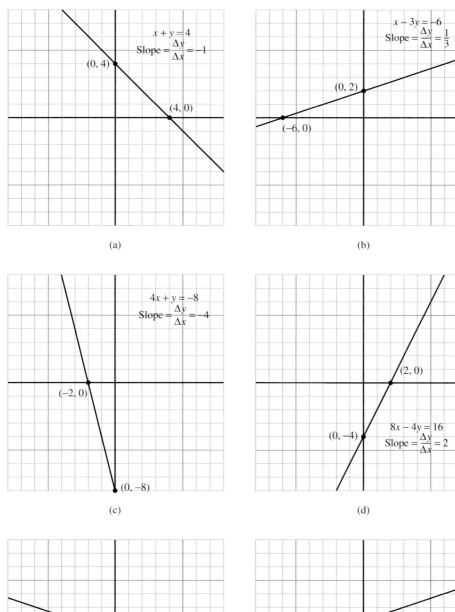

(a)

(b)

(c)

(d)

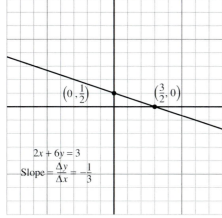

(e)

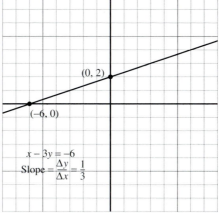

(f)

SLOPE-INTERCEPT FORM

If a linear equation contains two variables, it can be written like this:

$$y = mx + b$$

y and x are the variables. m and b are constants. When the equation is written in this form, it is said to be written in the *slope–intercept form* because m is the slope and b is the vertical or y-intercept.

☞ **KEY POINT** *Any linear equation containing two variables can be written in the slope–intercept form.*

EXAMPLE 13-3

Given the equation $3y - 2x = 9$, find the slope and the y-intercept.

SOLUTION   Using the rules of algebra, we can change to slope–intercept form:

$$3y - 2x = 9$$
$$3y = 2x + 9$$
$$y = \frac{2}{3}x + 3$$

The slope is $\frac{2}{3}$ and the y-intercept is 3. A slope of $\frac{2}{3}$ means that if x is increased 3 units, y increases 2 units. The y-intercept coordinates are $(0, 3)$. With this informa-

FIGURE 13–13
Defining the slope and y-intercept for the equation $y = \frac{2}{3}x + 3$.

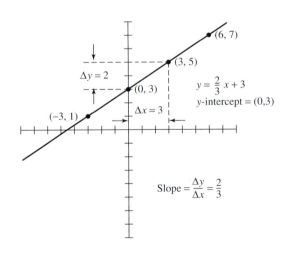

CHAPTER 13

tion we can graph the equation. In Figure 13–13 we first plot the point $(0,3)$, the y-intercept. Next we find a second point. The slope is $\frac{2}{3}$, which means that if we change x from 0 to 3, y will increase from 3 to 5. This point is $(3,5)$ and is plotted. A third point can be plotted by changing x from 3 to 6 and changing y from 5 to 7. This point is $(6,7)$. Starting at the y-intercept, we could have plotted the point $(-3,1)$. This point results from changing x from 0 to -3, which causes a change in y from 3 to 1.

PRACTICE PROBLEMS 13–4

Change the following equations into the slope–intercept form. Determine the slope and y-intercept. Graph the equation.

1. $4x - 3y = -15$
3. $3x + 3y = 5$
5. $3y - 7x = -12$

7. $6x - 3y = 4$

2. $3y - 4x = 18$
4. $6y - 5x = 24$
6. $4x - y = 5$

8. $4x + 3 = \dfrac{y}{2}$

SOLUTIONS

1. $y = \frac{4}{3}x + 5$. Slope $= \frac{4}{3}$, y-intercept $= (0,5)$. The graph of the equation is shown in Figure 13–14(a).

2. $y = \frac{4}{3}x + 6$. Slope $= \frac{4}{3}$, y-intercept $= (0,6)$. The graph of the equation is shown in Figure 13–14(b).

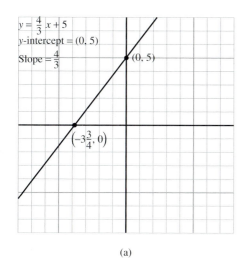

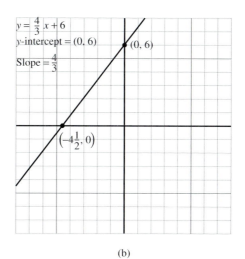

FIGURE 13–14

(a)

(b)

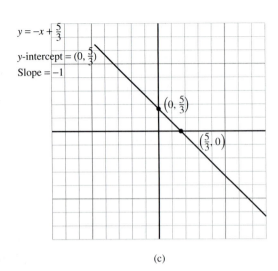

$y = -x + \frac{5}{3}$

y-intercept $= (0, \frac{5}{3})$

Slope $= -1$

$(0, \frac{5}{3})$

$(\frac{5}{3}, 0)$

(c)

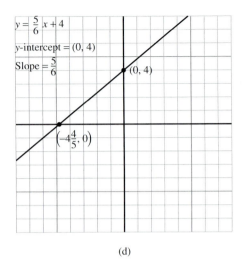

$y = \frac{5}{6}x + 4$

y-intercept $= (0, 4)$

Slope $= \frac{5}{6}$

$(0, 4)$

$(-4\frac{4}{5}, 0)$

(d)

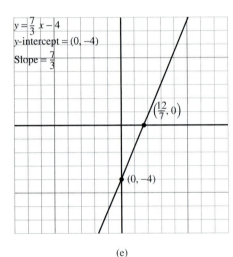

$y = \frac{7}{3}x - 4$

y-intercept $= (0, -4)$

Slope $= \frac{7}{3}$

$(\frac{12}{7}, 0)$

$(0, -4)$

(e)

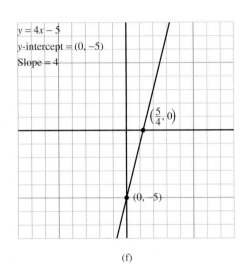

$y = 4x - 5$

y-intercept $= (0, -5)$

Slope $= 4$

$(\frac{5}{4}, 0)$

$(0, -5)$

(f)

FIGURE 13–14 *(cont.)*

3. $y = -x + \frac{5}{3}$. Slope $= -1$, y-intercept $= (0, \frac{5}{3})$. The graph of the equation is shown in Figure 13–14(c).

5. $y = \frac{7}{3}x - 4$. Slope $= \frac{7}{3}$, y-intercept $= (0, -4)$. The graph of the equation is shown in Figure 13–14(e).

4. $y = \frac{5}{6}x + 4$. Slope $= \frac{5}{6}$, y-intercept $= (0, 4)$. The graph of the equation is shown in Figure 13–14(d).

6. $y = 4x - 5$. Slope $= 4$, y-intercept $= (0, -5)$. The graph of the equation is shown in Figure 13–14(f).

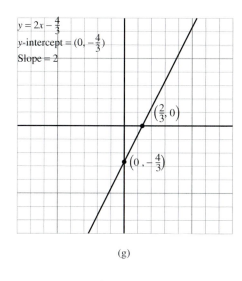

(g)

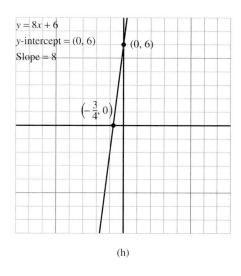

(h)

FIGURE 13–14

(cont.)

7. $y = 2x - \frac{4}{3}$. Slope = 2, y-intercept = $(0, -\frac{4}{3})$. The graph of the equation is shown in Figure 13–14(g).

8. $y = 8x + 6$. Slope = 8, y-intercept = $(0, 6)$. The graph of the equation is shown in Figure 13–14(h).

Additional practice problems are at the end of the chapter.

SELF-TEST 13–1

1. Plot the equation $5x + 3y = 15$ on linear graph paper. Identify the x- and y-intercepts.

2. Plot the equation $3x - y = 6$ on linear graph paper. Identify the x- and y-intercepts. Determine the slope.

3. Change the equation $2x - 3y = 6$ into slope–intercept form. Determine the slope. Identify the y-intercept. Graph the equation.

Answers to Self-test 13–1 are at the end of the chapter.

INTERPRETING GRAPHS   13–5

In Figure 13–15 we have plotted current versus voltage for a resistor with values of voltage from 0 to 20 V. The resulting graph is a straight line. We can determine the resistance by using the techniques learned previously. Notice that I is plotted along the y-axis and V is plotted along the x-axis. Relating this graph to a system of rectangular coordinates, we get

$$\text{Slope} = \frac{\Delta y}{\Delta x} = \frac{\Delta I}{\Delta V} = \frac{4 \text{ mA}}{4 \text{ V}} = 1 \text{ mS}$$

FIGURE 13–15
Plotting current
versus voltage
for values from
0 to 20 volts.

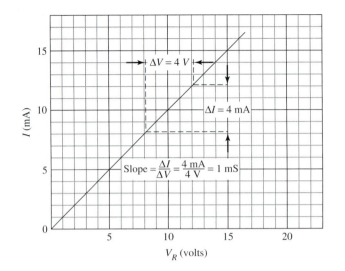

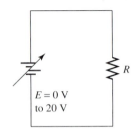

The slope is equal to the conductance. That is, the slope defines the conductance characteristic of the curve. To find R,

$$R = \frac{1}{G} = \frac{1}{1 \text{ mS}} = 1 \text{ k}\Omega$$

The graph, then, is a graph of current versus voltage for a 1-kΩ resistor. With such a graph we can easily find the resulting current for any value of voltage plotted. Because the graph is linear, the slope is the same anywhere along the graph. Therefore, any convenient place along the graph can be chosen to determine the slope.

We see from the graph in Figure 13–15 that resistors are linear devices. Some devices such as diodes and transistors are nonlinear. That is, their resistance characteristics are not straight-line functions. Figure 13–16 is the graph of a forward-biased silicon diode. The resulting graph of current versus voltage is nonlinear. When the graph is nonlinear, the slope of the line depends on where we choose to measure. The most accurate method of determining the slope of such a graph is to first select a point on the graph and then construct a tangent to that point. From geometry a *tangent* is defined as a line that touches a curve at only one point. In Figure 13–17 we have constructed a tangent to the curve where $V_f = 0.5$ V. The slope of this line is the same as the slope of the curve at that point.

$$\text{Slope} = \frac{\Delta I}{\Delta V} = \frac{8 \text{ mA}}{0.3 \text{ V}} = 26.7 \text{ mS}$$

The resistance determined from the slope is symbolized r_d (lowercase r signifies a dynamic or ac resistance).

$$r_d = \frac{1}{\text{slope}} = \frac{1}{26.7 \text{ mS}} = 37.5 \ \Omega$$

In addition to the method discussed previously, r_d can be found by the $\Delta I/\Delta V$ method if we consider that small increments of change along the curve will be nearly a straight

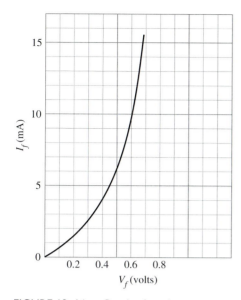

FIGURE 13–16   Graph of nonlinear resistance.

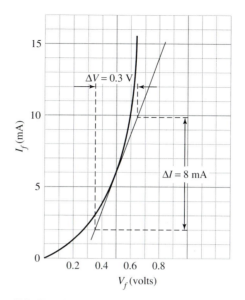

FIGURE 13–17   Finding the slope of nonlinear resistance in Figure 13–16.

line. Notice that r_d changes drastically as V_{dc} changes. This tells us that, when we are required to find r_d, we must specify where on the curve the measurement is to be made.

At the point in question, there are really two resistances. If we simply plot the point, we have a voltage and some resulting current. Since we are examining one point, we consider this to be a dc resistance. This dc resistance is symbolized R_D. The general equation is

$$R_D = \frac{V_{dc}}{I_{dc}}$$

Solving our problem for R_D, we get

$$R_D = \frac{V_{dc}}{I_{dc}} = \frac{0.5\text{ V}}{6\text{ mA}} = 83.3\ \Omega$$

☞ **KEY POINT:** Non-linear devices can be characterized by two resistances whose equations are:

Dynamic resistance $= r_d = \dfrac{\Delta V}{\Delta I}$

DC resistance $= R_D = \dfrac{V}{I}$

Both values depend on the point on the curve where measurement is made.

Figure 13–18 shows the relationship between V_C and V_R in a dc circuit consisting of resistance and capacitance. One time constant equals the product of R and C. Time

FIGURE 13–18
Universal time
constant chart.

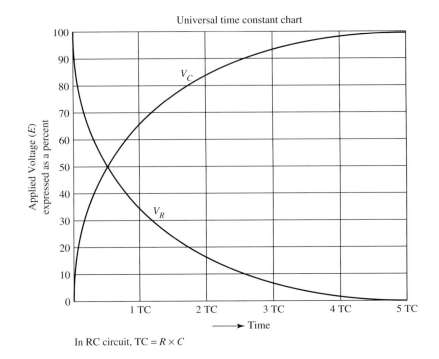

Universal time constant chart

In RC circuit, TC = $R \times C$

constants are plotted along the horizontal axis and the applied voltage (E) is plotted along the vertical axis. When power is applied, the voltage drop across the capacitor (V_C) will increase from zero to the applied voltage, E, in 5 TC. The applied voltage is expressed as a percentage because no matter what the applied voltage is, it takes 5 TC for the capacitor charge to reach that value. The curve shows that V_C will equal approximately 63% of E (whatever value E is) after 1 TC, approximately 87% of E after 2 TC, approximately 95% of E after 3 TC, approximately 99% of E after 4 TC, and equal to E after 5 TC. During that time, V_R decreases from E to zero in 5 TC. At any time, $E = V_C + V_R$. The curves of V_C and V_R are called *exponential curves*. Exponential curves will be analyzed in Chapters 24 and 26.

EXAMPLE 13–4

In a series *RC* circuit connected to a dc source, if the applied voltage is 50 V, find V_C and V_R after 1.5 TC.

SOLUTION From the curve we see that the capacitor has charged to 78% of E in 1.5 TC. V_R is 22% of E. Then

$$V_C = 78\% \times 50 \text{ V} = 39 \text{ V}$$
$$V_R = 22\% \times 50 \text{ V} = 11 \text{ V}$$

EXAMPLE 13–5

In Figure 13–18, (a) how many time constants are required for the capacitor to charge to 60% of E, and (b) if the applied voltage is 60 V, how many time constants are required to charge the capacitor to 48 V?

SOLUTION   (a) From the curve we see that it takes approximately 0.9 TC for the capacitor to charge to 60% of E. (b) We must first find what percent 48 V is of 60 V.

$$\frac{48 \text{ V}}{60 \text{ V}} \times 100 = 80\%$$

Approximately 1.6 TC are required to charge the capacitor to 48 V (80% of E).

PRACTICE PROBLEMS 13–5

1. Refer to Figure 13–19. Find the slope. Find R.
3. Refer to Figure 13–16. Find R_D and r_d if $V_f = 0.4$ V.

5. Refer to Figure 13–18. Let $E = 10$ V. Find V_C and V_R at 0.7 TC and at 2 TC.
7. In Figure 13–18: (a) How many TC are required for the capacitor to charge to 20% of E? (b) If $E = 5$ V, how many TC are required to charge the capacitor to 4.5 V?

2. Refer to Figure 13–20. Find the slope. Find R.
4. Refer to Figure 13–16. If ΔV_f is measured from 0.1 to 0.5 V, what is ΔI_f? What is the slope? What is the value of r_d?
6. Refer to Figure 13–18. Let $E = 50$ V. Find V_C and V_R after 1 TC and 2.3 TC.

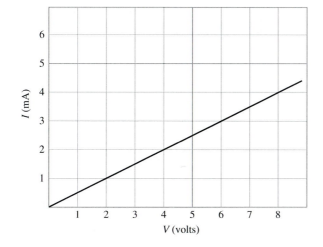

FIGURE 13–19

FIGURE 13–20

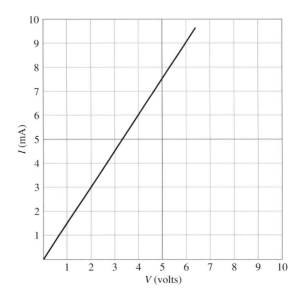

SOLUTIONS

1. Slope $= 500\ \mu\text{S}$, $R = 2\ \text{k}\Omega$
3. $R_D \approx 100\ \Omega$, $r_d \approx 55\ \Omega$

2. Slope $= 1.5\ \text{mS}$, $R = 667\ \Omega$
4. $\Delta I_f \approx 5.3\ \text{mA}$, slope $\approx 13.3\ \text{mS}$,
   $r_d \approx 75\ \Omega$

5. At 0.7 TC, $V_C \approx 5\ \text{V}$ and $V_R \approx 5\ \text{V}$;
   at 2 TC, $V_C \approx 8.4\ \text{V}$ and $V_R \approx 1.6\ \text{V}$
7. (a) ≈ 0.2 TC; (b) ≈ 2.6 TC

6. At 1 TC, $V_C \approx 33\ \text{V}$ and $V_R \approx 17\ \text{V}$
   At 2.3 TC, $V_C \approx 44\ \text{V}$ and $V_R \approx 6\ \text{V}$

Additional practice problems are at the end of the chapter.

13–6   PLOTTING CURVES

Refer again to Figure 13–15. A voltage source was connected in series with the resistor. As we vary the applied voltage in the circuit, the current changes. In this circuit the variable V is called the *independent* variable. I is the *dependent* variable. Its value depends on the value of V selected. In plotting curves it is standard practice to plot the independent variable along the x-axis. This leaves the dependent variable to be plotted along the y-axis. Whenever we are required to plot a curve, we must determine the position of the variables in this manner.

EXAMPLE 13–6

In a resistive circuit, plot a curve of V versus I where $R = 10\ \text{k}\Omega$. Vary V from 0 to 10 V.

V (volts)	I (μA)
2	200
5	500
8	800

(a)

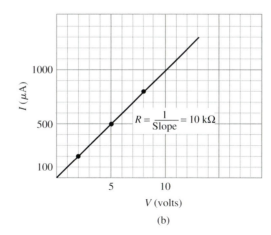

FIGURE 13–21
Graph of current versus voltage where $R = 10 \text{ k}\Omega$: (a) Given data; (b) Resulting curve of current versus voltage.

SOLUTION Select an appropriate scale. It would be convenient to use a scale of 1 V/div for voltage and 100 μA/div for current. This is just one of many scales that could be selected. Since this is a resistive circuit, the curve will be linear. We could draw the curve by using two points, but we will use three. The third point is used to check for errors. Using Ohm's law, we can construct a data table of variable values. Figure 13–21(a) is such a table. The resulting curve is shown in Figure 13–21(b).

Let's check our work. The reciprocal of the slope should equal 10 kΩ.

$$\text{Slope} = \frac{\Delta I}{\Delta V}$$

$$\text{Let } \Delta V = 2 \text{ V}$$

$$\text{Then } \Delta I = 200 \text{ μA}$$

$$\frac{\Delta I}{\Delta V} = \frac{200 \text{ μA}}{2 \text{ V}} = 100 \text{ μS}$$

$$R = \frac{1}{G} = \frac{1}{100 \text{ μS}} = 10 \text{ k}\Omega$$

EXAMPLE 13–7

Plot a curve of current versus voltage from the given data in Figure 13–22(a) on the next page. V is the independent variable.

SOLUTION Select an appropriate scale. 10 V/div for voltage and 10 mA/div for current were chosen for our solution. The resulting curve is shown in Figure 13–22(b).

FIGURE 13–22
Graph of
current versus
voltage where R
is nonlinear: (a)
Given data;
(b) Resulting
curve of current
versus voltage.

V (volts)	I (mA)
0	0
8	10
20	20
39	30
60	40
85	50

(a)

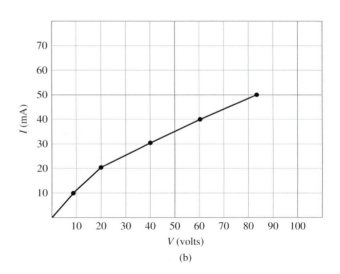

(b)

PRACTICE PROBLEMS 13–6

1. Plot a curve of current versus voltage for $R = 20\ \text{k}\Omega$ for values of voltage from 0 to 50 V.

2. Plot a curve of current versus voltage from the data in Figure 13–23.

SOLUTIONS

1. See Figure 13–24.

2. See Figure 13–25.

Additional practice problems are at the end of the chapter.

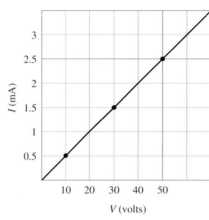

V (volts)	I (mA)
0	0
10	0.5
30	1.5
50	2.5

V (volts)	I (mA)
0	0
0.2	0.2
0.3	0.4
0.4	0.65
0.5	1.0
0.6	1.5
0.7	2.4

FIGURE 13–23

FIGURE 13–24

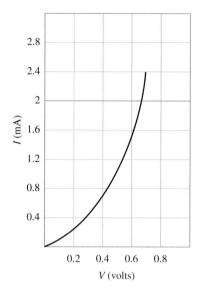

FIGURE 13–25

SELF-TEST 13–2

1. Refer to Figure 13–26. Find R.

2. Refer to Figure 13–16. Find R_D and r_d if $V_f = 0.4$ V.

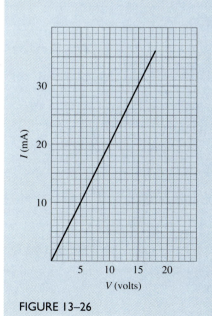

FIGURE 13–26

CHAPTER 13 AT A GLANCE

PAGE	KEY POINTS AND DEFINITIONS	EXAMPLE
385	*Key Point:* The graph of any linear equation is a straight line.	

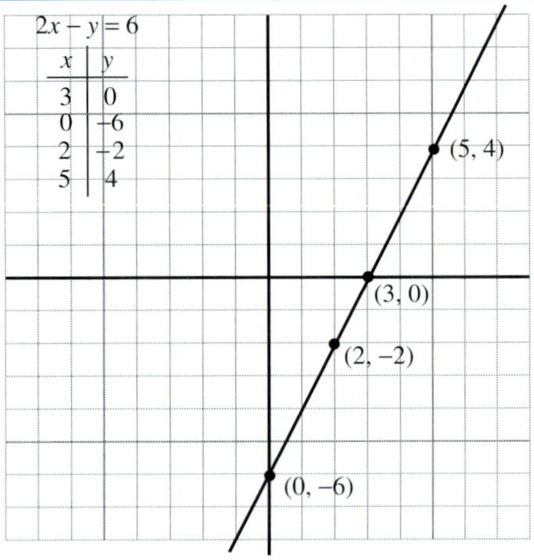

FIGURE 13–28

385   The *x-intercept* is the point where the line crosses the *x*-axis. The *y-intercept* is the point where the line crosses the *y*-axis.

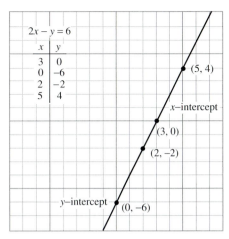

FIGURE 13–29

390   *Key Point:* The *slope* of a line is defined as the change in *y* that results from an *increase* in *x*.

Slope $= \Delta y/\Delta x$.

394   *Key Point:* Any linear equation containing two variables can be written in the slope–intercept form.

$y = mx + b$. *m* is the slope. *b* is the *y*-intercept.

402   When plotting curves, the *independent* variable is plotted along the *x*-axis and the *dependent* variable is plotted along the *y*-axis.

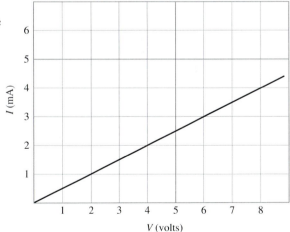

FIGURE 13–30

Key Point: Non-linear devices can be characterized by two resistances whose equations are:

Dynamic resistance $= r_d = \dfrac{\Delta V}{\Delta I}$

DC resistance $= R_D = \dfrac{V}{I}$

Both values depend on the point on the curve where measurements are made.

See Figure 13–17.
At $V_f = 0.5$ volts:

$$r_d = \frac{\Delta V}{\Delta I} = \frac{0.3 \text{ V}}{8 \text{ mA}} = 37.5 \ \Omega$$

$$R_D = \frac{V}{I} = \frac{0.5 \text{ V}}{6 \text{ mA}} = 83.3 \ \Omega$$

(Both of the above calculations are based on values on the graph where $V_f = 0.5$ volts.)

END OF CHAPTER PROBLEMS 13–1

1. Plot and label the following points on the graph of Figure 13–31:
   (a) $(0,6)$ (b) $(-5,0)$ (c) $(3,-5)$
   (d) $(4,4)$ (e) $(-2,-5)$ (f) $(-4,3.5)$

2. Plot and label the following points on the graph of Figure 13–31:
   (a) $(-5,-5)$ (b) $(-4,4)$ (c) $(6,3)$
   (d) $(0,-4)$ (e) $(4,0)$ (f) $(2,-4.5)$

3. Plot and label the following points on the graph of Figure 13–32:
   (a) $(30,-20)$ (b) $(-40,25)$
   (c) $(30,30)$ (d) $(-10,-30)$
   (e) $(0,45)$ (f) $(-30,0)$
   (g) $(7,-18)$ (h) $(-23,47)$

4. Plot and label the following points on the graph of Figure 13–32:
   (a) $(20,-30)$ (b) $(10,10)$
   (c) $(-40,15)$ (d) $(-20,-20)$
   (e) $(0,-25)$ (f) $(40,0)$
   (g) $(-13,22)$ (h) $(23,-8)$

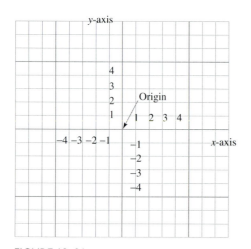

FIGURE 13–31

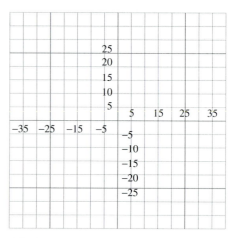

FIGURE 13–32

END OF CHAPTER PROBLEMS 13–2

On a system of rectangular coordinates, plot the following equations. Identify the x- and y-intercepts. Plot at least three points.

1. $x + y = 4$
2. $2y = 4x + 12$
3. $2x - y = 6$
4. $x + 2y = -4$
5. $x + 3y = 3$
6. $x + 2y = 6$
7. $2x + 3y = 8$
8. $x - 4y = 6$
9. $x = -4y + 6$
10. $3x = 2y - 7$
11. $-4x - 3y = 24$
12. $3x - 5y = -10$
13. $-x + 2y = -7$
14. $4x - 5y = -12$
15. $6 + 3x = 2y$
16. $12 + 4x = 6y$
17. $5 + 2y = 7x$
18. $7 + 3y = 8x$
19. $21 - 7y = 36x$
20. $28 - 4y = 7x$

END OF CHAPTER PROBLEMS 13–3

Plot the following equations on linear graph paper. Identify the x- and y-intercepts. Determine the slope.

1. $2x + y = 6$
2. $4x - 8y = 12$
3. $2y = 5x + 5$
4. $3x = 2y + 9$
5. $3x + 5y = 15$
6. $-4x - 6y = 36$
7. $2x - 7y = -14$
8. $6x + 3y = -24$
9. $-3x + 4y = -24$
10. $3y = 7x + 21$
11. $x - 2 = \dfrac{y}{5}$
12. $x + 2 = \dfrac{y}{4}$
13. $y + 3 = -2x$
14. $y + 3 = x$

END OF CHAPTER PROBLEMS 13–4

Change the following equations into the slope–intercept form. Determine the slope and the y-intercept. Graph the equation.

1. $x + 3y = 3$
2. $x - 5y = 10$
3. $3x + 4y = 12$
4. $y - x = -6$
5. $x - 4y = 12$
6. $4x + 3y = 12$
7. $5y - 2x = 10$
8. $x - y = -6$
9. $3x - 7y = 21$
10. $8x - 6y = 12$
11. $x + 3y = 6$
12. $x + 2y = 8$
13. $2x - 3y = 8$
14. $2x - 5y = 10$
15. $2y = x - 6$
16. $3y = x - 12$
17. $3x - 8 = 2y$
18. $4x - 16 = 3y$
19. $3x + y = -9$
20. $x + 3y = 12$

END OF CHAPTER PROBLEMS 13–5

1. Refer to Figure 13–33. Find R_{dc} and r_{ac} where $V = 3$ V.
2. Refer to Figure 13–33. Find R_{dc} and r_{ac} where $V = 5$ V.
3. Refer to Figure 13–34. If ΔV_f is measured from 0.4 to 0.6 V, what is ΔI_f? What is the slope? What is the value of r_d?
4. Refer to Figure 13–34. If ΔV_f is measured from 0.2 to 0.6 V, what is ΔI_f? What is the slope? What is the value of r_d?

FIGURE 13–33

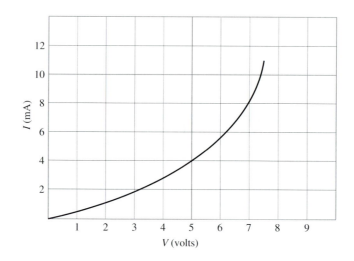

FIGURE 13–34

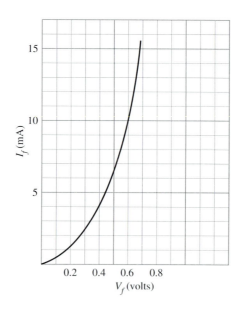

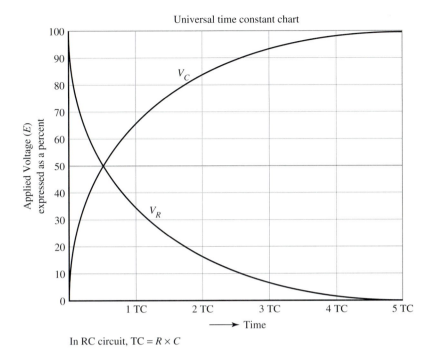

Universal time constant chart

FIGURE 13–35

Applied Voltage (E) expressed as a percent

V_C

V_R

1 TC   2 TC   3 TC   4 TC   5 TC

Time

In RC circuit, TC = $R \times C$

Refer to Figure 13–35.

5. Let $E = 100$ V. Find V_C and V_R at
   (a) 0.5 TC and (b) 3 TC.
7. Let $E = 20$ V. Find V_C and V_R at
   (a) 1.5 TC and (b) 4 TC.
9. (a) How many time constants will
   it take for the capacitor to charge
   to 80% of E? (b) If E equals 50 V,
   how many time constants will it
   take for the capacitor to charge to
   20 V?
11. Refer to Figure 13–36(a) on the next
    page. Find the slope. Find R.

6. Let $E = 80$ V. Find V_C and V_R at
   (a) 1 TC and (b) 3.5 TC.
8. Let $E = 50$ V. Find V_C and V_R at
   (a) 1 TC and (b) 3 TC.
10. (a) How many time constants will
    it take for the capacitor to charge
    to 50% of E? (b) If E equals 10 V,
    how many time constants will it
    take for the capacitor to charge to
    7 V?
12. Refer to Figure 14–36(b). Find the
    slope. Find R.

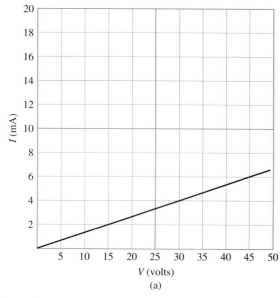

FIGURE 13–36

END OF CHAPTER PROBLEMS 13–6

Plot a curve of I versus V_R for the following:

1. $R = 100 \ \Omega$ for values of V_R from 0 to 50 V.
2. $R = 4 \ k\Omega$ for values of V_R from 0 to 20 V.
3. $R = 1 \ k\Omega$ for values of V_R from 0 to 100 V.
4. $R = 2 \ k\Omega$ for values of V_R from 0 to 80 V.

Plot a curve of I versus V:

5. From the data in Figure 13–37.
6. From the data in Figure 13–38.
7. From the data in Figure 13–39.
8. From the data in Figure 13–40.

V (volts)	I (mA)
0	0
1	15
2	23
3	30
4	34
5	38
6	42

FIGURE 13–37

V (mV)	I (mA)
0	0
40	1
80	2
120	3.2
200	6
240	8
300	11

FIGURE 13–38

V (volts)	I (mA)
0	0
10	20
20	30
30	40
40	44
50	48
70	48

FIGURE 13–39

V (volts)	I (mA)
1	0.5
2	2
3	5
4	6
5	6.5

FIGURE 13–40

ANSWERS TO SELF-TESTS

1. See Figure 13–41.   2. See Figure 13–42.   3. See Figure 13–43.

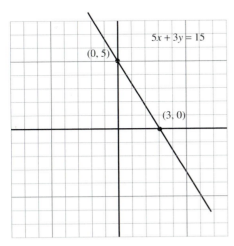

FIGURE 13–41

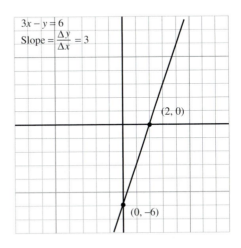

FIGURE 13–42

FIGURE 13–43

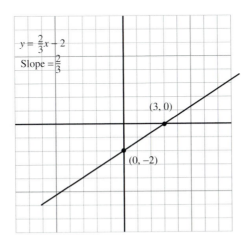

1. $R = 500\ \Omega$

2. $R_D \simeq 95\ \Omega$, $r_d \simeq 54\ \Omega$

3. $V_C \simeq 22.5$ V, $V_R \simeq 2.5$ V

4. See Figure 13–44.

5. See Figure 13–45.

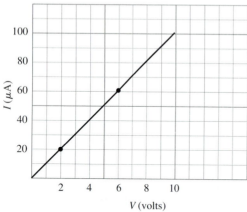

FIGURE 13–44

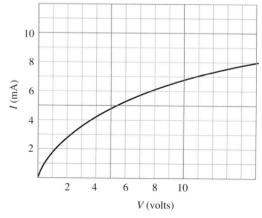

FIGURE 13–45

Simultaneous Linear Equations

Introduction

As we saw in Chapter 13, every linear equation in two unknowns has an unlimited number of solutions. If we have two equations, each equation still has an unlimited number of solutions. However, if the two equations are plotted on the same system of rectangular coordinates, they will intersect at some common point provided they (1) are not parallel lines and (2) are not the same line. The coordinates of this common point will satisfy both equations. When such equations have this common point of intersection, they are said to be *simultaneous*.

In the chapters on circuit analysis we used various laws and theorems in solving circuit problems. In this chapter we introduce another method of solving complex circuit problems. When there is more than one path for current, we can write an equation in terms of the voltages, currents, and resistances for each path. Because parts of each path will be common, we can solve for the unknown values using *simultaneous linear equations*. In this chapter we will show the various methods that can be used to solve sets of linear equations simultaneously, including graphical solutions.

Chapter Objectives

In this chapter you will learn how to:

1. Solve simultaneous linear equations graphically.
2. Solve simultaneous linear equations using the addition or subtraction method.
3. Solve simultaneous linear equations using the substitution method.
4. Solve simultaneous linear equations using determinants.
5. Use simultaneous linear equations to solve complex circuit problems.

GRAPHICAL SOLUTION   14-1

Using the methods for graphing linear equations we learned in Chapter 13, let's plot two equations on the same system of rectangular coordinates.

EXAMPLE 14-1

Plot the equations $x - 2y = 4$ and $3x + 2y = 4$ on the same system of rectangular coordinates. Find the point of intersection.

SOLUTION   The plot of the equations is shown in Figure 14–1. The point of intersection is at $(2, -1)$. This point is common to both equations. Because the lines intersect, there is a common solution, which in this case is $(2, -1)$. Furthermore, this is the only solution. $x = 2$ and $y = -1$ are the only values common to both equations.

FIGURE 14–1
Plot of the
equations
$x - 2y = 4$ and
$3x + 2y = 4$.

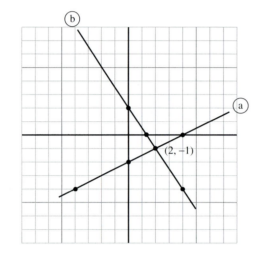

$x - 2y = 4$			$3x + 2y = 4$	
x	y		x	y
4	0		$1\frac{1}{3}$	0
0	−2		0	2
−4	−4		4	−4

PRACTICE PROBLEMS 14–1

Plot the following equations on the same system of rectangular coordinates. Find the coordinates of the point of intersection.

1. $x + y = 5$
   $x - y = 3$

2. $x + y = 3$
   $2x + y = 7$

3. $x + 2y = 7$
   $2x + 2y = 10$

4. $\dfrac{x}{4} + \dfrac{y}{3} = \dfrac{7}{12}$

   $\dfrac{x}{2} - \dfrac{y}{4} = \dfrac{1}{4}$

5. $x - y = 1.5$
   $0.9375x + y = 3.75$

SOLUTIONS

1. $(4, 1)$. The graph is shown in Figure 14–2.

2. $(4, -1)$. The graph is shown in Figure 14–3.

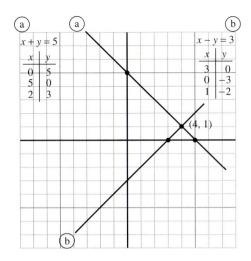

FIGURE 14–2

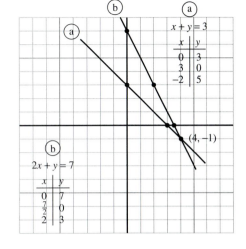

FIGURE 14–3

3. $(3,2)$. The graph is shown in Figure 14–4.

5. $(2.6, 1.1)$. The graph is shown in Figure 14–6.

4. $(1,1)$. The graph is shown in Figure 14–5.

In graphical analysis all answers are approximations because we cannot read exact values from graphs. We were able to determine accurately the answers to the first four problems because the point of intersection was a whole number and the graphs were scaled in whole numbers. In problem 5 we had to approximate the answer.

Additional practice problems are at the end of the chapter.

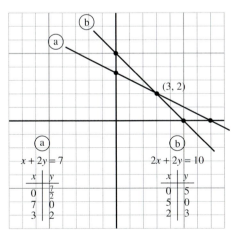

FIGURE 14–4

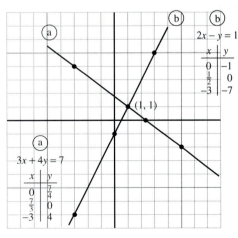

FIGURE 14–5

FIGURE 14–6

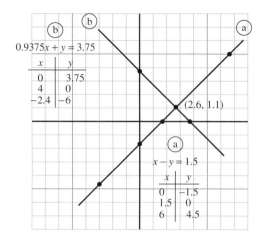

14–2   SOLUTION BY ADDITION OR SUBTRACTION

Solving the two equations in Example 14–1 graphically resulted in $x = 2$ and $y = -1$. Let's solve the same problem by using the addition or subtraction method. This method uses rules of algebra we have already developed. We will change one or the other of the equations so that both x values, or both y values, have the same coefficient. If the signs of the coefficients are the same, we will subtract one equation from the other. If the signs are different, we will add the equations. The subtraction (or addition) results in a new equation that has only one unknown. We can now find the value of this unknown. This value is common to both original equations. The value of this unknown is then substituted back into one of the original equations and the other unknown is then found. Let's see how it works.

EXAMPLE 14–2

Solve the following set of equations for x and y by the addition or subtraction method:

(14–1) $$x - 2y = 4$$

(14–2) $$3x + 2y = 4$$

SOLUTION We can eliminate the y term by addition.

(14–1) $$x - 2y = 4$$

(14–2) $$\underline{3x + 2y = 4}$$

(14–3) $$\text{(add)} \quad 4x \qquad = 8$$

Solve Equation 14–3 for x.

$$4x = 8$$

$$x = 2$$

Substitute $x = 2$ in Equation 14–1 and solve for y.

$$x - 2y = 4 \qquad (14\text{–}1)$$

$$2 - 2y = 4$$

$$-2y = 2$$

$$y = -1$$

The common values of x and y are $x = 2$ and $y = -1$. Check Equation 14–1.

$$x - 2y = 4$$

$$2 - 2(-1) = 4$$

$$2 + 2 = 4$$

Check Equation 14–2.

$$3x + 2y = 4$$

$$3(2) + 2(-1) = 4$$

$$6 - 2 = 4$$

The solution is verified.

EXAMPLE 14–3

Solve the following set of equations for x and y by the addition or subtraction method:

$$2x - y = 4 \qquad (14\text{–}4)$$

$$x - 2y = -1 \qquad (14\text{–}5)$$

SOLUTION

Write 14–4. $\qquad 2x - y = 4$

Multiply 14–5 by 2. $\qquad \underline{2x - 4y = -2} \qquad (14\text{–}6)$

Subtract 14–6. $\qquad 0 + 3y = 6 \qquad (14\text{–}7)$

Solve Equation 14–7 for y:

$$3y = 6$$

$$y = 2$$

Substitute $y = 2$ in Equation 14–4 and solve for x.

$$2x - y = 4 \qquad (14\text{–}4)$$

$$2x - 2 = 4$$

$$2x = 6$$

$$x = 3$$

The common values of x and y are $x = 3$ and $y = 2$. Check Equation 14–4.

(14–4)
$$2x - y = 4$$
$$2(3) - 2 = 4$$
$$6 - 2 = 4$$

Check Equation 14–5.

(14–5)
$$x - 2y = -1$$
$$3 - 2(2) = -1$$
$$3 - 4 = -1$$

PRACTICE PROBLEMS 14–2

Solve the following sets of equations for x and y by the addition or subtraction method.

1. $x + y = 5$
   $x - y = 3$

2. $x + y = 3$
   $2x + y = 7$

3. $x + 2y = 7$
   $2x + 2y = 10$

4. $\dfrac{x}{4} + \dfrac{y}{3} = \dfrac{7}{12}$

   $\dfrac{x}{2} - \dfrac{y}{4} = \dfrac{1}{4}$

5. $x - y = 1.5$
   $x + y = 3.75$

SOLUTIONS

1. $x = 4, y = 1$
2. $x = 4, y = -1$
3. $x = 3, y = 2$
4. $x = 1, y = 1$
5. $x = 2.625, y = 1.125$

Additional practice problems are at the end of the chapter.

14–3   SOLUTION BY SUBSTITUTION

Let's work our example problems again, but this time we will solve the equations by substitution. This involves finding the value of x (or y) in one equation and then substituting that value back into the other equation. This eliminates one of the unknowns, and we can then find the value of the other. Example 14–4 shows how it works.

EXAMPLE 14–4

Solve the following set of equations for x and y by using the substitution method:

$$x - 2y = 4 \qquad (14\text{–}1)$$

$$3x + 2y = 4 \qquad (14\text{–}2)$$

SOLUTION   Solve Equation 14–1 for x.

$$x - 2y = 4 \qquad (14\text{–}1)$$

$$x = 4 + 2y \qquad (14\text{–}8)$$

Substitute the value of x in Equation 14–8 back into Equation 14–2 and then solve the equation for y.

$$3x + 2y = 4 \qquad (14\text{–}2)$$

$$3(4 + 2y) + 2y = 4$$

$$12 + 6y + 2y = 4$$

$$8y = -8$$

$$y = -1$$

Now substitute -1 for y in either equation and solve for x. (We will choose Equation 14–1.)

$$x - 2y = 4 \qquad (14\text{–}1)$$

$$x - 2(-1) = 4$$

$$x + 2 = 4$$

$$x = 2$$

The solution is $x = 2$ and $y = -1$. We could check the solutions as we did in the previous section.

EXAMPLE 14–5

Solve the following set of equations for x and y by using the substitution method:

$$2x - y = 4 \qquad (14\text{–}4)$$

$$x - 2y = -1 \qquad (14\text{–}5)$$

SOLUTION   Solve for x in Equation 14–4.

$$2x - y = 4 \qquad (14\text{–}4)$$

$$2x = y + 4$$

$$x = \frac{y + 4}{2} \qquad (14\text{–}9)$$

Substitute the value of x in Equation 14–9 for the value of x in Equation 14–5.

(14–5)
$$x - 2y = -1$$

$$\frac{y + 4}{2} - 2y = -1$$

Multiply both sides by 2.

$$y + 4 - 4y = -2$$

$$-3y = -6$$

$$y = 2$$

Now substitute this value for y in Equation 14–4.

(14–4)
$$2x - y = 4$$

$$2x - 2 = 4$$

$$2x = 6$$

$$x = 3$$

The solution is $x = 3$ and $y = 2$. We could check the solutions as we did in the previous section.

PRACTICE PROBLEMS 14–3

Solve the sets of equations in Practice Problems 14–2 by using the substitution method.

SOLUTIONS: The solutions are the same as for Practice Problems 14–2.

Additional practice problems are at the end of the chapter.

14-4   SECOND-ORDER DETERMINANTS

Another way to solve simultaneous equations is by using *determinants*. Determinants are arrays of numbers. These numbers correspond to the coefficients and constants of the different equations.

Consider the following general equations with two unknowns:

(14–10)
$$a_1x + b_1y = k_1$$

(14–11)
$$a_2x + b_2y = k_2$$

a_1, b_1, a_2, and b_2 are the coefficients of the unknowns. k_1 and k_2 are the constants. Notice that the equations are set up the same way we set them up when we were using the addition or subtraction method. After the terms of the equations have been arranged in this way, we can find the values for x and y by using the following equations:

$$x = \frac{\begin{vmatrix} k_1 & b_1 \\ k_2 & b_2 \end{vmatrix}}{\begin{vmatrix} a_1 & b_1 \\ a_2 & b_2 \end{vmatrix}} = \frac{k_1 b_2 - k_2 b_1}{a_1 b_2 - a_2 b_1}$$

$$y = \frac{\begin{vmatrix} a_1 & k_1 \\ a_2 & k_2 \end{vmatrix}}{\begin{vmatrix} a_1 & b_1 \\ a_2 & b_2 \end{vmatrix}} = \frac{a_1 k_2 - a_2 k_1}{a_1 b_2 - a_2 b_1}$$

Let's analyze these equations. The array of four values in the denominator is called the *determinant of the denominator*. This determinant has two rows and two columns and is called a *second-order determinant*. The numbers a_1, a_2, b_1, and b_2 are called the *elements* of the determinants. $a_1 b_2 - a_2 b_1$ is called the *expansion* of the determinants. The numbers a_1 and b_2 are the *principal diagonal*. The numbers a_2 and b_1 are the *secondary diagonal*. The expansion of the determinant results from the product of the elements in the principal diagonal ($a_1 b_2$) minus the product of the elements in the secondary diagonal ($a_2 b_1$). This process is illustrated as follows:

$$\begin{vmatrix} a_1 & b_1 \\ a_2 & b_2 \end{vmatrix} = a_1 b_2 - a_2 b_1$$

This is the denominator for both unknowns.

The determinant for the numerator of the unknown, x, is written

$$\begin{vmatrix} k_1 & b_1 \\ k_2 & b_2 \end{vmatrix} = k_1 b_2 - k_2 b_1$$

Here we have replaced a_1 and a_2, the coefficients of the first unknown (x), with the constants k_1 and k_2 from the right side of the equations. The determinant for the numerator of the unknown, y, is written

$$\begin{vmatrix} a_1 & k_1 \\ a_2 & k_2 \end{vmatrix} = a_1 k_2 - a_2 k_1$$

Here we have replaced b_1 and b_2, the coefficients of the second unknown (y), with the constants from the right side of the equations.

   ☞ **KEY POINT:** The steps for solving simultaneous equations using *determinants* are:

1. Arrange the terms of the equations with the unknowns on the left side and the constants on the right side.

2. Arrange the unknowns so that they appear in the same order for each equation.
3. Use the coefficients of the unknowns to form an array of numbers called the *determinant of the denominator*.
4. For each unknown, create an array of numbers similar to the above array, except for its coefficients, substitute the constants from the right side of the equations; this is called the *determinant of the numerator*.
5. The value of each unknown can then be calculated by dividing its *determinant of the numerator* by the common *determinant of the denominator*.
6. To check your answers, substitute the solutions into the original equations and confirm the equalities.

EXAMPLE 14–6

Solve for x and y in the following set of equations by using determinants. (This is the same set of equations used in Example 14–2.)

(14–12) $$x - 2y = 4$$

(14–13) $$3x + 2y = 4$$

SOLUTION   Determine the numbers to be used in the following arrays:

$$a_1 = 1, \qquad b_1 = -2, \qquad k_1 = 4$$
$$a_2 = 3, \qquad b_2 = 2, \qquad k_2 = 4$$

Now substitute these known values in our general equations for x and y.

$$x = \frac{\begin{vmatrix} 4 & -2 \\ 4 & 2 \end{vmatrix}}{\begin{vmatrix} 1 & -2 \\ 3 & 2 \end{vmatrix}} = \frac{8 - (-8)}{2 - (-6)} = \frac{16}{8} = 2$$

$$y = \frac{\begin{vmatrix} 1 & 4 \\ 3 & 4 \end{vmatrix}}{\begin{vmatrix} 1 & -2 \\ 3 & 2 \end{vmatrix}} = \frac{4 - 12}{2 - (-6)} = \frac{-8}{8} = -1$$

The solution is $x = 2$, $y = -1$.

Notice that the denominators are identical. This means that we need to calculate the denominator only once. This method may be less laborious than either the addition or subtraction method or the substitution method in problem-solving situations.

EXAMPLE 14–7

Solve for x and y in the following set of equations. (This is the same set of equations used in Example 14–5.)

$$2x - y = 4$$

$$x - 2y = -1$$

SOLUTION Determine the numbers to be used in the following arrays:

$$a_1 = 2, \qquad b_1 = -1, \qquad k_1 = 4$$
$$a_2 = 1, \qquad b_2 = -2, \qquad k_2 = -1$$

$$x = \frac{\begin{vmatrix} 4 & -1 \\ -1 & -2 \end{vmatrix}}{\begin{vmatrix} 2 & -1 \\ 1 & -2 \end{vmatrix}} = \frac{-8 - 1}{-4 - (-1)} = \frac{-9}{-3} = 3$$

$$y = \frac{\begin{vmatrix} 2 & 4 \\ 1 & -1 \end{vmatrix}}{\begin{vmatrix} 2 & -1 \\ 1 & -2 \end{vmatrix}} = \frac{-2 - 4}{-3} = \frac{-6}{-3} = 2$$

The solution is $x = 3$, $y = 2$.

PRACTICE PROBLEMS 14–4

Solve the sets of equations in Practice Problems 14–2 by using determinants.

SOLUTIONS: The solutions are the same as for Practice Problems 14–2.

Additional practice problems are at the end of the chapter.

THIRD-ORDER DETERMINANTS   14–5

Let's now consider three equations and three unknowns. This determinant has three rows and three columns and is called a *third-order determinant.* Our general equations look like this:

$$a_1 x + b_1 y + c_1 z = k_1$$
$$a_2 x + b_2 y + c_2 z = k_2$$
$$a_3 x + b_3 y + c_3 z = k_3$$

The determinants for the three unknowns look like this:

$$x = \frac{\begin{vmatrix} k_1 & b_1 & c_1 \\ k_2 & b_2 & c_2 \\ k_3 & b_3 & c_3 \end{vmatrix}}{\begin{vmatrix} a_1 & b_1 & c_1 \\ a_2 & b_2 & c_2 \\ a_3 & b_3 & c_3 \end{vmatrix}} = \frac{k_1 b_2 c_3 + b_1 c_2 k_3 + c_1 k_2 b_3 - c_1 b_2 k_3 - k_1 c_2 b_3 - b_1 k_2 c_3}{a_1 b_2 c_3 + b_1 c_2 a_3 + c_1 a_2 b_3 - c_1 b_2 a_3 - a_1 c_2 b_3 - b_1 a_2 c_3} \qquad (14\text{–}14)$$

$$(14\text{-}15) \quad y = \frac{\begin{vmatrix} a_1 & k_1 & c_1 \\ a_2 & k_2 & c_2 \\ a_3 & k_3 & c_3 \\ a_1 & b_1 & c_1 \\ a_2 & b_2 & c_2 \\ a_3 & b_3 & c_3 \end{vmatrix}}{} = \frac{a_1k_2c_3 + k_1c_2a_3 + c_1a_2k_3 - c_1k_2a_3 - a_1c_2k_3 - k_1a_2c_3}{a_1b_2c_3 + b_1c_2a_3 + c_1a_2b_3 - c_1b_2a_3 - a_1c_2b_3 - b_1a_2c_3}$$

$$(14\text{-}16) \quad z = \frac{\begin{vmatrix} a_1 & b_1 & k_1 \\ a_2 & b_2 & k_2 \\ a_3 & b_3 & k_3 \\ a_1 & b_1 & c_1 \\ a_2 & b_2 & c_2 \\ a_3 & b_3 & c_3 \end{vmatrix}}{} = \frac{a_1b_2k_3 + b_1k_2a_3 + k_1a_2b_3 - k_1b_2a_3 - a_1k_2b_3 - b_1a_2k_3}{a_1b_2c_3 + b_1c_2a_3 + c_1a_2b_3 - c_1b_2a_3 - a_1c_2b_3 - b_1a_2c_3}$$

Various methods are used to find values of x, y, and z. Here is one of them: Rewrite columns 1 and 2 to the right of the determinant. For the unknown, x, the numerator would look like this:

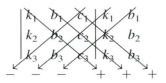

Diagonal lines have been drawn through the elements wherever there are three elements in the diagonal group. Where there are fewer than three elements in the diagonal group, that diagonal group is not used. The result is three diagonal groups running down from left to right and three diagonal groups running down from right to left. The products of each diagonal group running down from left to right are added. The products of the groups that run down from right to left are subtracted. This procedure is used for each numerator and each denominator. Since all the denominators are alike, however, the denominators need to be determined only once.

EXAMPLE 14–8

Solve for the unknowns in the following set of equations by using determinants:

$$x + 2y + 3z = 14$$
$$2x + y + 2z = 10$$
$$3x + 4y - 3z = 2$$

SOLUTION Let's first write down the coefficients:

$$a_1 = 1, \quad b_1 = 2, \quad c_1 = 3, \quad k_1 = 14$$
$$a_2 = 2, \quad b_2 = 1, \quad c_2 = 2, \quad k_2 = 10$$
$$a_3 = 3, \quad b_3 = 4, \quad c_3 = -3, \quad k_3 = 2$$

Next let's find the denominator since it will be common to all solutions. Referring to Equation 14–14 and replacing the literal numbers with real numbers, we get

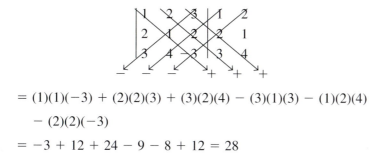

$$= (1)(1)(-3) + (2)(2)(3) + (3)(2)(4) - (3)(1)(3) - (1)(2)(4)$$

$$- (2)(2)(-3)$$

$$= -3 + 12 + 24 - 9 - 8 + 12 = 28$$

Using the same method, let's find x.

$$x = \frac{}{28}$$

$$\frac{(14)(1)(-3) + (2)(2)(2) + (3)(10)(4) - (3)(1)(2) - (14)(2)(4) - (2)(10)(-3)}{28}$$

$$= \frac{-42 + 8 + 120 - 6 - 112 + 60}{28} = \frac{28}{28} = 1$$

Now let's find y the same way.

$$y = \frac{}{28}$$

$$= \frac{(1)(10)(-3) + (14)(2)(3) + (3)(2)(2) - (3)(10)(3) - (1)(2)(2) - (14)(2)(-3)}{28}$$

$$= \frac{-30 + 84 + 12 - 90 - 4 + 84}{28} = \frac{56}{28} = 2$$

Now we solve for z and are finished.

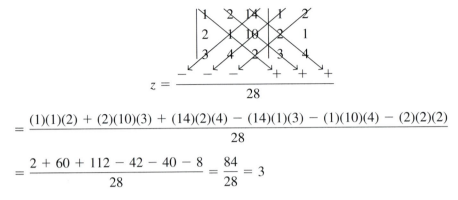

$$z = \frac{}{28}$$

$$= \frac{(1)(1)(2) + (2)(10)(3) + (14)(2)(4) - (14)(1)(3) - (1)(10)(4) - (2)(2)(2)}{28}$$

$$= \frac{2 + 60 + 112 - 42 - 40 - 8}{28} = \frac{84}{28} = 3$$

PRACTICE PROBLEMS 14–5

Use determinants to solve for the unknowns in the following sets of equations:

1. $x + 2y + z = 6$
   $2x - y + 3z = -13$
   $3x - 2y + 3z = -16$

2. $3x + 2y = 12$
   $4x + 2z = 16$
   $4y + 3z = 24$

(Even though only two of the three unknowns appear in each equation, we must use all three in our solution. When the third unknown is missing, we assign a value of zero to that coefficient. For example, $c_1 = 0$ in the first equation of problem 2.)

SOLUTIONS

1. $x = 1.25, y = 4.25, z = -3.75$

2. $x = 2, y = 3, z = 4$

Additional practice problems are at the end of the chapter.

SELF-TEST 14–1

Solve the following sets of equations (a) graphically, (b) by the addition or subtraction method, (c) by the substitution method, and (d) by using determinants.

1. $4x + y = 9$
   $x + y = 6$
3. $3x - 5y = 11$
   $x + 8y = -6$

2. $2x + 4y = 4$
   $x - y = 5$

Answers to Self-test 14–1 are at the end of the chapter.

In Chapter 10 we learned how to use various circuit theorems to solve complex circuit problems. An alternative method is solution by simultaneous linear equations.

As a first example of solution by simultaneous equations, let's solve for the currents in Figure 14–7. In this example we will solve the problem using all three methods: by addition and subtraction, by substitution, and by determinants.

EXAMPLE 14–9

Find the currents and the voltage drops in the circuit of Figure 14–7.

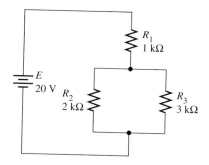

FIGURE 14–7
Circuit used to find currents and voltage drops in Example 14–9.

SOLUTION This circuit can easily be solved by using Ohm's law and Kirchhoff's laws. This solution would yield

$$R_T = R_1 + R_2 \| R_3 = 1 \text{ k}\Omega + 1.2 \text{ k}\Omega = 2.2 \text{ k}\Omega$$

$$I_T = \frac{E}{R_T} = \frac{20 \text{ V}}{2.2 \text{ k}\Omega} = 9.09 \text{ mA}$$

$$V_1 = IR_1 = 9.09 \text{ mA} \times 1 \text{ k}\Omega = 9.09 \text{ V}$$

$$V_2 = V_3 = IR_x = 9.09 \text{ mA} \times 1.2 \text{ k}\Omega = 10.9 \text{ V}$$

$$(R_x = R_2 \| R_3)$$

$$I_2 = \frac{V_2}{R_2} = \frac{10.9 \text{ V}}{2 \text{ k}\Omega} = 5.45 \text{ mA}$$

$$I_3 = \frac{V_3}{R_3} = \frac{10.9 \text{ V}}{3 \text{ k}\Omega} = 3.64 \text{ mA}$$

$$I_1 = I_T = 9.09 \text{ mA}$$

Now let's work the same problem using simultaneous linear equations. Let's first establish the polarities of the voltage drops across each resistor as in Figure 14–8.

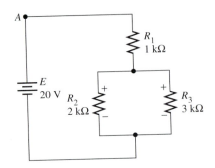

FIGURE 14–8
Circuit in Figure
14–7 with
polarities
shown.

Using Kirchhoff's voltage law, let's start at point A and move counterclockwise around the loop containing E, R_1, and R_2.

$$E - V_2 - V_1 = 0$$

(14–17)   or $$E = V_1 + V_2$$

Starting at point A and moving counterclockwise around the loop containing E, R_1, and R_3, we get

$$E - V_3 - V_1 = 0$$

(14–18)   or $$E = V_1 + V_3$$

If we express these equations in terms of IR drops, then Equation 14–17 would look like this:

$$E = I_1 R_1 + I_2 R_2$$

Substituting known values gives

(14–19) $$20\text{ V} = 1\text{k}I_1 + 2\text{k}I_2$$

Equation 14–18 would look like this:

$$E = I_1 R_1 + I_3 R_3$$

Substituting known values gives

(14–20) $$20\text{ V} = 1\text{k}I_1 + 3\text{k}I_3$$

Now we have two equations that we can solve simultaneously.

(14–19) $$20\text{ V} = 1\text{k}I_1 + 2\text{k}I_2$$

(14–20) $$20\text{ V} = 1\text{k}I_1 + 3\text{k}I_3$$

In Equations 14–19 and 14–20 we have three unknowns. We can simplify the solution by reducing the number of unknowns to two. Since $I_1 = I_2 + I_3$, then $I_3 = I_1 - I_2$. Let's substitute $(I_1 - I_2)$ for I_3 in Equation 14–20.

$$20\text{ V} = 1\text{k}I_1 + 3\text{k}I_3$$
$$= 1\text{k}I_1 + 3\text{k}(I_1 - I_2)$$
$$= 1\text{k}I_1 + 3\text{k}I_1 - 3\text{k}I_2$$

(14–21) $$= 4\text{k}I_1 - 3\text{k}I_2$$

Now let's look at Equations 14–19 and 14–21. We have two equations and two unknowns and can now solve the problem.

$$20 \text{ V} = 1\text{k}I_1 + 2\text{k}I_2 \qquad (14\text{–}19)$$

$$20 \text{ V} = 4\text{k}I_1 - 3\text{k}I_2 \qquad (14\text{–}21)$$

SOLUTION BY ADDITION AND SUBTRACTION   To eliminate one unknown, we will multiply Equation 14–19 by 4.

$$80 \text{ V} = 4\text{k}I_1 + 8\text{k}I_2 \qquad (14\text{–}22)$$

Subtracting Equation 14–22 from Equation 14–21 gives

$$20 \text{ V} = 4\text{k}I_1 - 3\text{k}I_2 \qquad (14\text{–}21)$$

$$\underline{80 \text{ V} = 4\text{k}I_1 + 8\text{k}I_2} \qquad (14\text{–}22)$$

$$-60 \text{ V} = 0 \quad - 11\text{k}I_2$$

Solving for I_2, we get

$$I_2 = \frac{-60 \text{ V}}{-11\text{k}} = 5.45 \text{ mA}$$

Now let's plug the known value of I_2 back into Equation 14–21 and solve for I_1.

$$20 \text{ V} = 4\text{k}I_1 - 3\text{k}I_2$$

$$= 4\text{k}I_1 - 3\text{k}(5.45 \text{ mA})$$

$$= 4\text{k}I_1 - 16.4 \text{ V}$$

$$36.4 \text{ V} = 4\text{k}I_1$$

$$\frac{36.4 \text{ V}}{4\text{k}} = 9.09 \text{ mA} = I_1$$

Since

$$I_1 = I_2 + I_3$$

then

$$I_3 = I_1 - I_2$$

$$I_3 = 9.09 \text{ mA} - 5.45 \text{ mA} = 3.64 \text{ mA}$$

Knowing I_1, I_2, and I_3, we can solve for V_1, V_2, and V_3:

$$V_1 = I_1 R_1 = 9.09 \text{ mA} \times 1 \text{ k}\Omega = 9.09 \text{ V}$$

$$V_2 = I_2 R_2 = 5.45 \text{ mA} \times 2 \text{ k}\Omega = 10.9 \text{ V}$$

$$V_3 = I_3 R_3 = 3.64 \text{ mA} \times 3 \text{ k}\Omega = 10.9 \text{ V}$$

Verifying, we get

$$E = V_1 + V_2 = 9.09 \text{ V} + 10.9 \text{ V} = 20 \text{ V}$$

(14–19) $$20 \text{ V} = 1k I_1 + 2k I_2$$

(14–21) $$20 \text{ V} = 4k I_1 - 3k I_2$$

In Equation 14–19, solve for one of the variables. Let's use I_1.

$$1k I_1 = 20 \text{ V} - 2k I_2$$

$$I_1 = \frac{20 \text{ V} - 2k I_2}{1k} = 20 \text{ mA} - 2 I_2$$

Substituting this value of I_1 in Equation 14–21, we get

$$20 \text{ V} = 4k(20 \text{ mA} - 2 I_2) - 3k I_2 = 80 \text{ V} - 8k I_2 - 3k I_2$$

$$20 \text{ V} = 80 \text{ V} - 11k I_2$$

$$11k I_2 = 60 \text{ V}$$

$$I_2 = 5.45 \text{ mA}$$

Substituting this value for I_2 in Equation 14–19 and solving for I_1,

$$20 \text{ V} = 1k I_1 + 2k(5.45 \text{ mA}) = 1k I_1 + 10.9 \text{ V}$$

$$20 \text{ V} - 10.9 \text{ V} = 1k I_1$$

$$I_1 = \frac{9.1 \text{ V}}{1k} = 9.10 \text{ mA}$$

I_3 and the voltage drops can be found as before.

SOLUTION BY DETERMINANTS

(14–19) $$20 \text{ V} = 1k I_1 + 2k I_2$$

(14–21) $$20 \text{ V} = 4k I_1 - 3k I_2$$

In the array,

$$a_1 = 1k, \quad b_1 = 2k, \quad \text{and} \quad k_1 = 20$$

$$a_2 = 4k, \quad b_2 = -3k, \quad \text{and} \quad k_2 = 20$$

$$I_1 = \frac{k_1 b_2 - k_2 b_1}{a_1 b_2 - a_2 b_1} = \frac{-60.0 \times 10^3 - 40.0 \times 10^3}{-11.0 \times 10^6} = 9.09 \text{ mA}$$

$$I_2 = \frac{a_1 k_2 - a_2 k_1}{a_1 b_2 - a_2 b_1} = \frac{20.0 \times 10^3 - 80.0 \times 10^3}{-11.0 \times 10^6} = 5.45 \text{ mA}$$

I_3 and the voltage drops can be found as before.

GRAPHICAL SOLUTION   The problem may also be solved graphically. Although this method is seldom used in solving electrical problems such as this, it is useful in some applications and is presented here. Figure 14–9 is the graphical solution.

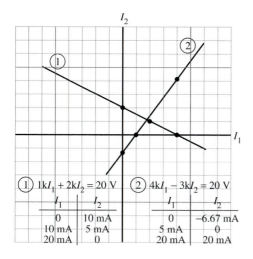

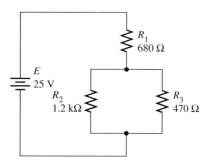

① $1k I_1 + 2k I_2 = 20$ V		② $4k I_1 - 3k I_2 = 20$ V	
I_1	I_2	I_1	I_2
0	10 mA	0	−6.67 mA
10 mA	5 mA	5 mA	0
20 mA	0	20 mA	20 mA

FIGURE 14–9   Graphical solution. Notice that while other solutions yield exact values or I_1 and I_2, the graphical solution provides only approximate values. That is, the intersection occurs where $I_1 \approx 9$ mA and $I_2 \approx 5$ mA.

FIGURE 14–10

PRACTICE PROBLEMS 14–6

Solve for the currents and voltage drops in Figure 14–10 by using simultaneous linear equations.

SOLUTION

The two basic equations should look like this:

$$E = V_1 + V_2 \quad \text{and} \quad E = V_1 + V_3$$

Substituting known values, we get

$$25 \text{ V} = 680 I_1 + 1.2 k I_2 \quad (14\text{–}23)$$

and

$$= 680 I_1 + 470 I_3 \quad (14\text{–}24)$$

Letting $I_3 = I_1 - I_2$, Equation 14–24 becomes

$$25 \text{ V} = 680 I_1 + 470 I_3$$
$$= 680 I_1 + 470(I_1 - I_2)$$
$$= 680 I_1 + 470 I_1 - 470 I_2$$
$$= 1150 I_1 - 470 I_2 \quad (14\text{–}25)$$

Now we need to multiply (or divide) Equations 14–23 and 14–25 by some factor or factors to make the coefficient of I_1 or I_2 equal in both equations. As we know, there are an infinite number of factors we could use. We choose to multiply Equation 14–25 by 2.55 so that the coefficient of I_2 in both equations equals 1.2.

$$25 \text{ V} = 1150 I_1 - 470 I_2 \quad (14\text{–}25)$$
$$25 \text{ V}(2.55) = 1150(2.55) I_1 - 470(2.55) I_2$$
$$63.8 \text{ V} = 2930 I_1 - 1200 I_2 \quad (14\text{–}26)$$

Putting Equations 14–23 and 14–26 together and adding, we get

$$25 \text{ V} = 680I_1 + 1200I_2 \quad (14\text{–}23)$$

$$63.8 \text{ V} = 2930I_1 - 1200I_2 \quad (14\text{–}26)$$

$$88.8 \text{ V} = 3610I_1$$

$$\frac{88.8 \text{ V}}{3610} = I_1 = 24.6 \text{ mA}$$

In this problem

$$I_1 = I_2 + I_3$$

$$I_3 = I_1 - I_2$$

Then

$$I_3 = 24.6 \text{ mA} - 6.92 \text{ mA} = 17.7 \text{ mA}$$

Substituting 24.6 mA for I_1 in Equation 14–23, we get

$$25 \text{ V} = 680(24.6 \text{ mA}) + 1200I_2$$

$$25 \text{ V} = 16.7 \text{ V} + 1200I_2$$

$$8.3 \text{ V} = 1200I_2$$

$$\frac{8.3 \text{ V}}{1200} = I_2 = 6.92 \text{ mA}$$

Using these currents, we get

$$V_1 = I_1R_1 = 24.6 \text{ mA} \times 680 \text{ } \Omega = 16.7 \text{ V}$$

$$V_2 = I_2R_2 = 6.92 \text{ mA} \times 1.2 \text{ k}\Omega = 8.3 \text{ V}$$

$$V_3 = I_3R_3 = 17.7 \text{ mA} \times 470 \text{ } \Omega = 8.32 \text{ V}$$

Let's solve the same problem using determinants:

$$25 \text{ V} = 680I_1 + 1200I_2 \quad (14\text{–}23)$$

$$25 \text{ V} = 1150I_1 - 470I_2 \quad (14\text{–}25)$$

In the array

$$a_1 = 680, \qquad b_1 = 1200, \qquad k_1 = 25$$

$$a_2 = 1150, \qquad b_2 = -470, \qquad k_2 = 25$$

$$I_1 = \frac{\begin{vmatrix} 25 & 1200 \\ 25 & -470 \end{vmatrix}}{\begin{vmatrix} 680 & 1200 \\ 1150 & -470 \end{vmatrix}}$$

$$= \frac{-11.8 \times 10^3 - 30 \times 10^3}{-320 \times 10^3 - 1.38 \times 10^6}$$

$$I_1 = \frac{-41.8 \times 10^3}{-1.70 \times 10^6} = 24.6 \text{ mA}$$

$$I_2 = \frac{\begin{vmatrix} 680 & 25 \\ 1150 & 25 \end{vmatrix}}{\begin{vmatrix} 680 & 1200 \\ 1150 & -470 \end{vmatrix}}$$

$$= \frac{17.0 \times 10^3 - 28.8 \times 10^3}{-320 \times 10^3 - 1.38 \times 10^6}$$

$$= \frac{-11.8 \times 10^3}{-1.7 \times 10^3} = 6.91 \text{ mA}$$

These values for I_1 and I_2 agree with the values found using the substitution method. The calculations to find I_3 and the voltage drops are the same as when using the substitution method.

Differences between voltage drops or currents using the various methods are due to rounding. You should understand these methods of solution before proceeding to the more complex problems.

Additional practice problems are at the end of the chapter.

APPLICATIONS—CIRCUITS WITH TWO SOURCES

In Chapter 12 we used Thévenin's theorem and Norton's theorem to solve problems where circuits contained more than one source. In this section, we will show how to use simultaneous linear equations to find the currents and voltage drops in multi-source circuits.

Let's find the currents and voltage drops in the circuit in Figure 14–11 using simultaneous linear equations.

We first have to determine the direction of current in the different branches. If the direction of current is not obvious, we simply assume a direction. If our assumption is wrong, when we solve for current, its magnitude will be correct but it will be negative.

Let's assume that the direction of current results from E_2. The resulting polarities would be as shown in Figure 14–12. There are three loops. One loop includes E_1, R_1, E_2, and R_3. A second loop includes E_1, R_1, and R_2, and a third loop includes E_2, R_2, and R_3. Even though there are three loops, we need use only two. Let's use the loops that contain E_2.

$$E_2 - V_2 - V_3 = 0$$

$$E_2 = V_2 + V_3 \tag{14–27}$$

$$E_2 - V_1 + E_1 - V_3 = 0$$

$$E_2 + E_1 = V_1 + V_3 \tag{14–28}$$

E_1 and E_2 are both included in the left member of Equation 14–28 only because they are both known. Substituting known values, we get

$$E_2 = V_2 + V_3 \tag{14–27}$$

$$40 \text{ V} = V_2 + V_3 \tag{14–29}$$

$$E_1 + E_2 = V_1 + V_3 \tag{14–28}$$

$$60 \text{ V} = V_1 + V_3 \tag{14–30}$$

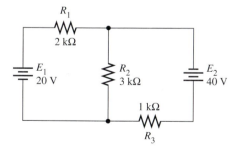

FIGURE 14–11 Circuit used to find currents and voltage drops using simultaneous linear equations.

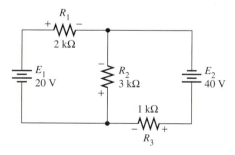

FIGURE 14–12 Circuit in Figure 14–11 with polarities shown.

SIMULTANEOUS LINEAR EQUATIONS

Expressing Equations 14–29 and 14–30 in terms of *IR* drops, we get

(14–29)
$$40 \text{ V} = V_2 + V_3$$
$$= I_2 R_2 + I_3 R_3$$

(14–31)
$$= 3kI_2 + 1kI_3$$

(14–30)
$$60 \text{ V} = V_1 + V_3$$
$$= I_1 R_1 + I_3 R_3$$

(14–32)
$$= 2kI_1 + 1kI_3$$

Let's look at Equations 14–31 and 14–32 together.

(14–31)
$$40 \text{ V} = 3000I_2 + 1000I_3$$

(14–32)
$$60 \text{ V} = 2000I_1 + 1000I_3$$

If we assume that the current through R_3 is the total current, then $I_3 = I_1 + I_2$, which makes $I_1 = I_3 - I_2$. Substituting $I_3 - I_2$ for I_1 in Equation 14–32 to eliminate one unknown yields

(14–32)
$$60 \text{ V} = 2000I_1 + 1000I_3$$
$$= 2000(I_3 - I_2) + 1000I_3$$
$$= 2000I_3 - 2000I_2 + 1000I_3$$

(14–33)
$$= -2000I_2 + 3000I_3$$

Now let's put Equations 14–31 and 14–33 together and solve.

(14–31)
$$40 \text{ V} = 3000I_2 + 1000I_3$$

(14–33)
$$60 \text{ V} = -2000I_2 + 3000I_3$$

SOLUTION BY ADDITION AND SUBTRACTION   Multiplying Equation 14–31 by 3 causes the coefficient of I_3 in both equations to be equal.

$$40 \text{ V}(3) = 3000I_2(3) + 1000I_3(3)$$

(14–34)
$$120 \text{ V} = 9000I_2 + 3000I_3$$

Now we can put Equations 14–33 and 14–34 together and solve by subtracting.

(14–34)
$$120 \text{ V} = 9000I_2 + 3000I_3$$

(14–33)
$$\underline{60 \text{ V} = -2000I_2 + 3000I_3}$$
$$60 \text{ V} = 11{,}000I_2$$

$$\frac{60 \text{ V}}{11{,}000 \text{ }\Omega} = 5.45 \text{ mA} = I_2$$

Substituting 5.45 mA for I_2 in Equation 14–34 yields

$$120 \text{ V} = 9000(5.45 \text{ mA}) + 3000I_3$$

$$120 \text{ V} = 49.1 \text{ V} + 3000I_3$$

$$70.9 \text{ V} = 3000I_3$$

$$\frac{70.9 \text{ V}}{3000 \text{ }\Omega} = I_3 = 23.6 \text{ mA}$$

If

$$I_3 = 23.6 \text{ mA} \qquad \text{and} \qquad I_2 = 5.45 \text{ mA}$$

then

$$I_1 = I_3 - I_2 = 23.6 \text{ mA} - 5.45 \text{ mA} = 18.2 \text{ mA}$$

Referring to Figure 14–11 and solving for the various voltage drops, we get

$$V_1 = I_1R_1 = 18.2 \text{ mA} \times 2 \text{ k}\Omega = 36.4 \text{ V}$$

$$V_2 = I_2R_2 = 5.45 \text{ mA} \times 3 \text{ k}\Omega = 16.4 \text{ V}$$

$$V_3 = I_3R_3 = 23.6 \text{ mA} \times 1 \text{ k}\Omega = 23.6 \text{ V}$$

To verify, we apply Kirchhoff's voltage law to the loops.

SOLUTION BY SUBSTITUTION

$$40 \text{ V} = 3000I_2 + 1000I_3 \tag{14–31}$$

$$60 \text{ V} = -2000I_2 + 3000I_3 \tag{14–33}$$

Solving Equation 14–31 for I_2, we get

$$I_2 = \frac{40 \text{ V} - 1000I_3}{3000} = 13.3 \text{ mA} - 3.33 \times 10^{-1}I_3$$

Substituting this value for I_2 in Equation 14–33 yields

$$60 \text{ V} = -2000(13.3 \text{ mA} - 3.33 \times 10^{-1}I_3) + 3000I_3$$

$$= -26.7 \text{ V} + 667I_3 + 3000I_3 = -26.7 \text{ V} + 3670I_3$$

$$86.7 \text{ V} = 3670I_3$$

$$I_3 = 23.6 \text{ mA}$$

Substituting this value back into Equation 14–33 and solving for I_2,

$$60 \text{ V} = -2000I_2 + 3000(23.6 \times 10^{-3}) = -2000I_2 + 70.8 \text{ V}$$

$$-10.8 \text{ V} = -2000I_2$$

$$I_2 = 5.45 \text{ mA}$$

I_1 and the voltage drops would be the same as those calculated when the solution was by addition and subtraction (some differences may exist due to rounding).

SOLUTION BY DETERMINANTS

(14–31)
$$40 \text{ V} = 3000I_2 + 1000I_3$$

(14–33)
$$60 \text{ V} = -2000I_2 + 3000I_3$$

In the array,

$$a_1 = 3000, \qquad b_1 = 1000, \qquad k_1 = 40$$
$$a_2 = -2000, \qquad b_2 = 3000, \qquad k_2 = 60$$

$$I_2 = \frac{k_1 b_2 - k_2 b_1}{a_1 b_2 - a_2 b_1} = \frac{120 \times 10^3 - 60 \times 10^3}{9.00 \times 10^6 - (-2 \times 10^6)} = 5.45 \text{ mA}$$

$$I_3 = \frac{a_1 k_2 - a_2 k_1}{a_1 b_2 - a_2 b_1} = \frac{180 \times 10^3 - (-80 \times 10^3)}{9.00 \times 10^6 - (-2 \times 10^6)} = 23.6 \text{ mA}$$

Values for I_1 and the voltage drops may be calculated as before.

PRACTICE PROBLEMS 14–7

Use simultaneous equations or determinants to find the currents and voltage drops of the circuits in Figure 14–13(a), (b), and (c).

FIGURE 14–13

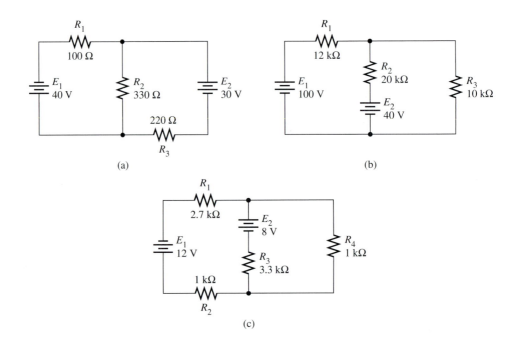

(a)

(b)

(c)

Remember, your method of solution may be different from the author's because you may not assume the same direction of currents or you may not use the same two loops.

(a) Refer to Figure 14–13(a). This solution uses the loops that contain E_1, R_1, and R_2; and E_1, E_2, R_1, and R_3. Assume that the direction of current results from E_1. Starting at the positive terminal of E_1 and moving in a clockwise direction gives

$$E_1 - V_1 - V_2 = 0$$

$$E_1 = V_1 + V_2$$

$$40 \text{ V} = V_1 + V_2$$

$$= 100I_1 + 330I_2 \quad (14\text{--}35)$$

$$E_1 - V_1 + E_2 - V_3 = 0$$

$$E_1 + E_2 = V_1 + V_3$$

$$40 \text{ V} + 30 \text{ V} = V_1 + V_3$$

$$70 \text{ V} = V_1 + V_3$$

$$= 100I_1 + 220I_3 \quad (14\text{--}36)$$

If we assume that I_1 is the total current, then $I_2 = I_1 - I_3$, and we can substitute $I_1 - I_3$ for I_2 in Equation 14–35:

$$40 \text{ V} = 100I_1 + 330I_2 \quad (14\text{--}35)$$

$$= 100I_1 + 330(I_1 - I_3)$$

$$= 430I_1 - 330I_3 \quad (14\text{--}37)$$

Multiplying Equation 14–36 by 4.3 yields

$$301 \text{ V} = 430I_1 + 946I_3 \quad (14\text{--}38)$$

Subtracting Equation 14–38 from Equation 14–37 yields

$$40 \text{ V} = 430I_1 - 330I_3 \quad (14\text{--}37)$$

$$\underline{301 \text{ V} = 430I_1 + 946I_3 \quad (14\text{--}38)}$$

$$-261 \text{ V} = 0 - 1276I_3$$

$$261 \text{ V} = 1276I_3$$

Then

$$I_3 = \frac{261 \text{ V}}{1276 \text{ }\Omega} = 205 \text{ mA}$$

Substituting in Equation 14–36 gives

$$70 \text{ V} = 100I_1 + 220(205 \text{ mA})$$

$$70 \text{ V} = 100I_1 + 45.1 \text{ V}$$

$$24.9 \text{ V} = 100I_1$$

$$I_1 = 249 \text{ mA}$$

Substituting in Equation 14–35 gives

$$40 \text{ V} = 100I_1 + 330I_2 \quad (14\text{--}35)$$

$$40 \text{ V} = 100(249 \text{ mA}) + 330I_2$$

$$40 \text{ V} = 24.9 \text{ V} + 330I_2$$

$$15.1 \text{ V} = 330I_2$$

$$I_2 = 45.8 \text{ mA}$$

$$V_1 = I_1R_1 = 249 \text{ mA} \times 100 \text{ }\Omega$$

$$= 24.9 \text{ V}$$

$$V_2 = I_2R_2 = 45.8 \text{ mA} \times 330 \text{ }\Omega$$

$$= 15.1 \text{ V}$$

$$V_3 = I_3R_3 = 205 \text{ mA} \times 220 \text{ }\Omega$$

$$= 45.1 \text{ V}$$

Apply Kirchhoff's law to verify the solution.

(b) Refer to Figure 14–13(b). This solution uses the loop that includes E_1, E_2, R_1 and R_2; and the loop that includes E_2, R_2 and R_3. Assume that the current in each loop results from E_1. Starting at the negative terminal of E_1 and moving clockwise, for the first loop we get

$$-E_1 + V_1 + V_2 + E_2 = 0$$

$$-100 \text{ V} + V_1 + V_2 + 40 \text{ V} = 0$$

$$V_1 + V_2 = 60 \text{ V}$$

$$12k I_1 + 20k I_2 = 60 \text{ V} \quad (14\text{–}38)$$

For the second loop, start at the negative terminal of E_2 and move clockwise:

$$-E_2 - V_2 - V_3 = 0$$

$$-40 \text{ V} - V_2 + V_3 = 0$$

$$-V_2 + V_3 = 40 \text{ V}$$

$$-20k I_2 + 10k I_3 = 40 \text{ V} \quad (14\text{–}39)$$

If we assume that I_1 is the total current, then $I_1 = I_2 + I_3$ and we can substitute $I_2 + I_3$ for I_1 in Equation 14–38.

$$12k(I_2 + I_3) + 20k I_2 = 60 \text{ V}$$

$$12k I_2 + 12k I_3 + 20k I_2 = 60 \text{ V}$$

$$32k I_2 + 12k I_3 = 60 \text{ V} \quad (14\text{–}40)$$

Using Equations 14–39 and 14–40, we will solve for the currents using the substitution method.

$$-20k I_2 + 10k I_3 = 40 \text{ V} \quad (14\text{–}39)$$

$$32k I_2 + 12k I_3 = 60 \text{ V} \quad (14\text{–}40)$$

Solving for I_2 in Equation 14–39, we get

$$I_2 = \frac{40 \text{ V} - 10k I_3}{-20k}$$

$$= -2 \text{ mA} + 0.5 I_3$$

Substituting this value into Equation 14–40,

$$32k(-2 \text{ mA} + 0.5 I_3) + 12k I_3 = 60 \text{ V}$$

$$-64 \text{ V} + 16k I_3 + 12k I_3 = 60 \text{ V}$$

$$28k I_3 = 124 \text{ V}$$

$$I_3 = 4.43 \text{ mA}$$

Substituting this value of I_3 back into Equation 14–39,

$$-20k I_2 + 10k(4.43 \text{ mA}) = 40 \text{ V}$$

$$-20k I_2 + 44.3 \text{ V} = 40 \text{ V}$$

$$-20k I_2 = -4.3 \text{ V}$$

$$I_2 = 215 \text{ } \mu A$$

Knowing I_2 and I_3, we can solve for the rest of the unknowns.

$$I_1 = I_2 + I_3 = 215 \text{ } \mu A + 4.43 \text{ mA}$$

$$= 4.65 \text{ mA}$$

$$V_1 = I_1 R_1 = 4.65 \text{ mA} \times 12 \text{ k}\Omega$$

$$= 55.7 \text{ V}$$

$$V_2 = I_2 R_2 = 215 \text{ } \mu A \times 20 \text{ k}\Omega$$

$$= 4.30 \text{ V}$$

$$V_3 = I_3 R_3 = 4.43 \text{ mA} \times 10 \text{ k}\Omega$$

$$= 44.3 \text{ V}$$

Apply Kirchhoff's laws to verify solutions.

(c) Refer to Figure 14–13(c). This solution uses the loop that includes E_1, E_2, R_1, R_2, and R_3 and the loop that includes E_2, R_3, and R_4. Assume that the current in each loop results from E_2. Starting at the positive terminal of E_2 and moving counterclockwise, for the first loop we get

$$E_2 - I_3 R_3 - I_1 R_2 + E_1 - I_1 R_1 = 0$$

$$8 \text{ V} - 3.3k I_3 - 1k I_1$$
$$+ 12 \text{ V} - 2.7k I_1 = 0$$

$$20 \text{ V} = 3.7k I_1 + 3.3k I_3 \quad (14\text{–}41)$$

(Because the current is the same through R_1 and R_2, we call that current I_1.)

For the second loop, starting again at the positive terminal of E_2 and moving counterclockwise,

$$E_2 - I_3R_3 - I_4R_4 = 0$$

$$8\text{ V} - 3.3kI_3 - 1kI_4 = 0$$

$$8\text{ V} = 3.3kI_3 + 1kI_4 \quad (14\text{-}42)$$

If we assume that I_3 is the total current, then $I_3 = I_1 + I_4$ or $I_4 = I_3 - I_1$, and we can substitute $I_3 - I_1$ for I_4 in Equation 14–42.

$$8\text{ V} = 3.3kI_3 + 1k(I_3 - I_1)$$

$$= 3.3kI_3 + 1kI_3 - 1kI_1$$

$$8\text{ V} = 4.3kI_3 - 1kI_1 \quad (14\text{-}43)$$

Using Equations 14–41 and 14–43, we will solve for the currents using determinants.

$$20\text{ V} = 3.7kI_1 + 3.3kI_3 \quad (14\text{-}41)$$

$$8\text{ V} = -1kI_1 + 4.3kI_3 \quad (14\text{-}43)$$

In the array

$$a_1 = 3.7k, \quad b_1 = 3.3k, \quad k_1 = 20\text{ V}$$

$$a_2 = -1k, \quad b_2 = 4.3k, \quad k_2 = 8\text{ V}$$

$$I_1 = \frac{k_1b_2 - k_2b_1}{a_1b_2 - a_2b_1}$$

$$= \frac{20\text{ V} \times 4.3k - 8\text{ V} \times 3.3k}{3.7k \times 4.3k - (-1k \times 3.3k)}$$

$$= \frac{86.0 \times 10^3 - 26.4 \times 10^3}{15.9 \times 10^6 + 3.3 \times 10^6}$$

$$= \frac{59.6 \times 10^3}{19.2 \times 10^6} = 3.10\text{ mA}$$

$$I_3 = \frac{a_1k_2 - a_2k_1}{a_1b_2 - a_2b_1}$$

$$= \frac{3.7k \times 8 - (-1k \times 20)}{3.7k \times 4.3k - (-1k \times 3.3k)}$$

$$= \frac{29.6 \times 10^3 + 20 \times 10^3}{15.9 \times 10^6 + 3.3 \times 10^6}$$

$$I_3 = \frac{49.6 \times 10^3}{19.2 \times 10^6} = 2.58\text{ mA}$$

We assumed that I_3 was total current. Then

$$I_4 = I_3 - I_1 = 2.58\text{ mA} - 3.1\text{ mA}$$

$$= -520\ \mu\text{A}$$

The negative value for I_4 tells us that we assumed the wrong direction for current through R_4. I_1 is the total current, not I_3. The values calculated are correct. We just have to reverse the direction of current through R_4.

$$V_1 = I_1R_1 = 3.1\text{ mA} \times 2.7\text{ k}\Omega$$

$$= 8.37\text{ V}$$

$$V_2 = I_2R_2 = 3.1\text{ mA} \times 1\text{ k}\Omega$$

$$= 3.1\text{ V}$$

$$V_3 = I_3R_3 = 2.58\text{ mA} \times 3.3\text{ k}\Omega$$

$$= 8.51\text{ V}$$

$$V_4 = I_4R_4 = 520\ \mu\text{A} \times 1\text{ k}\Omega$$

$$= 0.52\text{ V}$$

Apply Kirchhoff's law to verify the solutions.

Additional practice problems are at the end of the chapter.

1. Use the addition or subtraction method to solve for the currents and voltage drops of the circuit in Figure 14–14.

2. Use the substitution method to solve for the currents and voltage drops of the circuit in Figure 14–15.

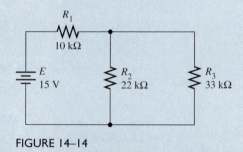

FIGURE 14–14

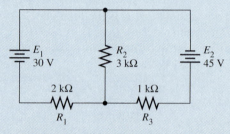

FIGURE 14–15

Answers to Self-test 14–2 are at the end of the chapter.

CHAPTER 14 AT A GLANCE

PAGE	METHOD OF SOLUTION	EXAMPLE
418	Sets of equations may be solved by addition or subtraction. In this set, add and solve for x. Knowing x we can then solve for y.	$x - 2y = 4$ (add) $\underline{3x + 2y = 4}$ $4x = 8$ $x = 2$ Then $y = -1$.
420	Sets of equations may be solved by substitution. Substitute $2y + 4$ for x in (2) and solve for y. Knowing y we can solve for x in either (1) or (2).	$x - 2y = 4$ (1) $3x + 2y = 4$ (2) Solve (1) for x $x = 2y + 4$ $3(2y + 4) + 2y = 4$ $y = -1$ Then $x = 2$.
423	Key Point: The steps for solving simultaneous equations using *determinants* are: 1. Arrange the terms of the equations with the unknowns on the left side and the constants on the right side.	See Example 14–6.

2. Arrange the unknowns so that they appear in the same order for each equation.
3. Use the coefficients of the unknowns to form an array of numbers called the *determinant of the denominator.*
4. For each unknown, create an array of numbers similar to the above array, except for its coefficients, substitute the constants from the right side of the equations; this is called the *determinant of the numerator.*
5. The value of each unknown can then be calculated by dividing its *determinant of the numerator* by the common *determinant of the denominator.*
6. To check your answers, substitute the solutions into the original equations and confirm the equalities.

END OF CHAPTER PROBLEMS 14–1

Solve the following sets of equations graphically:

1. $2x - y = 4$
   $2y = x + 1$

2. $3x + 3y = 5$
   $x - y = 5$

3. $5x + 2y = 2$
   $3x - y = 10$

4. $x + 2y = 6$
   $2x - 4y = 8$

5. $8x - 4y = 12$
   $x + 2y = 4$

6. $3x - y = 6$
   $x + y = 4$

7. $2x - 5y = 1$
   $3x - 8y = 2$

8. $3x - 2y = 3$
   $2x + y = 4$

9. $2x - 3y = 4$
   $x + 2y = 8$

10. $3x + 4y = 12$
    $4x - 3y = -12$

11. $\dfrac{x}{3} + \dfrac{y}{2} = \dfrac{5}{12}$

    $\dfrac{x}{4} - \dfrac{y}{3} = \dfrac{5}{24}$

12. $\dfrac{x}{2} + \dfrac{y}{5} = \dfrac{7}{20}$

    $\dfrac{x}{3} - \dfrac{y}{5} = -\dfrac{4}{15}$

13. $\dfrac{x}{2} + \dfrac{y}{3} = 3$

    $\dfrac{x}{2} - \dfrac{y}{3} = 1$

14. $x + 3y = 9$
    $2x - 3y = 9$

END OF CHAPTER PROBLEMS 14–2

Solve the following equations for *x* and *y* by using the addition or subtraction method:

1. $2x - y = 4$
   $2y = x + 1$
2. $3x + 3y = 5$
   $x - y = 5$
3. $5x + 2y = 2$
   $3x - y = 10$
4. $x + 2y = 6$
   $2x - 4y = 8$
5. $8x - 4y = 12$
   $x + 2y = 4$
6. $3x - y = 6$
   $x + y = 4$
7. $2x - 5y = 1$
   $3x - 8y = 2$
8. $3x - 2y = 3$
   $2x + y = 4$
9. $\dfrac{x}{2} + \dfrac{y}{3} = 3$

   $\dfrac{x}{2} - \dfrac{y}{3} = 1$
10. $3x + 4y = 12$
    $4x - 3y = -12$
12. $\dfrac{x}{2} + \dfrac{y}{5} = \dfrac{7}{20}$

    $\dfrac{x}{3} - \dfrac{y}{5} = -\dfrac{4}{15}$
11. $2x - 3y = 4$
    $x + 2y = 8$
13. $\dfrac{x}{3} + \dfrac{y}{2} = \dfrac{5}{12}$

    $\dfrac{x}{4} - \dfrac{y}{3} = \dfrac{5}{24}$
14. $x + 3y = 9$
    $2x - 3y = 9$

END OF CHAPTER PROBLEMS 14–3

Solve the following sets of equations using the substitution method:

1. $x + 2y = 4$
   $2x - y = 2$
2. $3x - y = 6$
   $2x - y = 2$
3. $4x - 2y = 3$
   $3x + 2y = 5$
4. $2x - 5y = 4$
   $3x - 3y = 9$
5. $5x + 3y = 5$
   $3x - 4y = 6$
6. $4x + 5y = 6$
   $5x + 3y = 4$
7. $x + y = 10$
   $4x - 4y = 0$
8. $x - y = 6$
   $3x + 3y = 0$
9. $3x - 3y = -5$
   $-x - y = 4$
10. $5x - y = -4$
    $-x - y = 5$

END OF CHAPTER PROBLEMS 14–4

Solve the following sets of equations using determinants:

1. $x + 2y = 4$
   $2x - y = 2$
2. $3x - y = 6$
   $2x - y = 2$
3. $4x - 2y = 3$
   $3x + 2y = 5$
4. $2x - 5y = 4$
   $3x + 4y = 8$

5. $5x + 3y = 5$
   $3x - 4y = 6$
6. $4x + 5y = 6$
   $5x + 3y = 4$
7. $x + y = 10$
   $4x - 4y = 0$
8. $x - y = 6$
   $3x + 3y = 0$
9. $3x - 3y = -5$
   $-x - y = 4$
10. $5x - y = -4$
    $-x - y = 5$

END OF CHAPTER PROBLEMS 14–5

Use determinants to solve for the unknowns in the following sets of equations:

1. $x + 2y + 3z = 14$
   $2x + y + 2z = 10$
   $3x + 4y - 3z = 2$
2. $x - y - z = -22$
   $2x - y + 2z = 22$
   $3x + 3y - z = 22$
3. $x - 2y + 4z = 4$
   $x + 3y - 2z = 6$
   $3x + y + 3z = 5$
4. $3x + 2y + 3z = 8$
   $4x - 4y - z = 4$
   $x + y - 3z = 2$
5. $x + y + z = 12$
   $2x - 3y - 4z = 8$
   $-3x + 2y + 2z = 5$
6. $10x - 15y + 20z = 12$
   $8x + 20y - 30z = 20$
   $15x + 20y + 25z = 40$

END OF CHAPTER PROBLEMS 14–6

Solve for the currents and voltage drops in the following problems by using simultaneous equations:

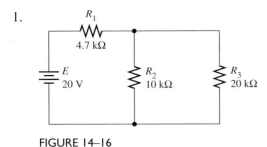

FIGURE 14–16   FIGURE 14–17

END OF CHAPTER PROBLEMS 14–7

Solve for the currents and voltage drops in the following problems by using simultaneous equations:

1. See Figure 14–18. Let $E_1 = 20$ V, $E_2 = 40$ V, $R_1 = 3$ kΩ, $R_2 = 2$ kΩ, and $R_3 = 1$ kΩ.
2. See Figure 14–18. Let $E_1 = 50$ V, $E_2 = 15$ V, $R_1 = 27$ kΩ, $R_2 = 56$ kΩ, and $R_3 = 33$ kΩ.
3. See Figure 14–19. Let $E_1 = 100$ V, $E_2 = 40$ V, $R_1 = 12$ kΩ, $R_2 = 20$ kΩ, and $R_3 = 10$ kΩ.
4. See Figure 14–19. Let $E_1 = 25$ V, $E_2 = 50$ V, $R_1 = 470$ Ω, $R_2 = 820$ Ω, and $R_3 = 330$ Ω.

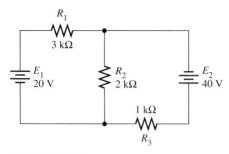

FIGURE 14–18

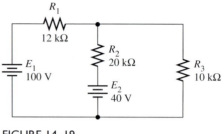

FIGURE 14–19

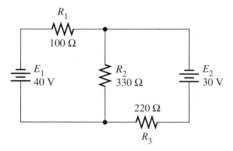

FIGURE 14–20

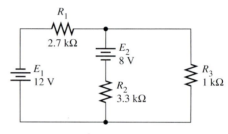

FIGURE 14–21

5. See Figure 14–20. Let $E_1 = 40$ V, $E_2 = 30$ V, $R_1 = 100$ Ω, $R_2 = 330$ Ω, and $R_3 = 220$ Ω.

6. See Figure 14–20. Let $E_1 = 10$ V, $E_2 = 15$ V, $R_1 = 270$ kΩ, $R_2 = 100$ kΩ, and $R_3 = 120$ kΩ.

7. See Figure 14–21. Let $E_1 = 12$ V, $E_2 = 8$ V, $R_1 = 2.7$ kΩ, $R_2 = 3.3$ kΩ, and $R_3 = 1$ kΩ.

8. See Figure 14–21. Let $E_1 = 80$ V, $E_2 = 50$ V, $R_1 = 68$ kΩ, $R_2 = 22$ kΩ, and $R_3 = 47$ kΩ.

ANSWERS TO SELF-TESTS

SELF-TEST 14–1

1. (a) See Figure 14–22 for graphical solution.
   (b) Addition–subtraction method:

   $$4x + y = 9 \quad (1)$$
   $$x + y = 6 \quad (2)$$

   Multiply (2) by 4: $\quad 4x + 4y = 24$
   Rewrite (1): $\qquad\quad \underline{4x + y = 9}$
   Subtract: $\qquad\qquad\quad 3y = 15$
   $$y = 5$$

   Substitute in (1).

   $$4x + 5 = 9$$
   $$4x = 4$$
   $$x = 1$$

   The solution is $x = 1$, $y = 5$.
   (c) Substitution method: Solve for x in (2).

   $$x + y = 6$$
   $$x = 6 - y$$

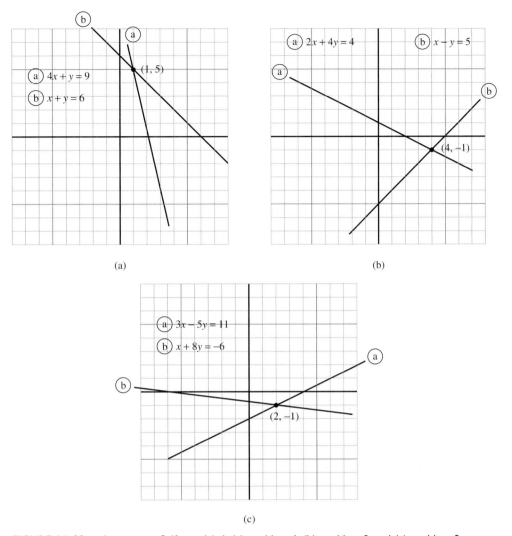

FIGURE 14–22 Answers to Self-test 14–1: (a) problem 1, (b) problem 2, and (c) problem 3.

Substitute for x in (1).

$$4(6 - y) + y = 9$$
$$24 - 4y + y = 9$$
$$-3y = -15$$
$$y = 5$$

Substitute for y in (1).

$$4x + 5 = 9$$
$$4x = 4$$
$$x = 1$$

The solution is $x = 1$, $y = 5$.

2. (a) See Figure 14–22(b) for graphical solution.

   (b) Addition–subtraction method:

   $$2x + 4y = 4 \quad (1)$$
   $$x - y = 5 \quad (2)$$

   Multiply (2) by 2: $\quad 2x - 2y = 10$
   Rewrite (1): $\qquad\quad \underline{2x + 4y = 4}$
   Subtract: $\qquad\qquad\quad -6y = 6$
   $$y = -1$$

   Substitute for y in (1).

   $$2x + 4(-1) = 4$$
   $$2x - 4 = 4$$
   $$2x = 8$$
   $$x = 4$$

   The solution is $x = 4$, $y = -1$.

   (c) Substitution method: Solve for x in (2).

   $$x - y = 5$$
   $$x = y + 5$$

   Substitute for x in (1).

   $$2(y + 5) + 4y = 4$$
   $$2y + 10 + 4y = 4$$
   $$6y = -6$$
   $$y = -1$$

   Substitute for y in (2).

   $$x - (-1) = 5$$
   $$x + 1 = 5$$
   $$x = 4$$

   The solution is $x = 4$, $y = -1$.

3. (a) See Figure 14–22(c) for graphical solution.

   (b) Addition–subtraction method:

   $$3x - 5y = 11 \quad (1)$$
   $$x + 8y = -6 \quad (2)$$

   Multiply (2) by 3: $3x + 24y = -18$
   Rewrite (1): $\qquad\quad \underline{3x - 5y = 11}$
   Subtract: $\qquad\qquad\quad 29y = -29$
   $$y = -1$$

   Substitute for y in (2).

   $$x + 8(-1) = -6$$
   $$x = 2$$

   The solution is $x = 2$, $y = -1$.

   (c) Substitution method: Solve for x in (2).

   $$x + 8y = -6$$
   $$x = -6 - 8y$$

   Substitute for x in (1).

   $$3(-6 - 8y) - 5y = 11$$
   $$-18 - 24y - 5y = 11$$
   $$-29y = 29$$
   $$y = -1$$

   Substitute for y in (2).

   $$x + 8(-1) = -6$$
   $$x = 2$$

   The solution is $x = 2$, $y = -1$.

1. $I_1 = 647\ \mu A$, $I_2 = 388\ \mu A$, $I_3 = 259\ \mu A$, $V_1 = 6.47\ V$, $V_2 = V_3 = 8.53\ V$

2. $I_1 = 23.2\ mA$, $I_2 = 5.47\ mA$, $I_3 = 28.6\ mA$, $V_1 = 46.4\ V$, $V_2 = 16.4\ V$, $V_3 = 28.6\ V$

Math for AC Electronics

The "AC" in "AC Electronics" stands for *alternating current*. A simple AC circuit is a group of electrical components connected in a closed loop and powered by a voltage source that periodically reverses its polarity. The current in this simple circuit would alternate direction as the polarity of the voltage changes. A practical example of a complex AC circuit is the electrical utility service in the United States that provides 120 volt and 240 volt AC power to both businesses and homes. In the electrical wiring in our homes, electrical current reverses its direction 120 times every second because an electrical generator at some distant utility plant is reversing the polarity of its output voltage 120 times every second.

The changing current in AC circuits causes energy to be temporarily stored in capacitors and inductors. The periodic storing and releasing of electrical energy in AC circuits makes AC calculations a bit more difficult than DC circuit calculations. To help solve AC circuit problems we need a few more math tools besides algebra.

In Chapter 15, we explain complex numbers, which are a combination of imaginary and real numbers. The concept of imaginary numbers helps us deal with Ohm's law calculations when some of the AC power is stored in capacitors or inductors.

In Chapter 16, we discuss the right triangle and some basic trigonometric ratios. When we graph AC voltages and currents, the real and imaginary parts form the sides of a right triangle.

In Chapter 17, we discuss the law of sines and the law of cosines which can be used in triangles that are not right triangles. This chapter does not involve AC circuits because our AC circuit analysis involves only right triangles. The law of sines and the law of cosines are useful in solving some physics problems.

In Chapter 18, we introduce AC fundamentals. In order to fully understand AC circuits, one must know the math concepts related to sine waves, radians, frequency, and phase shifts.

In Chapter 19, we show how to use trigonometry functions to find phase angles in AC circuits. When capacitors and inductors are subjected to a changing voltage, they store and then release electrical energy, which causes a time delay between the current and voltage waveforms. In the frequency domain, the time delay results in a phase shift.

In Chapter 20, we use trigonometry functions to solve problems in parallel and series–parallel AC circuits. We also learn how to transform a series circuit to an equivalent parallel circuit and vice-versa.

In Chapter 21, we use algebra, trigonometry, and graphing to analyze electrical filters, which are devices that allow AC energy at some frequencies to pass through while blocking other frequencies.

Complex Numbers

Introduction

Complex numbers are numbers that have a *real number part* and an *imaginary number part*. All previous chapters have dealt exclusively with *real numbers*. In this chapter, we introduce *imaginary numbers* because they allow us to accurately describe the relationship between voltages and currents in AC circuits.

When we have capacitance and inductance, in addition to resistance, in ac circuits, we are dealing with both energy users (resistance) and energy storers (capacitance and inductance). Circuit resistance converts energy to some other form, usually heat or light or both. Capacitance and inductance *store* energy. However, the calculations for voltage drops, current, and power dissipated are the same whether the component has the property of resistance, capacitance, or inductance. Thus, there are two kinds of power dissipation that we calculate: a real part, due to circuit resistance, and an imaginary part, due to circuit capacitance and/or inductance.

When imaginary parts exist, circuit currents and voltages do not lie along the same plane. They may be displaced by as little as a few degrees, in circuits that contain mostly real parts, to as much as 180° in circuits that are made up mostly of imaginary parts. In all such cases we say the variables are *out of phase*. The numbers that represent these variables have both real and imaginary parts and are called *complex* numbers.

This chapter, then, leads us into our study of ac circuits and those circuit variables that are out of phase. We will learn how to add, subtract, multiply, and divide complex numbers and also how to graph them. A thorough understanding of complex numbers is essential as we continue in the following chapters to develop the techniques used in solving ac circuit problems.

Chapter Objectives

In this chapter you will learn how to:

1. Solve problems that contain imaginary numbers.
2. Add, subtract, multiply, and divide complex numbers.
3. Graph complex numbers.

What is the result if we find the square root of a negative number? What is $\sqrt{-36}$? The answer cannot be 6 or -6 because the square of either of these numbers results in 36. We cannot take the square root of a negative number and come up with a real-number answer. We can find a solution to the problem in the following manner:

$$\sqrt{-36} = \sqrt{(36)(-1)} = \sqrt{36} \cdot \sqrt{-1} = 6\sqrt{-1}$$

A number in this form is called an *imaginary* number.

> ☞ **KEY POINT** Imaginary numbers are numbers that result from taking the square root of negative numbers. In mathematics, the letter i is used to represent $\sqrt{-1}$.
>
> $$\sqrt{-36} = \sqrt{36}i = 6i$$

The letter i in electronics denotes current. Therefore, we use the letter j to represent $\sqrt{-1}$. We place the j in front of the number to make it easier to recognize the presence of an imaginary number.

$$\sqrt{-36} = \sqrt{(36)(-1)} = \sqrt{36} \cdot \sqrt{-1} = j6$$

Whenever we see a number preceded by the letter j, we know that the number is multiplied by $\sqrt{-1}$.

The word *imaginary* is used simply as a means of separating these numbers from real numbers. Imaginary numbers exist, and they are very useful in analyzing ac circuits, as we will see in later chapters.

> ☞ **KEY POINT** A *vector* is a line drawn from the origin of a rectangular coordinate system to some given point. In electronics we call a vector a *phasor.*

The word phasor (*pha*se vec*tor*) is more descriptive than vector because we are usually describing and/or calculating phase relations between voltages and currents by the use of vectors/phasors.

In Chapter 13 we developed a system of rectangular coordinates. We can develop a similar set of coordinates for plotting imaginary and complex numbers because *the operator j causes a vector to be rotated 90°*. On a system of rectangular coordinates, 0° is plotted along the positive x-axis. Positive rotation is counterclockwise, as shown in Figure 15–1. We start with a point 4 units long, plotted along the positive x-axis. The line drawn from the origin to this point is the phasor.

The operator j rotates the phasor 90°. Thus, if we "j" our point, we are at 90° or are at the positive y-axis. Our phasor is now at $j4$. $j4$ tells us that the phasor is 4 units long and falls along the positive y-axis. (The operator j is always written in front of the number to help distinguish j as an operator from j as a literal number.) We can "j" the phasor again, which causes another 90° rotation to 180°. $j \cdot j = j^2$. 180° falls along the negative x-axis, so we see that j^2 is also -1. Our phasor is now at $j^2 4$ or -4. Rotating

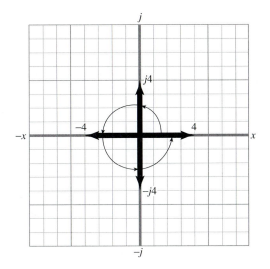

FIGURE 15–1
A phasor 4
units long is
plotted on a
system of
rectangular
coordinates.

another 90° to 270° yields $j \cdot j \cdot j$ or j^3. This point is also $-j$ since the phasor falls along the negative y-axis. $j \cdot j \cdot j = j^3 = -j$. Our phasor is at $j^3 4$ or $-j4$. (Notice that we could also arrive at $-j$ by rotating the phasor *clockwise* from the x-axis.)

We can complete the circle by j-ing again for another 90° rotation. $j \cdot j \cdot j \cdot j = j^4 = 1$. The circle is complete and we are back to the positive x-axis.

Remembering our basic relationship, that $j = \sqrt{-1}$, our graphic illustration helps us see the following relationships:

$$j = \sqrt{-1}$$
$$j^2 = \sqrt{-1} \cdot \sqrt{-1} = -1$$
$$j^3 = \sqrt{-1} \cdot \sqrt{-1} \cdot \sqrt{-1} = -1\sqrt{-1} = -j$$
$$j^4 = \sqrt{-1} \cdot \sqrt{-1} \cdot \sqrt{-1} \cdot \sqrt{-1} = (-1)(-1) = 1$$

Let's use these relationships to work the following problems.

EXAMPLE 15–1

Find $\sqrt{-144}$.

SOLUTION

$$\sqrt{-144} = j12$$

Imaginary numbers can be added or subtracted together, just as real numbers can be added or subtracted.

EXAMPLE 15–2

Find $\sqrt{-36} + \sqrt{-36}$.

$$j6 + j6 = j12$$

Multiplication and division of real and imaginary numbers are performed as a simple multiplication of two numbers. The operator j is treated as though it were a literal number.

EXAMPLE 15–3

Find $\sqrt{-36} \cdot \sqrt{36}$.

SOLUTION

$$j6 \cdot 6 = j36$$

EXAMPLE 15–4

Find $\sqrt{-36} \cdot \sqrt{-36}$.

SOLUTION

$$j6 \cdot j6 = j^2 36 = (-1) \cdot 36 = -36$$

EXAMPLE 15–5

Find $j^2 6 \cdot j5$.

SOLUTION

$$j^2 6 = (-1)6 = -6$$
$$j^2 6 \cdot j5 = -6 \cdot j5 = -j30$$

EXAMPLE 15–6

Find $\dfrac{\sqrt{-36}}{3}$.

SOLUTION

$$\frac{\sqrt{-36}}{3} = \frac{j6}{3} = j2$$

EXAMPLE 15–7

Find $\dfrac{j^2 36}{j9}$.

SOLUTION

$$\frac{j^2 36}{j9} = j4$$

EXAMPLE 15–8

Find $\dfrac{j24}{j^2 6}$.

SOLUTION

$$\frac{j24}{j^2 6} = \frac{j24}{-6} = -j4$$

PRACTICE PROBLEMS 15–1

Perform the indicated operations:

1. $\sqrt{-81} + \sqrt{-81}$
2. $\sqrt{-36} + \sqrt{-49}$
3. $\sqrt{-49} - \sqrt{-16}$
4. $j^2 6 \cdot j4$
5. $\dfrac{\sqrt{81}}{\sqrt{-9}}$
6. $\dfrac{j^3 72}{j8}$

SOLUTIONS

1. $j9 + j9 = j18$
2. $j6 + j7 = j13$
3. $j7 - j4 = j3$
4. $j^3 24 = -j24$
5. $\dfrac{9}{j3} = -j3$
6. $j^2 9 = -9$

Additional practice problems are at the end of the chapter.

COMPLEX NUMBERS   15–2

☞ **KEY POINT** *Complex* numbers are combinations of real and imaginary numbers.

$4 + j3$ is a complex number. 4 is the real-number part and $j3$ is the imaginary-number part. These two parts cannot be added together.

EXAMPLE 15–9

Find $\sqrt{36} + \sqrt{-36}$.

$$\sqrt{36} = 6$$
$$\sqrt{-36} = j6$$
$$\sqrt{36} + \sqrt{-36} = 6 + j6$$

EXAMPLE 15–10

Find $j6 + j^2 12 + j^3 2$.

SOLUTION

$$j^2 12 = -12$$
$$j^3 2 = -j2$$

Then

$$j6 + j^2 12 + j^3 2 = j6 - 12 - j2 = -12 + j4$$

15-2-1 Graphing Complex Numbers

Consider the complex number $2 + j4$. A look at this number tells us that there is a real part that is 2 units long and an imaginary part that is 4 units long. The point plots as $(2,4)$ on a system of rectangular coordinates. The resulting phasor is shown in Figure 15–2.

In the number $-4 - j6$, the real-number part is -4 and the imaginary part is $-j6$. The resulting phasor is also plotted in Figure 15–2.

FIGURE 15–2
Plotting
phasors.

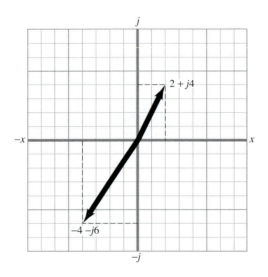

PRACTICE PROBLEMS 15–2

Perform the indicated operations:

1. $j^3 + j^2 3$
2. $j^2 4 + j^3 5 - j^2 6 + j^3$
3. $-j4 + j^3 2 + j^2 4 + 3$
4. $4 + j^2 3 - j^2 4 + j^3 6$

Graph the following complex numbers:

5. $6 + j3$
6. $3 - j6$
7. $-4 + j3$
8. $-3 - j5$
9. $0 - j4$

SOLUTIONS

1. $j^3 + j^2 3 = -j + (-1)3 = -3 - j$
2. $j^2 4 + j^3 5 - j^2 6 + j^3 = (-1)4 + (-j)5 - (-1)6 + (-j)$
   $$= -4 - j5 + 6 - j = 2 - j6$$
3. $-j4 + j^3 2 + j^2 4 + 3 = -j4 + (-j)2 + (-1)4 + 3$
   $$= -j4 - j2 - 4 + 3 = -1 - j6$$
4. $4 + j^2 3 - j^2 4 + j^3 6 = 4 + (-1)3 - (-1)4 + (-j)6$
   $$= 4 - 3 + 4 - j6 = 5 - j6$$

See Figure 15–3 for solutions to problems 5 through 9.

Additional practice problems are at the end of the chapter.

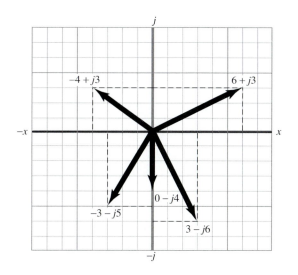

FIGURE 15–3

☞ **KEY POINT** To add (or subtract) complex numbers, first add (or subtract) the real parts, then add (or subtract) the imaginary parts.

EXAMPLE 15–11

Find the sum of $6 + j3$ and $4 - j7$.

SOLUTION

$$\begin{array}{r} 6 + j3 \\ 4 - j7 \\ \hline 10 - j4 \end{array}$$

$$(6 + j3) + (4 - j7) = 10 - j4$$

Some calculators will add, subtract, multiply, or divide complex numbers. Consult the user's manual for your calculator for this feature.

EXAMPLE 15–12

Find $(25 - j16) - (-15 + j12)$.

SOLUTION

$$\begin{array}{r} 25 - j16 \\ (-)-15 + j12 \\ \hline 40 - j28 \end{array}$$

$$(25 - j16) - (-15 + j12) = 40 - j28$$

PRACTICE PROBLEMS 15–3

Perform the following additions and subtractions.

1. $(8 + j3) + (7 + j2)$
2. $(10 + j2) + (20 - j14)$
3. $(15 - j4) + (14 + j7)$
4. $(-20 - j6) + (-15 - j10)$
5. $(10 + j10) - (15 + j6)$
6. $(22 + j16) - (17 - j12)$
7. $(6 - j14) - (15 + j8)$
8. $(-40 - j20) - (-25 - j30)$

SOLUTIONS

1. $15 + j5$
2. $30 - j12$
3. $29 + j3$
4. $-35 - j16$

5. $-5 + j4$

6. $5 + j28$

7. $-9 - j22$

8. $-15 + j10$

Additional practice problems are at the end of the chapter.

MULTIPLICATION AND DIVISION OF COMPLEX NUMBERS 15–4

☞ **KEY POINT** We multiply complex numbers in the same way we multiply algebraic terms.

EXAMPLE 15–13

Find the product of $(2 + j3)(2 + j4)$.

SOLUTION

$$(2 + j3)(2 + j4) = 4 + j6 + j8 + j^2 12$$
$$= 4 + j14 + (-1)12 = -8 + j14$$

☞ **KEY POINT** To divide complex numbers, multiply the numerator and denominator by the conjugate of the denominator. This will remove the imaginary part from the denominator.

The conjugate of a complex number is the same number, but the sign of the imaginary part is changed. The conjugate of $2 + j3$ is $2 - j3$. The conjugate of $4 - j5$ is $4 + j5$, and so on.

EXAMPLE 15–14

Perform the following division. Round answers to three places.

$$\frac{2 + j3}{4 - j6}$$

SOLUTION The conjugate of the denominator is $4 + j6$. If we multiply by $\frac{4 + j6}{4 + j6}$, we are really multiplying by 1 since any number divided by itself is 1. We perform this multiplication in order to simplify the denominator.

$$\frac{2 + j3}{4 - j6} \times \frac{4 + j6}{4 + j6} = \frac{(2 + j3)(4 + j6)}{(4 - j6)(4 + j6)}$$

$$= \frac{8 + j12 + j12 + j^2 18}{16 - j24 + j24 - j^2 36} = \frac{-10 + j24}{16 - j^2 36}$$

$$= \frac{-10 + j24}{52} = -\frac{10}{52} + j\frac{24}{52}$$

$$= -0.192 + j0.462$$

PRACTICE PROBLEMS 15–4

Perform the following multiplications. Round answers to three places.

1. $(3 + j4)(3 + j6)$ 2. $(3 + j2)(3 - j3)$
3. $(5 + j4)(5 - j4)$ 4. $(-2 + j5)(3 - j6)$

Perform the following divisions. Round answers to three places.

5. $\dfrac{2 + j3}{3 - j2}$ 6. $\dfrac{6 - j2}{2 + j4}$

7. $\dfrac{-6 + j4}{4 + j9}$ 8. $\dfrac{7 - j6}{4 - j3}$

9. $\dfrac{4 - j3}{-3 + j5}$ 10. $\dfrac{4 - j3}{-3 - j5}$

Remember, only the sign of the *imaginary* part is changed when finding the conjugate of the denominator.

SOLUTIONS

1. $-15 + j30$ 2. $15 - j3$
3. $41 + j0 = 41$ 4. $24 + j27$
5. $0 + j1 = j$ 6. $0.2 - j1.4$
7. $0.124 + j0.722$ 8. $1.84 - j0.120$
9. $-0.794 - j0.324$ 10. $0.0882 + j0.853$

Additional practice problems are at the end of the chapter.

SELF-TEST 15–1

Perform the indicated operations. Round answers to three places.

1. $j^2 + j^2 4$ 2. $j^3 4 - j6$

3. $\dfrac{\sqrt{-81}}{\sqrt{9}}$ 4. $\dfrac{\sqrt{121}}{\sqrt{-25}}$

5. $\dfrac{j^2 56}{j^3 8}$ 6. $(6 - j3)(3 + j7)$

7. $(3 - j6)(4 - j2)$

8. $(4 + j3)(4 - j3)$

9. $\dfrac{3 - j2}{1 + j2}$

10. $\dfrac{2 + j5}{2 - j2}$

Graph the following complex numbers.

11. $6 - j4$

12. $-4 + j6$

Answers to Self-test 15–1 are at the end of the chapter.

CHAPTER 15 AT A GLANCE

PAGE	KEY POINTS	EXAMPLE
452	*Key Point:* Imaginary numbers are numbers that result from taking the square root of negative numbers.	$\sqrt{-36} = 6\sqrt{-1}$
452	*Key Point:* A *vector* is a line drawn from the origin to some given point. In electronics we call a vector a *phasor.* Graphically, the operator *j* rotates a phasor 90°.	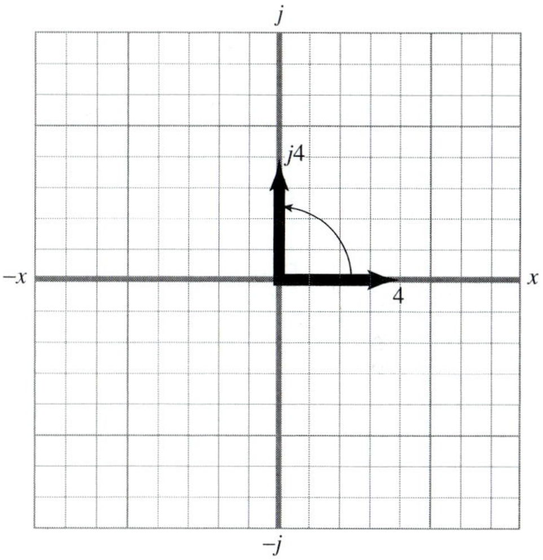 FIGURE 15–4
455	*Key Point:* Complex numbers are numbers that are part real and part imaginary.	$4 + j3$
458	*Key Point:* To add (or subtract) complex numbers, • first add (or subtract) the real parts, • then add (or subtract) the imaginary parts.	$(-9 + j6) + (12 - j2)$ $= 3 + (j6 - j2)$ $= 3 + j4$

| 459 | *Key Point:* Complex numbers are multiplied the same way we multiply algebraic terms. | $(2 + j3)(2 + j4)$ $= 4 + j6 + j8 + j^2 12$ $= 4 + j14 + (-1)12$ $= -8 + j14$ |

| 459 | *Key Point:* To divide complex numbers, multiply the numerator and denominator by the conjugate of the denominator. This will remove the imaginary part from the denominator. | |

$$\frac{2 + j3}{4 - j6} \times \frac{4 + j6}{4 + j6}$$

$$= \frac{(2 + j3)(4 + j6)}{(4 - j6)(4 + j6)}$$

$$= \frac{8 + j12 + j12 + j^2 18}{16 - j24 + j24 - j^2 36}$$

$$= \frac{-10 + j24}{16 - j^2 36}$$

$$= \frac{-10 + j24}{52} = -\frac{10}{52} + j\frac{24}{52}$$

$$= -0.192 + j0.462$$

END OF CHAPTER PROBLEMS 15–1

Perform the indicated operations:

1. $\sqrt{-49} + \sqrt{-36}$
3. $\sqrt{-64} - \sqrt{-49}$
5. $\sqrt{-25} \cdot \sqrt{-25}$
7. $\sqrt{-9} \cdot \sqrt{-36}$
9. $\sqrt{25} \cdot \sqrt{-49}$
11. $\sqrt{7.02} \cdot \sqrt{-6.2}$
13. $j^2 6 \cdot j3$
15. $j^3 4 \cdot j3$
17. $\dfrac{\sqrt{-144}}{\sqrt{16}}$
19. $\dfrac{\sqrt{100}}{\sqrt{-25}}$
21. $\dfrac{j^3 42}{j6}$
23. $\dfrac{j^2 100}{j^3 5}$

2. $\sqrt{-16} + \sqrt{-81}$
4. $\sqrt{-81} - \sqrt{-25}$
6. $\sqrt{-144} \cdot \sqrt{-144}$
8. $\sqrt{-121} \cdot \sqrt{-64}$
10. $\sqrt{-100} \cdot \sqrt{16}$
12. $\sqrt{12} \cdot \sqrt{-18}$
14. $j^2 4 \cdot j6$
16. $j^3 5 \cdot j6$
18. $\dfrac{\sqrt{-64}}{\sqrt{9}}$
20. $\dfrac{\sqrt{144}}{\sqrt{-25}}$
22. $\dfrac{j^3 63}{j9}$
24. $\dfrac{j^2 64}{j^3 6}$

END OF CHAPTER PROBLEMS 15–2

Perform the indicated operations:

1. $j^3 4 + j^2 4$
3. $j^3 8 + j^2 4$

2. $j^3 6 + j^2 5$
4. $j^3 8 + j^2 5$

5. $j^3 4.5 - j^2 3.1$
6. $j^3 7.3 - j^2 12$
7. $5 + j7 - j^2 6$
8. $4 + j9 - j^2 6$
9. $8.5 + j^3 4.2 + j^2 6.6$
10. $2.75 - j^3 4.55 + j^2 9.2$
11. $-j2 + j^3 5 + j^2 6 + 5$
12. $-j12 + j^3 2 + j^2 3 + 8$
13. $8 + j^2 5 - j^2 7 + j^3 6$
14. $3 + j^2 10 - j^2 2 + j^3 9$
15. $j^2 + j^3 + j + 1$
16. $j^2 4 + j^3 2 + j2 + 2$
17. $-j4 + 2 - j^2 4 - j^3 4$
18. $-j6 + 6 - j^2 3 - j^3 8$
19. $j^3 3 + j^2 4 - j3 + 4$
20. $j^3 4 + j^2 5 - j2 + 5$

Graph the following complex numbers.

21. $0 + j6$
22. $0 + j8$
23. $8 + j0$
24. $5 + j0$
25. $4 + j4$
26. $6 + j6$
27. $3 - j4$
28. $5 - j6$
29. $-4 + j3$
30. $-4 + j$
31. $-20 - j40$
32. $-60 - j80$
33. $-50 + j10$
34. $-100 + j60$
35. $3 \text{ k} + j4 \text{ k}$
36. $5 \text{ k}\Omega - j4 \text{ k}\Omega$
37. $-300 \ \Omega + j400 \ \Omega$
38. $-600 \ \Omega + j800 \ \Omega$
39. $-50 \text{ k}\Omega - j70 \text{ k}\Omega$
40. $20 \text{ k}\Omega - j30 \text{ k}\Omega$

END OF CHAPTER PROBLEMS 15–3

Perform the following additions and subtractions. Verify your answers using the calculator.

1. $(10 + j15) + (5 + j12)$
2. $(40 + j16) + (10 + j12)$
3. $(12 + j10) + (13 + j15)$
4. $(14 + j6) + (8 + j6)$
5. $(14 + j6) + (13 - j14)$
6. $(7 + j8) + (15 - j12)$
7. $(25 + j30) + (5 - j20)$
8. $(15 + j10) + (10 - j50)$
9. $(16 - j4) + (4 + j6)$
10. $(7 - j20) + (7 + j10)$
11. $(12 - j16) + (8 + j8)$
12. $(14 - j20) + (6 + j12)$
13. $(4 - j3) + (6 - j3)$
14. $(2 - j5) + (4 - j10)$
15. $(10 - j6) + (7 - j4)$
16. $(7 - j16) + (8 - j18)$
17. $(-4 - j5) + (-6 - j7)$
18. $(-10 - j15) + (-12 - j18)$
19. $(-15 - j10) + (-6 - j10)$
20. $(-12 - j12) + (-9 - j15)$
21. $(10 + j15) - (5 + j12)$
22. $(40 + j16) - (10 + j20)$
23. $(12 + j10) - (13 + j15)$
24. $(22 + j6) - (8 + j6)$
25. $(14 + j6) - (13 - j14)$
26. $(8 + j8) - (15 - j15)$
27. $(25 + j30) - (5 - j20)$
28. $(15 + j10) - (20 - j40)$
29. $(16 - j4) - (4 + j6)$
30. $(7 - j20) - (7 + j24)$
31. $(12 - j16) - (8 + j8)$
32. $(14 - j20) - (10 + j16)$
33. $(4 - j3) - (6 - j3)$
34. $(8 - j5) - (4 - j7)$
35. $(10 - j6) - (7 - j4)$
36. $(7 - j16) - (10 - j14)$
37. $(-4 - j5) - (-6 - j7)$
38. $(-10 - j20) - (-12 - j12)$
39. $(-15 - j10) - (-6 - j10)$
40. $(-12 - j12) - (-9 - j16)$

END OF CHAPTER PROBLEMS 15–4

Perform the following multiplications and divisions. Verify your answers using the calculator.

1. $(5 + j2)(3 - j2)$
2. $(6 - j2)(2 - j6)$
3. $(2 + j3)(2 - j3)$
4. $(4 - j3)(4 + j3)$
5. $(4 - j3)(4 - j3)$
6. $(6 + j2)(6 + j4)$
7. $(1 - j5)(3 - j3)$
8. $(6 - j4)(3 - j4)$
9. $(4 + j5)(6 + j2)$
10. $(2 + j2)(6 + j6)$
11. $(5 - j3)(6 + j1)$
12. $(2 - j2)(5 + j5)$
13. $\dfrac{5 + j3}{1 + j4}$
14. $\dfrac{1 + j}{1 - j}$
15. $\dfrac{4 + j4}{3 - j}$
16. $\dfrac{4 - j5}{3 - j7}$
17. $\dfrac{1 + j2}{5 + j2}$
18. $\dfrac{2 - j3}{3 + j4}$
19. $\dfrac{2 - j5}{-3 + j4}$
20. $\dfrac{3 - j6}{-2 + j5}$
21. $\dfrac{-8 - j2}{-4 - j5}$
22. $\dfrac{-3 - j6}{-3 - j7}$
23. $\dfrac{3 + j4}{4 + j5}$
24. $\dfrac{1 + j6}{5 + j3}$

ANSWERS TO SELF-TEST 15–1

1. -5
2. $-j10$
3. $j3$
4. $-j2.2$
5. $-j7$
6. $39 + j33$
7. $0 - j30$
8. 25
9. $-0.2 - j1.6$
10. $-0.75 + j1.75$
11. See Figure 15–5.
12. See Figure 15–5.

FIGURE 15–5

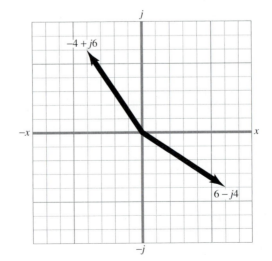

The Right Triangle

16

Introduction

Let's review some basic trigonometry definitions. Trigonometry is the study of triangles and the relationships between the lengths of their sides and the magnitudes of their angles. A triangle is a three-sided figure with three angles. The sum of the magnitudes of the three angles of any triangle equals 180 degrees. Angles are grouped into three classes, *acute* angles, *right* angles, and *obtuse* angles. *Acute* angles are angles that are between 0 and 90 degrees. *Right* angles are 90 degree angles. *Obtuse* angles are angles that are greater than 90 degrees but less than 180 degrees.

There are three kinds of triangles: those that contain an obtuse angle and two acute angles, those that contain three acute angles, and those that contain a right angle and two acute angles. Examples of the three triangles are shown in Figure 16–1 on the next page. In any triangle, the sum of the three angles equals 180°.

We are particularly interested in the right triangle because almost all ac circuit behavior can be evaluated mathematically by use of the right triangle. The effects of inductance and capacitance in ac circuits are at right angles or 90° from the effects of circuit resistance. The result is that we are often asked to compute circuit variables for which voltages or currents are acted on by forces that are displaced by 90°.

In this chapter we will introduce the student to the right triangle and to the trigonometric functions. The skills developed here in problem solving will be used throughout the following five chapters.

Chapter Objectives

In this chapter you will learn how to:

1. Identify the sides and angles of triangles.
2. Recognize the relationship between sides and angles.
3. Use the Pythagorean theorem to find the unknown sides of right triangles.
4. Use trigonometric functions to find the angles and sides of right triangles.
5. Use trigonometric tables.
6. Use the calculator to solve problems using trig functions.

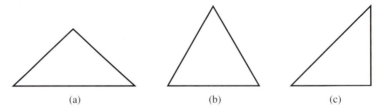

FIGURE 16–1   (a) Triangle with one obtuse and two acute angles;
(b) triangle with three acute angles; and (c) triangle with one right angle
and two acute angles.

16-1   SIDES AND ANGLES

Figure 16–2 is a right triangle shown in two common positions. For purposes of discussion, the triangles are labeled so that *A* denotes angle *A*, *B* denotes angle *B*, and *C* denotes the 90° or right angle. Side *a* lies opposite angle *A*, side *b* lies opposite angle *B*, and side *c* lies opposite angle *C*. In a right triangle the side opposite the right angle is called the *hypotenuse*. In Figure 16–2, side *c* is the hypotenuse.

Let's develop some basic relationships between the sides and angles in angles of right triangles such as the ones in Figure 16–2.

> ☞   **KEY POINT**   In any right triangle: (1) the sum of
> the acute angles equals 90°, (2) the hypotenuse is the
> side of greatest length, and (3) the greater side lies op-
> posite the greater angle.

When discussing angle *A*, side *b* is the *adjacent* side (the side next to angle *A*) and side *a* is the opposite side. When discussing angle *B*, side *a* is the adjacent side and side *b* is the opposite side. Side *c* is always called the hypotenuse.

Let's look at the right triangles in Figure 16–3. In Figure 16–3(a), angle *A* is 30°; therefore, angle *B* must equal 60° since the sum of angles *A* and *B* must equal 90°. The greater side lies opposite the greater angle. Therefore, since angle *B* is greater than angle *A*, side *b* is greater than side *a*. Since the hypotenuse is always the side of greatest length, our comparison of length is limited to the relationship between sides *a* and *b*. In Figure 16–3(b), angle *A* is the greater angle; therefore, side *a* is greater than side *b*.

FIGURE 16–2
Right triangles.

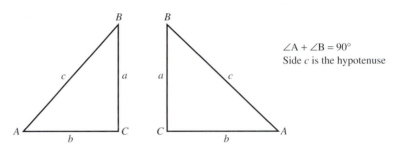

$\angle A + \angle B = 90°$
Side *c* is the hypotenuse

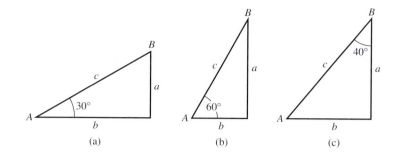

FIGURE 16–3

(a)   (b)   (c)

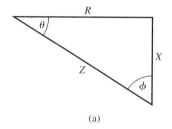
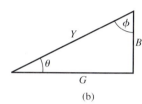
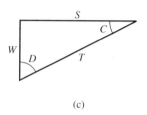

FIGURE 16–4

(a)   (b)   (c)

Finally, in Figure 16–3(c), B is 40°; therefore, angle A must be 50°, making side a the greater side.

Not all triangles are labeled as in Figure 16–3 and not all triangles are shown in this position. Notice in Figure 16–4(a) that the hypotenuse is labeled Z and the angles are θ (the Greek letter *theta*) and ϕ (the Greek letter *phi*). Side R lies opposite angle ϕ and is adjacent to angle θ. Side X lies opposite angle θ and is adjacent to angle ϕ. In Figure 16–4(b), side Y is the hypotenuse. Side G lies opposite angle ϕ and is adjacent to angle θ. Side B lies opposite angle θ and is adjacent to angle ϕ.

In Figure 16–4(c), side T is the hypotenuse. Side W lies opposite angle C and is adjacent to angle D. Side S lies opposite angle D and is adjacent to angle C. You should be able to identify the relationships among sides and angles regardless of the position of the triangle.

PRACTICE PROBLEMS 16–1

1. In the triangles in Figure 16–5, identify which of sides a and b is greater. Determine the size of the unknown angles.

2. Refer to Figure 16–6(a).
   (a) Which side is adjacent to angle θ?
   (b) Which side lies opposite angle ϕ?
   (c) Identify the hypotenuse.

3. Refer to Figure 16–6(b).
   (a) Which side lies opposite angle B?
   (b) Side m is adjacent to which angle?
   (c) Identify the hypotenuse.

4. Refer to Figure 16–6(c).
   (a) Side G is opposite which angle?
   (b) Which side is adjacent to angle θ?
   (c) Identify the hypotenuse.

FIGURE 16–5

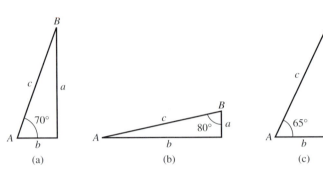

(a) (b) (c)

FIGURE 16–6

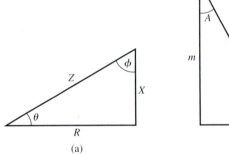
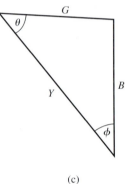

(a) (b) (c)

SOLUTIONS

1. In Figure 16–5(a), angle B is 20°. Side a is greater than side b. In (b), angle A is 10°; therefore, side b is greater. In (c), angle B is 25°; therefore, side a is greater.

2. (a) Side R is adjacent to angle θ.
   (b) Side R is opposite angle ϕ.
   (c) Side Z is the hypotenuse.

3. (a) Side m lies opposite angle B.
   (b) Side m is adjacent to angle A.
   (c) Side l is the hypotenuse.

4. (a) Side G is opposite angle ϕ.
   (b) Side G is adjacent to angle θ.
   (c) Side Y is the hypotenuse.

Additional practice problems are at the end of the chapter.

16–2   PYTHAGOREAN THEOREM

If two sides of a right triangle are known, the third side can be found by using the Pythagorean theorem, which is stated as follows:

☞ **KEY POINT** *In a right triangle, the square of the hypotenuse is equal to the sum of the squares of the other two sides.*

(16–1)
$$c^2 = a^2 + b^2$$

Taking the square root of both sides, we get

$$c = \sqrt{a^2 + b^2}$$

(16–2)

EXAMPLE 16–1

Suppose sides a and b in Figure 16–7 are of length 6 and 8, respectively. What is the length of the hypotenuse?

SOLUTION

$$c = \sqrt{a^2 + b^2} = \sqrt{6^2 + 8^2} = \sqrt{36 + 64} = \sqrt{100} = 10$$

Using the calculator,

$$6 \boxed{x^2} \boxed{+} 8 \boxed{x^2} \boxed{=} \boxed{\sqrt{x}} \qquad \text{Answer: } 10$$

EXAMPLE 16–2

If side $a = 5$ and side $c = 8$ in Figure 16–7, what is the value of side b?

SOLUTION

$$c^2 = a^2 + b^2$$

(16–1)

$$b^2 = c^2 - a^2$$

$$b = \sqrt{c^2 - a^2}$$

(16–3)

$$b = \sqrt{8^2 - 5^2} = \sqrt{64 - 25} = \sqrt{39} = 6.24$$

$$8 \boxed{x^2} \boxed{-} 5 \boxed{x^2} \boxed{=} \boxed{\sqrt{x}} \qquad \text{Answer: } 6.24$$

EXAMPLE 16–3

If side $b = 75$ and side $c = 150$ in Figure 16–7, what is the value of side a?

SOLUTION

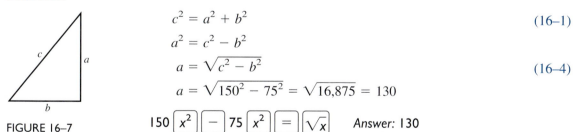

$$c^2 = a^2 + b^2$$

(16–1)

$$a^2 = c^2 - b^2$$

$$a = \sqrt{c^2 - b^2}$$

(16–4)

$$a = \sqrt{150^2 - 75^2} = \sqrt{16{,}875} = 130$$

FIGURE 16–7

$$150 \boxed{x^2} \boxed{-} 75 \boxed{x^2} \boxed{=} \boxed{\sqrt{x}} \qquad \text{Answer: } 130$$

In an ac circuit the sides of the triangle could represent circuit resistance (R), reactance (X), and impedance (Z), whose unit of measure is the ohm (Ω). Z is the hypotenuse; therefore, the equation is

$$Z^2 = R^2 + X^2$$

(16–5)

EXAMPLE 16–4

Refer to Figure 16–8. Find X if $R = 68$ kΩ and $Z = 80$ kΩ. Which of angles θ and ϕ is greater?

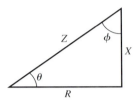

FIGURE 16–8

SOLUTION

$$Z^2 = R^2 + X^2$$
$$X^2 = Z^2 - R^2$$
$$X = \sqrt{Z^2 - R^2} = \sqrt{(80 \text{ k}\Omega)^2 - (68 \text{ k}\Omega)^2} = 42.1 \text{ k}\Omega \quad \phi > \theta$$

80 [EE] 3 [x^2] [$-$] 68 [EE] 3 [x^2] [$=$] [$\sqrt{x}$]

Answer: 42.1 $\times$ 10^3

The triangle could also represent circuit conductance (G), susceptance (B), and admittance (Y), whose unit of measure is the siemen (S). Y is the hypotenuse; therefore, the equation is

(16–6)

$$Y^2 = G^2 + B^2$$

EXAMPLE 16–5

Refer to Figure 16–9. Find Y where $G = 400$ μS and $B = 600$ μS.

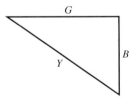

FIGURE 16–9

SOLUTION

$$Y^2 = G^2 + B^2$$
$$Y = \sqrt{G^2 + B^2} = \sqrt{(400 \text{ }\mu\text{S})^2 + (600 \text{ }\mu\text{S})^2} = 721 \text{ }\mu\text{S}$$

400 [EE] [+/−] 6 [x^2] [$+$] 600 [EE] [+/−] 6 [x^2]

[$=$] [$\sqrt{x}$]

Answer: 721 $\times$ 10^{-6}

FIGURE 16–10

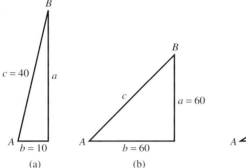

(a)   (b)   (c)

PRACTICE PROBLEMS 16–2

1. Find the length of the unknown side in the triangles in Figure 16–10(a), (b), and (c). Determine whether angle A or angle B is greater.

3. In Figure 16–12, find the unknown side and determine whether θ or ϕ is the greater angle when:
   (a) $B = 3$ mS and $G = 2$ mS
   (b) $Y = 400$ μS and $G = 150$ μS
   (c) $Y = 1.3$ mS and $B = 600$ μS

2. In Figure 16–11, find the unknown side and determine whether θ or ϕ is the greater angle when:
   (a) $Z = 17.3$ kΩ and $R = 10$ kΩ
   (b) $Z = 800$ Ω and $X = 500$ Ω
   (c) $R = 6.8$ kΩ and $X = 3$ kΩ

SOLUTIONS

1. For triangle (a), use Equation 16–4.

$$a = \sqrt{c^2 - b^2} = \sqrt{40^2 - 10^2} = 38.7$$

Angle A is greater than angle B because side a is greater than side b. Remember, the greater angle lies opposite the greater side. For triangle (b), use Equation 16–2.

$$c = \sqrt{a^2 + b^2} = \sqrt{60^2 + 60^2} = 84.9$$

Angles A and B are equal because the sides are equal. For triangle (c), use Equation 16–3.

$$b = \sqrt{c^2 - a^2} = \sqrt{14^2 - 8^2} = 11.5$$

Side b is greater than side a; therefore, angle B is greater than angle A.

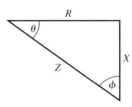

FIGURE 16–11

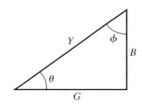

FIGURE 16–12

2. (a) $X = 14.1$ kΩ and θ is the greater angle.
   (b) $R = 624$ Ω and ϕ is the greater angle.
   (c) $Z = 7.43$ kΩ and ϕ is the greater angle.

3. (a) $Y = 3.61$ mS and θ is the greater angle.
   (b) $B = 371$ μS and θ is the greater angle.
   (c) $G = 1.15$ mS and ϕ is the greater angle.

Additional practice problems are at the end of the chapter.

16-3 TRIGONOMETRIC FUNCTIONS

The Pythagorean theorem has its limitations. We must know two sides in order to find the third side, and we can't determine the actual size of the angles.

We could find the length of the sides *and* the angles by using graph paper, a protractor, and a rule. Let's draw a line 3 units long, as in Figure 16–13. Let this be side *a*. Another line is drawn at right angles to side *a* and is 4 units in length (side *b*). If we complete the triangle and measure the length of side *c* (the hypotenuse), we will find that side *c* is 5 units as measured with the rule. $\angle A$ would equal approximately 37° and $\angle B$ would equal approximately 53° as determined by the protractor.

If we doubled sides *a* and *b*, as in Figure 16–14, so that side *a* is 6 units long and side *b* is 8 units long, how long would the hypotenuse be? What would $\angle A$ and $\angle B$ equal? If the length of *a* and *b* is doubled, the length of the hypotenuse would also double. This could be demonstrated either graphically (as in Figure 16–14) or by using the Pythagorean theorem. Again using the protractor, we could measure $\angle A$ and $\angle B$. They would still equal approximately 37° and 53°. They did not change because the relative lengths of the sides did not change. In Figure 16–14 the ratio of side *a* to side *b* was 3 to 4. When we doubled the length of the sides, the ratio of side *a* to side *b* was 6 to 8, which in both cases is 0.75 to 1. 3 to 4 = 0.75 and 6 to 8 = 0.75. As long as the ratio of side *a* to side *b* is 0.75 to 1, that is, whenever side *a* is 0.75 times as long as side *b*, $\angle A$ is 37°. We could expand on this by selecting other lengths of sides *a* and *b* and measuring the angles and the hypotenuse. In this manner, we could determine the approximate value for angles *A* and *B* for any ratio of side *a* to side *b*. We must consider the angles measured with the protractor as approximate since most protractors are not precision instruments.

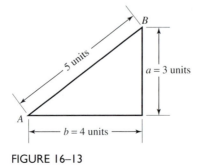

FIGURE 16–13

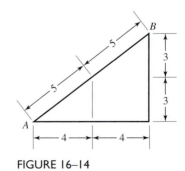

FIGURE 16–14

It is rarely necessary to solve problems graphically, however, because mathematicians have prepared tables of these and other ratios for us to use. These tables are called *trigonometric tables*.

Since a triangle has three sides, there are three possible trigonometric ratios. In Figure 16–15 these ratios would be side a to side b (as previously discussed), side a to side c, and side b to side c. These trigonometric ratios are called *functions* and are identified as *sine, cosine,* and *tangent*. (Actually, there are three additional functions, secant, cosecant, and cotangent, which are reciprocals of the sine, cosine, and tangent functions. But we will confine our discussion to the sine, cosine, and tangent functions since they are the only ones we need to solve for angles and sides of a right triangle.)

$$\text{sine of an angle} = \frac{\text{opposite side}}{\text{hypotenuse}}$$

(16–7)

$$\sin \theta = \frac{a}{c}$$

$$\text{cosine of an angle} = \frac{\text{adjacent side}}{\text{hypotenuse}}$$

(16–8)

$$\cos \theta = \frac{b}{c}$$

$$\text{tangent of an angle} = \frac{\text{opposite side}}{\text{adjacent side}}$$

(16–9)

$$\tan \theta = \frac{a}{b}$$

The sine function expresses the ratio of the length of the opposite side to the length of the hypotenuse. The equation would read $\sin \theta = \frac{a}{c}$. The cosine function expresses the ratio of the length of the adjacent side to the length of the hypotenuse. The equation would read $\cos \theta = \frac{b}{c}$. The tangent function expresses the ratio of the length of the opposite side to the length of the adjacent side. The equation would read $\tan \theta = \frac{a}{b}$.

FIGURE 16–15

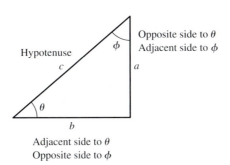

Opposite side to θ
Adjacent side to ϕ

Hypotenuse
c

a

ϕ

θ

b

Adjacent side to θ
Opposite side to ϕ

FIGURE 16–16

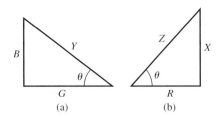

FIGURE 16–17

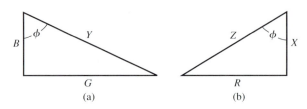

PRACTICE PROBLEMS 16–3

1. Refer to Figure 16–16(a). Write the equations for the three functions with reference to the given angle.

2. Refer to Figure 16–16(b). Write the equations for the three functions with reference to the given angle.

3. Refer to Figure 16–17(a). Write the equations for the three functions with reference to the given angle.

4. Refer to Figure 16–17(b). Write the equations for the three functions with reference to the given angle.

SOLUTIONS

1. In Figure 16–16(a), side G is the adjacent side, side B is the opposite side, and side Y is the hypotenuse when considering the angle θ. The sine function expresses the ratio of the opposite side to the hypotenuse; therefore, $\sin \theta = B/Y$. The cosine function expresses the ratio of the adjacent side to the hypotenuse; therefore, $\cos \theta = G/Y$. Finally, $\tan \theta = B/G$ since the tangent expresses the ratio of the opposite side to the adjacent side.

2. In Figure 16–16(b), side R is the adjacent side, side X is the opposite side, and side Z is the hypotenuse when considering the angle θ. Therefore, $\sin \theta = X/Z$, $\cos \theta = R/Z$, and $\tan \theta = X/R$.

3. If you determined in Figure 16–17(a) that the adjacent side is side B and the opposite side is side G, you are right. Then, $\sin \phi = G/Y$, $\cos \phi = B/Y$, and $\tan \phi = G/B$.

4. You should have determined in Figure 16–17(b) that the adjacent side is side X and the opposite side is side R. Then, $\sin \phi = R/Z$, $\cos \phi = X/Z$, and $\tan \phi = R/X$.

Additional practice problems are at the end of the chapter.

Refer to Figure 16–18.

1. $\angle\phi = 47.3°$. Find θ.
2. If $\angle\theta = 37.5°$, which side is longer?
3. If $Z = 73.6$ kΩ and $R = 47$ kΩ, find X.
4. In problem 3, which angle is greater?
5. Write the equations for the three functions with reference to $\angle\theta$.

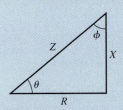

FIGURE 16–18

Answers to Self-test 16–1 are at the end of the chapter.

TRIGONOMETRIC TABLES   16–4

Now that we understand what a trig function is, let's see how we can use trig tables to find unknown angles or ratios.

Table 16–1 is a partial table of the trig functions. Appendix B is a complete table of trig functions.

If we wanted to look up the trigonometric ratios for the angles between 30.0 and 45.0 degrees, we would first find the angle in the *lefthand* column and then use the column headings to identify the sine, tangent, cotangent, and cosine for that angle. If we wanted to look up the trigonometric ratios for the angles between 45.0 and 60.0 degrees, we would first find the angle in the *righthand* column and then use the column footings to identify the sine, tangent, cotangent, and cosine for that angle. It is important to notice that the column footings are in the reverse order from the column headings! By using the column headings for the angles between 30.0 and 45.0 degrees and column footings for the angles between 45.0 and 60.0 degrees, we can print twice as much data on each page.

Let's take a look at Table 16–1. It shows angles from 30° to 60° and the sine, cosine, tangent, and cotangent functions. What is the tangent of a 40° angle? First, we find 40° and then we find the *tan* column. Tan 40° = 0.8391. What is tan 50°? Tan 50° = 1.1918. What is the sine of 35°? This time we find 35° and locate the sine under the column marked *sin*. Sin 35° = 0.57358. What is the cosine of 54°? The cosine column is marked *cos*. Cos 54° = 0.58779. What is the sine of 33.7°? We note that each degree is divided into tenths of a degree. First, we find 33° and then move down to 0.7. This is 7 tenths or 33.7°. Sin 33.7° = 0.55484.

Table 16-1   Trigonometric Functions (Partial Table)

↓ANGLE°	SIN	TAN	COT	COS	
30.0	.50000	.57735	1.7321	.86603	60.0
.1	.50151	.57968	1.7251	.86515	.9
.2	.50302	.58201	1.7182	.86427	.8
.3	.50453	.58435	1.7113	.86340	.7
.4	.50603	.58670	1.7045	.86251	.6
.5	.50754	.58905	1.6977	.86163	.5
.6	.50904	.59140	1.6909	.86074	.4
.7	.51054	.59376	1.6842	.85985	.3
.8	.51204	.59612	1.6775	.85895	.2
.9	.51354	.59849	1.6709	.85806	.1
31.0	.51504	.60086	1.6643	.85717	59.0
.1	.51653	.60324	1.6577	.85627	.9
.2	.51803	.60562	1.6512	.85536	.8
.3	.51952	.60801	1.6447	.85446	.7
.4	.52101	.61040	1.6383	.85355	.6
.5	.52250	.61280	1.6319	.85264	.5
.6	.52399	.61520	1.6255	.85173	.4
.7	.52547	.61761	1.6191	.85081	.3
.8	.52697	.62003	1.6128	.84989	.2
.9	.52844	.62245	1.6066	.84897	.1
32.0	.52992	.62487	1.6003	.84805	58.0
.1	.53140	.62730	1.5941	.84712	.9
.2	.53288	.62973	1.5880	.84619	.8
.3	.53435	.63217	1.5818	.84526	.7
.4	.53583	.63462	1.5757	.84433	.6
.5	.53730	.63707	1.5697	.84339	.5
.6	.53877	.63953	1.5637	.84245	.4
.7	.54024	.64199	1.5577	.84151	.3
.8	.54171	.64446	1.5517	.84057	.2
.9	.54317	.64693	1.5458	.83962	.1
33.0	.54464	.64941	1.5399	.83867	57.0
.1	.54610	.65189	1.5340	.83772	.9
.2	.54756	.65438	1.5282	.83676	.8
.3	.54902	.65688	1.5224	.83581	.7
.4	.55048	.65938	1.5166	.83485	.6
.5	.55194	.66189	1.5108	.83389	.5
.6	.55339	.66440	1.5051	.83292	.4
.7	.55484	.66692	1.4994	.83195	.3
.8	.55630	.66944	1.4938	.83098	.2
.9	.55775	.67197	1.4882	.83001	.1
	COS	COT	TAN	SIN	ANGLE° ↑

Table 16-1 *(cont.)*

↓ANGLE°	SIN	TAN	COT	COS	
34.0	.55919	.67451	1.4826	.82904	56.0
.1	.56064	.67705	1.4770	.82806	.9
.2	.56208	.67960	1.4715	.82708	.8
.3	.56353	.68215	1.4659	.82610	.7
.4	.56497	.68471	1.4605	.82511	.6
.5	.56641	.68728	1.4550	.82413	.5
.6	.56784	.68985	1.4496	.82314	.4
.7	.56928	.69243	1.4442	.82214	.3
.8	.57071	.69502	1.4388	.82115	.2
.9	.57215	.69761	1.4335	.82015	.1
35.0	.57358	.70021	1.4281	.81915	55.0
.1	.57501	.70281	1.4229	.81815	.9
.2	.57643	.70542	1.4176	.81714	.8
.3	.57786	.70804	1.4124	.81614	.7
.4	.57928	.71066	1.4071	.81513	.6
.5	.58070	.71329	1.4019	.81412	.5
.6	.58212	.71593	1.3968	.81310	.4
.7	.58354	.71857	1.3916	.81208	.3
.8	.58496	.72122	1.3865	.81106	.2
.9	.58637	.72388	1.3814	.81004	.1
36.0	.58779	.72654	1.3764	.80902	54.0
.1	.58920	.72921	1.3713	.80799	.9
.2	.59061	.73189	1.3663	.80696	.8
.3	.59201	.73457	1.3613	.80593	.7
.4	.59342	.73726	1.3564	.80489	.6
.5	.59482	.73996	1.3514	.80386	.5
.6	.59622	.74267	1.3465	.80282	.4
.7	.59763	.74538	1.3416	.80178	.3
.8	.59902	.74810	1.3367	.80073	.2
.9	.60042	.75082	1.3319	.79968	.1
37.0	.60182	.75355	1.3270	.79864	53.0
.1	.60321	.75629	1.3222	.79758	.9
.2	.60460	.75904	1.3175	.79653	.8
.3	.60599	.76180	1.3127	.79547	.7
.4	.60738	.76456	1.3079	.79441	.6
.5	.60876	.76733	1.3032	.79335	.5
.6	.61015	.77010	1.2985	.79229	.4
.7	.61153	.77289	1.2938	.79122	.3
.8	.61291	.77568	1.2892	.79016	.2
.9	.61429	.77848	1.2846	.78908	.1
	COS	COT	TAN	SIN	ANGLE° ↑

Table 16–1 (cont.)

↓ANGLE°	SIN	TAN	COT	COS	
38.0	.61566	.78129	1.2799	.78801	52.0
.1	.61704	.78410	1.2753	.78694	.9
.2	.61841	.78692	1.2708	.78586	.8
.3	.61978	.78975	1.2662	.78478	.7
.4	.62115	.79259	1.2617	.78369	.6
.5	.62251	.79544	1.2572	.78261	.5
.6	.62388	.79829	1.2527	.78152	.4
.7	.62524	.80115	1.2482	.78043	.3
.8	.62660	.80402	1.2437	.77934	.2
.9	.62796	.80690	1.2393	.77824	.1
39.0	.62932	.80978	1.2349	.77715	51.0
.1	.63068	.81268	1.2305	.77605	.9
.2	.63203	.81558	1.2261	.77494	.8
.3	.63338	.81849	1.2218	.77384	.7
.4	.63473	.82141	1.2174	.77273	.6
.5	.63608	.82434	1.2131	.77162	.5
.6	.63742	.82727	1.2088	.77051	.4
.7	.63877	.83022	1.2045	.76940	.3
.8	.64011	.83317	1.2002	.76828	.2
.9	.64145	.83613	1.1960	.76717	.1
40.0	.64279	.83910	1.1918	.76604	50.0
.1	.64412	.84208	1.1875	.76492	.9
.2	.64546	.84507	1.1833	.76380	.8
.3	.64679	.84806	1.1792	.76267	.7
.4	.64812	.85107	1.1750	.76154	.6
.5	.64945	.85408	1.1708	.76041	.5
.6	.65077	.85710	1.1667	.75927	.4
.7	.65210	.86014	1.1626	.75813	.3
.8	.65342	.86318	1.1585	.75700	.2
.9	.65474	.86623	1.1544	.75585	.1
41.0	.65606	.86929	1.1504	.75471	49.0
.1	.65738	.87236	1.1463	.75356	.9
.2	.65869	.87543	1.1423	.75241	.8
.3	.66000	.87852	1.1383	.75126	.7
.4	.66131	.88162	1.1343	.75011	.6
.5	.66262	.88473	1.1303	.74896	.5
.6	.66393	.88784	1.1263	.74780	.4
.7	.66523	.89097	1.1224	.74664	.3
.8	.66653	.89410	1.1184	.74548	.2
.9	.66783	.89725	1.1145	.74431	.1
	COS	COT	TAN	SIN	ANGLE° ↑

Table 16-1 *(cont.)*

↓ANGLE°	SIN	TAN	COT	COS	
42.0	.66913	.90040	1.1106	.74314	48.0
.1	.67043	.90357	1.1067	.74198	.9
.2	.67172	.90674	1.1028	.74080	.8
.3	.67301	.90993	1.0990	.73963	.7
.4	.67430	.91313	1.0951	.73846	.6
.5	.67559	.91633	1.0913	.73728	.5
.6	.67688	.91955	1.0875	.73610	.4
.7	.67816	.92277	1.0837	.73491	.3
.8	.67944	.92601	1.0799	.73373	.2
.9	.68072	.92926	1.0761	.73254	.1
43.0	.68200	.93252	1.0724	.73135	47.0
.1	.68327	.93578	1.0686	.73016	.9
.2	.68455	.93906	1.0649	.72897	.8
.3	.68582	.94235	1.0612	.72777	.7
.4	.68709	.94565	1.0575	.72657	.6
.5	.68835	.94896	1.0538	.72537	.5
.6	.68962	.95229	1.0501	.72417	.4
.7	.69088	.95562	1.0464	.72294	.3
.8	.69214	.95897	1.0428	.72176	.2
.9	.69340	.96232	1.0392	.72055	.1
44.0	.69466	.96569	1.0355	.71934	46.0
.1	.69591	.96907	1.0319	.71813	.9
.2	.69717	.97246	1.0283	.71691	.8
.3	.69842	.97586	1.0247	.71569	.7
.4	.69966	.97927	1.0212	.71447	.6
.5	.70091	.98270	1.0176	.71325	.5
.6	.70215	.98613	1.0141	.71203	.4
.7	.70339	.98958	1.0105	.71080	.3
.8	.70463	.99304	1.0070	.70957	.2
.9	.70587	.99652	1.0035	.70834	.1
45.0	.70711	1.00000	1.0000	.70711	45.0
COS	COT	TAN	SIN	ANGLE° ↑	

PRACTICE PROBLEMS 16–4

Use the trig table to find the following. Verify your answers by using a calculator.

1. sin 42°
2. cos 58°
3. tan 38°
4. tan 45°
5. sin 45°
6. cos 45°

7. sin 31.6°

8. cos 48.4°

9. tan 50.1°

10. cos 55.6°

11. sin 55.6°

12. tan 31.9°

SOLUTIONS

1. 0.66913

2. 0.52992

3. 0.78129

4. 1.0000

5. 0.70711

6. 0.70711

7. 0.52399

8. 0.66393

9. 1.1960

10. 0.56497

11. 0.82511

12. 0.62245

Additional practice problems are at the end of the chapter.

16–5   INVERSE TRIG FUNCTIONS

Suppose we know the length of the sides and need to find the angles. One of the basic relationships is

$$\sin \theta = \frac{\text{opposite side}}{\text{hypotenuse}}$$

If $\theta = 30°$, then $\sin \theta = \sin 30° = 0.5$. This tells us that if the angle is 30° the ratio of the opposite side to the hypotenuse is 0.5.

Now let's turn it around and say, "Given some ratio of sides, what is the angle?" For example, suppose the ratio of the opposite side to the hypotenuse is 0.74314. What is the angle? The solution requires the use of inverse trig functions. *Inverse trig functions* are written *arcsin, arccos,* and *arctan.* Or they are written $\sin^{-1}$, $\cos^{-1}$, and $\tan^{-1}$. Most of the time we will use $\sin^{-1}$, $\cos^{-1}$, and $\tan^{-1}$.

In our example the ratio was opposite side to hypotenuse, which indicates a sine function. Therefore, our inverse trig function is $\sin^{-1}$. $\sin^{-1} 0.74314 = \angle\theta$. The expression $\sin^{-1} 0.74314$ asks the question, "What is the angle whose sine is 0.74314?" To find the answer, we follow along the sin column in Table 16–1 until we find 0.74314, and then we find the angle that corresponds to that value. In this case, the angle is 48°.

$$\sin^{-1} 0.74314 = 48°$$

Using the Calculator

All the problems in this chapter have angles measured in degrees. When using a calculator in this chapter make sure the calculator is in the "degree mode" and not in the "radian mode."

Angles can be measured in degrees or radians, as you will learn in the chapter on AC Fundamentals. Measuring angles in degrees or radians is similar to measuring lengths using inches or centimeters; both methods are valid but you must be consistent when performing calculations.

With the TI calculator, to find the sine of an angle you must first enter the angle and then press the $\boxed{\text{sin}}$ key. With the Casio, to find the sine of an angle you first press $\boxed{\text{sin}}$, then enter the angle, and then press $\boxed{=}$.

With the TI calculator, to find the arcsine of an angle you must first enter the ratio and then press the $\boxed{\text{sin}^{-1}}$ key, (the second function of the $\boxed{\text{sin}}$ key). With the Casio, to find the arcsine of an angle you first press $\boxed{\text{sin}^{-1}}$, (the second function of the $\boxed{\text{sin}}$ key), then enter the ratio, and then press $\boxed{=}$.

EXAMPLE 16–6

Find the angle whose cosine is 0.90183.

SOLUTION

$$\cos^{-1} 0.90183 = 25.6°$$

$.90183$ $\boxed{\cos^{-1}}$ *Answer:* **25.6°**

EXAMPLE 16–7

Find the angle whose tangent is 1.4994.

SOLUTION

$$\tan^{-1} 1.4994 = 56.3°$$

1.4994 $\boxed{\tan^{-1}}$ *Answer:* **56.3°**

PRACTICE PROBLEMS 16–5

Use the trig table in Appendix B to find the following angles. Verify your answers by using a calculator.

1. $\sin^{-1} 0.86603$
2. $\sin^{-1} 0.70711$
3. $\sin^{-1} 0.35184$
4. $\cos^{-1} 0.83581$
5. $\tan^{-1} 1.7321$
6. $\cos^{-1} 0.25207$
7. $\cos^{-1} 0.5$
8. $\cos^{-1} 0.86603$
9. $\tan^{-1} 1.0$
10. $\tan^{-1} 0.57735$
11. $\tan^{-1} 1.2437$
12. $\sin^{-1} 0.22325$

SOLUTIONS

1. $\sin^{-1} 0.86603 = 60°$
3. $\sin^{-1} 0.35184 = 20.6°$
5. $\tan^{-1} 1.7321 = 60°$
7. $\cos^{-1} 0.5 = 60°$
9. $\tan^{-1} 1.0 = 45°$
11. $\tan^{-1} 1.2437 = 51.2°$

2. $\sin^{-1} 0.70711 = 45°$
4. $\cos^{-1} 0.83581 = 33.3°$
6. $\cos^{-1} 0.25207 = 75.4°$
8. $\cos^{-1} 0.86603 = 30°$
10. $\tan^{-1} 0.57735 = 30°$
12. $\sin^{-1} 0.22325 = 12.9°$

Additional practice problems are at the end of the chapter.

16-6 TRIGONOMETRIC EQUATIONS

Now let's see how we can use both trig functions and inverse trig functions to find the unknown sides and angles of a right triangle. Consider the triangle in Figure 16–19. Let's assume that side b equals 25 and angle $\theta = 40°$. How long are sides a and c and what is the angle ϕ? First, let's write the equations of the three functions.

(16–7)
$$\sin \theta = \frac{\text{opposite side}}{\text{hypotenuse}} = \frac{a}{c}$$

(16–8)
$$\cos \theta = \frac{\text{adjacent side}}{\text{hypotenuse}} = \frac{b}{c}$$

(16–9)
$$\tan \theta = \frac{\text{opposite side}}{\text{adjacent side}} = \frac{a}{b}$$

As with any linear equation, if we know two of the quantities, we can manipulate the equation to solve for the unknown. In our problem we know the length of side b and the angle θ. An examination of the equations shows that to find side c, the hypotenuse, we would use the cosine function. To find the opposite side, we would use the tangent function. To find the angle ϕ, we would simply subtract the angle θ from 90°. $\phi = 90° - 40°$. Let's find the hypotenuse first.

(16–8)
$$\cos \theta = \frac{b}{c}$$

FIGURE 16–19

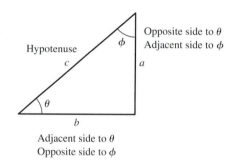

Hypotenuse
c
ϕ
a
Opposite side to θ
Adjacent side to ϕ
θ
b
Adjacent side to θ
Opposite side to ϕ

$$c = \frac{b}{\cos \theta}$$

(16–10)

☞ **KEY POINT** When rearranging the equations of the functions to solve for an unknown side or angle, *sin θ* (or *cos θ* or *tan θ*) is one quantity and cannot be separated.

That is, we could not move "cos" and leave θ as one member of the equation. And we could not move "θ" and leave cos as one member. Cos θ would move from one member to the other. In Equation 16–8, since c is unknown, we solve for c by multiplying both members by c and dividing both members by cos θ. Plugging in known values, we get

$$c = \frac{25}{\cos 40°} = \frac{25}{0.766} = 32.6$$

Even though we learned how to use the trig tables, most of our problem solving is done with the calculator. Since the calculator has a trig table stored in its memory, we need the table in the book only as a reference. Since three-place accuracy in answers is almost always close enough, we will continue that practice here. Using the calculator,

25 $\boxed{\div}$ 40 $\boxed{\text{cos}}$ $\boxed{=}$ *Answer:* $c = $ **32.6**

Now let's find side a. Since we now know θ, side b, and the hypotenuse, we have a choice. We can use the tangent function or we can use the sine function. We will do it both ways.

$$\tan \theta = \frac{a}{b}$$

(16–9)

Solve for a:

$$a = b \tan \theta$$

(16–11)

$$a = 25 \tan 40° = 25 \times 0.839 = 21.0$$

With the calculator,

25 $\boxed{\times}$ 40 $\boxed{\text{tan}}$ $\boxed{=}$ *Answer:* $a = $ **21.0**

Or

$$\sin \theta = \frac{a}{c}$$

(16–7)

Solve for a:

$$a = c \sin \theta$$

(16–12)

$$a = 32.6 \sin 40° = 32.6 \times 0.643 = 21.0$$

Using the calculator,

$$32.6 \boxed{\times} 40 \boxed{\sin} \boxed{=} \qquad \textit{Answer: } a = 21.0$$

$$\angle\phi = 90° - \angle\theta = 90° - 40° = 50°$$

EXAMPLE 16–8

Side $a = 150$ and side $b = 300$ in Figure 16–19. Find the hypotenuse and $\angle\theta$.

SOLUTION   Only the tangent function contains one unknown.

$$\tan\theta = \frac{a}{b}$$

$$\tan\theta = \frac{150}{300} = 0.5$$

Now that we know that $\tan\theta = 0.5$, our next step is to find the angle whose tangent is 0.5.

$$\tan^{-1} 0.5 = \theta = 26.6°$$

We were given sides a and b and we found $\angle\theta$. We can find the hypotenuse by using the Pythagorean theorem:

$$c = \sqrt{a^2 + b^2} = \sqrt{1.125 \times 10^5} = 335$$

Another solution uses the sine function:

$$\sin\theta = \frac{a}{c}$$

$$c = \frac{a}{\sin\theta} = \frac{150}{\sin 26.6°} = 335$$

A third solution uses the cosine function:

$$\cos\theta = \frac{b}{c}$$

$$c = \frac{b}{\cos\theta} = \frac{300}{\cos 26.6°} = 335$$

All solutions yield the same result, and they are all valid.

EXAMPLE 16–9

Refer to Figure 16–20. Find R, Z, and $\angle\phi$ if $X = 2.83 \text{ k}\Omega$ and $\theta = 33.4°$.

FIGURE 16–20

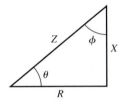

SOLUTION

$$\tan \theta = \frac{X}{R}$$

$$R = \frac{X}{\tan \theta} = \frac{2.83 \text{ k}\Omega}{\tan 33.4°} = \frac{2.83 \text{ k}\Omega}{0.659} = 4.29 \text{ k}\Omega$$

Using the calculator,

2.83 [EE] 3 [÷] 33.4 [tan] [=]

Answer: R = 4.29 kΩ

$$\sin \theta = \frac{X}{Z}$$

$$Z = \frac{X}{\sin \theta} = \frac{2.83 \text{ k}\Omega}{\sin 33.4°} = 5.14 \text{ k}\Omega$$

Using the calculator,

2.83 [EE] 3 [÷] 33.4 [sin] [=]

Answer: Z = 5.14 kΩ

$$\phi = 90° - \theta = 90° - 33.4° = 56.6°$$

EXAMPLE 16–10

Refer to Figure 16–21. Find B, θ, and ϕ if $Y = 670 \ \mu S$ and $G = 430 \ \mu S$.

SOLUTION Find θ.

$$\cos \theta = \frac{G}{Y}$$

$$\cos \theta = \frac{430 \ \mu S}{670 \ \mu S} = 0.642$$

$$\cos^{-1} 0.642 = \theta = 50.1°$$

FIGURE 16–21

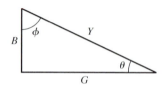

Using the calculator,

$$430 \boxed{\text{EE}} \boxed{+/-} 6 \boxed{\div} 670 \boxed{\text{EE}} \boxed{+/-} 6 \boxed{=}$$

$$\boxed{\cos^{-1}} \boxed{\text{STO}}$$

Answer: $\theta = 50.1°$ and is stored in memory.

$$\phi = 90° - \theta = 90° - 50.1° = 39.9°$$

Remember, we stored θ, so

$$90 \boxed{-} \boxed{\text{RCL}} \boxed{=} \qquad \textit{Answer:} \phi = 39.9°$$

We will use the sine function to find B.

$$\sin \theta = \frac{B}{Y}$$

$$Y \sin \theta = B$$

$$670 \ \mu S \ \sin 50.1° = 670 \ \mu S \times 0.767 = 514 \ \mu S$$

Using the calculator,

$$670 \boxed{\text{EE}} \boxed{+/-} 6 \boxed{\times} \boxed{\text{RCL}} \boxed{\sin} \boxed{=}$$

Answer: $B = 514 \ \mu S$

PRACTICE PROBLEMS 16–6

1. Refer to Figure 16–19.
   (a) $a = 25.5$, $\theta = 20°$. Find b, c, and ϕ.
   (b) $b = 1400$, $\theta = 65.4°$. Find a, c, and ϕ.
   (c) $a = 350$, $b = 500$. Find θ and c.
   (d) $a = 7.6$, $c = 11.2$. Find θ and b.
   (e) $a = 120$, $\theta = 45°$. Find b and c.
   (f) $\phi = 40°$, $c = 850$. Find a and b.

2. Refer to Figure 16–20.
   (a) $X = 700 \ \Omega$, $R = 900 \ \Omega$. Find Z and θ.
   (b) $\theta = 68.3°$, $X = 30 \ k\Omega$. Find R and Z.
   (c) $Z = 73.6 \ k\Omega$, $R = 56 \ k\Omega$. Find X and θ.
   (d) $\theta = 27°$, $R = 300 \ \Omega$. Find X and Z.
   (e) $Z = 12.3 \ k\Omega$, $X = 8.77 \ k\Omega$. Find R and θ.
   (f) $\theta = 30°$, $Z = 3 \ k\Omega$. Find X and R.

3. Refer to Figure 16–21.

   (a) $G = 2.8$ mS, $B = 3.65$ mS. Find
       Y and θ.
   (b) $\theta = 30°$, $B = 250$ μS. Find G
       and Y.
   (c) $G = 10.3$ mS, $Y = 15.8$ mS. Find
       B and θ.

   (d) $\theta = 52°$, $Y = 650$ μS. Find G
       and B.
   (e) $B = 140$ μS, $Y = 270$ μS. Find
       G and θ.
   (f) $\theta = 45°$, $G = 100$ μS. Find B and Y.

SOLUTIONS

1. (a) $b = 70.1$, $c = 74.6$, $\phi = 70°$
   (b) $a = 3060$, $c = 3360$, $\phi = 24.6°$
   (c) $c = 610$, $\theta = 35°$
   (d) $b = 8.23$, $\theta = 42.7°$
   (e) $b = 120$, $c = 170$
   (f) $a = 651$, $b = 546$

2. (a) $Z = 1.14$ kΩ, $\theta = 37.9°$
   (b) $R = 11.9$ kΩ, $Z = 32.3$ kΩ
   (c) $X = 47.8$ kΩ, $\theta = 40.5°$
   (d) $X = 153$ Ω, $Z = 337$ Ω
   (e) $R = 8.62$ kΩ, $\theta = 45.5°$
   (f) $R = 2.6$ kΩ, $X = 1.5$ kΩ

3. (a) $Y = 4.60$ mS, $\theta = 52.5°$
   (b) $G = 433$ μS, $Y = 500$ μS
   (c) $B = 12.0$ mS, $\theta = 49.3°$
   (d) $B = 512$ μS, $G = 400$ μS
   (e) $G = 231$ μS, $\theta = 31.2°$
   (f) $B = 100$ μS, $Y = 141$ μS

Additional practice problems are at the end of the chapter.

SELF-TEST 16–2

1. Find the sine, cosine, and tangent
   of 62.4°.
3. Find $\cos^{-1} 0.75927$.
5. Refer to Figure 16–22. Find b and θ if
   $a = 27$ and $c = 42.3$.
7. Refer to Figure 16–24. Find B and Y if
   $G = 220$ μS and $\theta = 38.2°$.

2. Find $\sin^{-1} 0.74548$.

4. Find $\tan^{-1} 1.42815$.
6. Refer to Figure 16–23. Find Z and θ if
   $R = 2.7$ kΩ and $X = 3.85$ kΩ.

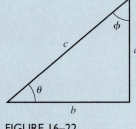

FIGURE 16–22

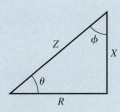

FIGURE 16–23

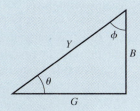

FIGURE 16–24

Answers to Self-test 16–2 are at the end of the chapter.

CHAPTER 16 AT A GLANCE

PAGE	KEY POINTS, THEOREMS, AND FUNCTIONS	EXAMPLE

466   *Key Point:* In any right triangle: (1) the sum of the acute angles equals 90°, (2) the hypotenuse is the side of greatest length, and (3) the greater side lies opposite the greater angle.

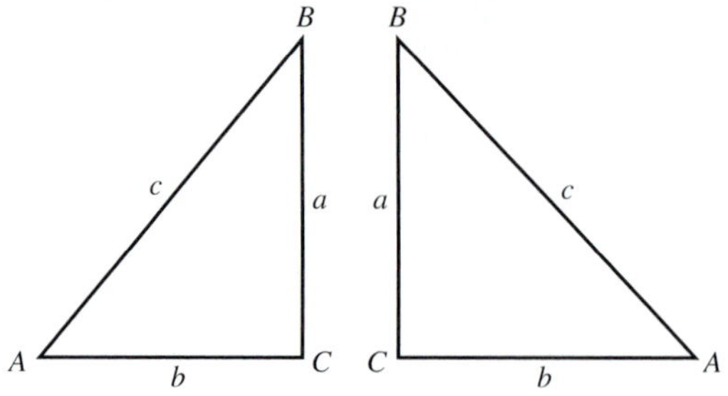

FIGURE 16–25

468   *Key Point:* The Pythagorean theorem states that *in a right triangle, the square of the hypotenuse is equal to the sum of the squares of the other two sides.*

$$c^2 = a^2 + b^2$$

473   Three trigonometric ratios, called *functions,* are the *sine, cosine,* and *tangent* functions. The *sine* function expresses the ratio of the length of the opposite side to the hypotenuse. The *cosine* function expresses the ratio of the length of the adjacent side to the hypotenuse. The *tangent* function expresses the ratio of the length of the opposite side to the adjacent side.

$$\sin \theta = \frac{X}{Z}$$

$$\cos \theta = \frac{R}{Z}$$

$$\tan \theta = \frac{X}{R}$$

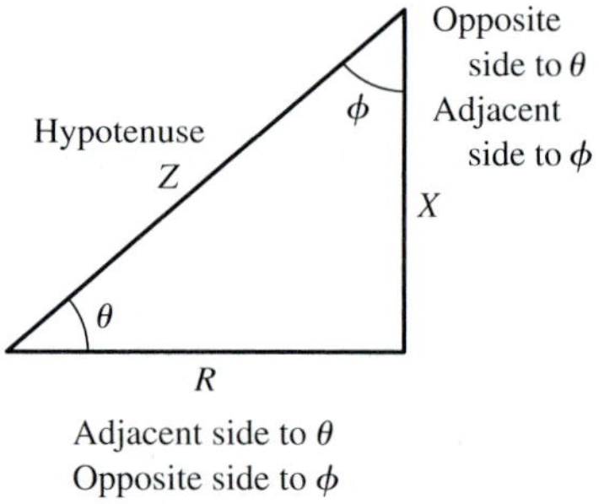

FIGURE 16–26

480   *Key Point:* Inverse trig functions are used to find the ratio of sides when the angle is known. Three *inverse trig functions* are *arcsin, arccos,* and *arctan.* They are usually written sin^{-1}, cos^{-1}, and tan^{-1}.

$$\sin^{-1} 0.74314 = 48.0°$$
$$\cos^{-1} 0.74314 = 42.0°$$
$$\tan^{-1} 0.74314 = 36.6°$$

483   *Key Point:* When rearranging the equations of the functions to solve for an unknown side or angle, $\sin \theta$ (or $\cos \theta$ or $\tan \theta$) is one quantity and cannot be separated.

$$\cos \theta = \frac{R}{Z}$$

$$Z = \frac{R}{\cos \theta}$$

END OF CHAPTER PROBLEMS 16–1

1. Refer to Figure 16–23:
   (a) If $\angle\theta = 35°$, find $\angle\phi$.
   (b) Which side (R or X) is greater?
   (c) Which side lies opposite $\angle\phi$?
   (d) Which side is adjacent to $\angle\theta$?
3. Refer to Figure 16–23:
   (a) Find Z when $R = 2.7$ kΩ and $X = 4$ kΩ.
   (b) Find R when $Z = 600$ Ω and $X = 350$ Ω.
   (c) Find X when $Z = 120$ kΩ and $R = 68$ kΩ.
5. Refer to Figure 16–24:
   (a) Find Y when $G = 1.83$ mS and $B = 2.37$ mS.
   (b) Find G when $Y = 1$ mS and $B = 600$ μS.
   (c) Find B when $Y = 460$ μS and $G = 175$ μS.

2. Refer to Figure 16–24:
   (a) If $\angle\phi = 70°$, find $\angle\theta$.
   (b) Which side (G or B) is greater?
   (c) Which side lies opposite $\angle\theta$? Which side is adjacent to $\angle\phi$?
4. Refer to Figure 16–23:
   (a) Find Z when $R = 20$ kΩ and $X = 30$ kΩ.
   (b) Find R when $Z = 600$ kΩ and $X = 300$ kΩ.
   (c) Find X when $Z = 11.7$ kΩ and $R = 9.1$ kΩ.
6. Refer to Figure 16–24:
   (a) Find Y when $G = 3.8$ mS and $B = 4.35$ mS.
   (b) Find G when $Y = 170$ μS and $B = 120$ μS.
   (c) Find B when $Y = 10.7$ mS and $G = 4.37$ mS.

END OF CHAPTER PROBLEMS 16–2

Refer to Figure 16–26:

1. Let $X = 14$ and $R = 22$. Find Z. Which of angles θ and ϕ is greater?
3. Let $X = 250$ and $Z = 350$. Find R.
5. Let $R = 73.2$ and $Z = 100$. Find X.

2. Let $X = 75$ and $R = 65$. Find Z. Which of angles θ and ϕ is greater?
4. Let $X = 8.93$ and $Z = 14.2$. Find R.
6. Let $R = 600$ and $Z = 800$. Find X.

Refer to Figure 16–27 on the next page:

7. Let $R = 680$ Ω and $X = 700$ Ω. Find Z. Which of angles θ and ϕ is greater?
9. Let $R = 470$ kΩ and $X = 560$ kΩ. Find Z. Which of angles θ and ϕ is greater?
11. Let $R = 75$ kΩ and $X = 39$ kΩ. Find Z. Which of angles θ and ϕ is greater?
13. Let $R = 4.7$ kΩ and $Z = 8.27$ kΩ. Find X. Which of angles θ and ϕ is greater?

8. Let $R = 330$ Ω and $X = 400$ Ω. Find Z. Which of angles θ and ϕ is greater?
10. Let $R = 680$ kΩ and $X = 900$ kΩ. Find Z. Which of angles θ and ϕ is greater?
12. Let $R = 22$ kΩ and $X = 18$ kΩ. Find Z. Which of angles θ and ϕ is greater?
14. Let $R = 8.2$ kΩ and $Z = 10.5$ kΩ. Find X. Which of angles θ and ϕ is greater?

15. Let $R = 100$ kΩ and $Z = 180$ kΩ. Find X. Which of angles $θ$ and $φ$ is greater?

16. Let $R = 390$ kΩ and $Z = 600$ kΩ. Find X. Which of angles $θ$ and $φ$ is greater?

17. Let $X = 43$ kΩ and $Z = 50.8$ kΩ. Find R. Which of angles $θ$ and $φ$ is greater?

18. Let $X = 82$ kΩ and $Z = 110$ kΩ. Find R. Which of angles $θ$ and $φ$ is greater?

19. Let $X = 47$ Ω and $Z = 102$ Ω. Find R. Which of angles $θ$ and $φ$ is greater?

20. Let $X = 33$ Ω and $Z = 60.5$ Ω. Find R. Which of angles $θ$ and $φ$ is greater?

Refer to Figure 16–28:

21. Let $G = 450$ μS and $B = 700$ μS. Find Y. Which of angles $θ$ and $φ$ is greater?

22. Let $G = 825$ μS and $B = 600$ μS. Find Y. Which of angles $θ$ and $φ$ is greater?

23. Let $G = 3.37$ mS and $B = 2.78$ mS. Find Y. Which of angles $θ$ and $φ$ is greater?

24. Let $G = 5.11$ mS and $B = 4.33$ mS. Find Y. Which of angles $θ$ and $φ$ is greater?

25. Let $G = 37$ μS and $Y = 42.7$ μS. Find B. Which of angles $θ$ and $φ$ is greater?

26. Let $G = 30.3$ μS and $Y = 60.5$ μS. Find B. Which of angles $θ$ and $φ$ is greater?

27. Let $G = 6.67$ mS and $Y = 8.07$ mS. Find B. Which of angles $θ$ and $φ$ is greater?

28. Let $G = 8$ μS and $Y = 12$ μS. Find B. Which of angles $θ$ and $φ$ is greater?

29. Let $B = 256$ μS and $Y = 612$ μS. Find G. Which of angles $θ$ and $φ$ is greater?

30. Let $B = 23.3$ μS and $Y = 48.5$ μS. Find G. Which of angles $θ$ and $φ$ is greater?

31. Let $B = 17.9$ mS and $Y = 22.3$ mS. Find G. Which of angles $θ$ and $φ$ is greater?

32. Let $B = 33.3$ μS and $Y = 55.6$ μS. Find G. Which of angles $θ$ and $φ$ is greater?

END OF CHAPTER PROBLEMS 16–3

1. Refer to Figure 16–20: Write the equation for the three functions with references to (a) $∠θ$ and (b) $∠φ$.

2. Refer to Figure 16–21: Write the equation for the three functions with reference to (a) $∠θ$ and (b) $∠φ$.

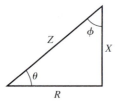

FIGURE 16–27

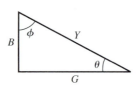

FIGURE 16–28

END OF CHAPTER PROBLEMS 16–4

Use the Trigonometric Functions Table in Appendix B to find the following. Verify your answers using the calculator.

1. $\sin 27.3°$
3. $\sin 63.6°$
5. $\cos 78.3°$
7. $\cos 23.7°$
9. $\tan 67.3°$
11. $\tan 40.3°$

2. $\sin 53.2°$
4. $\sin 37.5°$
6. $\cos 41.7°$
8. $\cos 58.9°$
10. $\tan 12.7°$
12. $\tan 75.7°$

END OF CHAPTER PROBLEMS 16–5

Use the Trigonometric Functions Table in Appendix B to find the following. Verify your answers using the calculator.

1. $\sin^{-1} 0.39715$
3. $\sin^{-1} 0.87036$
5. $\cos^{-1} 0.55630$
7. $\cos^{-1} 0.78043$
9. $\tan^{-1} 1.4715$

2. $\sin^{-1} 0.95106$
4. $\sin^{-1} 0.45865$
6. $\cos^{-1} 0.45088$
8. $\cos^{-1} 0.56641$
10. $\tan^{-1} 1.1303$

END OF CHAPTER PROBLEMS 16–6

Refer to Figure 16–19:

1. $a = 25$, $b = 40$. Find θ, ϕ, and c.
3. $a = 200$, $\phi = 35°$. Find θ, b, and c.
5. $c = 17.3$, $\theta = 40°$. Find a and b.

2. $a = 67.5$, $b = 43$. Find θ, ϕ, and c.
4. $a = 6.35$, $\phi = 54°$. Find θ, b, and c.
6. $c = 680$, $\theta = 72°$. Find a and b.

Refer to Figure 16–27:

7. $X = 17.6 \text{ k}\Omega$, $\theta = 22.3°$. Find R and Z.
9. $X = 650 \ \Omega$, $\theta = 22.3°$. Find R and Z.
11. $R = 5.6 \text{ k}\Omega$, $X = 10 \text{ k}\Omega$. Find Z and θ.
13. $R = 27 \text{ k}\Omega$, $X = 19.7 \text{ k}\Omega$. Find Z and θ.
15. $R = 20 \text{ k}\Omega$, $\theta = 30°$. Find X and Z.
17. $R = 270 \text{ k}\Omega$, $\theta = 37.5°$. Find X and Z.
19. $Z = 40 \text{ k}\Omega$, $\theta = 67.5°$. Find R and X.

8. $X = 35.7 \text{ k}\Omega$, $\theta = 53°$. Find R and Z.
10. $X = 18.7 \text{ k}\Omega$, $\theta = 54.3°$. Find R and Z.
12. $R = 100 \text{ k}\Omega$, $X = 250 \text{ k}\Omega$. Find Z and θ.
14. $R = 120 \ \Omega$, $X = 180 \ \Omega$. Find Z and θ.
16. $R = 4.7 \text{ k}\Omega$, $\theta = 60°$. Find X and Z.
18. $R = 120 \text{ k}\Omega$, $\theta = 60.3°$. Find X and Z.
20. $Z = 6.2 \text{ k}\Omega$, $\theta = 25.4°$. Find R and X.

21. $Z = 785 \ \Omega$, $\theta = 40.3°$. Find
    R and X.
22. $Z = 310 \ k\Omega$, $\theta = 63.5°$. Find
    R and X.
23. $R = 12 \ k\Omega$, $Z = 22 \ k\Omega$. Find
    X and θ.
24. $R = 39 \ k\Omega$, $Z = 60 \ k\Omega$. Find
    X and θ.
25. $R = 1.8 \ k\Omega$, $Z = 2.55 \ k\Omega$. Find
    X and θ.
26. $R = 56 \ k\Omega$, $Z = 93.7 \ k\Omega$. Find
    X and θ.
27. $X = 17.5 \ k\Omega$, $Z = 38 \ k\Omega$. Find
    R and θ.
28. $X = 450 \ \Omega$, $Z = 800 \ \Omega$. Find
    R and θ.
29. $X = 15.3 \ k\Omega$, $Z = 26.8 \ k\Omega$. Find
    R and θ.
30. $X = 800 \ \Omega$, $Z = 1 \ k\Omega$. Find
    R and θ.

Refer to Figure 16–28:

31. $G = 60 \ \mu S$, $\theta = 25°$. Find B and Y.
32. $G = 6.7 \ mS$, $\theta = 68°$. Find B and Y.
33. $G = 135 \ \mu S$, $\theta = 65°$. Find B and Y.
34. $G = 3.45 \ mS$, $\theta = 37.5°$. Find
    B and Y.
35. $G = 12.5 \ mS$, $B = 8.73 \ mS$.
    Find Y and θ.
36. $G = 440 \ \mu S$, $B = 660 \ \mu S$. Find
    Y and θ.
37. $G = 12.7 \ \mu S$, $B = 17.3 \ \mu S$.
    Find Y and θ.
38. $G = 256 \ \mu S$, $B = 330 \ \mu S$. Find
    Y and θ.
39. $B = 35 \ mS$, $\theta = 10.7°$. Find G and Y.
40. $B = 800 \ \mu S$, $\theta = 60°$. Find G and Y.
41. $B = 3.32 \ mS$, $\theta = 27.4°$. Find
    G and Y.
42. $B = 4.73 \ mS$, $\theta = 38.7°$. Find
    G and Y.
43. $G = 6.35 \ mS$, $Y = 9.35 \ mS$. Find
    B and θ.
44. $G = 300 \ \mu S$, $Y = 500 \ \mu S$. Find
    B and θ.
45. $G = 2.33 \ mS$, $Y = 3.30 \ mS$. Find
    B and θ.
46. $G = 170 \ \mu S$, $Y = 300 \ \mu S$. Find
    B and θ.
47. $Y = 70 \ \mu S$, $\theta = 32°$. Find G and B.
48. $Y = 600 \ \mu S$, $\theta = 60°$. Find G and B.
49. $Y = 6.57 \ mS$, $\theta = 55°$. Find G and B.
50. $Y = 200 \ \mu S$, $\theta = 30°$. Find G and B.
51. $B = 7.5 \ mS$, $Y = 11.5 \ mS$. Find
    G and θ.
52. $B = 6.35 \ mS$, $Y = 11 \ mS$. Find
    G and θ.
53. $B = 170 \ \mu S$, $Y = 320 \ \mu S$. Find
    G and θ.
54. $B = 450 \ \mu S$, $Y = 600 \ \mu S$. Find
    G and θ.

ANSWERS TO SELF-TESTS

SELF-TEST 16–1

1. $\theta = 42.7°$

2. Side R is longer.

3. $X = 56.6 \ k\Omega$

4. $\angle\theta$

5. $\sin \theta = \dfrac{X}{Z}$, $\cos \theta = \dfrac{R}{Z}$,

   $\tan \theta = \dfrac{X}{R}$

1. $\sin 62.4° = 0.88620$,
    $\cos 62.4° = 0.46330$,
    $\tan 62.4° = 1.91282$

2. 48.2°

3. 40.6°

4. 55.0°

5. $b = 32.6$, $\theta = 39.7°$

6. $Z = 4.70 \text{ k}\Omega$, $\theta = 55°$

7. $B = 173 \ \mu\text{S}$, $Y = 280 \ \mu\text{S}$

Trigonometric Identities

17

Introduction

We learned in Chapter 16 how to solve triangles when one of the angles is a right angle (90°). In this chapter we will consider triangles other than right triangles.

Let's review a few things we have learned about angles and triangles. Recall that there are three kinds of angles: acute angles (angles that are less than 90°), right angles (90° angles), and obtuse angles (angles that are greater than 90°). In triangles, the sum of the three angles must equal 180°. When we dealt with the right triangle, one angle was 90°. The other two angles, then, had to be acute angles since the sum of the two had to be 90°. Triangles that do not contain a right angle are called *oblique* triangles. An oblique triangle may include an obtuse angle, depending on its shape. In such a case, the other two angles must be acute, since the sum of the three must equal 180°.

In solving right triangles, we used the Pythagorean theorem and/or the trig functions sine, cosine, and tangent. When the triangle is oblique, we must use additional methods of solution. For such triangles, various trigonometric identities have been derived to help in the solutions. In this chapter, we will learn how to use some of these identities in problem solving.

Chapter Objectives

In this chapter you will learn how to:

1. Use the law of sines to find the sides and angles of oblique triangles.
2. Use the law of cosines to find the sides and angles of oblique triangles.

Several important relationships have been developed between the trigonometric functions and the Pythagorean theorem. These relationships are called trigonometric *identities*. These identities make it possible to simplify complex trigonometric problems so that they can be easily solved. Let's see how we can express the Pythagorean theorem in terms of trig functions: $a^2 + b^2 = c^2$, where a and b are the sides and c is the hypotenuse of the right triangle in Figure 17–1. Dividing both sides by c^2 we get

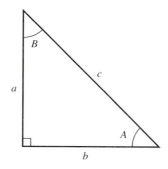

FIGURE 17–1
Right triangle.

$$\frac{a^2}{c^2} + \frac{b^2}{c^2} = \frac{c^2}{c^2}, \qquad \frac{c^2}{c^2} = 1$$

$$\frac{a^2}{c^2} + \frac{b^2}{c^2} = 1$$

Using trig functions,

$$\sin A = \frac{a}{c} \quad \text{and} \quad \cos A = \frac{b}{c}$$

Then

$$\frac{b^2}{c^2} = (\cos A)^2$$

[$(\cos A)^2$ is usually written as $\cos^2 A$] and

$$\frac{a^2}{c^2} = \sin^2 A$$

Therefore,

$$\sin^2 A + \cos^2 A = 1$$

From this identity, other identities have been derived: $\tan^2 A + 1 = \sec^2 A$ and $1 + \cot^2 A = \csc^2 A$. In addition, as stated previously, many other identities have been derived to help in the solution of complex problems. In this chapter we are going to limit ourselves to two of the more useful identities, the law of cosines and the law of sines.

LAW OF SINES 17–1

The law of sines (or sine rule) is that in any triangle, the ratio of the sine of one angle to the length of the opposite side is the same for all three angles and sides. Stated mathematically:

$$\text{Law of sines is:} \quad \frac{\sin A}{a} = \frac{\sin B}{b} = \frac{\sin C}{c}$$

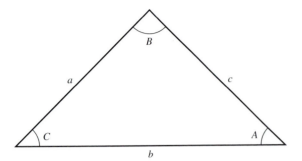

FIGURE 17–2   Oblique triangle.

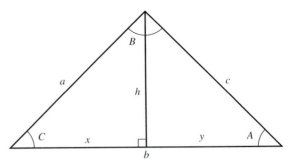

FIGURE 17–3   Oblique triangle with line h used to form two right triangles.

This can also be written as: $\dfrac{a}{\sin A} = \dfrac{b}{\sin B} = \dfrac{c}{\sin C}$

Using some basic concepts of algebra and the right triangle from previous chapters, we can derive the law of sines.

Consider the oblique triangle in Figure 17–2. We have labeled the sides a, b, and c. These sides lie opposite the angles A, B, and C. We need to develop relationships between the angles and sides so that when we know some of them we can find the others. To do this, let's draw a line from angle B to side b such that we form two right triangles as in Figure 17–3. We have labeled this line h. Now that we have right triangles, we can apply the trig functions that we learned in previous chapters.

$$\sin \theta = \frac{\text{opposite side}}{\text{hypotenuse}} \qquad \text{(general equation)}$$

From the two triangles of Figure 17–3, we can say

$$\sin C = \frac{h}{a} \quad \text{and} \quad \sin A = \frac{h}{c}$$

Moving a and c from the right member to the left member of the equations, we get

$$a \sin C = h \quad \text{and} \quad c \sin A = h$$

If $a \sin C = h$ and $c \sin A = h$, then

(17–1)
$$a \sin C = c \sin A$$

Rearranging the equation,

(17–2)
$$\frac{\sin C}{c} = \frac{\sin A}{a}$$

Using the same triangles, we could also prove that

(17–3)
$$\frac{\sin C}{c} = \frac{\sin B}{b}$$

and we could prove that

$$\frac{\sin A}{a} = \frac{\sin B}{b}$$

(17–4)

It follows then that we can combine Equations 17–2, 17–3, and 17–4:

$$\frac{\sin A}{a} = \frac{\sin B}{b} = \frac{\sin C}{c}$$

(17–5)

These equations are the trigonometric identities called the *law of sines.* These equations can be used where (1) two angles and a side opposite one of them are given or (2) two sides and an angle opposite one of them are given. In Example 17–1, two angles and a side are given.

EXAMPLE 17–1

Refer to Figure 17–2. Using the law of sines, find angle A and sides a and c where angle B is 27°, angle C is 64°, and side b is 700.

SOLUTION   Let's first find angle A. $A + B + C = 180°$.

$$A = 180° - B - C = 180° - 27° - 64° = 89°$$

To find side a using the law of sines, we could use Equation 17–4. We couldn't use Equation 17–2 because there are too many unknowns (sides a and c). Let's solve Equation 17–4 for a.

$$a = \frac{b \sin A}{\sin B} = \frac{700 \sin 89°}{\sin 27°} = \frac{700 \times 0.9998}{0.4540} = 1542$$

We will use Equation 17–3 to find side c. Let's rearrange the equation and solve for c.

$$c = \frac{b \sin C}{\sin B} = \frac{700 \times \sin 64°}{\sin 27°} = \frac{700 \times 0.8988}{0.4540} = 1386$$

The law of sines can be used to solve problems where two sides and an angle opposite one are given. This is sometimes referred to as the *ambiguous case* because there are three possible results, depending on the information given:

1. There is one solution. This is the case if the given angle is greater than 90° (making the other two angles acute angles), or if the side opposite the given angle is longer than the other given side (see Ex. 17–2 and 17–3).
2. There are two solutions. This is the case if the given angle is less than 90° and the side opposite the given angle is shorter than the other given side (see Ex. 17–4).
3. There is no solution. This is the case if, in solving a problem, we arrive at a value for the sine of an angle that is greater than one. If we calculate a value greater

than one for the sine, that tells us that there is a problem with the data given (see Ex. 17–5).

EXAMPLE 17–2

Refer to Figure 17–2. Using the law of sines, find angles B and C and side c where angle $A = 125°$, side $a = 10$, and side $b = 8$.

SOLUTION   Equation 17–4 contains only one unknown (angle A), so let's use that equation and solve for B.

$$\sin B = \frac{b \sin A}{a} = \frac{8 \sin 125°}{10} = \frac{8 \times 0.8192}{10} = 0.6553$$

$$B = \sin^{-1} 0.6553 = 40.94°$$

$$C = 180° - A - B = 180° - 125° - 40.94° = 14.06°$$

Now that we know angle C, we can use Equation 17–2 to find side c (we could also have used Equation 17–3).

$$c = \frac{a \sin C}{\sin A} = \frac{10 \sin 14.06°}{\sin 125°} = \frac{10 \times 0.2429}{0.8192} = 2.966$$

EXAMPLE 17–3

In an oblique triangle, side $a = 70$, side $b = 40$, and angle $A = 65°$. Find the unknown side and angles.

SOLUTION   Let's use Equation 17–4 and solve for the unknown, angle B.

$$\sin B = \frac{b \sin A}{a} = \frac{40 \sin 65°}{70} = \frac{40 \times 0.9063}{70} = 0.5179$$

$$\sin^{-1} 0.5179 = B = 31.19°$$

$$C = 180° - A - B = 180° - 65° - 31.19° = 83.81°$$

Now we can solve for side c using either Equation 17–2 or 17–3. We will use 17–2.

$$c = \frac{a \sin C}{\sin A} = \frac{70 \sin 83.81°}{\sin 65°} = \frac{70 \times 0.5179}{0.9063} = 76.79$$

EXAMPLE 17–4

In an oblique triangle, side $b = 10$, side $c = 15$, and angle $B = 30°$. Find the unknown side and angles.

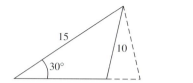 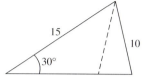

FIGURE 17-4.   Two triangles that can be constructed from the data given in Example 17–4.

SOLUTION   This is a case where it is extremely helpful to draw a diagram of the problem to gain insight about the possible solutions (see Figure 17–4). Rearranging Equation 17–3 and solving for angle C, we get

$$\sin C = \frac{c \sin B}{b} = \frac{15 \sin 30°}{10} = \frac{15 \times 0.5}{10} = 0.75$$

$$\sin^{-1} 0.75 = C = 48.59°$$

$$A = 180° - B - C = 180° - 30° - 48.59° = 101.4°$$

But angle C could also be a quadrant II angle since the problem meets the conditions listed above for two solutions. That is, the given angle is an acute angle, and the side opposite the given angle is smaller than the other given side. Then angle C could also be

$$C = 180° - 48.59° = 131.4°$$

which would make angle A:

$$A = 180° - 30° - 131.4° = 18.6°$$

There are two possible answers:

$$C = 48.59° \quad \text{and} \quad A = 101.4°$$

or

$$C = 131.4° \quad \text{and} \quad A = 18.6°$$

Let's find side a for the first case. To avoid compounding possible errors, it's better to use equations that contain given values whenever possible, so let's use Equation 17–4.

$$a = \frac{b \sin A}{\sin B} = \frac{10 \sin 101.4°}{\sin 30°} = \frac{10 \times 0.9803}{0.5} = 19.60$$

and for the second case

$$a = \frac{b \sin A}{\sin B} = \frac{10 \sin 18.6°}{\sin 30°} = \frac{10 \times 0.3190}{0.5} = 6.379$$

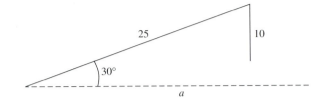

FIGURE 17–5
Diagram for
data given in
Example 17–5.

EXAMPLE 17–5

If the data from the previous problem was somehow misstated, then we could have angles and sides for an unrealizable triangle. For example, if the data for the previous problem was given as side $b = 10$, side $c = 25$, and the angle $B = 30$ degrees, then applying the law of sines we would have:

$$\sin C = (c \sin B)/b = (25 \sin 30°)/10 = (25 \times 0.5)/10 = 1.25$$

This can not be true because the sine function has a range of zero to one and this answer is greater than one. If we tried to draw this triangle, we would have Figure 17–5.

As we can see, this is not a triangle because side b does not touch side a. To make it a triangle, we would have to reduce side c to 20 or less so that $\sin c$ becomes 1 or less. We could also increase the length of side b or decrease the angle B until the end of side b touches side a so that we have a complete triangle.

PRACTICE PROBLEMS 17–1

Refer to Figure 17–2. Find the unknown sides and angles in the following problems using the law of sines.

1. $b = 45$, $B = 72°$, and $C = 53°$
2. $b = 110$, $B = 110°$, and $C = 15°$
3. $c = 50$, $A = 35°$, and $C = 70°$.
4. $a = 100$, $b = 70$, and $C = 80°$.
5. $a = 200$, $b = 150$, and $A = 65°$
6. $a = 200$, $b = 300$, and $A = 20°$
7. $a = 25$, $b = 40$, and $A = 35°$
8. $a = 25$, $b = 15$, and $A = 40°$
9. $b = 150$, $c = 178$, and $C = 80°$
10. $b = 8.5$, $c = 6$, and $C = 32°$

SOLUTIONS

1. $a = 38.76$
   $c = 37.79$
   $A = 55°$
3. $a = 30.52$
   $b = 51.40$
   $B = 75°$
5. $c = 210.1$
   $B = 42.82°$
   $C = 72.18°$

2. $a = 95.89$
   $c = 30.30$
   $A = 55°$
4. $c = 84.60$
   $B = 43.58°$
   $C = 56.42°$
6. Two solutions:
   (1) $c = 453.8$ (2) $c = 110.3$
       $B = 30.87°$ $B = 149.1°$
       $C = 129.1°$ $C = 10.87°$

7. Two solutions:
    (1) $c = 42.70$ (2) $c = 22.84$
        $B = 66.60°$ $B = 113.4°$
        $C = 78.40°$ $C = 31.60°$

8. $c = 34.56$
   $B = 22.69°$
   $C = 117.3°$

9. $a = 125.4$
   $A = 43.91°$
   $B = 56.09°$

10. Two solutions:
    (1) $a = 11.17$ (2) $a = 3.23$
        $A = 99.35°$ $A = 16.6°$
        $B = 48.65°$ $B = 131.4°$

Additional practice problems are at the end of the chapter.

LAW OF COSINES   17–2

The Law of Sines can be used when we know:
> Two angles and any side
> Or
> Two sides and an angle opposite one of the known sides.

The Law of Cosines must be used when we know only:
> Three sides
> Or
> Two sides and the included angle.

    Let's look again at Figure 17–3. This time let's use the Pythagorean theorem and write equations concerning the right triangle with sides x, h, and a.

$$h^2 = a^2 - x^2$$

Next, write the equation using the right triangle with sides y, h, and c.

$$h^2 = c^2 - y^2$$

If $a^2 - x^2 = h^2$ and $c^2 - y^2 = h^2$, then

$$a^2 - x^2 = c^2 - y^2 \qquad (17–6)$$

Now we can manipulate Equation 17–6 so that we can solve for a^2 in terms of b^2 and c^2 in the original triangle. Solving for a^2, we get

$$a^2 = c^2 - y^2 + x^2 \qquad (17–7)$$

Since $b = x + y$ in the triangle, then $x = b - y$. Now substitute $b - y$ for x and we get

$$a^2 = c^2 - y^2 + (b - y)^2 \qquad (17–8)$$
$$a^2 = c^2 - y^2 + b^2 - 2by + y^2 = c^2 + b^2 - 2by$$

Using trig functions, $y = c \cos A$. Then

$$a^2 = c^2 + b^2 - 2bc \cos A \qquad (17–9)$$

In a similar manner, we could prove that

(17–10)

$$b^2 = a^2 + c^2 - 2ac \cos B$$

and

(17–11)

$$c^2 = a^2 + b^2 - 2ab \cos C$$

Equations 17–9, 17–10, and 17–11 are called the *law of cosines*. These equations and the law of sines will enable us to solve any triangle for which there is a solution.

EXAMPLE 17–6

Refer to Figure 17–2. Using the law of cosines, find the unknown angles when $a = 400$, $b = 300$, and $c = 250$.

SOLUTION   Let's rearrange Equation 17–9 and solve for angle A.

$$-2bc \cos A = a^2 - b^2 - c^2$$

$$\cos A = \frac{a^2 - b^2 - c^2}{-2bc} = \frac{400^2 - 300^2 - 250^2}{-2 \times 300 \times 250} = -0.050$$

$$\cos^{-1}(-0.05) = A = 92.87°$$

Cos A was negative because the angle is in the second quadrant (greater than 90° but less than 180°). Using the calculator:

TI: 400 $\boxed{x^2}$ $\boxed{-}$ 300 $\boxed{x^2}$ $\boxed{-}$ 250 $\boxed{x^2}$ $\boxed{=}$ $\boxed{\div}$

$\boxed{(}$ 2 $\boxed{+/-}$ $\boxed{\times}$ 300 $\boxed{\times}$ 250 $\boxed{)}$ $\boxed{=}$ $\boxed{\cos^{-1}}$

Answer: A = **92.87°**

For the Casio calculator, the keystrokes are the same except that the minus sign must be entered before the parentheses in the denominator, rather than after the number 2.

Now that we know one angle, we can find the other angles by using either the law of sines or the law of cosines. Using the law of sines is less rigorous, so let's rearrange Equation 17–2 and find angle C.

$$\sin C = \frac{c \sin A}{a} = \frac{250 \sin 92.87°}{400} = \frac{250 \times 0.9988}{400} = 0.6242$$

$$\sin^{-1} 0.6242 = C = 38.62°$$

$$\text{angle } B = 180° - A - C = 180° - 38.62° - 92.87° = 48.51°$$

EXAMPLE 17–7

Refer to Figure 17–2. Find the unknown sides and angles when $b = 60$, $c = 80$, and angle $A = 75°$.

SOLUTION Solve for side a using Equation 17–9.

$$a = \sqrt{b^2 + c^2 - 2bc \cos A}$$
$$= \sqrt{60^2 + 80^2 - 2 \times 60 \times 80 \times \cos 75°}$$
$$= \sqrt{7515} = 86.69$$

Using the calculator,

TI: 60 $\boxed{x^2}$ $\boxed{+}$ 80 $\boxed{x^2}$ $\boxed{-}$ $\boxed{(}$ 2 $\boxed{\times}$ 60 $\boxed{\times}$ 80

$\boxed{\times}$ 75 $\boxed{\cos}$ $\boxed{)}$ $\boxed{=}$ $\boxed{\sqrt{x}}$

Answer: $a = $ **86.69**

For the Casio calculator the keystrokes are similar; the square root key is pressed first and then the three terms are enclosed by parentheses.

Next find angle B using the law of sines.

$$\sin B = \frac{b \sin A}{a} = \frac{60 \sin 75°}{86.69} = 0.6685$$

$$\sin^{-1} 0.6685 = B = 41.95°$$

$$\text{angle } C = 180° - A - B = 180° - 75° - 41.95° = 63.05°$$

PRACTICE PROBLEMS 17–2

Refer to Figure 17–6. Find the missing sides and angles using the law of cosines and the law of sines where:

1. Side $a = 10$, $b = 15$, and $c = 20$
2. Side $a = 12$, $b = 10$, and $c = 15$
3. Side $b = 55$, $c = 70$, and angle $A = 40°$
4. Side $b = 140$, $c = 100$, and angle $A = 85°$
5. Side $a = 30$, $c = 40$, and angle $B = 75°$
6. Side $a = 85$, $c = 115$, and angle $B = 110°$

FIGURE 17–6

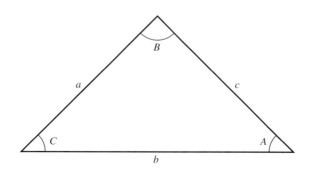

1. $A = 28.96°$, $B = 46.57°$, $C = 104.5°$
2. $A = 52.89°$, $B = 41.65°$, $C = 85.46°$
3. $a = 45.02$, $B = 51.75°$, $C = 88.25°$
4. $a = 164.8$, $B = 57.81°$, $C = 37.19°$
5. $b = 43.35$, $A = 41.95°$, $C = 63.05°$
6. $b = 164.7$, $A = 29.00°$, $C = 41.00°$

Additional practice problems are at the end of the chapter.

SELF-TEST 17–1

Refer to Figure 17–6. Find the unknown sides and angles from the given values, using the law of sines and the law of cosines, where:

1. Side $c = 310$, $A = 55°$, and $C = 80°$
2. Side $a = 65$, $b = 50$, and $A = 100°$
3. Side $a = 50$, $b = 100$, and $c = 60$
4. Side $b = 25$, $c = 15$, and $A = 110°$

Answers to Self-test 17–1 are at the end of the chapter.

CHAPTER 17 AT A GLANCE

PAGE	TRIGONOMETRIC IDENTITIES	EQUATIONS
495	The law of sines can be used when we know: Two angles and any side Or Two sides and an angle opposite one of the known sides.	$\dfrac{\sin A}{a} = \dfrac{\sin B}{b} = \dfrac{\sin C}{c}$ $(\sin 27°)/700 = (\sin 64°)/c$ (Ex. 17–1) $(\sin 125°)/10 = (\sin B)/8$ (Ex. 17–2)
501	The law of cosines must be used when we know: Three sides Or Two sides and the included angle.	$a^2 = c^2 + b^2 - 2bc \cos A$ $b^2 = a^2 + c^2 - 2ac \cos B$ $c^2 = a^2 + b^2 - 2ab \cos C$ $\cos A = (400^2 - 300^2 - 250^2)/$ $(-2 \times 300 \times 250)$ (Ex. 17–6) $a = \sqrt{60^2 + 80^2 - 2 \times 60 \times 80 \times \cos 75°}$ $= 86.69$ (Ex. 17–7)

END OF CHAPTER PROBLEMS 17–1

Refer to Figure 17–6. Find the unknown sides and angles using the law of sines when:

1. Side $c = 35$, $A = 50°$, $C = 70°$
2. Side $a = 7.5$, $A = 25°$, $B = 80°$
3. Side $b = 60$, $c = 100$, $C = 35°$
4. Side $b = 45$, $c = 78$, $C = 110°$
5. Side $a = 15$, $b = 22$, $A = 28°$
6. Side $a = 45$, $b = 60$, $A = 35°$

7. Side $a = 10, A = 30°, B = 70°$
8. Side $a = 300, A = 115°, B = 25°$
9. Side $b = 160, c = 100, C = 22°$
10. Side $b = 17, c = 12, C = 44°$
11. Side $a = 10, b = 20, A = 25°$
12. Side $a = 40, b = 70, A = 30°$
13. Side $b = 55, B = 50°, C = 105°$
14. Side $b = 20, B = 60°, C = 105°$
15. Side $a = 120, b = 65, A = 95°$
16. Side $a = 170, b = 210, A = 40°$
17. Side $b = 6, c = 4.7, C = 34°$
18. Side $b = 95, c = 70, C = 15°$
19. Side $c = 14, A = 125°, C = 45°$
20. Side $c = 65, A = 25°, C = 40°$
21. Side $c = 350, A = 30°, C = 110°$
22. Side $c = 38, A = 105°, C = 32°$
23. Side $b = 40, c = 65, B = 28°$
24. Side $b = 650, c = 900, B = 42°$
25. Side $b = 350, c = 275, B = 100°$
26. Side $b = 55, c = 40, B = 80°$
27. Side $b = 17, B = 95°, C = 22°$
28. Side $b = 40, B = 60°, C = 60°$
29. Side $b = 95, B = 20°, C = 20°$
30. Side $b = 200, B = 80°, C = 45°$
31. Side $b = 250, c = 200, B = 70°$
32. Side $b = 70, c = 40, B = 105°$
33. Side $b = 150, c = 100, B = 120°$
34. Side $b = 16, c = 12, B = 75°$
35. Side $b = 20, c = 33, B = 33°$
36. Side $b = 70, c = 100, B = 33°$
37. Side $c = 110, A = 47°, C = 95°$
38. Side $c = 220, A = 25°, C = 100°$
39. Side $c = 4.3, A = 75°, C = 45°$
40. Side $c = 150, A = 35°, C = 70°$

END OF CHAPTER PROBLEMS 17–2

Refer to Figure 17–2. From the given information, find the unknown sides and angles using the law of cosines and the law of sines when:

1. Side $a = 70, b = 130, c = 90$
2. Side $a = 100, b = 200, c = 200$
3. Side $a = 250, b = 100, c = 200$
4. Side $a = 14, b = 14, c = 20$
5. Side $b = 60, c = 80, A = 95°$
6. Side $b = 45, c = 100, A = 80°$
7. Side $b = 12, c = 36, A = 80°$
8. Side $b = 20, c = 70, A = 102°$
9. Side $a = 500, c = 400, B = 100°$
10. Side $a = 130, c = 95, B = 55°$
11. Side $a = 400, b = 300, c = 150$
12. Side $a = 500, b = 350, c = 200$
13. Side $a = 35, b = 40, c = 25$
14. Side $a = 8, b = 9, c = 6$
15. Side $a = 8.4, c = 10, B = 30°$
16. Side $a = 14, c = 12, B = 40°$
17. Side $a = 40, c = 62, B = 105°$
18. Side $a = 90, c = 110, B = 65°$
19. Side $b = 600, c = 400, A = 95°$
20. Side $b = 120, c = 160, A = 60°$
21. Side $b = 10, c = 15, A = 120°$
22. Side $b = 35, c = 55, A = 63°$
23. Side $a = 45, b = 45, c = 60$
24. Side $a = 125, b = 105, c = 140$
25. Side $a = 170, b = 140, c = 190$
26. Side $a = 50, b = 35, c = 20$
27. Side $a = 425, b = 525, c = 300$
28. Side $a = 27, b = 40, c = 40$
29. Side $a = 5.35, b = 8.5, c = 8.5$
30. Side $a = 1100, b = 950, c = 1400$

ANSWERS TO SELF-TEST 17–1

1. Side $a = 257.9, b = 222.6,$
   $B = 45°$
2. Side $c = 33.75, B = 49.25°,$
   $C = 30.75°$
3. Angle $A = 22.33°,$
   $B = 130.5°, C = 27.13°$
4. Side $a = 33.26, B = 44.93°,$
   $C = 25.07°$

ac Fundamentals

18

Introduction

An automobile or an airplane traveling in a particular direction has a *linear velocity* measured in miles per hour or kilometers per hour. Other moving objects may have a linear velocity measured in feet per second or centimeters per second. This linear velocity is commonly referred to as the *speed* at which the object is traveling.

Electrical generators have *angular* motion or *angular velocity*. The speed of rotation of the armature of a generator, the angular velocity, is measured in revolutions per minute or, more commonly, in revolutions per second (cycles per second). We can also measure angular velocity in degrees per second or in radians per second. The measurement of angular velocity in degrees or radians per second is necessary in electronics for us to understand the characteristics of alternating voltages and currents.

When alternating values of current and voltage are generated, the value of current or voltage changes sinusoidally. That is, the variation has the characteristics of a sine wave, and we can determine the value of current or voltage at any instant by using the sine function. Since the sine wave repeats itself periodically, we can easily compute the angular velocity.

Because all generated waveforms in electronics are basically sine waves, an understanding of how to compute angular velocity, how to relate time to frequency, and how to distinguish among rms, peak, and instantaneous values of current and voltage is a necessary part of our studies.

Chapter Objectives

In this chapter you will learn how to:

1. Identify the parts of a system of rectangular coordinates.
2. Generate angles on a system of rectangular coordinates.
3. Convert angles from degrees to radians and from radians to degrees.
4. Calculate angular velocity.
5. Calculate instantaneous values of voltage and current.
6. Convert between effective, peak, and peak-to-peak values of voltage and current.
7. Determine instantaneous values of voltage and current when there is an angle of lead or lag.

18-1-1 Generating Angles

To develop an understanding of ac concepts, we need to discuss some basic mathematical and physical ideas. In Figure 18–1, two lines are drawn at right angles to each other. The lines are xx_1 and yy_1. Such a figure has been defined as a system of rectangular coordinates. The origin is the point where the lines cross. A line extending outward from the origin is called a *vector*. Recall from previous chapters that we call a rotating vector a *phasor*. Hereafter, we will use the word *phasor* whenever we are discussing rotating vectors.

Let's consider a line or phasor that extends along the x-axis as in Figure 18–2. If the phasor were rotated either clockwise or counterclockwise until it were again along the x-axis, the phasor would have gone through 360° or a complete circle. It is standard practice to consider the x-axis as 0° of the circle and to rotate the phasor counterclockwise. The y-axis is displaced 90° from the x-axis, the x_1-axis is displaced 180° from the x-axis, and the y_1-axis is displaced 270°. Thus, counterclockwise rotation is considered positive rotation. Clockwise rotation is considered negative rotation. From this we can say that y_1 is displaced 270° or $-90°$ from the x-axis or 0°.

An angle is in its standard position when its initial side is on the positive x-axis and its vertex is at the origin. The x-y coordinates divide the system into fourths or quadrants, as labeled in Figure 18–1. Note that in the first quadrant both x and y are positive. In the second quadrant, x is negative and y is positive. In the third quadrant, both x and y are negative, and in the fourth quadrant, x is positive and y is negative.

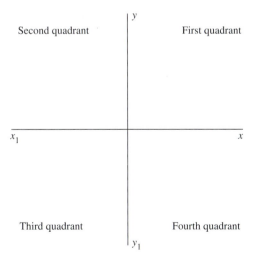

FIGURE 18–1   System of rectangular coordinates.

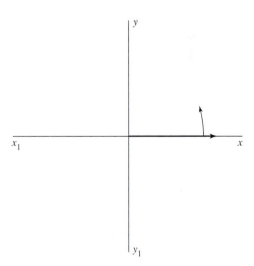

FIGURE 18–2   System of rectangular coordinates with phasor drawn along the x-axis.

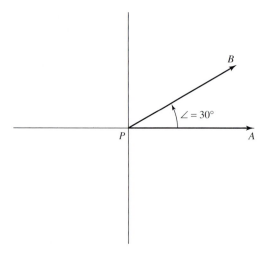

FIGURE 18–3   Phasor rotated through 30°.

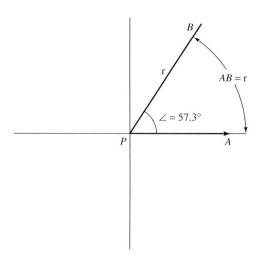

FIGURE 18–4   Phasor rotated through 57.3° or 1 radian.

In Figure 18–3 a phasor has been rotated through 30°. The angle is formed by first extending a line from point *P*, the origin, to point *A*. In this position the line is in its *initial* position. We then rotate the line through 30° to point *B*. The line is now in its *terminal* position. Since the resulting angle has its vertex at the origin, the phasor has a length equal to the radius of the circle. Notice that we rotated the line counterclockwise.

Consider Figure 18–4. The line from point *P* to any point on the circumference is the radius. Let's rotate the line again.

☞ **KEY POINT**   When the length of arc *AB* equals the length of the radius, the angle is equal to 1 *radian*. In degrees, this angle equals 57.3°. So

(18–1)
$$1 \text{ radian} = 57.3°$$

In our studies of ac circuits, we will measure angles in either degrees or radians.[1]

If the line in Figure 18–4 were rotated through 360°, a complete circle would be drawn, as illustrated in Figure 18–5. From such a circle, the relationship between degrees and radians can be determined. In a circle there are 360° or 6.28 radians. It follows then that there are 3.14 radians in 180°. The greek letter π (*pi*) is used to denote 3.14 rad. π radians = 3.14 radians.

(18–2)
$$180° = \pi \text{ radians}$$

_____

[1]In some professions the angle is also measured in *grads*. A grad is $\frac{1}{100}$ of a right angle, or, said another way, 90° = 100 grads. We will use only degrees and radians in our discussions of angles.

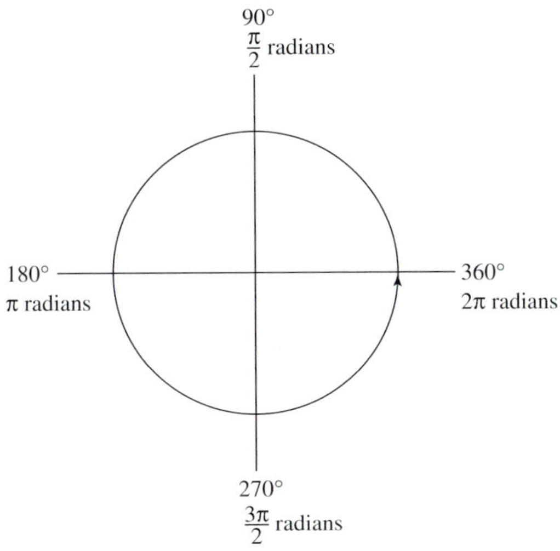

FIGURE 18–5
Phasor rotated
through 360° or
2π radians.

We determined previously that there are 57.3° in 1 radian. Rearranging Equation 18–2 is another way of finding the number of degrees in 1 radian.

$$1 \text{ rad} = \frac{180°}{\pi \text{ rad}} = 57.3° \qquad (18\text{–}3)$$

Since we can use Equation 18–3 to find the number of degrees in 1 radian of rotation, then, to convert from radians to degrees for any angle, we get

$$\text{angle in degrees} = \frac{180°}{\pi \text{ rad}} \times \text{number of radians} \qquad (18\text{–}4)$$

and to convert from degrees to radians,

$$\text{angle in radians} = \frac{\pi \text{ rad}}{180°} \times \text{number of degrees} \qquad (18\text{–}5)$$

EXAMPLE 18–1

Convert an angle of 85° to radians.

SOLUTION To convert from degrees to radians, we use Equation 18–5.

$$\text{angle in radians} = \frac{\pi}{180°} \times \text{angle in degrees} \qquad (18\text{–}5)$$

$$= \frac{\pi}{180°} \times 85° = 1.48 \text{ rad}$$

Most calculators will convert between degrees, radians, and grads by activating the DRG key. With the calculator in degrees mode and 85 displayed (85°), change to radians mode using the DRG key. 1.48 should be displayed. Consult the user's manual for the key combination needed by your calculator.

EXAMPLE 18–2

Convert an angle of 3.42 rad to degrees.

SOLUTION To convert from radians to degrees, use Equation 18–4.

(18–4)
$$\text{angle in degrees} = \frac{180°}{\pi} \times \text{angle in radians}$$

$$= \frac{180°}{\pi} \times 3.42 = 196°$$

18-1-2 Generating Sine Waves

Let's consider the system of rectangular coordinates shown in Figure 18–6(a). If we start at 0° (the positive x-axis) and rotate in a counterclockwise direction, we will have drawn a complete circle. This rotation could represent one complete cycle of a generated sine wave of voltage or current. The radius of the circle could equal the maximum or peak value of the sine wave. Suppose this represents a voltage waveform whose maximum voltage is 100 V. $V = 100\ V_{pk}$. As we rotate our phasor from 0°, the radius V_{pk} is the hypotenuse of a right triangle. The vertical component, v, falls along the y-axis and is the instantaneous value of voltage.

FIGURE 18–6
(a) V_{pk} equals the radius of the circle and is the hypotenuse of a right triangle;
(b) Graph of the circle created in (a) by rotating the phasor through 360°.

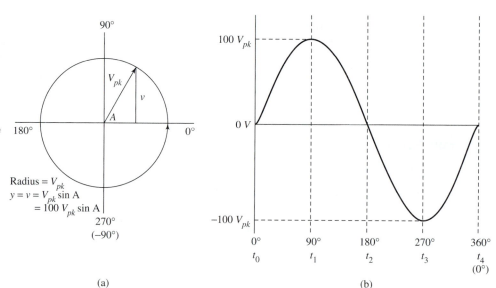

(a)

(b)

The instantaneous voltage is increasing from 0 V at 0° to 100 V at 90°. This is shown graphically in Figure 18–6(b). t_0 on the graph corresponds to 0° on our system of rectangular coordinates. As we increase to 90° ($t = t_1$), we can calculate the voltage generated at any instant by using the equation $v = V_{pk} \sin \angle A$. (Lowercase v and i are used to denote instantaneous values of voltage and current.)

Continuing on from 90° to 180° (t_2), we could calculate instantaneous values and plot our graph using our equation for v ($v = V_{pk} \sin A$). At this point we have completed one half-cycle or one-half revolution and are back to 0 V. As we continue on past 180°, and using the equation $v = V_{pk} \sin \angle A$, v is negative because $\sin \angle A$ is negative. We can also show graphically why this is so. Look at the angles generated in Figure 18–7. Remember that

$$\sin A = \frac{\text{opposite side}}{\text{hypotenuse}}$$

The side opposite the angle A is the value in the y direction, which we have labeled v. The hypotenuse is the radius vector, which has been labeled V_{pk}. The radius phasor is always considered to be positive. The opposite side, v, is positive in the first and second quadrants because the y-axis is positive. When we are in the third or fourth quadrant, the y-axis is negative; therefore, v is negative.

A look at our graph in Figure 18–6(b) shows that, as we complete the last half-cycle, v goes from 0 to -100 V as we rotate from 180° to 270°. v changes from a value equal to -100 V to 0 V as we rotate from 270° back to 0°. When ac voltages and currents are generated, this is the waveform produced. This waveform results from the constant rotation of a generator or from the natural oscillation of an electronics circuit. This sine curve could also represent the back and forth movement of a pendulum or the ripple created when a rock is dropped into still water.

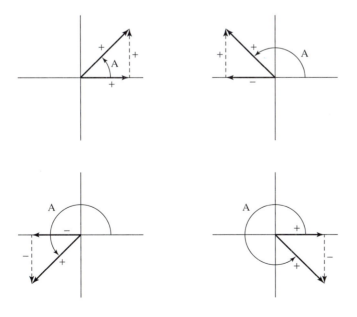

FIGURE 18–7
Position of phasor at various times.

In Figure 18–6(b), at time zero (t_0) the amplitude of the waveform is zero. Notice that time is plotted along the x-axis and amplitude is plotted along the y-axis. This is standard notation; t_0 on our graph corresponds to 0° on our system of rectangular coordinates. The maximum amplitude in the y direction corresponds to the radius of our circle.

18-1-3 Frequency and Time

The number of times per second that we generate the sine wave is called *frequency*. The time it takes to generate one complete sine wave is called the *period*. We can express the relationship between frequency and time as an equation:

(18–6)
$$f = \frac{1}{T}$$

where f is the frequency in cycles per second and T is the period in seconds. Frequency and time are reciprocals of each other. The unit of frequency is the *hertz*. One hertz (Hz) equals one cycle per second. The unit of time is the second (s).

EXAMPLE 18–3

If the frequency of rotation is 250 Hz, what is the period?

SOLUTION   Solve for T:

(18–6)
$$f = \frac{1}{T}$$

(18–7)
$$T = \frac{1}{f}$$

$$T = \frac{1}{250 \text{ Hz}} = 4 \text{ ms}$$

If there are 250 cycles per second (250 Hz), the time for one cycle (the period) is 4 ms.

EXAMPLE 18–4

If it takes 100 μS to complete one cycle, what is the frequency?

SOLUTION

$$f = \frac{1}{T}$$

$$f = \frac{1}{100 \ \mu s} = 10 \text{ kHz}$$

If the period is 100 μs, we will generate 10,000 cycles in one second, or 10 kHz.

When using the calculator, conversions between time and frequency are easily solved using the reciprocal key.

1. $f = 15$ kHz. Find T.
2. $f = 500$ Hz. Find T.
3. $T = 400$ μs. Find f.
4. $T = 16.7$ ms. Find f.
5. Convert $145°$ to radians.
6. Convert $300°$ to radians.
7. Convert 2 rad to degrees.
8. Convert 5.73 rad to degrees.

SOLUTIONS

1. 66.7 μs
2. 2 ms
3. 2.5 kHz
4. 59.9 Hz
5. 2.53 rad
6. 5.24 rad
7. 115°
8. 328°

Additional practice problems are at the end of the chapter.

ANGULAR VELOCITY 18–2

Let's generate an angle by rotating our radius phasor in Figure 18–3 at a constant speed. In angular motion this speed is called *angular velocity*. Angular velocity is measured in radians per second (rad/s) or in degrees per second. The symbol for angular velocity is ω (the Greek letter *omega*). In radian measure, ω equals 2π times the frequency

$$\omega = 2\pi \, \frac{\text{radians}}{\text{cycle}} \, f \, \frac{\text{cycles}}{\text{second}} = 2\pi f \frac{\text{radians}}{\text{second}} \qquad (18\text{–}8)$$

If the angle is measured in degrees, ω equals $360°$ times the frequency.

$$\omega = 360 \, \frac{\text{degrees}}{\text{cycle}} \, f \, \frac{\text{cycles}}{\text{second}} = 360f \frac{\text{degrees}}{\text{second}} \qquad (18\text{–}9)$$

Angular velocity is more commonly measured in radians per second.

When the frequency is 1 Hz (one cycle per second), the angular velocity is 2π rad/s or $360°$/s. If the frequency is 10 Hz, the angular velocity, in radians or degrees, is

$$\omega = 2\pi f$$

$$\omega = 2\pi \times 10 \text{ Hz} = 20\pi \text{ rad/s}$$

$$= 62.8 \text{ rad/s}$$

or $\qquad \omega = 360° \, f = 360° \times 10 \text{ Hz} = 3600°/\text{s}$

The total angle swept by a phasor in a given time (t) is ωt. In radian measure, the equation would be

(18–10)

$$\omega t = 2\pi \frac{\text{radians}}{\text{cycle}} f \frac{\text{cycles}}{\text{second}} t \text{ seconds} = 2\pi ft \text{ radians}$$

In degrees, the equation would be

(18–11)

$$\omega t = 360 \frac{\text{degrees}}{\text{cycle}} f \frac{\text{cycles}}{\text{second}} = t \text{ seconds} = 360 ft \text{ degrees}$$

EXAMPLE 18–5

Find the angle generated after 5 ms when the frequency is 500 Hz. Express your answer in degrees and radians.

SOLUTION

(18–10)

$$\omega t = 2\pi ft$$
$$= 2\pi \times 500 \times 5 \text{ ms} = 15.7 \text{ rad}$$
$$= 57.3° \times 15.7 \text{ rad} = 900°$$

EXAMPLE 18–6

Find the frequency if an angle of 2.35 rad is generated in 150 μs.

SOLUTION

(18–10)

$$\omega t = 2\pi ft$$
$$f = \frac{\omega t}{2\pi t} = \frac{2.35 \text{ rad}}{2\pi \times 150 \text{ }\mu s} = 2.49 \text{ kHz}$$

EXAMPLE 18–7

How much time is required to generate an angle of 75° when the frequency is 1 kHz?

SOLUTION

(18–11)

$$\omega t = 360° ft$$
$$t = \frac{\omega t}{360° f} = \frac{75°}{360 \times 1 \text{ kHz}} = 208 \text{ }\mu s$$

1. If the frequency is 2.5 kHz, find the angle in both degrees and radians after (a) 50 μs, (b) 100 μs, (c) 200 μs, and (d) 1 ms.
3. What is the frequency if an angle of 0.56 rad is generated in 55 μs?
5. How much time is required to generate an angle of 1.7 rad when the frequency is 25 kHz?

2. Find the frequency if the angular velocity is (a) 670 rad/s, (b) 2000 rad/s, (c) 7200°/s, and (d) 12,000°/s.
4. What is the frequency if an angle of 285° is generated in 200 μs?
6. How much time is required to generate an angle of 400° when the frequency is 1.5 kHz?

SOLUTIONS

1. (a) 0.785 rad or 45°   (b) 1.57 rad or 90°
   (c) 3.14 rad or 180°   (d) 15.7 rad or 900°
3. 1.62 kHz
5. 10.8 μs

2. (a) 107 Hz   (b) 318 Hz
   (c) 20 Hz   (d) 33.3 Hz

4. 3.96 kHz
6. 741 μs

Additional practice problems are at the end of the chapter.

SELF-TEST 18–1

1. $f = 50$ kHz. Find T.
3. Convert 0.628 rad to degrees.

2. Convert 160° to radians.
4. If the frequency is 400 Hz, find the angle generated after 750 μs. Express your answer in both degrees and radians.

5. What is the frequency if an angle of 120° is generated in 200 μs?

6. How much time is required to generate an angle of 1.3 rad when the frequency is 5 kHz?

Answers to Self-test 18–1 are at the end of the chapter.

INSTANTANEOUS VALUES OF VOLTAGE AND CURRENT   18–3

In Figure 18–8, at t_0 the voltage is 0 V and at 90° the voltage is at its maximum value, which, in this example, is 100 V. If the frequency is 1 kHz, how long does it take to

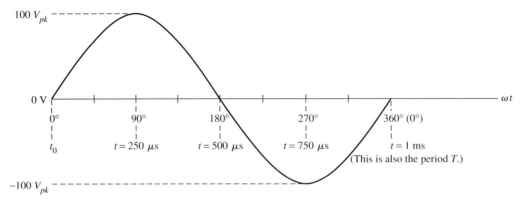

FIGURE 18-8   Sine wave of voltage.

reach this 100 V? To solve the problem, the first thing we have to do is find the period, the time it takes to complete one cycle:

$$(18\text{--}7) \qquad T = \frac{1}{f}$$

$$T = \frac{1}{1 \text{ kHz}} = 1 \text{ ms}$$

Since 90° is one-quarter of a complete cycle; then it must take one-quarter the time for a complete cycle to reach 90°:

$$\text{time to complete } 90° = \frac{\text{period}}{4} = \frac{1 \text{ ms}}{4} = 250 \ \mu s$$

During the time from t_0 to $t = 250 \ \mu s$, V is increasing from 0 to 100 V. From $t = 250 \ \mu s$ to $t = 750 \ \mu s$, the amplitude is changing from $+100$ to -100 V. The voltage is 0 V as we pass through 180°, and $t = T/2$ or 500 μs. From $t = 750 \ \mu s$, the amplitude is changing from -100 to $+100$ V. Zero volts is reached when we have completed the cycle, which is 360° or 1 ms. This brings us back to 0°. The cycle then repeats itself 1000 times per second. This change in voltage is not a linear change but a *sinusoidal* change. That is, the amplitude at any instant is a function of the sine of the generated angle. We have already determined that V_{pk} is the hypotenuse of a right triangle, and the opposite side is the magnitude of the voltage at some instant in time (see Figure 18-9). If we display the sine wave on a system of rectangular coordinates, the resulting equation for finding v at any instant is

$$(18\text{--}12) \qquad v = V_{pk} \sin \omega t$$

When working with an emf, our equation would be

$$(18\text{--}13) \qquad e = E_{pk} \sin \omega t$$

The equation for instantaneous circuit current would be

$$(18\text{--}14) \qquad i = I_{pk} \sin \omega t$$

Before starting any problem that involves trigonometric functions, you must decide if you want to use the radian or degree mode of the calculator. If all angles are given in

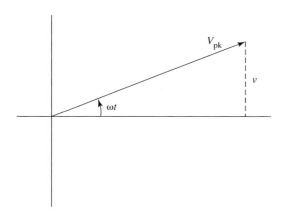

FIGURE 18–9
Diagram to
show how the
instantaneous
voltage (v)
depends on the
peak voltage
(V_{pk}) and the
angle (ωt) of the
sine wave.

$$\sin \omega t = v/V_{pk}$$

degrees, then set the calculator to the degree mode. If all angles are given in radians, then set the calculator to the radian mode. If angles are given in both degrees and radians, then you must decide what calculator mode will be easier and convert all angles to the mode you select.

EXAMPLE 18–8

$e = 100V_{pk} \sin \omega t$. If $f = 1\,\text{kHz}$, find e after   (a) $50\,\mu s$,   (b) $250\,\mu s$, and   (c) $600\,\mu s$.

SOLUTION

$$e = E_{pk} \sin \omega t \qquad\qquad (18\text{–}13)$$

$$= 100\,V_{pk} \sin 2\pi f t$$

(a) $\qquad\qquad e = 100\,V_{pk} \sin (2\pi \times 1\,\text{kHz} \times 50\,\mu s)$

$$= 100\,V_{pk} \sin 0.314\,\text{rad}$$

$$= 100\,V_{pk} \times 0.309 = 30.9\,V_t$$

We will use the subscript "t" to remind us that we found an instantaneous value of voltage (or current). $e = 30.9\,V_t$.

When we use $\omega = 2\pi f$, the calculator must be in the radians mode to find $\sin \omega t$; then

TI: $100\ \boxed{\times}\ \boxed{(}\ 2\ \boxed{\times}\ \boxed{\pi}\ \boxed{\times}\ 1\ \boxed{\text{EE}}\ 3\ \boxed{\times}$

$50\ \boxed{\text{EE}}\ \boxed{+/-}\ 6\ \boxed{)}\ \boxed{\sin}\ \boxed{=}$ Answer: $e = 30.9\,V_t$

Casio: $100\ \boxed{\sin}\ \boxed{(}\ 2\ \boxed{\pi}\ \boxed{\times}\ 1\ \boxed{\text{EXP}}\ 3\ \boxed{\times}\ 50$

$\boxed{\text{EXP}}\ \boxed{(-)}\ 6\ \boxed{)}\ \boxed{=}$

Notice that in all three problems, (a), (b), and (c), $2\pi f$ is the same. When this happens, we can save keystrokes by storing that product before we start.

TI: 2 $\boxed{\times}$ $\boxed{\pi}$ $\boxed{\times}$ I $\boxed{\text{EE}}$ 3 $\boxed{=}$ $\boxed{\text{STO}}$ I

Answer: $2\pi f = 6.28$ kHz

Casio: 2 $\boxed{\pi}$ $\boxed{\times}$ I $\boxed{\text{EXP}}$ 3 $\boxed{\text{STO}}$ $\boxed{\text{A}}$

Now we solve the problem:

TI: 100 $\boxed{\times}$ $\boxed{(}$ $\boxed{\text{RCL}}$ I $\boxed{\times}$ 50 $\boxed{\text{EE}}$ $\boxed{+/-}$ 6 $\boxed{)}$

$\boxed{\sin}$ $\boxed{=}$ Answer: $e = 30.9\,V_t$

Casio: 100 $\boxed{\sin}$ $\boxed{(}$ $\boxed{\text{RCL}}$ $\boxed{\text{A}}$ $\boxed{\times}$ 50 $\boxed{\text{EXP}}$ $\boxed{(-)}$

6 $\boxed{)}$ $\boxed{=}$

(b) $e = 100\ V_{pk} \sin (2\pi \times 1 \text{ kHz} \times 250\ \mu s)$

TI: 100 $\boxed{\times}$ $\boxed{(}$ $\boxed{\text{RCL}}$ I $\boxed{\times}$ 250 $\boxed{\text{EE}}$ $\boxed{+/-}$ 6 $\boxed{)}$

$\boxed{\sin}$ $\boxed{=}$ Answer: $e = 100\,V_t$

Casio: 100 $\boxed{\sin}$ $\boxed{(}$ $\boxed{\text{RCL}}$ $\boxed{\text{A}}$ $\boxed{\times}$ 250 $\boxed{\text{EXP}}$ $\boxed{(-)}$

6 $\boxed{)}$ $\boxed{=}$

We had determined previously that e would equal E_{pk} 250 μs after t_0.

(c) $e = 100\ V_{pk} \sin (2\pi \times 1 \text{ kHz} \times 600\ \mu s)$

TI: 100 $\boxed{\times}$ $\boxed{(}$ $\boxed{\text{RCL}}$ I $\boxed{\times}$ 600 $\boxed{\text{EE}}$ $\boxed{+/-}$ 6 $\boxed{)}$

$\boxed{\sin}$ $\boxed{=}$ Answer: $e = -58.8\,V_t$

Casio: 100 $\boxed{\sin}$ $\boxed{(}$ $\boxed{\text{RCL}}$ $\boxed{\text{A}}$ $\boxed{\times}$ 600 $\boxed{\text{EXP}}$ $\boxed{(-)}$

6 $\boxed{)}$ $\boxed{=}$

e is negative because 600 μs puts us in the third quadrant.

PRACTICE PROBLEMS 18-3

1. $V = 60 \text{ V}_{\text{pk}}$, $f = 500$ Hz. Find v after
   (a) 100 μs, (b) 250 μs, (c) 500 μs,
   (d) 800 μs, (e) 1.2 ms, and (f) 1.6 ms.

2. $I = 30 \text{ mA}_{\text{pk}}$, $f = 250$ Hz. Find i after
   (a) 400 μs, (b) 1 ms, (c) 1.7 ms,
   (d) 2.1 ms, (e) 3 ms, and (f) 11.3 ms.

SOLUTIONS

$$e = 60 \text{ V}_{\text{pk}} \sin (2\pi \times 500 \times t)$$

$$i = 30 \text{ mA}_{\text{pk}} \sin (2\pi \times 500 \times t)$$

1. (a) 18.5 V_t; (b) 42.4 V_t; (c) 60.0 V_t;
   (d) 35.3 V_t; (e) -35.3 V_t; (f) -57.1 V_t

2. (a) 17.6 mA_t; (b) 30.0 mA_t;
   (c) 13.6 mA_t; (d) -4.69 mA_t;
   (e) -30.0 mA_t; (f) -26.7 mA_t

Additional practice problems are at the end of the chapter.

RMS VALUES OF VOLTAGE AND CURRENT   18-4

In most AC circuits, the electrical power is of more interest than the peak voltage or peak current. For example, all voltages and currents related to household appliances are measured and specified according to their *rms* values, not their peak values (120 volt residential service refers to the *rms* voltage supplied to homes in the United States). *rms* stands for "Root-Mean-Squared" and its mathematical value can be derived by taking the square root of the average (mean) of the squares of the instantaneous values. Almost all AC test equipment measures the *rms* values of voltages and currents; only special test equipment can measure the peak values. When peak AC voltages or currents are specified, we always use a subscript, such as "pk", to distinguish the value from the *rms* value. When there is no subscript, we can assume the voltage or current is the *rms* value.

In determining currents and voltage drops in ac circuits, *rms* values of current and voltage are used. The rms value of a current or voltage is the value that converts the same energy as does a dc value. For example, in a dc circuit, 100 mA flows through a resistance of 10 Ω. The power dissipated is 100 mW. For an ac current to dissipate 100 mW in that same resistor, the rms current would also have to be 100 mA.

The rms value of an ac voltage or current is found by taking the square root of the average (mean) of the squares of the instantaneous values. Because of the method of calculation, this value is also called the *effective value*. This rms or effective value is the same for all sine waves and equals 0.707 of the maximum or peak value.

$$I_{\text{rms}} = I_{\text{eff}} = 0.707 \, I_{\text{pk}} = \frac{I_{\text{pk}}}{\sqrt{2}} \qquad (18\text{–}15)$$

$$V_{\text{rms}} = V_{\text{eff}} = 0.707 \, V_{\text{pk}} \qquad (18\text{–}16)$$

If we solve for peak values, we get

$$(18\text{--}17) \qquad I_{pk} = 1.414\ I_{rms} = \sqrt{2}\ I_{rms}$$

$$(18\text{--}18) \qquad V_{pk} = 1.414\ V_{rms}$$

Since peak-to-peak values of voltage and current equal twice the peak values ($V_{p\text{-}p} = 2\ V_{pk}$), then

$$(18\text{--}19) \qquad V_{p\text{-}p} = 2.828\ V_{rms} = 2\sqrt{2}\ V_{rms}$$

$$(18\text{--}20) \qquad I_{p\text{-}p} = 2.828\ I_{rms}$$

Solving for effective values, we get

$$V_{eff} = V_{rms} = 0.3535\ V_{p\text{-}p} = \frac{V_{p\text{-}p}}{2\sqrt{2}}$$

$$I_{eff} = I_{rms} = 0.3535\ I_{p\text{-}p}$$

When ac voltages or currents are given, values are always rms values. Thus, if we see an ac voltage written as 120 V, we know that it is 120 V_{rms}. If the voltage or current is either a peak or a peak-to-peak value, the proper subscript must be used. Therefore, ten volts rms is written 10 V. Ten volts peak is written 10 V_{pk}. Ten volts peak-to-peak is written 10 $V_{p\text{-}p}$.

EXAMPLE 18–9

Convert 12 V_{pk} to rms and p-p.

SOLUTION

$$V_{rms} = 0.707\ V_{pk}$$

$$V_{rms} = 0.707 \times 12\ V_{pk} = 8.48\ V$$

$$V_{p\text{-}p} = 2\ V_{pk}$$

$$V_{p\text{-}p} = 2 \times 12\ V_{pk} = 24\ V_{p\text{-}p}$$

EXAMPLE 18–10

Convert 20 mA to pk and p-p.

SOLUTION

$$I_{pk} = 1.414\ I_{rms}$$

$$= 1.414 \times 20\ mA = 28.3\ mA_{pk}$$

$$I_{p\text{-}p} = 2.828\ I_{rms}$$

$$= 2.828 \times 20\ mA = 56.6\ mA_{p\text{-}p}$$

EXAMPLE 18-11

Convert 20 V_{p-p} to rms and pk.

SOLUTION

$$V_{rms} = 0.3535\ V_{p-p}$$
$$= 0.3535 \times 20\ V_{p-p} = 7.07\ V$$

$$V_{pk} = \frac{V_{p-p}}{2}$$
$$= \frac{20\ V_{p-p}}{2} = 10\ V$$

PRACTICE PROBLEMS 18-4

1. An ac voltage is 30 V. Convert to pk and p-p.
2. An ac current is 150 mA. Convert to pk and p-p.
3. An ac voltage is 50 V_{pk}. Convert to p-p and rms.
4. An ac current is 70 mA_{pk}. Convert to p-p and rms.
5. An ac voltage is 200 V_{p-p}. Convert to pk and rms.
6. An ac current is 1.5 A_{p-p}. Convert to pk and rms.

SOLUTIONS

1. 30 V = 42.4 V_{pk} = 84.9 V_{p-p}
3. 50 V_{pk} = 100 V_{p-p} = 35.4 V
5. 200 V_{p-p} = 100 V_{pk} = 70.7 V

2. 150 mA = 212 mA_{pk} = 424 mA_{p-p}
4. 70 mA_{pk} = 140 mA_{p-p} = 49.5 mA
6. 1.5 A_{p-p} = 750 mA_{pk} = 530 mA

Additional practice problems are at the end of the chapter.

ANGLE OF LEAD OR LAG   18-5

When we have a purely resistive ac circuit, the applied voltage and circuit current equal zero at the same time. They also reach their peak values at the same time. In such a circuit we say that the voltage and current are *in phase*. The voltage drops across the various resistances are also in phase with the current and applied voltage.

> ☞ **KEY POINT** Whenever we have capacitance or inductance in an ac circuit, the current and voltage are *out of phase*. When the phase angle, *theta* (θ), is known, we can calculate the instantaneous values of current and voltage.

The applied voltage either leads or lags the current. If the applied voltage *leads* the current, this means that E reaches its values of 0 V and E_{pk} at an earlier time than does the current. The phase difference can be anywhere from 0° to 90°.

Because the applied voltage is common to each branch, *voltage is the reference in parallel circuits.* In parallel circuits, then, the current either leads or lags the voltage depending on whether the circuit is resistive and capacitive or resistive and inductive. The general equations for instantaneous values of current and voltage in parallel circuits are

(18–21)
$$e = E_{pk} \sin \omega t$$

(18–22)
$$i = I_{pk} \sin (\omega t \pm \theta)$$

Use + if current leads voltage in a parallel circuit. Use − if current lags voltage in a parallel circuit.

We use the greek letter theta (θ) to denote phase angle.

Because current is the same throughout the circuit, *current is the reference in series circuits.* In series circuits, then, the voltage either leads or lags the current. The general equations for instantaneous values of current and voltage in series circuits are

(18–23)
$$i = I_{pk} \sin \omega t$$

(18–24)
$$e = E_{pk} \sin (\omega t \pm \theta)$$

Use + if voltage leads current in a series circuit. Use − if voltage lags current in a series circuit.

Suppose that E leads I by 30° in a series RL circuit as illustrated in Figure 18–10. We could design and construct such a circuit and then look at the current and voltage waveforms on a dual-trace oscilloscope. If we properly calibrate the time base, we can measure the phase angle by noting the time difference between E and I. Because this is a series circuit, current is the reference. The equations for E and I at any instant are

(18–25)
$$i = I_{pk} \sin \omega t$$

(18–26)
$$e = E_{pk} \sin (\omega t \pm \theta)$$

$$e = E_{pk} \sin (\omega t + 30°)$$

FIGURE 18–10
Sine wave of voltage and current. Voltage leads by some angle θ.

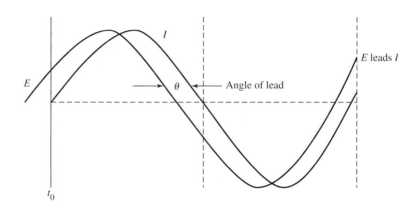

EXAMPLE 18–12

$I = 50$ mA$_{pk}$, $E = 100$ V$_{pk}$, and the frequency is 2 kHz. Using current as the reference, compute the instantaneous values of i and e, 40 μs after t_0 when I leads E by 60°.

FIGURE 18–11

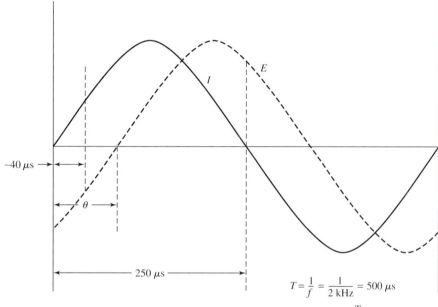

$$T = \frac{1}{f} = \frac{1}{2 \text{ kHz}} = 500 \ \mu s$$

$$\text{Time for } 180° = \frac{T}{2} = 250 \ \mu s$$

SOLUTION   If we draw the waveforms as in Figure 18–11 with I leading E by 60°, we see that at 40 μs I will be positive and close to one-half its maximum value, or approximately 25 mA. E is negative and moving in a positive direction. Let's estimate its value at about −40 V. (Our freehand graph isn't accurate, but it gets us in the ball park and gives us a visual display of the conditions at a specific time.)

$$\omega t = 360° ft = 360 \times 2 \text{ kHz} \times 40 \ \mu s = 28.8°$$

TI: 360 $\boxed{\times}$ 2 $\boxed{\text{EE}}$ 3 $\boxed{\times}$ 40 $\boxed{\text{EE}}$ $\boxed{\pm}$ 6 $\boxed{=}$

Answer: $\omega t = 28.8°$

Casio: Use $\boxed{\text{EXP}}$ instead of $\boxed{\text{EE}}$.

We found ωt in degrees because the phase angle was given in degrees. Knowing ωt, we can find the instantaneous current using Equation 18–25.

$$i = I_{pk} \sin \omega t = 50 \text{ mA}_{pk} \sin 28.8° = 24.1 \text{ mA}_t$$

TI: 50 $\boxed{\text{EE}}$ $\boxed{\pm}$ 3 $\boxed{\times}$ 28.8 $\boxed{\sin}$ $\boxed{=}$

Answer: $i = 24.1$ mA$_t$

Casio: 50 $\boxed{\text{EXP}}$ $\boxed{(-)}$ 3 $\boxed{\times}$ $\boxed{\sin}$ 28.8 $\boxed{=}$

E lags I by 60°. Looking at Figure 18–11, we can see that, when the current is at 28.8° (approximately 40 μs from t_0), E is moving from its maximum negative value toward 0 (which would occur 60° after I is zero) and is at an angle of $-31.2°$. Knowing the phase angle, we can compute e using Equation 18–26.

$$e = E_{pk} \sin (\omega t \pm \theta) = 100 \text{ V}_{pk} \sin (28.8° - 60°)$$
$$= 100 \text{ V}_{pk} \sin (-31.2°) = -51.8 \text{ V}_t$$

TI: 100 $\boxed{\times}$ $\boxed{(}$ 28.8 $\boxed{-}$ 60 $\boxed{)}$ $\boxed{\sin}$ $\boxed{=}$

Answer: $e = -51.8 \text{V}_t$

Casio: 100 $\boxed{\times}$ $\boxed{\sin}$ $\boxed{(}$ 28.8 $\boxed{-}$ 60 $\boxed{)}$ $\boxed{=}$

EXAMPLE 18–13

$I = 20 \text{ mA}_{pk}$, $E = 20 \text{ V}_{pk}$, $f = 12 \text{ kHz}$, and E leads I by 45°. Compute i and e 15 μs after t_0. Use voltage as the reference.

SOLUTION   A graph "ball parking" E and I is drawn in Figure 18–12 on the next page. The time for 90° is about 20 μs. 15 μs puts us approximately three-quarters of the distance between 0° and 90°, so ωt is about 70°. Its actual value is

$$\omega t = 360° \, ft = 360° \times 12 \text{ kHz} \times 15 \text{ } \mu s = 64.8°$$

TI: 360 $\boxed{\times}$ 12 $\boxed{EE}$ 3 $\boxed{\times}$ 15 $\boxed{EE}$ $\boxed{\pm}$ 6 $\boxed{=}$

Answer: $\omega t = 64.8°$

Casio: Use $\boxed{EXP}$ instead of $\boxed{EE}$.

E leads I by 45° ($\theta = 45°$). 45° is one-half the distance between 0° and 90°. The graph shows that, at 15 μs after t_0, E is increasing and is near its peak positive value. Using Equation 18–21,

$$e = E_{pk} \sin \omega t = 20 \text{ V}_{pk} \sin 64.8° = 18.1 \text{ V}_t$$

TI: 20 $\boxed{\times}$ 64.8 $\boxed{\sin}$ $\boxed{=}$ Answer: $e = 18.1 \text{ V}_t$

Casio: 20 $\boxed{\times}$ $\boxed{\sin}$ 64.8 $\boxed{=}$

I has crossed zero and is increasing toward its peak positive value. Using Equation 18–22,

$$i = I_{pk} \sin (\omega t \pm \theta) = 20 \text{ mA} \sin (64.8° - 45°)$$
$$= 20 \text{ mA} \sin 19.8° = 6.77 \text{ mA}_t$$

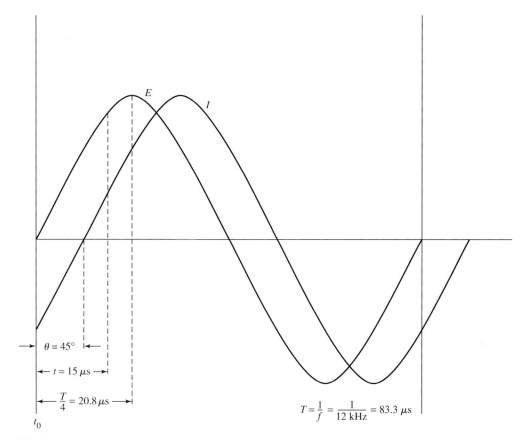

FIGURE 18–12

$$\text{TI: } 20 \boxed{\text{EE}} \boxed{\pm} 3 \boxed{\times} \boxed{(} 64.8 \boxed{-} 45 \boxed{)}$$

$$\boxed{\sin} \boxed{=} \quad \text{Answer: } i = 6.77 \text{ mA}_t$$

$$\text{Casio: } 20 \boxed{\text{EXP}} \boxed{(-)} 3 \boxed{\times} \boxed{\sin} \boxed{(} 64.8 \boxed{-} 45$$

$$\boxed{)} \boxed{=}$$

PRACTICE PROBLEMS 18–5

1. $I = 100$ mA$_{pk}$, $E = 30$ V$_{pk}$, and $f = 500$ Hz. Use current as the reference. I leads E by 40°. Find i and e after (a) 50 μs, (b) 225 μs, (c) 800 μs, and (d) 1.2 ms.

2. $I = 12$ mA$_{pk}$, $E = 15$ V$_{pk}$, and $f = 7.5$ kHz. Use voltage as the reference. I leads E by 60°. Find i and e after (a) 10 μs, (b) 40 μs, (c) 100 μs, and (d) 150 μs.

1. (a) $i = 15.6$ mA$_t$, $e = -15.5$ V$_t$
   (b) $i = 64.9$ mA$_t$, $e = 0.262$ V$_t$
   (c) $i = 58.8$ mA$_t$, $e = 29.1$ V$_t$
   (d) $i = -58.8$ mA$_t$, $e = 2.09$ V$_t$

2. (a) $i = 12$ mA$_t$, $e = 6.81$ V$_t$
   (b) $i = 2.49$ mA$_t$, $e = 14.3$ V$_t$
   (c) $i = -6.00$ mA$_t$, $e = -15.0$ V$_t$
   (d) $i = 11.6$ mA$_t$, $e = 10.6$ V$_t$

Additional practice problems are at the end of the chapter.

SELF-TEST 18–2

1. If $V = 25$ V$_{pk}$ and $f = 8$ kHz, find v after 25 μs.

3. Convert 25 V to peak values and peak-to-peak values.
5. I leads E by 25° in an ac circuit. If $E = 50$ V and the frequency is 25 kHz, find e after 7 μs. Use current as the reference.

2. What is the instantaneous value of current after 2 ms if the frequency is 400 Hz and I_{pk} is 20 mA?
4. Convert 60 mA$_{p-p}$ to peak values and rms values.

Answers to Self-test 18–2 are at the end of the chapter.

CHAPTER 18 AT A GLANCE

PAGE	KEY POINTS, DEFINITIONS, AND EQUATIONS	EXAMPLE
508	*Key Point:* When the length of an arc equals the length of the radius of a circle, the angle is equal to 1 radian. One radian equals 57.3°.	

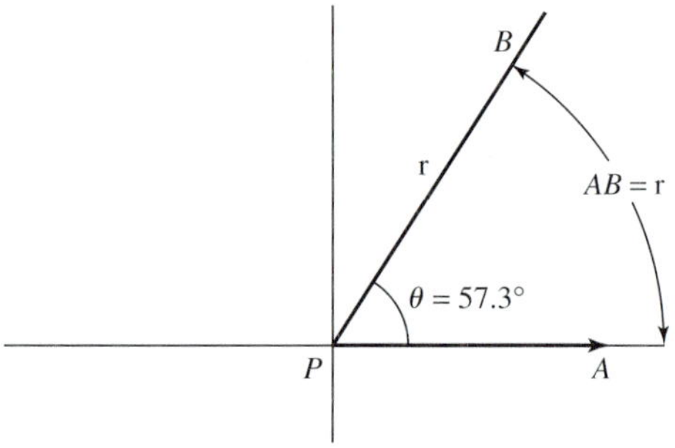

FIGURE 18–13

509	To convert from radians to degrees: $$\frac{180°}{\pi \text{ rad}} \times \text{number of radians}$$	angle $= 1.77$ rad $$\frac{180°}{\pi} \times 1.77 = 101°$$

509   To convert from degrees to radians:

$$\frac{\pi \text{ rad}}{180°} \times \text{number of degrees}$$

angle $= 37°$

$$\frac{\pi}{180} \times 37 = 0.646 \text{ rad}$$

512   The *frequency* (f) is the number of times per second that we generate a sine wave. The *period* (T) is the time it takes to generate one complete sine wave.

$f = 1/T$
$T = 1/f$

513   *Angular velocity* (ω) is the angular motion or speed of a rotating vector or phasor. Angular velocity is measured in radians per second or in degrees per second.

$\omega = 2\pi f$
$\omega = 360° f$

514   The generated angle at any instant, t, can be calculated as

angle in radians $= 2\pi ft = \omega t$
angle in degrees $= 360° ft = \omega t$

Given: $f = 2$ kHz $t = 100 \; \mu s$
$\omega t = 1.26$ rad
$\omega t = 72.0°$

516   In ac circuits, instantaneous values of current and voltage may be found using the following equations:

$i = I_{pk} \sin \omega t$
$e = E_{pk} \sin \omega t$

Let $I_{pk} = 40$ mA, $E_{pk} = 12$ V, and $f = 5$ kHz
Find i and e after 70 μs.
$i = 40 \text{ mA}_{pk} \sin \omega t$
   $= 32.4 \text{ mA}_t$
$e = 12 \text{ V}_{pk} \sin \omega t$
   $= 9.71 \text{ V}_t$

519   The *rms* or *effective* value of an ac voltage or current is the value that converts the same energy as does a dc value. The rms value equals 0.707 of the maximum or peak of a sine wave.

$I_{rms} = I_{eff} = 0.707 \, I_{pk} = \dfrac{I_{pk}}{\sqrt{2}}$
$V_{rms} = V_{eff} = 0.707 \, V_{pk}$
$I_{pk} = 1.414 \, I_{rms} = \sqrt{2} \, I_{rms}$
$V_{pk} = 1.414 \, V_{rms}$

521   *Key Point:* Whenever we have capacitance or inductance in an ac circuit, the current and voltage are out of phase. When the phase angle, *theta* (θ), is known, we can calculate the instantaneous values of current and voltage.

When current is the reference,
$i = I_{pk} \sin \omega t$
$e = E_{pk} \sin (\omega t \pm \theta)$
When voltage is the reference,
$e = E_{pk} \sin \omega t$
$i = I_{pk} \sin (\omega t \pm \theta)$

END OF CHAPTER PROBLEMS 18–1

1. $f = 600$ Hz. Find T.
2. $f = 4.75$ kHz. Find T.
3. $f = 1.2$ kHz. Find T.
4. $f = 250$ Hz. Find T.
5. $f = 1.76$ MHz. Find T.
6. $f = 20$ kHz. Find T.
7. $T = 400$ μs. Find f.
8. $T = 80$ ms. Find f.
9. $T = 1.33$ ms. Find f.
10. $T = 66.7$ μs. Find f.
11. $T = 25$ μs. Find f.
12. $T = 200$ ms. Find f.
13. Convert $25°$ to radians.
14. Convert $30°$ to radians.
15. Convert $90°$ to radians.
16. Convert $55°$ to radians.
17. Convert $180°$ to radians.
18. Convert $150°$ to radians.
19. Convert $75°$ to radians.
20. Convert $330°$ to radians.
21. Convert $600°$ to radians.
22. Convert $850°$ to radians.
23. Convert 0.146 rad to degrees.
24. Convert 1.45 rad to degrees.
25. Convert 0.5 rad to degrees.
26. Convert 3.75 rad to degrees.
27. Convert $\pi/2$ rad to degrees.
28. Convert $\dfrac{5\pi}{2}$ rad to degrees.
29. Convert 1.75 rad to degrees.
30. Convert 5.75 rad to degrees.
31. Convert 10 rad to degrees.
32. Convert 12 rad to degrees.

END OF CHAPTER PROBLEMS 18–2

1. Find the angular velocity in both degrees per second and radians per second if the frequency is (a) 60 Hz, (b) 150 Hz, (c) 200 Hz, (d) 300 Hz, (e) 600 Hz, (f) 800 Hz, and (g) 1.25 kHz.

2. Find the angular velocity in both degrees per second and radians per second if the frequency is (a) 20 Hz, (b) 40 Hz, (c) 120 Hz, (d) 180 Hz, (e) 300 Hz, (f) 500 Hz, (g) 1 kHz, and (h) 1.2 kHz.

3. Find the frequency if the angular velocity is (a) 12 rad/s, (b) 50 rad/s, (c) 400 rad/s, (d) 900 rad/s, (e) 2500 rad/s, (f) 5000 rad/s, and (g) 7000 rad/s.

4. Find the frequency if the angular velocity is (a) 18 rad/s, (b) 42 rad/s, (c) 300 rad/s, (d) 800 rad/s, (e) 3500 rad/s, (f) 4500 rad/s, and (g) 8000 rad/s.

5. Find the frequency if the angular velocity is (a) 10,000°/s, (b) 15,000°/s, (c) 50,000°/s, (d) 120,000°/s, (e) 200,000°/s, (f) 7.35×10^5 degrees/s, and (g) 1.65×10^6 degrees/s.

6. Find the frequency if the angular velocity is (a) 4000°/s, (b) 20,000°/s, (c) 75,000°/s, (d) 2.5×10^5 degrees/s, (e) 6.45×10^5 degrees/s, (f) 1×10^6 degrees/s, and (g) 3.25×10^6 degrees/s.

7. If the frequency is 500 Hz, find the angle generated in both degrees and radians after (a) 100 μs, (b) 250 μs, (c) 1 ms, and (d) 2.8 ms.

8. If the frequency is 35 kHz, find the angle generated in both degrees and radians after (a) 25 μs, (b) 120 μs, (c) 1.5 ms, and (d) 5 ms.

9. If the frequency is 2.55 kHz, find the angle generated in both degrees and radians after (a) 100 μs, (b) 250 μs, (c) 1 ms, and (d) 2.8 ms.

10. If the frequency is 1.75 kHz, find the angle generated in both degrees and radians after (a) 25 μs, (b) 120 μs, (c) 1.5 ms, and (d) 5 ms.

11. What is the frequency when an angle of (a) 1.8 rad is generated in 20 μs, (b) 3.25 rad is generated in 500 μs, (c) 8.73 rad is generated in 150 μs, and (d) 17.6 rad is generated in 45 μs?

12. What is the frequency when an angle of (a) 0.25 rad is generated in 10 μs, (b) 4.6 rad is generated in 90 μs, (c) 27 rad is generated in 17.5 μs, and (d) 30 rad is generated in 120 ms?

13. What is the frequency when an angle of (a) 145° is generated in 2.3 ms, (b) 310° is generated in 500 μs, (c) 1100° is generated in 50 μs, and (d) 15,000° is generated in 25 μs?

14. What is the frequency when an angle of (a) 75° is generated in 25 μs, (b) 500° is generated in 200 μs, (c) 3000° is generated in 100 μs, and (d) 14,000° is generated in 12 ms?

15. If the frequency is 5 kHz, how much time is required to generate an angle of (a) 1 rad, (b) 0.125 rad, (c) 4.73 rad, (d) 20°, (e) 90°, and (f) 300°?

16. If the frequency is 125 Hz, how much time is required to generate an angle of (a) 0.35 rad, (b) 2.55 rad, (c) 6.6 rad, (d) 400°, (e) 225°, and (f) 270°?

17. If the frequency is 15 kHz, how much time is required to generate an angle of (a) 1 rad, (b) 0.125 rad, (c) 4.73 rad, (d) 20°, (e) 90°, and (f) 300°?

18. If the frequency is 7.5 kHz, how much time is required to generate an angle of (a) 0.35 rad, (b) 2.55 rad, (c) 5.8 rad, (d) 60°, (e) 180°, and (f) 300°.

END OF CHAPTER PROBLEMS 18–3

1. $E = 25$ V$_{pk}$, $f = 2.5$ kHz. Find e after (a) 10 μs, (b) 25 μs, (c) 150 μs, (d) 225 μs, (e) 300 μs, and (f) 375 μs.

2. $E = 15$ V$_{pk}$, $f = 18$ kHz. Find e after (a) 5 μs, (b) 12 μs, (c) 25 μs, (d) 50 μs, (e) 80 μs, and (f) 120 μs.

3. $E = 12$ V$_{pk}$, $f = 12.7$ kHz. Find e after (a) 10 μs, (b) 25 μs, (c) 150 μs, (d) 225 μs, (e) 300 μs, and (f) 375 μs.

4. $E = 40$ V$_{pk}$, $f = 5.75$ kHz. Find e after (a) 5 μs, (b) 12 μs, (c) 25 μs, (d) 50 μs, (e) 80 μs, and (f) 120 μs.

5. $I = 700$ μA$_{pk}$, $f = 50$ kHz. Find i after (a) 5 μs, (b) 10 μs, (c) 15 μs, (d) 20 μs, (e) 25 μs, and (f) 30 μs.

6. $I = 50$ mA$_{pk}$, $f = 2$ kHz. Find i after (a) 40 μs, (b) 150 μs, (c) 250 μs, (d) 350 μs, (e) 500 μs, and (f) 600 μs.

7. $I = 3.45$ mA$_{pk}$, $f = 800$ Hz. Find i after (a) 5 μs, (b) 10 μs, (c) 15 μs, (d) 20 μs, (e) 25 μs, and (f) 30 μs.

8. $I = 450$ μA$_{pk}$, $f = 8.55$ kHz. Find i after (a) 40 μs, (b) 150 μs, (c) 250 μs, (d) 350 μs, (e) 500 μs, and (f) 600 μs.

END OF CHAPTER PROBLEMS 18–4

Convert the following voltages and currents to pk and p-p values:

1. (a) 65 V, (b) 110 V
2. (a) 400 mV, (b) 32.5 V
3. (a) 220 V, (b) 440 V
4. (a) 200 V, (b) 400 V
5. (a) 600 μA, (b) 1.2 mA
6. (a) 150 μA, (b) 4.75 mA
7. (a) 37.8 mA, (b) 3 A
8. (a) 750 mA, (b) 7.5 A

Convert the following peak values of voltage and current to p-p and rms values:

9. (a) $350 \, \mu V_{pk}$, (b) $400 \, mV_{pk}$
10. (a) $900 \, \mu V_{pk}$, (b) $55 \, mV_{pk}$
11. (a) $80 \, V_{pk}$, (b) $200 \, V_{pk}$
12. (a) $650 \, mV_{pk}$, (b) $18.5 \, V_{pk}$
13. (a) $200 \, \mu A_{pk}$, (b) $9.45 \, mA_{pk}$
14. (a) $600 \, \mu A_{pk}$, (b) $16.5 \, mA_{pk}$
15. (a) $500 \, mA_{pk}$, (b) $1.75 \, A_{pk}$
16. (a) $250 \, mA_{pk}$, (b) $1.4 \, A_{pk}$

Convert the following peak-to-peak values of voltage and current to pk and rms:

17. (a) $320 \, \mu V_{p-p}$, (b) $70 \, mV_{p-p}$
18. (a) $500 \, \mu V_{p-p}$, (b) $45 \, mV_{p-p}$
19. (a) $17 \, V_{p-p}$, (b) $283 \, V_{p-p}$
20. (a) $35 \, V_{p-p}$, (b) $100 \, V_{p-p}$
21. (a) $50 \, \mu A_{p-p}$, (b) $700 \, \mu A_{p-p}$
22. (a) $65 \, \mu A_{p-p}$, (b) $725 \, \mu A_{p-p}$
23. (a) $9.38 \, mA_{p-p}$ (b) $650 \, mA_{p-p}$
24. (a) $25 \, mA_{p-p}$ (b) $280 \, mA_{p-p}$

END OF CHAPTER PROBLEMS 18–5

1. $I = 4 \, mA_{pk}$, $E = 10 \, V_{pk}$, and $f =$ 20 kHz. Use current as the reference. E leads I by 25°. Find i and e after (a) 5 μs, (b) 12 μs, (c) 25 μs, and (d) 40 μs.

2. $I = 150 \, mA_{pk}$, $E = 50 \, V$, and $f =$ 3 kHz. Use voltage as the reference. E leads I by 45°. Find i and e after (a) 30 μs, (b) 70 μs, (c) 180 μs, and (d) 400 μs.

3. $I = 25 \, mA_{pk}$, $E = 20 \, V$, and $f =$ 100 Hz. Use voltage as the reference. I leads E by 35°. Find i and e after (a) 1 ms, (b) 2.5 ms, (c) 3.2 ms, and (d) 8.3 ms.

4. $I = 600 \, \mu A_{pk}$, $E = 9 \, V_{pk}$, and $f =$ 5 kHz. Use current as the reference. I leads E by 30°. Find i and e after (a) 20 μs, (b) 50 μs, (c) 160 μs, and (d) 350 μs.

5. $I = 600 \, \mu A_{pk}$, $E = 80 \, V_{pk}$, and $f = 3.75$ kHz. Use current as the reference. I leads E by 55°. Find i and e after (a) 10 μs, (b) 40 μs, (c) 120 μs, and (d) 200 μs.

6. $I = 1.2 \, mA_{pk}$, $E = 12 \, V_{pk}$, and $f =$ 18 kHz. Use current as the reference. I leads E by 38.5°. Find i and e after (a) 25 μs, (b) 40 μs, (c) 100 μs, and (d) 500 μs.

7. $I = 25 \, mA_{pk}$, $E = 65 \, V_{pk}$, and $f =$ 2.35 kHz. Use voltage as the reference. I leads E by 40°. Find i and e after (a) 25 μs, (b) 70 μs, (c) 200 μs, and (d) 600 μs.

8. $I = 800 \, \mu A_{pk}$, $E = 15 \, V_{pk}$, and $f =$ 10 kHz. Use voltage as the reference. I leads E by 50°. Find i and e after (a) 7.5 μs, (b) 25 μs, (c) 70 μs, and (d) 120 μs.

ANSWERS TO SELF-TESTS

SELF-TEST 18–1

1. $T = 20 \, \mu s$
2. 2.79
3. 36.0°
4. 1.88 rad $= 108°$
5. 1.67 kHz
6. 41.4 μs

SELF-TEST 18–2

1. 23.8 V
2. -19.0 mA
3. $35.4 \, V_{pk} = 70.7 \, V_{p-p}$
4. $30 \, mA_{pk} = 21.2 \, mA$
5. 30.8 V

ac Circuit Analysis: Series Circuits

19

Introduction

All electrical circuits contain the properties of resistance, capacitance, and inductance. We didn't consider the effects of capacitance and inductance in the dc circuits in Chapters 10, 11, and 12 because in dc circuits their effects are negligible. In those chapters we used Kirchhoff's laws and Ohm's law to solve series circuit problems where the only measurable circuit property was resistance and the electrical energy source was some dc source.

In reality, most circuits we work with in electronics contain ac sources and measurable values of inductance and capacitance as well as resistance. Furthermore, the effects of inductance and capacitance are displaced from the effects of circuit resistance by 90°. These effects cause the applied voltage and circuit current to be out of phase. We learned how to solve problems when one angle was a right triangle in the chapters on trig. In this chapter we will use trig functions and the Pythagorean theorem to find the phase angle, the various voltage drops, and the current in ac series circuit problems.

When the effects of both inductance and capacitance are measurable in series circuits, series resonance occurs. *Resonance* is that frequency where the effects of capacitance and inductance cancel each other and the phase angle is 0°. The concept of series resonance is discussed, and the computation of series resonance and the conditions that occur at series resonance are also covered in this chapter.

Chapter Objectives

In this chapter you will learn how to:

1. Use trigonometric functions to find phase angles, reactances, and impedances in series *RC* and *RL* circuits.
2. Express phasors in polar and rectangular form.
3. Express answers to series *RC* and *RL* circuit problems in both polar and rectangular forms.

4. Add and subtract impedances in both polar and rectangular forms.
5. Find the equivalent series circuit of ac circuits containing all three circuit properties: capacitance, inductance, and resistance.
6. Determine the resonant frequency of series *RCL* circuits.

19–1   SERIES RC CIRCUITS—FINDING VOLTAGES AND PHASE ANGLES

Capacitance in ac circuits causes the current to lead the applied voltage. In a purely capacitive circuit ($R = 0\ \Omega$), current leads the voltage by 90°. At the other extreme, if the circuit is purely resistive, current and applied voltage are in phase. Let's see what happens in a circuit that contains both resistance and capacitance. Consider the circuit in Figure 19–1.

☞ **KEY POINT**   In a series circuit, we use *I* as the reference. In an *RC* circuit, current leads, so V_C and X_C are plotted along the negative *y*-axis. **E** (or **Z**) is the hypotenuse, and θ is the phase angle.

We have drawn a series *RC* circuit in Figure 19–1(a). As shown in Figure 19–1(b), the current leads the voltage drop across the capacitor by 90°. V_R is the side adjacent to the angle θ and V_C is the opposite side.

EXAMPLE 19–1

Refer to Figure 19–1(a). Let $V_R = 10$ V and $V_C = -12$ V. Find **E** and θ.

SOLUTION   Refer to Figure 19–2 on the next page.

$$\tan \theta = \frac{\text{opposite side}}{\text{adjacent side}} = \frac{V_C}{V_R} = \frac{-12\ \text{V}}{10\ \text{V}} = -1.2$$

$$\tan^{-1}(-1.2) = \theta = -50.2°$$

FIGURE 19–1
An *RC* series circuit with phasor diagrams.

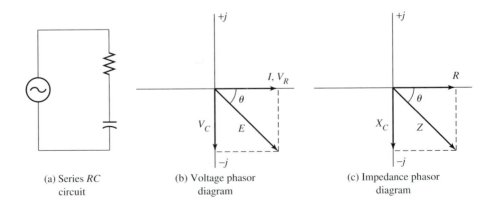

(a) Series *RC* circuit   (b) Voltage phasor diagram   (c) Impedance phasor diagram

FIGURE 19–2

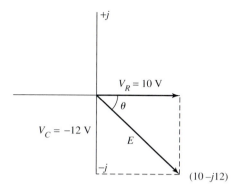

For the problems in this chapter, let's set up the calculator in engineering mode and display 2 digits to the right of the decimal point.

TI: [eng] [fix] 2

Casio: [mode] [mode] [mode] [mode] 1 2

Using the calculator,

TI: 12 [+/−] [÷] 10 [=] [tan⁻¹] [sto] 1

Casio: [tan⁻¹] [(] [−] 12 [÷] 10 [)] [=] [sto] [A]

Answer: $\theta = -50.2°$

Store θ because we will use it for the next part of the problem. (Remember, V_C is negative because we plotted V_C along the negative y-axis.) We can now find **E** by using either the sine or cosine function. We will use the sine function.

$$\sin \theta = \frac{\text{opposite side}}{\text{hypotenuse}} = \frac{V_C}{\mathbf{E}}$$

$$\mathbf{E} = \frac{V_C}{\sin \theta} = \frac{-12 \text{ V}}{\sin (-50.2°)} = \frac{-12 \text{ V}}{-0.768} = 15.6 \text{ V}$$

TI: 12 [+/−] [÷] [RCL] 1 [sin] [=]

Casio: [−] 12 [÷] [sin] [RCL] [A] [=]

Answer: E = 15.6 V

EXAMPLE 19-2

Refer to Figure 19–1(a). Let $\mathbf{E} = 25$ V and $V_R = 18$ V. Find V_C and θ.

SOLUTION

$$\cos \theta = \frac{V_R}{\mathbf{E}} = \frac{18 \text{ V}}{25 \text{ V}} = 0.72$$

$$\cos^{-1} 0.72 = \theta = -43.9°$$

From the given information we know that θ is negative because we know that we are operating in the fourth quadrant.

$$\sin \theta = \frac{V_C}{\mathbf{E}}$$

$$V_C = \mathbf{E} \sin \theta = 25 \text{ V} \sin (-43.9°) = -17.3 \text{ V}$$

PRACTICE PROBLEMS 19-1

Refer to Figure 19–1 for the following problems:

1. $V_R = 6.8$ V, $V_C = -10$ V. Find $\mathbf{E}$ and θ.
2. $V_R = 30$ V, $\mathbf{E} = 50$ V. Find V_C and θ.
3. $V_R = 10$ V, $\theta = -60°$. Find V_C and $\mathbf{E}$.
4. $\mathbf{E} = 20$ V, $\theta = -35°$. Find V_R and V_C.
5. $\mathbf{E} = 50$ V, $V_C = -40$ V. Find V_R and θ.
6. $V_C = -11$ V, $\theta = -65°$. Find $\mathbf{E}$ and V_R.

SOLUTIONS

1. $\mathbf{E} = 12.1$ V; $\theta = -55.8°$
2. $V_C = -40$ V; $\theta = -53.1°$
3. $V_C = -17.3$ V; $\mathbf{E} = 20$ V
4. $V_R = 16.4$ V; $V_C = -11.5$ V
5. $V_R = 30$ V; $\theta = -53.1°$
6. $\mathbf{E} = 12.1$ V; $V_R = 5.13$ V

Additional practice problems are at the end of the chapter.

19-2   SERIES RC CIRCUITS—RESISTANCE, REACTANCE, AND IMPEDANCE

In Figure 19–1(c), since V_R and I are in phase, R is plotted along the x-axis. X_C, the capacitive reactance, must be plotted along the negative y-axis to show the 90° phase difference between capacitive values and resistive values. The capacitive reactance is determined by the following equation:

(19–1)

$$X_C = \frac{1}{2\pi fC}$$

EXAMPLE 19–3

Refer to Figure 19–1(c). Let $R = 2$ kΩ, $C = 20$ nF, and $f = 3$ kHz. Find X_C, Z, and θ.

SOLUTION

$$X_C = \frac{1}{2\pi fC} = \frac{1}{2\pi \times 3 \text{ kHz} \times 20 \text{ nF}} = 2.65 \text{ k}\Omega$$

Using the calculator,

TI: 2 $\boxed{\times}$ $\boxed{\pi}$ $\boxed{=}$ $\boxed{\text{STO}}$ 1 $\boxed{\times}$ 3 $\boxed{\text{EE}}$ 3 $\boxed{\times}$ 20

$\boxed{\text{EE}}$ $\boxed{+/-}$ 9 $\boxed{=}$ $\boxed{1/x}$

Casio: 2 $\boxed{\pi}$ $\boxed{\text{STO}}$ $\boxed{\text{A}}$ $\boxed{\times}$ 3 $\boxed{\text{EXP}}$ 3 $\boxed{\times}$ 20 $\boxed{\text{EXP}}$

$\boxed{(-)}$ 9 $\boxed{=}$ $\boxed{x^{-1}}$ $\boxed{=}$ Answer: $X_C = 2.65$ kΩ

We stored 2π since we will use it often in future problems. X_C will be negative because it is plotted along the negative y-axis (see Figure 19–2).

$$\tan \theta = \frac{\text{opposite side}}{\text{adjacent side}} = \frac{X_C}{R} = \frac{-2.65 \text{ k}\Omega}{2 \text{ k}\Omega} = -1.33$$

$$\tan^{-1}(-1.33) = \theta = -53.0°$$

The calculator already displays X_C. We need to change its sign and then continue solving the problem:

TI: $\boxed{+/-}$ $\boxed{\div}$ 2 $\boxed{\text{EE}}$ 3 $\boxed{=}$ $\boxed{\tan^{-1}}$

Casio: $\boxed{\tan^{-1}}$ $\boxed{(}$ $\boxed{\text{Ans}}$ $\boxed{\div}$ 2 $\boxed{\text{EXP}}$ 3 $\boxed{)}$ $\boxed{=}$

Answer: $\theta = -53.0°$

θ is negative because we are in the fourth quadrant.

$$\cos \theta = \frac{R}{Z}$$

$$Z = \frac{R}{\cos \theta} = \frac{2 \text{ k}\Omega}{\cos(-53°)} = 3.32 \text{ k}\Omega$$

Now θ is displayed and we need to divide 2 kΩ by the cosine of θ. Dividing by cosine θ is the same as multiplying by the reciprocal of cosine θ.

$$\frac{2 \text{ k}\Omega}{\cos \theta} = 2 \text{ k}\Omega \times \frac{1}{\cos \theta}$$

So

$$\text{TI: } \boxed{\cos}\ \boxed{1/x}\ \boxed{\times}\ 2\ \boxed{\text{EE}}\ 3\ \boxed{=}$$

Casio: $2\ \boxed{\text{EXP}}\ 3\ \boxed{\div}\ \boxed{\cos}\ \boxed{\text{Ans}}\ \boxed{=}$

Answer: $Z = 3.32\ k\Omega$

EXAMPLE 19–4

Refer to Figure 19–1(c). Let $R = 10\ k\Omega$, $C = 50\ nF$, $f = 500\ Hz$, and $\mathbf{E} = 10\ V$. Find X_C, Z, θ, V_R, V_C, and I.

SOLUTION

$$X_C = \frac{1}{2\pi f C} = \frac{1}{2\pi \times 500 \times 50\ nF} = 6.37\ k\Omega$$

$$\tan\theta = \frac{\text{opposite side}}{\text{adjacent side}} = \frac{X_C}{R} = \frac{-6.37\ k\Omega}{10\ k\Omega} = -0.637$$

$$\tan^{-1}(-0.637) = -32.5°$$

$$\sin\theta = \frac{X_C}{Z}$$

$$Z = \frac{X_C}{\sin\theta} = \frac{-6.37\ k\Omega}{\sin(-32.5°)} = 11.9\ k\Omega$$

With the calculator,

$$\text{TI: } \boxed{\text{RCL}}\ 1\ \boxed{\times}\ 500\ \boxed{\times}\ 50\ \boxed{\text{EE}}\ \boxed{+/-}\ 9\ \boxed{-}\ \boxed{1/x}$$

Casio: $1\ \boxed{\div}\ \boxed{(}\ \boxed{\text{RCL}}\ \boxed{\text{A}}\ \boxed{\times}\ 500\ \boxed{\times}\ 50\ \boxed{\text{EXP}}$

$\boxed{(-)}\ 9\ \boxed{)}\ \boxed{=}$ Answer: $X_C = 6.37\ k\Omega$

Remember, we had previously stored 2π. Keep X_C displayed but record its value before going to the next step.

$$\text{TI: } \boxed{+/-}\ \boxed{\div}\ 10\ \boxed{\text{EE}}\ 3\ \boxed{=}\ \boxed{\tan^{-1}}$$

Casio: $\boxed{\tan^{-1}}\ \boxed{(}\ \boxed{-}\ \boxed{\text{Ans}}\ \boxed{\div}\ 10\ \boxed{\text{EXP}}\ 3\ \boxed{)}\ \boxed{=}$

Answer: $\theta = -32.5°$

Change the sign of X_C and then divide by the resistance, $10\ k\Omega$, to get $\tan\theta$, which is -0.637. The arctan gives us θ. Keep θ displayed, but record its value before going to the next step.

TI: $\boxed{\sin}$ $\boxed{1/x}$ $\boxed{\times}$ $\boxed{6.37}$ $\boxed{+/-}$ $\boxed{\text{EE}}$ $\boxed{3}$ $\boxed{=}$

Casio: $\boxed{-}$ $\boxed{6.37}$ $\boxed{\text{EXP}}$ $\boxed{3}$ $\boxed{\div}$ $\boxed{\sin}$ $\boxed{\text{Ans}}$ $\boxed{=}$

Answer: $Z = 11.9 \text{ k}\Omega$

Find $\sin \theta$, take its reciprocal, and multiply by X_C to find the impedance, Z.
Now that we know Z, we can find I.

$$I = \frac{E}{Z}$$

$$= \frac{10 \text{ V}}{11.9 \text{ k}\Omega} = 844 \ \mu\text{A}$$

To find V_C and V_R, we can use Ohm's law:

$$V_C = IX_C = 844 \ \mu\text{A} \times 6.37 \text{ k}\Omega = 5.37 \text{ V}$$

$$V_R = IR = 844 \ \mu\text{A} \times 10 \text{ k}\Omega = 8.44 \text{ V}$$

We could also find V_C and V_R by using trig functions since we know $\mathbf{E}$ and θ.

$$V_R = \mathbf{E} \cos \theta = 10 \text{ V} \cos (-32.5°) = 8.44 \text{ V}$$

$$V_C = \mathbf{E} \sin \theta = 10 \text{ V} \sin (-32.5°) = -5.37 \text{ V}$$

(Differences in the values of V_R and V_C in the two methods are a result of rounding the value of Z to three places.)

EXAMPLE 19–5

Refer to Figure 19–3. $R = 5 \text{ k}\Omega$, $C = 10 \text{ nF}$, $Z = 8.3 \text{ k}\Omega$, and $\mathbf{E} = 10 \text{ V}$. Find I, θ, f, X_C, V_C, and V_R.

SOLUTION

$$I = \frac{E}{Z} = \frac{10 \text{ V}}{8.3 \text{ k}\Omega} = 1.2 \text{ mA}$$

Since we know R and Z, we can find X_C and θ. We will use the cosine function to find θ and the sine function to find X_C.

FIGURE 19–3

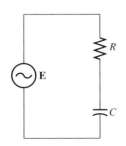

$$\cos \theta = \frac{R}{Z} = \frac{5 \text{ k}\Omega}{8.3 \text{ k}\Omega} = 0.602$$

$$\cos^{-1} 0.602 = \theta = -53.0°$$

$$\sin \theta = \frac{X_C}{Z}$$

$$X_C = Z \sin \theta = 8.3 \text{ k}\Omega \times \sin (-53.0°) = -6.62 \text{ k}\Omega$$

(Remember, mathematically the value of X_C is negative because we are operating in the fourth quadrant. Electrically, X_C is treated as a positive value.)

Now we can find the frequency because we know C and X_C. If we solve for f in the equation

$$X_C = \frac{1}{2\pi f C}$$

we get

$$f = \frac{1}{2\pi C X_C} = \frac{1}{2\pi \times 10 \text{ nF} \times 6.62 \text{ k}\Omega} = 2.4 \text{ kHz}$$

We can find V_R and V_C by using Ohm's law or by using trig functions. We should solve both ways to check our work. We will use Ohm's law here.

$$V_R = IR = 1.2 \text{ mA} \times 5 \text{ k}\Omega = 6.02 \text{ V}$$

$$V_C = IX_C = 1.2 \text{ mA} \times 6.62 \text{ k}\Omega = 7.98 \text{ V}$$

PRACTICE PROBLEMS 19–2

Refer to Figure 19–3 for the following problems and find X_C, Z, and θ.

1. $R = 5.6 \text{ k}\Omega$, $C = 50 \text{ nF}$, and $f = 2.5 \text{ kHz}$
2. $R = 120 \text{ }\Omega$, $C = 0.2 \text{ }\mu\text{F}$, and $f = 4.85 \text{ kHz}$
3. $R = 39 \text{ k}\Omega$, $C = 400 \text{ pF}$, and $f = 12 \text{ kHz}$

Refer to Figure 19–3 for the following problems and find X_C, Z, θ, V_R, V_C, and I.

4. $R = 470 \text{ }\Omega$, $C = 0.2 \text{ }\mu\text{F}$, $f = 1 \text{ kHz}$, and $\mathbf{E} = 9 \text{ V}$
5. $R = 2 \text{ k}\Omega$, $C = 30 \text{ nF}$, $f = 3 \text{ kHz}$, and $\mathbf{E} = 20 \text{ V}$
6. $R = 27 \text{ k}\Omega$, $C = 100 \text{ pF}$, $f = 35 \text{ kHz}$, and $\mathbf{E} = 15 \text{ V}$

Refer to Figure 19–3 for the following problems and find I, θ, f, X_C, V_R, and V_C.

7. $R = 12 \text{ k}\Omega$, $C = 200 \text{ pF}$, $Z = 20 \text{ k}\Omega$, and $\mathbf{E} = 15 \text{ V}$
8. $R = 680 \text{ }\Omega$, $C = 1 \text{ }\mu\text{F}$, $Z = 800 \text{ }\Omega$, and $\mathbf{E} = 25 \text{ V}$
9. $R = 3 \text{ k}\Omega$, $C = 50 \text{ nF}$, $Z = 4.5 \text{ k}\Omega$, and $\mathbf{E} = 30 \text{ V}$

1. $X_C = 1.27 \text{ k}\Omega$, $Z = 5.74 \text{ k}\Omega$,
   $\theta = -12.8°$
2. $X_C = 164 \ \Omega$, $Z = 203 \ \Omega$,
   $\theta = -53.8°$
3. $X_C = 33.2 \text{ k}\Omega$, $Z = 51.2 \text{ k}\Omega$,
   $\theta = -40.4°$
4. $X_C = 796 \ \Omega$, $Z = 924 \ \Omega$, $\theta = -59.4°$,
   $V_R = 4.58 \text{ V}$, $V_C = 7.75 \text{ V}$,
   $I = 9.74 \text{ mA}$
5. $X_C = 1.77 \text{ k}\Omega$, $Z = 2.67 \text{ k}\Omega$,
   $\theta = -41.5°$, $V_R = 15 \text{ V}$, $V_C = -13.2 \text{ V}$,
   $I = 7.49 \text{ mA}$
6. $X_C = 45.5 \text{ k}\Omega$, $Z = 52.9 \text{ k}\Omega$,
   $\theta = -59.3°$, $V_R = 7.66 \text{ V}$,
   $V_C = -12.9 \text{ V}$, $I = 284 \ \mu\text{A}$
7. $I = 750 \ \mu\text{A}$, $\theta = -53.1°$,
   $f = 49.7 \text{ kHz}$, $X_C = 16 \text{ k}\Omega$,
   $V_R = 9 \text{ V}$, $V_C = -12 \text{ V}$
8. $I = 31.3 \text{ mA}$, $\theta = -31.8°$, $f = 378 \text{ Hz}$,
   $X_C = 421 \ \Omega$, $V_R = 21.3 \text{ V}$,
   $V_C = -13.2 \text{ V}$
9. $I = 6.67 \text{ mA}$, $\theta = -48.2°$,
   $f = 949 \text{ Hz}$, $X_C = 3.35 \text{ k}\Omega$,
   $V_R = 20 \text{ V}$, $V_C = -22.4 \text{ V}$

Additional practice problems are at the end of the chapter.

SERIES RL CIRCUITS—FINDING VOLTAGES AND PHASE ANGLES   19-3

☞ **KEY POINT** In a series circuit, we use I as the reference. In an RL circuit, voltage leads, so V_L and X_L are plotted along the positive y-axis. **E** (or **Z**) is the hypotenuse, and θ is the phase angle.

We have drawn a series RL circuit in Figure 19–4(a). As shown in Figure 19–4(b), the voltage drop across the inductor leads the current by 90°. V_R is the side adjacent to the angle θ and V_L is the opposite side.

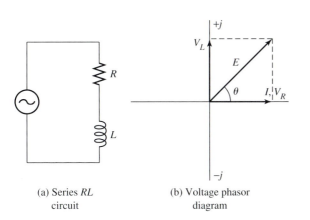

(a) Series RL circuit

(b) Voltage phasor diagram

(c) Impedance phasor diagram

FIGURE 19–4
An RL series circuit with phasor diagrams.

EXAMPLE 19–6

Let $V_R = 15$ V and $V_L = 10$ V in Figure 19–4(a). Find **E** and θ.

FIGURE 19–5

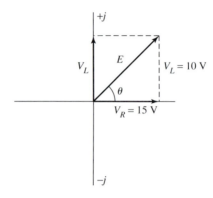

SOLUTION   Refer to Figure 19–5.

$$\tan \theta = \frac{V_L}{V_R} = \frac{10 \text{ V}}{15 \text{ V}} = 0.667$$

$$\tan^{-1} 0.667 = \theta = 33.7°$$

E can be found by using either the sine function or the cosine function. Let's use the cosine function.

$$\cos \theta = \frac{V_R}{\mathbf{E}}$$

$$\mathbf{E} = \frac{V_R}{\cos \theta} = \frac{15 \text{ V}}{\cos 33.7°} = 18 \text{ V}$$

EXAMPLE 19–7

Let **E** = 10 V and $V_L = 5$ V in Figure 19–4(a). Find V_R and θ.

SOLUTION

$$\sin \theta = \frac{V_L}{\mathbf{E}} = \frac{5 \text{ V}}{10 \text{ V}} = 0.5$$

$$\sin^{-1} 0.5 = \theta = 30°$$

Now that we know θ, we can use either the cosine function or the tangent function to find V_R. Let's use the tangent function.

$$\tan \theta = \frac{V_L}{V_R}$$

$$V_R = \frac{V_L}{\tan \theta} = \frac{5 \text{ V}}{\tan 30°} = 8.66 \text{ V}$$

Refer to Figure 19–4 for the following problems.

1. $V_R = 12$ V, $V_L = 18$ V. Find **E** and θ.
3. $V_R = 35$ V, **E** = 50 V. Find V_L and θ.
5. $V_L = 15$ V, $\theta = 27°$. Find V_R and **E**.

2. $V_R = 27$ V, $\theta = 65°$. Find **E** and V_L.
4. $V_L = 6$ V, **E** = 10 V. Find V_R and θ.
6. **E** = 25 V, $\theta = 33°$. Find V_R and V_L.

SOLUTIONS

1. **E** = 21.6 V; $\theta = 56.3°$
3. $V_L = 35.7$ V; $\theta = 45.6°$
5. $V_R = 29.4$ V; **E** = 33 V

2. **E** = 63.9 V; $V_L = 57.9$ V
4. $V_R = 8$ V; $\theta = 36.9°$
6. $V_R = 21$ V; $V_L = 13.6$ V

Additional practice problems are at the end of the chapter.

SERIES RL CIRCUITS—RESISTANCE, REACTANCE, AND IMPEDANCE   19-4

In Figure 19–4(c), since V_R and I are in phase, R is plotted along the x-axis. X_L, the inductive reactance, must be plotted along the positive y-axis to show the 90° phase difference between inductive and resistive values. The inductive reactance is determined by the equation

$$X_L = 2\pi fL \tag{19–2}$$

EXAMPLE 19–8

Refer to Figure 19–4(a). Let $R = 2.7$ kΩ, $L = 200$ mH, and $f = 3$ kHz. Find X_L, θ, and Z.

SOLUTION

$$X_L = 2\pi fL = 2\pi \times 3 \text{ k}\Omega \times 200 \text{ mH} = 3.77 \text{ k}\Omega$$

Knowing R and X_L, we can use the tangent function to find θ.

$$\tan \theta = \frac{X_L}{R} = \frac{3.77 \text{ k}\Omega}{2.7 \text{ k}\Omega} = 1.4$$

$$\tan^{-1} 1.4 = \theta = 54.4°$$

We can use either the sine function or the cosine function to find Z. Let's use the sine function.

$$\sin \theta = \frac{X_L}{Z}$$

$$Z = \frac{X_L}{\sin \theta} = \frac{3.77 \text{ k}\Omega}{\sin 54.4°} = 4.64 \text{ k}\Omega$$

EXAMPLE 19–9

Refer to Figure 19–4. Let $R = 10 \text{ k}\Omega$, $L = 100 \text{ mH}$, $f = 20 \text{ kHz}$, and $\mathbf{E} = 15 \text{ V}$. Find X_L, θ, Z, V_R, V_L, and I.

SOLUTION

$$X_L = 2\pi fL = 2\pi \times 20 \text{ kHz} \times 100 \text{ mH} = 12.6 \text{ k}\Omega$$

Knowing R and X_L, we can find Z and θ.

$$\tan \theta = \frac{X_L}{R} = \frac{12.6 \text{ k}\Omega}{10 \text{ k}\Omega} = 1.26$$

$$\tan^{-1} 1.26 = \theta = 51.5°$$

$$\cos \theta = \frac{R}{Z}$$

$$Z = \frac{R}{\cos \theta} = \frac{10 \text{ k}\Omega}{\cos 51.5°} = 16.1 \text{ k}\Omega$$

We can find V_R and V_L by using Ohm's law.

$$I = \frac{\mathbf{E}}{Z} = \frac{15 \text{ V}}{16.1 \text{ k}\Omega} = 934 \text{ }\mu\text{A}$$

$$V_R = IR = 934 \text{ }\mu\text{A} \times 10 \text{ k}\Omega = 9.34 \text{ V}$$

$$V_L = IX_L = 934 \text{ }\mu\text{A} \times 12.6 \text{ k}\Omega = 11.7 \text{ V}$$

We can also find V_R and V_L by using trig functions.

$$V_R = \mathbf{E} \cos \theta = 15 \text{ V} \times \cos 51.5° = 9.34 \text{ V}$$

$$V_L = \mathbf{E} \sin \theta = 15 \text{ V} \times \sin 51.5° = 11.7 \text{ V}$$

EXAMPLE 19–10

Refer to Figure 19–4. Let $L = 30 \text{ mH}$, $Z = 12 \text{ k}\Omega$, $R = 7.94 \text{ k}\Omega$, and $V_L = 4.19 \text{ V}$. Find X_L, θ, E, I, f, and V_R.

SOLUTION Let's find θ since we know Z and R.

$$\cos \theta = \frac{R}{Z} = \frac{7.94 \text{ k}\Omega}{12 \text{ k}\Omega} = 0.662$$

$$\cos^{-1} 0.662 = \theta = 48.6°.$$

Having found θ, we can now find X_L, E, and V_R.

$$\sin \theta = \frac{X_L}{Z}$$

$$X_L = Z \sin \theta = 12 \text{ k}\Omega \sin 48.6° = 9.00 \text{ k}\Omega$$

$$\sin \theta = \frac{V_L}{E}$$

$$E = \frac{V_L}{\sin \theta} = \frac{4.19 \text{ V}}{\sin 48.6°} = 5.59 \text{ V}$$

$$\cos \theta = \frac{V_R}{E}$$

$$V_R = E \cos \theta = 5.59 \cos 48.6° = 3.70 \text{ V}$$

We can use Ohm's law to find I.

$$I = \frac{E}{Z} = \frac{5.59 \text{ V}}{12 \text{ k}\Omega} = 466 \text{ }\mu\text{A}$$

Solving for f in the equation $X_L = 2\pi fL$, we get

$$f = \frac{X_L}{2\pi L} = \frac{9 \text{ k}\Omega}{2\pi \times 30 \text{ mH}} = 47.7 \text{ kHz}$$

Using the calculator,

TI: 9 $\boxed{\text{EE}}$ 3 $\boxed{\div}$ $\boxed{(}$ 2 $\boxed{\times}$ $\boxed{\pi}$ $\boxed{\times}$ 30 $\boxed{\text{EE}}$ $\boxed{+/-}$

3 $\boxed{)}$ $\boxed{=}$

Answer: f = 47.7 kHz

For Casio use $\boxed{\text{EXP}}$ instead of $\boxed{\text{EE}}$.

PRACTICE PROBLEMS 19–4

Refer to Figure 19–6 for the following problems and find X_L, Z, and θ.

FIGURE 19–6

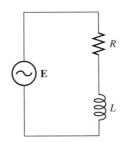

1. $R = 2.7\ k\Omega$, $L = 1\ H$, and
   $f = 600\ Hz$

2. $R = 25\ k\Omega$, $L = 200\ mH$, and
   $f = 15\ kHz$

Refer to Figure 19–6 for the following problems and find X_L, Z, θ, V_R, V_L, and I.

3. $R = 600\ \Omega$, $L = 400\ mH$, $f = 200\ Hz$,
   and $E = 20\ V$

4. $R = 3\ k\Omega$, $L = 50\ mH$, $f = 15\ kHz$,
   and $E = 9\ V$

Refer to Figure 19–6 for the following problems and find X_L, θ, f, E, V_R, and I.

5. $L = 100\ mH$, $Z = 7.3\ k\Omega$, $R = 5\ k\Omega$,
   and $V_L = 7.2\ V$

6. $L = 20\ mH$, $Z = 1\ k\Omega$, $R = 300\ \Omega$,
   and $V_L = 3.2\ V$

Refer to Figure 19–6 for the following problems and find X_L, L, θ, V_R, E, and I.

7. $R = 12\ k\Omega$, $Z = 22.7\ k\Omega$, $f = 5.3\ kHz$,
   and $V_L = 14\ V$

8. $R = 680\ \Omega$, $Z = 870\ \Omega$, $f = 1.35\ kHz$,
   and $V_L = 7.35\ V$

SOLUTIONS

1. $X_L = 3.77\ k\Omega$; $Z = 4.64\ k\Omega$; $\theta = 54.4°$
2. $X_L = 18.8\ k\Omega$; $Z = 31.3\ k\Omega$; $\theta = 37.0°$
3. $X_L = 503\ \Omega$; $\theta = 40°$; $Z = 783\ \Omega$;
   $V_R = 15.3\ V$; $V_L = 12.8\ V$;
   $I = 25.6\ mA$
4. $X_L = 4.71\ k\Omega$; $\theta = 57.5°$;
   $Z = 5.59\ k\Omega$; $V_R = 4.83\ V$;
   $V_L = 7.59\ V$; $I = 1.61\ mA$
5. $X_L = 5.32\ k\Omega$; $\theta = 46.8°$;
   $f = 8.47\ kHz$; $E = 9.88\ V$;
   $V_R = 6.77\ V$; $I = 1.35\ mA$
6. $X_L = 954\ \Omega$; $\theta = 72.5°$;
   $f = 7.59\ kHz$; $E = 3.35\ V$;
   $V_R = 1.01\ V$; $I = 3.35\ mA$
7. $X_L = 19.3\ k\Omega$; $L = 579\ mH$;
   $\theta = 58.1°$; $V_R = 8.72\ V$; $E = 16.5\ V$;
   $I = 727\ \mu A$
8. $X_L = 543\ \Omega$; $L = 64.0\ mH$;
   $\theta = 38.6°$; $V_R = 9.21\ V$; $E = 11.8\ V$;
   $I = 13.5\ mA$

Additional practice problems are at the end of the chapter.

SELF-TEST 19–1

Refer to Figure 19–7 on the next page.

1. $V_L = 3.3\ V$, $\theta = 65°$. Find V_R and E.

2. $R = 4.7\ k\Omega$, $L = 175\ mH$, $f = 5\ kHz$, and
   $E = 25\ V$. Find X_L, θ, Z, V_R, V_L, and I.

3. $L = 50\ mH$, $Z = 1.2\ k\Omega$, $R = 750\ \Omega$, and
   $V_L = 7.8\ V$. Find X_L, θ, f, E, V_R, and I.

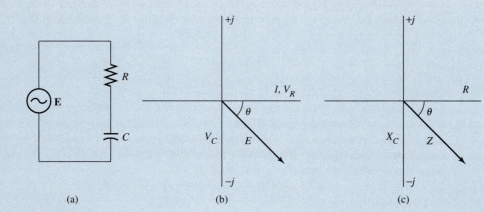

FIGURE 19–7

Refer to Figure 19–6.

4. $V_R = 17$ V, $\theta = -36°$. Find V_C and $\mathbf{E}$.

5. $R = 2$ kΩ, $C = 50$ nF, $f = 1$ kHz, and $\mathbf{E} = 10$ V. Find X_C, Z, θ, V_R, V_C, and I.

6. $R = 330$ Ω, $C = 1$ μF, $Z = 600$ Ω, and $\mathbf{E} = 9$ V. Find X_C, f, θ, V_R, V_C, and I.

Answers to Self-test 19–1 are at the end of the chapter.

POLAR TO RECTANGULAR CONVERSION 19–5

Any phasor may be identified by expressing the phasor in terms of its magnitude and direction or in terms of its horizontal and vertical components. For example, if we are given a phasor of magnitude 50 at an angle of $-30°$, we could represent the phasor graphically as in Figure 19–8. In equation form we could write $r = 50 \,/\!-30°$. When the magnitude and direction are given, we say that the quantity is in *polar form*. Polar

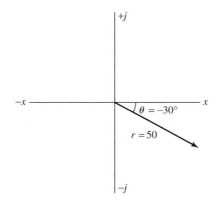

FIGURE 19–8
Phasor with polar coordinates shown.

form notation gives us the magnitude of the hypotenuse of a right triangle and the phase angle. Knowing the hypotenuse and phase angle, we can quickly determine the x and y values necessary to cause such a phasor.

$$x = r \cos \theta = 50 \cos (-30°) = 43.3$$

$$y = r \sin \theta = 50 \sin (-30°) = -25$$

In Chapter 15 we saw how the j operator could be used to indicate 90° rotation from the x-axis. $+j$ indicates 90° rotation in a positive direction and $-j$ indicates 90° rotation in a negative direction. $+j$, then, puts us at 90° and $-j$ puts us at 270° or $-90°$. We can use the j operator anytime we want to show 90° rotation. This means that we could write the x and y values above in this manner:

$$r = 43.3 - j25$$

We have expressed the magnitude of the hypotenuse in terms of its rectangular coordinates. A phasor expressed in these terms is said to be in *rectangular form*.

$$r_{\text{polar}} = 50 \; \underline{/-30°}$$

$$r_{\text{rect}} = 43.3 - j25$$

Either form describes the phasor.

Most scientific or engineering calculators will convert between polar and rectangular forms.

EXAMPLE 19-11

Convert $43.3 - j25$ from rectangular to polar form using the calculator.

SOLUTION   The calculator sequence is

TI: 43.3 $\boxed{X \leftrightarrow Y}$ 25 $\boxed{\pm}$ $\boxed{R \rightarrow P}$ *Answer:* 50

$\boxed{X \leftrightarrow Y}$ *Answer:* -30.0

The calculator stores the hypotenuse, 50, in the x register, and the angle, $-30°$, in the y register. We perform the second x-y interchange so that the angle is displayed.

Casio: $\boxed{\text{Pol (}}$ 43.3 $\boxed{,}$ $\boxed{-}$ 25 $\boxed{=}$ *Answer:* 50

$\boxed{\text{RCL}}$ $\boxed{\text{F}}$ *Answer:* -30.0

$$43.3 - j25 = 50 \; \underline{/-30°}$$

EXAMPLE 19-12

Convert $50 \; \underline{/-30°}$ from polar to rectangular form using the calculator.

SOLUTION   The calculator sequence is

TI: 50 $\boxed{X\leftrightarrow Y}$ 30 $\boxed{\pm}$ $\boxed{P\rightarrow R}$ *Answer:* 43.3

Casio: $\boxed{Rec\,(}$ 50 $\boxed{,}$ $\boxed{-}$ 30 $\boxed{=}$ *Answer:* 43.3

The real-number part, which is the side adjacent to the phase angle θ, is displayed. To display the imaginary part, the opposite side, we again use the *x-y* interchange.

TI: $\boxed{X\leftrightarrow Y}$ *Answer:* −25.0

Casio: $\boxed{RCL}$ $\boxed{F}$ *Answer:* −25.0

Polar and rectangular notation is used in ac circuit analysis to define the impedance (Z) and the applied voltage (E) in a series circuit. It is also used to define admittance (Y) and total current (I) in a parallel circuit.

19-5-1 Series RC and RL Circuits

Consider the circuit in Figure 19–9(a). We could express E in both polar and rectangular forms.

$$E \underline{/\theta} = V_R - jV_C$$

EXAMPLE 19–13

Let $V_R = 20$ V and $V_C = -30$ V in Figure 19–9(a). Find E and θ. Express E in both polar and rectangular forms.

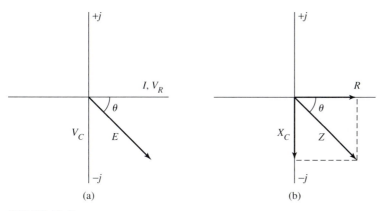

(a)   (b)

FIGURE 19–9

SOLUTION

$$E_{\text{rect}} = 20 \text{ V} - j30 \text{ V}$$

$$\tan \theta = \frac{V_C}{V_R} = \frac{-30 \text{ V}}{20 \text{ V}} = -1.5$$

$$\tan^{-1}(-1.5) = \theta = -56.3°$$

$$\sin \theta = \frac{V_C}{E}$$

$$E = \frac{V_C}{\sin \theta} = \frac{-30 \text{ V}}{\sin(-56.3°)} = 36.1 \text{ V}$$

$$E_{\text{polar}} = 36.1 \underline{/-56.3°} \text{ V}$$

In Figure 19–9(b) we show the relationship between X_C, R, and Z. Z in polar and rectangular forms is

$$Z \underline{/\theta} = R - jX_C$$

EXAMPLE 19–14

Let $R = 7.5 \text{ k}\Omega$ and $\theta = -27.5°$ in Figure 19–9(b). Express Z in both polar and rectangular forms.

SOLUTION

$$\cos \theta = \frac{R}{Z}$$

$$Z = \frac{R}{\cos \theta} = \frac{7.5 \text{ k}\Omega}{\cos(-27.5°)} = 8.46 \text{ k}\Omega$$

$$Z_{\text{polar}} = 8.46 \underline{/-27.5°} \text{ k}\Omega$$

$$X_C = Z \sin \theta = 8.46 \text{ k}\Omega \times \sin(-27.5°) = -3.9 \text{ k}\Omega$$

$$Z_{\text{rect}} = 7.5 \text{ k}\Omega - j3.9 \text{ k}\Omega$$

In the RL circuit of Figure 19–10 we could express E in both polar and rectangular forms.

$$E \underline{/\theta} = V_R + jV_L$$

EXAMPLE 19–15

Let $V_R = 15 \text{ V}$ and $V_L = 10 \text{ V}$ in Figure 19–10(a) on the next page. Find E and θ. Express E in both polar and rectangular forms.

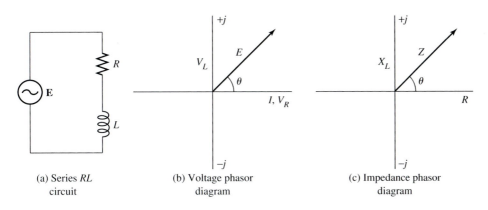

(a) Series *RL* circuit

(b) Voltage phasor diagram

(c) Impedance phasor diagram

FIGURE 19–10
An *RL* series
circuit with
phasor
diagrams.

SOLUTION

$$E_{\text{rect}} = 15\ \text{V} + j10\ \text{V}$$

$$\tan \theta = \frac{V_L}{V_R} = \frac{10\ \text{V}}{15\ \text{V}} = 0.667$$

$$\tan^{-1} 0.667 = \theta = 33.7°$$

$$\cos \theta = \frac{V_R}{E}$$

$$E = \frac{V_R}{\cos \theta} = \frac{15\ \text{V}}{\cos 33.7°} = 18.0\ \text{V}$$

$$E_{\text{polar}} = 18.0\ \underline{/33.7°}\ \text{V}$$

In Figure 19–10(c) we show the relationship between X_L, R, and Z. Z in polar and rectangular forms is

$$Z\ \underline{/\theta} = R + jX_L$$

EXAMPLE 19–16

Let $X_L = 27\ \text{k}\Omega$ and $\theta = 55°$ in Figure 19–10. Express Z in both polar and rectangular forms.

SOLUTION

$$\sin \theta = \frac{X_L}{Z}$$

$$Z = \frac{X_L}{\sin \theta} = \frac{27\ \text{k}\Omega}{\sin 55°} = 33\ \text{k}\Omega$$

$$Z_{\text{polar}} = 33\ \underline{/55°}\ \text{k}\Omega$$

$$R = Z \cos \theta = 33\ \text{k}\Omega \cos 55° = 18.9\ \text{k}\Omega$$

$$Z_{\text{rect}} = 18.9\ \text{k}\Omega + j27\ \text{k}\Omega$$

PRACTICE PROBLEMS 19–5

Express your answers to the following problems in (a) polar form and (b) rectangular form:

1. $R = 3.3$ kΩ
   $\theta = 30°$
2. $X_C = 12.8$ kΩ
   $\theta = -60°$
3. $Z = 17$ kΩ
   $X_L = 10$ kΩ
4. $R = 270$ Ω
   $X_L = 500$ Ω
5. $V_C = 8.6$ V
   $\theta = -35°$
6. $V_R = 7.3$ V
   $\theta = 47°$
7. $E = 15$ V
   $\theta = -25°$
8. $V_R = 10.7$ V
   $V_L = 6.3$ V

SOLUTIONS

1. $Z_{\text{polar}} = 3.81 \; \underline{/30°}$ kΩ
   $Z_{\text{rect}} = 3.3$ k$\Omega + j1.91$ kΩ
2. $Z_{\text{polar}} = 14.8 \; \underline{/-60°}$ kΩ
   $Z_{\text{rect}} = 7.39$ k$\Omega - j12.8$ kΩ
3. $Z_{\text{polar}} = 17 \; \underline{/36°}$ kΩ
   $Z_{\text{rect}} = 13.7$ k$\Omega + j10$ kΩ
4. $Z_{\text{polar}} = 568 \; \underline{/61.6°}$ Ω
   $Z_{\text{rect}} = 270$ $\Omega + j500$ Ω
5. $E_{\text{polar}} = 15 \; \underline{/-35°}$ V
   $E_{\text{rect}} = 12.3$ V $- j8.6$ V
6. $E_{\text{polar}} = 10.7 \; \underline{/47°}$ V
   $E_{\text{rect}} = 7.3$ V $+ j7.83$ V
7. $E_{\text{polar}} = 15 \; \underline{/-25°}$ V
   $E_{\text{rect}} = 13.6$ V $- j6.34$ V
8. $E_{\text{polar}} = 12.4 \; \underline{/30.5°}$ V
   $E_{\text{rect}} = 10.7$ V $+ j6.3$ V

Additional practice problems are at the end of the chapter.

19-6   ADDITION AND SUBTRACTION OF PHASORS

☞ **KEY POINT** Only phasors that lie along the same line can be added or subtracted directly in polar form. Phasors not on the same line must be added in rectangular form.

For example, in Figure 19–11 we have plotted a number of phasors on a system of rectangular coordinates. These phasors are labeled **A** through **G**. We could add phasors **A** and **B** algebraically. The resultant phasor would be 2 units long and would fall along the positive x-axis. Phasors **E** and **G** could be added together. The resultant phasor would be 2 units long and would be in the direction of phasor **E**. Since phasor **F** does not fall on the same plane as any other phasor, it cannot be added to any other phasor.

Consider the phasors labeled Z_1 and Z_2 in Figure 19–12. In polar form the impedances are

$$Z_1 = 1 \; \underline{/30°} \text{ k}\Omega$$

$$Z_2 = 1.5 \; \underline{/50°} \text{ k}\Omega$$

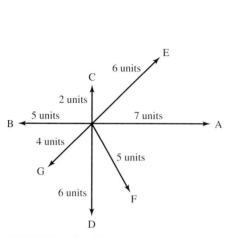

FIGURE 19–11 Phasors shown on a system of rectangular coordinates.

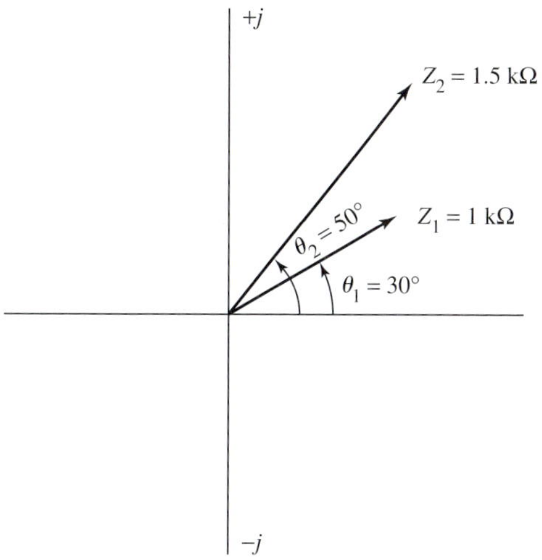

FIGURE 19–12 Two phasors on the same system of rectangular coordinates.

We cannot find Z_T by adding Z_1 and Z_2 because they don't lie along the same line. However, their rectangular components do fall along the same line. R_1, the side adjacent to θ_1, and R_2, the side adjacent to θ_2, fall on the same line. XL_1, the side opposite θ_1, and XL_2, the side opposite θ_2, fall on the same line. Therefore, if we change Z_1 and Z_2 to rectangular form, we can add the rectangular components. The result of this addition is Z_T in rectangular form. This is easily converted to polar form so that we know the magnitude of Z_T and the phase angle.

$$Z_1 = 1 \underline{/30°}\ k\Omega\ \ \ = 866\ \Omega + j500\ \Omega$$
$$Z_2 = 1.5 \underline{/50°}\ k\Omega = 964\ \Omega + j1150\ \Omega$$
$$Z_T = 1830\ \Omega + j1650\ \Omega$$
$$Z_T = 2.46 \underline{/42.0°}\ k\Omega$$

EXAMPLE 19–17

$Z_1 = 2 \underline{/30°}\ k\Omega$ and $Z_2 = 2 \underline{/-60°}\ k\Omega$. Find Z_T. Express your answer in polar form.

SOLUTION

$$Z_1 = 2 \underline{/30°}\ k\Omega\ \ \ = 1.73\ k\Omega + j1.00\ k\Omega$$
$$Z_2 = 2 \underline{/-60°}\ k\Omega = 1.00\ k\Omega - j1.73\ k\Omega$$
$$Z_T = 2.73\ k\Omega - j0.732\ k\Omega$$
$$Z_T = 2.83 \underline{/-15.0°}\ k\Omega$$

Find Z_T in the following problems. Express your answers in polar form.

1. $Z_1 = 17\ \underline{/60°}$ kΩ, $Z_2 = 30\ \underline{/20°}$ kΩ
2. $Z_1 = 270\ \underline{/-35°}$ Ω, $Z_2 = 1\ \underline{/65°}$ kΩ
3. $Z_1 = 100\ \underline{/-70°}$ kΩ,
   $Z_2 = 100\ \underline{/-40°}$ kΩ
4. $Z_1 = 48\ \underline{/20°}$ kΩ, $Z_2 = 70\ \underline{/-65°}$ kΩ
5. $Z_1 = 6.8\ \underline{/70°}$ kΩ, $Z_2 = 3\ \underline{/25°}$ kΩ

SOLUTIONS

1. $44.4\ \underline{/34.3°}$ kΩ
2. $990\ \underline{/49.4°}$ Ω
3. $193\ \underline{/-55.0°}$ kΩ
4. $88.3\ \underline{/-32.2°}$ kΩ
5. $9.17\ \underline{/56.6°}$ kΩ

Additional practice problems are at the end of the chapter.

19–7   EQUIVALENT SERIES CIRCUIT

☞ **KEY POINT**   The equivalent series circuit (ESC) of components in series is the resultant of the vector addition of the circuit R, C, and L.

Any passive ac network may be reduced to a resistance in series with some reactance. The reactance may be either capacitive or inductive depending on the component values. In some circuits the equivalent circuit is purely resistive. Circuits such as these will be discussed later in the chapter.

In a circuit such as the one in Figure 19–13(a), the total resistance is the sum of the individual resistances. $R_T = R_1 + R_2 = 12$ kΩ. Once we have found R_T, we can find Z.

$$Z_{\text{rect}} = 12\ \text{k}\Omega + j15\ \text{k}\Omega$$

$$Z_{\text{polar}} = 19.2\ \underline{/51.3°}\ \text{k}\Omega$$

FIGURE 19–13
(a) *RL* series circuit; (b) equivalent circuit.

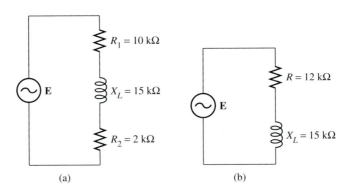

(a)   (b)

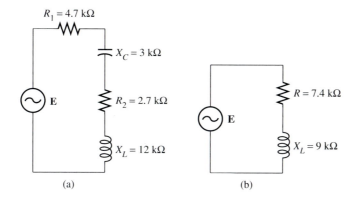

FIGURE 19–14
(a) *RCL* series
circuit;
(b) equivalent
circuit when X_L
is greater than
X_C.

(a)   (b)

The equivalent series circuit (ESC) is shown in Figure 19–13(b).

In a circuit such as the one in Figure 19–14, we could express Z as the sum of all the resistances and reactances. All the resistances are added together to get R_T. The reactances are added together to find the resultant reactance. Since X_C and X_L are 180° out of phase, the total reactance will be the difference between X_L and X_C.

$$Z_{rect} = R_1 + R_2 + jX_L - jX_C$$

$$= 7.4 \text{ k}\Omega + j12 \text{ k}\Omega - j3 \text{ k}\Omega$$

$$= 7.4 \text{ k}\Omega + j9 \text{ k}\Omega$$

$$Z_{polar} = 11.7 \underline{/50.6°} \text{ k}\Omega$$

The ESC is shown in Figure 19–14(b).

EXAMPLE 19–18

In Figure 19–14(a), let $f = 2.5$ kHz, $C = 10$ nF, $L = 600$ mH, $R_1 = 3.3$ kΩ, and $R_2 = 2$ kΩ. Find the ESC. Include component values.

SOLUTION

$$X_L = 2\pi f L = 9.42 \text{ k}\Omega$$

$$X_C = \frac{1}{2\pi f C} = 6.37 \text{ k}\Omega$$

$$Z_{rect} = 3.3 \text{ k}\Omega + 2 \text{ k}\Omega + j9.42 \text{ k}\Omega - j6.37 \text{ k}\Omega$$

$$= 5.3 \text{ k}\Omega + j3.06 \text{ k}\Omega$$

The ESC includes a resistance of 5.3 kΩ and an inductive reactance of 3.06 kΩ. The equivalent inductance is

$$L = \frac{X_L}{2\pi f} = \frac{3.06 \text{ k}\Omega}{2\pi \times 2.5 \text{ kHz}} = 195 \text{ mH}$$

The ESC is shown in Figure 19–15 on the next page.

FIGURE 19–15
Equivalent
circuit of Figure
19–14 showing
equivalent values
of R and L.

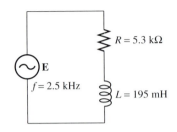
R = 5.3 kΩ

E

f = 2.5 kHz

L = 195 mH

PRACTICE PROBLEMS 19–7

1. Refer to Figure 19–13(a). Find the ESC if $R_1 = 27$ kΩ, $R_2 = 12$ kΩ, and $X_L = 30$ kΩ. Express Z in both polar and rectangular forms.

2. Refer to Figure 19–14(a). Let $R_1 = 1.5$ kΩ, $R_2 = 1.2$ kΩ, $X_L = 1.5$ kΩ, and $X_C = 2.7$ kΩ. Find the ESC. Express Z in both polar and rectangular forms.

3. Refer to Figure 19–14(a). Let $R_1 = 1.5$ kΩ, $R_2 = 2.7$ kΩ, $C = 50$ nF, $L = 800$ mH, and $f = 400$ Hz. Draw the ESC. Include component values.

4. Refer to Figure 19–16. Find Z_{rect}, Z_{polar}, V_R, V_C, V_L, and I. Draw the ESC.

SOLUTIONS

1. $Z_{polar} = 49.2 \,\underline{/37.6°}$ kΩ
   $Z_{rect} = 39$ kΩ $+ j30$ kΩ.
   See Figure 19–17.

3. $Z_{rect} = 4.2$ kΩ $- j5.95$ kΩ
   $Z_{polar} = 7.28 \,\underline{/-54.8°}$ kΩ.
   See Figure 19–19.
   In the ESC: $R = 4.20$ kΩ and $C = 66.9$ nF

2. $Z_{rect} = 2.7$ kΩ $- j1.2$ kΩ
   $Z_{polar} = 2.95 \,\underline{/-24.0°}$ kΩ. See Figure 19–18.

4. $Z_{rect} = 1$ kΩ $- j2.16$ kΩ
   $Z_{polar} = 2.38 \,\underline{/-65.2°}$ kΩ. See Figure 19–20.
   $V_R = 4.20$ V
   $V_C = 22.3$ V
   $V_L = 13.2$ V
   $I = 4.20$ mA

Additional practice problems are at the end of the chapter.

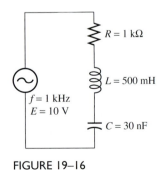
R = 1 kΩ

L = 500 mH

f = 1 kHz
E = 10 V

C = 30 nF

FIGURE 19–16

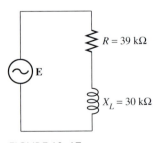
R = 39 kΩ

E

$X_L = 30$ kΩ

FIGURE 19–17

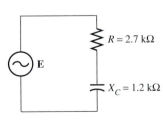
R = 2.7 kΩ

E

$X_C = 1.2$ kΩ

FIGURE 19–18

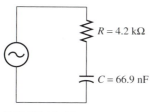

FIGURE 19–19

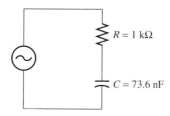

FIGURE 19–20

In circuits that include capacitance and inductance, there is a frequency where $X_L = X_C$. This frequency is called the *resonant frequency*. At this frequency (labeled f_r) the phase angle is 0°. The circuit is purely resistive because the resultant reactance is 0 Ω. The impedance is equal to R. If Z equals R, then Z must be at its minimum value. This makes I equal to its maximum value. So far we have listed the following conditions at resonance:

$$X_C = X_L$$

$$\theta = 0°$$

$$Z = R \text{ and is minimum}$$

$$I \text{ is maximum}$$

The voltage drop across R must equal the applied voltage. The voltage drop across C would equal the voltage drop across L.

$$V_R = E$$

$$V_C = V_L$$

The resonant frequency can be determined in the following manner: Since resonance is defined as that frequency where $X_C = X_L$, then by substitution we can write

$$\frac{1}{2\pi fC} = 2\pi fL$$

Solving for f, we get

$$f = f_r = \frac{1}{2\pi\sqrt{LC}}$$

EXAMPLE 19–19

In Figure 19–21(a) find f_r; at this frequency find V_R, V_L, V_C, X_L, X_C, Z, and I. Draw the ESC.

FIGURE 19–21

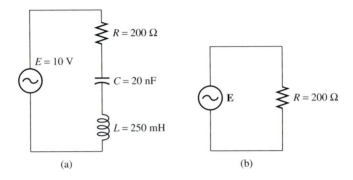

(a)

(b)

SOLUTION

$$f_r = \frac{1}{2\pi\sqrt{LC}}$$

$$= \frac{1}{2 \times \pi \times \sqrt{250 \times 10^{-3} \times 20 \times 10^{-9}}} = 2.25 \text{ kHz}$$

Using the calculator,

TI: 2 × π × (250 EE +/− 3 ×

20 EE +/− 9) √x = 1/x

Casio: 1 ÷ (2 × π × √ (250 EXP (−)

3 × 20 EXP (−) 9) =   Answer: f_r = 2.25 kHz

Knowing f_r, we can solve the rest of the problem by using Ohm's law or trig functions.

$$X_L = 2\pi f L = 3.54 \text{ k}\Omega$$

$$X_C = \frac{1}{2\pi f C} = 3.54 \text{ k}\Omega$$

If your values of X_L and X_C are not exactly equal it is because you rounded f_r to 2.25 kHz. To five places, $f_r = 2.2508$ kHz, which makes $X_C = X_L = 3.5355$ kΩ. Let's stick to our three-place accuracy and remember that some slight differences may be the result. If

$$X_C = X_L \cong 3.54 \text{ k}\Omega$$

then Z is

$$Z_{rect} = 200 \ \Omega + j3.54 \ k\Omega - j3.54 \ k\Omega$$

$$= 200 \ \Omega + j0$$

$$Z_{polar} = 200 \ \underline{/0°} \ \Omega$$

$$I = \frac{E}{Z} = \frac{10 \ V}{200 \ \Omega} = 50 \ mA$$

$$V_R = IR = 10 \ V$$

$$V_C = IX_C = 177 \ V$$

$$V_L = IX_L = 177 \ V$$

The ESC is shown in Figure 19–21(b).

The fact that V_C and V_L are both greater than E does not mean we got something for nothing, and it does not mean that we have violated Kirchhoff's laws. Remember, V_C and V_L are 180° out of phase. They are not 177 V at the same time. Added vectorially, the resultant voltage drop is 0 V.

Let's see what conditions exist above and below resonance.

EXAMPLE 19–20

Assume that the operating frequency in Figure 19–21 is 2.5 kHz. Find X_L, X_C, V_R, V_L, V_C, θ, and I. Draw the ESC.

SOLUTION

$$X_C = \frac{1}{2\pi fC} = 3.18 \ k\Omega$$

$$X_L = 2\pi fL = 3.93 \ k\Omega$$

$$Z_{rect} = 200 \ \Omega + j3.93 \ k\Omega - j3.18 \ k\Omega$$

$$= 200 \ \Omega + j744 \ \Omega$$

$$Z_{polar} = 770 \ \underline{/75.0°} \ \Omega$$

$$I = \frac{E}{Z} = \frac{10 \ V}{770 \ \Omega} = 13.0 \ mA$$

$$V_L = 51.0 \ V$$

$$V_C = 41.3 \ V$$

$$V_R = 2.60 \ V$$

The ESC is shown in Figure 19–22 on the next page. The resultant reactance is 744 Ω and is inductive. This inductance is

$$L = \frac{X_L}{2\pi f} = 47.4 \ mH$$

FIGURE 19–22

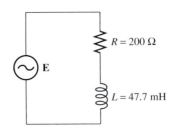

EXAMPLE 19–21

Change the operating frequency to 2 kHz in Figure 19–21. Find X_L, X_C, V_R, V_L, V_C, Z, θ, and I. Draw the ESC.

SOLUTION

$$X_L = 2\pi fL = 3.14 \text{ k}\Omega$$

$$X_C = \frac{1}{2\pi fC} = 3.98 \text{ k}\Omega$$

$$Z_{\text{rect}} = 200 \ \Omega + j3.14 \text{ k}\Omega - j3.98 \text{ k}\Omega$$

$$= 200 \ \Omega - j840 \ \Omega$$

$$Z_{\text{polar}} = 863 \ \underline{/-76.6°} \ \Omega$$

$$I = \frac{E}{Z} = \frac{10 \text{ V}}{863 \ \Omega} = 11.6 \text{ mA}$$

$$V_R = IR = 2.32 \text{ V}$$

$$V_C = IX_C = 46.1 \text{ V}$$

$$V_L = IX_L = 36.4 \text{ V}$$

The ESC is shown in Figure 19–23. The resultant reactance is 840 Ω and is capacitive. This capacitance is

$$C = \frac{1}{2\pi fX_C} = 94.7 \text{ nF}$$

FIGURE 19–23

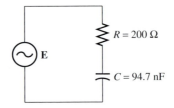

The examples above show that at frequencies above resonance the equivalent circuit is RL. Below resonance, the equivalent circuit is RC. The impedance increased as we moved away from resonance, and the current decreased.

PRACTICE PROBLEMS 19–8

Refer to Figure 19–21.

1. $R = 500\ \Omega$, $C = 30\ nF$, $L = 100\ mH$, and $E = 25\ V$. Find f_r. Determine X_L, X_C, V_R, V_C, V_L, Z, θ, I, and the ESC at (a) $f = f_r$, (b) $f = 3.5\ kHz$, and (c) $f = 2.7\ kHz$.

2. $R = 1\ k\Omega$, $C = 500\ pF$, $L = 50\ mH$, and $E = 15\ V$. Find f_r. Determine X_L, X_C, V_R, V_C, V_L, Z, θ, I, and the ESC at (a) $f = f_r$, (b) $f = 33\ kHz$, and (c) $f = 29\ kHz$.

SOLUTIONS

1. $f_r = 2.91\ kHz$
   (a) $X_L = 1.83\ k\Omega$, $X_C = 1.83\ k\Omega$,
   $Z_{rect} = 500\ \Omega + j0$,
   $Z_{polar} = 500\ \underline{/0°}\ \Omega$, $I = 50\ mA$,
   $V_R = 25\ V$, $V_C = -91.3\ V$, and
   $V_L = 9.13\ V$. The ESC consists of
   $500\ \Omega$ of resistance.
   (b) $X_L = 2.20\ k\Omega$, $X_C = 1.52\ k\Omega$,
   $Z_{rect} = 500\ \Omega + j683\ \Omega$,
   $Z_{polar} = 847\ \underline{/53.8°}\ \Omega$,
   $I = 29.5\ mA$, $V_R = 14.8\ V$,
   $V_L = 64.9\ V$, and $V_C = -44.8\ V$.
   The ESC consists of $500\ \Omega$ of
   resistance and $31.1\ mH$ of
   inductance.
   (c) $X_L = 1.70\ k\Omega$, $X_C = 1.96\ k\Omega$,
   $Z_{rect} = 500\ \Omega - j268\ \Omega$,
   $Z_{polar} = 567\ \underline{/-28.2°}\ \Omega$,
   $I = 44.1\ mA$, $V_R = 22.0\ V$,
   $V_L = 74.7\ V$, and $V_C = -86.4\ V$.
   The ESC consists of $500\ \Omega$ of
   resistance and $220\ nF$ of
   capacitance.

2. $f_r = 31.8\ kHz$
   (a) $X_L = 10\ k\Omega$, $X_C = 10\ k\Omega$,
   $Z_{rect} = 1\ k\Omega + j0$,
   $Z_{polar} = 1\ \underline{/0°}\ k\Omega$, $I = 15\ mA$,
   $V_R = 15\ V$, $V_L = 150\ V$, and
   $V_C = -150\ V$. The ESC consists of
   $1\ k\Omega$ of resistance.
   (b) $X_L = 10.4\ k\Omega$, $X_C = 9.65\ k\Omega$,
   $Z_{rect} = 1\ k\Omega + j722\ \Omega$,
   $Z_{polar} = 1.23\ \underline{/35.8°}\ k\Omega$,
   $I = 12.2\ mA$, $V_R = 12.2\ V$, $V_L =$
   $126\ V$, and $V_C = -117\ V$. The ESC
   consists of $1\ k\Omega$ of resistance and
   $3.48\ mH$ of inductance.
   (c) $X_L = 9.11\ k\Omega$, $X_C = 11\ k\Omega$,
   $Z_{rect} = 1\ k\Omega - j1.87\ k\Omega$,
   $Z_{polar} = 2.12\ \underline{/-61.8°}\ k\Omega$,
   $I = 7.09\ mA$, $V_R = 7.09\ V$,
   $V_L = 64.6\ V$, and $V_C = -77.8\ V$.
   The ESC consists of $1\ k\Omega$ of
   resistance and $2.94\ nF$ of
   capacitance.

Additional practice problems are at the end of the chapter.

SELF-TEST 19–2

1. $X_L = 11.3\ k\Omega$, $\theta = 27.5°$. Find Z in both polar and rectangular forms.

2. $Z_1 = 2.7\ \underline{/17.5°}\ k\Omega$, $Z_2 = 4\ \underline{/-58°}\ k\Omega$. Find Z_T in both polar and rectangular forms.

3. Refer to Figure 19–24. Let $R_1 = 7.5$ kΩ, $R_2 = 4.7$ kΩ, $X_L = 20$ kΩ, and $X_C = 3.8$ kΩ. Find the ESC. Express Z in both polar and rectangular forms.

4. Refer to Figure 19–25. Let $R = 1.2$ kΩ, $C = 10$ nF, $L = 500$ mH, $E = 20$ V, and $f = 2.1$ kHz. Determine X_L, X_C, V_R, V_L, V_C, Z, θ, I, and the ESC.

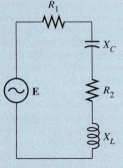

FIGURE 19–24   FIGURE 19–25

Answers to Self-test 19–2 are at the end of the chapter.

CHAPTER 19 AT A GLANCE

PAGE	KEY POINTS AND EQUATIONS	EXAMPLE
532	*Key Point:* In a series circuit, we use I as the reference. In an RC circuit, current leads, so V_C and X_C are plotted along the negative y-axis. **E** (or **Z**) is the hypotenuse, and θ is the phase angle. (See Figure 19–26.)	FIGURE 19–26
539	*Key Point:* In a series circuit, we use I as the reference. In an RL circuit, voltage leads, so V_L and X_L are plotted along the positive y-axis. **E** (or **Z**) is the hypotenuse, and θ is the phase angle. (See Figure 19–27.)	FIGURE 19–27

| 545 | Phasors are quantities that have both magnitude and direction. They can be represented in either polar form or rectangular form. | $r_{polar} = 50 \underline{/-30°}$
 $r_{rect} = 43.3 - j25$ |

550   Only phasors that lie along the same line can be added or subtracted directly in polar form. Phasors not on the same line must be added in rectangular form.

Add $Z_1 = 1 \underline{/30°}$ kΩ
and $Z_2 = 1.5 \underline{/50°}$ kΩ.
$Z_1 = 1 \underline{/30°}$ kΩ $= 866\ Ω + j500\ Ω$
$Z_2 = 1.5 \underline{/50°}$ kΩ $= 964\ Ω + j1150\ Ω$
$\qquad\qquad Z_r = 1830\ Ω + j1650\ Ω$
$\qquad\qquad Z_r = 2.46 \underline{/42°}$ kΩ

552   The equivalent series circuit (ESC) of components in series is the resultant of the vector addition of the circuit R, C, and L.

Find the ESC of the circuit in Figure 19–14.

$$\begin{aligned} ESC &= 7.4\ kΩ + j12\ kΩ \\ &\quad - j3kΩ \\ &= 7.4\ kΩ + j9\ kΩ \\ &= 11.7 \underline{/50.6°}\ kΩ \end{aligned}$$

555   Resonance in a series RCL circuit can be found using the equation

$$f_r = \frac{1}{2\pi\sqrt{LC}}$$

Find f_r if $L = 250$ mH and $C = 20$ nF.

$$f_r = \frac{1}{2\pi \times \sqrt{250\ mH \times 20\ nF}}$$
$$= 2.25\ kHz$$

END OF CHAPTER PROBLEMS 19–1

Refer to Figure 19–28 for the following problems:

1. $V_R = 20$ V, $V_C = -20$ V. Find **E** and θ.
2. $V_R = 1.8$ V, $V_C = -1$ V. Find **E** and θ.
3. $V_R = 12$ V, $V_C = -8$ V. Find **E** and θ.
4. $V_R = 55$ V, $V_C = -45$ V. Find **E** and θ.
5. $V_R = 6.35$ V, **E** $= 10$ V. Find V_C and θ.
6. $V_R = 21.3$ V, **E** $= 40$ V. Find V_C and θ.
7. $V_R = 21$ V, **E** $= 40$ V. Find V_C and θ.
8. $V_R = 6$ V, **E** $= 12$ V. Find V_C and θ.

FIGURE 19–28

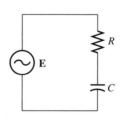

9. $V_R = 15$ V, $\theta = -40°$. Find **E** and V_C.

10. $V_R = 10$ V, $\theta = -70°$. Find **E** and V_C.

11. $V_R = 5.5$ V, $\theta = -25°$. Find **E** and V_C.

12. $V_R = 20$ V, $\theta = -60°$. Find **E** and V_C.

13. **E** $= 18$ V, $\theta = -35.6°$. Find V_R and V_C.

14. **E** $= 9$ V, $\theta = -40°$. Find V_R and V_C.

15. **E** $= 70$ V, $\theta = -27°$. Find V_R and V_C.

16. **E** $= 5$ V, $\theta = -50°$. Find V_R and V_C.

17. **E** $= 30$ V, $V_C = -20$ V. Find V_R and θ.

18. **E** $= 12$ V, $V_C = -3$ V. Find V_R and θ.

19. **E** $= 50$ V, $V_C = -40$ V. Find V_R and θ.

20. **E** $= 8$ V, $V_C = -6$ V. Find V_R and θ.

21. $V_C = -5.5$ V, $\theta = -32°$. Find **E** and V_R.

22. $V_C = -20$ V, $\theta = -55°$. Find **E** and V_R.

23. $V_C = -30$ V, $\theta = -70°$, Find **E** and V_R.

24. $V_C = -3$ V, $\theta = -20°$. Find **E** and V_R.

END OF CHAPTER PROBLEMS 19–2

Refer to Figure 19–28 for the following problems and find X_C, Z, and θ:

1. $R = 10$ kΩ, $C = 50$ nF, $f = 500$ Hz
2. $R = 4.7$ kΩ, $C = 10$ nF, $f = 4.5$ kHz
3. $R = 700$ Ω, $C = 1$ μF, $f = 200$ Hz
4. $R = 18$ kΩ, $C = 200$ pF, $f = 40$ kHz
5. $R = 2$ kΩ, $C = 20$ nF, $f = 2.5$ kHz
6. $R = 1$ kΩ, $C = 1$ nF, $f = 250$ kHz
7. $R = 27$ kΩ, $C = 5$ nF, $f = 1.55$ kHz
8. $R = 12$ kΩ, $C = 30$ nF, $f = 600$ Hz
9. $R = 3.3$ kΩ, $C = 0.1$ μF, $f = 2.3$ kHz
10. $R = 680$ Ω, $C = 2$ μF, $f = 100$ Hz

Refer to Figure 19–28 for the following problems and find X_C, Z, θ, V_R, V_C, and I:

11. $R = 2$ kΩ, $C = 40$ nF, $f = 3$ kHz, **E** $= 30$ V
12. $R = 25$ kΩ, $C = 500$ pF, $f = 10$ kHz, **E** $= 5$ V
13. $R = 10$ kΩ, $C = 100$ pF, $f = 100$ kHz, **E** $= 10$ V
14. $R = 200$ Ω, $C = 2$ μF, $f = 250$ Hz, **E** $= 20$ V
15. $R = 3.3$ kΩ, $C = 20$ nF, $f = 1.5$ kHz, **E** $= 30$ V
16. $R = 560$ Ω, $C = 0.25$ μF, $f = 1$ kHz, **E** $= 50$ V
17. $R = 10$ kΩ, $C = 500$ pF, $f = 6.5$ kHz, **E** $= 20$ V
18. $R = 68$ Ω, $C = 400$ nF, $f = 5$ kHz, **E** $= 10$ V
19. $R = 470$ Ω, $C = 200$ nF, $f = 2$ kHz, **E** $= 100$ V
20. $R = 2.2$ kΩ, $C = 40$ nF, $f = 2$ kHz, **E** $= 35$ V

Refer to Figure 19–28 for the following problems and find I, θ, f, X_C, V_R, and V_C:

21. $R = 8.2$ kΩ, $C = 200$ pF, Z $= 12$ kΩ, **E** $= 40$ V
22. $R = 1$ kΩ, $C = 40$ nF, Z $= 2$ kΩ, **E** $= 9$ V
23. $R = 750$ Ω, $C = 10$ nF, Z $= 1.2$ kΩ, **E** $= 15$ V
24. $R = 30$ kΩ, $C = 200$ pF, Z $= 47$ kΩ, **E** $= 40$ V

25. $R = 2.7$ kΩ, $C = 30$ pF, $Z = 3.5$ kΩ,
    $\mathbf{E} = 30$ V
26. $R = 18$ kΩ, $C = 2$ nF, $Z = 25$ kΩ,
    $\mathbf{E} = 5$ V
27. $R = 56$ kΩ, $C = 25$ nF, $Z = 64.4$ kΩ,
    $\mathbf{E} = 7.5$ V
28. $R = 2.7$ kΩ, $C = 1$ μF, $Z = 2.75$ kΩ,
    $\mathbf{E} = 9$ V
29. $R = 810$ Ω, $C = 1$ μF, $Z = 1.14$ kΩ,
    $\mathbf{E} = 12$ V
30. $R = 1.8$ kΩ, $C = 0.5$ μF,
    $Z = 1.97$ kΩ, $\mathbf{E} = 15$ V

END OF CHAPTER PROBLEMS 19–3

Refer to Figure 19–29 for the following problems:

1. $V_R = 8$ V, $V_L = 10$ V. Find
   $\mathbf{E}$ and θ.
2. $V_R = 22$ V, $V_L = 35$ V. Find
   $\mathbf{E}$ and θ.
3. $V_R = 14$ V, $V_L = 20$ V. Find
   $\mathbf{E}$ and θ.
4. $V_R = 38.7$ V, $V_L = 24.3$ V. Find
   $\mathbf{E}$ and θ.
5. $V_R = 6.7$ V, $\theta = 27°$. Find
   $\mathbf{E}$ and V_L.
6. $V_R = 15$ V, $\theta = 50°$. Find
   $\mathbf{E}$ and V_L.
7. $V_R = 27.4$ V, $\theta = 67.4°$. Find
   $\mathbf{E}$ and V_L.
8. $V_R = 9.45$ V, $\theta = 34.4°$. Find
   $\mathbf{E}$ and V_L.
9. $V_R = 8.35$ V, $\mathbf{E} = 12$ V. Find
   V_L and θ.
10. $V_R = 15$ V, $\mathbf{E} = 40$ V. Find
    V_L and θ.
11. $V_R = 43.7$ V, $\mathbf{E} = 60$ V. Find
    V_L and θ.
12. $V_R = 5.36$ V, $\mathbf{E} = 15$ V. Find
    V_L and θ.
13. $V_L = 15$ V, $\mathbf{E} = 30$ V. Find
    V_R and θ.
14. $V_L = 30$ V, $\mathbf{E} = 45$ V. Find
    V_R and θ.
15. $V_L = 10$ V, $\mathbf{E} = 17$ V. Find
    V_R and θ.
16. $V_L = 27.5$ V, $\mathbf{E} = 45.5$ V. Find
    V_R and θ.
17. $V_L = 3$ V, $\theta = 62°$. Find
    $\mathbf{E}$ and V_R.
18. $V_L = 10$ V, $\theta = 47.5°$. Find
    $\mathbf{E}$ and V_R.
19. $V_L = 15.5$ V, $\theta = 30°$. Find
    $\mathbf{E}$ and V_R.
20. $V_L = 67.4$ V, $\theta = 63.4°$. Find
    $\mathbf{E}$ and V_R.
21. $\mathbf{E} = 9$ V, $\theta = 58°$. Find
    V_R and V_L.
22. $\mathbf{E} = 22$ V, $\theta = 32.5°$. Find
    V_R and V_L.
23. $\mathbf{E} = 50$ V, $\theta = 45°$. Find
    V_R and V_L.
24. $\mathbf{E} = 40$ V, $\theta = 63°$. Find
    V_R and V_L.

FIGURE 19–29

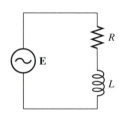

END OF CHAPTER PROBLEMS 19–4

Refer to Figure 19–29 and find X_L, Z, and θ in the following problems:

1. $R = 3.9$ kΩ, $L = 600$ mH, $f = 1$ kHz
2. $R = 12$ kΩ, $L = 100$ mH, $f = 20$ kHz
3. $R = 500$ Ω, $L = 40$ mH, $f = 1.5$ kHz
4. $R = 33$ kΩ, $L = 2$ mH, $f = 2.5$ kHz
5. $R = 27$ kΩ, $L = 250$ mH, $f = 8.5$ kHz
6. $R = 56$ Ω, $L = 750$ μH, $f = 17.5$ kHz
7. $R = 1.8$ kΩ, $L = 10$ mH, $f = 20$ kHz
8. $R = 910$ Ω, $L = 500$ μH, $f = 455$ kHz

Refer to Figure 19–29 and find X_L, Z, θ, V_R, V_L, and I in the following problems:

9. $R = 1$ kΩ, $L = 500$ mH, $f = 250$ Hz, $\mathbf{E} = 15$ V
10. $R = 22$ kΩ, $L = 150$ mH, $f = 10$ kHz, $\mathbf{E} = 5$ V
11. $R = 300$ Ω, $L = 150$ mH, $f = 500$ Hz, $\mathbf{E} = 50$ V
12. $R = 10$ kΩ, $L = 40$ mH, $f = 10$ kHz, $\mathbf{E} = 25$ V
13. $R = 27$ kΩ, $L = 1.35$ H, $f = 2.35$ kHz, $\mathbf{E} = 20$ V
14. $R = 150$ kΩ, $L = 2.5$ H, $f = 12.7$ kHz, $\mathbf{E} = 9$ V
15. $R = 100$ Ω, $L = 500$ μH, $f = 25$ kHz, $\mathbf{E} = 12$ V
16. $R = 470$ Ω, $L = 65$ mH, $f = 1.55$ kHz, $\mathbf{E} = 18$ V

Refer to Figure 19–29 and find X_L, θ, f, $\mathbf{E}$, V_R, and I in the following problems:

17. $R = 4.7$ kΩ, $Z = 6$ kΩ, $L = 650$ mH, $V_L = 7.5$ V
18. $R = 68$ kΩ, $Z = 100$ kΩ, $L = 2$ H, $V_L = 12$ V
19. $R = 18$ kΩ, $Z = 32$ kΩ, $L = 75$ mH, $V_L = 3.7$ V
20. $R = 820$ Ω, $Z = 1.6$ kΩ, $L = 800$ mH, $V_L = 8.7$ V
21. $R = 22$ kΩ, $Z = 35.7$ kΩ, $L = 800$ mH, $V_L = 17$ V
22. $R = 33$ kΩ, $Z = 47.3$ kΩ, $L = 760$ mH, $V_L = 7.8$ V
23. $R = 750$ Ω, $Z = 1.77$ kΩ, $L = 100$ mH, $V_L = 27$ V
24. $R = 560$ Ω, $Z = 940$ Ω, $L = 200$ mH, $V_L = 15.6$ V

Refer to Figure 19–29 and find X_L, L, θ, V_R, $\mathbf{E}$, and I in the following problems:

25. $R = 6.8$ kΩ, $Z = 11$ kΩ, $f = 4.3$ kHz, $V_L = 6$ V
26. $R = 910$ Ω, $Z = 1.25$ kΩ, $f = 2$ kHz, $V_L = 13.6$ V
27. $R = 2.2$ kΩ, $Z = 4.8$ kΩ, $f = 7.5$ kHz, $V_L = 4.5$ V
28. $R = 680$ Ω, $Z = 900$ Ω, $f = 25$ kHz, $V_L = 14$ V
29. $R = 47$ kΩ, $Z = 83$ kΩ, $f = 5.5$ kHz, $V_L = 20$ V
30. $R = 22$ kΩ, $Z = 30.5$ kΩ, $f = 2.3$ kHz, $V_L = 6.7$ V
31. $R = 3.9$ kΩ, $Z = 6.75$ kΩ, $f = 12$ kHz, $V_L = 8.5$ V
32. $R = 5.6$ kΩ, $Z = 8.32$ kΩ, $f = 9.5$ kHz, $V_L = 11.3$ V

END OF CHAPTER PROBLEMS 19–5

In problems 1 through 20, find Z. Express your answers in both *polar* and *rectangular* forms:

1. $X_L = 23$ kΩ
   $\theta = 48°$
2. $X_L = 17.5$ kΩ
   $\theta = 63.2°$
3. $X_C = 740$ Ω
   $\theta = -51°$
4. $X_C = 17.5$ kΩ
   $\theta = -32°$
5. $R = 820$ Ω
   $\theta = -17°$
6. $R = 22$ kΩ
   $\theta = 72°$
7. $Z = 6.35$ kΩ
   $\theta = -45°$
8. $Z = 38.6$ kΩ
   $\theta = 36°$
9. $X_C = 30$ kΩ
   $R = 40$ kΩ
10. $X_C = 5.85$ kΩ
    $R = 7.5$ kΩ
11. $X_L = 120$ kΩ
    $R = 180$ kΩ
12. $X_L = 17.5$ kΩ
    $R = 15$ kΩ
13. $X_C = 3$ kΩ
    $Z = 4.6$ kΩ
14. $X_C = 450$ Ω
    $Z = 560$ Ω
15. $X_L = 3.75$ kΩ
    $Z = 5.35$ kΩ
16. $X_L = 800$ Ω
    $Z = 1$ kΩ
17. $X_L = 18$ kΩ
    $R = 10$ kΩ
18. $X_L = 46$ kΩ
    $R = 56$ kΩ
19. $X_C = 1.45$ kΩ
    $R = 1.8$ kΩ
20. $X_C = 18$ kΩ
    $R = 27$ kΩ

In problems 21 through 40, find **E**. Express your answers in both *polar* and *rectangular* forms:

21. $V_L = 12.7$ V
    $V_R = 10$ V
22. $V_L = 800$ mV
    $V_R = 600$ mV
23. $V_C = -6.25$ V
    $V_R = 4.32$ V
24. $V_C = -17$ V
    $V_R = 27$ V
25. $V_C = -5.7$ V
    $E = 12$ V
26. $V_C = -10$ V
    $E = 12$ V
27. $V_L = 8.5$ V
    $E = 15$ V
28. $V_L = 24$ V
    $E = 35$ V
29. $E = 15$ V
    $\theta = -40°$
30. $E = 25$ V
    $\theta = 65.5°$
31. $V_L = 7.93$ V
    $\theta = 34.6°$
32. $V_L = 12$ V
    $\theta = 73°$
33. $V_C = -64$ V
    $\theta = -28°$
34. $V_C = -28$ V
    $\theta = -56°$
35. $V_R = 2.6$ V
    $E = 5$ V
36. $V_R = 550$ mV
    $E = 1$ V
37. $V_L = 16$ V
    $E = 30$ V
38. $V_L = 22$ V
    $E = 40$ V
39. $V_C = -6$ V
    $E = 10$ V
40. $V_C = -20$ V
    $E = 30$ V

END OF CHAPTER PROBLEMS 19–6

Find Z_T in the following problems. Express your answers in polar form.

1. $Z_1 = 2 \underline{/20°}$ kΩ, $Z_2 = 3 \underline{/65°}$ kΩ
3. $Z_1 = 11 \underline{/-20°}$ kΩ,
   $Z_2 = 15 \underline{/-30°}$ kΩ
5. $Z_1 = 7.3 \underline{/-25°}$ kΩ, $Z_2 = 3 \underline{/40°}$ kΩ

7. $Z_1 = 78 \underline{/38°}$ kΩ, $Z_2 = 50 \underline{/-65°}$ kΩ
9. $Z_1 = 270 \underline{/30°}$ kΩ, $Z_2 = 150 \underline{/30°}$ kΩ
11. $Z_1 = 7.5 \underline{/-90°}$ kΩ,
    $Z_2 = 13.5 \underline{/22°}$ kΩ

2. $Z_1 = 3 \underline{/60°}$ kΩ, $Z_2 = 4.5 \underline{/30°}$ kΩ
4. $Z_1 = 600 \underline{/-35°}$ Ω, $Z_2 = 1 \underline{/-45°}$ kΩ

6. $Z_1 = 1.73 \underline{/-60°}$ kΩ,
   $Z_2 = 3.5 \underline{/-35°}$ kΩ

8. $Z_1 = 100 \underline{/70°}$ Ω, $Z_2 = 400 \underline{/-20°}$ Ω
10. $Z_1 = 5 \underline{/90°}$ kΩ, $Z_2 = 4 \underline{/0°}$ kΩ
12. $Z_1 = 10 \underline{/0°}$ kΩ, $Z_2 = 18 \underline{/-75°}$ kΩ

END OF CHAPTER PROBLEMS 19–7

1. Refer to Figure 19–30. $R_1 = 560$ Ω, $R_2 = 1.2$ kΩ, and $X_L = 3$ kΩ. Draw the ESC. Express Z in both polar and rectangular forms.
3. Refer to Figure 19–31. $R_1 = 470$ Ω, $R_2 = 680$ Ω, $X_C = 2$ kΩ, and $X_L = 3.3$ kΩ. Draw the ESC. Express Z in both polar and rectangular forms.
5. Refer to Figure 19–31. $R_1 = 1.2$ kΩ, $R_2 = 2.2$ kΩ, $C = 100$ nF, $L = 800$ mH, and $f = 1.2$ kHz. Draw the ESC. Include component values.
7. Refer to Figure 19–32. $R = 200$ Ω, $L = 200$ mH, $C = 350$ nF, $f = 500$ Hz, and $E = 25$ V. Find Z_{rect}, Z_{polar}, V_R, V_C, V_L, and I. Draw the ESC.

2. Refer to Figure 19–30. $R_1 = 1.5$ kΩ, $R_2 = 1.2$ kΩ, and $X_L = 2$ kΩ. Draw the ESC. Express Z in both polar and rectangular forms.
4. Refer to Figure 19–31. $R_1 = 2.2$ kΩ, $R_2 = 3.3$ kΩ, $X_C = 6.1$ kΩ, and $X_L = 1.3$ kΩ. Draw the ESC. Express Z in both polar and rectangular forms.
6. Refer to Figure 19–31. $R_1 = 150$ Ω, $R_2 = 330$ Ω, $C = 0.2$ μF, $L = 350$ mH, and $f = 700$ Hz. Draw the ESC. Include component values.
8. Refer to Figure 19–32. $R = 500$ Ω, $L = 400$ mH, $C = 5$ nF, $f = 3$ kHz, and $E = 20$ V. Find Z_{rect}, Z_{polar}, V_R, V_C, V_L, and I. Draw the ESC.

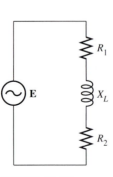

FIGURE 19–30

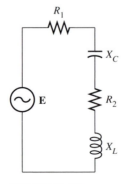

FIGURE 19–31

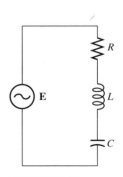

FIGURE 19–32

END OF CHAPTER PROBLEMS 19-8

Refer to Figure 19-32.

1. $R = 470 \ \Omega$, $C = 5$ nF, $L = 750$ mH, and $E = 10$ V. Find f_r. Determine X_L, X_C, V_R, V_C, V_L, Z, θ, I, and the ESC at (a) $f = f_r$, (b) $f = 2.3$ kHz, and (c) $f = 2.7$ kHz.

2. $R = 1.2$ kΩ, $C = 1$ nF, $L = 45$ mH, and $E = 20$ V. Find f_r. Determine X_L, X_C, V_R, V_C, V_L, Z, θ, I, and the ESC at (a) $f = f_r$, (b) $f = 20$ kHz, and (c) $f = 25$ kHz.

3. $R = 300 \ \Omega$, $C = 500$ nF, $L = 1$ H, and $E = 25$ V. Find f_r. Determine X_L, X_C, V_R, V_C, V_L, Z, θ, I, and the ESC at (a) $f = f_r$, (b) $f = 200$ Hz, and (c) $f = 300$ Hz.

4. $R = 560 \ \Omega$, $C = 50$ nF, $L = 200$ mH, and $E = 10$ V. Find f_r. Determine X_L, X_C, V_R, V_C, V_L, Z, θ, I, and the ESC at (a) $f = f_r$, (b) $f = 1$ kHz, and (c) $f = 2$ kHz.

ANSWERS TO SELF-TESTS

SELF-TEST 19-1

1. $\mathbf{E} = 3.64$ V; $V_R = 1.54$ V

2. $X_L = 5.5$ kΩ; $\theta = 49.5°$; $Z = 7.23$ kΩ; $V_R = 16.2$ V; $V_L = 19$ V; $I = 3.46$ mA

3. $X_L = 937 \ \Omega$; $\theta = 51.3°$; $f = 2.98$ kHz; $\mathbf{E} = 9.99$ V; $V_R = 6.24$ V; $I = 8.33$ mA

4. $V_C = -12.4$ V; $\mathbf{E} = 21$ V

5. $X_C = 3.18$ kΩ; $Z = 3.76$ kΩ; $\theta = 57.9°$; $V_R = 5.32$ V; $V_C = -8.47$ V; $I = 2.66$ mA

6. $X_C = 501 \ \Omega$; $f = 318$ Hz; $\theta = -56.6°$; $V_R = 4.95$ V; $V_C = -7.52$ V; $I = 15$ mA

SELF-TEST 19-2

1. $Z_{rect} = 21.7$ k$\Omega + j11.3$ kΩ $Z_{polar} = 24.5 \ \underline{/27.5°}$ kΩ

2. $Z_{rect} = 4.69$ k$\Omega - j2.58$ kΩ $Z_{polar} = 5.36 \ \underline{/-28.8°}$ kΩ

3. $Z_{rect} = 12.2$ k$\Omega + j16.2$ kΩ $Z_{polar} = 20.3 \ \underline{/53.0°}$ kΩ The ESC consists of 12.2 kΩ of resistance and 16.2 kΩ of inductive reactance.

4. $X_C = 7.58$ kΩ, $X_L = 6.60$ kΩ, $Z_{polar} = 1.55 \ \underline{/-39.3°}$ kΩ, $I = 12.9$ mA; $V_R = 15.5$ V, $V_C = -97.8$ V, $V_L = 85.1$ V. The ESC consists of 1.2 kΩ of resistance and 77.2 nF of capacitance.

ac Circuit Analysis: Parallel Circuits

20

Introduction

Continuing our study of ac circuits, in this chapter we will solve parallel ac circuit problems using trig functions and our calculator. We learned in our study of dc circuits in Chapters 10 and 11 that the current is the same through all components connected in series and that the voltage is the same across all parallel branches. These concepts are also true in ac circuits. In ac circuits, resistance, reactance, and impedance are the series circuit parameters. Conductance, susceptance, and admittance are the parallel circuit parameters.

In the chapters on dc circuits, we discussed the duality between current and conductance in parallel circuits and voltage drops and resistance in series circuits. This duality holds true in ac circuits between current and admittance in parallel circuits and voltage and impedance in series circuits. Students should think admittance when dealing with parallel circuits, just as they think impedance when working with series circuits. Familiarity with conductance, susceptance, and admittance is necessary as you further your studies into electronic circuits and systems, especially in the area of microwave circuits and systems.

In this chapter you will also learn how to reduce complex circuits to their equivalent series or parallel circuits in order to simplify their solutions.

Chapter Objectives

In this chapter you will learn how to:

1. Analyze *RC* and *RL* parallel circuits to determine currents, phase angle, and admittance in both polar and rectangular forms.
2. Find the equivalent impedance of impedances in parallel.
3. Convert from parallel to equivalent series circuits and from series to equivalent parallel circuits.
4. Reduce complex circuits to equivalent series circuits or to equivalent parallel circuits.

In parallel circuits the voltage drops across the various branches are equal.

> 👉 **KEY POINT**   In a parallel circuit, we use E as the reference. In an RC circuit, current leads, so I_C and B_C are plotted along the positive y-axis. I_T (or Y) is the hypotenuse and θ is the phase angle.

As shown in Figure 20–1(b), the current through the capacitor leads the voltage drop across it by 90°. This puts the total current, I_T, in the first quadrant. I_R is the side adjacent to the angle θ, and I_C is the opposite side.

EXAMPLE 20–1

In the circuit of Figure 20–1(a), let $I_R = 7$ mA and $I_C = 10$ mA. Find I_T and θ. Express I_T in both rectangular and polar forms.

SOLUTION   Refer to Figure 20–2 on the next page.

$$I_{rect} = 7 \text{ mA} + j10 \text{ mA}$$

$$\tan \theta = \frac{\text{opposite side}}{\text{adjacent side}}$$

FIGURE 20–1

(a)

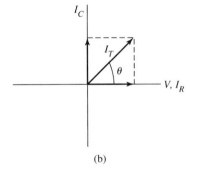

(b)

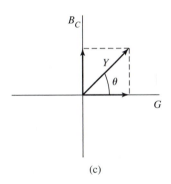

(c)

FIGURE 20–2

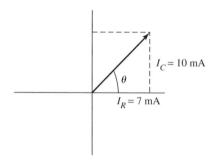

$$\tan \theta = \frac{I_C}{I_R} = \frac{10 \text{ mA}}{7 \text{ mA}} = 1.43$$

$$\tan^{-1} 1.43 = 55.0°$$

We can now find I_T by using either the sine or cosine function. We will use the cosine function.

$$\cos \theta = \frac{\text{adjacent side}}{\text{hypotenuse}}$$

$$\cos \theta = \frac{I_R}{I_T}$$

$$I_T = \frac{I_R}{\cos \theta} = \frac{7 \text{ mA}}{\cos 55.0°} = 12.2 \text{ mA}$$

$$I_{\text{polar}} = 12.2 \underline{/55.0°} \text{ mA}$$

$$I_{\text{rect}} = 7 \text{ mA} + j10 \text{ mA}$$

Using the calculator and converting from rectangular form to polar form:

TI: 7 $\boxed{\text{EE}}$ $\boxed{\pm}$ 3 $\boxed{\text{X}\leftrightarrow\text{Y}}$ 10 $\boxed{\text{EE}}$ $\boxed{\pm}$ 3 $\boxed{\text{R}\rightarrow\text{P}}$

Casio: $\boxed{\text{Pol (}}$ 7 $\boxed{\text{EXP}}$ $\boxed{(-)}$ 3 $\boxed{,}$ 10 $\boxed{\text{EXP}}$ $\boxed{(-)}$ 3 $\boxed{=}$

Answer: 12.2 × 10^{-3}

TI: $\boxed{\text{X}\leftrightarrow\text{Y}}$ Answer: 55 $I_{\text{polar}} = 12.2 \underline{/55.0°}$ mA

Casio: $\boxed{\text{RCL}}$ $\boxed{\text{F}}$

EXAMPLE 20–2

In Figure 20–1(a) let $I_T = 300 \mu A$ and $I_C = 100 \mu A$. Find I_R and θ. Express I_T in both rectangular and polar forms.

SOLUTION

$$\sin \theta = \frac{I_C}{I_T} = \frac{100 \ \mu A}{300 \ \mu A} = 0.333$$

$$\sin^{-1} 0.333 = 19.5°$$

$$\cos \theta = \frac{I_R}{I_T}$$

$$I_R = I_T \cos \theta = 300 \ \mu A \cos 19.5° = 283 \ \mu A$$

$$I_{\text{rect}} = 283 \ \mu A + j100 \ \mu A$$

$$I_{\text{polar}} = 300 \ \underline{/19.5°} \ \mu A$$

PRACTICE PROBLEMS 20–1

Refer to Figure 20–1(a). From the information given in problems 1 through 7, express I_T in rectangular and polar forms.

1. $I_T = 30$ mA and $\theta = 40°$
2. $I_R = 3.5$ mA and $I_T = 6$ mA
3. $I_C = 40 \ \mu A$ and $I_T = 80 \ \mu A$
4. $I_T = 500$ mA and $\theta = 63.2°$
5. $I_R = 200 \ \mu A$ and $I_C = 300 \ \mu A$
6. $I_R = 750 \ \mu A$ and $\theta = 37.8°$
7. $I_C = 6.75$ mA and $\theta = 64.7°$

SOLUTIONS

1. $I_{\text{rect}} = 23$ mA $+ j19.3$ mA
   $I_{\text{polar}} = 30 \ \underline{/40°}$ mA
2. $I_{\text{rect}} = 3.5$ mA $+ j4.87$ mA
   $I_{\text{polar}} = 6 \ \underline{/54.3°}$ mA
3. $I_{\text{rect}} = 69.3 \ \mu A + j40 \ \mu A$
   $I_{\text{polar}} = 80 \ \underline{/30°} \ \mu A$
4. $I_{\text{rect}} = 225$ mA $+ j446$ mA
   $I_{\text{polar}} = 500 \ \underline{/63.2°}$ mA
5. $I_{\text{rect}} = 200 \ \mu A + j300 \ \mu A$
   $I_{\text{polar}} = 361 \ \underline{/56.3°} \ \mu A$
6. $I_{\text{rect}} = 750 \ \mu A + j582 \ \mu A$
   $I_{\text{polar}} = 949 \ \underline{/37.8°} \ \mu A$
7. $I_{\text{rect}} = 3.19$ mA $+ j6.75$ mA
   $I_{\text{polar}} = 7.47 \ \underline{/64.7°}$ mA

Additional practice problems are at the end of the chapter.

RC CIRCUIT ANALYSIS— CIRCUIT PARAMETERS 20–2

We learned in the chapter on Ohm's law that conductance and resistance are reciprocals. Susceptance and reactance are reciprocals, as are admittance and impedance. Conductance, susceptance, and admittance are parallel circuit parameters, just as resistance, reactance, and impedance are series circuit parameters.

The phase relationship between conductance, G, capacitive susceptance, B_C, and admittance, Y, is shown in Figure 20–1(c). The capacitive susceptance is determined by the equation

$$B_C = 2\pi f C$$

(20–1)

EXAMPLE 20–3

Refer to Figure 20–1(a). Let $R = 3.3$ kΩ, $C = 75$ nF, and $f = 500$ Hz. Find G, B_C, Y, and θ. Express Y in rectangular and polar forms.

SOLUTION

$$B_C = 2\pi f C = 2\pi \times 500 \text{ Hz} \times 75 \text{ nF} = 236 \ \mu\text{S}$$

$$G = \frac{1}{R} = \frac{1}{3.3 \text{ k}\Omega} = 303 \ \mu\text{S}$$

$$Y_{\text{rect}} = 303 \ \mu\text{S} + j236 \ \mu\text{S}$$

$$\tan \theta = \frac{\text{opposite side}}{\text{adjacent side}} = \frac{B_C}{G} = \frac{236 \ \mu\text{S}}{303 \ \mu\text{S}} = 0.779$$

$$\tan^{-1} 0.779 = 37.9°$$

$$\sin \theta = \frac{B_C}{Y}$$

$$Y = \frac{B_C}{\sin \theta} = \frac{236 \ \mu\text{S}}{\sin 37.9°} = 384 \ \mu\text{S}$$

$$Y_{\text{polar}} = 384 \ \underline{/37.9°} \ \mu\text{S}$$

EXAMPLE 20–4

Refer to Figure 20–1(a). Let $R = 680$ Ω, $C = 100$ nF, $f = 5.5$ kHz, and $I_T = 25$ mA. Find G, B_C, Y, θ, I_R, I_C, and V. Express Y in rectangular and polar forms.

SOLUTION

$$B_C = 2\pi f C = 2\pi \times 5.5 \text{ kHz} \times 100 \text{ nF} = 3.46 \text{ mS}$$

$$G = \frac{1}{R} = \frac{1}{680 \ \Omega} = 1.47 \text{ mS}$$

$$Y_{\text{rect}} = 1.47 \text{ mS} + j3.46 \text{ mS}$$

$$\tan \theta = \frac{B_C}{G} = \frac{3.46 \text{ mS}}{1.47 \text{ mS}} = 2.35$$

$$\tan^{-1} 2.35 = 66.9°$$

$$\sin \theta = \frac{B_C}{Y}$$

CHAPTER 20

$$Y = \frac{B_C}{\sin \theta} = \frac{3.46 \text{ mS}}{\sin 66.9°} = 3.76 \text{ mS}$$

$$Y_{\text{polar}} = 3.76 \underline{/66.9°} \text{ mS}$$

$$I_{\text{polar}} = 25 \underline{/66.9°} \text{ mA}$$

$$I_R = I_T \cos \theta = 25 \text{ mA} \cos 66.9° = 9.79 \text{ mA}$$

$$I_C = I_T \sin \theta = 25 \text{ mA} \sin 66.9° = 23.0 \text{ mA}$$

$$V = \frac{I_T}{Y} = 6.66 \text{ V}$$

PRACTICE PROBLEMS 20–2

Refer to Figure 20–1(a) for problems 1 through 4. Find Y_{rect}, Y_{polar}, I_{rect}, and V.

1. $R = 15 \text{ k}\Omega$, $C = 10 \text{ nF}$, $f = 1.5 \text{ kHz}$, $I_T = 500 \ \mu\text{A}$
2. $R = 1.8 \text{ k}\Omega$, $C = 200 \text{ nF}$, $f = 350 \text{ Hz}$, $I_T = 2 \text{ mA}$
3. $R = 6.8 \text{ k}\Omega$, $C = 50 \text{ nF}$, $f = 300 \text{ Hz}$, $I_T = 10 \text{ mA}$
4. $R = 47 \text{ k}\Omega$, $C = 200 \text{ pF}$, $f = 15 \text{ kHz}$, $I_T = 50 \ \mu\text{A}$
5. Refer to Figure 20–1(a). Let $B_c = 345 \ \mu\text{S}$, $C = 50 \text{ nF}$, $\theta = 42.3°$, $I_R = 1.65 \text{ mA}$. Find R, f, I_T, and V.

SOLUTIONS

1. $Y_{\text{rect}} = 66.7 \ \mu\text{S} + j94.2 \ \mu\text{S}$
   $Y_{\text{polar}} = 115 \underline{/54.7°} \ \mu\text{S}$
   $I_{\text{rect}} = 289 \ \mu\text{A} + j408 \ \mu\text{A}$
   $V = 4.33 \text{ V}$
2. $Y_{\text{rect}} = 556 \ \mu\text{S} + j440 \ \mu\text{S}$
   $Y_{\text{polar}} = 709 \underline{/38.4°} \ \mu\text{S}$
   $I_{\text{rect}} = 1.57 \text{ mA} + j1.24 \text{ mA}$
   $V = 2.82 \text{ V}$
3. $Y_{\text{rect}} = 147 \ \mu\text{S} + j94.2 \ \mu\text{S}$
   $Y_{\text{polar}} = 175 \underline{/32.7°} \ \mu\text{S}$
   $I_{\text{rect}} = 8.42 \text{ mA} + j5.40 \text{ mA}$
   $V = 57.3 \text{ V}$
4. $Y_{\text{rect}} = 21.3 \ \mu\text{S} + j18.8 \ \mu\text{S}$
   $Y_{\text{polar}} = 28.4 \underline{/41.5°} \ \mu\text{S}$
   $I_{\text{rect}} = 37.4 \ \mu\text{A} + j33.2 \ \mu\text{A}$
   $V = 1.76 \text{ V}$
5. $G = 379 \ \mu\text{S}$. Then $R = 2.64 \text{ k}\Omega$;
   $f = 1.10 \text{ kHz}$; $I_T = 2.23 \text{ mA}$;
   $V = 4.36 \text{ V}$.

Additional practice problems are at the end of the chapter.

RL CIRCUIT ANALYSIS—CIRCUIT CURRENTS   20–3

☞ **KEY POINT** In a parallel circuit, we use E as the reference. In an *RL* circuit, voltage leads, so I_L and B_L are plotted along the negative y-axis. I_T (or Y) is the hypotenuse and θ is the phase angle.

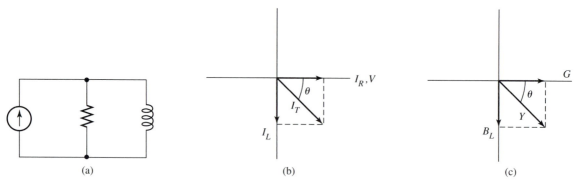

FIGURE 20–3

As shown in Figure 20–3(b), the voltage drop across the inductor leads the current through it by 90°. This puts the total current I_T in the fourth quadrant. I_R is the side adjacent to the angle θ, and I_L is the opposite side.

EXAMPLE 20–5

In the circuit of Figure 20–3(a), let $I_R = 30$ mA, and $I_L = 20$ mA. Find I and θ. Express I_T in rectangular and polar forms.

SOLUTION Refer to Figure 20–4.

$$\tan \theta = \frac{\text{opposite side}}{\text{adjacent side}}$$

$$\tan \theta = \frac{I_L}{I_R} = \frac{-20 \text{ mA}}{30 \text{ mA}} = -0.667$$

$$\tan^{-1}(-0.667) = -33.7°$$

$$\sin \theta = \frac{\text{opposite side}}{\text{hypotenuse}} = \frac{I_L}{I_T}$$

FIGURE 20–4

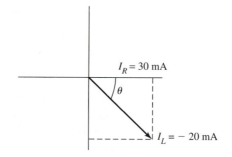

$$I_T = \frac{I_L}{\sin \theta} = \frac{-20 \text{ mA}}{\sin (-33.7°)} = 36.1 \text{ mA}$$

$$I_{\text{rect}} = 30 \text{ mA} - j20 \text{ mA}$$

$$I_{\text{polar}} = 36.1 \; \underline{/-33.7°} \text{ mA}$$

Using the calculator to find I_{polar}:

TI: 30 [EE] [±] 3 [X↔Y] 20 [±] [EE] [±] 3 [R→P]

Casio: [Pol (] 30 [EXP] [(−)] 3 [,] [−] 20 [EXP] [(−)] 3 [=]

Answer: 36.1 × 10^{-3}

TI: [X↔Y]   *Answer:* −33.7,

$I_{\text{polar}} = 36.1 \; \underline{/-33.7°}$ mA

Casio: [RCL] [F]

EXAMPLE 20–6

In Figure 20–3(a), let I_T = 7.3 mA and I_R = 3.6 mA. Find I_L and θ. Express I in both rectangular and polar forms.

SOLUTION

$$\cos \theta = \frac{I_R}{I_T} = \frac{3.6 \text{ mA}}{7.3 \text{ mA}} = 0.493$$

$$\cos^{-1} 0.493 = -60.5°$$

$$I_{\text{polar}} = 7.3 \; \underline{/-60.5°} \text{ mA}$$

$$I_L = I_T \sin \theta = 7.3 \text{ mA} \times \sin (-60.5°) = -6.35 \text{ mA}$$

$$I_{\text{rect}} = 3.6 \text{ mA} - j6.35 \text{ mA}$$

PRACTICE PROBLEMS 20–3

Refer to Figure 20–3(a). From the information given, find I_T in both rectangular and polar forms.

1. I_R = 400 μA, I_L = 300 μA
2. I_R = 17 mA, I_T = 25 mA
3. I_R = 4.5 mA, θ = −65°
4. I_T = 50 mA, θ = −36°
5. I_L = 50 μA, θ = −17°

SOLUTIONS

1. I_{rect} = 400 μA − j300 μA
   I_{polar} = 500 $\underline{/-36.9°}$ μA

2. I_{rect} = 17 mA − j18.3 mA
   I_{polar} = 25 $\underline{/-47.2°}$ mA

3. $I_{rect} = 4.5$ mA $- j9.65$ mA
   $I_{polar} = 10.6 \underline{/-65°}$ mA
5. $I_{rect} = 164$ μA $- j50$ μA
   $I_{polar} = 171 \underline{/-17°}$ μA

4. $I_{rect} = 40.5$ mA $- j29.4$ mA
   $I_{polar} = 50 \underline{/-36°}$ mA

Additional practice problems are at the end of the chapter.

| 20–4 | RL CIRCUIT ANALYSIS— CIRCUIT PARAMETERS |

The phase relationship between conductance, inductive susceptance, B_L, and admittance is shown in Figure 20–3(c). The inductive susceptance is determined by the equation

(20–2)
$$B_L = \frac{1}{2\pi fL}$$

EXAMPLE 20–7

Refer to Figure 20–3(a). Let $R = 6.8$ kΩ, $L = 200$ mH, $f = 4$ kHz, and $I_T = 10$ mA. Find Y and I_T in both rectangular and polar forms.

SOLUTION

$$G = \frac{1}{R} = \frac{1}{6.8 \text{ k}\Omega} = 147 \text{ μS}$$

$$B_L = \frac{1}{2\pi fL} = \frac{1}{2\pi \times 4 \text{ kHz} \times 200 \text{ mH}} = 199 \text{ μS}$$

$$Y_{rect} = 147 \text{ μS} - j199 \text{ μS}$$

$$\tan \theta = \frac{B_L}{G} = \frac{-199 \text{ μS}}{147 \text{ μS}} = -1.35$$

$$\tan^{-1}(-1.35) = -53.5°$$

$$\sin \theta = \frac{B_L}{Y}$$

$$Y = \frac{B_L}{\sin \theta} = \frac{-199 \text{ μS}}{\sin(-53.5°)} = 247 \text{ μS}$$

$$Y_{polar} = 248 \underline{/-53.5°} \text{ μS}$$

$$I_{polar} = 10 \underline{/-53.5°} \text{ mA}$$

$$I_R = I_T \cos \theta = 10 \text{ mA} \times \cos(-53.5°) = 5.94 \text{ mA}$$

$$I_L = I_T \sin \theta = 10 \text{ mA} \times \sin(-53.5°) = -8.04 \text{ mA}$$

$$I_{rect} = 5.94 \text{ mA} - j8.04 \text{ mA}$$

We could have found I_R and I_L by using Ohm's law:

$$I_R = \frac{I_T G}{Y} = \frac{10 \text{ mA} \times 147 \text{ }\mu\text{S}}{247 \text{ }\mu\text{S}} = 5.94 \text{ mA}$$

$$I_L = \frac{I_T B_L}{Y} = \frac{10 \text{ mA} \times 199 \text{ }\mu\text{S}}{247 \text{ }\mu\text{S}} = 8.04 \text{ mA}$$

Differences in the answers are due to rounding.

EXAMPLE 20–8

Refer to Figure 20–5. Let $R = 12 \text{ k}\Omega$, $L = 100 \text{ mH}$, $\theta = -40°$, and $I_R = 50 \text{ }\mu\text{A}$. Find f. Find Y and I_T in both rectangular and polar forms.

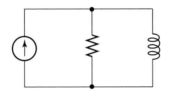

FIGURE 20–5

SOLUTION

$$G = \frac{1}{R} = \frac{1}{12 \text{ k}\Omega} = 83.3 \text{ }\mu\text{S}$$

$$B_L = G \tan \theta = 83.3 \text{ }\mu\text{S} \times \tan (-40°) = -69.9 \text{ }\mu\text{S}$$

$$Y_{\text{rect}} = 83.3 \text{ }\mu\text{S} - j69.9 \text{ }\mu\text{S}$$

$$\cos \theta = \frac{G}{Y}$$

$$Y = \frac{G}{\cos \theta} = \frac{83.3 \text{ }\mu\text{S}}{\cos (-40°)} = 109 \text{ }\mu\text{S}$$

$$Y_{\text{polar}} = 109 \text{ }\underline{/-40°} \text{ }\mu\text{S}$$

$$f = \frac{1}{2\pi B_L L} = \frac{1}{2\pi \times 69.9 \text{ }\mu\text{S} \times 100 \text{ mH}} = 22.8 \text{ kHz}$$

$$I_L = I_R \tan \theta = 50 \text{ }\mu\text{A} \times \tan (-40°) = -42.0 \text{ }\mu\text{A}$$

$$I_{\text{rect}} = 50 \text{ }\mu\text{A} - j42.0 \text{ }\mu\text{A}$$

$$I_T = \frac{I_R}{\cos \theta} = \frac{50 \text{ }\mu\text{A}}{\cos (-40°)} = 65.3 \text{ }\mu\text{A}$$

$$I_{\text{polar}} = 65.3 \text{ }\underline{/-40°} \text{ }\mu\text{A}$$

PRACTICE PROBLEMS 20–4

Refer to Figure 20–5. Find Y and I_T in both rectangular and polar forms in problems 1 through 6.

1. $R = 1.2 \text{ k}\Omega$, $L = 50 \text{ mH}$, $f = 3 \text{ kHz}$, $I_T = 400 \text{ }\mu\text{A}$
2. $R = 15 \text{ k}\Omega$, $L = 20 \text{ mH}$, $f = 100 \text{ kHz}$, $I_T = 1 \text{ mA}$
3. $R = 470 \text{ }\Omega$, $L = 400 \text{ mH}$, $f = 200 \text{ Hz}$, $I_T = 20 \text{ mA}$
4. $R = 5.6 \text{ k}\Omega$, $L = 100 \text{ mH}$, $f = 9 \text{ kHz}$, $I_T = 50 \text{ mA}$
5. $R = 7.5 \text{ k}\Omega$, $L = 200 \text{ mH}$, $\theta = -56.5°$, $I_R = 1.85 \text{ mA}$
6. $L = 10 \text{ mH}$, $f = 50 \text{ kHz}$, $\theta = -35°$, and $I_L = 60 \text{ }\mu\text{A}$.

SOLUTIONS

1. $Y_{\text{rect}} = 833 \text{ }\mu\text{S} - j1.06 \text{ mS}$
   $Y_{\text{polar}} = 1.35 \underline{/-51.9°} \text{ mS}$
   $I_{\text{rect}} = 247 \text{ }\mu\text{A} - j315 \text{ }\mu\text{A}$
   $I_{\text{polar}} = 400 \underline{/-51.9°} \text{ }\mu\text{A}$
3. $Y_{\text{rect}} = 2.13 \text{ mS} - j1.99 \text{ mS}$
   $Y_{\text{polar}} = 2.91 \underline{/-43.1°} \text{ mS}$
   $I_{\text{rect}} = 14.6 \text{ mA} - j13.7 \text{ mA}$
   $I_{\text{polar}} = 20 \underline{/-43.1°} \text{ mA}$
5. $Y_{\text{rect}} = 133 \text{ }\mu\text{S} - j201 \text{ }\mu\text{S}$
   $Y_{\text{polar}} = 242 \underline{/-56.5°} \text{ }\mu\text{S}$
   $I_{\text{rect}} = 1.85 \text{ mA} - j2.80 \text{ mA}$
   $I_{\text{polar}} = 3.07 \underline{/-56.5°} \text{ mA}$

2. $Y_{\text{rect}} = 66.7 \text{ }\mu\text{S} - j79.6 \text{ }\mu\text{S}$
   $Y_{\text{polar}} = 104 \underline{/-50.0°} \text{ }\mu\text{S}$
   $I_{\text{rect}} = 642 \text{ }\mu\text{A} - j767 \text{ }\mu\text{A}$
   $I_{\text{polar}} = 1 \underline{/-50.0°} \text{ mA}$
4. $Y_{\text{rect}} = 179 \text{ }\mu\text{S} - j177 \text{ }\mu\text{S}$
   $Y_{\text{polar}} = 251 \underline{/-44.7°} \text{ }\mu\text{S}$
   $I_{\text{rect}} = 35.5 \text{ mA} - j35.2 \text{ mA}$
   $I_{\text{polar}} = 50 \underline{/-44.7°} \text{ mA}$
6. $Y_{\text{rect}} = 455 \text{ }\mu\text{S} - j318 \text{ }\mu\text{S}$
   $Y_{\text{polar}} = 555 \underline{/-35°} \text{ }\mu\text{S}$
   $I_{\text{rect}} = 85.7 \text{ }\mu\text{A} - j60 \text{ }\mu\text{A}$
   $I_{\text{polar}} = 105 \underline{/-35°} \text{ }\mu\text{A}$

Additional practice problems are at the end of the chapter.

SELF-TEST 20–1

1. Refer to Figure 20–6. Let $R = 10 \text{ k}\Omega$, $f = 4.3 \text{ kHz}$, $C = 2.5 \text{ nF}$, and $I_T = 5 \text{ mA}$. Find Y_{rect}, Y_{polar}, I_{rect}, I_{polar}, and V.
2. Refer to Figure 20–6. Let $R = 27 \text{ k}\Omega$, $f = 800 \text{ Hz}$, $C = 10 \text{ nF}$, and $I_T = 500 \text{ }\mu\text{A}$. Find Y_{rect}, Y_{polar}, I_{rect}, I_{polar}, and V.
3. Refer to Figure 20–5. Let $R = 120 \text{ k}\Omega$, $f = 50 \text{ kHz}$, $L = 0.5 \text{ H}$, and $I_T = 70 \text{ }\mu\text{A}$. Find Y_{rect}, Y_{polar}, I_{rect}, and I_{polar}.
4. Refer to Figure 20–5. Let $R = 68 \text{ k}\Omega$, $f = 1.5 \text{ kHz}$, $L = 5.3 \text{ H}$, and $I_T = 2 \text{ mA}$. Find Y_{rect}, Y_{polar}, I_{rect}, and I_{polar}.

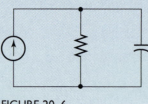

FIGURE 20–6

5. Refer to Figure 20–5. Let $R = 27$ kΩ,
$L = 100$ mH, $f = 30$ kHz, and $I_T = 2$ mA.
Find Y and I_T in both rectangular and
polar forms.

Answers to Self-test 20–1 are at the end of the chapter.

EQUIVALENT CIRCUITS—PHASORS 20–5

We determined in Chapter 15 that addition or subtraction of phasors must be done in rectangular form. When we multiply or divide phasors, the multiplications and divisions can be done in *either* form, but they are simpler to do in polar form. Multiplication of phasors usually occurs when simplifying complex circuits in which impedances are in parallel. Such circuits will be analyzed in Chapter 21. Division of phasors occurs in the simplification of complex circuits and when making impedance–admittance conversions.

When multiplying impedances, we will put the impedances in polar form. The magnitudes are multiplied together. The angles are *added algebraically.* When dividing impedances, the magnitudes are divided. The angles are *subtracted.*

> ☞ **KEY POINT** When multiplying phasors, put them in polar form. Multiply the magnitudes and add the angles. When dividing phasors, divide the magnitudes and subtract the angles.

EXAMPLE 20–9

Find the circuit impedance when the circuit admittance is 300 $\underline{/-30°}$ μS.

SOLUTION Impedance and admittance are reciprocals.

$$Z = \frac{1}{Y}$$

$$Z = \frac{1}{300 \, \underline{/-30°} \, \mu S}$$

The rule is that we divide in polar form; therefore, nothing further need be done to solve for Z.

$$Z = 3.33 \, \underline{/30°} \text{ k}\Omega$$

The phase angle is positive because $(-)-30° = 30°$. For some students the process of subtraction is simplified by first moving the phase angle to the numerator, changing its sign, and then adding.

$$Z = \frac{1\ \underline{/30^\circ}}{300\ \mu S} = 3.33\ \underline{/30^\circ}\ k\Omega$$

EXAMPLE 20–10

Two impedances Z_1 and Z_2 are in parallel. $Z_1 = 15\ \underline{/60^\circ}\ k\Omega$ and $Z_2 = 10\ \underline{/-60^\circ}\ k\Omega$. Find Z_T.

SOLUTION

$$Z_T = \frac{Z_1 Z_2}{Z_1 + Z_2}$$

We will simplify the denominator first. Remembering that impedances must be added in rectangular form, our first step is to convert Z_1 and Z_2 to rectangular form.

$$Z_{1(\text{polar})} = 15\ \underline{/60^\circ}\ k\Omega$$
$$R_1 = 15\ k\Omega\ \cos\theta = 7.5\ k\Omega$$
$$X_L = 15\ k\Omega\ \sin\theta = 13.0\ k\Omega$$
$$Z_{2(\text{polar})} = 10\ \underline{/-60^\circ}\ k\Omega$$
$$R_2 = 10\ k\Omega\ \cos(-60^\circ) = 5\ k\Omega$$
$$X_C = 10\ k\Omega\ \sin(-60^\circ) = -8.66\ k\Omega$$
$$Z_{1(\text{rect})} = 7.5\ k\Omega + j13.0\ k\Omega$$
$$Z_{2(\text{rect})} = \underline{5\ k\Omega - j8.66\ k\Omega}$$
$$Z_1 + Z_2 = 12.5\ k\Omega + j4.33\ k\Omega$$
$$Z_T = \frac{15\ \underline{/60^\circ}\ k\Omega \times 10\ \underline{/-60^\circ}\ k\Omega}{12.5\ k\Omega + j4.33\ k\Omega}$$

The division is done in polar form; therefore, we must change the result of $Z_1 + Z_2$ back into polar form.

$$(Z_1 + Z_2)_{\text{rect}} = 12.5\ k\Omega + j4.33\ k\Omega$$
$$\tan\theta = \frac{X_L}{R} = \frac{4.33\ k\Omega}{12.5\ k\Omega} = 0.347$$
$$\tan^{-1} 0.347 = 19.1^\circ$$
$$Z = \frac{X_L}{\sin\theta} = \frac{4.33\ k\Omega}{\sin 19.1^\circ} = 13.2\ k\Omega$$
$$(Z_1 + Z_2)_{\text{polar}} = 13.2\ \underline{/19.1^\circ}\ k\Omega$$

Using the calculator,

TI: 12.5 $\boxed{\text{EE}}$ 3 $\boxed{X \leftrightarrow Y}$ 4.33 $\boxed{\text{EE}}$ 3 $\boxed{R \rightarrow P}$

Casio: $\boxed{\text{Pol (}}$ 12.5 $\boxed{\text{EXP}}$ 3 $\boxed{,}$ 4.33 $\boxed{\text{EXP}}$ 3 $\boxed{=}$

Answer: 13.2×10^3

TI: $\boxed{X \leftrightarrow Y}$ *Answer:* 19.1

Casio: $\boxed{RCL}$ $\boxed{F}$

$$Z_1 + Z_2 = 13.2 \; \underline{/19.1°} \text{ k}\Omega \qquad \text{in polar form}$$

$$Z_T = \frac{15 \; \underline{/60°} \text{ k}\Omega \times 10 \; \underline{/-60°} \text{ k}\Omega}{13.2 \; \underline{/19.1°} \text{ k}\Omega}$$

Bringing 19.1° from the denominator results in an angle of $-19.1°$ added to the existing angles.

$$Z_T = \frac{15 \text{ k}\Omega \times 10 \text{ k}\Omega \; \underline{/60° + (-)60° + (-)19.1°}}{13.2 \text{ k}\Omega}$$

$$= \frac{15 \text{ k}\Omega \times 10 \text{ k}\Omega \; \underline{/-19.1°}}{13.2 \text{ k}\Omega} = 11.3 \; \underline{/-19.1°} \text{ k}\Omega$$

PRACTICE PROBLEMS 20–5

1. Find Z_T if $Y_T = 200 \; \underline{/40°} \; \mu\text{S}$.
2. Find Y_T if $Z_T = 43.7 \; \underline{/-27°} \text{ k}\Omega$.
3. Z_1 and Z_2 are in parallel. Find Z_T if $Z_1 = 2 \; \underline{/20°} \text{ k}\Omega$ and $Z_2 = 4 \; \underline{/-50°} \text{ k}\Omega$.
4. Z_1 and Z_2 are in parallel. Find Z_T if $Z_1 = 20 \; \underline{/-25°} \text{ k}\Omega$ and $Z_2 = 10 \; \underline{/-60°} \text{ k}\Omega$.

SOLUTIONS

1. $Z_T = 5 \; \underline{/-40°} \text{ k}\Omega$
2. $Y_T = 22.9 \; \underline{/27°} \; \mu\text{S}$
3. $Z_T = 1.59 \; \underline{/-1.86°} \text{ k}\Omega$
4. $Z_T = 6.95 \; \underline{/-48.5°} \text{ k}\Omega$

Additional practice problems are at the end of the chapter.

EQUIVALENT CIRCUITS— PARALLEL-SERIES CONVERSIONS   20–6

In solving complex circuit problems, it is often necessary to reduce circuits to equivalent parallel circuits (EPC) or to equivalent series circuits (ESC). The parallel circuit problems we have been solving can easily be changed to equivalent series circuits. Figure 20–7 shows a parallel circuit with the values of conductance, susceptance, and frequency given. (See Example 20–3 for the original circuit values.) We determined that $Y_{\text{polar}} = 384 \; \underline{/37.9°} \; \mu\text{S}$. The equivalent series circuit would have an impedance equal to the reciprocal of this admittance.

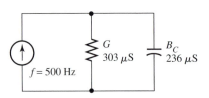

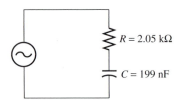

FIGURE 20–7   Parallel circuit with values of conductance, susceptance, and frequency given.

FIGURE 20–8   Equivalent series circuit of Figure 20–7.

$$Z_{polar} = \frac{1}{Y_{polar}} = \frac{1}{384 \underline{/37.9°} \ \mu S} = 2.60 \underline{/-37.9°} \ k\Omega$$

Knowing Z_{polar}, we can determine the values of R and X_C that make up the impedance.

$$R = Z \cos \theta = 2.60 \ k\Omega \times \cos (-37.9°) = 2.05 \ k\Omega$$

$$X_C = Z \sin \theta = 2.60 \ k\Omega \times \sin (-37.9°) = -1.60 \ k\Omega$$

$$Z_{rect} = 2.05 \ k\Omega - j1.60 \ k\Omega$$

The frequency in Example 20–3 was 500 Hz. Therefore,

$$C = \frac{1}{2\pi f X_C} = \frac{1}{2\pi \times 500 \ Hz \times 1.6 \ k\Omega} = 199 \ nF$$

The equivalent series circuit consists of a resistance of 2.05 kΩ and a capacitance of 199 nF, as shown in Figure 20–8.

EXAMPLE 20–11

Find the equivalent series circuit and the branch currents of the parallel circuit in Figure 20–9.

SOLUTION

$$G = \frac{1}{R} = \frac{1}{12 \ k\Omega} = 83.3 \ \mu S$$

$$B_C = 2\pi f C = 2\pi \times 3.5 \ kHz \times 100 \ nF = 2.20 \ mS$$

$$B_L = \frac{1}{2\pi f L} = \frac{1}{2\pi \times 3.5 \ kHz \times 20 \ mH} = 2.27 \ mS$$

FIGURE 20–9

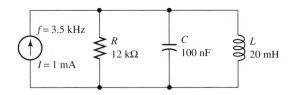

$$Y_{rect} = 83.3 \ \mu S + j2.2 \ mS - j2.27 \ mS = 83.3 \ \mu S - j74.5 \ \mu S$$

$$Y_{polar} = 112 \ \underline{/-41.8°} \ \mu S$$

For the ESC,

$$Z_{polar} = \frac{1}{Y_{polar}} = \frac{1}{112 \ \underline{/41.8°} \ \mu S} = 8.94 \ \underline{/41.8°} \ k\Omega$$

$$R = Z \cos \theta = 8.94 \ k\Omega \times \cos 41.8° = 6.67 \ k\Omega$$

$$X_L = Z \sin \theta = 8.94 \ k\Omega \times \sin 41.8° = 5.96 \ k\Omega$$

$$L = \frac{X_L}{2\pi f} = \frac{5.96 \ k\Omega}{2\pi \times 3.5 \ kHz} = 271 \ mH$$

The ESC is shown in Figure 20–10. In the original circuit,

$$V = \frac{I_T}{Y} = \frac{1 \ mA}{112 \ \mu S} = 8.94 \ V$$

$$I_R = VG = 8.94 \ V \times 83.3 \ \mu S = 745 \ \mu A$$

$$I_C = VB_C = 8.94 \ V \times 2.2 \ mS = 19.7 \ mA$$

$$I_L = VB_L = 8.94 \ V \times 2.27 \ mS = 20.3 \ mA$$

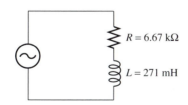

FIGURE 20–10

PRACTICE PROBLEMS 20–6

Find the branch currents Y, Z, V, and the ESC in the following problems:

1. Refer to Figure 20–11 on the next page. $f = 1$ kHz, $C = 20$ nF, $R = 10$ kΩ, and $I_T = 10$ mA.
2. Refer to Figure 20–11. $f = 2.5$ kHz, $C = 0.5 \ \mu F$, $R = 82 \ \Omega$, and $I_T = 300 \ \mu A$.
3. Refer to Figure 20–12 on the next page. $f = 15$ kHz, $L = 300$ mH, $R = 22$ kΩ, and $I_T = 2$ mA.
4. Refer to Figure 20–12. $f = 2.5$ kHz, $L = 200$ mH, $R = 4.7$ kΩ, and $I_T = 1$ mA.
5. Refer to Figure 20–9. $f = 3.7$ kHz, $C = 100$ nF, $L = 20$ mH, $R = 12$ kΩ, and $I_T = 100 \ \mu A$.
6. Refer to Figure 20–9. $f = 7.5$ kHz, $C = 5$ nF, $L = 50$ mH, $R = 20$ kΩ, and $I_T = 15$ mA.

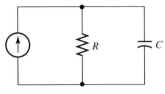

FIGURE 20–11

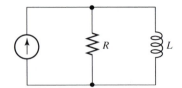

FIGURE 20–12

SOLUTIONS

1. $I_R = 6.23$ mA, $I_C = 7.82$ mA, $Y = 161 \underline{/51.5°}$ µS, $Z = 6.23 \underline{/-51.5°}$ kΩ, and $V = 62.3$ V. In the ESC, $R = 3.88$ kΩ and $C = 32.7$ nF.

2. $I_R = 252$ µA, $I_C = 163$ µA, $Y = 14.5 \underline{/32.8°}$ mS, $Z = 68.9 \underline{/-32.8°}$ Ω, and $V = 20.7$ mV. In the ESC, $R = 58.0$ Ω and $C = 1.71$ µF.

3. $I_R = 1.58$ mA, $I_L = 1.23$ mA, $Y = 57.6 \underline{/-37.9°}$ µS, $Z = 17.4 \underline{/37.9°}$ kΩ, and $V = 34.7$ V. In the ESC, $R = 13.7$ kΩ and $L = 113$ mH.

4. $I_R = 556$ µA, $I_L = 831$ µA, $Y = 383 \underline{/-56.2°}$ µS, $Z = 2.61 \underline{/56.2°}$ kΩ, and $V = 2.61$ V. In the ESC, $R = 1.45$ kΩ and $L = 138$ mH.

5. $I_R = 43.2$ µA, $I_C = 1.2$ mA, $X_L = 1.11$ mA, $V = 518$ mV, $Y = 193 \underline{/64.4°}$ µS, and $Z = 5.18 \underline{/-64.4°}$ kΩ. In the ESC, $R = 2.24$ kΩ and $C = 9.20$ nF.

6. $I_R = 3.84$ mA, $I_C = 18.1$ mA, $I_L = 32.6$ mA, $V = 76.8$ V, $Y = 195 \underline{/-75.2°}$ µS, and $Z = 5.12 \underline{/75.2°}$ kΩ. In the ESC, $R = 1.31$ kΩ and $L = 105$ mH.

SELF-TEST 20–2

1. Find Z_T, where $Z_1 = 7 \underline{/25°}$ kΩ and $Z_2 = 10 \underline{/-60°}$ kΩ.

2. Refer to Figure 20–11. Find the branch currents, Y, Z, V, and the ESC, where $R = 1.8$ kΩ, $C = 200$ nF, $f = 350$ Hz, and $I_T = 200$ µA.

3. Refer to Figure 20–12. Find the branch currents, Y, Z, V, and the ESC, where $R = 22$ kΩ, $f = 15$ kHz, $L = 300$ mH, and $I_T = 20$ mA.

4. Refer to Figure 20–9. Find the branch currents, Y, Z, V, and the ESC, where $R = 10$ kΩ, $L = 2$ mH, $C = 500$ pF, $f = 150$ kHz, and $I_T = 500$ µA.

Answers to Self-test 20–2 are at the end of the chapter.

AC NETWORKS—TRANSFORMS 20-7

As we get into more complex ac circuit problems, many series–parallel conversions become necessary. The methods shown previously may be shortened by the use of equations that we will call *transforms*. These transforms allow us to equate series resistance and reactance to their parallel equivalent conductance and susceptance, or vice versa.

To convert from a series circuit to its equivalent parallel circuit, the equations are

$$G = \frac{R}{R^2 + X^2} \quad \text{and} \quad B = \frac{X}{R^2 + X^2} \tag{20-3}$$

To convert from a parallel circuit to its equivalent series circuit, the equations are

$$R = \frac{G}{G^2 + B^2} \quad \text{and} \quad X = \frac{B}{G^2 + B^2} \tag{20-4}$$

These equations yield absolute values. Signs must be assigned to reactive and susceptive values, depending on circuit components, when problem solving.

Consider the circuit in Figure 20–7 again. In that circuit, the conductance, G, is 303 μS and the susceptance, B, is 236 μS. Let's convert to its equivalent series circuit using Equation 20–4.

$$R = \frac{G}{G^2 + B_C^2} = \frac{303 \times 10^{-6}}{(303 \times 10^{-6})^2 + (236 \times 10^{-6})^2} = 2.05 \text{ k}\Omega$$

TI: 303 $\boxed{\text{EE}}$ $\boxed{+/-}$ 6 $\boxed{\div}$ $\boxed{(}$ 303 $\boxed{\text{EE}}$ $\boxed{+/-}$ 6 $\boxed{x^2}$

$\boxed{+}$ 236 $\boxed{\text{EE}}$ $\boxed{+/-}$ 6 $\boxed{x^2}$ $\boxed{)}$ $\boxed{\text{STO}}$ $\boxed{=}$

Answer: R = 2.05 kΩ

Casio: 303 $\boxed{\text{EXP}}$ $\boxed{(-)}$ 6 $\boxed{\div}$ $\boxed{(}$ 303 $\boxed{\text{EXP}}$ $\boxed{(-)}$ 6 $\boxed{x^2}$

$\boxed{+}$ 236 $\boxed{\text{EXP}}$ $\boxed{(-)}$ 6 $\boxed{x^2}$ $\boxed{)}$ $\boxed{=}$

Store the denominator because it is used in the next part.

$$X_C = \frac{B_C}{G^2 + B_C^2} = \frac{236 \times 10^{-6}}{(303 \times 10^{-6})^2 + (236 \times 10^{-6})^2} = 1.60 \text{ k}\Omega$$

TI: 236 $\boxed{\text{EE}}$ $\boxed{+/-}$ 6 $\boxed{\div}$ $\boxed{\text{RCL}}$ $\boxed{=}$

Answer: X_C = 1.60 kΩ

With the Casio you can edit the previous calculation: just change the 303 in the numerator to 236 and press the = key.

In problems of this type, the denominators are the same in the two equations, so store the denominator the first time you compute it to save time in solving the second equation.

The equivalent series circuit in rectangular form for this problem would be

$$Z = R - jX_C = 2.05 \text{ k}\Omega - j1.60 \text{ k}\Omega$$

and the circuit of Figure 20–8 results.

EXAMPLE 20–12

Consider the circuit in Figure 20–13. (It's like the circuit described in Example 20–11.) $X_L = 3.77$ kΩ and the impedance in rectangular form is

$$Z = 2.7 \text{ k}\Omega + j3.77 \text{ k}\Omega$$

Let's convert to the equivalent parallel circuit using Equation 20–3.

SOLUTION

$$G = \frac{R}{R^2 + X_L{}^2} = \frac{2.7 \times 10^3}{(2.7 \times 10^3)^2 + (3.77 \times 10^3)^2} = 126 \ \mu S$$

$$B_L = \frac{X_L}{R^2 + X_L{}^2} = \frac{3.77 \times 10^3}{(2.7 \times 10^3)^2 + (3.77 \times 10^3)^2} = 175 \ \mu S$$

$$L = \frac{1}{2\pi f B_L} = \frac{1}{6.28 \times 3 \times 10^3 \times 175 \times 10^{-6}} = 303 \text{ mH}$$

The equivalent parallel circuit in rectangular form is

$$Y = G - jB_L = 126 \ \mu S - j175 \ \mu S$$

The equivalent parallel circuit is drawn in Figure 20–14.

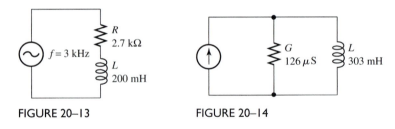

FIGURE 20–13   FIGURE 20–14

PRACTICE PROBLEMS 20–7

Find Y_{rect} and C or L in the equivalent parallel circuit for the following problems.

1. In a series RC circuit where R = 4.7 kΩ, C = 0.05 μF, and f = 500 Hz.

2. In a series RL circuit where R = 27 kΩ, L = 200 mH, and f = 15 kHz.

Find Z_{rect} and C or L in the equivalent series circuit for the following problems.

3. In a parallel RC circuit where G = 66.7 μS, C = 500 pF, and f = 30 kHz.

4. In a parallel RL circuit where G = 500 μS, L = 500 mH, and f = 500 Hz.

SOLUTIONS

1. $Y = 75.1 \ \mu S + j102 \ \mu S$, $C = 32.4$ nF
3. $Z = 5.00 \text{ k}\Omega - j7.07 \text{ k}\Omega$, $C = 750$ pF

2. $Y = 24.9 \ \mu S - j17.4 \ \mu S$, $L = 610$ mH
4. $Z = 763 \ \Omega + j972 \ \Omega$, $L = 309$ mH

Additional practice problems are at the end of the chapter.

Let's see how we might use these transforms to solve complex circuit problems.

EXAMPLE 20–13

Find the equivalent series circuit and the phase angle of the circuit in Figure 20–15.

SOLUTION First, let's convert the series RL circuit to its equivalent parallel circuit.

$$X_L = 2\pi fL = 6.28 \times 10 \text{ kHz} \times 100 \text{ mH} = 6.28 \text{ k}\Omega$$

$$G = \frac{R}{R^2 + X_L^2} = \frac{100}{100^2 + (6.28 \times 10^3)^2} = 2.53 \text{ } \mu S$$

$$B_L = \frac{X_L}{R^2 + X_L^2} = \frac{6.28 \times 10^3}{100^2 + (6.28 \times 10^3)^2} = 159 \text{ } \mu S$$

Next solve for B_C:

$$B_C = 2\pi fC = 6.28 \times 10 \text{ kHz} \times 3 \text{ nF} = 188 \text{ } \mu S$$

The parallel circuit in rectangular form is

$$Y = G + jB_C - jB_L = 2.53 \text{ } \mu S + j188 \text{ } \mu S - j159 \text{ } \mu S$$

$$= 2.53 \text{ } \mu S + j29.0 \text{ } \mu S$$

The equivalent parallel circuit is R and C and is shown in Figure 20–16. Use Equation 20–4 to find the equivalent series circuit.

$$R = \frac{G}{G^2 + B_C^2} = \frac{2.53 \times 10^{-6}}{(2.53 \times 10^{-6})^2 + (29.3 \times 10^{-6})^2} = 2.92 \text{ k}\Omega$$

$$X_C = \frac{B_C}{G^2 + B_C^2} = \frac{29.3 \times 10^{-6}}{(2.53 \times 10^{-6})^2 + (29.3 \times 10^{-6})^2} = 33.8 \text{ k}\Omega$$

The equivalent series circuit in rectangular form is

$$Z = 2.92 \text{ k}\Omega - j33.8 \text{ k}\Omega$$

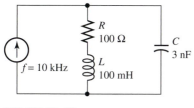

FIGURE 20–15

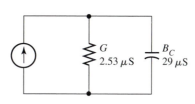

FIGURE 20–16

FIGURE 20–17

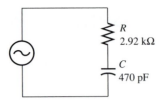

The equivalent series capacitance is

$$C = \frac{1}{2\pi f X_C} = \frac{1}{6.28 \times 10 \times 10^3 \times 34.2 \times 10^3} = 470 \text{ pF}$$

The phase angle is

$$\tan \theta = \frac{X_C}{R} = \frac{33.8 \text{ k}\Omega}{2.99 \text{ k}\Omega} = 11.6$$

$$\tan^{-1} 11.6 = 85.1°$$

The ESC is shown in Figure 20–17.

This is a circuit encountered whenever L and C are in parallel. The inductance usually contains some measurable series resistance because of the resistance of the wire that makes up the inductor. The result, then, of L and C in parallel is really a circuit such as the one in Figure 20–15, and it is this circuit configuration you must consider when finding the various circuit variables.

EXAMPLE 20–14

Find the ESCs for the circuit in Figure 20–18.

SOLUTION   Convert the RL series circuit (call it branch X) to its EPC using Equation 20–3.

$$G_X = \frac{R_2}{R_2^2 + X_L^2} = \frac{200}{200^2 + (10 \times 10^3)^2} = 2.00 \text{ }\mu\text{S}$$

$$B_L = \frac{X_L}{R_2^2 + X_L^2} = \frac{10 \times 10^3}{200^2 + (10 \times 10^3)^2} = 100 \text{ }\mu\text{S}$$

FIGURE 20–18

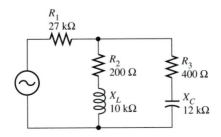

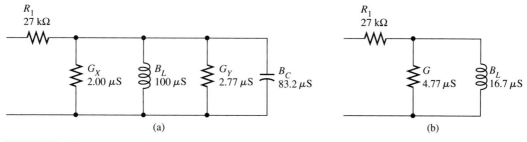

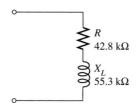

FIGURE 20–19

Next convert the RC series circuit (call it branch Y) to its EPC using Equation 20–3.

$$G_Y = \frac{R_3}{R_3{}^2 + X_C{}^2} = \frac{400}{400^2 + (12 \times 10^3)^2} = 2.77 \ \mu S$$

$$B_C = \frac{X_C}{R_3{}^2 + X_C{}^2} = \frac{12 \times 10^3}{400^2 + (12 \times 10^3)^2} = 83.2 \ \mu S$$

Figure 20–19(a) is the circuit at this point. The EPC is

$$Y = G_X + G_Y + jB_C - jB_L$$

$$= 2.00 \ \mu S + 2.77 \ \mu S + j83.2 \ \mu S - j100 \ \mu S$$

$$= 4.77 \ \mu S - j16.7 \ \mu S$$

At this point the circuit looks like Figure 20–19(b).

Next find the equivalent series circuit of the parallel portion of Figure 20–18.

$$R_{eq} = \frac{G}{G^2 + B_L{}^2} = \frac{4.77 \times 10^{-6}}{(4.77 \times 10^{-6})^2 + (16.7 \times 10^{-6})^2}$$

$$= 15.8 \ k\Omega$$

$$X_L = \frac{B_L}{G^2 + B_L{}^2} = \frac{16.7 \times 10^{-6}}{(4.77 \times 10^{-6})^2 + (16.7 \times 10^{-6})^2}$$

$$= 55.3 \ k\Omega$$

The equivalent series circuit in rectangular form is

$$Z = R_1 + R_{eq} + jX_L = 42.8 \ k\Omega + j55.3 \ k\Omega$$

and is drawn in Figure 20–20.

FIGURE 20–20

PRACTICE PROBLEMS 20–8

Find the equivalent series circuit for the following problems. Express answers in rectangular form.

1. Refer to Figure 20–15. Let $R = 2\ \text{k}\Omega$, $X_L = 3\ \text{k}\Omega$, and $X_C = 5\ \text{k}\Omega$.
2. Refer to Figure 20–21.
3. Refer to Figure 20–22.
4. Refer to Figure 20–23. In addition to the ESC, find the component values and the phase angle.

SOLUTIONS

1. The *RL* branch conversion is

$$Y = 154\ \mu S - j231\ \mu S$$

The equivalent parallel circuit in rectangular form is

$$Y = 154\ \mu S + j200\ \mu S - j231\ \mu S$$

$$= 154\ \mu S - j30.8\ \mu S$$

Converting to the ESC, we get

$$Z = 6.25\ \text{k}\Omega + j1.25\ \text{k}\Omega$$

The ESC is resistive and inductive.

2. The *RL* branch conversion is

$$Y = 2.00\ \mu S - 100\ \mu S$$

The *RC* branch conversion is

$$Y = 2.77\ \mu S + j83.2\ \mu S$$

The resultant equivalent parallel circuit is

$$Y = G_1 + G_2 + jB_C - jB_L$$

$$= 2.00\ \mu S + 2.77\ \mu S$$

$$+ j83.2\ \mu S - j100\ \mu S$$

$$= 4.77\ \mu S - j16.7\ \mu S$$

Converting to the ESC, we get

$$Z = 15.8\ \text{k}\Omega + j55.3\ \text{k}\Omega$$

The circuit is resistive and inductive.

FIGURE 20–21

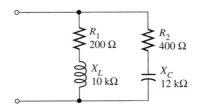

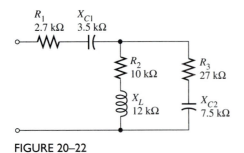

FIGURE 20–22

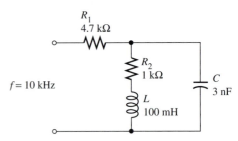

FIGURE 20–23

3. The RL branch conversion is

$$Y_L = G_2 - jB_L = 41.0 \ \mu S - j49.2 \ \mu S$$

The RC branch conversion is

$$Y_C = G_3 + jB_C = 34.4 \ \mu S + j9.55 \ \mu S$$

The EPC of the two branches in rectangular form is

$$Y = G_2 + G_3 + jB_C - jB_L$$

$$= 41.0 \ \mu S + 34.4 \ \mu S + j9.55 \ \mu S$$

$$- j49.2 \ \mu S$$

$$= 75.4 \ \mu S - j39.6 \ \mu S$$

Equation 20–4 yields an ESC of $10.4 \ k\Omega + j5.47 \ k\Omega$. The ESC for the entire circuit is

$$Z = R_1 + R_{eq} + jX_L - jX_C$$

$$= 2.7 \ k\Omega + 10.4 \ k\Omega + j5.47 \ k\Omega$$

$$- j3.5 \ k\Omega$$

$$= 13.1 \ k\Omega + j1.97 \ k\Omega$$

The equivalent series circuit is resistive and inductive.

4. In the RL branch $X_L = 2\pi fL = 6.28 \ k\Omega$ and $R = 4.7 \ k\Omega$. Equation 20–3 yields an EPC of $24.7 \ \mu S - j155 \ \mu S$. In the capacitive branch $B_C = 2\pi fC = 188 \ \mu S$.

$$Y_T = G + jB_C - jB_L$$

$$= 24.7 \ \mu S - j155 \ \mu S + j188 \ \mu S$$

$$= 24.7 \ \mu S + j33.3 \ \mu S$$

(this is the EPC of the two branches) Equation 20–4 yields an ESC of $14.4 \ k\Omega - j19.4 \ k\Omega$.

$$Z_{ckt} = R_1 + R_{eq} - jX_C$$

$$= 4.7 \ k\Omega + j14.4 \ k\Omega - j19.4 \ k\Omega$$

$$= 19.1 \ k\Omega - j19.4 \ k\Omega$$

The circuit is resistive and capacitive. $R = 19.1 \ k\Omega$.

$$C = \frac{1}{2\pi fX_C} = 821 \ pF$$

$$\text{phase angle} = \tan^{-1} \frac{X_C}{R}$$

$$= \tan^{-1} \frac{-19.4 \ k\Omega}{19.1 \ k\Omega}$$

$$= -45.5°$$

Additional practice problems are at the end of the chapter.

SELF-TEST 20–3

1. Refer to Figure 20–24. Let $f = 7$ kHz, $L = 0.3$ H, $C = 2$ nF, and $R = 1 \ k\Omega$. Find the ESC, the component values, and the phase angle.
2. Refer to Figure 20–22. Let $R_1 = 1 \ k\Omega$, $X_{C1} = 5 \ k\Omega$, $R_2 = 10 \ k\Omega$, $R_3 = 15 \ k\Omega$, $X_L = 2 \ k\Omega$, and $X_{C2} = 7.5 \ k\Omega$. Find the ESC resistance and reactance. Express the answer in rectangular form.

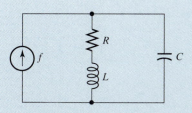

FIGURE 20–24

Answers to Self-test 20–3 are at the end of the chapter.

CHAPTER 20 AT A GLANCE

PAGE	KEY POINTS AND EQUATIONS	EXAMPLE
569	*Key Point:* In a parallel circuit, we use E as the reference. In an RC circuit, current leads, so I_C and B_C are plotted along the positive y-axis. I_T (or Y) is the hypotenuse and θ is the phase angle.	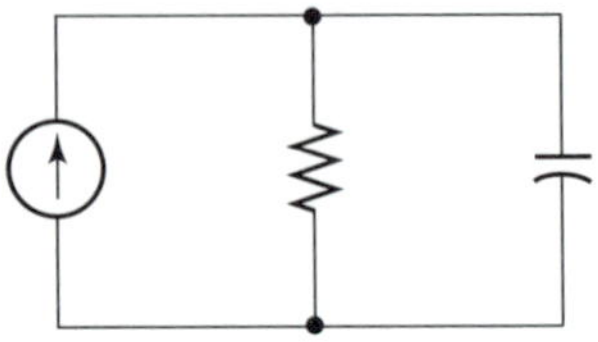 FIGURE 20–25
573	*Key Point:* In a parallel circuit, we use E as the reference. In an RL circuit, voltage leads, so I_L and B_L are plotted along the negative y-axis. I_T (or Y) is the hypotenuse and θ is the phase angle.	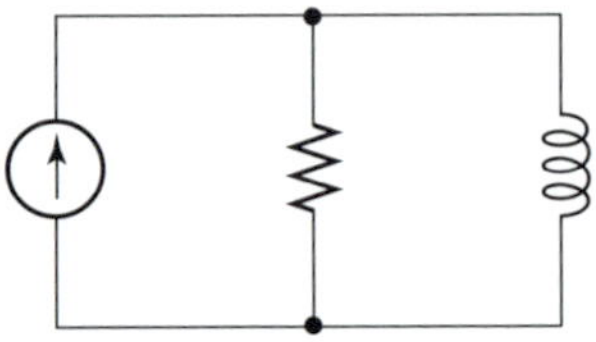 FIGURE 20–26
579	*Key Point:* When multiplying phasors, put in polar form. Multiply the magnitudes and add the angles.	$3\,\angle 40^\circ\ \text{k}\Omega \times 4\,\angle -25.3^\circ\ \text{k}\Omega = 12\,\angle 14.7^\circ\ \text{k}\Omega$
579	When dividing phasors, divide the magnitudes and subtract the angles.	$3\,\angle 40^\circ\ \text{k}\Omega \div 4\,\angle -25.3^\circ\ \text{k}\Omega = 750\,\angle 65.3^\circ\ \Omega$
585	To convert from a series circuit to its equivalent parallel circuit, the equations are $G = \frac{R}{R^2 + X^2}$ and $B = \frac{X}{R^2 + X^2}$	Given: $R = 2.7\ \text{k}\Omega$ and $X_L = 3.77\ \text{k}\Omega$. $G = \frac{R}{R^2 + X_L^2} = \frac{2.7 \times 10^3}{(2.7 \times 10^3)^2 + (3.77 \times 10^3)^2} = 126\ \mu\text{S}$ $B_L = \frac{X_L}{R^2 + X_L^2} = \frac{3.77 \times 10^3}{(2.7 \times 10^3)^2 + (3.77 \times 10^3)^2} = 175\ \mu\text{S}$

585   To convert from a parallel circuit to its equivalent series circuit, the equations are

$$R = \frac{G}{G^2 + B^2} \quad \text{and} \quad X = \frac{B}{G^2 + B^2}$$

Given: $G = 303 \ \mu S$ and $B_C = 236 \ \mu S$.

$$X_C = \frac{B_C}{G^2 + B_C^2}$$

$$= \frac{236 \times 10^{-6}}{(303 \times 10^{-6})^2 + (236 \times 10^{-6})^2}$$

$$= 1.60 \ k\Omega$$

$$R = \frac{G}{G^2 + B_C^2}$$

$$= \frac{303 \times 10^{-6}}{(303 \times 10^{-6})^2 + (236 \times 10^{-6})^2}$$

$$= 2.05 \ k\Omega$$

END OF CHAPTER PROBLEMS 20–1

Refer to Figure 20–27 and express I_T in *rectangular* and *polar* forms:

1. $I_T = 40$ mA and $\theta = 54°$
2. $I_T = 500 \ \mu A$ and $\theta = 40°$
3. $I_T = 2.75$ mA and $\theta = 67.4°$
4. $I_T = 7.65$ mA and $\theta = 35.5°$
5. $I_T = 175 \ \mu A$ and $\theta = 22°$
6. $I_T = 120 \ \mu A$ and $\theta = 65°$
7. $I_R = 350 \ \mu A$ and $I_C = 270 \ \mu A$
8. $I_R = 63.5$ mA and $I_C = 34$ mA
9. $I_R = 4.77$ mA and $I_C = 7.05$ mA
10. $I_R = 9.5$ mA and $I_C = 7.4$ mA
11. $I_R = 100 \ \mu A$ and $I_C = 100 \ \mu A$
12. $I_R = 1.45$ mA and $I_C = 850 \ \mu A$
13. $I_R = 1$ mA and $\theta = 33°$
14. $I_R = 4.5$ mA and $\theta = 52°$
15. $I_R = 640 \ \mu A$ and $\theta = 20.5°$
16. $I_R = 170 \ \mu A$ and $\theta = 53.7°$
17. $I_R = 25.3$ mA and $\theta = 62.0°$
18. $I_R = 47.3$ mA and $\theta = 40°$
19. $I_C = 45$ mA and $\theta = 71°$
20. $I_C = 33$ mA and $\theta = 22.5°$
21. $I_C = 5.44$ mA and $\theta = 59.5°$
22. $I_C = 12$ mA and $\theta = 45°$
23. $I_C = 275 \ \mu A$ and $\theta = 55.7°$
24. $I_C = 805 \ \mu A$ and $\theta = 35.5°$
25. $I_R = 7.5$ mA and $I_T = 12$ mA
26. $I_R = 25$ mA and $I_T = 50$ mA
27. $I_R = 650 \ \mu A$ and $I_T = 1$ mA
28. $I_R = 275 \ \mu A$ and $I_T = 300 \ \mu A$
29. $I_R = 2.55$ mA and $I_T = 3.15$ mA
30. $I_R = 120 \ \mu A$ and $I_T = 180 \ \mu A$
31. $I_C = 25$ mA and $I_T = 43$ mA
32. $I_C = 13.5$ mA and $I_T = 24$ mA
33. $I_C = 200 \ \mu A$ and $I_T = 400 \ \mu A$
34. $I_C = 900 \ \mu A$ and $I_T = 1.75$ mA
35. $I_C = 3.55$ mA and $I_T = 6.08$ mA
36. $I_C = 875 \ \mu A$ and $I_T = 1.25$ mA

FIGURE 20–27

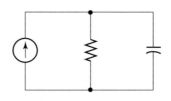

END OF CHAPTER PROBLEMS 20–2

Refer to Figure 20–27 for problems 1 through 10. Find Y_{rect}, Y_{polar}, I_{rect}, and V.

1. $R = 3.3\ k\Omega$, $C = 30\ nF$, $f = 2\ kHz$,
   $I_T = 4\ mA$
2. $R = 20\ k\Omega$, $C = 200\ nF$, $f = 100\ Hz$,
   $I_T = 1\ mA$
3. $R = 750\ \Omega$, $C = 5\ nF$, $f = 60\ kHz$,
   $I_T = 1\ mA$
4. $R = 15\ k\Omega$, $C = 500\ pF$, $f = 35\ kHz$,
   $I_T = 25\ mA$
5. $R = 6.8\ k\Omega$, $C = 10\ nF$, $f = 1.5\ kHz$,
   $I_T = 20\ mA$
6. $R = 27\ k\Omega$, $C = 100\ nF$, $f = 50\ Hz$,
   $I_T = 40\ \mu A$
7. $R = 10\ k\Omega$, $C = 30\ nF$, $f = 1\ kHz$,
   $I_T = 10\ mA$
8. $R = 82\ \Omega$, $C = 750\ nF$, $f = 3\ kHz$,
   $I_T = 300\ \mu A$
9. $R = 15\ k\Omega$, $C = 300\ pF$, $f = 35\ kHz$,
   $I_T = 2\ mA$
10. $R = 15\ k\Omega$, $C = 1\ nF$, $f = 15\ kHz$,
    $I_T = 7.5\ mA$
11. Refer to Figure 20–27. $B_C = 200\ \mu S$,
    $C = 25\ nF$, $\theta = 50°$, and $I_R = 3\ mA$.
    Find R, f, I_T, and V.
12. Refer to Figure 20–27. $B_C = 3.72\ mS$,
    $C = 200\ nF$, $\theta = 37.5°$, and $I_R =$
    $30\ \mu A$. Find R, f, I_T, and V.

END OF CHAPTER PROBLEMS 20–3

Refer to Figure 20–28 and express I_T in *rectangular* and *polar* forms.

1. $I_T = 3.8\ mA$ and $\theta = -30°$
2. $I_T = 7.3\ mA$ and $\theta = -48°$
3. $I_T = 210\ \mu A$ and $\theta = -39°$
4. $I_T = 420\ \mu A$ and $\theta = -52.5°$
5. $I_T = 1.34\ mA$ and $\theta = -23.4°$
6. $I_T = 850\ \mu A$ and $\theta = -37.8°$
7. $I_R = 12\ mA$ and $I_L = -8\ mA$
8. $I_R = 18\ mA$ and $I_L = -22\ mA$
9. $I_R = 30\ mA$ and $I_L = -40\ mA$
10. $I_R = 550\ \mu A$ and $I_L = -400\ \mu A$
11. $I_R = 150\ \mu A$ and $I_L = -180\ \mu A$
12. $I_R = 2.35\ mA$ and $I_L = -3.73\ mA$
13. $I_R = 2\ mA$ and $\theta = -28.5°$
14. $I_R = 3\ mA$ and $\theta = -65°$
15. $I_R = 1.3\ mA$ and $\theta = -73°$
16. $I_R = 2.4\ mA$ and $\theta = -20°$
17. $I_R = 37.6\ \mu A$ and $\theta = -27.4°$
18. $I_R = 350\ \mu A$ and $\theta = -65.5°$
19. $I_L = -730\ \mu A$ and $\theta = -45°$
20. $I_L = -600\ \mu A$ and $\theta = -53.3°$
21. $I_L = -9\ mA$ and $\theta = -40°$
22. $I_L = -8\ mA$ and $\theta = -40°$
23. $I_L = -475\ \mu A$ and $\theta = -43.5°$
24. $I_L = -370\ \mu A$ and $\theta = -38.8°$
25. $I_L = -20\ mA$ and $I_T = 27\ mA$
26. $I_L = -7\ mA$ and $I_T = 15\ mA$
27. $I_L = -150\ \mu A$ and $I_T = 300\ \mu A$
28. $I_L = -225\ \mu A$ and $I_T = 300\ \mu A$
29. $I_L = -925\ \mu A$ and $I_T = 1.83\ mA$
30. $I_L = -830\ \mu A$ and $I_T = 1.23\ mA$
31. $I_R = 4.7\ mA$ and $I_T = 6.7\ mA$
32. $I_R = 3.35\ mA$ and $I_T = 6.35\ mA$
33. $I_R = 825\ \mu A$ and $I_T = 1.35\ mA$
34. $I_R = 670\ \mu A$ and $I_T = 970\ \mu A$
35. $I_R = 73.8\ mA$ and $I_T = 100\ mA$
36. $I_R = 87.5\ mA$ and $I_T = 100\ mA$

FIGURE 20–28

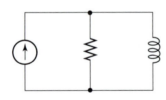

END OF CHAPTER PROBLEMS 20–4

Refer to Figure 20–28. Find Y and I_T in both *polar* and *rectangular* forms.

1. $R = 20 \text{ k}\Omega$, $L = 20 \text{ mH}$, $f = 200 \text{ kHz}$, $I_T = 40 \text{ mA}$
2. $R = 2.2 \text{ k}\Omega$, $L = 400 \text{ mH}$, $f = 600$ Hz, $I_T = 500 \text{ }\mu\text{A}$
3. $R = 10 \text{ k}\Omega$, $L = 60 \text{ mH}$, $f = 20 \text{ kHz}$, $I_T = 200 \text{ mA}$
4. $R = 2.7 \text{ k}\Omega$, $L = 5 \text{ mH}$, $f = 65 \text{ kHz}$, $I_T = 15 \text{ mA}$
5. $R = 4.7 \text{ k}\Omega$, $L = 100 \text{ mH}$, $f = 10 \text{ kHz}$, $I_T = 100 \text{ }\mu\text{A}$
6. $R = 18 \text{ k}\Omega$, $L = 40 \text{ mH}$, $f = 50 \text{ kHz}$, $I_T = 1 \text{ mA}$
7. $R = 27 \text{ k}\Omega$, $L = 2.5 \text{ H}$, $f = 2 \text{ kHz}$, $I_T = 400 \text{ }\mu\text{A}$
8. $R = 3.9 \text{ k}\Omega$, $L = 150 \text{ mH}$, $f = 3 \text{ kHz}$, $I_T = 2.7 \text{ mA}$
9. $R = 750 \text{ }\Omega$, $L = 600 \text{ mH}$, $f = 400 \text{ Hz}$, $I_T = 30 \text{ mA}$
10. $R = 1 \text{ k}\Omega$, $L = 400 \text{ mH}$, $f = 300 \text{ Hz}$, $I_T = 12.5 \text{ mA}$

END OF CHAPTER PROBLEMS 20–5

1. Find Z_T if $Y_T = 3.72 \text{ } \underline{/27°} \text{ mS}$.
2. Find Z_T if $Y_T = 45 \text{ } \underline{/-62°} \text{ } \mu\text{S}$.
3. Find Y_T if $Z_T = 6.32 \text{ } \underline{/-40°} \text{ k}\Omega$.
4. Find Y_T if $Z_T = 12.7 \text{ } \underline{/70°} \text{ k}\Omega$.

Find Z_T in the following problems by using the equation $Z_T = \dfrac{Z_1 Z_2}{Z_1 + Z_2}$, where Z_1 and Z_2 are in parallel.

5. $Z_1 = 700 \text{ } \underline{/70°} \text{ k}\Omega$, $Z_2 = 1.5 \text{ } \underline{/-25°} \text{ k}\Omega$
6. $Z_1 = 12 \text{ } \underline{/-65°} \text{ k}\Omega$, $Z_2 = 7.5 \text{ } \underline{/-40°} \text{ k}\Omega$
7. $Z_1 = 35 \text{ } \underline{/20°} \text{ k}\Omega$, $Z_2 = 10 \text{ } \underline{/-40°} \text{ k}\Omega$
8. $Z_1 = 5.6 \text{ } \underline{/45°} \text{ k}\Omega$, $Z_2 = 12 \text{ } \underline{/-70°} \text{ k}\Omega$
9. $Z_1 = 3.3 \text{ } \underline{/0°} \text{ k}\Omega$, $Z_2 = 4.2 \text{ } \underline{/-90°} \text{ k}\Omega$
10. $Z_1 = 12 \text{ } \underline{/0°} \text{ k}\Omega$, $Z_2 = 20 \text{ } \underline{/90°} \text{ k}\Omega$

END OF CHAPTER PROBLEMS 20–6

Find the branch currents, Y, Z, V, and the ESC in the following problems:

1. Refer to Figure 20–27. $f = 35 \text{ kHz}$, $C = 400 \text{ pF}$, $R = 15 \text{ k}\Omega$, and $I_T = 2 \text{ mA}$.
2. Refer to Figure 20–27. $f = 15 \text{ kHz}$, $C = 1 \text{ nF}$, $R = 12 \text{ k}\Omega$, and $I_T = 7.5 \text{ mA}$.
3. Refer to Figure 20–27. $f = 3 \text{ kHz}$, $C = 150 \text{ nF}$, $R = 470 \text{ }\Omega$, and $I_T = 40 \text{ mA}$.
4. Refer to Figure 20–27. $f = 200 \text{ Hz}$, $C = 150 \text{ nF}$, $R = 4.7 \text{ k}\Omega$, and $I_T = 100 \text{ }\mu\text{A}$.
5. Refer to Figure 20–28. $f = 150 \text{ kHz}$, $L = 60 \text{ mH}$, $R = 47 \text{ k}\Omega$, and $I_T = 250 \text{ }\mu\text{A}$.
6. Refer to Figure 20–28. $f = 500 \text{ Hz}$, $L = 3 \text{ H}$, $R = 10 \text{ k}\Omega$, and $I_T = 500 \text{ }\mu\text{A}$.
7. Refer to Figure 20–28. $f = 5 \text{ kHz}$, $L = 30 \text{ mH}$, $R = 680 \text{ }\Omega$, and $I_T = 5 \text{ mA}$.
8. Refer to Figure 20–28. $f = 10 \text{ kHz}$, $L = 15 \text{ mH}$, $R = 1.5 \text{ k}\Omega$, and $I_T = 100 \text{ }\mu\text{A}$.
9. Refer to Figure 20–29 on the next page. $f = 25 \text{ kHz}$, $L = 30 \text{ mH}$, $C = 300 \text{ pF}$, $R = 10 \text{ k}\Omega$, and $I_T = 20 \text{ mA}$.
10. Refer to Figure 20–29. $f = 6 \text{ kHz}$, $L = 26.5 \text{ mH}$, $C = 150 \text{ nF}$, $R = 250 \text{ }\Omega$, and $I_T = 250 \text{ }\mu\text{A}$.

FIGURE 20–29

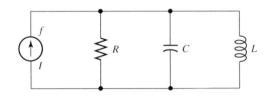

11. Refer to Figure 20–29. $f = 15$ kHz, $L = 4$ mH, $C = 30$ nF, $R = 6.8$ kΩ, and $I_T = 400$ μA.

12. Refer to Figure 20–29. $f = 250$ Hz, $L = 70$ mH, $C = 6$ μF, $R = 820$ Ω, and $I_T = 150$ μA.

END OF CHAPTER PROBLEMS 20–7

Find Y_{rect} and C or L in the equivalent parallel circuit for the following problems:

1. In a series RC circuit where $R = 680$ Ω, $C = 50$ nF, and $f = 4$ kHz.
2. In a series RC circuit where $R = 1.2$ kΩ, $C = 100$ nF, and $f = 1.2$ kHz.
3. In a series RC circuit where $R = 15$ kΩ, $C = 100$ pF, and $f = 120$ kHz.
4. In a series RC circuit where $R = 9.1$ kΩ, $C = 1$ nF, and $f = 30$ kHz.
5. In a series RL circuit where $R = 2.2$ kΩ, $L = 100$ mH, and $f = 2.65$ kHz.
6. In a series RL circuit where $R = 5.6$ kΩ, $L = 1.5$ H, and $f = 600$ Hz.
7. In a series RL circuit where $R = 15$ kΩ, $L = 200$ mH, and $f = 10$ kHz.
8. In a series RL circuit where $R = 5.6$ kΩ, $L = 0.5$ H, and $f = 2$ kHz.

Find Z_{rect} and C or L in the equivalent series circuit for the following problems:

9. In a parallel RC circuit where $G = 256$ μS, $C = 2$ nF, and $f = 10$ kHz.
10. In a parallel RC circuit where $G = 556$ μS, $C = 40$ nF, and $f = 1.55$ kHz.
11. In a parallel RC circuit where $G = 1.22$ mS, $C = 100$ nF, and $f = 3.25$ kHz.
12. In a parallel RC circuit where $G = 833$ μS, $C = 600$ nF, and $f = 800$ Hz.
13. In a parallel RL circuit where $G = 14.7$ mS, $L = 200$ mH, and $f = 100$ Hz.
14. In a parallel RL circuit where $G = 4.55$ mS, $L = 40$ mH, and $f = 2.5$ kHz.
15. In a parallel RL circuit where $G = 110$ μS, $L = 30$ mH, and $f = 8.1$ kHz.
16. In a parallel RL circuit where $G = 50$ μS, $L = 250$ mH, and $f = 7.5$ kHz.

END OF CHAPTER PROBLEMS 20–8

For problems 1 to 8, find the ESC. Express answers in rectangular form.

1. Refer to Figure 20–30 on the next page. Let $R = 500$ Ω, $X_L = 10$ kΩ, and $X_C = 20$ kΩ.
2. Refer to Figure 20–30. Let $R = 150$ Ω, $X_L = 250$ Ω, and $X_C = 150$ Ω.

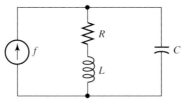

FIGURE 20–30

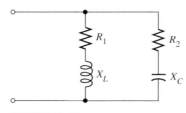

FIGURE 20–31

3. Refer to Figure 20–31. Let $R_1 =$ 15 kΩ, $R_2 = 10$ kΩ, $X_L = 5$ kΩ, and $X_C = 5$ kΩ.

4. Refer to Figure 20–31. Let $R_1 =$ 10 kΩ, $R_2 = 4.7$ kΩ, $X_L = 5$ kΩ, and $X_C = 8$ kΩ.

5. Refer to Figure 20–32. Let $R_1 =$ 5.6 kΩ, $R_2 = 27$ kΩ, $X_L = 35$ kΩ, and $X_C = 20$ kΩ.

6. Refer to Figure 20–32. Let $R_1 =$ 12 kΩ, $R_2 = 5$ kΩ, $X_L = 10$ kΩ, and $X_C = 25$ kΩ.

7. Refer to Figure 20–33. Let $R_1 =$ 1.2 kΩ, $R_2 = 12$ kΩ, $R_3 = 10$ kΩ, $X_L = 12$ kΩ, $X_{C1} = 10$ kΩ, and $X_{C2} = 3$ kΩ.

8. Refer to Figure 20–33. Let $R_1 =$ 2.7 kΩ, $R_2 = 12$ kΩ, $R_3 = 8.2$ kΩ, $X_L = 10$ kΩ, $X_{C1} = 5.2$ kΩ, and $X_{C2} = 7.5$ kΩ.

For problems 9 to 16, find the ESC. Include component values and the phase angle.

9. Refer to Figure 20–30. Let $f = 1$ kHz, $R = 1$ kΩ, $L = 500$ mH, and $C = 3$ nF.

10. Refer to Figure 20–30. Let $f = 5$ kHz, $R = 2$ kΩ, $L = 200$ mH, and $C = 10$ nF.

11. Refer to Figure 20–31. Let $R_1 =$ 500 Ω, $R_2 = 200$ Ω, $L = 250$ mH, $C = 3$ nF, and $f = 5$ kHz.

12. Refer to Figure 20–31. Let $R_1 =$ 1 kΩ, $R_2 = 470$ Ω, $L = 500$ mH, $C = 500$ pF, and $f = 10$ kHz.

13. Refer to Figure 20–33. $R_1 = 3.3$ kΩ, $R_2 = 20$ kΩ, $R_3 = 6.8$ kΩ, $C_1 = 20$ nF, $C_2 = 8$ nF, $L = 3.5$ H, and $f = 1$ kHz.

14. Refer to Figure 20–33. $R_1 = 1$ kΩ, $R_2 = 100$ Ω, $R_3 = 100$ Ω, $C_1 = 1$ μF, $C_2 = 1.2$ μF, $L = 1.6$ H, and $f = 100$ Hz.

15. Refer to Figure 20–32. $R_1 = 5.6$ kΩ, $R_2 = 2$ kΩ, $L = 150$ mH, $C = 10$ nF, and $f = 6.7$ kHz.

16. Refer to Figure 20–32. $R_1 = 220$ Ω, $R_2 = 470$ Ω, $L = 40$ mH, $C = 150$ nF, and $f = 2.7$ kHz.

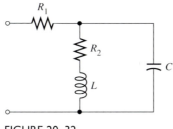

FIGURE 20–32

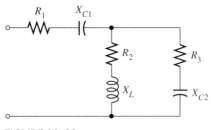

FIGURE 20–33

ANSWERS TO SELF-TESTS

SELF-TEST 20–1

1. $Y_{rect} = 100 \ \mu S + j67.5 \ \mu S$, $Y_{polar} = 121 \ \underline{/34°} \ \mu S$, $I_{rect} = 4.14 \ mA + j2.8 \ mA$, $I_{polar} = 5 \ \underline{/34°} \ mA$, $V = 41.3 \ V$

2. $Y_{rect} = 37.0 \ \mu S + j50.3 \ \mu S$, $Y_{polar} = 62.4 \ \underline{/53.6°} \ \mu S$, $I_{rect} = 296 \ \mu A + j403 \ \mu A$, $I_{polar} = 500 \ \underline{/53.6°} \ \mu A$, $V = 8.01 \ V$

3. $Y_{rect} = 8.33 \ \mu S - j6.37 \ \mu S$, $Y_{polar} = 10.5 \ \underline{/-37.4°} \ \mu S$, $I_{rect} = 55.6 \ \mu A - j42.5 \ \mu A$, $I_{polar} = 70 \ \underline{/-37.4°} \ \mu A$

4. $Y_{rect} = 14.7 \ \mu S - j20 \ \mu S$, $Y_{polar} = 24.8 \ \underline{/-53.7°} \ \mu S$, $I_{rect} = 1.18 \ mA - j1.61 \ mA$, $I_{polar} = 2 \ \underline{/-53.7°} \ mA$

5. $Y_{rect} = 37.0 \ \mu S - j53.1 \ \mu S$, $Y_{polar} = 64.7 \ \underline{/-55.1°} \ \mu S$, $I_{rect} = 1.14 \ mA - j1.64 \ mA$, $I_{polar} = 2 \ \underline{/-55.1°} \ mA$

SELF-TEST 20–2

1. $5.51 \ \underline{/-8.31°} \ k\Omega$

2. $I_R = 157 \ \mu A$, $I_C = 124 \ \mu A$, $Y_{polar} = 709 \ \underline{/38.4°} \ \mu S$, $Z_{polar} = 1.41 \ \underline{/-38.4°} \ k\Omega$, and $V = 282 \ mV$. In the ESC, $R = 1.11 \ k\Omega$ and $C = 519 \ nF$.

3. $I_R = 15.8 \ mA$, $I_L = -12.3 \ mA$, $Y_{polar} = 57.6 \ \underline{/-37.9°} \ \mu S$, $Z_{polar} = 17.4 \ \underline{/37.9°} \ k\Omega$, and $V = 347 \ V$. In the ESC, $R = 13.7 \ k\Omega$ and $L = 113 \ mH$.

4. $I_R = 430 \ \mu A$, $I_C = 2.03 \ mA$, $I_L = -2.28 \ mA$, $Y_{polar} = 116 \ \underline{/-30.7°} \ \mu S$, $Z_{polar} = 8.60 \ \underline{/30.7°} \ k\Omega$, and $V = 4.30 \ V$. In the ESC, $R = 7.40 \ k\Omega$ and $L = 4.65 \ mH$.

SELF-TEST 20–3

1. The ESC in rectangular form is $Z = 29.8 \ k\Omega - j65.8 \ k\Omega$. $C = 346 \ pF$ and the phase angle is $-65.6°$.

2. The EPC of the *RL* branch is $Y = 96.2 \ \mu S - j19.2 \ \mu S$. The EPC of the *RC* branch is $Y = 53.3 \ \mu S + j26.7 \ \mu S$. This yields a total admittance of $Y = 149 \ \mu S + j7.44 \ \mu S$. The ESC of the two branches is $Z = 6.69 \ k\Omega - j334\Omega$. The circuit ESC in rectangular form is $Z = 7.69 \ k\Omega - j5.33 \ k\Omega$.

Filters

21

Introduction

An electronic filter is a circuit that allows some AC signal frequencies to pass through with very little loss of power while greatly reducing signal power at other frequencies. The filters in this chapter are constructed with resistors, capacitors, and inductors. Filters are often tested by applying a constant amplitude voltage at the input of the filter and measuring the filter output voltage as the frequency of the input signal is changed.

In the preceding chapters on ac circuit analysis we calculated voltage drops and currents in many different kinds of circuits. We have seen how the distributions of these voltage drops and currents depend on component values and frequency. If we consider that component values remain fixed in a circuit, then the voltage drops and currents are frequency dependent only.

Typically, in an electronic circuit or system, the output voltage or current is affected by circuits that are combinations of series and parallel components. Because these circuits are often complex, the analysis is easier if we use circuit theorems and reduce the circuits to their equivalent series or parallel circuits.

In this chapter, then, we will see how frequency affects the output voltage of an *RL*, *RC*, or *RCL* circuit. We will also take the output from across different components and note how the voltage changes with frequency.

Chapter Objectives

In this chapter you will learn how to:

1. Identify high-pass and low-pass filters, calculate their cutoff frequencies, and determine the conditions that exist at cutoff.
2. Use circuit theorems to analyze various kinds of band-pass filters.
3. Calculate the various parameters in a series resonant circuit.

Consider the circuit in Figure 21–1. The circuit output voltage is developed across the capacitor. Therefore, V_o must vary as V_C varies. Without considering exact values of R and C, let's determine, in general, what happens to V_R and V_C as the frequency changes.

Because X_C is greatest at low frequencies, V_C will be greatest at low frequencies. If we plotted a curve of output voltage versus frequency, we would get a curve like the one in Figure 21–2. The curve shows that at very low frequencies $V_o = V_{in}$. As the frequency increases, X_C decreases and so does V_o.

Again referring to Figure 21–1, as we increase frequency from some low value, we will eventually reach a frequency where $X_C = R$. This frequency is called the *cutoff frequency* (also called the *break frequency* or the *corner frequency*). Because $X_C = R$ at the cutoff frequency (f_{co}), other conditions also exist: $\theta = -45°$, $V_C = V_R = 0.707\ V_{in}$, $I = 0.707\ I_{max}$, and $P = 0.5\ P_{max}$.

EXAMPLE 21–1

In Figure 21–1, let $V_{in} = 10$ V, $R = 10$ kΩ, and $C = 10$ nF. Find f_{co}. Find V_o, V_R, I, P, and θ, at f_{co}.

SOLUTION

At f_{co}:
$$Z = R - jX_C = 10\text{ k}\Omega - j10\text{ k}\Omega = 14.1\ \underline{/-45°}\text{ k}\Omega$$

$$I = \frac{V}{Z} = \frac{10\text{ V}}{14.1\text{ k}\Omega} = 0.707\text{ mA}$$

$$V_o = V_C = IX_C = 7.07\text{ V}$$

or
$$V_o = V_{in} \sin \theta = 7.07\text{ V}$$

$$V_R = IR = 7.07\text{ V}$$

or
$$V_R = V_{in} \cos \theta = 7.07\text{ V}$$

$$P = IV_o = 0.707\text{ mA} \times 7.07\text{ V} = 5\text{ mW}$$

$$\theta = -45°$$

$$f_{co} = \frac{1}{2\pi RC} = 1.59\text{ kHz}$$

FIGURE 21–1
Low-pass filter with output taken across the capacitor.

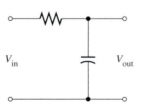

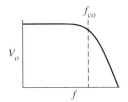

FIGURE 21–2 Curve of output voltage versus frequency for the low-pass filter in Figure 21–1.

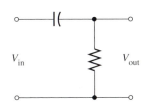

FIGURE 21–3 High-pass filter with output taken across the resistor.

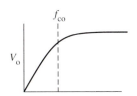

FIGURE 21–4 Curve of output voltage versus frequency for the high-pass filter in Figure 21–3.

A circuit such as the one in Example 21–1 is called a *low-pass filter*. That is, this circuit passes the low frequencies and rejects the high frequencies. Frequencies that result in output voltages that are less than 70.7% of maximum output are considered lost due to their low amplitude.

If we reversed the position of the components as in Figure 21–3, we would have a *high-pass filter*. The curve of output voltage versus frequency for this circuit is shown in Figure 21–4. The relationships discussed in reference to Figure 21–1 are also true here except that the output voltage is taken from across R. This causes V_o to *increase* with frequency. However, the cutoff frequency is the same, and all other conditions at f_{co} are the same.

PRACTICE PROBLEMS 21–1

1. Refer to Figure 21–1. Let $V_{in} = 20$ V and $R = 3$ kΩ. Find I_{max}, P_{max}, and $V_{o(max)}$. Find X_C, I, P, and V_o at f_{co}.

2. Refer to Figure 21–3. Let $V_{in} = 10$ V, $R = 2.2$ kΩ, and $C = 200$ nF. Find f_{co}, I, V_o, and P at f_{co}.

SOLUTIONS

1. $I_{max} = 6.67$ mA, $P_{max} = 133$ mW, $V_{o(max)} = 20$ V. At f_{co}, $I = 4.71$ mA, $P = 66.7$ mW, and $V_o = 14.1$ V.

2. $f_{co} = 362$ Hz, $I = 3.21$ mA, $V_o = 7.07$ V, and $P = 22.7$ mW.

Additional practice problems are at the end of the chapter.

THÉVENIN'S AND NORTON'S THEOREMS ⬛ 21–2

The use of Thévenin's theorem can simplify the solutions of some ac circuits. For example, consider the circuit in Figure 21–5(a). Suppose we wanted to find V_C. If we used trig functions, the problem would be fairly complex. Using Thévenin's theorem reduces the problem to a simple circuit quite easily. If we consider that C is the load,

FIGURE 21–5
(a) Series–
parallel *RC*
circuit;
(b) finding V_{OC}
where *C* is the
load;
(c) resulting
circuit to find
R_{TH};
(d) Thévenin's
equivalent
circuit.

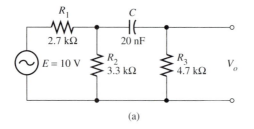

(a)

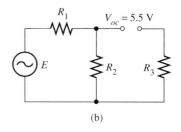

(b)

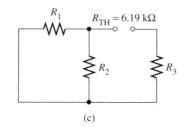

(c)

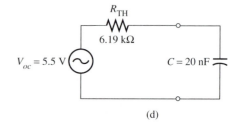

(d)

then V_{OC} and R_{TH} can be found. (See Chapter 13 for a review of Thévenin's theorem.) With the load removed as in Figure 21–5(b), $V_{OC} = V_2$ since there is no drop across R_3.

$$V_{OC} = V_2 = \frac{ER_2}{R_1 + R_2} = \frac{10 \text{ V} \times 3.3 \text{ k}\Omega}{2.7 \text{ k}\Omega + 3.3 \text{ k}\Omega} = 5.50 \text{ V}$$

Shorting the source results in an R_{TH} of 6.19 kΩ. The circuit is shown in Figure 21–5(c).

$$R_{TH} = R_3 + (R_1 \parallel R_2) = 4.7 \text{ k}\Omega + 1.49 \text{ k}\Omega = 6.19 \text{ k}\Omega$$

Thévenin's equivalent circuit with the load connected is shown in Figure 21–5(d). This equivalent circuit allows us to find the cutoff frequency of the circuit. f_{co} is the frequency where $X_C = R_{TH}$, so

(21–1)
$$f_{co} = \frac{1}{2\pi C R_{TH}}$$

$$= \frac{1}{2\pi \times 20 \text{ nF} \times 6.19 \text{ k}\Omega} = 1.29 \text{ kHz}$$

The output voltage is taken from across R_3. This results in a *high-pass* filter. The circuit will pass all frequencies above 1.29 kHz.

We can find V_o at f_{co} by using Thévenin's theorem. We know from the equivalent circuit that X_C must equal 6.19 kΩ at f_{co}. Therefore, $V_C = V_{OC} \sin 45° = 3.89$ V. Applying Ohm's law and Kirchhoff's current law, we get

$$I_C = \frac{V_C}{X_C} = \frac{3.89 \text{ V}}{6.19 \text{ k}\Omega} = 629 \text{ }\mu\text{A}$$

$$I_C = I_3 \qquad (C \text{ and } R_3 \text{ are in series})$$

$$V_3 = I_3 R_3 = 628 \text{ }\mu\text{A} \times 4.7 \text{ k}\Omega = 2.96 \text{ V}$$

$$V_o = V_3 = 2.96 \text{ V}$$

In a circuit such as the one in Figure 21–6(a), V_o and f_{co} can be found easily by using Norton's theorem. In Figure 21–6(b), the load is replaced with a short circuit. The short circuit current (I_{SC}) would be

$$I_{SC} = \frac{12 \text{ V}}{2.7 \text{ k}\Omega} = 4.44 \text{ mA}$$

(The short circuit reduces the resistance of the parallel circuit to 0 Ω.) G_N is found from the circuit in Figure 21–6(c).

$$G_N = G_1 + G_2 = 370 \ \mu\text{S} + 303 \ \mu\text{S} = 673 \ \mu\text{S}$$

The equivalent circuit with the load connected is shown in Figure 21–6(d).

The circuit is a low-pass filter. C provides a low-impedance path for current at high frequencies. f_{co} may be calculated from the equivalent circuit:

$$f_{co} = \frac{G_N}{2\pi C} \tag{21-2}$$

$$f_{co} = \frac{673 \ \mu\text{S}}{2\pi \times 20 \text{ nF}} = 5.36 \text{ kHz}$$

The output voltage at f_{co} can be determined by first determining $V_{o(max)}$. $V_{o(max)}$ would occur where $G_N \gg B_C$.

$$V_{o(max)} = \frac{I_{SC}}{G_N} = 6.60 \text{ V}$$

At f_{co}

$$V_o = 0.707 \ V_{o(max)} = 4.67 \text{ V}$$

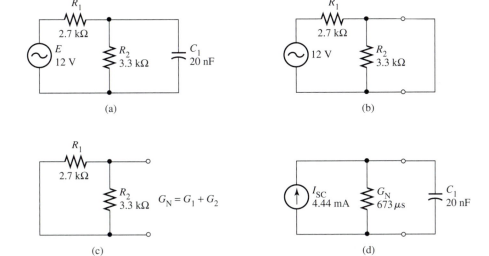

(a)

(b)

(c)

(d)

FIGURE 21–6
(a) Series–parallel RC circuit;
(b) finding I_{SC} where C is the load;
(c) resulting circuit used to find G_N;
(d) Norton's equivalent circuit.

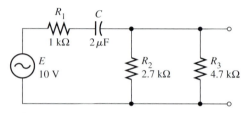

FIGURE 21–7   High-pass filter.

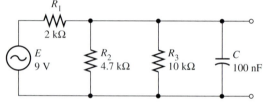

FIGURE 21–8   Low-pass filter.

PRACTICE PROBLEMS 21–2

1. Refer to Figure 21–5. Let $R_1 = 1$ kΩ, $R_2 = 2$ kΩ, $R_3 = 3$ kΩ, $C = 100$ nF, and $E = 10$ V. Thévenize the circuit. Find f_{co} and V_o at f_{co}.
2. Refer to Figure 21–7. Thévenize the circuit. Find f_{co} and V_o at f_{co}.
3. Refer to Figure 21–6. Let $R_1 = 2$ kΩ, $R_2 = 2.7$ kΩ, $C = 500$ pF, and $E = 12$ V. Nortonize the circuit. Find f_{co} and V_o at f_{co}.
4. Refer to Figure 21–8. Nortonize the circuit. Find f_{co} and V_o at f_{co}.

SOLUTIONS

1. $V_{oc} = 6.67$ V, $R_{TH} = 3.67$ kΩ, $f_{co} = 434$ Hz, V_o at $f_{co} = 3.86$ V
2. $V_{oc} = 10$ V, $R_{TH} = 2.71$ kΩ, $f_{co} = 29.4$ Hz, V_o at $f_{co} = 4.46$ V
3. $I_{SC} = 6$ mA, $G_N = 870$ μS, $f_{co} = 277$ kHz, V_o at $f_{co} = 4.87$ V
4. $I_{SC} = 4.5$ mA, $G_N = 813$ μS, $f_{co} = 1.29$ kHz, V_o at $f_{co} = 3.92$ V

Additional practice problems are at the end of the chapter.

SELF-TEST 21–1

1. Refer to Figure 21–9 on the next page. $V_{in} = 25$ V, $R = 6042.2$ kΩ, and $C = 200$ nF. Find f_{co}. Find I, V_o, and P at f_{co}.

2. Refer to Figure 21–5. Let $R_1 = 4.7$ kΩ, $R_2 = 1$ kΩ, $R_3 = 1.8$ kΩ, $C = 250$ nF, and $E = 12$ V. Thévenize the circuit. Find f_{co}.

3. Refer to Figure 21–7. Let $R_1 = 3.3$ kΩ, $R_2 = 1.8$ kΩ, $R_3 = 4.7$ kΩ, $C = 500$ nF, and $E = 12$ V. Thévenize the circuit. Find f_{co}, and V_o at f_{co}.

4. Refer to Figure 21–6. Let $R_1 = 680$ Ω, $R_2 = 2.7$ kΩ, $C = 5$ nF, and $E = 20$ V. Nortonize the circuit. Find f_{co}, and V_o at f_{co}.

5. Refer to Figure 21–8. Let $R_1 = 680$ Ω, $R_2 = 2.7$ kΩ, $R_3 = 6.8$ kΩ, $C = 10$ nF, and $E = 15$ V. Nortonize the circuit. Find f_{co}, and V_o at f_{co}.

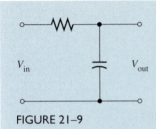

FIGURE 21–9

Answers to Self-test 21–1 are at the end of the chapter.

BAND-PASS FILTERS

Band-pass filters are filters that pass a band of frequencies. That is, the filter rejects both high frequencies and low frequencies but passes a range of frequencies between. Figure 21–10 shows a curve of such a filter. Various circuit configurations could produce such a curve, but, in general, the curve usually results from both series and parallel capacitances in the circuit. The curve has two cutoff frequencies. f_1 is the lower cutoff frequency and f_2 is the upper cutoff frequency.

Consider the circuit in Figure 21–11. This circuit could be the equivalent circuit of an amplifier or passive network. At low frequencies the capacitive reactances will be quite large compared to the circuit resistances. Because C_2 is connected in parallel, its high reactance (low susceptance) will have negligible effect on the circuit. C_1, though, is connected in series, thus resulting in low output voltages at these low frequencies. We can better see the low-frequency circuit if we Thévenize it. Since C_2 will have little effect on the circuit at low frequencies, we can treat it as an open circuit. With C_2 open-circuited, the new circuit is shown in Figure 21–12.

Even though the output is taken across R_2, for the low frequency equivalent circuit, we Thévenize at C_1 because we are interested in the relationship between X_{C1} and the resistance that C_1 "sees." If we Thévenize at C_1, the resulting equivalent resistance, R_{TH}, is $R_1 + R_2$. The Thévenin's equivalent circuit is shown in Figure 21–13. At f_1, $X_{C1} = R_{TH}$.

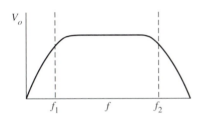

FIGURE 21–10   Output voltage versus frequency of a band-pass filter.

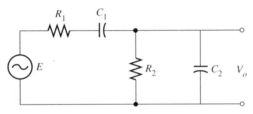

FIGURE 21–11   Typical equivalent circuit at the output of an amplifier.

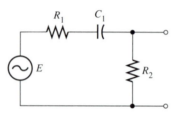

FIGURE 21–12   Low-frequency
circuit. Effects of C_2 are negligible.

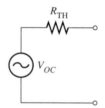

FIGURE 21–13
Low-frequency
circuit Thévenized.

(21–3)

$$f_1 = \frac{1}{2\pi R_{TH}C_1}$$

At high frequencies we can ignore the low reactance of C_1. The high susceptance of C_2 (low reactance) causes a reduction in output voltage at high frequencies. Because we are interested in the relationship between X_{C2} and the resistance that C_2 "sees," and because the components at high frequencies are in parallel, we will use Norton's equivalent circuit. G_N in the equivalent circuit results from R_1 and R_2 in parallel. $G_N = G_1 + G_2$. The resulting circuit at high frequencies is shown in Figure 21–14. Norton's equivalent circuit is shown in Figure 21–15. At f_2, $G_N = B_C$.

(21–4)

$$f_2 = \frac{G_N}{2\pi C_2}$$

In the mid-frequency range (the range of frequencies between f_1 and f_2) we ignore the effects of both capacitances. The reactance of C_1 is negligible compared to $R_{TH}(R_1 + R_2)$. The susceptance of C_2 is negligible compared to G_N $(G_1 + G_2)$. The mid-frequency circuit consists of R_1 and R_2, as in Figure 21–16. Even though we consider that the circuit passes a band or range of frequencies, a *middle* frequency (f_m) may be found by using the following equation:

$$f_m = \sqrt{f_1 f_2}$$

Theoretically, this middle frequency is at the mid-point of the mid-frequency range.

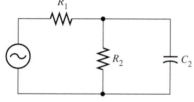

FIGURE 21–14   High-frequency
circuit. Effects of C_1 are negligible.

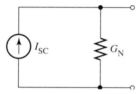

FIGURE 21–15   High-
frequency circuit
Nortonized.

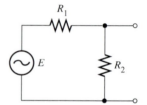

FIGURE 21–16   Mid-
frequency circuit. Effects
of circuit capacitances are
negligible.

EXAMPLE 21-2

Refer to Figure 21–11. Let $R_1 = 2$ kΩ, $R_2 = 3$ kΩ, $C_1 = 1$ μF, $C_2 = 500$ pF, and $E = 20$ V. Find f_1, f_2, V_o at f_m and V_o at f_{co}.

SOLUTION

$$f_1 = \frac{1}{2\pi R_{TH} C_1}$$

$$R_{TH} = R_1 + R_2 = 2 \text{ k}\Omega + 3 \text{ k}\Omega = 5 \text{ k}\Omega$$

$$f_1 = \frac{1}{2\pi \times 5 \text{ k}\Omega \times 1 \text{ } \mu F} = 31.8 \text{ Hz}$$

$$f_2 = \frac{G_N}{2\pi C_2} \qquad\qquad (21\text{--}5)$$

$$G_N = G_1 + G_2 = 500 \text{ } \mu S + 333 \text{ } \mu S = 833 \text{ } \mu S$$

$$f_2 = \frac{833 \text{ } \mu S}{2\pi \times 500 \text{ pF}} = 265 \text{ kHz}$$

At f_m, the effects of C_1 and C_2 are negligible so we treat the circuit as a simple voltage divider where the output is taken across R_2.

$$V_o = \frac{ER_2}{R_1 + R_2} = \frac{20 \text{ V} \times 3 \text{ k}\Omega}{5 \text{ k}\Omega} = 12 \text{ V}$$

At f_1 and f_2 the phase angle is 45°. At these frequencies, $V_o = 0.707$ times $V_{o(max)}$. $V_{o(max)}$ is the output voltage in the mid-frequency range.

$$V_o \text{ at } f_1 = V_o \text{ at } f_2 = 12 \text{ V} \times 0.707 = 8.49 \text{ V}$$

EXAMPLE 21-3

Refer to Figure 21–17 on the next page. Find f_1 and f_2.

SOLUTION
Ignoring C_2 and treating C_1 as the load, we can determine the value of f_1:

$$R_{TH} = R_3 + R_1 \| R_2 = 5.24 \text{ k}\Omega$$

$$f_1 = \frac{1}{2\pi C_1 R_{TH}} = 30.4 \text{ Hz}$$

Treating C_1 as a short circuit and assuming that C_2 is the load, we can determine the value of f_2:

$$G_N = G_1 + G_2 + G_3 = 1.47 \text{ mS} + 370 \text{ } \mu S + 213 \text{ } \mu S = 2.05 \text{ mS}$$

$$f_2 = \frac{G_N}{2\pi C_2} = 163 \text{ kHz}$$

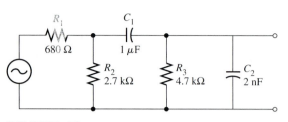

FIGURE 21–17

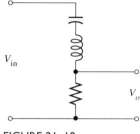

FIGURE 21–18

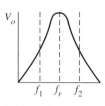

FIGURE 21–19

Consider the circuit in Figure 21–18. We recognize a series *LCR* bandpass circuit. We have determined previously that the resonant frequency (f_r) is that frequency where $X_L = X_C$. Other conditions at resonance are $\theta = 0°$, $V_{in} = V_R = V_o$, $V_C = V_L$, $Z = R$, I is max, and P is max. The curve of output voltage versus frequency is shown in Figure 21–19.

As we increase the frequency above resonance, X_L increases and X_C decreases. Eventually, we reach a frequency where the difference between X_L and X_C equals R. This is the *upper cutoff frequency, f_2*.

As we decrease the frequency below resonance, X_C increases and X_L decreases. Eventually we reach a frequency where the difference between X_C and X_L equals R. This is the *lower cutoff frequency, f_1*. In *RCL* circuits the range of frequencies between f_1 and f_2 is called the *bandwidth* (BW).

$$BW = f_2 - f_1$$

$$f_1 = f_r - \frac{BW}{2}$$

$$f_2 = f_r + \frac{BW}{2}$$

Bandpass (and band-reject) filters are usually designed to be very selective. They will pass (or reject) a narrow band of frequencies. The highest quality filters have the narrowest bandwith. This selectivity characteristic is referred to as Q (for quality) and is defined as:

$$Q = \frac{\text{resonant frequency}}{\text{bandwidth}} = \frac{f_r}{BW}$$

EXAMPLE 21–4

In Figure 21–18, let $V_{in} = 3$ V, $R = 100\ \Omega$, $f_r = 1$ kHz, and $Q = 20$. Find the bandwidth, f_1, f_2, I_{max}, P_{max}, and V_R at resonance, and V_R, I, and P at f_1 and f_2.

SOLUTION

$$BW = \frac{f_r}{Q} = \frac{1\ \text{kHz}}{20} = 50\ \text{Hz}$$

$$f_1 = f_r - \frac{BW}{2} = 1 \text{ kHz} - 25 \text{ Hz} = 975 \text{ Hz}$$

$$f_2 = f_r + \frac{BW}{2} = 1 \text{ kHz} + 25 \text{ Hz} = 1025 \text{ Hz}$$

$V_R = V_{in} = 3$ V at resonance. At other frequencies, V_R would be less than 3 V. Therefore,

$$I_{max} = \frac{V_R}{R} = \frac{3 \text{ V}}{100 \ \Omega} = 30 \text{ mA}$$

$$P_{max} = V_R I = 3 \text{ V} \times 30 \text{ mA} = 90 \text{ mW}$$

At f_1, $\theta = -45°$. At f_2, $\theta = 45°$. Then, at either f_1 or f_2,

$$V_R = V_{in} \cos 45° = 3 \text{ V} \times 0.707 = 2.12 \text{ V}$$

$$I = \frac{V_R}{R} = \frac{2.12 \text{ V}}{100 \ \Omega} = 21.2 \text{ mA}$$

$$P = I V_R = 21.2 \text{ mA} \times 2.12 \text{ V} = 45.0 \text{ mW}$$

This circuit (Figure 21–18) is a *band-pass* filter. The circuit passes the band of frequencies between f_1 and f_2 and rejects all others.

EXAMPLE 21–5

In Figure 21–18, let $L = 40$ mH, $C = 200$ nF, and $R = 30 \ \Omega$. Find f_r, f_1, f_2, BW, and Q.

SOLUTION

$$f_r = \frac{1}{2\pi\sqrt{LC}} = 1.78 \text{ kHz}$$

$$Q = \frac{X_L}{R} = \frac{2\pi f_r L}{R} = \frac{447 \ \Omega}{30 \ \Omega} = 14.9$$

$$BW = \frac{f_r}{Q} = \frac{1.78 \text{ kHz}}{14.9} = 119$$

$$f_1 = f_r - \frac{BW}{2} = 1.78 \text{ kHz} - 60 \text{ Hz} = 1.72 \text{ kHz}$$

$$f_2 = f_r + \frac{BW}{2} = 1.78 \text{ kHz} + 60 \text{ Hz} = 1.84 \text{ kHz}$$

The circuit in Figure 21–20 is a *band-reject* filter. The curve of output voltage versus frequency is shown in Figure 21–21. The parallel *LC* circuit offers a maximum impedance to the circuit at resonance. Therefore, circuit current is minimum and V_o is minimum. The frequencies between f_1 and f_2 are rejected.

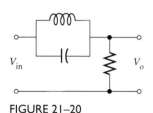

FIGURE 21–20

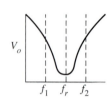

FIGURE 21–21

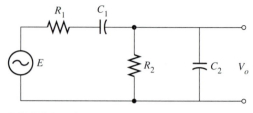

FIGURE 21–22

PRACTICE PROBLEMS 21–3

1. Refer to Figure 21–22. Let $R_1 =$ 4.7 kΩ, $R_2 = 1.8$ kΩ, $C_1 = 100$ nF, $C_2 = 1$nF, and $E = 20$ V. Find f_1, f_2, V_o at f_m, and V_o at f_{co}.

2. Refer to Figure 21–17. Let $R_1 =$ 1.5 kΩ, $R_2 = 3.3$ kΩ, $R_3 = 2.7$ kΩ, $C_1 = 200$ nF, and $C_2 = 700$ pF. Find f_1 and f_2.

3. Refer to Figure 21–18. $V_{in} = 9$ V, $R =$ 200 Ω, $f_r = 800$ Hz, and $Q = 20$. Find BW, f_1, and f_2. Find V_R, P, and I at f_r, f_1, and f_2.

4. Refer to Figure 21–18. $L = 200$ mH, $C = 200$ nF, and $R = 100$ Ω. Find f_r, f_1, f_2, BW, and Q.

SOLUTIONS

1. $f_1 = 245$ Hz, $f_2 = 122$ kHz, V_o at $f_m =$ 5.54 V, V_o at $f_{co} = 3.92$ V

2. $f_1 = 213$ kHz, $f_2 = 305$ kHz

3. BW $= 40$ Hz, $f_1 = 780$ Hz, $f_2 =$ 820 Hz. At f_r, $V_R = 9$ V, $I = 45$ mA, and $P = 405$ mW. At f_{co}, $V_R = 6.36$ V, $I = 31.8$ mA, and $P = 202$ mW.

4. $f_r = 796$ Hz, $f_1 = 756$ Hz, $f_2 = 836$ Hz, BW $= 79.6$ Hz, $Q = 10$

Additional practice problems are at the end of the chapter.

SELF-TEST 21–2

1. Refer to Figure 21–22. Let $R_1 = 2$ kΩ, $R_2 = 1.2$ kΩ, $C_1 = 200$ nF, $C_2 = 2$ nF, and $E = 9$ V. Find f_1, f_2, V_o at f_m, and V_o at f_{co}.

2. Refer to Figure 21–17. Let $R_1 = 3.9$ kΩ, $R_2 = 4.7$ kΩ, $R_3 = 2.2$ kΩ, $C_1 = 100$ nF, and $C_2 = 2$ nF. Find f_1 and f_2.

3. Refer to Figure 21–18. $L = 200$ mH, $C = 2$ nF, and $R = 1$ kΩ. Find f_r, f_1, f_2, BW, and Q.

Answers to Self-test 21–2 are at the end of the chapter.

CHAPTER 21 AT A GLANCE

PAGE	DEFINITIONS	EXAMPLE
600	The *cutoff frequency* is that frequency where the reactance equals the resistance.	$R = X$

601 — A *low-pass* filter passes low frequencies and rejects those frequencies above the cutoff frequency.

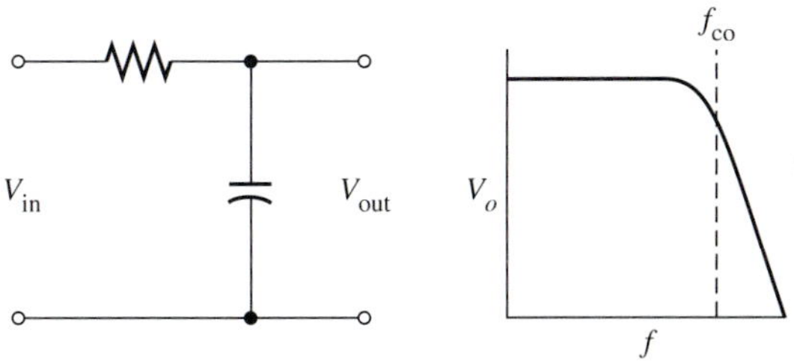

FIGURE 21–23

601 — A *high-pass* filter passes high frequencies and rejects those frequencies below the cutoff frequency.

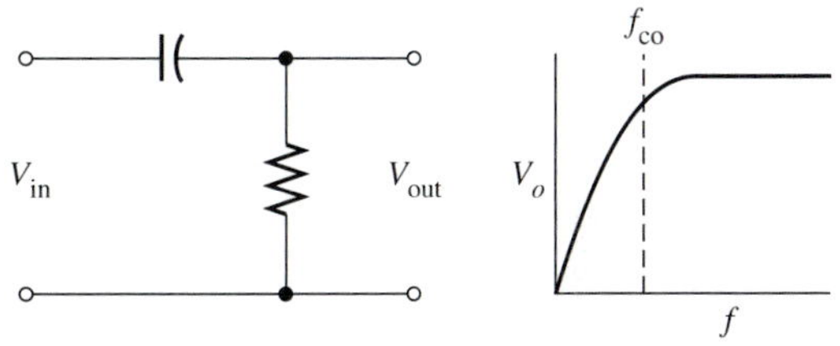

FIGURE 21–24

605 — A *band-pass* filter passes that band of frequencies between the lower and upper cutoff frequencies.

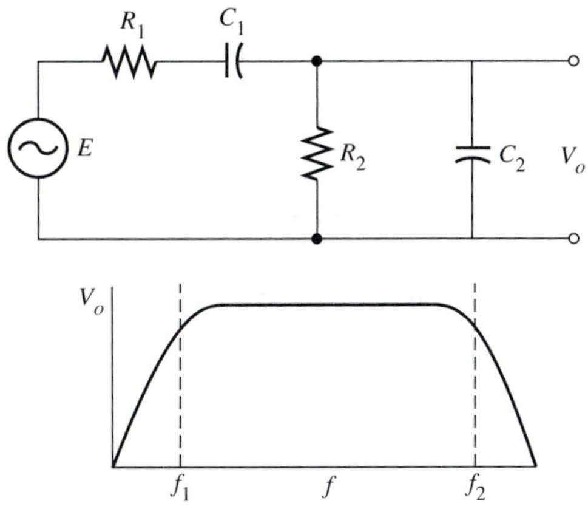

FIGURE 21–25

609 — A *band-reject* filter rejects that band of frequencies between the cutoff frequencies and passes all other frequencies.

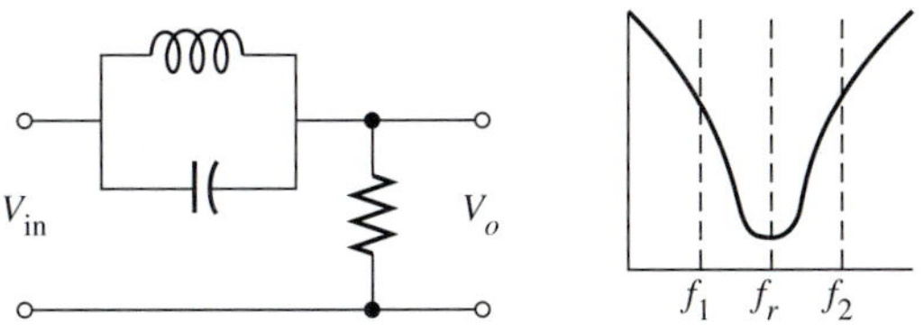

FIGURE 21–26

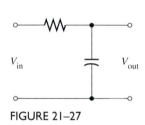

FIGURE 21–27

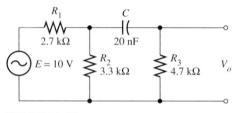

FIGURE 21–28

END OF CHAPTER PROBLEMS 21–1

Refer to Figure 21–27.

1. Let $V_{in} = 15$ V and $R = 470$ Ω. Find I_{max}, P_{max}, and $V_{o(max)}$. Find X_C, I, P, and V_o at f_{co}.
3. Let $V_{in} = 300$ mV and $R = 5.6$ kΩ. Find I_{max}, P_{max}, and $V_{o(max)}$. Find X_C, I, P, and V_o at f_{co}.
5. Let $V_{in} = 20$ V, $R = 1.2$ kΩ, and $C = 500$ nF. Find f_{co}. Find I, V_o, and P at f_{co}.
7. Let $V_{in} = 40$ V, $R = 2.2$ kΩ, and $C = 200$ nF. Find f_{co}. Find I, V_o, and P at f_{co}.

2. Let $V_{in} = 20$ V and $R = 5.6$ kΩ. Find I_{max}, P_{max}, and $V_{o(max)}$. Find X_C, I, P, and V_o at f_{co}.
4. Let $V_{in} = 12$ V and $R = 1.8$ kΩ. Find I_{max}, P_{max}, and $V_{o(max)}$. Find X_C, I, P, and V_o at f_{co}.
6. Let $V_{in} = 25$ V, $R = 300$ Ω, and $C = 2$ μF. Find f_{co}. Find I, V_o, and P at f_{co}.
8. Let $V_{in} = 600$ mV, $R = 100$ Ω, and $C = 1$ μF. Find f_{co}. Find I, V_o, and P at f_{co}.

END OF CHAPTER PROBLEMS 21–2

1. Refer to Figure 21–28. Let $R_1 = 1.8$ kΩ, $R_2 = 3.3$ kΩ, $R_3 = 1$ kΩ, $C = 500$ nF, and $E = 10$ V. Thévenize the circuit and find f_{co} and V_o at f_{co}.
3. Refer to Figure 21–29. Thévenize the circuit and find f_{co} and V_o at f_{co}.

5. Refer to Figure 21–30. Let $R_1 = 2.7$ kΩ, $R_2 = 3.9$ kΩ, $R_3 = 5.6$ kΩ, $C = 1$ μF, and $E = 10$ V. Thévenize the circuit. Find f_{co} and V_o at f_{co}.

2. Refer to Figure 21–28. Let $R_1 = 3.3$ kΩ, $R_2 = 1.2$ kΩ, $R_3 = 3$ kΩ, $C = 1$ μF, and $E = 15$ V. Thévenize the circuit and find f_{co} and V_o at f_{co}.
4. Refer to Figure 21–29. Let $R_1 = 5.6$ kΩ, $R_2 = 2.7$ kΩ, $C = 1$ μF, and $E = 12$ V. Thévenize the circuit and find f_{co} and V_o at f_{co}.
6. Refer to Figure 21–30. Let $R_1 = 750$ Ω, $R_2 = 2$ kΩ, $R_3 = 1.5$ kΩ, $C = 800$ nF, and $E = 20$ V. Thévenize the circuit. Find f_{co} and V_o at f_{co}.

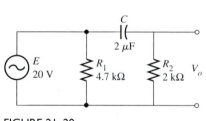

FIGURE 21–29

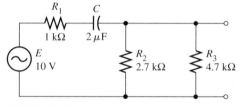

FIGURE 21–30

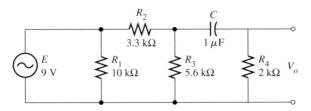

FIGURE 21–31

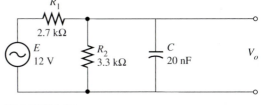

FIGURE 21–32

7. Refer to Figure 21–31. Thévenize the circuit and find f_{co}.

8. Refer to Figure 21–31. Let $R_1 =$ 7.5 kΩ, $R_2 = 2.7$ kΩ, $R_3 = 4.7$ kΩ, $R_4 = 3.9$ kΩ, $C = 500$ nF, and $E = 15$ V. Thévenize the circuit and find f_{co}.

9. Refer to Figure 21–32. Let $R_1 =$ 470 Ω, $R_2 = 1.2$ kΩ, $C = 50$ nF, and $E = 15$ V. Nortonize the circuit and find f_{co} and V_o at f_{co}.

10. Refer to Figure 21–32. Let $R_1 =$ 2.2 kΩ, $R_2 = 3.3$ kΩ, $C = 100$ nF, and $E = 20$ V. Nortonize the circuit and find f_{co} and V_o at f_{co}.

11. Refer to Figure 21–33. Let $R_1 =$ 2 kΩ, $R_2 = 4.7$ kΩ, $R_3 = 5.6$ kΩ, $C = 1$ nF, and $E = 10$ V. Nortonize the circuit and find f_{co} and V_o at f_{co}.

12. Refer to Figure 21–33. Let $R_1 =$ 3 kΩ, $R_2 = 2.7$ kΩ, $R_3 = 3.3$ kΩ, $C = 2$ nF, and $E = 5$ V. Nortonize the circuit and find f_{co} and V_o at f_{co}.

END OF CHAPTER PROBLEMS 21–3

Refer to Figure 21–34.

1. Let $R_1 = 680$ Ω, $R_2 = 1.2$ kΩ, $C_1 =$ 500 nF, $C_2 = 2$ nF, and $E = 9$ V. Find f_1, f_2, V_o at f_m, and V_o at f_{co}.

2. Let $R_1 = 2.7$ kΩ, $R_2 = 2.2$ kΩ, $C_1 =$ 100 nF, $C_2 = 100$ pF, and $E = 10$ V. Find f_1, f_2, V_o at f_m, and V_o at f_{co}.

3. Let $R_1 = 3.3$ kΩ, $R_2 = 2$ kΩ, $C_1 =$ 2 μF, $C_2 = 1$ nF, and $E = 15$ V. Find f_1, f_2, V_o at f_m, and V_o at f_{co}.

4. Let $R_1 = 1.8$ kΩ, $R_2 = 1.2$ kΩ, $C_1 =$ 750 nF, $C_2 = 500$ pF, and $E = 5$ V. Find f_1, f_2, V_o at f_m, and V_o at f_{co}.

5. Let $R_1 = 2.2$ kΩ, $R_2 = 3.9$ kΩ, $C_1 =$ 1.2 μF, $C_2 = 20$ nF, and $E = 2$ V. Find f_1, f_2, V_o at f_m, and V_o at f_{co}.

6. Let $R_1 = 5.6$ kΩ, $R_2 = 3.3$ kΩ, $C_1 =$ 0.5 μF, $C_2 = 100$ nF, and $E =$ 600 mV. Find f_1, f_2, V_o at f_m, and V_o at f_{co}.

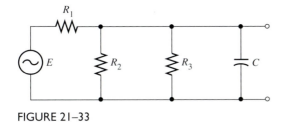

FIGURE 21–33

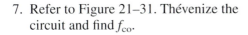

FIGURE 21–34

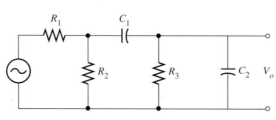

FIGURE 21-35

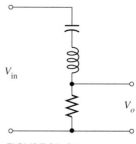

FIGURE 21-36

7. Let $R_1 = 330\ \Omega$, $R_2 = 820\ \Omega$, $C_1 = 200$ nF, $C_2 = 500$ pF, and $E = 3$ V. Find f_1, f_2, V_o at f_m, and V_o at f_{co}.

8. Let $R_1 = 910\ \Omega$, $R_2 = 2\ k\Omega$, $C_1 = 700$ nF, $C_2 = 1$ nF, and $E = 7.5$ V. Find f_1, f_2, V_o at f_m, and V_o at f_{co}.

Refer to Figure 21–35.

9. Let $R_1 = 5.6\ k\Omega$, $R_2 = 2.7\ k\Omega$, $R_3 = 1.8\ k\Omega$, $C_1 = 1\ \mu F$, and $C_2 = 2$ nF. Find f_1 and f_2.

10. Let $R_1 = 6.8\ k\Omega$, $R_2 = 2\ k\Omega$, $R_3 = 4.7\ k\Omega$, $C_1 = 1\ \mu F$, and $C_2 = 2$ nF. Find f_1 and f_2.

11. Let $R_1 = 1.2\ k\Omega$, $R_2 = 4.3\ k\Omega$, $R_3 = 6.8\ k\Omega$, $C_1 = 2\ \mu F$, and $C_2 = 800$ pF. Find f_1 and f_2.

12. Let $R_1 = 470\ \Omega$, $R_2 = 560\ \Omega$, $R_3 = 1\ k\Omega$, $C_1 = 1\ \mu F$, and $C_2 = 1$ nF. Find f_1 and f_2.

Refer to Figure 21–36.

13. $V_{in} = 20$ V, $R = 1\ k\Omega$, $f_r = 2$ kHz, and $Q = 25$. Find BW, f_1, and f_2. Find V_R, I, and P at f_r, f_1, and f_2.

14. $V_{in} = 10$ V, $R = 810\ \Omega$, $f_r = 1$ kHz, and $Q = 10$. Find BW, f_1, and f_2. Find V_R, I, and P at f_r, f_1, and f_2.

15. $V_{in} = 5$ V, $R = 120\ \Omega$, $f_r = 5.5$ kHz, and $Q = 14$. Find BW, f_1, and f_2. Find V_R, I, and P at f_r, f_1, and f_2.

16. $V_{in} = 200$ mV, $R = 300\ \Omega$, $f_r = 2.73$ kHz, and $Q = 20$. Find BW, f_1, and f_2. Find V_R, I, and P at f_r, f_1, and f_2.

17. $L = 100$ mH, $C = 250$ nF, and $R = 45\ \Omega$. Find f_r, f_1, f_2, BW, and Q.

18. $L = 500$ mH, $C = 2$ nF, and $R = 470\ \Omega$. Find f_r, f_1, f_2, BW, and Q.

19. $L = 20$ mH, $C = 100$ nF, and $R = 75\ \Omega$. Find f_r, f_1, f_2, BW, and Q.

20. $L = 500$ mH, $C = 1$ nF, and $R = 100\ \Omega$. Find f_r, f_1, f_2, BW, and Q.

ANSWERS TO SELF-TESTS

SELF-TEST 21–1

1. $f_{co} = 362$ Hz, $I = 8.03$ mA, $V_o = 17.7$ V, $P = 142$ mW

2. $V_{oc} = 2.11$ V, $R_{TH} = 2.62\ k\Omega$, $f_{co} = 243$ Hz

3. $V_{oc} = 12$ V, $R_{TH} = 4.6\ k\Omega$, $f_{co} = 69.2$ Hz, V_o at $f_{co} = 2.40$ V

4. $I_{SC} = 29.4$ mA, $G_N = 1.84$ mS, $f_{co} = 58.6$ kHz, V_o at $f_{co} = 11.3$ V

5. $I_{SC} = 22.1$ mA, $G_N = 1.99$ mS, $f_{co} = 31.6$ kHz, V_o at $f_{co} = 7.85$ V

SELF-TEST 21–2

1. $f_1 = 249$ Hz, $f_2 = 106$ kHz, V_o at $f_m = 3.38$ V, V_o at $f_{co} = 2.39$ V

2. $f_1 = 367$ Hz, $f_2 = 73.5$ kHz

3. $f_r = 7.96$ kHz, $f_1 = 7.56$ kHz, $f_2 = 8.36$ kHz, BW $= 796$ Hz, $Q = 10$

Logarithms in Electronics

Logarithms are exponents. They are usually used to represent quantities of power. Examples are:

1. Audio sound leveler. The human ear responds to sound in a logarithmic fashion. Audio amplifiers use a logarithmic scale to measure sound and noise output.
2. Radio frequency (RF) transmitters and receivers use a logarithmic scale to measure RF energy. Radar transmitters can send out megawatt pulses of energy and RF receivers can detect signals that have less than a microwatt of electrical energy.

Logarithms are usually used in the electronics field when discussing radio frequency (RF) power levels. Chapter 22 defines logarithms, shows how to convert between logarithmic expressions and exponential expressions, shows how to multiply or divide using logarithms, shows how to raise a number to a power using logarithms, and describes the differences between common and natural logarithms.

Chapter 23 shows how to solve equations involving both common and natural logarithms. It also shows how to use the calculator to help solve equations. Chapter 24 explains how to apply logarithmic equations to solve practical electronics problems.

Logarithms

22

Introduction

Logarithms are used in solving many electrical and electronics problems. The use of logarithms simplifies the solution of some complex problems. Their use reduces multiplication to addition, division to subtraction, raising to a power to simple multiplication, and extracting square roots to simple division.

Multiplication, division, or raising a number to some power can be accomplished with a calculator. However, in some problem-solving situations an understanding of the solution to such problems using logarithms is necessary. The following chapters dealing with logarithmic equations and applications of logarithms require this understanding.

Chapter Objectives

In this chapter you will learn how to:

1. Identify logarithmic expressions and exponential expressions.
2. Convert between logarithmic and exponential forms.
3. Find common logarithms of numbers using a calculator.
4. Multiply and divide numbers by using logarithms.
5. Raise a number to some power by using logarithms.
6. Find natural logarithms of numbers using a calculator.

COMMON LOGARITHMS · 22-1

Consider the expression $10^2 = 100$. Recall from previous chapters that 10 is called the base, 2 is the exponent, and 100 is its numerical value. The *logarithm* of the expression is the exponent 2.

EXAMPLE 22–1

The logarithm is the exponent.

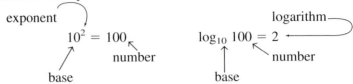

$$\underset{\text{base}}{1}0^{\underset{\text{exponent}}{2}} = \underset{\text{number}}{100} \qquad \log_{\underset{\text{base}}{10}} \underset{\text{number}}{100} = \underset{\text{logarithm}}{2}$$

☞ **KEY-POINT** The common logarithm of a number (N) is an exponent (x) where $10^x = N$. The general equation is:

$$\log_{10} N = x \text{ where } 10^x = N$$

That is, the logarithm is the power to which 10 (the base) must be raised in order for it to equal 100 (the number). Simply stated, *a logarithm is an exponent.*

Ten is the base of this system of logarithms. When 10 is the base, the system is called a system of *common logarithms.* In the system of logarithms, this is one of the two bases used in problem solving. The other base is 2.718. This system is referred to as a system of *natural logarithms.* Natural logarithms will be discussed later in the chapter.

When dealing with common logarithms, the logarithmic expression for $10^2 = 100$ would be $\log_{10} 100 = 2$. This is read, "The log, to the base 10, of 100 is 2." When using common logs, it is not necessary to indicate the base because it is assumed to be 10. The expression "log 100 = 2" is more often used. Notice that we have expressed the quantity two different ways. The first expression, $10^2 = 100$, is called an *exponential expression.* We have expressed the quantity in *exponential form.* The second expression, log 100 = 2, is called a *logarithmic expression.* We have written the quantity in *logarithmic form.*

Let's find the log of 1000. First, we express the quantity in exponential form. To do this, we merely find the power of 10 that equals 1000. In exponential form, $10^3 = 1000$. The logarithm is the exponent to which 10 is raised and equals 3. In logarithmic form, log 1000 = 3.

Find the log of 0.0001. Again, we first express the quantity in exponential form: $10^{-4} = 0.0001$. The logarithm is the exponent and equals −4. In logarithmic form, log 0.0001 = −4.

PRACTICE PROBLEMS 22–1

Find the logarithms of the following numbers. Express your answers first in logarithmic form and then in exponential form.

1. 0.00001
2. 0.01
3. 10,000
4. 1,000,000

SOLUTIONS

Logarithmic form	Exponential form
1. $\log 0.00001 = -5$	$10^{-5} = 0.00001$
2. $\log 0.01 = -2$	$10^{-2} = 0.01$
3. $\log 10{,}000 = 4$	$10^4 = 10{,}000$
4. $\log 1{,}000{,}000 = 6$	$10^6 = 1{,}000{,}000$

Additional practice problems are at the end of the chapter.

FINDING THE LOGARITHMS OF NUMBERS BETWEEN 1 AND 10   22-2

What about the logs of numbers that are not multiples of 10? Let's consider numbers between 1 and 10. Since $10^0 = 1$ and $10^1 = 10$, then numbers between 1 and 10 would have logs or exponents between 0 and 1. For example (to three significant figures) $5.32 = 10^{0.726}$ and $7.63 = 10^{0.883}$. In logarithmic form, $\log 5.32 = 0.726$ and $\log 7.63 = 0.883$.

We can find the logarithms of all numbers between 1 and 10 in a *table of common logarithms*. Your calculator also contains a table of common logarithms. The table in your calculator contains the logarithms of *all* numbers. The calculator has made tables of common logarithms obsolete in almost all problem-solving situations because it is faster and easier to use.

EXAMPLE 22-2

Find $\log 4.23$.

SOLUTION

$$\log 4.23 = 0.626$$

To find logarithms using the calculator, let's first clear and set up to display 3 digits to the right of the decimal point.

$\boxed{\text{AC}}\ \boxed{\text{FIX}}\ 3\ 4.23\ \boxed{\text{log}}$ *Answer:* $\log 4.23 = 0.626$

In exponential form,

$$10^{0.626} = 4.23$$

10^x is usually the second function of the LOG key; so

$0.626\ \boxed{10^x}$ *Answer:* $10^{0.626} = 4.23$

EXAMPLE 22–3

Find log 5.03.

SOLUTION

$$\log 5.03 = 0.702$$

In exponential form,

$$10^{0.702} = 5.03$$

PRACTICE PROBLEMS 22–2

Using the calculator, find the logs of the following numbers to three significant figures. Express your answers in both logarithmic and exponential forms.

1. 1.76
2. 2.03
3. 9.83
4. 4.35
5. 3.16

SOLUTIONS

Logarithmic form	*Exponential form*
1. $\log 1.76 = 0.246$	$10^{0.246} = 1.76$
2. $\log 2.03 = 0.307$	$10^{0.307} = 2.03$
3. $\log 9.83 = 0.993$	$10^{0.993} = 9.83$
4. $\log 4.35 = 0.638$	$10^{0.638} = 4.35$
5. $\log 3.16 = 0.500$	$10^{0.500} = 3.16$

Additional practice problems are at the end of the chapter.

22–3 FINDING THE LOGARITHMS OF NUMBERS GREATER THAN 10

We can see that if we wanted to find the logarithm of 50, for example, the exponent (or log) would be greater than 1 but less than 2. (The number is greater than 10 but less than 100.) The log will consist of a whole number and a decimal fraction.

☞ **KEY POINT** Only numbers that are multiples of 10 will have logarithms that are whole numbers.

Notice that in Practice Problems 22–2 the logs of the numbers were all decimal fractions. This was because all the numbers were between 1 and 10. Remember, the log of 1 is 0 and the log of 10 is 1. If problems contained numbers between 1 and 10, each logarithm would be a number between 0 and 1.

If problems contained numbers between 10 and 100, the logarithms would be numbers between 1 and 2 because

$$\log 10 = 1 \quad \text{and} \quad \log 100 = 2$$

If the numbers were between 100 and 1000, the logarithms would be numbers between 2 and 3 because

$$\log 100 = 2 \quad \text{and} \quad \log 1000 = 3$$

If the numbers were between 1000 and 10,000, the logarithms would be numbers between 3 and 4, and so on.

What is the log of 50? Log 50 = 1.699. What is the log of 500? 5000? 50,000? Log 500 = 2.699; log 5000 = 3.699; log 50,000 = 4.699. Notice that in each case the fractional part of the logarithm is 0.699. The whole-number part depends on the multiples of 10 contained in the number. For example, let's express the number 5000 in scientific notation: $5000 = 5 \times 10^3$. We know from previous discussions that $\log 10^3 = \log 1000 = 3$. Log 5 = 0.699. If we express $5 \times 10^3 = 5000$ by using the laws of exponents, $10^{0.699+3} = 10^{3.699} = 5000$. When finding logarithms of numbers, we are finding a fractional part that results from finding the log of a number between 1 and 10 and a whole-number part that is the power of 10. What is log 276,000? Log 276,000 = 5.441. When we consider that $276,000 = 2.76 \times 10^5$, we really found the log of 2.76 and the log of 10^5. Log 2.76 = 0.441 and $\log 10^5 = 5$. $10^5 \times 10^{0.441} = 10^{5.441}$. The logarithm is the exponent to the base 10. We found the power to which 10 is raised so that its value is 276,000.

EXAMPLE 22–4

Find log 37,500.

SOLUTION

$$\log 37,500 = 4.574$$

In exponential form,

$$10^{4.574} = 37,500$$

PRACTICE PROBLEMS 22–3

Find the logarithms of the following numbers. Express your answers in both logarithmic and exponential forms. Set up your calculator to display three significant figures to the right of the decimal point (FIX 3).

1. 67.3
3. 742
5. 8730

2. 67,300
4. 103,000

Logarithmic form	Exponential form
1. log 67.3 = 1.828	$10^{1.828} = 67.3$
2. log 67,300 = 4.828	$10^{4.828} = 67,300$
3. log 742 = 2.870	$10^{2.870} = 742$
4. log 103,000 = 5.013	$10^{5.013} = 103,000$
5. log 8730 = 3.941	$10^{3.941} = 8730$

Additional practice problems are at the end of the chapter.

22-4   FINDING THE LOGARITHMS OF NUMBERS LESS THAN 1

If problems contained numbers between 1 and 0.1, the logarithms would be numbers between 0 and −1. If the numbers were between 0.1 and 0.01, the logarithms would be numbers between −1 and −2, and so on. Therefore, we can make the following statement:

☞ **KEY POINT**   For any number less than 1, the logarithm (or exponent) is a negative number.

Find the log of 0.0275. Log 0.0275 = −1.561. In exponential form, $10^{-1.561} = 0.0275$. Using scientific notation gives $0.0275 = 2.75 \times 10^{-2}$. Log 2.75 = 0.439 and log $10^{-2} = -2$. The resulting exponent is −2 + 0.439 = −1.561.

EXAMPLE 22-5

Find log 0.432.

SOLUTION

$$\log 0.432 = -0.365$$

In exponential form,

$$10^{-0.365} = 0.432$$

PRACTICE PROBLEMS 22-4

Find the logs of the following numbers. Express your answers in both logarithmic and exponential forms. Set up your calculator to display three significant figures to the right of the decimal point.

1. 0.0463
2. 0.000935
3. 0.00205
4. 0.127
5. 0.00000812

Logarithmic form	*Exponential form*
1. $\log 0.0463 = -1.334$	$10^{-1.334} = 0.0463$
2. $\log 0.000935 = -3.029$	$10^{-3.029} = 0.000935$
3. $\log 0.00205 = -2.688$	$10^{-2.688} = 0.00205$
4. $\log 0.127 = -0.896$	$10^{-0.896} = 0.127$
5. $\log 0.00000812 = -5.090$	$10^{-5.090} = 0.00000812$

Additional practice problems are at the end of the chapter.

SELF-TEST 22–1

Find the logarithms of the following numbers. Express your answers in both logarithmic and exponential forms.

1. 7
2. 98.3
3. 0.00420
4. 7,650,000
5. 5762
6. 0.273
7. 0.00017
8. 0.00000983
9. 1786
10. 0.0933

Answers to Self-test 22–1 are at the end of the chapter.

MULTIPLICATION OF NUMBERS BY USING LOGARITHMS   22-5

Numbers can be multiplied by using logarithms. When we covered exponents in Chapter 2, we learned that, when multiplying like bases, the exponents are added. For example,

$$10^2 \times 10^3 = 10^{2+3} = 10^5$$

$$y^3 \times y^4 = y^{3+4} = y^7$$

$$10^{2.34} \times 10^{1.72} = 10^{4.06}$$

Since this is true, when multiplying numbers together we could change the numbers by using logarithms so that the bases are alike. Then we could add the logs (exponents) together. The number in exponential form would be the answer. For instance, suppose we wanted to multiply 37×763. We would first change the numbers to base 10. Log 37 = 1.568 and log 763 = 2.883. In exponential form,

$$37 = 10^{1.568} \qquad \text{and} \qquad 763 = 10^{2.883}$$

Then

$$37 \times 763 = 10^{1.568} \times 10^{2.883} = 10^{4.451}$$

$$10^{4.451} = 2.82 \times 10^4$$

Since logarithms are exponents, we can make the following statement:

☞ **KEY POINT** Multiplication of numbers, using logarithms, is reduced to addition. The general equation is:

$$\log ab = \log a + \log b$$

To find the product of numbers using logs:

1. Find the log of each number.
2. Add the logs (exponents).
3. Find the value of 10 raised to the summed exponents.

When we found the product of 37×763, we could have written the problem in this way:

$$\log (37 \times 763) = \log 37 + \log 763$$

The general equation is

$$\log ab = \log a + \log b$$

Remember, the logs are the exponents to the base 10 and, according to the laws of exponents, are added.

EXAMPLE 22–6

Find the product of 7600×230. Round the answer to three places.

SOLUTION

$$\log (7600 \times 230) = \log 7600 + \log 230 = 3.881 + 2.362 = 6.243$$

$$7600 \times 230 = 10^{3.881} \times 10^{2.362} = 10^{6.243} = 1.75 \times 10^6$$

EXAMPLE 22–7

Find the product of 1760×0.0273. Round the answer to three places.

SOLUTION

$$\log (1760 \times 0.0273) = \log 1760 + \log 0.0273$$

$$= 3.246 + (-1.564) = 1.682$$

$$1760 \times 0.0273 = 10^{3.246} \times 10^{-1.564} = 10^{1.682} = 4.8 \times 10$$

EXAMPLE 22–8

Find the product of 0.0783×0.0000342.

SOLUTION

$$\log (0.0783 \times 0.0000342) = \log 0.0783 + \log 0.0000342$$
$$= -1.106 + (-4.466) = -5.572$$
$$10^{-5.572} = 2.68 \times 10^{-6}$$

PRACTICE PROBLEMS 22–5

Find the product of the following numbers by using logs. Round your answers to three digits.

1. 76×2730
2. 0.783×6500
3. 0.00376×0.107
4. $7,600,000 \times 9800$
5. $10,700 \times 0.0083$

SOLUTIONS

1. $\log (76 \times 2730) = \log 76 + \log 2730$
   $$= 1.881 + 3.436$$
   $$= 5.317$$
   $$10^{5.317} = 2.07 \times 10^5$$

2. $\log (0.783 \times 6500) = \log 0.783$
   $$+ \log 6500$$
   $$= -0.106 + 3.813$$
   $$= 3.707$$
   $$10^{3.707} = 5.09 \times 10^3$$

3. $\log (0.00376 \times 0.107) = \log 0.00376$
   $$+ \log 0.107$$
   $$= -2.425 +$$
   $$(-0.971)$$
   $$= -3.395$$
   $$10^{-3.395} = 4.02 \times 10^{-4}$$

4. $\log (7,600,000 \times 9800) = \log$
   $$7,600,000$$
   $$+ \log 9800$$
   $$= 6.881 +$$
   $$3.991$$
   $$= 10.872$$
   $$10^{10.872} = 7.45 \times 10^{10}$$

5. $\log (10,700 \times 0.0083) = \log 10,700$
   $$+ \log 0.0083$$
   $$= 4.029 +$$
   $$(-2.081)$$
   $$= 1.948$$
   $$10^{1.948} = 8.88 \times 10$$

Additional practice problems are at the end of the chapter.

Division of numbers by using logarithms is possible by again observing the laws of exponents. When like bases are divided, the exponents are subtracted. For example,

$$\frac{10^7}{10^4} = 10^{7-4} = 10^3$$

$$\frac{8^2}{8^5} = 8^{2-5} = 8^{-3}$$

$$\frac{a^5}{a^{-3}} = a^{5-(-3)} = a^{5+3} = a^8$$

$$\frac{10^{4.384}}{10^{2.732}} = 10^{4.384-2.732} = 10^{1.652}$$

Let's find the quotient of $\frac{6270}{220}$ by using logs. To do this:

1. Find the log of both dividend and divisor.
2. Subtract the log of the divisor from the log of the dividend.
3. Change to exponential form.

We would solve the problem this way:

$$\log\left(\frac{6270}{220}\right) = \log 6270 - \log 220 = 3.797 - 2.342 = 1.455$$

$$10^{1.455} = 2.85 \times 10$$

As with multiplication, finding the logs of the numbers gives us the opportunity to add or subtract exponents of like bases:

$$\frac{6270}{220} = \frac{10^{3.797}}{10^{2.342}} = 10^{3.797-2.342} = 10^{1.455} = 2.85 \times 10$$

EXAMPLE 22-9

Find the quotient of $\frac{1.73}{96,700}$.

SOLUTION

$$\log\left(\frac{1.73}{96,700}\right) = \log 1.73 - \log 96,700 = 0.238 - 4.985$$

$$= -4.747$$

$$10^{-4.747} = 1.79 \times 10^{-5}$$

EXAMPLE 22–10

Find the quotient of $\dfrac{0.0735}{0.00034}$.

SOLUTION

$$\log\left(\frac{0.0735}{0.00034}\right) = \log 0.0735 - \log 0.00034 = -1.134 - (-3.469)$$

$$= 2.335$$

$$10^{2.335} = 2.16 \times 10^2$$

PRACTICE PROBLEMS 22–6

Find the quotients of the following problems by using logs. Round your answers to three places.

1. $\dfrac{76{,}000}{42}$

2. $\dfrac{207}{56{,}300}$

3. $\dfrac{1670}{0.0073}$

4. $\dfrac{0.000426}{0.862}$

5. $\dfrac{0.0000942}{6040}$

SOLUTIONS

1. $\log\left(\dfrac{76{,}000}{42}\right) = \log 76{,}000 - \log 42$
$$= 4.881 - 1.623$$
$$= 3.258$$
$$10^{3.258} = 1.81 \times 10^3$$

2. $\log\left(\dfrac{207}{56{,}300}\right) = \log 207 - \log 56{,}300$
$$= 2.316 - 4.751$$
$$= -2.435$$
$$10^{-2.435} = 3.68 \times 10^{-3}$$

3. $\log\left(\dfrac{1670}{0.0073}\right) = \log 1670 - \log 0.0073$
$$= 3.223 - (-2.137)$$
$$= 5.359$$
$$10^{5.359} = 2.29 \times 10^5$$

4. $\log\left(\dfrac{0.000426}{0.862}\right) = \log 0.000426$
$$- \log 0.862$$
$$= -3.371 - (-0.0645)$$
$$= -3.306$$
$$10^{-3.306} = 4.94 \times 10^{-4}$$

5. $\log\left(\dfrac{0.0000942}{6040}\right) = \log 0.0000942$
$$- \log 6040$$
$$= -4.026 - 3.781$$
$$= -7.807$$
$$10^{-7.807} = 1.56 \times 10^{-8}$$

Additional practice problems are at the end of the chapter.

RAISING A NUMBER TO A POWER BY USING LOGARITHMS

☞ **KEY POINT** Raising a number to a power is a process of multiplication. In general, to find a^x using logarithms,

$$\log a^x = x(\log a)$$

Let's find 4^5 by using logarithms. To solve problems of this type:

1. Find the log of the number.
2. Multiply the log by the exponent.
3. Change to exponential form.

We can express the problem this way: $\log 4^5 = 5 \log 4$. Remember, we are going to multiply the log of the number by the exponent.

$$5 \log 4 = 5 \times 0.602 = 3.01$$
$$10^{3.01} = 1.02 \times 10^3$$
$$4^5 = (10^{0.602})^5 = 10^{(0.602)(5)} = 10^{3.01} = 1.02 \times 10^3$$

EXAMPLE 22–11

What does 0.0745^3 equal?

SOLUTION

$$\log 0.0745^3 = 3 \log 0.0745 = (3)(-1.128) = -3.384$$
$$10^{-3.384} = 4.13 \times 10^{-4}$$
$$0.0745^3 = (10^{-1.128})^3 = 10^{-3.384} = 4.13 \times 10^{-4}$$

EXAMPLE 22–12

What does $74.6^{-2.5}$ equal?

SOLUTION

$$\log 74.6^{-2.5} = -2.5 \log 74.6 = -2.5 \times 1.873 = -4.682$$
$$10^{-4.682} = 2.08 \times 10^{-5}$$
$$74.6^{-2.5} = (10^{1.873})^{-2.5} = 10^{-4.682} = 2.08 \times 10^{-5}$$

Find the answers to the following problems by using logarithms. Round to three places.

1. 6.7^5

2. $2.3^{-3.6}$

3. 0.00403^4

4. $0.6^{4.2}$

5. 50^3

SOLUTIONS

1. $\log 6.7^5 = 5 \log 6.7 = 5 \times 0.826$
   $ = 4.130$
   $10^{4.130} = 1.35 \times 10^4$

2. $\log 2.3^{-3.6} = -3.6 \log 2.3 = -3.6$
   $\phantom{\log 2.3^{-3.6} = -3.6 \log 2.3} \times 0.362$
   $\phantom{\log 2.3^{-3.6}} = -1.302$
   $10^{-1.302} = 4.99 \times 10^{-2}$

3. $\log 0.00403^4 = 4 \log 0.00403 = 4$
   $ \times (-2.395)$
   $ = -9.579$
   $10^{-9.579} = 2.64 \times 10^{-10}$

4. $\log 0.6^{4.2} = 4.2 \log 0.6 = 4.2$
   $\phantom{\log 0.6^{4.2} = 4.2 \log 0.6} \times (-0.222)$
   $\phantom{\log 0.6^{4.2}} = -0.932$
   $10^{-0.932} = 0.117 = 1.17 \times 10^{-1}$

5. $\log 50^3 = 3 \log 50 = 3 \times 1.699$
   $ = 5.097$
   $10^{5.097} = 1.25 \times 10^5$

Additional practice problems are at the end of the chapter.

SELF-TEST 22–2

Find the answers to the following problems by using logarithms. Round to three places.

1. 463×0.027

2. $\dfrac{34.3}{8700}$

3. $7^{4.6}$

4. $\dfrac{0.0063}{2.73}$

5. 0.463^4

6. $97{,}000 \times 170$

7. 0.0128×0.000055

8. $27.3^{2.6}$

9. $\dfrac{43.2}{0.000772}$

10. 0.00193×1930

Answers to Self-test 22–2 are at the end of the chapter.

NATURAL LOGARITHMS 22–8

Common logarithms are exponents where 10 is the base. *Natural logarithms* are exponents where e is the base. The numerical value of e, rounded to three digits, is 2.72. In electronics, the instantaneous charge or discharge of a capacitor or the instantaneous current in a coil can be determined by equations using natural logarithms.

22-8-1 Finding Natural Logarithms of Numbers

Just as we have tables of common logarithms, we have tables of natural logarithms. We used the abbreviation "log" for common logarithms. We use the abbreviation "ln" for natural logarithms.

As with common logarithms, the calculator stores a table of natural logarithms. What is the natural log of 5? Using the calculator,

$$5 \boxed{\text{Ln}} \qquad \textit{Answer:} \ln 5 = 1.609$$

In exponential form, $e^{1.6094} = 5$. e^x is usually the second function of the $\boxed{\text{Ln}}$ key, so

$$1.609 \boxed{e^x} \qquad \textit{Answer:} e^{1.609} = 5$$

Notice that the change from logarithmic form to exponential form is the same for natural logs as for common logs. The only difference is in the base. Note the following examples:

Logarithmic form	*Exponential form*
$\log 5 = 0.699$	$10^{0.699} = 5$
$\ln 5 = 1.609$	$e^{1.609} = 5$
$\log 1 = 0$	$10^0 = 1$
$\ln 1 = 0$	$e^0 = 1$
$\log 10 = 1$	$10^1 = 10$
$\ln e = 1$	$e^1 = e$

Find $\ln 45$. $\ln 45 = 3.807$. In exponential form, $e^{3.807} = 45$. Find the ln of 0.832. $\ln 0.832 = -0.184$. In exponential form, $e^{-0.184} = 0.832$. As with common logarithms, the natural logarithms of numbers greater than 1 are *positive,* and the natural logarithms of numbers less than 1 are *negative.*

If we know the exponent to the base e, we can find the number by using the calculator, just as we found numbers when we were using base 10. What does $e^{3.434}$ equal? $e^{3.434} = 31.0$. What does e^2 equal? $e^2 = 7.39$.

PRACTICE PROBLEMS 22-8

Find the natural logarithms of the following numbers by using the calculator.

1. 7.4
2. 300
3. 0.8
4. 3.4
5. 100

Find the number whose natural logarithm is:

6. 1.946   7. 3
8. 4.489   9. 0.0953
10. 1.435

SOLUTIONS

1. $\ln 7.4 = 2.001$ 2. $\ln 300 = 5.704$
3. $\ln 0.8 = -0.223$ 4. $\ln 3.4 = 1.224$
5. $\ln 100 = 4.605$ 6. $e^{1.946} = 7.00$
7. $e^3 = 20.1$ 8. $e^{4.489} = 89.0$
9. $e^{0.0953} = 1.10$ 10. $e^{1.435} = 4.20$

Additional practice problems are at the end of the chapter.

SELF-TEST 22–3

Find the natural logarithm of the following numbers. Convert your answers to exponential form.

1. 14   2. 0.0073
3. 740   4. 1073
5. 10   6. 0.00001
7. 0.463   8. 0.0673
9. 4.73   10. 73

Answers to Self-test 22–3 are at the end of the chapter.

CHAPTER 22 AT A GLANCE

PAGE	KEY POINTS AND DEFINITIONS	EXAMPLE
620	*Key Point:* The common logarithm of a number (N) is an exponent (x) where $10^x = N$. The general equations is $\log_{10} N = x$ where $10^x = N$	$\log_{10} 1000 = 3$ $10^3 = 1000$
6225	*Key Point:* Only numbers that are multiples of 10 will have logarithms that are whole numbers.	$\log 10 = 1$ $\log 100 = 2$
624	*Key Point:* For any number less than 1, the logarithm is a negative number.	$\log 1 = 0$ $\log 0.1 = -1$
626	*Key Point:* Multiplication of numbers, using logarithms, is reduced to addition.	$\log ab = \log a + \log b$

| 630 | *Key Point:* Raising a number to a power is a process of multiplication. | $\log a^x = x(\log a)$ |
| 631 | *Natural logarithms* are exponents where e is the base. $e = 2.718$. | $\ln 10 = 2.30$ |

END OF CHAPTER PROBLEMS 22–1

Find the logarithms of the following numbers. Express answers in both logarithmic and exponential forms.

1. 1000	2. 100,000
3. 1	4. 10
5. 10,000,000	6. 100
7. 0.1	8. 0.001
9. 0.000001	10. 0.0001

END OF CHAPTER PROBLEMS 22–2

Find the logarithms of the following numbers accurate to three significant figures. Express answers in both logarithmic and exponential forms.

1. 3.42	2. 6.45
3. 4.05	4. 8.33
5. 6.44	6. 4.73
7. 9.45	8. 5.85
9. 2.50	10. 9.66

END OF CHAPTER PROBLEMS 22–3

Find the logarithms of the following numbers accurate to three significant figures. Express answers in both logarithmic and exponential forms. Set up your calculator to display three significant figures to the right of the decimal point.

1. 14	2. 476
3. 743	4. 7650
5. 1070	6. 6070
7. 43.4	8. 64.7
9. 10,000	10. 500,000
11. 23,400	12. 44,500
13. 47,700	14. 275
15. 14.6	16. 3.84
17. 37,000	18. 201,000
19. 2,000,000	20. 94.7

END OF CHAPTER PROBLEMS 22–4

Find the logarithms of the following numbers accurate to three significant figures. Express answers in both logarithmic and exponential forms. Set up your calculator to display three significant figures to the right of the decimal point.

1. 0.00377
2. 0.000448
3. 0.037
4. 0.000837
5. 0.746
6. 0.532
7. 0.407
8. 0.003
9. 0.000614
10. 0.000476
11. 0.49
12. 0.00899
13. 0.00814
14. 0.00719
15. 0.301
16. 0.000601
17. 0.975
18. 0.00055
19. 0.0275
20. 0.00475

END OF CHAPTER PROBLEMS 22–5

Find the product of the following numbers by using logarithms. Round answers to three places:

1. 27×56
2. 83×920
3. $107 \times 243,000$
4. 110×5070
5. 43×0.143
6. 53×0.243
7. 0.043×0.0106
8. 0.00274×0.0776
9. $0.0000176 \times 0.000357$
10. 0.000555×0.00677
11. 256×0.00246
12. 670×0.0378
13. 78×0.473
14. 48.7×0.00842
15. 4730×0.000506
16. 2780×0.000425
17. 760×7360
18. 0.146×0.0556
19. 2560×0.00457
20. 2650×0.00735

END OF CHAPTER PROBLEMS 22–6

Find the quotients of the following numbers by using logarithms. Round answers to three places:

1. $\dfrac{1760}{43}$
2. $\dfrac{6530}{3.25}$
3. $\dfrac{65}{467}$
4. $\dfrac{27}{630}$
5. $\dfrac{0.00273}{172}$
6. $\dfrac{0.000825}{50}$
7. $\dfrac{4300}{0.15}$
8. $\dfrac{5430}{0.335}$
9. $\dfrac{0.0675}{0.00103}$
10. $\dfrac{0.000565}{0.093}$

11. $\dfrac{0.00467}{73.2}$

12. $\dfrac{0.00675}{82.3}$

13. $\dfrac{0.147}{0.0432}$

14. $\dfrac{0.473}{0.00563}$

15. $\dfrac{4730}{0.443}$

16. $\dfrac{7430}{257}$

17. $\dfrac{0.445}{635}$

18. $\dfrac{0.655}{625}$

19. $\dfrac{0.00275}{4.35}$

20. $\dfrac{0.000375}{64.5}$

END OF CHAPTER PROBLEMS 22–7

Express the following numbers by using logarithms. Round answers to three places:

1. 3.4^3
2. 6.26^8
3. $6^{-1.3}$
4. $34^{-2.43}$
5. $0.43^{2.5}$
6. 0.007^3
7. $6^{4.3}$
8. $4^{3.45}$
9. $0.27^{-4.4}$
10. $0.063^{-2.1}$
11. $8.7^{6.3}$
12. $8.67^{5.3}$
13. $3.4^{3.2}$
14. $4.75^{4.3}$
15. $6.35^{-2.3}$
16. $1.73^{6.25}$
17. $0.045^{3.6}$
18. $0.00863^{4.3}$
19. $0.00735^{-2.5}$
20. $0.00545^{-3.6}$

END OF CHAPTER PROBLEMS 22–8

Find the natural logarithms of the following numbers. Express your answers in both logarithmic and exponential forms. Round answers to three places.

1. 2
2. 8.78
3. 0.932
4. 0.00273
5. 60
6. 43
7. 0.073
8. 0.545
9. 247
10. 675
11. 0.00146
12. 0.000541
13. 1100
14. 1200
15. 0.00027
16. 0.000249
17. 8.43
18. 15.3
19. 0.346
20. 0.000072
21. 3.67
22. 94.7
23. 0.107
24. 0.0058
25. 176
26. 700
27. 0.293
28. 0.472
29. 873
30. 1600

Find the number whose natural logarithm is:

31. 2.34	32. 1.20
33. 3.77	34. 4.5
35. 4.25	36. 3.75
37. 5.3	38. 6.2
39. 1.55	40. 5.22
41. 0.43	42. 0.532
43. −0.0045	44. −0.078
45. 0.15	46. 0.335
47. −0.315	48. −0.61
49. −0.142	50. −0.741

ANSWERS TO SELF-TESTS

SELF-TEST 22–1

Logarithmic form	Exponential form
1. log 7 = 0.845	$10^{0.845} = 7$
2. log 98.3 = 1.993	$10^{1.993} = 98.3$
3. log 0.00420 = −2.377	$10^{-2.377} = 0.00420$
4. log 7,650,000 = 6.884	$10^{6.884} = 7,650,000$
5. log 5762 = 3.761	$10^{3.761} = 5762$
6. log 0.273 = −0.564	$10^{-0.564} = 0.273$
7. log 0.00017 = −3.770	$10^{-3.770} = 0.00017$
8. log 0.00000983 = −5.007	$10^{-5.007} = 0.00000983$
9. log 1786 = 3.252	$10^{3.252} = 1786$
10. log 0.0933 = −1.030	$10^{-1.030} = 0.0933$

SELF-TEST 22–2

1. $\log (463 \times 0.027) = \log 463 + \log 0.027 = 2.666 + (-1.569) = 1.097$
   $10^{1.097} = 1.25 \times 10$

2. $\log \left(\dfrac{34.3}{8700}\right) = \log 34.3 - \log 8700 = 1.535 - 3.94 = -2.404$
   $10^{-2.404} = 3.94 \times 10^{-3}$

3. $\log 7^{4.6} = 4.6 \log 7 = 4.6 \times 0.845 = 3.887$
   $10^{3.887} = 7.72 \times 10^{3}$

4. $\log \left(\dfrac{0.0063}{2.73}\right) = \log 0.0063 - \log 2.73 = -2.201 - 0.436 = -2.637$
   $10^{-2.637} = 2.31 \times 10^{-3}$

5. $\log 0.463^{4} = 4 \log 0.463 = 4 \times (-0.334) = -1.338$
   $10^{-1.338} = 4.60 \times 10^{-2}$

6. $\log (97,000 \times 170) = \log 97,000 + \log 170 = 4.987 + 2.23 = 7.217$
   $10^{7.217} = 1.65 \times 10^{7}$

7. $\log (0.0128 \times 0.000055) = \log 0.0128 + \log 0.000055 = -1.893 + (-4.26) = -6.152$
$$10^{-6.152} = 7.04 \times 10^{-7}$$

8. $\log 27.3^{2.6} = 2.6 \log 27.3 = 2.6 \times 1.436 = 3.734$
$$10^{3.734} = 5.42 \times 10^{3}$$

9. $\log \dfrac{43.2}{0.000772} = \log 43.2 - \log 0.000772 = 1.635 - (-3.112) = 4.748$
$$10^{4.748} = 5.60 \times 10^{4}$$

10. $\log (0.00193 \times 1930) = \log 0.00193 + \log 1930 = -2.714 + 3.286 = 0.571$
$$10^{0.571} = 3.72$$

SELF-TEST 22–3

Logarithmic form	Exponential form
1. $\ln 14 = 2.639$	$e^{2.639} = 14$
2. $\ln 0.0073 = -4.92$	$e^{-4.92} = 0.0073$
3. $\ln 740 = 6.607$	$e^{6.607} = 740$
4. $\ln 1073 = 6.978$	$e^{6.978} = 1073$
5. $\ln 10 = 2.303$	$e^{2.303} = 10$
6. $\ln 0.00001 = -11.51$	$e^{-11.51} = 0.00001$
7. $\ln 0.463 = -0.770$	$e^{-0.770} = 0.463$
8. $\ln 0.0673 = -2.699$	$e^{-2.699} = 0.0673$
9. $\ln 4.73 = 1.554$	$e^{1.554} = 4.73$
10. $\ln 73 = 4.29$	$e^{4.29} = 73$

Logarithmic Equations

23

Introduction

This chapter continues our study of logarithms. Because of the logarithmic nature of the way the human ear perceives changes in sound, we must develop circuits to deal with this characteristic. Whenever we analyze circuits, we work with equations. If we are working with amplifiers and developing curves and collecting data that describe the amplifiers' characteristics, we will be working with some sort of logarithmic equation. In this chapter we will work with the kinds of logarithmic equations you are most likely to encounter in your electronics studies.

Chapter Objectives

In this chapter you will learn how to:

1. Write equations in both logarithmic and exponential forms.
2. Solve equations involving both common and natural logarithms.
3. Change equations from logarithmic to exponential form and from exponential to logarithmic form.
4. Use the calculator to help solve logarithmic equations.

LOGARITHMIC AND EXPONENTIAL FORMS 23-1

We have discussed logarithmic and exponential expressions in previous chapters. Now let's write them again in the form of simple equations.

Logarithmic	Exponential	
(1) $\log N = x$	$10^x = N$	(23–1)
(2) $\ln a = b$	$e^b = a$	(23–2)

In Equation 23–1, the exponent is x, the number is N, and the base is 10. In Equation 23–2, the exponent is b, the number is a, and the base is e.

> ☞ **KEY POINT** Logarithmic equations can either be solved in logarithmic form or in exponential form.

Consider the following equation:

$$\log 73 = t$$

This equation is in logarithmic form. Seventy-three is the number and t is the exponent. The equation is solved when we find log 73.

> ☞ **KEY POINT** Logarithmic equations can be solved directly when the exponent is unknown.

$$\log 73 = t$$
$$t = 1.86$$

In the following logarithmic equation, the number is unknown.

$$\log y = 4.73$$

We can't solve this equation in its present form because we don't know the value of y. However, the equation is easily solved if we change to exponential form.

$$\log y = 4.73 \quad \text{(logarithmic form)}$$
$$10^{4.73} = y \quad \text{(exponential form)}$$
$$y = 5.37 \times 10^4$$

> ☞ **KEY POINT** Change from logarithmic to exponential form when the number is unknown.

PRACTICE PROBLEMS 23–1

Change the following equations from logarithmic form to exponential form:

1. $\log 7600 = a$
2. $\log x = -3.74$
3. $\log y = 0.742$
4. $\log 0.00432 = x$

Change the following equations from exponential form to logarithmic form:

5. $10^{4.2} = y$
6. $10^t = 0.043$
7. $10^y = 173$
8. $10^{-2.43} = y$

1. $10^a = 7600$
2. $10^{-3.74} = x$
3. $10^{0.742} = y$
4. $10^x = 0.00432$
5. $\log y = 4.2$
6. $\log 0.043 = t$
7. $\log 173 = y$
8. $\log y = -2.43$

Additional practice problems are at the end of the chapter.

LOGARITHMIC EQUATIONS: COMMON LOGS 23-2

Let's consider some of the equations we might encounter in problem solving that involve the use of common logs.

EXAMPLE 23-1

$x = \log 250$. Solve for x.

$$x = 2.40$$

EXAMPLE 23-2

$\log a = 7.83$. Solve for a.

SOLUTION Change to exponential form:

$$10^{7.83} = a$$

$$a = 6.76 \times 10^7$$

TI: 7.83 $\boxed{10^x}$ *Answer:* 6.76 × 10⁷

Casio: $\boxed{10^x}$ 7.83 $\boxed{=}$ *Answer:* 6.76 × 10⁷

☞ **KEY POINT** We solve logarithmic equations using rules developed in Chapter 7: We may perform any mathematical operation on one side of an equation provided we perform the same operation on the other side.

EXAMPLE 23-3

$3^x = 104$. Solve for x.

SOLUTION Take the log of both sides to get rid of the exponent:

$$x \log 3 = \log 104$$

$$x = \frac{\log 104}{\log 3} = \frac{2.02}{0.477}$$

$$= 4.23$$

(We could have found log 3 and log 104 first and then solved for x.)

TI: 104 $\boxed{\log}$ $\boxed{\div}$ 3 $\boxed{\log}$ $\boxed{=}$ *Answer:* **4.23**

Casio: $\boxed{\log}$ 104 $\boxed{\div}$ $\boxed{\log}$ 3 $\boxed{=}$ *Answer:* **4.23**

EXAMPLE 23–4

$x^{3.6} = 85$. Solve for x.

SOLUTION   Take the log of both sides to get rid of the exponent:

$$3.6 \log x = \log 85$$

$$\log x = \frac{\log 85}{3.6} = \frac{1.93}{3.6} = 0.536$$

Change to exponential form:

$$10^{0.536} = x$$

$$x = 3.44$$

TI: 85 $\boxed{\log}$ $\boxed{\div}$ 3.6 $\boxed{=}$ $\boxed{10^x}$ *Answer:* **3.44**

Casio: $\boxed{10^x}$ $\boxed{(}$ $\boxed{\log}$ 85 $\boxed{\div}$ 3.6 $\boxed{)}$ $\boxed{=}$ *Answer:* **3.44**

EXAMPLE 23–5

$6.3^4 = x$. Solve for x.

SOLUTION   Take the log of both sides to get rid of the exponent:

$$4 \log 6.3 = \log x$$

$$4(0.799) = \log x$$

$$3.20 = \log x$$

Change to exponential form:

$$10^{3.20} = x$$

$$x = 1.58 \times 10^3$$

PRACTICE PROBLEMS 23–2

Solve for the unknown in the following problems:

1. $\log 473 = x$
2. $\log 0.0783 = x$
3. $\log t = 2.74$
4. $\log x = -3.74$
5. $\log a = 0.274$
6. $\log 10{,}700 = y$
7. $4^x = 256$
8. $x^{2.7} = 56$
9. $3.6^3 = x$
10. $x^{-1.4} = 0.073$
11. $3.4^{-3} = x$
12. $0.087^x = 4$

1. $x = 2.67$
2. $x = -1.11$
3. $t = 5.5 \times 10^2$
4. $x = 1.82 \times 10^{-4}$
5. $a = 1.88$
6. $y = 4.03$
7. $x = 4.00$
8. $x = 4.44$
9. $x = 46.7$
10. $x = 6.49$
11. $x = 2.54 \times 10^{-2}$
12. $x = 5.68 \times 10^{-1}$

Additional practice problems are at the end of the chapter.

MORE COMMON LOGARITHMIC EQUATIONS 23–3

EXAMPLE 23–6

$3^{1-x} = 12$. Solve for x.

SOLUTION   Take the log of both sides to get rid of the exponent:

$$(1 - x) \log 3 = \log 12$$

$$1 - x = \frac{\log 12}{\log 3} = \frac{1.08}{0.477} = 2.26$$

$$1 - 2.26 = x$$

$$x = -1.26$$

EXAMPLE 23–7

$27^{x/3} = 18$. Solve for x.

SOLUTION   Take the log of both sides:

$$\frac{x}{3} \log 27 = \log 18$$

$$\frac{x}{3} = \frac{\log 18}{\log 27} = \frac{1.255}{1.431} = 0.877$$

$$x = 3 \times 0.877 = 2.63$$

EXAMPLE 23–8

$4 = \log \dfrac{x}{100}$. Solve for x.

Change to exponential form:

$$10^4 = \frac{x}{100}$$

$$x = 100 \times 10^4$$

$$x = 10^6$$

EXAMPLE 23–9

$8 = \log \dfrac{200}{x}$. Solve for x.

SOLUTION   Change to exponential form:

$$10^8 = \frac{200}{x}$$

$$10^8 x = 200$$

$$x = \frac{200}{10^8}$$

$$x = 2.00 \times 10^{-6}$$

EXAMPLE 23–10

$65 = 10 \log \dfrac{x}{0.03}$. Solve for x.

SOLUTION

$$\frac{65}{10} = \log \frac{x}{0.03}$$

$$6.5 = \log \frac{x}{0.03}$$

Change to exponential form:

$$10^{6.5} = \frac{x}{0.03}$$

$$0.03 \times 10^{6.5} = x$$

$$x = 9.49 \times 10^4$$

PRACTICE PROBLEMS 23–3

Solve for x in the following problems:

1. $4^{3-x} = 128$

2. $5.4^{x+2} = 200$

3. $53^{x/2} = 35$

4. $3.6 = \log \dfrac{x}{25}$

5. $7.3 = \log \dfrac{75}{x}$

6. $x = \log \dfrac{36}{2}$

7. $43 = 10 \log \dfrac{x}{12}$

8. $85 = 20 \log \dfrac{40}{x}$

SOLUTIONS

1. $x = -0.500$

2. $x = 1.14$

3. $x = 1.79$

4. $x = 9.95 \times 10^4$

5. $x = 3.76 \times 10^{-6}$

6. $x = 1.26$

7. $x = 2.39 \times 10^5$

8. $x = 2.25 \times 10^{-3}$

Additional practice problems are at the end of the chapter.

SELF-TEST 23–1

Solve for x in the following problems:

1. $x = \log 400$

2. $4^{3.5} = x$

3. $12^{-4} = x$

4. $x^{1.7} = 40$

5. $5^{x+2} = 625$

6. $9^{1-x} = 36$

7. $65^{x/3} = 10$

8. $3^{4/x} = 0.506$

9. $5 = \log \dfrac{x}{40}$

10. $14 = \log \dfrac{200}{x}$

11. $\log x^4 = 2$

12. $\log 3^x = 0.93$

Answers to Self-test 23–1 are at the end of the chapter.

LOGARITHMIC EQUATIONS: NATURAL LOGS | 23–4

☞ **KEY POINT** Equations involving natural logarithms are solved just like equations involving common logarithms.

Here are some problems involving the use of natural logs.

EXAMPLE 23–11

$e^{1.7} = y$. Solve for y.

SOLUTION $y = 5.47$

EXAMPLE 23–12

$e^y = 83$. Solve for y.

SOLUTION   Take the ln of both sides:

$$y \ln e = \ln 83$$

$$y(1) = \ln 83 \qquad \text{(remember: } \ln e = 1\text{)}$$

$$y = 4.42$$

EXAMPLE 23–13

$89 = e^{-x}$. Solve for x.

SOLUTION

$$\ln 89 = -x \ln e$$

$$4.49 = -x$$

$$x = -4.49$$

EXAMPLE 23–14

$\ln x = 1.7$. Solve for x.

SOLUTION   Change to exponential form:

$$e^{1.7} = x$$

$$x = 5.47$$

PRACTICE PROBLEMS 23–4

Solve for the unknown in the following problems:

1. $e^{3.4} = x$ 2. $e^x = 49$
3. $32 = e^t$ 4. $y = e^{4.5}$
5. $e^y = 58$ 6. $e^x = 150$
7. $\ln x = 3$ 8. $\ln x = -2.7$

SOLUTIONS

1. $x = 30.0$ 2. $x = 3.89$
3. $t = 3.47$ 4. $y = 90.0$

5. $y = 4.06$ 6. $x = 5.01$
7. $x = 20.1$ 8. $x = 6.72 \times 10^{-2}$

Additional practice problems are at the end of the chapter.

MORE NATURAL LOGARITHMIC EQUATIONS   23–5

EXAMPLE 23–15

$x = 20(1 - e^{-0.5})$. Solve for x.

SOLUTION

$$x = 20(1 - 0.606)$$
$$= 20(0.393)$$
$$= 7.87$$

EXAMPLE 23–16

$30 = x(1 - e^{-4})$. Solve for x.

SOLUTION

$$30 = x(1 - 0.0183)$$
$$30 = x(0.982)$$
$$x = 30.6$$

EXAMPLE 23–17

$5 = 15(1 - e^{-1/x})$. Solve for x.

SOLUTION

$$0.333 = 1 - e^{-1/x}$$
$$0.333 - 1 = -e^{-1/x}$$
$$-0.667 = -e^{-1/x}$$
$$0.667 = e^{-1/x}$$
$$\ln 0.667 = -\frac{1}{x} \ln e$$
$$-0.405 = \frac{-1}{x}$$
$$0.405x = 1$$
$$x = 2.47$$

Solve for the unknown in the following problems:

1. $x = 25(1 - e^{-1.75})$
2. $40 = 60(1 - e^{-x})$
3. $15 = 45(1 - e^{-2/3x})$
4. $0.45 = e^{-x/3.2}$
5. $0.175 = 1 - e^{-3x/4}$
6. $22 = x(1 - e^{-2.3})$

SOLUTIONS

1. $x = 20.7$
2. $x = 1.10$
3. $x = 1.64$
4. $x = 2.56$
5. $x = 0.256$
6. $x = 24.5$

SELF-TEST 23–2

Solve for the unknown in the following problems:

1. $x = \ln 40$
2. $\ln x = 14$
3. $e^x = 20$
4. $x = 1 - e^{-0.6}$
5. $36 = 100e^{-x}$
6. $x = 50(1 - e^{-2.75})$
7. $86 = x(1 - e^{-1.5})$
8. $40 = 80e^{-2/x}$
9. $0.45 = e^{-x/4}$
10. $0.2 = 1 - e^{-2/6x}$

Answers to Self-test 23–2 are at the end of the chapter.

CHAPTER 23 AT A GLANCE

PAGE	KEY POINTS AND RULES	EXAMPLE
640	*Key Point:* Logarithmic equations can be solved either in logarithmic form or in exponential form.	$\log N = x$ $10^x = N$
640	*Key Point:* Logarithmic equations can be solved directly when the exponent is unknown.	$x = \log 250$ $= 2.40$
640	*Key Point:* Change from logarithmic to exponential form when the number is unknown.	$\log a = 7.83$ $a = 10^{7.83}$ $= 6.76 \times 10^7$

641	*Key Point:* We solve logarithmic equations using rules developed in Chapter 7.	$3^x = 104$. Solve for x. Take the log of both sides. $x \log 3 = \log 104$ $$x = \frac{\log 104}{\log 3} = \frac{2.02}{0.477} = 4.23$$
645	*Key Point:* Equations involving natural logarithms are solved just like equations involving common logarithms.	$e^y = 83$. Solve for y. $y \ln e = \ln 83$ $y(1) = \ln 83$ $y = 4.42$

END OF CHAPTER PROBLEMS 23–1

Change the following equations from logarithmic form to exponential form:

1. $\log 340 = x$
2. $\log 5760 = x$
3. $\log 14 = x$
4. $\log 45 = x$
5. $\log 7.4 = x$
6. $\log 4.3 = x$
7. $\log 0.014 = y$
8. $\log 0.073 = y$
9. $\log 0.734 = y$
10. $\log 0.617 = y$
11. $\log -0.00178 = y$
12. $\log 0.00827 = y$
13. $\log x = -2.43$
14. $\log x = -0.64$
15. $\log x = -3.64$
16. $\log x = -5.74$
17. $\log x = -0.143$
18. $\log x = -0.00173$
19. $\log y = 1.44$
20. $\log y = 0.082$
21. $\log y = 2.83$
22. $\log y = 7.73$
23. $\log y = 4.14$
24. $\log y = 5.75$

Change the following equations from exponential form to logarithmic form:

25. $10^{3.6} = y$
26. $10^{2.6} = y$
27. $10^{4.73} = y$
28. $10^{5.66} = y$
29. $10^{2.73} = y$
30. $10^{1.03} = y$
31. $10^a = 732$
32. $10^a = 8.73$
33. $10^a = 47.3$
34. $10^a = 376$
35. $10^a = 276$
36. $10^a = 97.4$
37. $10^y = 0.043$
38. $10^y = 0.236$
39. $10^y = 0.177$
40. $10^y = 0.666$
41. $10^y = 0.00178$
42. $10^y = 0.00856$
43. $10^{-3.4} = x$
44. $10^{-3.02} = x$
45. $10^{-2.76} = x$
46. $10^{-7.45} = x$
47. $10^{-1.17} = x$
48. $10^{-5.14} = x$

END OF CHAPTER PROBLEMS 23–2

Find the unknown in the following problems:

1. $\log y = 7.36$
2. $\log y = 0.0476$
3. $\log x = 0.000736$
4. $\log x = 8.6$
5. $\log a = -4.23 \times 10^{-1}$
6. $\log a = -3.73$
7. $\log x = -3.64 \times 10^{-3}$
8. $\log x = -5.44 \times 10^{-2}$
9. $10^x = 462$
10. $10^x = 1700$
11. $10^x = 0.0173$
12. $10^x = 0.00306$
13. $10^y = 2.5 \times 10^{-3}$
14. $10^y = 7.6 \times 10^{-6}$
15. $10^y = 4.66 \times 10^4$
16. $10^y = 9.4 \times 10^3$
17. $10^{4.16} = x$
18. $10^{5.23} = x$
19. $10^{0.033} = x$
20. $10^{0.056} = x$
21. $10^{-3.6} = y$
22. $10^{-5.73} = y$
23. $10^{-0.173} = y$
24. $10^{-0.0066} = y$
25. $\log 746 = x$
26. $\log 317 = x$
27. $\log 0.00716 = x$
28. $\log 0.0000406 = x$
29. $7.26^4 = x$
30. $12.4^{5.3} = x$
31. $0.026^x = 7$
32. $0.173^x = 0.097$
33. $x^{4.3} = 7300$
34. $x^{-4.3} = 0.15$
35. $\log y^3 = 9$
36. $\log 5^x = 7$
37. $\log y = 7.36$
38. $\log y = 0.0576$
39. $x^7 = 14$
40. $x^{4.5} = 95$
41. $70^x = 12$
42. $200^x = 8000$

END OF CHAPTER PROBLEMS 23–3

Solve for the unknown in the following problems:

1. $6^{2-y} = 68$
2. $8^{1+y} = 70$
3. $75^{y/4} = 3.54$
4. $38^{y/3} = 8.73$
5. $1.7 = \log \dfrac{x}{42}$
6. $3 = \log \dfrac{x}{0.5}$
7. $6 = \log \dfrac{100}{x}$
8. $7 = \log \dfrac{0.43}{x}$
9. $3^{x-2} = 50$
10. $5^{x-1} = 50$
11. $16^{3/x} = 200$
12. $7^{2/x} = 0.963$
13. $6 = \log \dfrac{x}{100}$
14. $4 = \log \dfrac{x}{1000}$
15. $-4 = \log \dfrac{20}{x}$
16. $-5 = \log \dfrac{10}{x}$
17. $14^{x/2} = 43$
18. $8^{x/4} = 47$
19. $35^{3-x} = 470$
20. $15^{4-x} = 327$
21. $3 = \log \dfrac{0.005}{x}$
22. $5 = \log \dfrac{0.043}{x}$

23. $4.3 = \log \dfrac{y}{9}$

24. $8.7 = \log \dfrac{y}{4}$

25. $45^{x-2} = 3.75$

26. $30^{x-4} = 4.36$

27. $8.7^{x+2} = 150$

28. $5.3^{x+3} = 220$

29. $60^{y/3} = 1.78$

30. $100^{y/4} = 9.4$

END OF CHAPTER PROBLEMS 23–4

Solve for the unknown in the following problems:

1. $x = \ln 11$

2. $x = \ln 67$

3. $\ln 15 = x$

4. $\ln 0.835 = x$

5. $e^x = 200$

6. $e^{-x} = 0.93$

7. $\ln 0.0004 = x$

8. $\ln 0.446 = x$

9. $e^{-x} = 15$

10. $e^x = 34.7$

11. $e^{2.7} = x$

12. $e^{7.6} = x$

13. $46 = e^y$

14. $20 = e^y$

15. $e^{-4} = y$

16. $e^{-3.6} = y$

END OF CHAPTER PROBLEMS 23–5

Solve for the unknown in the following problems:

1. $0.625 = 1 - e^{-x}$

2. $x = 1 - e^{0.35}$

3. $50 = 80e^{-x}$

4. $75 = 60e^{-x}$

5. $x = 25(1 - e^{-0.75})$

6. $x = 100(1 - e^{-1.5})$

7. $30 = x(1 - e^{-3})$

8. $20 = x(1 - e^{-2})$

9. $10 = 25e^{-x/3}$

10. $5.7 = 10e^{-x/1.7}$

11. $3 = 9e^{-4/x}$

12. $7 = 12e^{-0.27/x}$

13. $0.176 = e^{-2x/3}$

14. $0.73 = e^{-x/4.7}$

15. $0.5 = e^{-2/x}$

16. $0.57 = e^{-3.2/x}$

17. $x = 1 - e^{-3}$

18. $1 - x = e^{-2.7}$

19. $0.376 = 1 - e^{-x}$

20. $0.873 = e^{-x}$

21. $e^{-x/2} = 0.76$

22. $1 - x = e^{-2.5}$

ANSWERS TO SELF-TESTS

SELF-TEST 23–1

1. $x = 2.60$

2. $x = 128$

3. $x = 4.82 \times 10^{-5}$

4. $x = 8.76$

5. $x = 2.00$

6. $x = -0.631$

7. $x = 1.65$

8. $x = -6.45$

9. $x = 4.00 \times 10^6$

10. $x = 2.00 \times 10^{-12}$

11. $x = 3.16$

12. $x = 1.95$

SELF-TEST 23–2

1. $x = 3.69$

2. $x = 1.20 \times 10^6$

3. $x = 3.00$

4. $x = 0.451$

5. $x = 1.02$

6. $x = 46.8$

7. $x = 111$

8. $x = 2.89$

9. $x = 3.19$

10. $x = 1.49$

Applications of Logarithms

24

Introduction

A parameter of many electronic circuits or systems is *gain*. The symbol for gain is *A*. Gain is the ratio of an output quantity to an input quantity. We may analyze a device in terms of power gain (A_p), voltage gain (A_v), current gain (A_i), or resistance gain (A_r). In ac circuits, gain will vary as the frequency varies. In this chapter we will calculate the gain of different kinds of circuits, note how the gain changes with frequency, and then plot curves of gain versus frequency so that we may have a visual display of these characteristics.

Chapter Objectives

In this chapter you will learn how to:

1. Compute both power and voltage gains of electrical systems.
2. Determine the frequency response of an electrical network or system.
3. Construct a Bode plot of the frequency response of an electrical network or system.
4. Compute the charge and discharge of a capacitor in a series *RC* circuit.

24-1   GAIN MEASUREMENTS

It has been demonstrated experimentally that the human ear responds logarithmically to changes in sound levels. It has also been found that it requires about a 1-decibel change in a sound level before we are aware that a change has taken place. Because of this logarithmic characteristic, power gains are usually expressed as the logarithm of the ratio of power levels.

$$A_{p(\text{bels})} = \log \frac{P_{\text{out}}}{P_{\text{in}}}$$

The unit of measure is the *bel* in honor of Alexander Graham Bell.

The bel is seldom used because it is considered too large a unit. Instead, the *decibel* is the unit of measure we normally use. A decibel (dB) equals $\frac{1}{10}$ bel. Therefore,

$$A_{p(dB)} = 10 \log \frac{P_{out}}{P_{in}} \tag{24-1}$$

It is common practice in industry to leave out the subscript dB. We can always assume that gain units are in decibels unless we are told otherwise.

Equation 24-1 can be solved for P_{out} or P_{in}.

Like Ohm's law, when we know any two of the variables, we can solve the equation for the third variable. Gain measurements are very important when studying RF systems and circuits. The cell phone industry is highly dependent on the correct analysis of RF power levels. The placement of cell phone antenna arrays in every city is determined by examining the RF power levels in each cell area to make sure that cell phone calls do not get dropped because of low signal power. Figure 24-1 shows a block diagram of an RF amplifier that would be a common component of a cell phone.

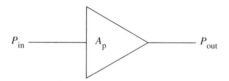

FIGURE 24-1
Block diagram
of an RF
amplifier. Power
gain = A_p

EXAMPLE 24-1

The power out of an amplifier is 30 W. The input power is 100 mW. What is the power gain in decibels?

SOLUTION

$$A_p = 10 \log \frac{P_{out}}{P_{in}}$$

$$A_p = 10 \log \frac{30 \text{ W}}{100 \text{ mW}} = 10 \log 300 = 24.8 \text{ dB}$$

EXAMPLE 24-2

The power gain of a power amplifier is 40 dB. If the input power is 50 μW, what is the output power?

$$A_p = 10 \log \frac{P_{out}}{P_{in}}$$

$$40 \text{ dB} = 10 \log \frac{P_{out}}{50 \ \mu\text{W}}$$

Divide both sides by 10:

$$4 = \log \frac{P_{out}}{50 \ \mu W}$$

Change to exponential form:

$$10^4 = \frac{P_{out}}{50 \ \mu W}$$

Multiply both sides by 50 μW:

$$50 \ \mu W \times 10^4 = P_{out}$$

$$P_{out} = 0.5 \ W$$

If we consider that

$$P_{out} = \frac{V_{out}^2}{R_{out}} \quad \text{and} \quad P_{in} = \frac{V_{in}^2}{R_{in}}$$

then we can write the equation for voltage gain in decibels.

$$A_v = 10 \log \left(\frac{V_{out}^2}{R_{out}} \div \frac{V_{in}^2}{R_{in}} \right)$$

$$= 10 \log \left(\frac{V_{out}^2}{V_{in}^2} \times \frac{R_{in}}{R_{out}} \right)$$

$$= 10 \log \frac{V_{out}^2}{V_{in}^2} + 10 \log \frac{R_{in}}{R_{out}}$$

If $R_{out} = R_{in}$, then

$$A_v = 10 \log \left(\frac{V_{out}}{V_{in}} \right)^2 + 10 \log 1 = 20 \log \frac{V_{out}}{V_{in}} + 0$$

(24–2)
$$A_{v(dB)} = 20 \log \frac{V_{out}}{V_{in}}$$

EXAMPLE 24–3

The output voltage of an amplifier is 1.73 V. The input voltage is 50 mV. Assuming that $R_{out} = R_{in}$, find the voltage gain in dB.

SOLUTION

$$A_v = 20 \log \frac{V_{out}}{V_{in}} = 20 \log \frac{1.73 \ V}{50 \ mV} = 20 \log 34.6 = 30.8 \ dB$$

TI: 1.73 $\boxed{\div}$ 50 $\boxed{EE}$ $\boxed{\pm}$ 3 $\boxed{=}$ $\boxed{\log}$ $\boxed{\times}$ 20 $\boxed{=}$

Answer: $A_v = 30.8$ dB

Casio: 20 $\boxed{\log}$ $\boxed{(}$ 1.73 $\boxed{\div}$ 50 $\boxed{EXP}$ $\boxed{(-)}$ 3 $\boxed{)}$ $\boxed{=}$

EXAMPLE 24–4

The voltage gain of a device is -6 dB. If the output voltage is 700 mV, find V_{in}.

SOLUTION

$$A_v = 20 \log \frac{V_{out}}{V_{in}}$$

$$-6 = 20 \log \frac{700 \text{ mV}}{V_{in}}$$

Divide by 20:

$$-0.3 = \log \frac{700 \text{ mV}}{V_{in}}$$

Change to exponential form:

$$10^{-0.3} = \frac{700 \text{ mV}}{V_{in}}$$

$$0.501 = \frac{700 \text{ mV}}{V_{in}}$$

Multiply by V_{in}:

$$0.501 \, V_{in} = 700 \text{ mV}$$

Divide by 0.501:

$$V_{in} = \frac{700 \text{ mV}}{0.501}$$

$$= 1.40 \text{ V}$$

24–1–1 Reference Levels

In the preceding examples the output powers and output voltages were given with reference to input values. Sometimes other reference levels are used. One common reference level is 6 mW. In other words, 0 dB equals 6 mW. Another reference level is 1 mW. When 1 mW is the reference, the decibel measurement is called a *dBm*. If a power gain is given in dBm, we assume that P_{in} is 1 mW.

EXAMPLE 24–5

The power gain of an amplifier is 33 dBm. Determine the output power.

$$A_p = 10 \log \frac{P_{out}}{P_{in}}$$

$$33 = 10 \log \frac{P_{out}}{1 \text{ mW}}$$

Divide by 10:

$$3.3 = \log \frac{P_{out}}{1 \text{ mW}}$$

Change to exponential form:

$$10^{3.3} = \frac{P_{out}}{1 \text{ mW}}$$

Multiply by 1 mW:

$$1 \text{ mW} \times 10^{3.3} = P_{out}$$

TI: 1 $\boxed{\text{EE}}$ $\boxed{\pm}$ 3 $\boxed{\times}$ 3.3 $\boxed{10^x}$ $\boxed{=}$

Answer: $P_{out} = 2.00$ W

Casio: 1 $\boxed{\text{EXP}}$ $\boxed{(-)}$ 3 $\boxed{10^x}$ 3.3 $\boxed{=}$

PRACTICE PROBLEMS 24–1

1. $P_{in} = 20$ mW, $P_{out} = 35$ W. Find $A_{p(dB)}$.
2. $A_p = 38$ dB, $P_{in} = 5$ mW. Find P_{out}.
3. $A_p = 25$ dB, $P_{out} = 15$ W. Find P_{in}.
4. $A_p = -3$ dB, $P_{out} = 1$ W. Find P_{in}.
5. $A_p = 15$ dBm. Find P_{out}.
6. $P_{out} = 10$ W. Find A_p in dBm.

Assume that $R_{out} = R_{in}$ for problems 7 through 10:

7. $V_{in} = 50$ μV, $V_{out} = 1$ V. Find $A_{v(dB)}$.
8. $A_v = 60$ dB, $V_{out} = 3$ V. Find V_{in}.
9. $A_v = 75$ dB, $V_{in} = 2$ mV. Find V_{out}.
10. $A_v = -10$ dB, $V_{in} = 2.7$ V. Find V_{out}.

SOLUTIONS

1. 32.4 dB
2. 31.5 W
3. 47.4 mW
4. 2.00 W
5. 31.6 mW
6. 40 dBm
7. 86.0 dB
8. 3.00 mV
9. 11.2 V
10. 854 mV

Additional practice problems are at the end of the chapter.

24-2-1 Bode Plot

A Bode plot is a useful tool for approximating the frequency response of a circuit or system. Let's construct a Bode plot of the frequency response characteristics of the RC circuit in Figure 24–2. Recall from ac circuit analysis that X_C varies inversely with frequency. Furthermore, the cutoff frequency f_{co} occurs where $X_C = R$. Because the output is taken from across R, $V_{out} = V_R$. We can compute the voltage gain of the circuit by using the equation learned previously.

$$A_{v(dB)} = 20 \log \frac{V_{out}}{V_{in}}$$

At high frequencies $V_{out} = V_{in}$. The voltage gain is

$$A_{v(dB)} = 20 \log \frac{10 \text{ V}}{10 \text{ V}} = 0 \text{ dB}$$

The maximum possible gain for this circuit is 0 dB.

The Bode plot is a plot of gain versus frequency. Semilog graph paper is used in such a plot. Frequency is the independent variable and is plotted along the logarithmic scale. A scale for A_v is chosen that will allow us to change the gain by at least -20 dB. Once our scale is chosen, a horizontal line is drawn across our graph at 0 dB as shown in Figure 24–3. Four-cycle semilog paper has been chosen with a minimum frequency of 1 Hz. This allows us to examine the gain at frequencies from 1 Hz to 10 kHz. The cutoff frequency is 100 Hz.

$$f_{co} = \frac{1}{2\pi CR} = 100 \text{ Hz}$$

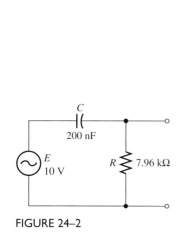

FIGURE 24–2

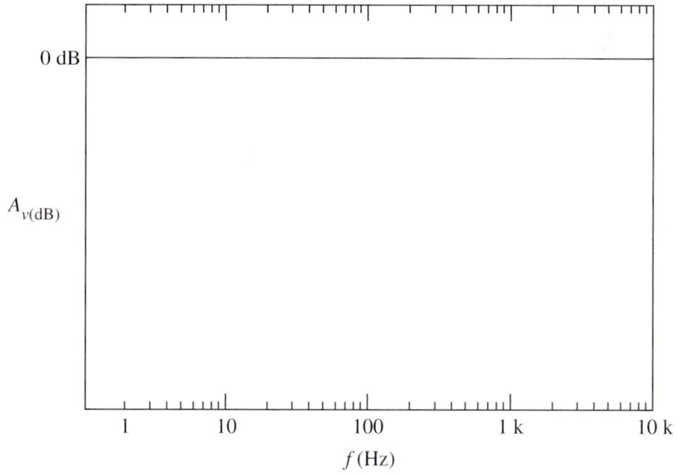

FIGURE 24–3

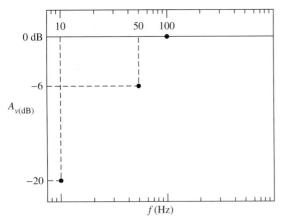

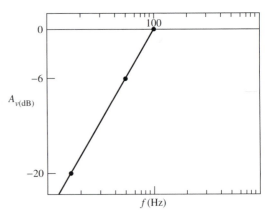

FIGURE 24–4   FIGURE 24–5

Figure 24–3 has been expanded in Figure 24–4 and the 100-Hz point has been plotted on our horizontal line. This becomes one point on our Bode plot.

> ☞ **KEY POINT** As a straight-line approximation, the gain will change at the rate of 6 *dB per octave* and 20 *dB per decade.*

An octave change in frequency is a halving or doubling of the frequency. A decade change is a change in frequency of one-tenth or ten times.

Because gain decreases as frequency decreases, a second point is plotted in Figure 24–4 where the gain is down 6 dB and the frequency is 50 Hz, one-half the cutoff frequency. A third point is plotted where the gain is down 20 dB and the frequency is 10 Hz, 0.1 times f_{co}. In Figure 24–5 a straight line is drawn to connect these three points. This is our Bode plot. All Bode plots are plotted in a like manner. When the gain decreases with frequency, the line slopes downward and to the *left* as in Figure 24–5. When the gain decreases as frequency *increases,* the line slopes downward and to the right.

PRACTICE PROBLEMS 24–2

1. Refer to Figure 24–2. Construct a Bode plot for $C = 100$ nF and $R = 2.7$ kΩ.

2. Refer to Figure 24–6. Construct a Bode plot of the frequency response.

SOLUTIONS

1. $f_{co} = 589$ Hz. The Bode plot is shown in Figure 24–7.

2. $f_{co} = 15.9$ kHz. See Figure 24–8 for the Bode plot.

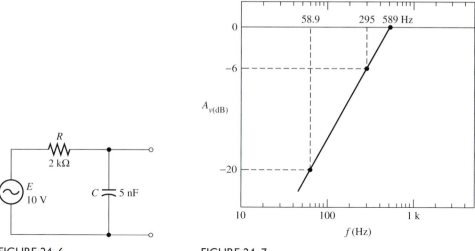

FIGURE 24–6   FIGURE 24–7

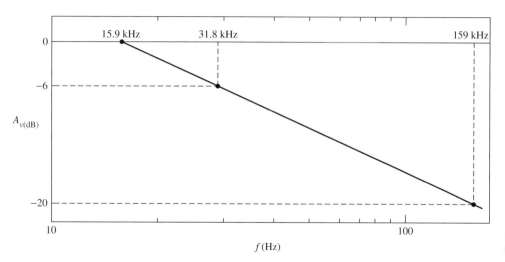

FIGURE 24–8

AMPLIFIER GAIN AND THE BODE PLOT   24–3

EXAMPLE 24–6

An amplifier has a mid-frequency gain of 25 dB. The lower cutoff frequency (symbolized f_1) is 90 Hz. The upper cutoff frequency (symbolized f_2) is 12 kHz. Construct a Bode plot of the frequency response.

SOLUTION   We will use five-cycle semilog graph paper for this example. Choose a practical scale for A_v and draw a horizontal line at 25 dB. Plot the lower cutoff frequency as before. One point is f_1 and 25 dB. $\dfrac{f_1}{2}$ and 19 dB is a second point (6 dB down). $\dfrac{f_1}{10}$ and 5 dB is a third point (20 dB down). The upper cutoff frequency is plotted by using the points f_2 and 25 dB, $2f_2$ and 19 dB, and $10f_2$ and 5 dB. The complete plot is shown in Figure 24–9.

FIGURE 24–9

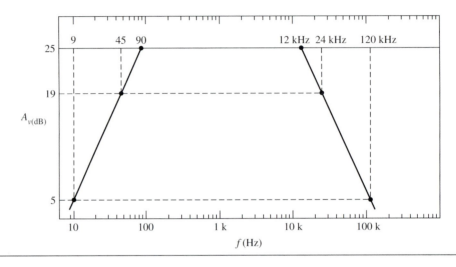

Consider the circuit in Figure 24–10.

FIGURE 24–10

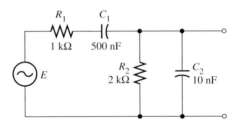

In circuits like this, the mid-frequency gain can be determined (if f_1 and f_2 are separated by at least one decade) by ignoring the effects of C_1 and C_2 and using the ratio of R_2 to the total resistance $R_1 + R_2$.

$$\frac{R_2}{R_1 + R_2} = \frac{V_{out}}{V_{in}}$$

The mid-frequency equivalent circuit is shown in Figure 24–11(a). (In practice there may be other factors affecting response, such as source impedance and cable capacitance.)

Mid-frequency gain would be determined by

(24–3)

$$A_v = 20 \log \frac{R_2}{R_1 + R_2}$$

CHAPTER 24

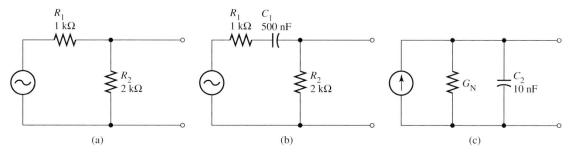

FIGURE 24–11   (a) Mid-frequency equivalent circuit of Figure 24–10; (b) low-frequency equivalent circuit of Figure 24–10; (c) high-frequency equivalent circuit of Figure 24–10.

In this circuit

$$A_v = 20 \log \frac{2 \text{ k}\Omega}{1 \text{ k}\Omega + 2 \text{ k}\Omega} = -3.52 \text{ dB}$$

TI: 2 EE 3 ÷ (1 EE 3 + 2 EE 3) =

log × 20 =   Answer: $A_v = -3.52$ dB

Casio: 20 log (2 EXP 3 ÷ (1 EXP 3 + 2

EXP 3)) =

This is the Bode limit on circuit gain. If the frequency is decreased, the reactance of C_1 becomes larger and some voltage drops across it. This causes a decrease in the output voltage. If the frequency is increased, the susceptance of C_2 increases, which again causes a decrease in output voltage. The series reactive components (C_1 in our circuit) affect the response as we lower frequency. The effect on the circuit due to C_2 is negligible, which is the case if f_1 and f_2 are at least a decade apart. The equivalent low-frequency circuit looks like the circuit in Figure 24–11(b).

At high frequencies, the reactance of C_1 becomes negligible (compared to R_T), but the susceptance of C_2 becomes a factor. Since C_2 is a parallel component, if we Nortonize the circuit of R_2, we get an equivalent circuit consisting of C_2, R_1, and R_2 all in parallel, as shown in Figure 24–11(c).

EXAMPLE 24–7

Find f_1 and f_2 in the circuit of Figure 24–10. Construct a Bode plot.

SOLUTION   We previously determined that the mid-frequency gain is -3.52 dB. Let's compute f_1 and f_2.

FIGURE 24–12

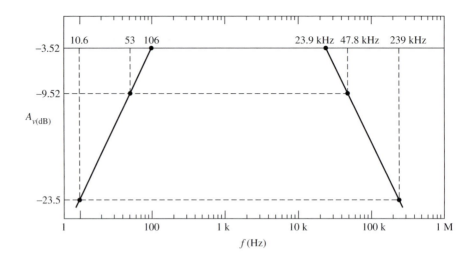

$$f_1 = \frac{1}{2\pi C_1(R_1 + R_2)} = \frac{1}{2 \times 3.14 \times 500 \times 10^{-9} \times 3 \times 10^3} = 106 \text{ Hz}$$

$$f_2 = \frac{G_1 + G_2}{2\pi C_2} = \frac{1.5 \times 10^{-3}}{2 \times 3.14 \times 10 \times 10^{-9}} = 23.9 \text{ kHz}$$

The Bode plot is constructed in Figure 24–12.

24-3-1 Frequency Response Curve

The Bode plot results in some error, particularly around the cutoff frequency. As we know, the gain is really down 3 dB at f_{co}. At frequencies other than cutoff, the error is less than 3 dB. The error is practically 0 one decade above and below f_{co}. The reduction in gain (from maximum gain) at any frequency can be found by using the following equations:

For an RC high-pass filter:

(24–4)
$$A_v = 20 \log \frac{1}{\sqrt{1 + \left(\dfrac{f_1}{f}\right)^2}}$$

where f_1 is the lower cutoff frequency and f is the frequency in question.

For an RC low-pass filter:

(24–5)
$$A_v = 20 \log \frac{1}{\sqrt{1 + \left(\dfrac{f}{f_2}\right)^2}}$$

where f_2 is the upper cutoff frequency and f is the frequency in question. The general shape of the frequency response curve results from plotting gain versus frequency

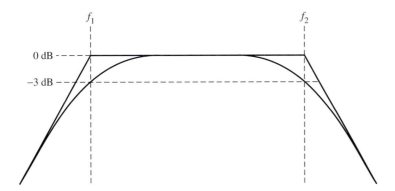

FIGURE 24–13

using these equations. The actual curve and the Bode plot are drawn together in Figure 24–13.

EXAMPLE 24–8

Refer to Figure 24–10. Find the gain at 50 Hz and at 100 kHz by constructing a Bode plot and finding those frequencies on the curve. Verify your answers by using Equations 24–4 and 24–5.

SOLUTION We previously determined that the mid-frequency gain (the maximum gain) is −3.52 dB. At 50 Hz

$$A_v = 20 \log \frac{1}{\sqrt{1 + \left(\frac{f_1}{f}\right)^2}} = 20 \log \frac{1}{\sqrt{1 + \left(\frac{106 \text{ Hz}}{50 \text{ Hz}}\right)^2}} = -7.4 \text{ dB}$$

TI: $106 \boxed{\div} 50 \boxed{=} \boxed{x^2} \boxed{+} 1 \boxed{=} \boxed{\sqrt{x}} \boxed{1/x}$

$\boxed{\log} \boxed{\times} 20 \boxed{=}$ *Answer:* $A_v = -7.4$ dB

Casio: $20 \boxed{\log} \boxed{(} 1 \boxed{\div} \boxed{\sqrt{}} \boxed{(} 1 \boxed{+} \boxed{(} 106$

$\boxed{\div} 50 \boxed{)} \boxed{x^2} \boxed{)} \boxed{)} \boxed{=}$

This tells us that we are down 7.4 dB from −3.52 dB. The resultant gain is

$$A_{v(dB)} = -3.52 \text{ dB} + (-7.4 \text{ dB}) = -10.9 \text{ dB}$$

At 100 kHz

$$A_v = 20 \log \frac{1}{\sqrt{1 + \left(\frac{f}{f_2}\right)^2}} = 20 \log \frac{1}{\sqrt{1 + \left(\frac{100 \text{ kHz}}{23.9 \text{ kHz}}\right)^2}}$$

$$= -12.7 \text{ dB}$$

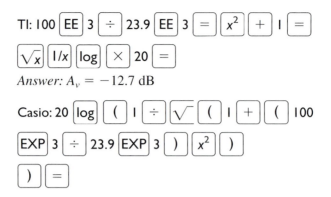

Answer: $A_v = -12.7$ dB

The resultant gain is

$$A_{v(dB)} = -3.52 \text{ dB} + (-12.7 \text{ dB}) = -16.2 \text{ dB}$$

The Bode plot is shown in Figure 24–12. An inspection of the Bode plot shows that our gain calculations at 50 Hz and at 100 kHz were correct.

PRACTICE PROBLEMS 24–3

1. An amplifier has a mid-frequency gain of 40 dB. $f_1 = 200$ Hz and $f_2 = 20$ kHz. Construct a Bode plot of the frequency response.

2. Refer to Figure 24–10. Let $R_1 = 330 \ \Omega$ and $R_2 = 680 \ \Omega$. Compute the gain at 100 Hz and 200 kHz. Construct a Bode plot of the frequency response. Do the computed gains at 100 Hz and 200 kHz fall on the Bode plot? If they do not, either the plot is wrong or your computations are wrong.

3. The mid-frequency gain of an amplifier is 27 dB. $f_1 = 100$ Hz and $f_2 = 15$ kHz. Find the gain at $f_1, f_2,$ 25 Hz, and 30 kHz.

SOLUTIONS

1. See Figure 24–14 for the Bode plot.

2. The mid-frequency gain is -3.44 dB, $f_1 = 315$ Hz and $f_2 = 71.6$ kHz. At 100 Hz, $A_v = -13.8$ dB (-3.44 dB + -10.4 dB). At 200 kHz, $A_v = -12.9$ dB (-3.44 dB + -9.44 dB). The Bode plot is constructed in Figure 24–15. An inspection of the plot shows that both 100 Hz and 200 kHz fall on the curve. This verifies the accuracy of our curve and our calculations.

FIGURE 24-14

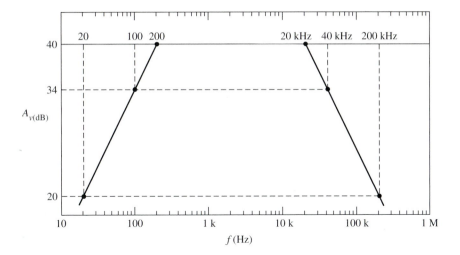

FIGURE 24-15

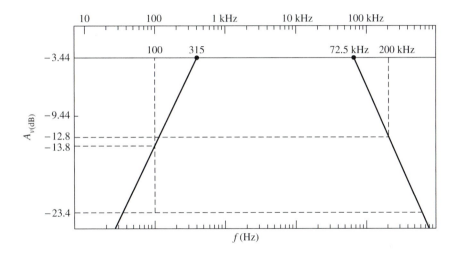

3. At f_1, $A_v = 24$ dB. At f_2, $A_v = 24$ dB.
   At 25 Hz, $A_v = 14.7$ dB. At 30 kHz,
   $A_v = 20.0$ dB.

Additional practice problems are at the end of the chapter.

RC CIRCUITS 24-4

In Chapter 13 we used a universal time constant curve to determine the charge on a capacitor at some time after power was applied to the circuit. In this section we will compute the charge or discharge of a capacitor by using logarithmic equations.

To determine the charge on a capacitor in a series RC circuit, the following equation is used:

(24–6)
$$v_C = E(1 - e^{-t/RC})$$

where v_C = charge on the capacitor

E = applied voltage

t = time in seconds

R = resistance in ohms

C = capacitance in farads

The equation for v_R is

(24–7)
$$v_R = Ee^{-t/RC}$$

When the capacitor is discharging, the equation is

(24–8)
$$v_C = Ee^{-t/RC}$$

EXAMPLE 24–9

In the circuit of Figure 24–16, find v_R and v_C 1 ms after the switch is closed.

SOLUTION

$$v_C = E(1 - e^{-t/RC}) = 70(1 - e^{-1ms/3.4 \times 10^{-3}})$$

TI: 1 $\boxed{\pm}$ $\boxed{EE}$ $\boxed{\pm}$ $\boxed{3}$ $\boxed{\div}$ $\boxed{(}$ $\boxed{680}$ $\boxed{\times}$ 5 $\boxed{EE}$ $\boxed{\pm}$

6 $\boxed{)}$ $\boxed{=}$ $\boxed{e^x}$ $\boxed{\pm}$ $\boxed{+}$ 1 $\boxed{=}$ $\boxed{\times}$ 70 $\boxed{=}$

Answer: 17.8 V

Casio: 70 $\boxed{(}$ 1 $\boxed{-}$ $\boxed{e^x}$ $\boxed{(}$ $\boxed{-}$ 1 $\boxed{EXP}$ $\boxed{(-)}$

3 $\boxed{\div}$ 3.4 $\boxed{EXP}$ $\boxed{(-)}$ 3 $\boxed{)}$ $\boxed{)}$ $\boxed{=}$

FIGURE 24–16

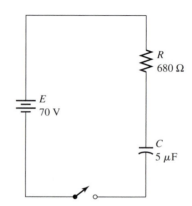

CHAPTER 24

An alternative method is

$$| \boxed{\pm} \boxed{\text{EE}} \boxed{\pm} 3 \boxed{\div} \boxed{(} 680 \boxed{\times} 5 \boxed{\text{EE}} \boxed{\pm} 6 \boxed{)} \boxed{=} \boxed{\text{STO}}$$

(store the exponent to which e will be raised)

Then

$$70 \boxed{\times} \boxed{(} | \boxed{-} \boxed{\text{RCL}} \boxed{e^x} \boxed{)} \boxed{=}$$

Answer: $v_C = 17.8$ V

v_R can be found by applying Kirchhoff's voltage law:

$$v_R = E - v_C = 70 \text{ V} - 17.8 \text{ V} = 52.2 \text{ V}$$

EXAMPLE 24–10

Refer to Figure 24–16. If $v_C = 40$ V, after the switch is closed, how much time has elapsed?

SOLUTION

$$v_C = E(1 - e^{-t/RC})$$
$$40 = 70(1 - e^{-t/3.4 \times 10^{-3}})$$

Divide by 70:

$$0.571 = 1 - e^{-t/3.4 \times 10^{-3}}$$

Subtract 1:

$$-0.429 = -e^{-t/3.4 \times 10^{-3}}$$

Change signs:

$$0.429 = e^{-t/3.4 \times 10^{-3}}$$

Take ln:

$$\ln 0.429 = -\frac{t}{3.4 \times 10^{-3}} \ln e$$

$$-0.847 = -\frac{t}{3.4 \times 10^{-3}} \qquad (\ln e = 1)$$

Multiply by -3.4×10^{-3}:

$$2.88 \times 10^{-3} = t$$

$$t = 2.88 \text{ ms}$$

PRACTICE PROBLEMS 24–4

Refer to Figure 24–16.

1. $t = 2$ ms. Find v_C and v_R.
2. $V_R = 10$ V. Find t.
3. $i = 20$ mA. Find t. (The equation for i is of the same form as Equation 24–7.)
4. $E = 50$ V, $R = 4.7$ kΩ, $v_C = 10$ V, and $t = 50$ μs. Find C.
5. In a circuit like Figure 24–16, $E = 50$ V, $C = 0.2$ μF, $v_R = 20$ V, and $t = 500$ ms. Find R.

SOLUTIONS

1. $v_C = 31.1$ V, $v_R = 38.9$ V
2. 6.62 ms
3. 5.57 ms
4. 47.7 nF
5. 2.73 MΩ

SELF-TEST 24–1

1. Refer to Figure 24–17. Let $C_1 = 1$ μF, $C_2 = 470$ pF, $R_1 = 4.7$ kΩ, and $R_2 = 2$ kΩ. Determine $A_{v(dB)}$ at 50 Hz and 200 kHz. Construct a Bode plot.
2. $A_p = 25$ dB, $P_{out} = 3$ W. Find P_{in}.
3. $A_p = 45$ dBm. Find P_{out}.
4. Refer to Figure 24–16. $E = 60$ V, $v_C = 20$ V, $C = 500$ nF, and $t = 3$ ms. Find R.

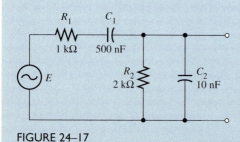

FIGURE 24–17

Answers to Self-test 24–1 are at the end of the chapter.

CHAPTER 24 AT A GLANCE

PAGE	KEY POINTS, DEFINITIONS, AND EQUATIONS	EXAMPLE
653	The decibel is the unit of measure we use when computing power gain. $$A_{p(dB)} = 10 \log \frac{P_{out}}{P_{in}}$$	$P_{in} = 100$ mW, $P_{out} = 30$ W. Find A_p. $$A_p = 10 \log \frac{P_{out}}{P_{in}}$$ $$= 10 \log \frac{30 \text{ W}}{100 \text{ mW}}$$ $$= 10 \log 300 = 24.8 \text{ dB}$$
654	The equation for voltage gain is $$A_{v(dB)} = 20 \log \frac{V_{out}}{V_{in}}$$	$V_{in} = 50$ mV, $V_{out} = 1.73$ V. Find A_v. $$A_v = 20 \log \frac{V_{out}}{V_{in}}$$
658	*Key Point:* As a straight-line approximation, the gain will change at the rate of 6 dB per octave and 20 dB per decade.	$$= 20 \log \frac{1.73 \text{ V}}{50 \text{ mV}}$$ $$= 20 \log 34.6 = 30.8 \text{ dB}$$
658	A Bode plot is a plot of gain versus frequency. As a straight-line approximation, the gain will change at the rate of 6 dB per octave and 20 dB per decade.	If the mid-frequency gain is 25 dB, f_1 is 90 Hz and f_2 is 12 kHz. Construct a Bode plot.

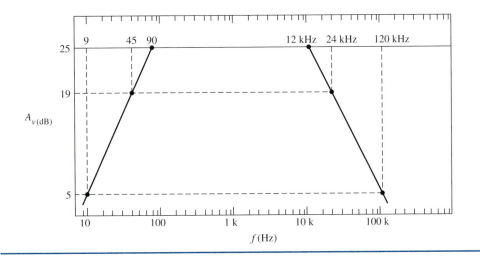

FIGURE 24–18

END OF CHAPTER PROBLEMS 24–1

1. $P_{in} = 30\ \mu W$, $P_{out} = 30$ W. Find A_p in dB.

2. $P_{in} = 1$ mW, $P_{out} = 3.16$ W. Find A_p in dB.

3. $P_{in} = 1$ mW, $P_{out} = 20$ W. Find A_p in dB.

4. $P_{in} = 100\ \mu W$, $P_{out} = 10$ W. Find A_p in dB.

5. $P_{in} = 2$ mW, $P_{out} = 50$ W. Find A_p in dB.

6. $P_{in} = 10$ mW, $P_{out} = 40$ W. Find A_p in dB.

7. $A_p = 35$ dB, $P_{out} = 12$ W. Find P_{in}.

8. $A_p = 30$ dB, $P_{out} = 2$ W. Find P_{in}.

9. $A_p = 20$ dB, $P_{out} = 100$ W. Find P_{in}.

10. $A_p = 40$ dB, $P_{out} = 1$ W. Find P_{in}.

11. $A_p = 32$ dB, $P_{out} = 5$ W. Find P_{in}.

12. $A_p = 50$ dB, $P_{out} = 40$ W. Find P_{in}.

13. $A_p = 22$ dB, $P_{in} = 10$ mW. Find P_{out}.

14. $A_p = 40$ dB, $P_{in} = 250$ mW. Find P_{out}.

15. $A_p = 38$ dB, $P_{in} = 500\ \mu W$. Find P_{out}.

16. $A_p = 27$ dB, $P_{in} = 100$ mW. Find P_{out}.

17. $A_p = 15$ dB, $P_{in} = 10$ mW. Find P_{out}.

18. $A_p = 45$ dB, $P_{in} = 70$ mW. Find P_{out}.

19. $A_p = 20$ dBm. Find P_{out}.

20. $A_p = 40$ dBm. Find P_{out}.

21. $A_p = 32$ dBm. Find P_{out}.

22. $A_p = 10$ dBm. Find P_{out}.

23. $A_p = 50$ dBm. Find P_{out}.

24. $A_p = 35$ dBm. Find P_{out}.

Assume $R_{out} = R_{in}$ for the following problems:

25. $V_{in} = 100\ \mu V$, $V_{out} = 2.7$ V. Find A_v in dB.

26. $V_{in} = 5.7$ V, $V_{out} = 3.8$ V. Find A_v in dB.

27. $V_{in} = 10$ mV, $V_{out} = 3$ V. Find A_v in dB.

28. $V_{in} = 600\ \mu V$, $V_{out} = 12$ V. Find A_v in dB.

29. $V_{in} = 50\ \mu V$, $V_{out} = 10$ V. Find A_v in dB.

30. $V_{in} = 2$ mV, $V_{out} = 12$ V. Find A_v in dB.

31. $A_v = 40$ dB, $V_{in} = 5$ mV. Find V_{out}.

32. $A_v = 70$ dB, $V_{in} = 20$ mV. Find V_{out}.

33. $A_v = 22.7$ dB, $V_{in} = 2$ mV. Find V_{out}.

34. $A_v = 37.5$ dB, $V_{in} = 10$ mV. Find V_{out}.

35. $A_v = 73$ dB, $V_{in} = 23$ mV. Find V_{out}.

36. $A_v = 44$ dB, $V_{in} = 5.7$ mV. Find V_{out}.

37. $A_v = 90$ dB, $V_{out} = 4$ V. Find V_{in}.

38. $A_v = 55$ dB, $V_{out} = 1.73$ V. Find V_{in}.

39. $A_v = 75$ dB, $V_{out} = 15$ V. Find V_{in}.

40. $A_v = 63$ dB, $V_{out} = 10$ V. Find V_{in}.

41. $A_v = 80$ dB, $V_{out} = 600$ mV. Find V_{in}.

42. $A_v = 32$ dB, $V_{out} = 400$ mV. Find V_{in}.

43. $A_v = -3$ dB, $V_{out} = 500$ mV. Find V_{in}.

44. $A_v = -3$ dB, $V_{out} = 20$ V. Find V_{in}.

45. $A_v = -6$ dB, $V_{out} = 1$ V. Find V_{in}.

46. $A_v = -6$ dB, $V_{out} = 6$ V. Find V_{in}.

47. $A_v = -3$ dB, $V_{out} = 5$ V. Find V_{in}.

48. $A_v = -3$ dB, $V_{out} = 9$ V. Find V_{in}.

49. $A_v = -30$ dB, $V_{out} = 100$ mV. Find V_{in}.

50. $A_v = 20$ dB, $V_{out} = 800\ \mu V$. Find V_{in}.

END OF CHAPTER PROBLEMS 24–2

Refer to Figure 24–19 and construct a Bode plot of the frequency response.

1. $C = 0.3\ \mu F$ and $R = 1\ k\Omega$
3. $C = 1\ \mu F$ and $R = 470\ \Omega$
5. $C = 300\ nF$ and $R = 3.9\ k\Omega$

2. $C = 0.5\ \mu F$ and $R = 330\ \Omega$
4. $C = 100\ nF$ and $R = 15\ k\Omega$
6. $C = 50\ nF$ and $R = 2.2\ k\Omega$

Refer to Figure 24–20 and construct a Bode plot of the frequency response.

7. $C = 50\ nF$ and $R = 2\ k\Omega$
9. $C = 200\ nF$ and $R = 750\ \Omega$
11. $C = 40\ nF$ and $R = 2\ k\Omega$

8. $C = 20\ nF$ and $R = 2.5\ k\Omega$
10. $C = 400\ nF$ and $R = 120\ \Omega$
12. $C = 120\ nF$ and $R = 1.8\ k\Omega$

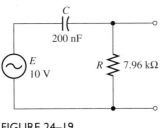

FIGURE 24–19

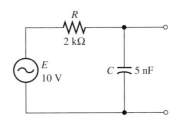

FIGURE 24–20

END OF CHAPTER PROBLEMS 24–3

Refer to Figure 24–17 and construct a Bode plot of the frequency response.

1. $C_1 = 250\ nF$, $C_2 = 40\ nF$, $R_1 = 680\ \Omega$, and $R_2 = 1.2\ k\Omega$.
3. $C_1 = 500\ nF$, $C_2 = 50\ nF$, $R_1 = 470\ \Omega$, and $R_2 = 1.8\ k\Omega$.
5. $C_1 = 200\ nF$, $C_2 = 30\ nF$, $R_1 = 3.3\ k\Omega$, and $R_2 = 2.2\ k\Omega$.
7. An amplifier has a mid-frequency gain of 20 dB. $f_1 = 50$ Hz and $f_2 = 18$ kHz. Construct a Bode plot of the frequency response. Find the gain at 40 Hz and 25 kHz.
9. An amplifier has a mid-frequency gain of 10 dB. $f_1 = 80$ Hz and $f_2 = 4$ kHz. Construct a Bode plot of the frequency response. Find the gain at 40 Hz and 12 kHz.

2. $C_1 = 400\ nF$, $C_2 = 25\ nF$, $R_1 = 2\ k\Omega$, and $R_2 = 2\ k\Omega$.
4. $C_1 = 1\ \mu F$, $C_2 = 400\ pF$, $R_1 = 1.8\ k\Omega$, and $R_2 = 2.5\ k\Omega$.
6. $C_1 = 450\ nF$, $C_2 = 2\ nF$, $R_1 = 820\ \Omega$, and $R_2 = 3.6\ k\Omega$.
8. An amplifier has a mid-frequency gain of 35 dB. $f_1 = 80$ Hz and $f_2 = 40$ kHz. Construct a Bode plot of the frequency response. Find the gain at 20 Hz and 100 kHz.
10. An amplifier has a mid-frequency gain of 40 dB. $f_1 = 20$ Hz and $f_2 = 70$ kHz. Construct a Bode plot of the frequency response. Find the gain at 10 Hz and 100 kHz.

11. Refer to Figure 24–16. If $R_1 = 470\ \Omega$, $R_2 = 680\ \Omega$, $C_1 = 1\ \mu F$, and $C_2 = 20$ nF, find the gain at 30 Hz and at 50 kHz.

12. Refer to problem 11. Find the gain at 300 Hz and at 30 kHz.

END OF CHAPTER PROBLEMS 24–4

Refer to Figure 24–16 for the following problems. Let $E = 100$ V, $R = 47$ kΩ, and $C = 50$ nF for problems 1 through 8.

1. $t = 1$ ms. Find v_R and v_C.
2. $t = 3$ ms. Find v_R and v_C.
3. $v_C = 65$ V. Find t.
4. $v_C = 50$ V. Find t.
5. $i = 2$ mA. Find t.
6. $i = 1$ mA. Find t.
7. $v_R = 10$ V. Find t.
8. $v_R = 60$ V. Find t.
9. $E = 25$ V, $C = 10$ nF, $v_R = 15$ V, and $t = 50\ \mu s$. Find R.
10. $E = 50$ V, $C = 200$ nF, $v_R = 20$ V, and $t = 200\ \mu s$. Find R.
11. $E = 100$ V, $R = 47$ kΩ, $v_C = 20$ V, and $t = 100\ \mu s$. Find C.
12. $E = 100$ V, $R = 2.7$ kΩ, $v_C = 50$ V, and $t = 70\ \mu s$. Find C.

ANSWERS TO SELF-TEST 24–1

1. At 50 Hz, $A_v = -11.4$ dB. At 200 kHz, $A_v = -12.8$ dB. See Figure 24–21 for the Bode plot.
2. $P_{in} = 9.49$ mV
3. $P_{out} = 31.6$ W
4. $R = 14.8$ kΩ

FIGURE 24–21

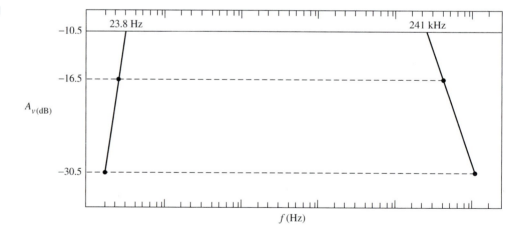

Math for Digital Electronics

Digital electronics is the branch of electronics that uses the ON and OFF states of transistor switches to perform specified tasks. Just as a wall switch can turn electrical lights ON or OFF by interrupting current, a transistor switch can interrupt the current in a digital circuit. Since transistors were invented over 50 years ago, their size and cost have decreased every year. The decreased cost has allowed engineers to design and develop hundreds of digital electronic circuit applications. Common examples are digital watches, TV remote controls, high definition TV, wireless telephones, calculators, and computers.

One of the keys to the understanding of digital electronics is understanding the mathematics that is the foundation upon which digital circuits are built.

In Chapter 25, students are introduced to computer number systems. Chapter 26 describes the mathematical properties of three basic digital circuits (AND, OR, and NOT) that are the elements of all other digital circuits. The interaction of these three digital functions can be described by the rules of Boolean Algebra. In Chapter 27, we study Karnaugh maps, which are a technique used to optimize digital circuits.

Computer Number Systems

25

Introduction

The basic tools for understanding computers are an understanding of the number systems used and an understanding of basic logic functions. Number systems are covered in this chapter and logic functions in Chapter 26.

We all know and understand the decimal number system, the number system we use daily, but the computer world makes use of the *binary* number system. All data are stored and manipulated inside the computer in binary. That is, within the computer all data are reduced to 1s and 0s. These 1s and 0s are stored in logic circuits called *registers* or they are stored in *memory*. The size of a register or memory location varies according to the kind of computer used. Microcomputers store 8 or 16 binary digits in each location. Large computers may store as many as 125 binary digits in each location.

We normally enter data into the computer in some number system other than binary because entering data in binary is too time-consuming and too prone to error. There are too many 1s and 0s. Data are entered into the computer by means of the decimal, octal, or hexadecimal number system. Octal and hexadecimal are used most often because they are more closely related to binary than is the decimal system.

In this chapter we will work with numbers in the binary, octal, decimal, and hexadecimal number systems. We use subscripts to denote the base we are working in. This must be done when working with different bases to avoid errors. For example, the number 11 is a valid number in all the bases we study, but it has a different value in each. If it is a binary number, we write it this way: 11_2. If it is an octal number, it is written this way: 11_8. The importance of indicating the base is apparent when you consider that 11_2 has a decimal value of 3, whereas 11_8 has a decimal value of 9!

Because all these number systems are used in the computer world in one way or another, it is necessary that we understand how they are related to each other.

Chapter Objectives

In this chapter you will learn how to:

1. Read and write numbers in the binary, octal, and hexadecimal number systems.
2. Convert between the decimal, binary, octal, and hexadecimal number systems.
3. Add numbers in the various number systems.
4. Subtract numbers in the various number systems using both the direct and complement methods.

25-1 | BINARY NUMBER SYSTEM

In Chapter 1, we talked about the decimal number system and how the value or *weight* of a digit depended on its position in the number. For example, in the number 7037, the 7 in the units position has a weight of 7, whereas the 7 in the thousands position has a weight of 7000. Furthermore, because our number system has ten digits (0 through 9), the next number after 9 causes a *carry* into the next position. In the decimal number system, we have no special symbol or character for numbers greater than 9, so we use combinations of these numbers in our *place value system* to indicate quantities greater than 9. Thus, the next number after 9 is 10. The one in the tens position has a weight of 10.

In powers of ten form in base 10, numbers in the units positions are raised to the zero power ($10^0 = 1$), numbers in the tens position are raised to the first power (10^1), numbers in the hundreds position are raised to the second power (10^2), and so on. Using the powers of ten form in the various positions, our number looks like this:

$$(7 \times 10^3) + (0 \times 10^2) + (3 \times 10^1) + (7 \times 10^0)$$

$$= 7000 + 0 + 30 + 7 = 7037$$

Those concepts, developed in Chapter 1 and reviewed here, apply to *all* number systems. The only difference is that other number systems have fewer (or more) characters.

In the binary number system, there are only two digits, 0 and 1. Digits in the units position can only have a value of 0 or 1 since that's all the digits there are. Numbers greater than 1 cause a carry into the next position. Just as in base 10, each position represents the base raised to a power. The units position has a power of 2^0, the next position is 2^1, and so on, as illustrated in Figure 25–1. In decimal, the weight or value of each position is a multiple of 10. In binary, the weight or value of each position is a multiple of 2.

Binary digits are called *bits* (a contraction of binary digits). Therefore, the digit in the units position is called the least significant *bit* (LSB), and so on, until the most significant bit (MSB) is reached.

FIGURE 25–1
Positional value
of the first five
positions for
the binary
number system.

MSB	4SB	3SB	2SB	LSB
2^4	2^3	2^2	2^1	2^0
16	8	4	2	1

25-1-1 Binary to Decimal Conversions

Converting from base 2 to base 10 is easy. If a 1 is present in a given position, the weight of that bit is added. If a 0 is present, the weight of that bit is not added. Consider the number 11010 in Example 25–1. The decimal equivalent is 26.

EXAMPLE 25–1

$11010_2 = (?)_{10}$

SOLUTION

$$(1 \times 2^4) + (1 \times 2^3) + (0 \times 2^2) + (1 \times 2^1) + (0 \times 2^0) =$$
$$16 \quad + \quad 8 \quad + \quad 0 \quad + \quad 2 \quad + \quad 0 \quad = 26$$

Because there are only 1s and 0s in binary, each bit position either equals the weight of that digit position or it equals 0. In Example 25–1, the values of the bit positions are added to get the decimal equivalent. In the 2^4 or 16's position, there is a 1. Therefore, that bit position converts to 16. In the 2^3 or 8's position, there is a 1. That bit position converts to 8. In the 2^2 or 4's position, there is a 0. That bit position converts to 0. In the 2's position, there is a 1. That bit position converts to 2. There is a 0 in the units position. If we add the numbers together, we get decimal 26.

> ☞ **RULE 25–1** In converting from binary to decimal, find the value or weight of the MSB. Work down to the LSB, adding the weight of that position if a 1 is present or a 0 if a 0 is present.

Table 25–1 on the next page is a conversion table for all the computer number systems.

EXAMPLE 25–2

Change the following binary numbers to decimal numbers:

    (a) 1100   (b) 10001   (c) 101011   (d) 111101

SOLUTIONS

    (a) $1100_2 = (1 \times 2^3) + (1 \times 2^2) + (0 \times 2^1) + (0 \times 2^0) =$
$$8 \quad + \quad 4 \quad + \quad 0 \quad + \quad 0 \quad = 12_{10}$$

(b) $10001_2 = (1 \times 2^4) + (0 \times 2^3) + (0 \times 2^2) + (0 \times 2^1) + (1 \times 2^0)$

$\qquad\qquad 16 \quad + \quad 0 \quad + \quad 0 \quad + \quad 0 \quad + \quad 1 \quad = 17_{10}$

(c) $101011_2 = (1 \times 2^5) + (0 \times 2^4) + (1 \times 2^3) + (0 \times 2^2) + (1 \times 2^1) +$

$\qquad (1 \times 2^0) = \quad 32 \quad + \quad 0 \quad + \quad 8 \quad + \quad 0 \quad + \quad 2 \quad + 1$

$$= 43_{10}$$

TABLE 25–1 Computer Numbering System.

Numbering system		Decimal		Binary						Octal		Hexa-decimal	
Base		10		2						8		16	
Position weight — Power		10^1	10^0	2^5	2^4	2^3	2^2	2^1	2^0	8^1	8^0	16^1	16^0
Position weight — Value		10	1	32	16	8	4	2	1	8	1	16	1
											(LSD)		(LSD)
		0	0	0	0	0	0	0	0	0	0	0	0
			1	0	0	0	0	0	1		1		1
			2	0	0	0	0	1	0		2		2
			3	0	0	0	0	1	1		3		3
			4	0	0	0	1	0	0		4		4
			5	0	0	0	1	0	1		5		5
			6	0	0	0	1	1	0		6		6
			7	0	0	0	1	1	1		7		7
			8	0	0	1	0	0	0	1	0		8
			9	0	0	1	0	0	1	1	1		9
		1	0	0	0	1	0	1	0	1	2		A
		1	1	0	0	1	0	1	1	1	3		B
		1	2	0	0	1	1	0	0	1	4		C
		1	3	0	0	1	1	0	1	1	5		D
		1	4	0	0	1	1	1	0	1	6		E
		1	5	0	0	1	1	1	1	1	7		F
		1	6	0	1	0	0	0	0	2	0	1	0
		1	7	0	1	0	0	0	1	2	1	1	1
		1	8	0	1	0	0	1	0	2	2	1	2
		1	9	0	1	0	0	1	1	2	3	1	3
		2	0	0	1	0	1	0	0	2	4	1	4
		2	1	0	1	0	1	0	1	2	5	1	5
		2	2	0	1	0	1	1	0	2	6	1	6
		2	3	0	1	0	1	1	1	2	7	1	7
		2	4	0	1	1	0	0	0	3	0	1	8
		2	5	0	1	1	0	0	1	3	1	1	9
		2	6	0	1	1	0	1	0	3	2	1	A
		2	7	0	1	1	0	1	1	3	3	1	B
		2	8	0	1	1	1	0	0	3	4	1	C
		2	9	0	1	1	1	0	1	3	5	1	D
		3	0	0	1	1	1	1	0	3	6	1	E
		3	1	0	1	1	1	1	1	3	7	1	F
		3	2	1	0	0	0	0	0	4	0	2	0

Example: $1_{(10)} = 1_{(2)} = 1_{(8)} = 1_{(16)}$

Example: $10_{(10)} = 1010_{(2)} = 12_{(8)} = 0A_{(16)}$

Example: $25_{(10)} = 11001_{(2)} = 31_{(8)} = 19_{(16)}$

(d) $111101_2 = (1 \times 2^5) + (1 \times 2^4) + (1 \times 2^3) + (1 \times 2^2) + (0 \times 2^1) +$

$(1 \times 2^0) = \quad 32 \quad + \quad 16 \quad + \quad 8 \quad + \quad 4 \quad + \quad 0 \quad + 1$

$= 61_{10}$

25-1-2 Decimal to Binary Conversions

Let's turn the process around and find the binary equivalent of a decimal number.

EXAMPLE 25-3

Convert 11_{10} to binary:

SOLUTION

Step 1. $\dfrac{11}{2} = 5 + 1$

Step 2. $\dfrac{5}{2} = 2 + 1$

Step 3. $\dfrac{2}{2} = 1 + 0$ $\quad$ LSB

Step 4. $\dfrac{1}{2} = 0 + 1 - \text{MSB} \rightarrow 1 \quad 0 \quad 1 \quad 1$

Since we are converting to base 2, we first divide the decimal number by 2 (step 1). The answer is 5 with a remainder of 1. This remainder is the LSB in the answer. The whole number left after this initial division (5) is again divided by 2 (step 2). The answer is 2 with a remainder of 1. This 1 is the bit in the 2^1 position. The whole number left (2) is divided by 2 (step 3). The result is 1 and a remainder of 0. This 0 is the bit in the 2^2 position. Step 4 is the final step because no further divisions are possible. The remainder in the final step is the MSB.

A second method of converting from decimal to binary is shown in the following example.

EXAMPLE 25-4

Convert 167 to binary.

SOLUTION The highest multiple of 2 contained in 167 is 128 (2^7). Therefore, we need a 128. So we start writing our binary number with a 1 in the MSB position, remembering that this "1" has a value of 128. Next we determine if we need a 64. If we do, we put a 1 in the next bit position. If we don't need a 64, we put a 0 in this position. Since $128 + 64$ is greater than 167, we must put a zero in this position. At this point we have 10, which is $128 + 0$.

FIGURE 25–2
Conversion of
167_{10} to binary.

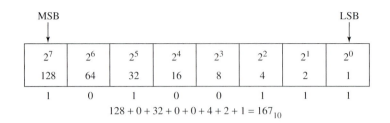

MSB							LSB
2^7	2^6	2^5	2^4	2^3	2^2	2^1	2^0
128	64	32	16	8	4	2	1
1	0	1	0	0	1	1	1

$$128 + 0 + 32 + 0 + 0 + 4 + 2 + 1 = 167_{10}$$

Binary No.	167	Original decimal number
1	-128	2^7 is the largest binary number that we can subtract from 167, so the MSB is in the 2^7 position.
	39	Remainder
0	-0	The 2^6 position must equal 0 (64 is greater than 39), so we must put a 0 in the 2^6 position.
1	-32	We can subtract 32 from 39, so there is a 1 in the 2^5 position.
	7	Remainder
0	-0	The 2^4 position must equal 0 (can't subtract 16 from 7).
0	-0	The 2^3 position must equal 0 (can't subtract 8 from 7).
1	-4	We can subtract 4 from 7, so there is a 1 in the 2^4 position.
	3	Remainder
1	-2	We can subtract 2 from 3, so there is a 1 in the 2^1 position.
1	1	Remainder: There is a 1 in the LSB position. (If there had been no remainder, the LSB would equal 0.)

This procedure is followed for each bit position. Can we use 32? Yes. We now have 101. Remember, we started at the MSB position and are working down to the LSB position. Can we use a 16? No; a 16 is too much. Put a zero in the 16 (2^4) position. We now have 1010. Can we use an 8? No; so put a zero in the 2^3 position. Now we have 10100. Can we use a 4? Yes. Our number is now 101001. Can we use a 2? Yes. We now have 1010011. Do we need a 1 or a 0 in the LSB position? We need a 1. Our binary number is 10100111. Any decimal number can be converted to binary in this manner. With practice, this method is faster than the method shown previously. The process is detailed in Figure 25–2.

EXAMPLE 25–5

Convert the following decimal numbers to binary numbers:

(a) 7   (b) 34   (c) 89   (d) 203

SOLUTIONS

(a) $\dfrac{7}{2} = 3 + 1$ ──────┐

$\dfrac{3}{2} = 1 + 1$ $\qquad\qquad\qquad\qquad$ $7_{10} = 111_2$

$\dfrac{1}{2} = 0 + 1 \rightarrow 1 \quad 1 \quad 1$

(b) $\dfrac{34}{2} = 17 + 0$ ──────────┐

$\dfrac{17}{2} = 8 + 1$

$\dfrac{8}{2} = 4 + 0$

$\dfrac{4}{2} = 2 + 0$ $\qquad\qquad\qquad$ $34_{10} = 100010_2$

$\dfrac{2}{2} = 1 + 0$

$\dfrac{1}{2} = 0 + 1 \rightarrow 1 \quad 0 \quad 0 \quad 0 \quad 1 \quad 0$

(c) $\dfrac{89}{2} = 44 + 1$ ──────────┐

$\dfrac{44}{2} = 22 + 0$

$\dfrac{22}{2} = 11 + 0$

$\dfrac{11}{2} = 5 + 1$ $\qquad\qquad\qquad$ $89_{10} = 1011001_2$

$\dfrac{5}{2} = 2 + 1$

$\dfrac{2}{2} = 1 + 0$

$\dfrac{1}{2} = 0 + 1 \rightarrow 1 \quad 0 \quad 1 \quad 1 \quad 0 \quad 0 \quad 1$

(d) $\dfrac{203}{2} = 101 + 1$ ─────────────┐

$\dfrac{101}{2} = 50 + 1$

$\dfrac{50}{2} = 25 + 0$

$\dfrac{25}{2} = 12 + 1$ $203_{10} = 11001011_2$

$\dfrac{12}{2} = 6 + 0$

$\dfrac{6}{2} = 3 + 0$

$\dfrac{3}{2} = 1 + 1$

$\dfrac{1}{2} = 0 + 1 \rightarrow 1$ 1   0   0   1   0   1   1

PRACTICE PROBLEMS 25–1

Convert the following binary numbers to decimal numbers:

1. 101_2 2. 111_2
3. 1001_2 4. 1110_2
5. 10110_2 6. 11001_2
7. 100101_2 8. 111011_2
9. 1010100_2 10. 1111001_2

Convert the following decimal numbers to binary numbers:

11. 4   12. 8
13. 10   14. 27
15. 39   16. 57
17. 78   18. 100
19. 150   20. 200

SOLUTIONS

1. 5   2. 7
3. 9   4. 14
5. 22   6. 25
7. 37   8. 59

9. 84
10. 121
11. 100_2
12. 1000_2
13. 1010_2
14. 11011_2
15. 100111_2
16. 111001_2
17. 1001110_2
18. 1100100_2
19. 10010110_2
20. 11001000_2

Additional practice problems are at the end of the chapter.

OCTAL NUMBER SYSTEM   25–2

Octal, or base 8, is an important computer number system. Since there are 8 digits in octal (0, 1, 2, 3, 4, 5, 6, 7), the digit of greatest value is 7, which is 1 less than the base. In counting, the sequence is 0 to 7. On the next count, we return to 0 in the units position (just as we do when we reach 9 in decimal) and a carry is generated. The resulting number is 10 and is read "one-zero." This number is equal to decimal 8 because the carry is equal to 8 (base 8). This process continues; the sequence goes from 0 to 7 and back to 0 again. Each time a count greater than 7 is reached, a carry is generated.

In the octal number system, digits in the units position have the indicated value. Digits in the next position have *place value*. That is, digits in this position indicate the number of 8s present. This, then, is the 8's position. The next position indicates the number of 64s present, and so on, as shown in Figure 25–3.

Again, the concept of *place value* must be understood so that we can work in this (or any) base. Consider the number 567_{10}. The five has a value of 500 because it is in the hundreds or 10^2 position ($5 \times 10^2 = 500$). The six has a value of 60 because it is in the tens position ($6 \times 10 = 60$). If the number were 567_8, the five would have a value of 320 in decimal because it is in the 8^2 or 64s position ($5 \times 8^2 = 320$). The six has a value of 48, in decimal, because it is in the eights (8^1) position, $6 \times 8 = 48$. In both bases the seven is in the units position and has a value of 7.

$$567_8 = (5 \times 8^2) + (6 \times 8) + 7 = 375_{10}$$

25–2–1 Octal to Decimal Conversions

Let's examine the number 1025 in octal. We might be tempted to read this as "one thousand twenty-five." Such language implies base 10. The number should be read "one-zero-two-five," which does not imply any base. We normally think in decimal,

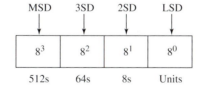

FIGURE 25–3
Positional value
of the first four
positions.

and decimal is the number system we use most; therefore, let's convert the number to base 10. There is a 1 in the 512's position + a 0 in the 64's position + a 2 in the 8's position + a 5 in the units position.

$$1025_8 = (1 \times 8^3) + (0 \times 8^2) + (2 \times 8^1) + (5 \times 8^0) = 533_{10}$$

$$1025_8 = 512 + 0 + 16 + 5 = 533_{10}$$

> ☞ **RULE 25-2** In converting from octal to decimal, find the weight of the digit in each position. Add the values of the digits in each position to determine the decimal equivalent.

EXAMPLE 25-6

Change the following octal numbers to decimal:

    (a) 73_8 (b) 432_8 (c) 600_8 (d) 1234_8

SOLUTIONS

(a) $73_8 = (7 \times 8^1) + (3 \times 8^0) = 56 + 3 = 59_{10}$
(b) $432_8 = (4 \times 8^2) + (3 \times 8^1) + (2 \times 8^0) = 256 + 24 + 2 = 282_{10}$
(c) $600_8 = (6 \times 8^2) + (0 \times 8^1) + (0 \times 8^0) = 384 + 0 + 0 = 384_{10}$
(d) $1234_8 = (1 \times 8^3) + (2 \times 8^2) + (3 \times 8^1) + (4 \times 8^0) = 512 + 128 + 24 + 4 = 668_{10}$

25-2-2 Decimal to Octal Conversions

Let's reverse the process. In Example 25–7, the number 189 is converted from decimal to octal.

EXAMPLE 25-7

$189_{10} = (?)_8$

SOLUTION

Step 1. $\dfrac{189}{8} = 23 + 5$

Step 2. $\dfrac{23}{8} = 2 + 7$

Step 3. $\dfrac{2}{8} = 0 + 2 -$ MSD $\rightarrow 2 \quad 7 \quad 5 \quad 189_{10} = 275_8$

Since we are converting to base 8, we first divide the decimal number by 8 (step 1). The answer is 23 with a remainder of 5. This remainder is the LSD (least significant digit) of our base 8 number. The whole number left after this initial division (23) is again divided by 8 to determine the value of the digit in the next (8^1) position. This division leaves a remainder of 7, which becomes the digit in the 8^1 position. The whole number resulting from this division (2) is again divided by 8, and the remainder (2) is the digit in the 8^2 position. This digit is the MSD (most significant digit) since no further division is possible.

EXAMPLE 25–8

Convert the following decimal numbers to octal numbers:

(a) 78   (b) 376   (c) 463

SOLUTIONS

(a) $\dfrac{78}{8} = 9 + 6$

$\dfrac{9}{8} = 1 + 1$

$\dfrac{1}{8} = 0 + 1 \rightarrow 1$ 1   6 $78_{10} = 116_8$

(b) $\dfrac{376}{8} = 47 + 0$

$\dfrac{47}{8} = 5 + 7$

$\dfrac{5}{8} = 0 + 5 \rightarrow 5$ 7   0 $376_{10} = 570_8$

(c) $\dfrac{1463}{8} = 182 + 7$

$\dfrac{182}{8} = 22 + 6$

$\dfrac{22}{8} = 2 + 6$

$\dfrac{2}{8} = 0 + 2 \rightarrow 2$ 6   6   7 $1463_{10} = 2667_8$

PRACTICE PROBLEMS 25–2

1. Convert the following numbers from base 8 to base 10:
   (a) 46_8 (b) 111_8 (c) 204_8 (d) 400_8 (e) 777_8
2. Convert the following numbers from base 10 to base 8:
   (a) 27_{10} (b) 89_{10} (c) 100_{10} (d) 256_{10} (e) 400_{10}

1. (a) 38_{10} (b) 73_{10} (c) 132_{10} (d) 256_{10} (e) 511_{10}
2. (a) 33_8 (b) 131_8 (c) 144_8 (d) 400_8 (e) 620_8

Additional practice problems are at the end of the chapter.

25-3 HEXADECIMAL NUMBER SYSTEM

Another important computer number system is the hexadecimal system (base 16). How many digits exist in hexadecimal? Sixteen, because the base is 16. What is the value of the largest digit in hexadecimal? Fifteen, one less than the base. However, 15 is a decimal number and requires *two digits* (15). In the hexadecimal number system, 16 *different* symbols must be used. Zero through 9 are used for the first ten, and A through F are used for the remaining six, as shown in Table 25–2. (Also refer to Table 25–1 on page 678 for the complete conversion table.)

Table 25–2   Hexadecimal to Decimal Conversions.

DECIMAL	HEXADECIMAL	DECIMAL	HEXADECIMAL
0	0	9	9
1	1	10	A
2	2	11	B
3	3	12	C
4	4	13	D
5	5	14	E
6	6	15	F
7	7	16	10
8	8		

Let's look at the number 567 again. Recall from the previous section that, in the number 567_{10}, the 5 is in the 10^2 position and so has a value of 500 ($5 \times 10^2 = 500$). Six is in the 10s position and so has a value of 60 ($6 \times 10 = 60$). In the number 567_{16}, the 5 is in the 16^2 position, so it has a value of 1280 in decimal ($5 \times 16^2 = 1280$). The 6 is in the 16 position and so has a value of 96 ($6 \times 16 = 96$). Again, the 7 is in the units position and so has a value of 7.

$$567_{16} = (5 \times 16^2) + (6 \times 16) + 7 = 1383$$

FIGURE 25–4
Positional value of a four-digit hexadecimal number.

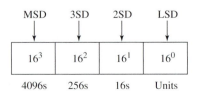

25-3-1 Hexadecimal to Decimal Conversions

The hexadecimal number 3AC2 may be converted to base 10 as follows (see Figure 25–4 on the previous page):

$$
\begin{array}{lr}
\text{Three 4096s, which equal} & 12{,}288 \\
+ \text{ ten 256s, which equal} & 2{,}560 \\
+ \text{ twelve 16s, which equal} & 192 \\
+ \text{ two 1's, which equal} & 2 \\
\hline
\text{SUM} & 15{,}042_{10}
\end{array}
$$

or

$$(3 \times 16^3) + (10 \times 16^2) + (12 \times 16^1) + (2 \times 16^0) = 15{,}042_{10}$$

☞ **RULE 25-3** In converting from hexadecimal to decimal, find the weight of the digit in each position. Add the values of the digits in each position to determine the decimal equivalent.

EXAMPLE 25–9

Convert the following hexadecimal numbers to decimal numbers:

(a) 72   (b) C29   (c) 12AB   (d) 1A2A.

SOLUTIONS

(a) $72_{16} = (7 \times 16^1) + (2 \times 16^0) = 112 + 2 = 114_{10}$
(b) $C29_{16} = (12 \times 16^2) + (2 \times 16^1) + (9 \times 16^0) = 3113_{10}$
(c) $12AB_{16} = (1 \times 16^3) + (2 \times 16^2) + (10 \times 16^1) + (11 \times 16^0) = 4779_{10}$
(d) $1A2A_{16} = (1 \times 16^3) + (10 \times 16^2) + (2 \times 16^1) + (10 \times 16^0) = 6698_{10}$

25-3-2 Decimal to Hexadecimal Conversions

The conversion of decimal numbers to hexadecimal is identical to conversion using other bases. In Example 25–10, the number 1324 is converted from decimal to hexadecimal. Since the hexadecimal number system has a base of 16, division is by 16. In working with remainders greater than 9, we substitute the characters A, B, C, D, E, and F as required.

EXAMPLE 25–10

$1324_{10} = (?)_{16}$

SOLUTION

Step 1. $\dfrac{1324}{16} = 82 + 12$ ⎯⎯⎯⎯⎯⎯⎯⎯⎯⎯⎯⎯⎯⎯⎯⎯⎮

Step 2. $\dfrac{82}{16} = 5 + 2$ ⎯⎯⎯⎯⎯⎯⎯⎯⎮ LSD

Step 3. $\dfrac{5}{16} = 0 + 5 - \text{MSD} \rightarrow 5 \quad 2 \quad \text{C} \qquad 1324_{10} = 52\text{C}_{16}$

EXAMPLE 25–11

Convert the following decimal numbers to hexadecimal numbers:

   (a) 672   (b) 1763   (c) 12,760

SOLUTIONS

(a) $\dfrac{672}{16} = 42 + 0$ ⎯⎯⎯⎯⎯⎯⎯⎮

$\dfrac{42}{16} = 2 + 10$

$\dfrac{2}{16} = 0 + 2 \rightarrow 2 \quad \text{A} \quad 0 \quad 672_{10} = 2\text{A}0_{16}$

(b) $\dfrac{1763}{16} = 110 + 3$ ⎯⎯⎯⎯⎯⎮

$\dfrac{110}{16} = 6 + 14$

$\dfrac{6}{16} = 0 + 6 \rightarrow 6 \quad \text{E} \quad 3 \quad 1763_{10} = 6\text{E}3_{16}$

(c) $\dfrac{12760}{16} = 797 + 8$ ⎯⎯⎯⎯⎯⎮

$\dfrac{797}{16} = 49 + 13$

$\dfrac{49}{16} = 3 + 1$

$\dfrac{3}{16} = 0 + 3 \rightarrow 3 \quad 1 \quad \text{D} \quad 8 \quad 12760_{10} = 31\text{D}8_{16}$

PRACTICE PROBLEMS 25–3

1. Convert the following hexadecimal numbers to decimal numbers:
   (a) D_{16} (b) $3F_{16}$ (c) $A4_{16}$ (d) $1CD_{16}$ (e) $10BE_{16}$
2. Convert the following decimal numbers to hexadecimal numbers:
   (a) 27_{10} (b) 85_{10} (c) 100_{10} (d) 256_{10} (e) 1500_{10}

SOLUTIONS

1. (a) 13_{10} (b) 63_{10} (c) 164_{10} 2. (a) $1B_{16}$ (b) 55_{16} (c) 64_{16}
   (d) 461_{10} (e) 4286_{10} (d) 100_{16} (e) $5DC_{16}$

Additional practice problems are at the end of the chapter.

BINARY TO OCTAL TO HEXADECIMAL CONVERSIONS   25-4

25-4-1 Binary to Octal Conversions

Many 1s and 0s are required when we use binary numbers to represent large quantities. If we convert from binary to octal, the number of digits required is reduced by a factor of 3 because one octal digit equals 3 bits. That is, all octal digits, 0 through 7, can be represented by 3 bits.

$$0 = 000 \qquad 4 = 100$$
$$1 = 001 \qquad 5 = 101$$
$$2 = 010 \qquad 6 = 110$$
$$3 = 011 \qquad 7 = 111$$

☞ **KEY POINT** To convert from binary to octal, separate the bits into 3-bit groups, starting with the LSB and moving left to the MSB.

EXAMPLE 25–12

Convert 110111100001_2 to octal:

SOLUTION

110	111	100	001	(binary number divided into groups of three)
6	7	4	1	(octal substitution for each 3-bit group)

EXAMPLE 25–13

Convert 11101001_2 to octal:

SOLUTION   011   101   001
                3   5   1 $11101001_2 = 351_8$

Notice that a zero was added to the leftmost group of bits to remind us that 3 bits represent one octal digit.

The procedure may be reversed to convert from octal to binary.

EXAMPLE 25–14

Convert 417_8 to binary:

SOLUTION   4   1   7 $417_8 = 100001111_2$
                100   001   111

EXAMPLE 25–15

Convert 362_8 to binary:

SOLUTION   3   6   2 $362_8 = 11110010_2$
                011   110   010

In each example, 3 bits are substituted for each octal digit. A zero in the left most bit position has no meaning and need not be written in the answer.

25-4-2 Binary to Hexadecimal Conversions

If we convert from binary to hexadecimal, the number of digits required is reduced by a factor of 4 because one hexadecimal digit equals 4 bits. That is, all hexadecimal digits (0 through F) can be represented by 4 bits.

$$0 = 0000 \qquad 8 = 1000$$
$$1 = 0001 \qquad 9 = 1001$$
$$2 = 0010 \qquad A = 1010$$
$$3 = 0011 \qquad B = 1011$$
$$4 = 0100 \qquad C = 1100$$
$$5 = 0101 \qquad D = 1101$$
$$6 = 0110 \qquad E = 1110$$
$$7 = 0111 \qquad F = 1111$$

☞ **KEY POINT** To convert from binary to hexadecimal, separate the bits into 4-bit groups, starting with the LSB and moving left to the MSB.

EXAMPLE 25–16

Convert 11011001101_2 to hexadecimal:

SOLUTION 1101 1001 1011 (binary number divided into 4-bit groups)
D   9   B   (hexadecimal substitution for each 4-bit group)

PRACTICE PROBLEMS 25–4

Convert the following binary numbers to (a) octal numbers and then (b) hexadecimal numbers:

1. 1101001101_2
2. 1000011001_2
3. 1100110001_2
4. 111001110101_2
5. 111101011000110_2

Convert the following numbers to binary numbers:

6. 73_8
7. 127_8
8. 617_8
9. 506_8
10. 1207_8
11. $A6_{16}$
12. $4C_{16}$
13. $C0D_{16}$
14. $BA1_{16}$
15. $F23_{16}$

SOLUTIONS

1. (a) 1515_8 (b) $34D_{16}$
2. (a) 1031_8 (b) 219_{16}
3. (a) 1461_8 (b) 331_{16}
4. (a) 7165_8 (b) $E75_{16}$
5. (a) 75306_8 (b) $7AC6_{16}$
6. 111011_2
7. 1010111_2
8. 110001111_2
9. 101000110_2
10. 1010000111_2
11. 10100110_2
12. 1001100_2
13. 110000001101_2
14. 101110100001_2
15. 111100100011_2

Additional practice problems are at the end of the chapter.

25-5-1 The Calculator

Conversion between number systems is possible with many calculators. However, use the calculator only after you have a thorough understanding of the concepts.

25-5-2 Converting between the Various Number Systems

The simplest octal to hexadecimal conversion is to first convert the number to binary and then make a conversion from binary to the other number system. We have converted from decimal to the other systems, and we have converted between binary, octal, and hexadecimal. Now let's practice some conversions among all the number systems we have studied. For example, let's convert 200_{10} to binary, octal, and hexadecimal.

EXAMPLE 25-17

$$200_{10} = \underline{\hspace{2cm}}_2 = \underline{\hspace{2cm}}_8 = \underline{\hspace{2cm}}_{16}$$

SOLUTION   There are many ways to solve this problem. We either could convert from 200_{10} to each of the other systems, or we could convert to base 2 and then convert to base 8 and base 16 from base 2. Another way, and probably the one requiring fewest steps, is to convert from base 10 to base 16, then from base 16 to base 2, and then from base 2 to base 8. Remember, fewer steps mean fewer chances for error.

$$\frac{200}{16} = 12 + 8 \longrightarrow$$

$$\frac{12}{16} = 0 + 12 \longrightarrow C \quad 8$$

$$200_{10} = C8_{16}$$

Convert to binary:

$$\begin{array}{cc} C & 8 \\ 1100 & 1000 \end{array} \quad C8_{16} = 11001000_2$$

Convert to octal:

$$\begin{array}{ccc} 011 & 001 & 000 \\ 3 & 1 & 0 \end{array} \quad 11001000_2 = 310_8$$

$$200_{10} = 11001000_2 = 310_8 = C8_{16}$$

EXAMPLE 25–18

$C27_{16} =$ _____ $_2 =$ _____ $_8 =$ _____ $_{10}$

SOLUTION In this example, we have a straight conversion to binary and then to octal. Convert to binary:

$$\begin{array}{cccc} C & 2 & 7 \\ 1100 & 0010 & 0111 \end{array} \quad C27_{16} = 110000100111_2$$

Convert to octal:

$$\begin{array}{cccc} 110 & 000 & 100 & 111 \\ 6 & 0 & 4 & 7 \end{array} \quad 110000100111_2 = 6047_8$$

The conversion to decimal can be made from any of the other systems, but conversion from base 16 requires fewer steps.

$$C27_{16} = (12 \times 16^2) + (2 \times 16) + (7 \times 16^0)$$
$$= 3072 + 32 + 7$$
$$= 3111_{10}$$
$$C27_{16} = 110000100111_2 = 6047_8 = 3111_{10}$$

You will find that some conversions in the following problems will exceed the limits of the calculator. Consult the user's manual and become familiar with the maximum numbers that can be stored and displayed in systems other than decimal.

PRACTICE PROBLEMS 25–5

1. $73_8 =$ _____ $_{16} =$ _____ $_2 =$ _____ $_{10}$
2. $273_8 =$ _____ $_{16} =$ _____ $_2 =$ _____ $_{10}$
3. $A3_{16} =$ _____ $_8 =$ _____ $_2 =$ _____ $_{10}$
4. $12E_{16} =$ _____ $_8 =$ _____ $_2 =$ _____ $_{10}$
5. $1101001_2 =$ _____ $_8 =$ _____ $_{16} =$ _____ $_{10}$
6. $10111100_2 =$ _____ $_8 =$ _____ $_{16} =$ _____ $_{10}$

SOLUTIONS

1. $73_8 = 3B_{16} = 111011_2 = 59_{10}$
2. $273_8 = BB_{16} = 10111011_2 = 187_{10}$
3. $A3_{16} = 243_8 = 10100011_2 = 163_{10}$
4. $12E_{16} = 456_8 = 100101110_2 = 302_{10}$
5. $1101001_2 = 151_8 = 69_{16} = 105_{10}$
6. $10111100_2 = 274_8 = BC_{16} = 188_{10}$

Additional practice problems are at the end of the chapter.

Convert the following numbers to decimal numbers:

1. 46_8
2. 276_8
3. $F6_{16}$
4. $C3A_{16}$
5. 1011011_2
6. 11001010_2

Convert the following decimal numbers to (a) binary, (b) octal, and (c) hexadecimal:

7. 10
8. 28
9. 187
10. 625

Convert the following to the bases indicated.

11. $93_{10} = $ _____ $_2 = $ _____ $_8 = $ _____ $_{16}$
12. $E1A_{16} = $ _____ $_2 = $ _____ $_8 = $ _____ $_{10}$
13. $11010010110_2 = $ _____ $_8 = $ _____ $_{10} = $ _____ $_{16}$
14. $573_8 = $ _____ $_2 = $ _____ $_{10} = $ _____ $_{16}$

Answers to Self-test 25–1 are at the end of the chapter.

25-6   DECIMAL AND OCTAL ADDITION

25-6-1   Adding Decimal Numbers

Add the numbers 23 and 45. When numbers are added, they are added in columns; one number is placed below the other. Either number may be placed first. In Example 25–19, the numbers 23 and 45 are placed in the position for addition. The digits are separated to emphasize positional differences. Eight is in the units position (3 + 5), and 6 is in the tens position (2 + 4). The answer is 68 (six 10s and eight 1s).

EXAMPLE 25–19

$$23 + 45 = ?$$

$$
\begin{array}{cc}
2 & 3 \\
4 & 5 \\
\hline
6 & 8 \\
\end{array}
$$

EXAMPLE 25–20

Add the numbers 64 and 87 in base 10:

SOLUTION

```
Carry   1    1
             6    4
             8    7
        ─────────────
        1  (15) (11)
            10    10
        ─────────────
        1    5    1     (remainders taken as decimal sum)
```
(subtract value of carry)

In Example 25–20, there is an 11 in the units position. This number is greater than 9, which tells us there must be a carry (10). The digit remaining in the units position is the difference of 10 and 11, or 1. The tens position now contains $6 + 8 +$ a carry of 1, or 15. Again, since the number is greater than 9, a carry is indicated. Subtracting 10 from 15, we get 5 with a carry of 1. The third position contains only the carry. Since we have been adding for years in decimal, our experience allows us to perform these operations automatically. We have taken time to discuss the arithmetic steps because this is the addition process in all bases.

EXAMPLE 25–21

Add the following decimal numbers using the procedure given in Example 25–20:

(a) $738 + 417$ (b) $9706 + 463$ (c) $1774 + 7268$

SOLUTIONS

```
(a)  1          1
          7     3        8
          4     1        7
        ─────────────────────
        1 (11)   5      (15)
           10            10           738 + 417 = 1155
        ─────────────────────
        1   1    5        5
```

```
(b)  1    1
          9     7   0    6
          4     6   3
        ─────────────────────
        1 (10) (11)  6   9          9706 + 463 = 10,169
           10   10
        ─────────────────────
        1   0    1   6   9
```

```
(c)  1    1    1
     1    7    7        4
     7    2    6        8
        ─────────────────────
     9  (10) (14)     (12)
         10   10       10           1774 + 7268 = 9042
        ─────────────────────
     9   0    4         2
```

25-6-2 Adding Octal Numbers

EXAMPLE 25–22

Add the octal numbers 736 and 215.

SOLUTION

```
(carry)                     1           1
                            7    3      6
                            2    1      5
(decimal sum)          1   (9)   5    (11)
(subtract value of carry)   8           8         736₈ + 215₈ = 1153₈
                       1    1    5      3
```

$$736_8 + 215_8 = 1153_8$$

Again, the LSD or units position is added first. The sum of 6 and 5 is 11. The number is greater than 7 (7 is the largest digit in base 8), which tells us there must be a carry. Subtracting 8 (the value of the carry) leaves 3 with a carry of 1. The 8^1 position now contains $3 + 1 + $ a carry of 1, or 5. Since 5 is less than 7, no carry exists. In the 8^2 position, $7 + 2$ is 9. Since the number is greater than 7, a carry is generated. Subtracting the carry (8) yields a remainder of 1 with a carry of 1. Since the 8^3 position contains only the carry, the answer is 1153_8.

EXAMPLE 25–23

Add the following octal numbers:

(a) 243
    172

(b) 764
    414

(c) 604
    777

SOLUTIONS

```
(a)  1                              (b)  1     1     1
     2    4    3                          7     6     4
     1    7    2                          4     1     4
     4  (11)   5                      1  (12)  (8)   (8)
          8                               8     8     8
     4    3    5  = 435₈              1    4     0     0  = 1400₈
```

```
(c)  1    1    1
     6    0    4
     7    7    7
 1  (14)  (8) (11)
     8    8    8
 1    6    0    3  = 1603₈
```

Add the following octal numbers.

1. 204
   316
2. 447
   173
3. 605
   726
4. 517
   567
5. 176
   761

SOLUTIONS

1. 522_8
2. 642_8
3. 1533_8
4. 1306_8
5. 1157_8

Additional practice problems are at the end of the chapter.

ADDING HEXADECIMAL NUMBERS 25–7

EXAMPLE 25–24

Add the hexadecimal numbers 7AF and 579.

SOLUTION

	(carry)	1	1	
		7	A	F
		5	7	9
(decimal sum)		(13)	(18)	(24)
(subtract value of carry)			16	16
		D	2	8

$= D28_{16}$

Adding the units position yields 24. Twenty-four is greater than the largest digit (F), which is equal to decimal 15. This indicates that a carry should exist. Subtracting the value of a carry (16) leaves a remainder of 8 with a carry of 1. In the 16^1 position, the sum of A (decimal 10) + 7 + a carry of 1 is 18. Again, a carry is generated, leaving a remainder of 2 with a carry of 1. In the 16^2 position, the carry +7 + 5 equals decimal 13. In hexadecimal, 13 is D; therefore, the answer is $D28_{16}$.

EXAMPLE 25–25

Add the following hexadecimal numbers:

(a) 4D3
    818
(b) 789
    C47
(c) F347
    E006

(a)
```
      4     D     3
      8     1     8
   _____
    (12)  (14)  (11)
      C     E     B  = CEB₁₆
```

(b)
```
   1 ←              1 ←
                7     8  │   9
                C     4  │   7
     _____
      1  │ (19)  (13) │ (16)
         └─ 16        └─ 16
     _____
      1     3     D     0  = 13D0₁₆
```

(c)
```
   1 ←
      │   F    3    4    7
      │   E    0    0    6
   _____
    1 │ (29)  3    4  (13)
      └─ 16
   _____
    1     D    3    4    D  = 1D34D₁₆
```

PRACTICE PROBLEMS 25–7

Add the following hexadecimal numbers:

1. C4
   1B

2. A4
   B7

3. A28
   6F9

4. 9B5
   4C8

5. FACE
   1701

1. DF_{16}

2. $15B_{16}$

3. 1121_{16}

4. $E7D_{16}$

5. $111CF_{16}$

Additional practice problems are found at the end of the chapter.

25–8   ADDING BINARY NUMBERS

EXAMPLE 25–26

Add the binary numbers 1101 and 1001.

SOLUTION

```
1 ←        1 ←      (carry)
    1  1  0    1
    1  0  0    1
(decimal sum)      (2) 1  1   (2)
(subtract value of carry)  2          2
    1   0  1  1    0  = 10110₂
```

$$1 \leftarrow \qquad 1 \leftarrow \quad \text{(carry)}$$

The method of addition is the same as that used with other bases. Notice that $1 + 1$ equals decimal 2. Since 2 is invalid in base 2, a carry is generated into the next position. We see from the example that a carry is generated whenever a sum greater than 1 results from the addition of numbers in a column.

EXAMPLE 25–27

Add the following binary numbers.

(a) 10
    11

(b) 111
    101

(c) 1001
    1010

(d) 11100
    01010

SOLUTIONS

(a)
```
1 ←
    1  0
    1  1
1  (2) 1
    2
1   0  1 = 101₂
```

(b)
```
1 ←  1 ←  1 ←
    1    1    1
    1    0    1
1  (3)  (2)  (2)
    2    2    2
1   1    0    0 = 1100₂
```

(c)
```
1 ←
    1  0  0  1
    1  0  1  0
1  (2) 0  1  1
    2
1   0  0  1  1 = 10011₂
```

(d)
```
1 ←  1 ←
    1    1  1  0  0
    0    1  0  1  0
1  (2)  (2) 1  1  0
    2    2
1   0    0  1  1  0 = 100110₂
```

From working with these and other examples of binary addition, we see that:

1. $0 + 0$ always equals 0.
2. $0 + 1$ always equals 1.
3. $1 + 1$ always equals 0 and a carry.
4. $1 + 1 +$ a carry always equals 1 and a carry.

PRACTICE PROBLEMS 25–8

Add the following binary numbers:

1. 1011
   1100

2. 1001
   1111

3. 10111
   10001

4. 11001
   11110

5. 101110
   111100

SOLUTIONS

1. 10111_2
2. 11000_2
3. 101000_2
4. 110111_2
5. 1101010_2

Additional practice problems are at the end of the chapter.

SELF-TEST 25–2

Add the following numbers:

1. AF_{16}
   36_{16}

2. $C4A_{16}$
   $17E_{16}$

3. $F4A9_{16}$
   $FE7A_{16}$

4. 73_8
   66_8

5. 63_8
   41_8

6. 73_8
   46_8

7. 101_2
   111_2

8. 1101_2
   1011_2

9. 10110_2
   11110_2

Answers to Self-test 25–2 are at the end of the chapter.

25-9-1 Subtracting Decimal Numbers

Subtraction in base 10 is so automatic for us that we may do it without thinking of the rules involved. Since these rules in base 10 are identical to those for other bases, and since we are more familiar with base 10 than with other bases, let's review the process of subtraction in base 10.

EXAMPLE 25-28

Subtract 25 from 43:

SOLUTION

$$
\begin{array}{rcl}
 & 3 & \text{borrow 1} \\
\text{(minuend)} & 4 \to (1) & 3 \\
\text{(subtrahend)} & 2 & 5 \\
\hline
\text{(difference)} & 1 & 8
\end{array}
$$

As you know, we start with the least significant position and subtract. Since we cannot subtract 5 from 3, we *borrow* from the next significant position of the minuend. We may now interpret the 3 as 13 and find the difference, which is 8. Moving to the next position, we must reduce the minuend by 1 because of the borrow. We now subtract the number in the subtrahend from the new minuend and the difference is 1. Remember, the borrow carries with it a weight equal to that of the base.

EXAMPLE 25-29

Subtract the following decimal numbers. Show the process of borrowing.

(a) 624   (b) 854   (c) 8534   (d) 705
   −276   −235   −6748   −528

SOLUTIONS

(a)
```
   5  (1)1
   6    2  (1)4
  −2    7    6
  ─────────────
   3    4    8
```

(b)
```
         4
      8  5  (1)4
     −2  3    5
     ───────────
      6  1    9
```

(c)
```
   7  (1)4  (1)2
   8    5    3  (1)4
  −6    7    4    8
  ───────────────────
   1    7    8    6
```

(d)
```
   6    9
   7  (1)0  (1)5
  −5    2    8
  ─────────────
   1    7    7
```

Notice in Example 25-29(d) that, since 8 is greater than 5, a borrow is indicated. However, the next significant digit is 0 and no borrow can take place (we can't borrow something from nothing). Therefore, we must borrow from the next significant digit

(7), giving the second position, a value of 10. Now we can borrow from that position to yield 15 in the least significant position.

25-9-2 Subtracting Octal Numbers

In Example 25–30, the number 267_8 is subtracted from 512_8. Notice that a borrow is necessary in the first position since 7 is greater than 2. The borrow creates a new number in the minuend, 12. (This number is read "one-two," *not* "twelve," because twelve is a decimal number. One-two in base 8 means $8 + 2$ in base 10.)

EXAMPLE 25–30

Subtract 267_8 from 512_8:

SOLUTION

$$
\begin{array}{r}
\overset{4}{\cancel{5}} \quad \overset{\nearrow 10}{\cancel{1}} \to (1)2 \\
2 \quad 6 \quad 7 \\
\hline
2 \quad 2 \quad 3 = 223_8
\end{array}
$$

(minuend) (subtrahend) (difference)

One-two minus 7 equals 3. If you think in base 10, then $(8 + 2) - 7$ equals 3. In the next position, 1 was borrowed from the number in the minuend, making its value 0. A borrow is again required; it makes the value of the minuend 10 (or $8 + 0$ in base 10). Subtracting 6 from 10 yields a difference of 2. In the next position, 1 was borrowed from the minuend, making its value 4. The difference in this position is 2.

EXAMPLE 25–31

Solve the following subtraction problems in base 8:

(a)   43
    (−)16

(b)   161
    (−)72

(c)   2172
    (−)1717

SOLUTIONS

(a)
$$
\begin{array}{r}
3 \\
\cancel{4} \quad (1)3 \\
(-)1 \quad 6 \\
\hline
2 \quad 5 = 25_8
\end{array}
$$

(b)
$$
\begin{array}{r}
0 \quad (1)5 \\
\cancel{1} \quad \cancel{6} \quad (1)1 \\
(-) \quad 7 \quad 2 \\
\hline
6 \quad 7 = 67_8
\end{array}
$$

(c)
$$
\begin{array}{r}
1 \quad\quad 6 \\
\cancel{2} \quad (1)1 \quad \cancel{7} \quad (1)2 \\
(-)1 \quad 7 \quad 1 \quad 7 \\
\hline
2 \quad 5 \quad 3 = 253_8
\end{array}
$$

Subtract the following octal numbers:

1.   73
     $(-)$56

2.   376
     $(-)$277

3.   673
     $(-)$176

4.   2133
     $(-)$1574

5.   6014
     $(-)$3247

SOLUTIONS

1. 15_8

2. 77_8

3. 475_8

4. 337_8

5. 2545_8

Additional practice problems are at the end of the chapter.

SUBTRACTING HEXADECIMAL NUMBERS | 25–10

Applying the same rules as in base 8, let's subtract the hexadecimal number 85E from the number C37.

EXAMPLE 25–32

$C37 - 85E = (?)_{16}$

SOLUTION

$$\begin{array}{cccc}
 & B & (1)2 & \\
\text{(minuend)} & \cancel{C} & \cancel{3} \longrightarrow & (1)7 \\
\text{(subtrahend)} & 8 & 5 & E \\
\hline
\text{(difference)} & 3 & D & 9 = 3D9_{16}
\end{array}$$

Notice that a borrow is necessary in the first position since E is greater than 7. One-seven (*not* seventeen) minus E equals 9 or, in base 10, $(16 + 7) - 14$ equals 9. (Remember, the borrow has a weight of 16.) In the next position, a borrow is required, so 5 is subtracted from 12 (one-two), which equals D. (In base 10, it would be $16 + 2 - 5 = 13$.) Finally, in the next position, subtracting 8 from B (remember that we borrowed 1, making C a B) yields 3 or, in base 10, $11 - 8 = 3$.

EXAMPLE 25–33

Subtract the following hexadecimal numbers:

(a)   A29
     $-$7BF

(b)   3A12
     $-$29A0

(c)   F1B2
     $-$ABC7

(a) 9 ↗ (1)1
     A̸   2̸ → (1)9
     7   B   F
     2   6   A = $26A_{16}$

(b)   9
     3   A̸ → (1)1   2
     2   9   A   0
     1   0   7   2 = 1072_{16}

(c) E ↗ (1)0 ↗(1)A
     F̸   1̸   B̸ → (1)2
     A   B   C   7
     4   5   E   B = $45EB_{16}$

PRACTICE PROBLEMS 25–10

Subtract the following hexadecimal numbers:

1.   A17
    (−)26B

2.   B27
    (−)6A4

3.   7A29
    (−)2BC9

4.   4A9B
    (−)100F

5.   FADE
    (−)2C3F

1. $7AC_{16}$
3. $4E60_{16}$
5. $CE9F_{16}$

2. 483_{16}
4. $3A8C_{16}$

Additional practice problems are at the end of the chapter.

25–11   SUBTRACTING BINARY NUMBERS

The process of subtraction in base 2 is similar to that in the other bases, as was discussed earlier in this chapter.

EXAMPLE 25–34

Subtract 0101 from 1110:

SOLUTION

$$
\begin{array}{llcccl}
\text{borrow} & & & 0 & \\
\text{(minuend)} & 1 & 1 & \not{1} \to (1)0 \\
\text{(subtrahend)} & 0 & 1 & 0 & & 1 \\
\hline
\text{(difference)} & 1 & 0 & 0 & & 1 = 1001_2
\end{array}
$$

EXAMPLE 25–35

Subtract 01101 from 10011:

SOLUTION

$$
\begin{array}{lcccccl}
\text{borrow} & 0 & & 1 & \\
\text{(minuend)} & 1 \to (\not{1})\not{0} \to (1)0 & & 1 & 1 \\
\text{(subtrahend)} & 0 & 1 & 1 & 0 & 1 \\
\hline
& 0 & 0 & 1 & 1 & 0 = 00110_2
\end{array}
$$

In Example 25–34, a borrow is necessary in the least significant position since the minuend is 0. The borrow has a weight of 2 because we are working in base 2. Note that in Example 25–35 a borrow is required in the 2^2 position. However, since the 2^3 position contains a 0, the borrow must come from the 2^4 position to the 2^3 position. A borrow is now possible from the 2^3 position, and the minuend in the 2^2 position is 10 (one-zero), resulting in a difference of 1. The new minuend in the next position (2^3) is now 1, and the difference is 0. Since our original borrow came from the next position (2^4), the minuend is 0 and the difference is 0.

PRACTICE PROBLEMS 25–11

Subtract the following binary numbers:

1. 1010
    (−)0101

2. 10110
    (−)01101

3. 110010
    (−)100111

4. 100111
    (−)001101

5. 1100110
    (−)1011001

SOLUTIONS

1. 101_2
3. 1011_2
5. 1101_2

2. 1001_2
4. 11010_2

Additional practice problems are at the end of the chapter.

COMPLEMENT METHOD OF SUBTRACTION

In the modern computer, subtraction is usually performed by addition of the complement of the subtrahend. This is done to simplify the design of the arithmetic section of the computer. To help us understand the complement method of subtraction in binary, we will first use the complement method of subtraction in decimal.

To subtract numbers in decimal, we must first find the nine's complement of the subtrahend. The nine's complement of a digit is the difference between 9 and that digit and can be found by a simple inspection procedure. Let's find the nine's complement of 235:

$$
\begin{array}{r}
999 \\
-235 \\
\hline
764 \quad \text{(nine's complement of 235)}
\end{array}
$$

The nine's complement may be converted to the ten's complement by adding 1. Find the ten's complement of 5789:

$$
\begin{array}{r}
9999 \\
-5789 \\
\hline
4210 \quad \text{(nine's complement)} \\
+1 \\
\hline
4211 \quad \text{(ten's complement)}
\end{array}
$$

EXAMPLE 25–36

Subtract 676 from 835 using the complement method.

SOLUTION   Find the ten's complement of the subtrahend; add this number to the minuend and drop the carry.

$$
\begin{array}{r}
999 \\
-676 \\
\hline
323 \quad \text{(nine's complement)} \\
+1 \\
\hline
324 \quad \text{(ten's complement)}
\end{array}
\qquad
\begin{array}{r}
835 \quad \text{(minuend)} \\
324 \quad \text{(ten's complement of 676)} \\
\hline
\cancel{1}159 \quad \text{(difference = 159)}
\end{array}
$$

EXAMPLE 25–37

Subtract 4678 from 6392 using the complement method.

SOLUTION

$$
\begin{array}{r}
9999 \\
-4678 \\
\hline
5321 \\
+1 \\
\hline
5322
\end{array}
\qquad
\begin{array}{r}
6392 \quad \text{(minuend)} \\
+\ 5322 \quad \text{(ten's complement of 4678)} \\
\hline
\cancel{1}1714 \quad \text{(difference = 1714)}
\end{array}
$$

25-12-1 One's Complement

The one's complement of a binary number is similar to the nine's complement of a decimal number. The binary number to be complemented is subtracted from a binary number made up of all 1's. In the following example the one's complement of 10110010 is found.

EXAMPLE 25–38

$$\begin{array}{r} 1\ 1\ 1\ 1\ 1\ 1\ 1\ 1 \\ -\ 1\ 0\ 1\ 1\ 0\ 0\ 1\ 0 \\ \hline 0\ 1\ 0\ 0\ 1\ 1\ 0\ 1 \end{array}$$ (one's complement of 10110010)

Careful inspection reveals that the one's complement of a binary number may be produced by changing all the ones in the number to zeros and changing all of the zeros to ones.

1 0 1 1 0 0 1 0 (invert this number to get the one's complement)
0 1 0 0 1 1 0 1

The computer is able to perform this operation easily by means of a circuit called an *inverter*. Notice that carries or borrows are not involved in this operation.

To subtract by using the one's complement method, the one's complement of the subtrahend is added to the minuend. The carry out of the highest bit position is then added to the *lowest bit position* (LBP). This is called *end-around carry*.

EXAMPLE 25–39

Find the difference between 10110011 and 01101101:

SOLUTION

		1	
(minuend)	10110011	10110011	
(subtrahend)	01101101	10010010	(one's complement)
		101000101	
		→1	(end-around carry)
		01000110	(difference)

EXAMPLE 25–40

Find the difference between 11011000 and 10110011:

SOLUTION

		1	
(minuend)	11011000	11011000	
(subtrahend)	10110011	01001100	(one's complement)
		100100100	
		→1	(end-around carry)
		00100101	(difference)

25-12-2 **Two's Complement**

☞ **KEY POINT** Computers subtract by using the two's complement method.

The two's complement of a number may be found by adding 1 to the one's complement. Let's find the two's complement of 10011101. First, we change all 0s to 1s and all 1's to 0's (one's complement). Then we add 1 as follows:

$$
\begin{array}{ll}
\text{10011101} & \quad 0\ \ 1\ \ 1\ \ 0\ \ 0\ \ 0\ \ 1\ \ 0 \quad \text{(one's complement)} \\
(+) & \quad \underline{\hspace{5.5cm} 1} \\
 & \quad 0\ \ 1\ \ 1\ \ 0\ \ 0\ \ 0\ \ 1\ \ 1 \quad \text{(two's complement)}
\end{array}
$$

To subtract, we obtain the two's complement of the subtrahend and then add to the minuend. A quick way to find the two's complement of a number is to copy all bits starting at the LSB and continuing through the first 1 bit. Then complement the remaining bits. This is the two's complement of the original binary number. Consider the number 10010100. To find the two's complement, copy all the bits starting at the LSB and continuing through the first 1 bit. This gives us the three least significant bits (100). Next, complement the remaining bits. The answer is 01101100.

EXAMPLE 25–41

Let's subtract 01001010 from 01100111:

SOLUTION

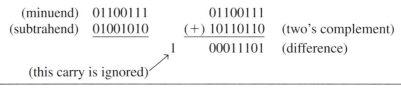

(minuend)   01100111   01100111
(subtrahend) 01001010   (+) 10110110   (two's complement)
                         1   00011101   (difference)

(this carry is ignored)

The carry resulting from the most significant bit position is ignored. There is no end-around carry as there is with the one's complement method. The computer obtains the one's complement by using an inverter. The two's complement is then obtained by starting the addition process with a carry, as shown in Example 25–42.

EXAMPLE 25–42

Subtract 00110110 from 01011011.

SOLUTION

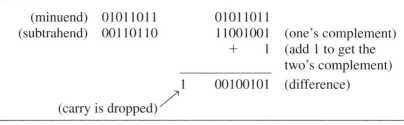

(minuend)   01011011   01011011
(subtrahend) 00110110   11001001   (one's complement)
                              +   1   (add 1 to get the
                                           two's complement)
                         1   00100101   (difference)

(carry is dropped)

The number of bit positions in the subtrahend must agree with the number of bit positions in the minuend. This is usually accomplished by providing a fixed number of storage cells in the computer. In the preceding examples, eight bit positions were used; therefore, eight storage cells were used to store the numbers. Eight cells were also used to store the difference, causing the carry out of the most significant bit position to be dropped because there was no place to store it. Each of the above eight-cell combinations is called a *register*. Example 25–43 illustrates the concept of registers with fixed numbers of bit positions.

EXAMPLE 25–43

Subtract 00001100 from 01001010 (Figure 25–5).

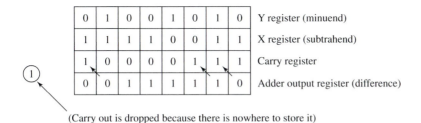

FIGURE 25–5

(Carry out is dropped because there is nowhere to store it)

SOLUTION   The minuend is placed in the Y register; the one's complement of the subtrahend is placed in the X register. The LSB is in the rightmost bit position. A 1 is placed in the LSB position of the carry register. All the other positions of the carry register store a zero at the start. If a carry is produced with the addition of a column, a 1 is stored in the *next* bit position of the carry register. Since the output register is limited to eight bit positions, the carry out of the MSB position is dropped.

PRACTICE PROBLEMS 25–12

For the following decimal numbers, find (a) the nine's complement and (b) the ten's complement:

1. 17
3. 88
5. 8976

2. 36
4. 276

For the following binary numbers, find (a) the one's complement and (b) the two's complement:

6. 1101
8. 100110
10. 11000011

7. 1001
9. 110001

Subtract the following binary numbers by using (a) the one's complement method and (b) the two's complement method:

11.　11001
　　　(−)01101

12.　101101
　　　(−)001011

13.　1100011
　　　(−)0011001

14.　10110010
　　　(−)00011101

15.　11100100
　　　(−)00011011

SOLUTIONS

1. (a) 82 　　　(b) 83
3. (a) 11 　　　(b) 12
5. (a) 1023 　　(b) 1024
7. (a) 0110 　　(b) 0111
9. (a) 001110 　(b) 001111
11. (a) 1

2. (a) 63 　　　(b) 64
4. (a) 723 　　(b) 724
6. (a) 0010 　　(b) 0011
8. (a) 011001 　(b) 011010
10. (a) 00111100 　(b) 00111101
(b) 1

11. (a) 1
　　　 11001
　　　 10010　(one's complement)
　　1　01011
　　　└──→ 1
　　　 01100

(b) 1
　　　 11001
　　　 10011　(two's complement)
　　1̸　01100

12. (a) 1
　　　 101101
　　　 110100　(one's complement)
　　1　100001
　　　└──→ 1
　　　 100010

(b) 1
　　　 101101
　　　 110101　(two's complement)
　　1̸　100010

13. (a) 1　　11
　　　 1100011
　　　 1100110　(one's complement)
　　1　1001001
　　　└──→ 1
　　　 1001010

(b) 1　　111
　　　 1100011
　　　 1100111　(two's complement)
　　1̸　1001010

14. (a) 1
　　　 10110010
　　　 11100010　(one's complement)
　　1　10010100
　　　└──→ 1
　　　 10010101

(b) 1　　11　1
　　　 10110010
　　　 11100011　(two's complement)
　　1̸　10010101

15. (a) 1   11   1
          11100100
          11100100   (one's complement)
       1 11001000
       └──────→ 1
          11001001

(b) 1   11   1
        11100100
        11100101   (two's complement)
     Ɨ 11001001

Additional practice problems are at the end of the chapter.

SIXTEEN'S AND EIGHT'S COMPLEMENT   **25-13**

When large signed binary numbers are added or subtracted, it is convenient to use the sixteen's complement method. The sixteen's complement is obtained by taking the fifteen's complement and adding one. The fifteen's complement is obtained by substituting each digit with the difference between its value and 15. For example, the sixteen's complement of 3B8 is

$$
\begin{array}{llll}
\text{FFF} & \text{C47} & \text{(fifteen's complement)} \\
(-)\text{3B8} & \underline{\quad 1} & \text{(plus 1)} \\
\text{C47} & \text{C48} & \text{(sixteen's complement)}
\end{array}
$$

In many systems, octal is used to represent large binary numbers. Therefore, it is convenient to use the eight's complement method for performing signed binary number additions and subtractions. The seven's complement is first obtained by substituting each octal digit with the difference between its value and 7. Then 1 is added to produce the eight's complement. For example, the eight's complement of 136_8 is

$$
\begin{array}{llll}
777 & 641 & \text{(seven's complement)} \\
(-)136 & \underline{\quad 1} & \text{(plus 1)} \\
641 & 642 & \text{(eight's complement)}
\end{array}
$$

The following subtraction is performed by using the two's complement method:

$$
\begin{array}{llll}
0011 \quad 1100 \quad 1011 \longrightarrow & 0011 \quad 1100 \quad 1011 & \text{(minuend)} \\
(-)0010 \quad 0110 \quad 0111 \longrightarrow & 1101 \quad 1001 \quad 1001 & \text{(two's complement)} \\
\text{(carry is dropped)} \longrightarrow & \text{Ɨ} \quad 0001 \quad 0110 \quad 0100 & \text{(difference)}
\end{array}
$$

The same problem may be solved by using the sixteen's complement method:

$$
\begin{array}{lll}
3\text{CB} & 3\text{CB} & \text{(minuend)} \\
(-)267 & \text{D99} & \text{(subtrahend in sixteen's complement form)} \\
& \text{Ɨ} \quad 164 & \text{(difference)}
\end{array}
$$

(carry is dropped) ⌐↑

If the above binary numbers are expressed in octal, the eight's complement method may be used:

$$1713_8 \qquad 1713 \quad \text{(minuend)}$$
$$(-)1147_8 \qquad \underline{6631} \quad \text{(subtrahend in eight's complement form)}$$
$$1 \quad 0544 \quad \text{(difference)}$$

(carry is dropped) ⟶

PRACTICE PROBLEMS 25–13

For the following octal numbers, find (a) the seven's complement and (b) the eight's complement:

1. 6
3. 56
5. 726

2. 12
4. 173

For the following hexadecimal numbers, find (a) the fifteen's complement and (b) the sixteen's complement.

6. A
8. 87
10. F209

7. 2C
9. 4A3

Perform the following subtractions: (a) in binary by using the two's complement method, (b) in octal by using the eight's complement method, and (c) in hexadecimal by using the sixteen's complement method.

11. 11010011
     (−)00111101

12. 10011101
     (−)00110011

13. 10111100
     (−)01001110

14. 11110000
     (−)00111100

15. 11001111
     (−)10000011

SOLUTIONS

1. (a) 1	(b) 2		2. (a) 65	(b) 66
3. (a) 21	(b) 22		4. (a) 604	(b) 605
5. (a) 051	(b) 052		6. (a) 5	(b) 6
7. (a) D3	(b) D4		8. (a) 78	(b) 79
9. (a) B5C	(b) B5D		10. (a) 0DF6	(b) 0DF7

11. (a)   11010011
         11000011   (two's complement)

$\cancel{1}$ 10010110_2

(b) 1101 0011 =   D3
    1100 0011 =   C3   (sixteen's complement)

$\cancel{1}$ 96_{16}

(c) 011 010 011 =   323
    111 000 011 =   703   (eight's complement)

$\cancel{1}$ 226_8

The subtrahend is 00111101. The two's complement is 11000011. Since we have already found the two's complement of 00111101 to be 11000011, we separate the two's complement into two 4-bit groups. Then 1100 0011 = C3 which is the sixteen's complement. To find the eight's complement, we must separate the subtrahend into three 3-bit groups. Since we have only 8 bits, we add a zero to the subtrahend in the original problem to get 000 111 101. The two's complement is 111 000 011. The eight's complement is 703.

12. (a)   10011101
         11001101   (two's complement)

$\cancel{1}$ 01101010_2

(b) 1001 1101 =   9D
    1100 1101 =   CD   (sixteen's complement)

$\cancel{1}$ $6A_{16}$

(c) 010 011 101 =   235
    111 001 101 =   715   (eight's complement)

$\cancel{1}$ 152_8

13. (a)   10111100
         10110010   (two's complement)

$\cancel{1}$ 01101110_2

(b) 1011 1100 =   BC
    1011 0010 =   B2   (sixteen's complement)

$\cancel{1}$ $6E_{16}$

(c) 010 111 100 =   274
    110 110 010 =   662   (eight's complement)

$\cancel{1}$ 156_8

14. (a)   11110000
         11000100   (two's complement)

$\cancel{1}$ 10110100_2

(b) 1111 0000 =   F0
    1100 0100 =   C4   (sixteen's complement)

$\cancel{1}$ $B4_{16}$

(c) 011 110 000 =   360
    111 000 100 =   704   (eight's complement)

$\cancel{1}$ 264_8

15. (a)   11001111
        01111101   (two's complement)
    $\cancel{1}$ 01001100

   (b) 1100   1111 =   CF
       0111   1101 =   7D   (sixteen's complement)
                  $\cancel{1}$ $4C_{16}$

   (c) 011   001   111 =   317
       101   111   101 =   575   (eight's complement)
                  $\cancel{1}$ 114_8

Additional practice problems are at the end of the chapter.

SELF-TEST 25–3

Work the following problems by using (a) the straight subtraction method and (b) the complement method:

1. 11101011_2
     $(-)01001101_2$

2. 11000110_2
     $(-)10011011_2$

3. 73_8
     $(-)46_8$

4. 603_8
     $(-)364_8$

5. $F6_{16}$
     $(-)3A_{16}$

6. $B03_{16}$
     $(-)8AE_{16}$

Answers to Self-test 25–3 are at the end of the chapter.

CHAPTER 25 AT A GLANCE

PAGE	KEY POINT OR RULE	EXAMPLE
677	*Rule 25–1:* In converting from binary to decimal, find the value or weight of the MSB. Work down to the LSB, adding the weight of that position if a 1 is present or a 0 if a 0 is present.	$101_2 = 1 \times 2^2$ $+ 0 \times 2^1$ $+ 1 \times 2^0 = 5_{10}$
684	*Rule 25–2:* In converting from octal to decimal, find the weight of the digit in each position. Add the values of the digits in each position to determine the decimal equivalent.	$473_8 = 4 \times 8^2 +$ $7 \times 8^1 +$ $3 \times 8^0 = 315_{10}$

687   *Rule 25–3:* In converting from hexadecimal to decimal, find the weight of the digit in each position. Add the values of the digits in each position to determine the decimal equivalent.

$$2BC_{16} = 2 \times 16^2$$
$$+ 11 \times 16^1$$
$$+ 12 \times 16^0 = 700_{10}$$

689   *Key Point:* To convert from binary to octal, separate the bits into 3-bit groups, starting with the LSB and moving left to the MSB.

$$10110110_2 = 266_8$$

691   *Key Point:* To convert from binary to hexadecimal, separate the bits into 4-bit groups, starting with the LSB and moving left to the MSB.

$$10011011_2 = 9B_{16}$$

708   *Key Point:* Computers subtract by using the two's complement method. The two's complement of the subtrahend is added to the minuend. The carry out of the MSB is ignored.

$$01100111 - 01001010$$
Solution
```
    01100111
(+)10110110   (two's complement)
1   00011101   (difference)
```

END OF CHAPTER PROBLEMS 25–1

Convert the following binary numbers to decimal numbers:

1. 111_2
2. 101_2
3. 1010_2
4. 1101_2
5. 1101_2
6. 1000_2
7. 10111_2
8. 11001_2
9. 100110_2
10. 111001_2
11. 111000_2
12. 101010_2
13. 1100101_2
14. 1001101_2
15. 1110111_2
16. 1111111_2
17. 11110000_2
18. 10101010_2
19. 11000011_2
20. 10011001_2

Convert the following decimal numbers to binary numbers:

21. 5
22. 5
23. 12
24. 14
25. 21
26. 31
27. 45
28. 55
29. 59
30. 60
31. 68
32. 84

33. 96

35. 135

37. 210

34. 117

36. 180

38. 254

END OF CHAPTER PROBLEMS 25–2

Convert the following numbers from octal to decimal:

1. 14_8

3. 77_8

5. 276_8

7. 4176_8

9. $11,714_8$

2. 25_8

4. 100_8

6. 676_8

8. 6237_8

10. $24,576_8$

Convert the following numbers from decimal to octal:

11. 20_{10}

13. 80_{10}

15. 360_{10}

17. 1417_{10}

19. 3789_{10}

12. 46_{10}

14. 103_{10}

16. 617_{10}

18. 5916_{10}

20. 6063_{10}

END OF CHAPTER PROBLEMS 25–3

Convert the following numbers from hexadecimal to decimal:

1. $1A_{16}$

3. $4C_{16}$

5. 200_{16}

7. $11AA_{16}$

9. $A02B_{16}$

2. $B1_{16}$

4. $1AC_{16}$

6. 500_{16}

8. $FADE_{16}$

10. $84AB_{16}$

Convert the following numbers from decimal to hexadecimal:

11. 22_{10}

13. 97_{10}

15. 512_{10}

17. 2700_{10}

19. 6075_{10}

12. 50_{10}

14. 127_{10}

16. 873_{10}

18. 5606_{10}

20. 8088_{10}

END OF CHAPTER PROBLEMS 25–4

Convert the following binary numbers to (a) octal numbers and then (b) hexadecimal numbers.

1. 11001010_2

3. 10011011_2

2. 10110110_2

4. 11100011_2

5. 10000001_2

6. 10101111_2

7. 110011001001_2

8. 110011001100_2

9. 101011000011_2

10. 110100011010_2

Convert the following numbers to binary numbers.

11. 17_8

12. 25_8

13. 44_8

14. 61_8

15. 200_8

16. 400_8

17. 560_8

18. 777_8

19. 1000_8

20. 1102_8

21. C_{16}

22. E_{16}

23. 24_{16}

24. 40_{16}

25. $4F_{16}$

26. $C4_{16}$

27. $E3_{16}$

28. $B7_{16}$

29. $19A_{16}$

30. $2EF_{16}$

END OF CHAPTER PROBLEMS 25–5

1. $75_8 =$ _____ $_2 =$ _____ $_{10} =$ _____ $_{16}$

2. $112_8 =$ _____ $_2 =$ _____ $_{10} =$ _____ $_{16}$

3. $140_8 =$ _____ $_2 =$ _____ $_{10} =$ _____ $_{16}$

4. $165_8 =$ _____ $_2 =$ _____ $_{10} =$ _____ $_{16}$

5. $2A_{16} =$ _____ $_2 =$ _____ $_8 =$ _____ $_{10}$

6. $6D_{16} =$ _____ $_2 =$ _____ $_8 =$ _____ $_{10}$

7. $1A7_{16} =$ _____ $_2 =$ _____ $_8 =$ _____ $_{10}$

8. $3EC_{16} =$ _____ $_2 =$ _____ $_8 =$ _____ $_{10}$

9. $100110_2 =$ _____ $_8 =$ _____ $_{10} =$ _____ $_{16}$

10. $101101_2 =$ _____ $_8 =$ _____ $_{10} =$ _____ $_{16}$

11. $111000101_2 =$ _____ $_8 =$ _____ $_{10} =$ _____ $_{16}$

12. $101111011011_2 =$ _____ $_8 =$ _____ $_{10} =$ _____ $_{16}$

13. $10_{10} =$ _____ $_2 =$ _____ $_8 =$ _____ $_{16}$

14. $200_{10} =$ _____ $_2 =$ _____ $_8 =$ _____ $_{16}$

15. $290_{10} =$ _____ $_2 =$ _____ $_8 =$ _____ $_{16}$

16. $433_{10} =$ _____ $_2 =$ _____ $_8 =$ _____ $_{16}$

END OF CHAPTER PROBLEMS 25–6

Add the following octal numbers:

1. 123
   234

2. 451
   116

3. 456
   317

4. 517
   126

5. 667
   107

6. 273
   706

7. 604
   617

8. 376
   412

9. 726
   161

10. 571
    707

END OF CHAPTER PROBLEMS 25–7

Add the following hexadecimal numbers:

1. 4A
   A3

2. D2
   1C

3. A07
   29E

4. B38
   398

5. 4AB
   8AB

6. C37
   8B9

7. ABF
   CDE

8. FED
   ABC

9. F0C6
   E9A6

10. BA09
    F6E4

END OF CHAPTER PROBLEMS 25–8

Add the following binary numbers:

1. 1100
   1001

2. 1011
   1111

3. 11100
   10111

4. 10011
   11101

5. 10011
   10110

6. 10111
   11101

7. 110011
   101111

8. 111001
   011101

9. 101110
   111011

10. 111101
    101111

END OF CHAPTER PROBLEMS 25–9

Subtract the following octal numbers:

1.   73
   (−)26

2.   64
   (−)40

3.   616
   (−)554

4.   624
   (−)267

5.   540
  (−)273

6.   405
  (−)267

7.   2004
  (−)1445

8.   3001
  (−)2653

9.   6334
  (−)4617

10.   5503
  (−)3573

END OF CHAPTER PROBLEMS 25–10

Subtract the following hexadecimal numbers:

1.   A6
  (−)8C

2.   CA
  (−)A9

3.   E06
  (−)A25

4.   C0A
  (−)2A9

5.   9A0
  (−)3AF

6.   8B4
  (−)4E8

7.   C00
  (−)103

8.   CA0
  (−)6A6

9.   B029
  (−)A04D

10.   E24A
  (−)174C

END OF CHAPTER PROBLEMS 25–11

Subtract the following binary numbers:

1.   1011
  (−)0110

2.   1101
  (−)0011

3.   1101
  (−)1011

4.   1101
  (−)0110

5.   11001
  (−)10111

6.   111011
  (−)010011

7.   10010
  (−)01111

8.   110110
  (−)100111

9.   110011
  (−)011101

10.   111001
  (−)011001

END OF CHAPTER PROBLEMS 25–12

Find the two's complement of:

1. 1010_2

2. 10110_2

3. 101101_2

4. 11001110_2

Use the one's complement method to find:

5. 11001_2
   $(-)01100_2$

6. 11100_2
   $(-)11011_2$

7. 101101_2
   $(-)001010_2$

8. 1101001_2
   $(-)0100110_2$

Use the two's complement method to find:

9. 10110_2
   $(-)01011_2$

10. 10011_2
   $(-)01000_2$

11. 1011001_2
   $(-)0010110_2$

12. 11100101_2
   $(-)10110010_2$

END OF CHAPTER PROBLEMS 25–13

Find the eight's complement of:

1. 64_8

2. 436_8

3. 635_8

4. 3732_8

Find the sixteen's complement of:

5. $A3_{16}$

6. $4C3_{16}$

7. $C1E_{16}$

8. $3E04_{16}$

Use the eight's complement method to find:

9. 63_8
   $(-)25_8$

10. 503_8
   $(-)236_8$

11. 626_8
   $(-)377_8$

12. 6026_8
   $(-)3662_8$

Use the sixteen's complement method to find:

13. $A2_{16}$
   $(-)8C_{16}$

14. $E4D_{16}$
   $(-)AAB_{16}$

15. $6A2_{16}$
   $(-)4FF_{16}$

16. $F40B_{16}$
   $(-)6BDD_{16}$

ANSWERS TO SELF-TESTS

SELF-TEST 25–1

1. 38

2. 190

3. 246

4. 3130

5. 91

6. 202

7. (a) 1010_2 (b) 12_8 (c) A_{16}
8. (a) 11100_2 (b) 34_8 (c) $1C_{16}$
9. (a) 10111011_2 (b) 273_8 (c) BB_{16}
10. (a) 1001110001_2 (b) 1161_8 (c) 271_{16}
11. (a) 1011101_2 (b) 135_8 (c) $5D_{16}$
12. (a) 111000011010_2 (b) 7032_8 (c) 3610_{10}
13. (a) 3226_8 (b) 1686_{10} (c) 696_{16}
14. (a) 101111011_2 (b) 379_{10} (c) $17B_{16}$

SELF-TEST 25–2

1. $E5_{16}$
2. $DC8_{16}$
3. $1F323_{16}$
4. 161_8
5. 124_8
6. 141_8
7. 1100_2
8. 11000_2
9. 110100_2

SELF TEST 25–3

1. 10011110_2
2. 101011_2
3. 25_8
4. 217_8
5. BC_{16}
6. 255_{16}

Boolean Algebra

26

Introduction

In today's world, technicians must have some background in digital logic circuits. In Chapter 25 we introduced the various kinds of number systems used in computers. In this chapter we will examine the basic building blocks of any logic system, the AND, OR, and NOT functions. We will learn how to relate between logic diagrams and logic expressions. We will simplify logic circuits and construct truth tables to verify the accuracy of our simplifications.

Chapter Objectives

In this chapter you will learn how to:

1. Identify the three basic logic operations.
2. Draw the logic symbols and construct truth tables for the three basic logic operations.
3. Write the Boolean expression for various logic circuits.
4. Use Boolean postulates and theorems to simplify logic expressions.

26-1   LOGIC GATES

26-1-1 Logic Variables

Boolean algebra is used in manipulating logic variables. A variable is either completely true or completely false; partly true or partly false values are not allowed. When a variable is not true, by implication it must be false. Conversely, if the variable is not false, it must be true. Because of this characteristic, Boolean algebra is ideally suited to variables that have two states or values, such as YES and NO, or for a number system that has two values, 1 and 0 (for example, the binary number system). For example, B (a variable) could represent the presence of Bob. B has two values: if Bob is present, B equals "true"; if Bob is absent, B equals "false." Note that Bob is not the variable; B is the variable that represents the presence of Bob.

FIGURE 26–1
SPST switch:
(a) closed;
(b) open.

(a)   (b)

A switch is ideally suited to represent the value of any two-state variable because it can only be "off" or "on." Consider the SPST switch in Figure 26–1. When the switch is in the closed position, it indicates that Bob is present (B = true). When it is in the open position, it then represents that Bob is absent (B = false).

A closed switch could also represent values such as true, yes, one (1), HIGH (H), go, and so on; an open switch could represent values such as false, no, zero (0), LOW (L), no go, and so on.

In digital circuits the TRUE state of a logic variable is usually indicated by the presence of 5 volts. The FALSE state is usually indicated by zero volts. There are a variety of terms that are used, instead of TRUE and FALSE, depending on the application. Table 26–1 shows examples of the synonyms for TRUE and FALSE used in industry.

26-1-2 Logic Operations

There are only three basic logic operations:

1. The *conjunction* (logical product), commonly called AND, symbolized by ($\cdot$). For example: $A \cdot B$.
2. The *disjunction* (logical sum), commonly called OR, symbolized by ($+$). For example: $A + B$.
3. The *negation,* commonly called NOT, symbolized by ($\overline{}$). For example: $\overline{A}$.

These operations are performed by logic circuits. All functions within a computer can be performed by combinations of these three basic logic operations.

26-1-3 The AND Function

> ☞ **KEY POINT** *For the AND function, the output is true only when ALL inputs are true. Otherwise, the output is false.*

Figure 26–2 on the next page shows the logic symbol and truth table for a two-input AND gate. A and B are the input variables. The function ($A \cdot B$) is produced when A is true AND B is true. That is, if both inputs are true, the output is true. If either input, or both, is false, the output is false. To show all the conditions that may exist at the input and output of a logic gate, truth tables are used. Since there are two inputs and

Table 26-1   Synonyms for TRUE and FALSE.

TRUE	T	YES	ON	CLOSED	HIGH	1	5 V
FALSE	F	NO	OFF	OPEN	LOW	0	0 V

FIGURE 26–2

Two-input AND
gate: (a) logic
symbol;
(b) truth table.

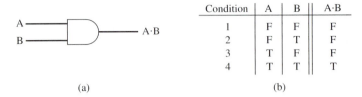

Condition	A	B	A·B
1	F	F	F
2	F	T	F
3	T	F	F
4	T	T	T

(a)

(b)

each input has two possible states (true and false), the number of possible conditions at the input would be 2 raised to the second power (2^2), or 4. Note in the truth table that condition 4 is the only one in which all the inputs are true so that the AND function is produced and a true output appears. If A is true AND B is true, the output will be true.

AND gates may have two or more inputs. Figure 26–3 shows a three-input AND gate and its truth table. The function (A · B · C) is produced when A is true AND B is true AND C is true. Ones (1) and zeros (0) are used for the values of the variables (true = 1 and false = 0). Because three variables are used, eight possible conditions exist ($2^3 = 8$). Condition 8 is the only one that will produce a true output (1) because all the input variables are true. Remember, in an AND gate, *all* the inputs must be true in order to produce a true output.

The (·) symbol used in the expression A · B · C is the AND operator and indicates logical multiplication. For example, condition 8 in Figure 26–3 can be interpreted as $1 \times 1 \times 1 = 1$, whereas condition 7 can be interpreted as $1 \times 1 \times 0 = 0$.

Using the same approach, we find that all other conditions (1 through 6) also produce a zero. It is important to remember that in the binary system a 2 cannot exist, only 1s and 0s. As with ordinary algebra, the operator symbol (·) can be omitted; thus, A · B · C = ABC and is read "A AND B AND C."

The AND function can be illustrated by the following analogy. The members of group A are Bob, Charley, and Dick. Note that the names in this group are combined by the conjunction "and." That means group A equals the presence of Bob AND Charley AND Dick. This may be symbolized as

$$A = BCD$$

A is true (group A is present) when B is true (Bob is present), AND C is true (Charley is present), AND D is true (Dick is present). A is not true if any one or more of the members are absent.

FIGURE 26–3

Three-input
AND gate:
(a) logic symbol;
(b) truth table.

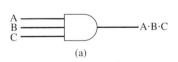

Condition	A	B	C	A·B·C
1	0	0	0	0
2	0	0	1	0
3	0	1	0	0
4	0	1	1	0
5	1	0	0	0
6	1	0	1	0
7	1	1	0	0
8	1	1	1	1

(a)

(b)

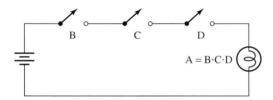

FIGURE 26–4   Switches connected in series to produce the AND function.

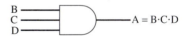

FIGURE 26–5   Logic symbol for the AND function BCD.

The circuit in Figure 26–4 can be used to produce the AND function of the above example. The light indicates that group A is present only when all members of the group are present. (All switches are closed.) A logic circuit producing this AND function is symbolized in Figure 26–5.

26-1-4 The OR Function

☞ **KEY POINT** For the OR function, the output is true when any one of the inputs is true. Otherwise, the output is false.

Figure 26–6 shows the logic symbol and truth table for a two-input OR gate. The function (A + B) is produced when A is true OR B is true. That is, if either input is true, the output is true. If both inputs are false, the output is false. Because there are two input variables, a total of four conditions is possible ($2^2 = 4$). Note in the truth table that conditions 2, 3, and 4 produce a true output because at least one of the input variables is true. Condition 1 produces a false output because neither of the input variables is true.

OR gates may have two or more inputs. Figure 26–7 on the next page shows a three-input OR gate and its truth table. Ones (1) and zeros (0) are used for the values of the variables. Because three variables are used, eight possible conditions occur ($2^3 = 8$). Conditions 2 through 8 produce a true (1) output because at least one of the input variables is true (1). Condition 1 is the only condition that does not produce a true output.

The + symbol used in the expression A + B + C is the OR operator and indicates logical addition. In logical addition $1 + 0 = 1$ and $1 + 1 = 1$. One OR one does not equal two! Condition 8 in Figure 26–7 may be interpreted as $1 + 1 + 1 = 1$, whereas condition 5 can be interpreted as $1 + 0 + 0 = 1$. Condition 1 produces a zero output because $0 + 0 + 0 = 0$.

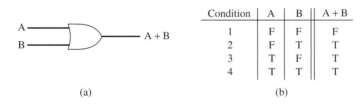

Condition	A	B	A + B
1	F	F	F
2	F	T	T
3	T	F	T
4	T	T	T

(a)   (b)

FIGURE 26–6
Two-input OR gate: (a) logic symbol; (b) truth table.

FIGURE 26–7
Three-input OR
gate: (a) logic
symbol;
(b) truth table.

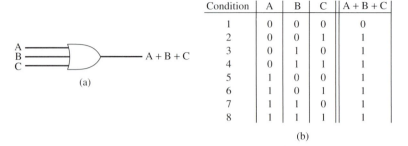

(a)

A + B + C

Condition	A	B	C	A + B + C
1	0	0	0	0
2	0	0	1	1
3	0	1	0	1
4	0	1	1	1
5	1	0	0	1
6	1	0	1	1
7	1	1	0	1
8	1	1	1	1

(b)

The OR function is illustrated by the following analogy. The members of group A are Bob, Charley, and Dick. A representative of this group could be Bob OR Charley OR Dick, or any combination of them. This expression may be symbolized as

$$R = B + C + D$$

where R is a representative of group A. R is true (a representative of group A is present) when B is true (Bob is present) OR C is true (Charley is present) OR D is true (Dick is present). Only one of the members needs to be present for group A to be represented. However, group A is also represented when more than one member is present. Group A will not be represented (false) when all members are absent.

The circuit in Figure 26–8 can be used to produce the OR function of the above example. The light (R) indicates that a representative of group A is present when one OR more members of the group are present. (At least one switch is closed.) A logic circuit producing this OR function is symbolized in Figure 26–9.

26-1-5 The NOT Function

☞ **KEY POINT** For the NOT function, the output is the complement of the input.

The concept of the NOT function can be illustrated by the circuit in Figure 26–10. When the switch is closed, the indicator lamp lights (true). Opening the switch will break the circuit and the lamp will go out (false). Only two conditions may exist.

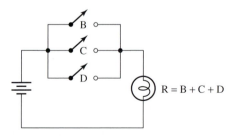

FIGURE 26–8   Switches connected in
parallel to produce the OR function.

R = B + C + D

FIGURE 26–9   Logic symbol for the
OR function B + C + D.

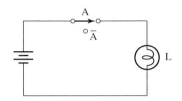

FIGURE 26–10   Representation of the NOT function.

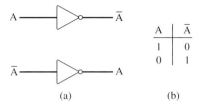

(a)   (b)

FIGURE 26–11   Inverter: (a) logic symbol; (b) truth table.

Either the lamp is on (true) or it is off (false). No other condition may exist. The conditions are *complements* of each other.

A logic circuit producing the NOT function is called an *inverter;* its symbol appears in Figure 26–11. An inverter converts the state or value of a variable to its complement. Thus, if variable A appears at the input, $\overline{A}$ (not A) is produced at the output; and, conversely, when $\overline{A}$ appears at the input, A is produced at the output. When performing the NOT function, a 1 will be changed to a 0, and vice versa, as shown in the truth table.

26-1-6 Boolean Expressions

The three basic logic functions discussed, AND, OR, and NOT, either individually or in various combinations, are the basic building blocks for all computer logic circuits. A few of these combinations will be illustrated by our group A analogy consisting of the presence of Bob (B), Charley (C), and Dick (D). Suppose we wish to describe a situation in which the group (A) is represented by the presence of Bob (B) and Charley (C), but not Dick (D). This situation could be expressed as $X = BC\overline{D}$, which states that Bob and Charley are present but Dick is NOT present; it is illustrated in Figure 26–12. Note the use of the inverter.

How would you describe a condition in which the group (A) is represented by at least two members—in other words, a majority? This situation could be expressed in Boolean algebra as $X = BC + BD + CD$ and is symbolized in Figure 26–13. The output is true if B AND C is true (Bob AND Charley are present), OR B AND D is true (Bob AND Dick are present), OR C AND D is true (Charley AND Dick are present).

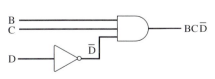

FIGURE 26–12   Logic circuit used to represent the expression $A = BC\overline{D}$.

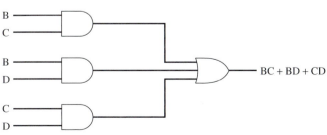

FIGURE 26–13   Logic circuit used to represent the expression $BC + BD + CD$.

FIGURE 26–14
Logic circuit
used to
represent the
expression
$\overline{B}C\overline{D}$.

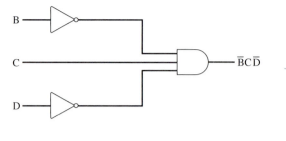

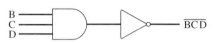

FIGURE 26–15   Logic circuit used to
represent the expression $\overline{BCD}$.

FIGURE 26–16   Logic circuit used to represent
the expression BC + D.

How could the group be represented by the presence of Charley and the absence of both Bob and Dick? One possible method would be $\overline{B}C\overline{D}$, as shown in Figure 26–14.

Suppose we wish to express the situation in which the entire group is NOT present. In other words, one or more members are absent. The Boolean expression would be $\overline{BCD}$ and is illustrated in Figure 26–15. Note that the entire group is affected by the inverter.

Finally, how would we express a situation in which the group is represented by either Bob and Charley or by Dick alone? One possible method is shown in Figure 26–16.

Logic diagrams are drawn to symbolize logic expressions. In the following problems, logic diagrams are to be interpreted or logic diagrams are to be drawn to symbolize logic expressions.

PRACTICE PROBLEMS 26–1

Draw the logic diagrams that represent the following Boolean expressions:

1. $B + \overline{C}$
2. $\overline{A}B$
3. $AB + \overline{C}$
4. $A(\overline{B} + C)$
5. $\overline{B + \overline{C} + D}$

6–10.   Write the Boolean expression represented by the logic circuits in Figure 26–18.

SOLUTIONS

1. In problem 1, Figure 26–17, the OR function is indicated; therefore, an OR gate is drawn. To realize the input $\overline{C}$, an inverter is used.

2. In problem 2, Figure 26–17, the AND function is indicated; therefore, an AND gate is drawn. To realize the input $\overline{A}$, an inverter is used.

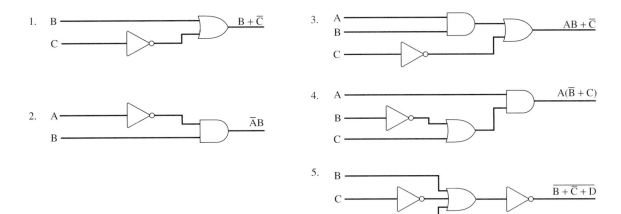

FIGURE 26–17   Answers to Practice Problems 26–1, problems 1 through 5.

3. In problem 3, Figure 26–17, both AND and OR functions are indicated. A and B are the inputs to the AND gate. The output of the AND gate (AB) and $\overline{C}$ are the inputs to the OR gate. An inverter is used to produce the input variable $\overline{C}$.

4. In problem 4, Figure 26–17, parentheses are used to indicate logical multiplication as in any algebraic expression. Therefore, a two-input AND gate is indicated. The inputs are A and $\overline{B}$ OR C. The input $\overline{B}$ OR C results from a two-input OR gate whose inputs are $\overline{B}$ and C.

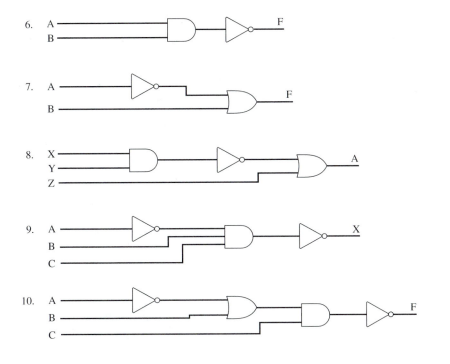

FIGURE 26–18
Logic circuits for Practice Problems 26–1, problems 6 through 10.

5. In problem 5, Figure 26–17, the entire output function is negated. The bar across the entire expression tells us this. Therefore, an inverter must be connected to the output of the OR gate. At the output of the OR gate is the expression $B + \overline{C} + D$. Because there are three variables, the gate is a three-input OR gate.

6. $F = \overline{AB}$. In problem 6 the AND function is indicated, followed by an inverter. Because the inverter is on the output side of the gate, the entire expression is complemented.

7. $F = \overline{A} + B$. In problem 7 the inverter complements one input to the OR gate and results in the output $\overline{A} + B$.

8. $A = \overline{XY} + Z$. The inverter complements the output of the AND gate in problem 8 to give the expression $\overline{XY}$. This expression is one input to the OR gate. The other input is Z. The output of the OR gate then is $\overline{XY} + Z$.

9. $X = \overline{\overline{ABC}}$. In problem 9 the three inputs to the AND gate are $\overline{A}$, B, and C. The output is $\overline{ABC}$. This output is complemented by the inverter and results in $\overline{\overline{ABC}}$ at its output.

10. $F = \overline{C(\overline{A} + B)}$. In problem 10 the output of the OR gate $(\overline{A} + B)$ is one input to the AND gate. C is the other input. The output of the AND gate is $C(\overline{A} + B)$. This output is complemented and results in $\overline{C(\overline{A} + B)}$ at the output of the inverter.

SELF-TEST 26–1

Problems **1** through **10**: Write the Boolean expressions for the logic circuits in Figure 26–19. Problems **11** through **20**: Draw the logic diagrams that represent the following Boolean expressions.

11. $A\overline{B}$

12. $\overline{A} + B$

13. $\overline{F + G}$

14. $\overline{C}(B + \overline{D})$

15. $A\overline{B} + C$

16. $AB + AC$

17. $(X + Z)(X + Y)$

18. $A\overline{B} + \overline{CD}$

19. $\overline{A(B + C)}$

20. $\overline{BC} + D$

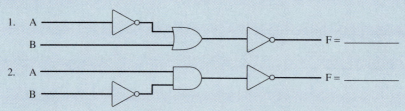

FIGURE 26–19   Logic circuits for problems 1 through 10 for Self-test 26–1.

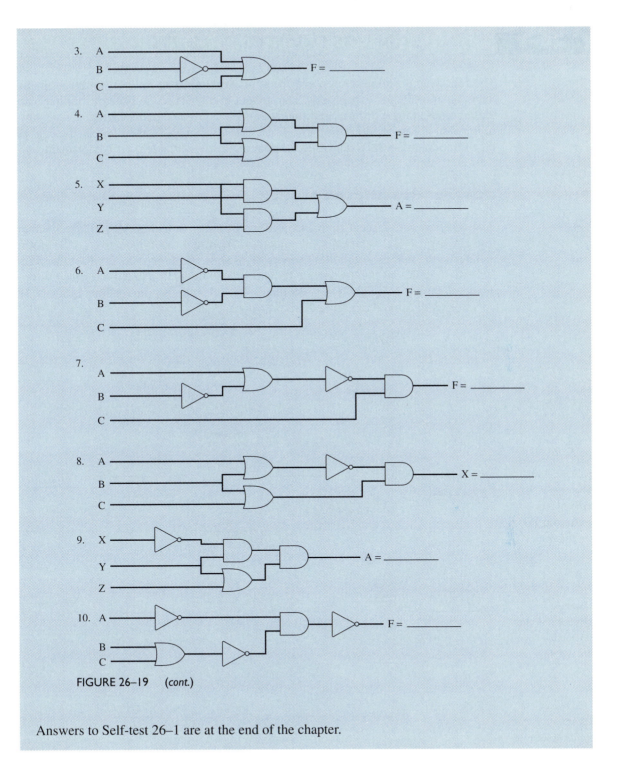

3. A
   B
   C
   F = _____

4. A
   B
   C
   F = _____

5. X
   Y
   Z
   A = _____

6. A
   B
   C
   F = _____

7. A
   B
   C
   F = _____

8. A
   B
   C
   X = _____

9. X
   Y
   Z
   A = _____

10. A
    B
    C
    F = _____

FIGURE 26–19   (cont.)

Answers to Self-test 26–1 are at the end of the chapter.

26–2 BOOLEAN POSTULATES AND THEOREMS

The following postulates, laws, and theorems are important in the simplification and manipulation of logic expressions. The student is encouraged to become familiar with these because they will be used throughout this chapter. Note that each postulate, law, or theorem is described in two parts. These are *duals* of each other; that is, the dual of the OR operator is the AND, while the dual of a given variable is its complement.

26–2–1 Postulates

Postulates are self-evident truths. Consider postulates 1a and 1b. A variable is either true (1) or false (0). Postulates 2a, 3a, and 4a represent the conjunctive (AND) form and define the function of the AND operator, as shown in Figure 26–20. Postulates 2b, 3b, and 4b represent the disjunctive (OR) form and define the function of the OR operator, as shown in Figure 26–21. Postulates 5a and 5b define the function of the NOT operator, as illustrated in Figure 26–22.

Postulates

1a. $A = 1$ (if $A \neq 0$)	1b. $A = 0$ (if $A \neq 1$)
2a. $0 \cdot 0 = 0$	2b. $0 + 0 = 0$
3a. $1 \cdot 1 = 1$	3b. $1 + 1 = 1$
4a. $1 \cdot 0 = 0$	4b. $1 + 0 = 1$
5a. $\overline{1} = 0$	5b. $\overline{0} = 1$

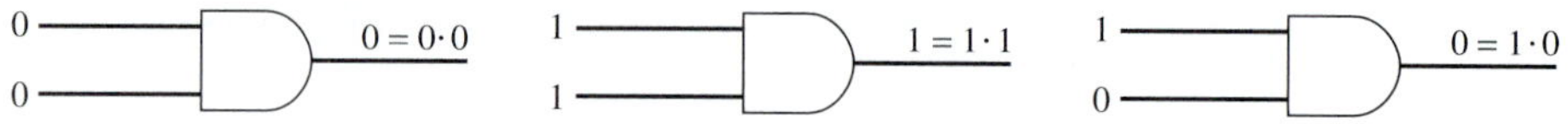

0
0
0 = 0·0
1
1
1 = 1·1
1
0
0 = 1·0

FIGURE 26–20 Logic symbolization of postulates 2a, 3a, and 4a.

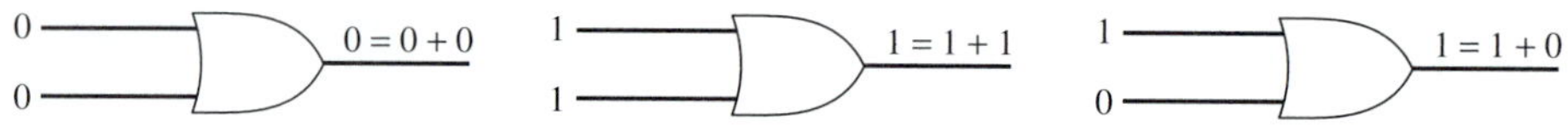

0
0
0 = 0 + 0
1
1
1 = 1 + 1
1
0
1 = 1 + 0

FIGURE 26–21 Logic symbolization of postulates 2b, 3b, and 4b.

1
$\overline{1} = 0$
0
$\overline{0} = 1$

FIGURE 26–22 Logic symbolization of postulates 5a and 5b.

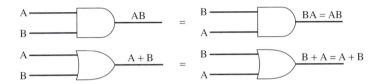

FIGURE 26–23
Logic symbols
used to
demonstrate
the
commutative
property.

26-2-2 Algebraic Properties

The following are properties of ordinary algebra that also apply to Boolean algebra. Remember, Boolean expressions contain variables having only two possible values.

Algebraic Properties
Commutative
  6a. $AB = BA$ 6b. $A + B = B + A$
Associative
  7a. $A(BC) = AB(C)$ 7b. $A + (B + C) = (A + B) + C$
Distributive
  8a. $A(B + C) = AB + AC$ 8b. $A + BC = (A + B)(A + C)$

The *commutative* property simply means that the circuit is not affected by the order or sequence of the variables. This is illustrated in Figure 26–23.

The *associative* property pertains to the parentheses. It shows that a sequence exclusively of AND functions (7a) or a sequence exclusively of OR functions (7b) is not affected by the placement of the parentheses, as indicated in Figure 26–24.

The *distributive* property for 8a and 8b may be proved by performing the algebraic multiplication or by factoring. Figure 26–25 on the next page shows that the

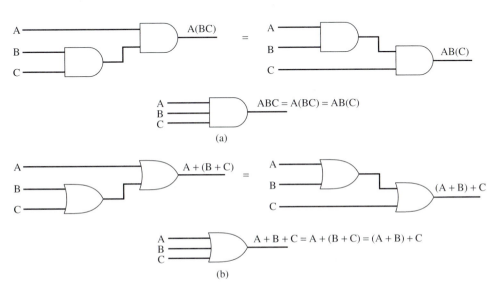

FIGURE 26–24
Logic symbols
and circuits
used to
demonstrate
the associative
property:
(a) logic circuit
to demonstrate
AND; (b) logic
circuit to
demonstrate
OR.

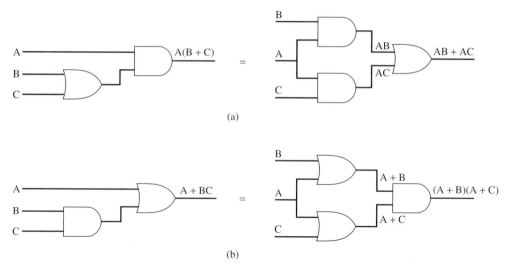

FIGURE 26–25   Logic symbols and circuits used to demonstrate the distributive property:
(a) logic circuit to demonstrate A(B + C) = AB + AC; (b) logic circuit to demonstrate A + BC =
(A + B)(A + C).

application of the distributive property produces two forms of an expression and, thus, two circuits may be realized for each.

26-2-3 Theorems

The following theorems define the application of the operators to variables:

> **Theorems**
> 9a. $A \cdot 0 = 0$ 9b. $A + 0 = A$
> 10a. $A \cdot 1 = A$ 10b. $A + 1 = 1$
> 11a. $A \cdot A = A$ 11b. $A + A = A$
> 12a. $A \cdot \overline{A} = 0$ 12b. $A + \overline{A} = 1$
> 13a. $\overline{\overline{A}} = A$ 13b. $A = \overline{\overline{A}}$

Theorems 9a, 10a, 11a, and 12a pertain to the AND function. These are symbolized in Figure 26–26. For theorem 9a, one of the variables is always a zero. Therefore, the output will be a zero regardless of the value of A. Theorem 10a tells us the output will be determined by the input variable A. If $A = 1$, then $1 \cdot 1 = 1$. If $A = 0$, then $1 \cdot 0 = 0$. For Theorem 11a, if the input variable $A = 1$, the output will be $1 \cdot 1 = 1$. If the input variable $A = 0$, the output will be $0 \cdot 0 = 0$.

For Theorem 12a, the output will always be 0, because, when the input variable $A = 1$, the other input variable will be 0 ($\overline{A}$), causing the output to be $1 \cdot 0 = 0$. The same output will be produced when the values of the input variables are reversed. Theorem 12a is called a *self-contradiction*.

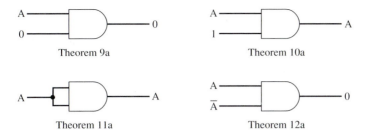

Theorem 9a

Theorem 10a

Theorem 11a

Theorem 12a

FIGURE 26–26
Logic symbolization of the AND function.

Theorem 13a expresses double negation. An expression that has been inverted twice equals its original value, as shown in Figure 26–27.

Theorems 9b, 10b, 11b, and 12b pertain to the OR function and are illustrated in Figure 26–28. For the OR function, one or more of the input variables must be true in order that a true output be present. For theorems 9b and 11b, the output is determined by the input variable A. For 10b and 12b, either A or $\overline{A}$ must always equal 1.

Consider the logic diagram in Figure 26–29 on the next page. The Boolean expression at its output is (AB)(BC). By applying various postulates, laws, and theorems, we can simplify the expression to ABC. Here's how we could do it.

1. (AB)(BC)   Original expression
2. ABBC   7a
3. ABC   11a

Step 1 shows the original expression. In step 2 we remove the parentheses. Our justification for this step is the associative property (7a). The simplified expression in step 3 results from applying theorem 11a. Any variable AND that same variable equals that variable (B · B = B). The simplified expression results in 3 two-input AND gates being replaced by 1 three-input AND gate, a substantial savings of both space and money to the designer.

Let's prove that the expressions are equal by using a truth table as in Figure 26–29(c). Notice that there are eight conditions. There are three input variables (A, B, and C) and each variable has two possible states ($2^3 = 8$). When we construct a truth table, it is important that we keep the order of the original statement and not skip any steps along the way. Thus, we list the variables first, then we list the function of the

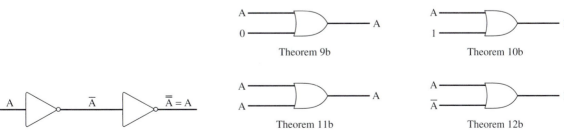

Theorem 9b

Theorem 10b

$\overline{\overline{A}} = A$

Theorem 11b

Theorem 12b

FIGURE 26–27   Logic symbolization of double negation.

FIGURE 26–28   Logic symbolization of the OR function.

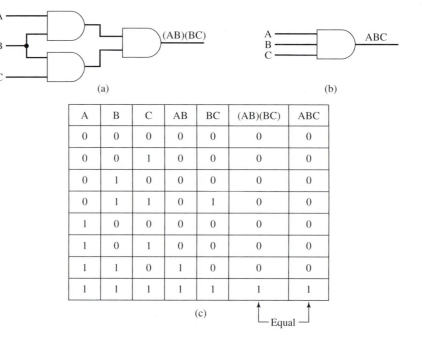

FIGURE 26–29
(a) Original circuit;
(b) simplified circuit; (c) truth table.

A	B	C	AB	BC	(AB)(BC)	ABC
0	0	0	0	0	0	0
0	0	1	0	0	0	0
0	1	0	0	0	0	0
0	1	1	0	1	0	0
1	0	0	0	0	0	0
1	0	1	0	0	0	0
1	1	0	1	0	0	0
1	1	1	1	1	1	1

(c)

└ Equal ┘

first two AND gates (AB and BC), and then we write the original expression and the simplified expression. The truth table shows that (AB)(BC) = ABC.

The process of using truth tables to prove the equality of two expressions is called *proof by perfect induction.* The student is encouraged to use truth tables to evaluate logic expressions or to prove the results of logic operations. Truth tables can also be used to define the functions of various logic circuits.

The distributive property (8a) states that A(B + C) = AB + AC. Let's prove that they are equal by constructing a truth table, as in Figure 26–30. First we list the input variables. Again, because there are three variables, eight conditions are possible. Next we list each intermediate step. B + C is the result of ORing the input variables B and C. The columns AB and AC are the result of ANDing the input variables A and B and the input variables A and C. Finally, there are columns showing the expressions A(B + C) and AB + AC.

Now let's do the same thing with the distributive property 8b: (A + B)(A + C). The logic diagrams and truth table are shown in Figure 26–31. Notice again how each part of each expression is listed in the truth table.

Let's see if it is possible to simplify the logic diagram in Figure 26–32 on p. 738. The step-by-step simplification and the justification for each step are shown by the following:

1. A + B + AC + AB   Original expression
2. A + AB + B + AC   6b
3. A(1 + B) + B + AC   8a
4. A(1) + B + AC   10b
5. A + B + AC   10a

736   CHAPTER 26

$$A(B + C) = AB + AC$$

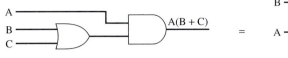

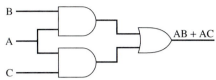

FIGURE 26–30
Logic diagrams
and truth tables
for A(B + C) =
AB + AC.

A	B	C	B + C	AB	AC	A(B + C)	AB + AC
0	0	0	0	0	0	0	0
0	0	1	1	0	0	0	0
0	1	0	1	0	0	0	0
0	1	1	1	0	0	0	0
1	0	0	0	0	0	0	0
1	0	1	1	0	1	1	1
1	1	0	1	1	0	1	1
1	1	1	1	1	1	1	1

└─ Equal ─┘

$$(A + B)(A + C) = A + BC$$

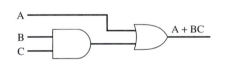

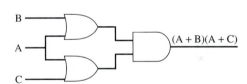

FIGURE 26–31
Logic diagrams
and truth table
for A + BC =
(A + B)(A +
C).

A	B	C	BC	A + B	A + C	A + BC	(A + B)(A + C)
0	0	0	0	0	0	0	0
0	0	1	0	0	1	0	0
0	1	0	0	1	0	0	0
0	1	1	1	1	1	1	1
1	0	0	0	1	1	1	1
1	0	1	0	1	1	1	1
1	1	0	0	1	1	1	1
1	1	1	1	1	1	1	1

└─ Equal ─┘

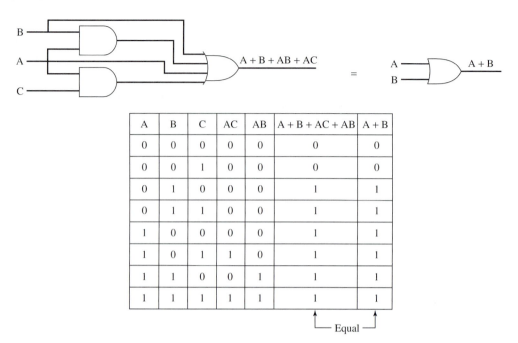

FIGURE 26–32 Logic symbols and truth table for A + B + AB + AC = A + B.

A	B	C	AC	AB	A + B + AC + AB	A + B
0	0	0	0	0	0	0
0	0	1	0	0	0	0
0	1	0	0	0	1	1
0	1	1	0	0	1	1
1	0	0	0	0	1	1
1	0	1	1	0	1	1
1	1	0	0	1	1	1
1	1	1	1	1	1	1

Equal

6. $A + AC + B$ 6b
7. $A(1 + C) + B$ 8a
8. $A(1) + B$ 10b
9. $A + B$ 10a

When many steps are involved in the simplification, there is usually more than one way to arrive at the final answer. The student may find alternative methods to the above simplification, but in each case the final simplification should be A + B. The original logic diagram, simplified logic diagram, and truth table are found in Figure 26–32.

PRACTICE PROBLEMS 26–2

For each logic diagram in Figure 26–33, (a) write the Boolean expression, (b) simplify the Boolean expression and justify each step, (c) draw the simplified logic diagram, and (d) construct a truth table to prove that the two expressions are equal.

SOLUTIONS

1. See Figure 26–34.
3. See Figure 26–36 (p. 740).

2. See Figure 26–35 (p. 740).
4. See Figure 26–37 (p. 741).

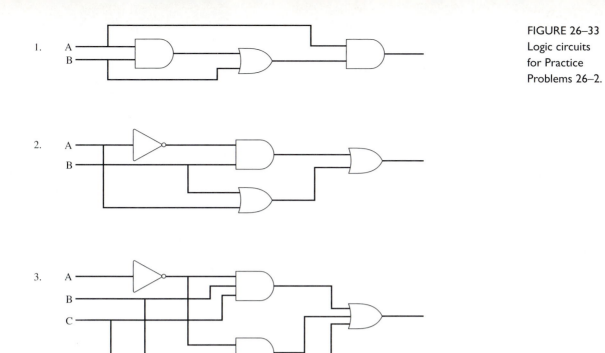

FIGURE 26–33
Logic circuits
for Practice
Problems 26–2.

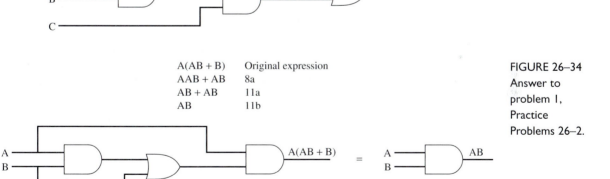

A(AB + B)	Original expression
AAB + AB	8a
AB + AB	11a
AB	11b

FIGURE 26–34
Answer to
problem 1,
Practice
Problems 26–2.

A	B	AB	AB + B	A(AB + B)
0	0	0	0	0
0	1	0	1	0
1	0	0	0	0
1	1	1	1	1

Equal

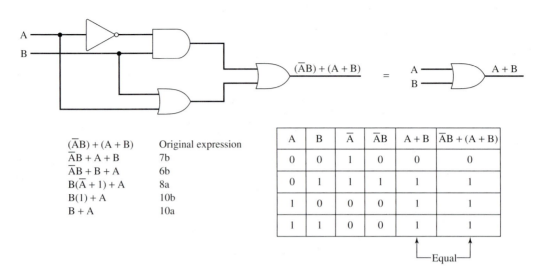

A	B	$\overline{A}$	$\overline{A}B$	$A + B$	$\overline{A}B + (A + B)$
0	0	1	0	0	0
0	1	1	1	1	1
1	0	0	0	1	1
1	1	0	0	1	1

$(\overline{A}B) + (A + B)$ — Original expression
$\overline{A}B + A + B$ — 7b
$\overline{A}B + B + A$ — 6b
$B(\overline{A} + 1) + A$ — 8a
$B(1) + A$ — 10b
$B + A$ — 10a

Equal

FIGURE 26–35   Answer to problem 2, Practice Problems 26–2.

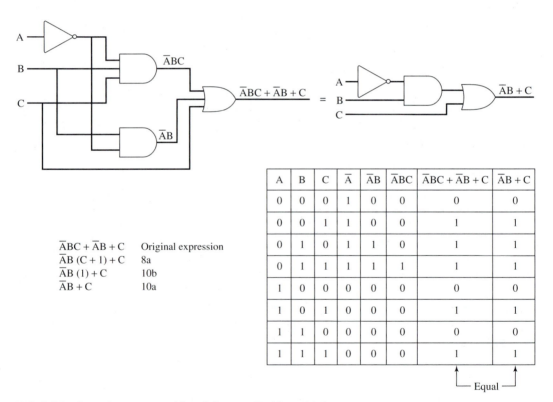

$\overline{A}BC + \overline{A}B + C$ — Original expression
$\overline{A}B (C + 1) + C$ — 8a
$\overline{A}B (1) + C$ — 10b
$\overline{A}B + C$ — 10a

A	B	C	$\overline{A}$	$\overline{A}B$	$\overline{A}BC$	$\overline{A}BC + \overline{A}B + C$	$\overline{A}B + C$
0	0	0	1	0	0	0	0
0	0	1	1	0	0	1	1
0	1	0	1	1	0	1	1
0	1	1	1	1	1	1	1
1	0	0	0	0	0	0	0
1	0	1	0	0	0	1	1
1	1	0	0	0	0	0	0
1	1	1	0	0	0	1	1

Equal

FIGURE 26–36   Answer to problem 3, Practice Problems 26–2.

A	B	C	AB	ABC	A + ABC
0	0	0	0	0	0
0	0	1	0	0	0
0	1	0	0	0	0
0	1	1	0	0	0
1	0	0	0	0	1
1	0	1	0	0	1
1	1	0	1	0	1
1	1	1	1	1	1

——— Equal ———

Expression	Reference
A + ABC	Original expression
A(1 + BC)	8a
A(1)	10b
A	10a

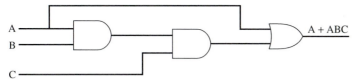

FIGURE 26–37   Answer to problem 4, Practice Problems 26–2.

DE MORGAN'S THEOREM
AND THE ABSORPTION THEOREM   26–3

26-3-1 De Morgan's Theorem

De Morgan's theorem provides us with an invaluable tool to use in simplifying Boolean expressions. By applying De Morgan's theorem, we can find the equivalent of a Boolean expression. For example, consider the logic diagram in Figure 26–38 on the next page. The Boolean expression at the output is $\overline{\overline{A}\,\overline{B}}$. We would not consider $\overline{\overline{A}\,\overline{B}}$ to be in its simplest form because of the bar extending over the entire expression. By applying De Morgan's theorem, the equivalent Boolean expression A + B is realized. This expression is much easier to understand and much more simple to construct.

Three steps are required in De Morganizing a Boolean expression: First, we replace all the OR operator symbols (+) with AND operator symbols (·) and all the AND operator symbols with OR operator symbols; next, we replace all variables with their complements; finally, we complement the entire expression. To De Morganize the expression $\overline{\overline{A}\,\overline{B}}$ above, we use the following steps:

· Step 1.   Replace the AND operator with the OR operator: $\overline{\overline{A} + \overline{B}}$.
· Step 2.   Complement each variable: $\overline{\overline{\overline{A}} + \overline{\overline{B}}} = \overline{A + B}$.
· Step 3.   Complement the entire expression: $\overline{\overline{A + B}} = A + B$.

The logic diagrams and truth table are found in Figure 26–39 on the next page.

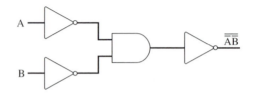

Let's determine De Morgan's equivalent of the expression $\overline{\overline{A} + \overline{B}}$.

- Step 1.   Replace the OR operator with the AND operator: $\overline{\overline{A} \cdot \overline{B}}$.
- Step 2.   Complement each variable: $\overline{\overline{\overline{A}} \cdot \overline{\overline{B}}} = \overline{AB}$.
- Step 3.   Complement the entire expression: $\overline{\overline{AB}} = AB$.

Therefore,

$$\overline{\overline{A} + \overline{B}} = AB$$

and is symbolized in Figure 26–40.

Let's find the De Morgan's equivalent of the expression in theorem 14a, $\overline{\overline{A}\,\overline{B}\,\overline{C}}$.

- Step 1. $\overline{A} + \overline{B} + \overline{C}$
- Step 2. $\overline{\overline{A}} + \overline{\overline{B}} + \overline{\overline{C}} = A + B + C$
- Step 3. $\overline{A + B + C}$

Therefore,

$$\overline{A}\,\overline{B}\,\overline{C} = \overline{A + B + C}$$

FIGURE 26–39
Logic diagrams
and truth table
for $\overline{\overline{AB}} =$
$A + B$.

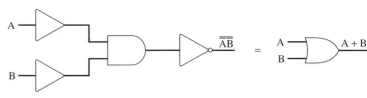

A	B	$\overline{A}$	$\overline{B}$	$\overline{AB}$	$\overline{\overline{AB}}$	A + B
0	0	1	1	1	0	0
0	1	1	0	0	1	1
1	0	0	1	0	1	1
1	1	0	0	0	1	1

⌐Equal⌐

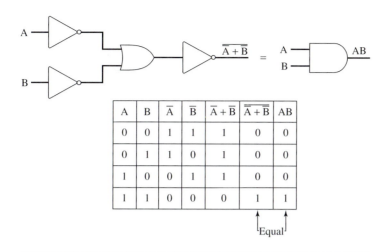

FIGURE 26–40
Logic diagrams
and truth table
for $\overline{\overline{A} + \overline{B}} =$
AB.

A	B	$\overline{A}$	$\overline{B}$	$\overline{A} + \overline{B}$	$\overline{\overline{A} + \overline{B}}$	AB
0	0	1	1	1	0	0
0	1	1	0	1	0	0
1	0	0	1	1	0	0
1	1	0	0	0	1	1

Equal

De Morgan's Theorem

14a. $\overline{A}\,\overline{B}\,\overline{C} = \overline{A + B + C}$ 14b. $\overline{A} + \overline{B} + \overline{C} = \overline{ABC}$

The symbols for the equivalent expressions above are given in Figure 26–41 on the next page.

Now let's try finding De Morgan's equivalent of the expression in theorem 14b, $\overline{A} + \overline{B} + \overline{C}$.

- Step 1. $\overline{\overline{A}} \cdot \overline{\overline{B}} \cdot \overline{\overline{C}}$
- Step 2. $\overline{\overline{A}} \cdot \overline{\overline{B}} \cdot \overline{\overline{C}} = A \cdot B \cdot C$
- Step 3. $\overline{A \cdot B \cdot C}$

Therefore,

$$\overline{A} + \overline{B} + \overline{C} = \overline{A \cdot B \cdot C}$$

The logic symbols for these expressions are given in Figure 26–42 on the next page.

The three-step solution process can be simplified. Consider the expression $\overline{\overline{A} + \overline{B}}$ that we simplified earlier. To simplify, we just break the bar and change the operator.

$$\overline{\overline{A} + \overline{B}} = \overline{\overline{A}} \cdot \overline{\overline{B}} = AB$$

This is an example of theorem 14a. When breaking the bar results in a double negation, we apply theorem 13a ($\overline{\overline{A}} = A$).

Let's simplify the Boolean expression $\overline{\overline{A} + B + \overline{C}}$.

$$\overline{\overline{A} + B + \overline{C}} \quad \text{Original expression}$$

$$\overline{\overline{A}} \cdot \overline{B} \cdot \overline{\overline{C}} \quad \text{14b}$$

$$A\overline{B}C \quad \text{13a}$$

Notice we broke the bar between each variable and changed each OR operator to AND. The logic diagrams and truth table are shown in Figure 26–43 (p. 745).

FIGURE 26–41
Logic diagrams
and truth table
for $\overline{A}\,\overline{B}\,\overline{C} =$
$\overline{A + B + C}$.

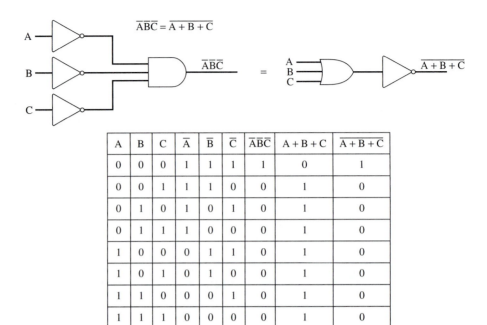

$\overline{A}\,\overline{B}\,\overline{C} = \overline{A + B + C}$

$\overline{A}\,\overline{B}\,\overline{C}$

$=$

$\overline{A + B + C}$

A	B	C	$\overline{A}$	$\overline{B}$	$\overline{C}$	$\overline{A}\,\overline{B}\,\overline{C}$	$A + B + C$	$\overline{A + B + C}$
0	0	0	1	1	1	1	0	1
0	0	1	1	1	0	0	1	0
0	1	0	1	0	1	0	1	0
0	1	1	1	0	0	0	1	0
1	0	0	0	1	1	0	1	0
1	0	1	0	1	0	0	1	0
1	1	0	0	0	1	0	1	0
1	1	1	0	0	0	0	1	0

└─── Equal ───┘

FIGURE 26–42
Logic diagrams
and truth
table for
$\overline{A} + \overline{B} + \overline{C} =$
$\overline{ABC}$.

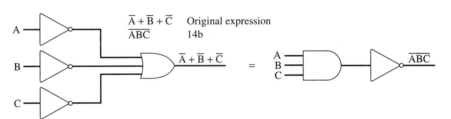

$\overline{A} + \overline{B} + \overline{C}$ Original expression
$\overline{ABC}$ 14b

$\overline{A} + \overline{B} + \overline{C}$

$=$

$\overline{ABC}$

A	B	C	$\overline{A}$	$\overline{B}$	$\overline{C}$	$\overline{A} + \overline{B} + \overline{C}$	ABC	$\overline{ABC}$
0	0	0	1	1	1	1	0	1
0	0	1	1	1	0	1	0	1
0	1	0	1	0	1	1	0	1
0	1	1	1	0	0	1	0	1
1	0	0	0	1	1	1	0	1
1	0	1	0	1	0	1	0	1
1	1	0	0	0	1	1	0	1
1	1	1	0	0	0	0	1	0

└─── Equal ───┘

A	B	C	$\overline{A}$	$\overline{B}$	$\overline{C}$	$\overline{A} + B + \overline{C}$	$\overline{\overline{A} + B + \overline{C}}$	$A\overline{B}C$
0	0	0	1	1	1	1	0	0
0	0	1	1	1	0	1	0	0
0	1	0	1	0	1	1	0	0
0	1	1	1	0	0	1	0	0
1	0	0	0	1	1	1	0	0
1	0	1	0	1	0	0	1	1
1	1	0	0	0	1	1	0	0
1	1	1	0	0	0	1	0	0

Equal

Let's simplify the expression $\overline{\overline{A} + \overline{B} + C}$. Again we use De Morgan's theorem since simplification requires removing the signs of negation (the bars). Let's start by breaking the bar between $\overline{A}$ and $\overline{B} + C$ and changing the operator.

$$\overline{\overline{A} + \overline{B} + C} \qquad \text{Original expression}$$

$$\overline{\overline{A}} \cdot \overline{\overline{B} + C} \qquad \text{14a}$$

$$A \cdot (B + C) \qquad \text{13a}$$

Or

$$A(B + C)$$

Notice the parentheses (sign of grouping) around B + C. We must treat $\overline{B + C}$ in the original expression as a single term (in algebra the bar is a sign of grouping). Therefore, when we De Morganize, we must include the sign of grouping around B + C to avoid an invalid expression. Without the sign of grouping, we get A · B + C or AB + C, which is not the same as A(B + C). We could have included the sign of grouping in the original expression:

$$\overline{\overline{A} + \overline{B} + C} = \overline{\overline{A} + (\overline{B} + C)}$$

It isn't necessary though because the bar $(\overline{B} + C)$ tells us to treat B + C as a single term or variable.

$$A(B + C)$$

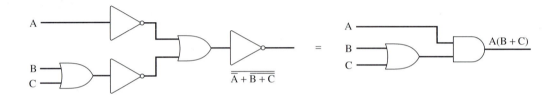

A	B	C	$\overline{A}$	B + C	$\overline{B + C}$	$\overline{A} + \overline{B + C}$	$\overline{\overline{A} + \overline{B + C}}$	A(B + C)
0	0	0	1	0	1	1	0	0
0	0	1	1	1	0	1	0	0
0	1	0	1	1	0	1	0	0
0	1	1	1	1	0	1	0	0
1	0	0	0	0	1	1	0	0
1	0	1	0	1	0	0	1	1
1	1	0	0	1	0	0	1	1
1	1	1	0	1	0	0	1	1

Equal

FIGURE 26–44 Logic diagrams and truth table for $\overline{\overline{A} + \overline{B + C}} = A(B + C)$.

We could have De Morganized $\overline{B + C}$ first:

$$\overline{\overline{A} + \overline{B + C}} \quad \text{Original expression} \qquad \overline{\overline{A} + \overline{B}\,\overline{C}} \qquad \text{14a}$$

Notice this did not affect the bar that extends over the entire expression since we were De Morganizing just $\overline{B + C}$. Next

$$\overline{\overline{A}} \cdot \overline{\overline{B}\,\overline{C}} \qquad \text{14a (again)} \qquad A \cdot \overline{\overline{B}\,\overline{C}} \qquad \text{13a}$$

$$A \cdot (\overline{\overline{B}} + \overline{\overline{C}}) \qquad \text{14b} \qquad A(B + C) \qquad \text{13a}$$

We arrive at the same answer by a much more difficult route. Notice the parentheses when theorem 14b was used. Only the expression BC was De Morganized. Therefore, we include the sign of grouping to indicate that B + C is a single variable (in this case one input to an AND gate). Figure 26–44 includes the logic diagrams and truth table.

26-3-2 Absorption Theorem

Sometimes variables appear in a certain pattern that can obviously be simplified. The *absorption* theorem shows how substitution may be used on some commonly derived patterns.

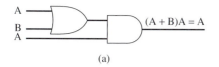

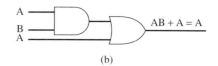

FIGURE 26–45
Logic
symbolization of
(a) theorem 15a
and (b) theorem
15b.

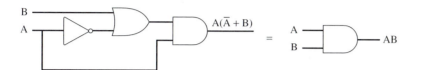

FIGURE 26–46
Logic
symbolization of
theorem 16a.

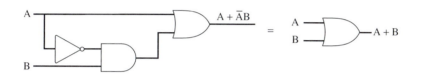

FIGURE 26–47
Logic
symbolization of
theorem 16b.

Absorption Theorem
15a. $A(A + B) = A$ 15b. $A + AB = A$
16a. $A(\overline{A} + B) = AB$ 16b. $A + \overline{A}B = A + B$

Theorems 15a and 15b show that the value of B is redundant. A true output occurs only when A is true, as shown in Figure 26–45.

Theorem 16a states that only when A AND B are true is the statement true, as shown in Figure 26–46. Recall that A and $\overline{A}$ cannot be true at the same time.

Theorem 16b is the dual of theorem 16a and is shown in Figure 26–47. For the expression to be true, either A must be true or B must be true. $\overline{A}$ is redundant.

PRACTICE PROBLEMS 26–3

For each logic diagram in Figure 26–48 (p. 748), (a) write the Boolean expression, (b) simplify the Boolean expression and justify each step, (c) draw the simplified logic diagram, and (d) construct a truth table to prove that the two expressions are equal.

SOLUTIONS

1. See Figure 26–49 (p. 749).
2. See Figure 26–50 (p. 749).
3. See Figure 26–51 (p. 750).
4. See Figure 26–52 (p. 750).
5. See Figure 26–53 (p. 751).

FIGURE 26–48
Practice
Problems 26–3.

1.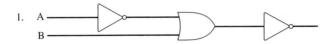

2.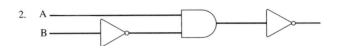

3.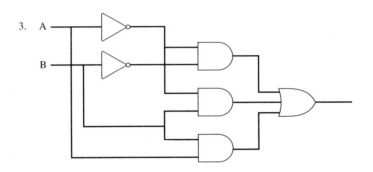

4.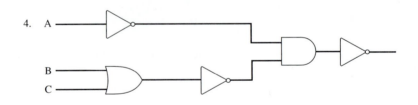

5.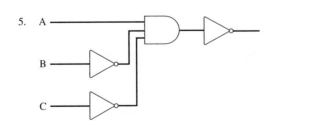

A	B	$\overline{A}$	$\overline{B}$	$\overline{A}+B$	$\overline{\overline{A}+B}$	$A\overline{B}$
0	0	1	1	1	0	0
0	1	1	0	1	0	0
1	0	0	1	0	1	1
1	1	0	0	1	0	0

└ Equal ┘

$\overline{\overline{A}+B}$ Original expression
$\overline{\overline{A}}\cdot\overline{B}$ 14a
$A\overline{B}$ 13a

FIGURE 26–49 Answer to problem 1, Practice Problems 26–3.

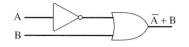

A	B	$\overline{A}$	$\overline{B}$	$A\overline{B}$	$\overline{A\overline{B}}$	$\overline{A}+B$
0	0	1	1	0	1	1
0	1	1	0	0	1	1
1	0	0	1	1	0	0
1	1	0	0	0	1	1

└ Equal ┘

$\overline{A\overline{B}}$ Original expression
$\overline{A}+\overline{\overline{B}}$ 14b
$\overline{A}+B$ 13b

FIGURE 26–50 Answer to problem 2, Practice Problems 26–3.

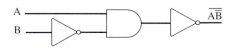

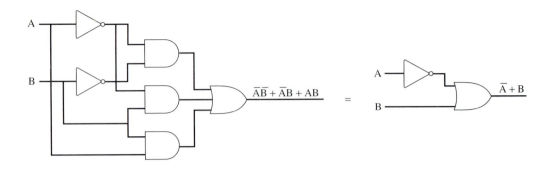

A	B	$\overline{A}$	$\overline{B}$	$\overline{A}\overline{B}$	$\overline{A}B$	AB	$\overline{A}\overline{B} + \overline{A}B + AB$	$\overline{A} + B$
0	0	1	1	1	0	0	1	1
0	1	1	0	0	1	0	1	1
1	0	0	1	0	0	0	0	0
1	1	0	0	0	0	1	1	1

└─ Equal ─┘

FIGURE 26–51   Answer to problem 3, Practice Problems 26–3.

$$\overline{\overline{A} \ \overline{(B + C)}}$$ Original expression
$$\overline{\overline{A}} + \overline{(B + C)}$$ 14b
$$A + (B + C)$$ 13a
$$A + B + C$$ 7b

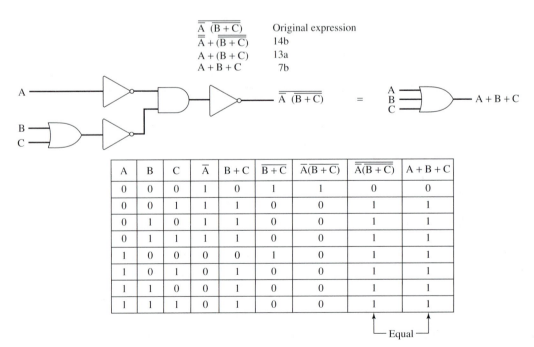

A	B	C	$\overline{A}$	B + C	$\overline{B + C}$	$\overline{A}(B + C)$	$\overline{\overline{A}(B + C)}$	A + B + C
0	0	0	1	0	1	1	0	0
0	0	1	1	1	0	0	1	1
0	1	0	1	1	0	0	1	1
0	1	1	1	1	0	0	1	1
1	0	0	0	0	1	0	1	1
1	0	1	0	1	0	0	1	1
1	1	0	0	1	0	0	1	1
1	1	1	0	1	0	0	1	1

└─ Equal ─┘

FIGURE 26–52   Answer to problem 4, Practice Problems 26–3.

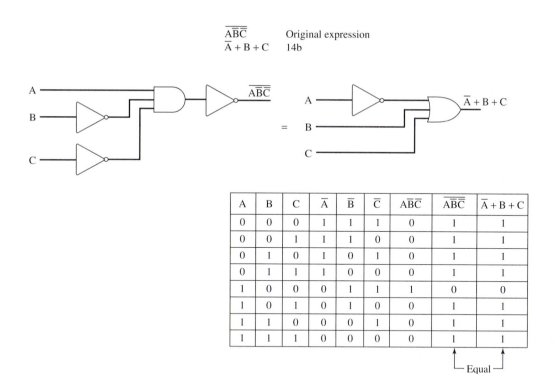

$$\overline{A\overline{B}\overline{C}} \qquad \text{Original expression}$$
$$\overline{A} + B + C \qquad 14b$$

A	B	C	$\overline{A}$	$\overline{B}$	$\overline{C}$	$A\overline{B}\,\overline{C}$	$\overline{A\overline{B}\,\overline{C}}$	$\overline{A} + B + C$
0	0	0	1	1	1	0	1	1
0	0	1	1	1	0	0	1	1
0	1	0	1	0	1	0	1	1
0	1	1	1	0	0	0	1	1
1	0	0	0	1	1	1	0	0
1	0	1	0	1	0	0	1	1
1	1	0	0	0	1	0	1	1
1	1	1	0	0	0	0	1	1

Equal

FIGURE 26–53   Answer to problem 5, Practice Problems 26–3.

APPLICATION PROBLEMS   26–4

Computers use thousands of logic circuits to perform their many functions. Each of these logic circuits can be separated into the basic logic circuits that have been presented in this chapter. We can write a Boolean equation for each of these circuits that will be a complete description of the logic functions that it performs. By combining all the Boolean equations for all the logic circuits in a computer, it is possible to completely describe the logical functions of a computer using Boolean equations. Emulators are software programs containing thousands of Boolean equations that describe complex logic circuits. Intel sells emulators for their microprocessors so that their customers can simulate the operations of Intel microprocessors on a computer. This type of simulation allows computer controlled testing in a fraction of the time it would take for bench testing.

Every logic circuit, no matter how complex, can be separated into simpler logic circuits with a single output. Every logic circuit in this chapter has a single output and we can write a Boolean equation for each of these circuits that takes the form:

Boolean expression using input variables = one logical output

In the following application problems we must analyze the English words to determine the inputs, the logical operators and the desired output. The logical operators are described using Boolean expressions. A Boolean equation has the logical output on one side of the equation and Boolean expressions on the other side of the equation.

The sequence of reasoning we can follow to solve these problems is to determine:

1. the logical output desired—this will become the output of the equivalent logic circuit
2. the input conditions, or signals—we can select alphabetic characters to represent these logic variables
3. the logical operations (AND, OR, NOT) that must be performed on the input variables in order to obtain the desired output
4. the Boolean equation

EXAMPLE 26–1

In a home, when a light switch is flipped to the ON position, a lamp will come on if its branch circuit breaker and the main circuit breaker are in the ON position. Write the Boolean Equation that represents the condition for the lamp to glow.

SOLUTION

First, we can identify that the logical output is that the lamp is glowing and we can specify the output variable to be:

$$L = \text{lamp is glowing}$$

Second, we can specify symbols to represent the input conditions necessary for the lamp to turn on.

$$M = \text{main circuit breaker is on}$$
$$B = \text{branch circuit breaker is on}$$
$$S = \text{light switch is on}$$

Third, we can determine, by analyzing the problem statement, that we must use the AND function for all three inputs. Fourth, we write the Boolean equation: $M \cdot B \cdot S = L$

EXAMPLE 26–2

The interior lights in an automobile are turned on when a door is opened or when the key is in the ignition and the interior light switch is put in the ON position. Write the Boolean equation that represents the conditions for the interior lights to be on.

SOLUTION

First, we can identify that the logical output is for the lights to be on. We can specify the output variable to be:

$$L = \text{lights ON}$$

Second, we can specify symbols to represent the input conditions necessary for the interior lights to glow.

$$D = \text{door is opened}$$
$$K = \text{key in ignition}$$
$$S = \text{switch in ON position}$$

Third, we must determine the logical operators by analyzing the problem statement. There are two conditions that can cause the light to come on. Either the door is opened *or* the key is in the ignition *and* the switch is on. By analyzing the use of the words *or* and *and* in the sentence, we can determine that both an AND and an OR operator are needed.

Fourth, we write the Boolean equation: $D + (K \cdot S) = L$.

PRACTICE PROBLEMS 26–4

1. A wall switch and an infrared (IR) detector that can sense the presence of a live body control the lights in a classroom. The wall switch has three positions—ON, AUTO, and OFF. In the ON position the lights are turned on, in the AUTO position the lights are turned on if the infrared detector senses the presence of a body, and in the OFF position the lights are always off. Write the Boolean equation that represents the conditions for the classroom lights to be on.

2. A single remote control can control both a TV and a VCR because of dual function keys. For example if the TV button is pressed and then the ON button, the TV will turn on; if the VCR button is pressed and then the ON button, the VCR will turn on. Write the Boolean equation that represents the conditions to control the TV and VCR.

3. A calculator has several dual function keys. The key labeled sin is used to calculate the sine of an angle. The same key can also calculate the inverse sine if the SHIFT key is pressed first. Write the Boolean equation that represents the conditions to perform each calculation.

4. The buttons on a computer mouse perform different functions, depending on the location of the mouse pointer. For example, when the mouse pointer is placed on the printer icon and the left mouse button is pressed, the printer will print a copy of the displayed document; if the mouse pointer is placed on the page icon and the left mouse button is pressed, the computer will display the document in the same format as it will be printed. Write the Boolean expression that represents the conditions to perform each function.

SOLUTIONS

1. $(\text{ON}) + ((\text{AUTO}) \cdot (\text{IR})) = (\text{Lights on})$

2. $(\text{TV pressed}) \cdot (\text{ON pressed}) = (\text{Turn TV on})$
   $(\text{VCR pressed}) \cdot (\text{ON pressed}) = (\text{Turn VCR on})$

3. $(\text{SHIFT}) \cdot (\text{SIN}) = $ (calculate inverse sine)
   $(\overline{\text{SHIFT}}) \cdot (\text{SIN}) = $ (calculate sine)

4. (Pointer on print icon) $\cdot$ (left mouse button) = (print document)
   (Pointer on page icon) $\cdot$ (left mouse button) = (display page)

SELF-TEST 26–2

Problems **1** through **5**: for the logic diagrams in Figure 26–54, (a) write the Boolean expression at the output, (b) simplify and justify each step, (c) draw the logic diagram of the simplified expression, and (d) construct a truth table to prove that the expressions are equal.

For the following Boolean expressions, (a) draw the logic diagram, (b) simplify and justify each step, (c) draw the logic diagram of the simplified expression, and (d) construct a truth table to prove that the expressions are equal.

6. $C(\overline{B + D})$

7. $\overline{A\overline{B} + \overline{C}}$

8. $\overline{\overline{A} + \overline{B}} + C$

9. $(AB + BC)(AC + BC)$

10. $\overline{\overline{AB} + A\overline{C} + \overline{BC}}$

1.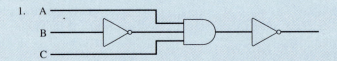

2.

3.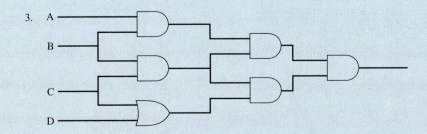

4.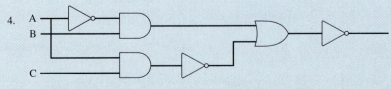

FIGURE 26–54   Problems 1 through 5, Self-test 26–2.

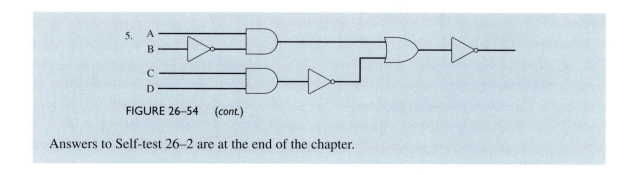

5.

FIGURE 26-54   (cont.)

Answers to Self-test 26–2 are at the end of the chapter.

CHAPTER 26 AT A GLANCE

PAGE	LOGIC FUNCTIONS, POSTULATES, AND THEOREMS	LOGIC SYMBOLS AND CIRCUITS
723	For the AND function, the output is true only when all inputs are true. Otherwise, the output is false.	

A
B
C   A·B·C
(a)

Condition	A	B	C	A·B·C
1	0	0	0	0
2	0	0	1	0
3	0	1	0	0
4	0	1	1	0
5	1	0	0	0
6	1	0	1	0
7	1	1	0	0
8	1	1	1	1

(b)

FIGURE 26–55   Three-input AND gate: (a) logic symbol; (b) truth table.

725	For the OR function, the output is true when any one of the inputs is true. Otherwise, the output is false.	

A
B
C   A + B + C
(a)

Condition	A	B	C	A + B + C
1	0	0	0	0
2	0	0	1	1
3	0	1	0	1
4	0	1	1	1
5	1	0	0	1
6	1	0	1	1
7	1	1	0	1
8	1	1	1	1

(b)

FIGURE 26–56   Three-input OR gate: (a) logic symbol; (b) truth table.

BOOLEAN ALGEBRA   **755**

726   For the NOT function, the output is the complement of the input.

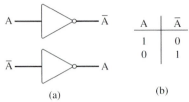

(a)

(b)

FIGURE 26–57   Inverter: (a) logic symbol; (b) truth table.

732   Postulates, laws, and theorems are used in the simplification and manipulation of logic expressions.

END OF CHAPTER PROBLEMS 26–1

Draw the logic diagrams that represent the following Boolean expressions:

1. $\overline{A + \overline{B}}$
2. $\overline{AB} + A\overline{C}$
3. $ABC + D$
4. $X(\overline{Y + Z})$
5. $\overline{A + B\overline{C}}$
6. $\overline{AB} + \overline{AC}$
7. $A\overline{B} + \overline{A}B$
8. $B(\overline{\overline{C} + D})$
9. $\overline{AB} + C + D$
10. $\overline{\overline{BCD} + E}$
11. $AB + A\overline{C} + B\overline{C}$
12. $\overline{(B + C)(\overline{C} + D)}$
13. $\overline{A\overline{B} + BC}$
14. $\overline{\overline{B}(A + \overline{C})}$
15. $\overline{\overline{AB}C + D}$
16. $\overline{A} + \overline{B} + C$
17. $(\overline{X} + Y)(Y + Z)$
18. $\overline{AC + \overline{D}}$
19. $\overline{\overline{AB} + \overline{CD}}$
20. $\overline{\overline{AB} + \overline{C}}$

Write the Boolean expression represented by the logic diagrams in Figures 26–58 through 26–60 (pp. 756–759).

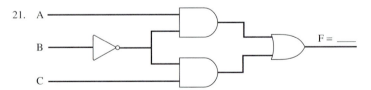

21.

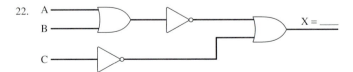

22.

FIGURE 26–58   Logic diagrams for End of Chapter Problems 26–1, problems 21 and 22.

23.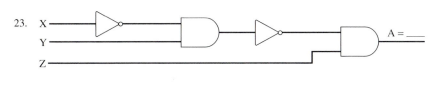

24.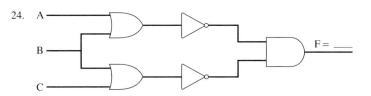

25.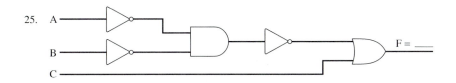

26.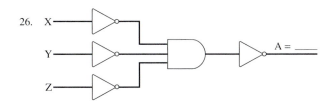

27. 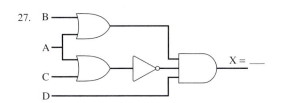   28.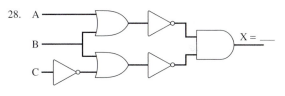

29. 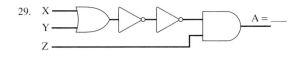   30.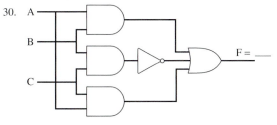

FIGURE 26–59   Logic diagrams for End of Chapter Problems 26–1, problems 23 through 30.

31. 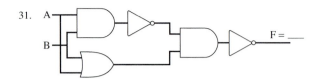 F = ___

32. 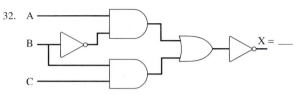 X = ___

33. 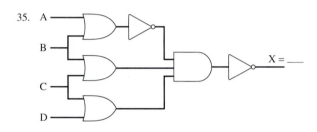 A = ___

34. 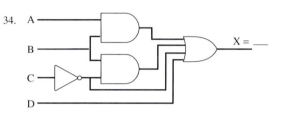 X = ___

35. 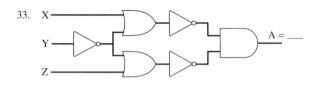 X = ___

36. 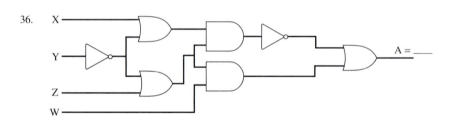 A = ___

37. 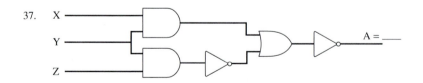 A = ___

38. 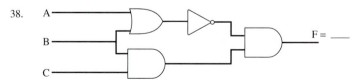 F = ___

FIGURE 26–60   Logic diagrams for End of Chapter Problems 26–1, problems 31 through 40.

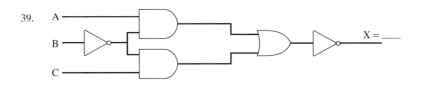

39. A
     B
     C
                                    X = ___

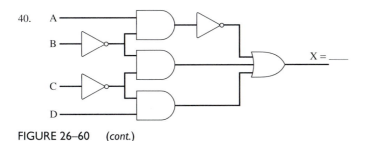

40. A
     B
     C
     D
                                    X = ___

FIGURE 26–60   *(cont.)*

END OF CHAPTER PROBLEMS 26–2

For each logic diagram in Figure 26–61, (a) write the Boolean expression, (b) simplify the Boolean expression and justify each step, (c) draw the simplified logic diagram, and (d) construct a truth table to prove that the two expressions are equal.

1. See Figure 26–61, circuit 1.
3. See Figure 26–61, circuit 3.
5. See Figure 26–61, circuit 5.

2. See Figure 26–61, circuit 2.
4. See Figure 26–61, circuit 4.
6. See Figure 26–61, circuit 6.

FIGURE 26–61
Circuits 1, 2, 3

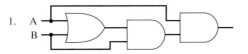

1. A
    B

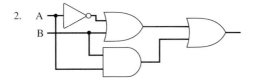

2. A
    B

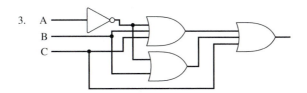

3. A
    B
    C

FIGURE 26–61
Circuits 4, 5,
and 6

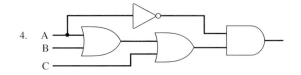

4.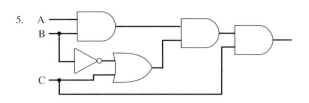

5.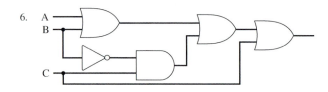

6.

END OF CHAPTER PROBLEMS 26–3

In each problem, (a) simplify the output expression and justify each step, (b) draw the simplified logic diagram, and (c) construct a truth table to prove that the expressions are equal.

1. See Figure 26–58, circuit for Problem 22.

2. See Figure 26–59, circuit for Problem 25.

3. See Figure 26–59, circuit for Problem 27.

4. See Figure 26–59, circuit for Problem 30.

5. See Figure 26–60, circuit for Problem 32.

6. See Figure 26–60, circuit for Problem 40.

7. $\overline{A + B\overline{\overline{C}}}$

8. $\overline{\overline{AB} + \overline{A}C}$

9. $\overline{(\overline{B + C})(C + D)}$

10. $\overline{\overline{\overline{B(\overline{A} + \overline{C})}}}$

11. $\overline{\overline{AB} + C\overline{D}}$

12. $\overline{A\overline{B} + C}$

END OF CHAPTER PROBLEMS 26–4

1. In some automobiles, in order to start the engine, the transmission must be in PARK and the ignition key must be turned to ON. Write the Boolean equation that represents the conditions for the engine to start.

2. To start a truck, the brake must be ON, the transmission must be in neutral, and the ignition key must be turned to ON. Write the Boolean equation that represents the conditions for the engine to start.

3. A computer needs three things to start up correctly. It needs power, the ON button must be pressed, and it needs operating system software. Write the Boolean equation that represents the conditions for the computer to start up.

4. A computer needs three things to start up correctly. It must be plugged into a power source, it must be turned on, and it must have operating system software. However, if there is a data disk in the floppy drive it will not start up. Write the Boolean equation that represents the conditions for the computer to start up correctly.

5. When using a word processor, you can scroll down a document by using the down arrow key, the page down key, or by placing the mouse pointer in the scroll bar and clicking the left mouse button. Write the Boolean equation that represents the condition to scroll down a document.

6. A "warm boot" can be initiated in some computers by pressing the "restart" button or by simultaneously pressing three keys—the Ctrl, Alt, and Del keys. Write the Boolean equation that represents the conditions to initiate a warm boot.

ANSWERS TO SELF-TESTS

SELF-TEST 26–1

1. $F = \overline{\overline{A} + B}$

2. $F = \overline{AB}$

3. $F = A + \overline{B} + C$

4. $F = (A + B)(B + C)$

5. $A = XY + XZ$

6. $F = \overline{A}\overline{B} + C$

7. $F = C(\overline{A + \overline{B}})$

8. $X = (\overline{A + B})(B + C)$

9. $A = (\overline{X}Y)(Y + Z)$

10. $F = \overline{\overline{A}(B + C)}$

11 through 20. See Figure 26–62.

11.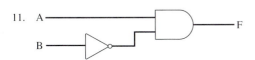

12.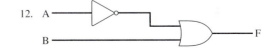

13.

14.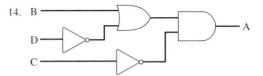

15.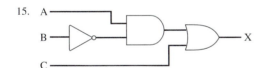

16.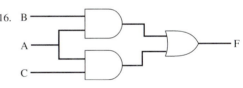

FIGURE 26–62   Answers to questions 11 through 20 for Self-test 26–1.

17.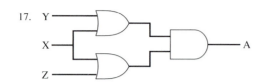

18.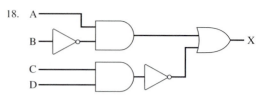

19.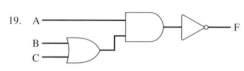

20.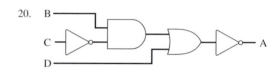

FIGURE 26–62 *(cont.)*

SELF-TEST 26–2

1. See Figure 26–63.
2. See Figure 26–64.
3. See Figure 26–65.
4. See Figure 26–66 (p. 764).
5. See Figure 26–67 (p. 765).
6. See Figure 26–68 (pp. 765–766).
7. See Figure 26–69 (p. 766).
8. See Figure 26–70 (p. 767).
9. See Figure 26–71 (pp. 767–768).
10. See Figure 26–72 (p. 768).

FIGURE 26–63
Logic diagrams
and truth table
for $\overline{A\overline{B}C} = \overline{A} +$
$B + \overline{C}$. Answer
to problem 1,
Self-test 26–2.

$\overline{A\overline{B}C}$ Original expression
$\overline{A} + B + \overline{C}$ 14b

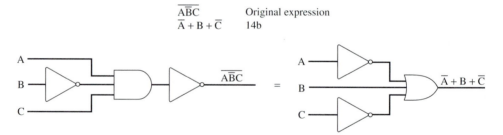

A	B	C	$\overline{A}$	$\overline{B}$	$\overline{C}$	$A\overline{B}C$	$\overline{A\overline{B}C}$	$\overline{A} + B + \overline{C}$
0	0	0	1	1	1	0	1	1
0	0	1	1	1	0	0	1	1
0	1	0	1	0	1	0	1	1
0	1	1	1	0	0	0	1	1
1	0	0	0	1	1	0	1	1
1	0	1	0	1	0	1	0	0
1	1	0	0	0	1	0	1	1
1	1	1	0	0	0	0	1	1

└─ Equal ─┘

$$\overline{\overline{A + \overline{B}} + C} \quad \text{Original expression}$$
$$\overline{\overline{\overline{A + \overline{B}}} \cdot \overline{C}} \quad 14a$$
$$(A + \overline{B})\overline{C} \quad 13a$$
$$\overline{C}(A + \overline{B}) \quad 6a$$

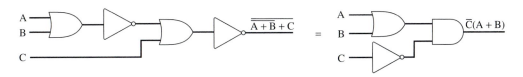

A	B	C	$\overline{C}$	$A + \overline{B}$	$\overline{A + \overline{B}}$	$\overline{A + \overline{B}} + C$	$\overline{\overline{A + \overline{B}} + C}$	$\overline{C}(A + \overline{B})$
0	0	0	1	0	1	1	0	0
0	0	1	0	0	1	1	0	0
0	1	0	1	1	0	0	1	1
0	1	1	0	1	0	1	0	0
1	0	0	1	1	0	0	1	1
1	0	1	0	1	0	1	0	0
1	1	0	1	1	0	0	1	1
1	1	1	0	1	0	1	0	0

└─ Equal ─┘

FIGURE 26–64 Logic diagram and truth table for $\overline{\overline{A + \overline{B}} + C} = \overline{C}(A + \overline{B})$. Answer to problem 2, Self-test 26–2.

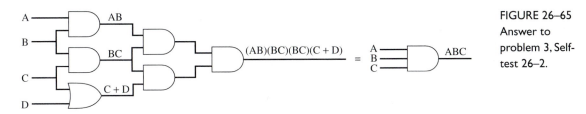

FIGURE 26–65 Answer to problem 3, Self-test 26–2.

(AB)(BC)(BC)(C + D)	Original expression
(AB)(BC)(C + D)	11b
ABC (C + D)	11a
ABCC + ABCD	8a
ABC + ABCD	11a
ABC (1 + D)	8a
ABC	10b

FIGURE 26–65
(cont.)

A	B	C	D	AB	BC	C + D	(AB)(BC)	(BC)(C + D)	(AB) (BC) (BC) (C + D)	ABC
0	0	0	0	0	0	0	0	0	0	0
0	0	0	1	0	0	1	0	0	0	0
0	0	1	0	0	0	1	0	0	0	0
0	0	1	1	0	0	1	0	0	0	0
0	1	0	0	0	0	0	0	0	0	0
0	1	0	1	0	0	1	0	0	0	0
0	1	1	0	0	1	1	0	1	0	0
0	1	1	1	0	1	1	0	1	0	0
1	0	0	0	0	0	0	0	0	0	0
1	0	0	1	0	0	1	0	0	0	0
1	0	1	0	0	0	1	0	0	0	0
1	0	1	1	0	0	1	0	0	0	0
1	1	0	0	1	0	0	0	0	0	0
1	1	0	1	1	0	1	0	0	0	0
1	1	1	0	1	1	1	1	1	1	1
1	1	1	1	1	1	1	1	1	1	1

└── Equal ──┘

$\overline{\overline{A}B + \overline{AC}}$ Original expression
$\overline{\overline{A}B} \cdot \overline{\overline{AC}}$ 14a
$(A + \overline{B}) AC$ 14b
$AAC + A\overline{B}C$ 8a
$AC + A\overline{B}C$ 11a
$AC (1 + \overline{B})$ 8a
$AC (1)$ 10b
AC 10a

A	B	C	$\overline{A}$	$\overline{A}B$	AC	$\overline{AC}$	$\overline{A}B + \overline{AC}$	$\overline{\overline{A}B + \overline{AC}}$
0	0	0	1	0	0	1	1	0
0	0	1	1	0	0	1	1	0
0	1	0	1	1	0	1	1	0
0	1	1	1	1	0	1	1	0
1	0	0	0	0	0	1	1	0
1	0	1	0	0	1	0	0	1
1	1	0	0	0	0	1	1	0
1	1	1	0	0	1	0	0	1

└── Equal ──┘

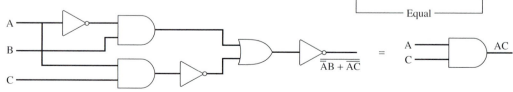

FIGURE 26–66   Answer to problem 4, Self-test 26–2.

$\overline{A\overline{B} + \overline{CD}}$ Original expression
$\overline{A\overline{B}} \cdot \overline{\overline{CD}}$ 14a
$\overline{A\overline{B}} \cdot CD$ 13a
$(\overline{A} + B)\, CD$ 14b
$CD\,(\overline{A} + B)$ 7a

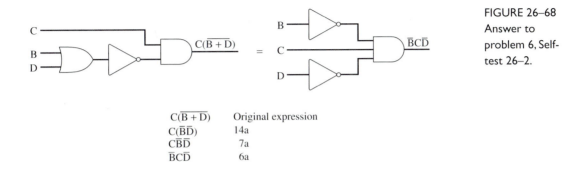

A	B	C	D	$\overline{A}$	$\overline{B}$	CD	$\overline{CD}$	$A\overline{B}$	$A\overline{B} + \overline{CD}$	$\overline{A\overline{B} + \overline{CD}}$	$\overline{A} + B$	$CD\,(\overline{A} + B)$
0	0	0	0	1	1	0	1	0	1	0	1	0
0	0	0	1	1	1	0	1	0	1	0	1	0
0	0	1	0	1	1	0	1	0	1	0	1	0
0	0	1	1	1	1	1	0	0	0	1	1	1
0	1	0	0	1	0	0	1	0	1	0	1	0
0	1	0	1	1	0	0	1	0	1	0	1	0
0	1	1	0	1	0	0	1	0	1	0	1	0
0	1	1	1	1	0	1	0	0	0	1	1	1
1	0	0	0	0	1	0	1	1	1	0	0	0
1	0	0	1	0	1	0	1	1	1	0	0	0
1	0	1	0	0	1	0	1	1	1	0	0	0
1	0	1	1	0	1	1	0	1	1	0	0	0
1	1	0	0	0	0	0	1	0	1	0	1	0
1	1	0	1	0	0	0	1	0	1	0	1	0
1	1	1	0	0	0	0	1	0	1	0	1	0
1	1	1	1	0	0	1	0	0	0	1	1	1

└─── Equal ───┘

FIGURE 26–67   Answer to problem 5, Self-test 26–2.

FIGURE 26–68
Answer to
problem 6, Self-
test 26–2.

$C(\overline{B + D})$ Original expression
$C(\overline{B}\overline{D})$ 14a
$C\overline{B}\overline{D}$ 7a
$\overline{B}C\overline{D}$ 6a

FIGURE 26–68
(cont.)

B	C	D	$\overline{B}$	$\overline{D}$	B + D	$\overline{B+D}$	$C(\overline{B+D})$	$\overline{B}C\overline{D}$
0	0	0	1	1	0	1	0	0
0	0	1	1	0	1	0	0	0
0	1	0	1	1	0	1	1	1
0	1	1	1	0	1	0	0	0
1	0	0	0	1	1	0	0	0
1	0	1	0	0	1	0	0	0
1	1	0	0	1	1	0	0	0
1	1	1	0	0	1	0	0	0

└ Equal ┘

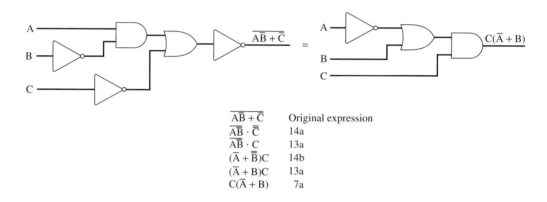

$\overline{A\overline{B}+\overline{C}}$	Original expression
$\overline{A\overline{B}} \cdot \overline{\overline{C}}$	14a
$\overline{A\overline{B}} \cdot C$	13a
$(\overline{A} + \overline{\overline{B}})C$	14b
$(\overline{A} + B)C$	13a
$C(\overline{A} + B)$	7a

A	B	C	$\overline{A}$	$\overline{B}$	$\overline{C}$	$A\overline{B}$	$A\overline{B} + \overline{C}$	$\overline{A\overline{B} + \overline{C}}$	$\overline{A} + B$	$C(\overline{A} + B)$
0	0	0	1	1	1	0	1	0	1	0
0	0	1	1	1	0	0	0	1	1	1
0	1	0	1	0	1	0	1	0	1	0
0	1	1	1	0	0	0	0	1	1	1
1	0	0	0	1	1	1	1	0	0	0
1	0	1	0	1	0	1	1	0	0	0
1	1	0	0	0	1	0	1	0	1	0
1	1	1	0	0	0	0	0	1	1	1

└─ Equal ─┘

FIGURE 26–69   Answer to problem 7, Self-test 26–2.

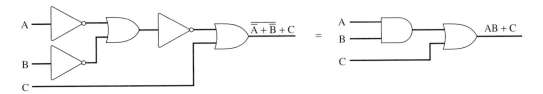

$$\overline{\overline{A} + \overline{B}} + C \qquad \text{Original expression}$$
$$\overline{\overline{\overline{A}} \cdot \overline{\overline{B}}} + C \qquad \text{14a}$$
$$AB + C \qquad \text{13a}$$

A	B	C	$\overline{A}$	$\overline{B}$	$\overline{A} + \overline{B}$	$\overline{\overline{A} + \overline{B}}$	$\overline{\overline{A} + \overline{B}} + C$	AB	AB + C
0	0	0	1	1	1	0	0	0	0
0	0	1	1	1	1	0	1	0	1
0	1	0	1	0	1	0	0	0	0
0	1	1	1	0	1	0	1	0	1
1	0	0	0	1	1	0	0	0	0
1	0	1	0	1	1	0	1	0	1
1	1	0	0	0	0	1	1	1	1
1	1	1	0	0	0	1	1	1	1

└— Equal —┘

FIGURE 26–70   Answer to problem 8, Self-test 26–2.

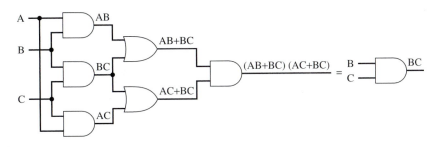

FIGURE 26–71
Answer to
problem 9, Self-
test 26–2.

(AB+BC) (AC+BC)	Original expression
(BC+AB) (BC+AC)	6b
BC + (AB) (AC)	8b
BC + AABC	6a
BC + ABC	11a
BC (1+A)	8a
BC(1)	11b
BC	10a

FIGURE 26–71
(cont.)

A	B	C	AB	BC	AC	AB+BC	AC+BC	(AB+BC)(AC+BC)
0	0	0	0	0	0	0	0	0
0	0	1	0	0	0	0	0	0
0	1	0	0	0	0	0	0	0
0	1	1	0	1	0	1	1	1
1	0	0	0	0	0	0	0	0
1	0	1	0	0	1	0	1	0
1	1	0	1	0	0	1	0	0
1	1	1	1	1	1	1	1	1

⬆ ———————— Equal ———————— ⬆

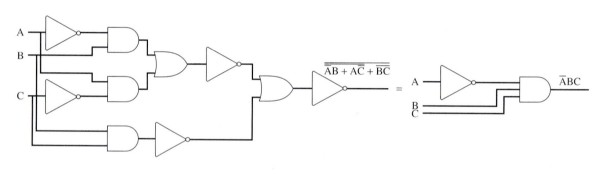

$\overline{\overline{A}B + A\overline{C} + \overline{BC}}$ Original expression
$(\overline{\overline{A}B + A\overline{C}})\overline{\overline{BC}}$ 14a
$\overline{A}\,\overline{B}BC + ABC\overline{C}$ 8a
$\overline{A}BC + 0$ 11a, 12a
$\overline{A}BC$ 9b

A	B	C	$\overline{A}$	$\overline{C}$	$\overline{A}B$	$A\overline{C}$	$\overline{A}B + A\overline{C}$	$\overline{\overline{A}B + A\overline{C}}$	BC	$\overline{BC}$	$\overline{\overline{A}B + A\overline{C}} + \overline{BC}$	$\overline{\overline{\overline{A}B + A\overline{C}} + \overline{BC}}$	$\overline{A}BC$
0	0	0	1	1	0	0	0	1	0	1	1	0	0
0	0	1	1	0	0	0	0	1	0	1	1	0	0
0	1	0	1	1	1	0	1	0	0	1	1	0	0
0	1	1	1	0	1	0	1	0	1	0	0	1	1
1	0	0	0	1	0	1	1	0	0	1	1	0	0
1	0	1	0	0	0	0	0	1	0	1	1	0	0
1	1	0	0	1	0	1	1	0	0	1	1	0	0
1	1	1	0	0	0	0	0	1	1	0	1	0	0

⬆ ——— Equal ——— ⬆

FIGURE 26–72   Answer to problem 10, Self-test 26–2.

Karnaugh Maps

27

Introduction

In Chapter 26, you learned how to simplify logic diagrams or logic expressions by using various laws, postulates, and theorems. In this chapter an alternative method of simplification is offered.

Chapter Objectives

In this chapter you will learn how to:

1. Interpret Karnaugh maps.
2. Simplify logic diagrams using Karnaugh maps.

TWO- AND THREE-VARIABLE EXPRESSIONS   27-1

27-1-1 Two-Variable Expression

When digital circuits are first designed, they may be more complex than necessary. By using Boolean algebra, theorems, and laws previously discussed, it is often possible to simplify logic expressions without changing the original logic functions. Simplified expressions are very desirable because they will result in less complex circuits with increased reliability. In addition, the smaller circuits result in increased speeds and greater economy.

Karnaugh maps provide a method by which a Boolean expression can be visualized and, more important, they represent a technique for simplification. A Karnaugh map is made up of a group of squares. The number of squares required depends on the number of variables. A two-variable expression requires four squares; three variables, eight squares; four variables, 16 squares; and so on. The number of squares required is equal to 2 raised to the number of variables. Karnaugh maps are seldom useful for more than five variables.

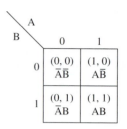

A	B	AB
0	0	0
0	1	0
1	0	0
1	1	1

(a)

A	B	AB
$\overline{A}$	$\overline{B}$	$\overline{A}\overline{B}$
$\overline{A}$	B	$\overline{A}B$
A	$\overline{B}$	$A\overline{B}$
A	B	AB

(b)

FIGURE 27–1   Truth tables for a two-input AND gate. (a) Standard truth table. (b) Truth table with 0s and 1s replaced with the variable and the complement of the variable.

FIGURE 27–2
Karnaugh map of the truth table in Figure 27–1.

To begin, consider the truth table in Figure 27–1. The standard truth table for a two-input AND gate is shown in Figure 27–1(a). Note that a true output (1) exists only when A is true AND B is true. The other three outputs are indicated as not true (0).

In Figure 27–1(b), a similar table is produced. In this table, 1 is replaced by the variable, and 0 is replaced by the complement of the variable. The variables are then ANDed. Again, there exist four possible combinations. These are $\overline{A}\overline{B} + \overline{A}B + A\overline{B} + AB$. When both tables of Figure 27–1 are combined, the result is the Karnaugh map illustrated in Figure 27–2.

Note that the squares represent each of the four possible conditions for the two variables. The top-left square corresponds to the condition where both A and B are logical 0, and so it is expressed as $\overline{A}\overline{B}$. The bottom-right square corresponds to the condition where both A and B equal a logical 1 and therefore can be expressed as AB. In the top-right square, A is equal to a logical 1, but B is equal to a logical 0, and it is expressed as $A\overline{B}$. In the bottom-left square, A is equal to 0 and B is equal to 1, and it is expressed as $\overline{A}B$.

Two Karnaugh maps, each having two variables, appear in Figure 27–3. Note that a 1 appears in the top-right and bottom-right squares of Figure 27–3(a). The 1s indicate the presence of the term that corresponds to that square. The Karnaugh map expression for Figure 27–3(a) is

$$\text{map} = A\overline{B} + AB \qquad \text{(refer to Figure 27–3(a))}$$

FIGURE 27–3
Two-variable Karnaugh map. (a) Map for the expression $A\overline{B}$ + AB. (b) Map for the expression $\overline{A}\,\overline{B}$ + $\overline{A}B$ + AB.

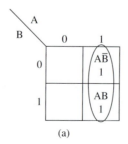

(a)

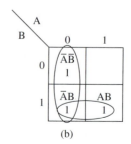

(b)

The logical expression for Figure 27–3(b) is

$$\text{map} = \overline{A}\,\overline{B} + \overline{A}B + AB \qquad \text{(refer to Figure 27–3(b))}$$

The empty squares represent the absence of a term and are omitted to make the map easier to read.

The rules for simplifying a two-variable Boolean expression using a Karnaugh map are as follows:

1. Two adjacent squares are combined to represent a single variable.
2. The resulting variable is common to both terms.

Returning to Figure 27–3(a), note that there are two adjacent 1s that may be combined to represent one variable. This is shown in Figure 27–4.

In the expression $A\overline{B} + AB$, A is the common variable for both terms. Therefore, $A\overline{B} + AB = A$, which is much simpler than the original expression. That this is true may be easily verified through the use of some of the laws and theorems previously discussed.

$A\overline{B} + AB$	original expression
$A(\overline{B} + B)$	algebraic property 8a
$A(1)$	theorem 12b
A	theorem 10a

Therefore,

$$A\overline{B} + AB = A$$

For Figure 27–3(b), you will note that there are two sets of adjacent 1s, as shown in Figure 27–5. The two *adjacent vertical* 1s give $\overline{A}\,\overline{B} + \overline{A}B$, which simplifies to $\overline{A}$, because $\overline{A}$ is common to both terms. The two *adjacent horizontal* 1s give $\overline{A}B + AB$, which simplifies to B because B is common to both terms. Therefore,

$$\overline{A}\,\overline{B} + \overline{A}B + AB = \overline{A} + B$$

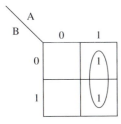

FIGURE 27–4   Two adjacent 1s combined to represent one variable.

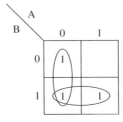

FIGURE 27–5 Karnaugh map showing two sets of adjacent squares. Map $= \overline{A}\,\overline{B} +$ $\overline{A}B + AB$.

You may verify this expression by using previous laws and theorems, as follows:

$$\overline{A}\overline{B} + \overline{A}B + AB \qquad \text{original expression}$$
$$\overline{A}(\overline{B} + B) + AB \qquad \text{theorem 8a}$$
$$\overline{A}(1) + AB \qquad \text{theorem 12b}$$
$$\overline{A} + AB \qquad \text{theorem 10a}$$
$$\overline{A} + B \qquad \text{theorem 16b}$$

Therefore,

$$\overline{A}\overline{B} + \overline{A}B + AB = \overline{A} + B$$

EXAMPLE 27–1

Write the Boolean expression for the map in Figure 27–6.

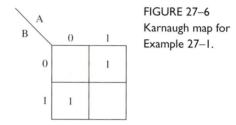

FIGURE 27–6
Karnaugh map for
Example 27–1.

SOLUTION   Map $= \overline{A}B + A\overline{B}$. All other boxes are omitted.

EXAMPLE 27–2

Draw the map for the expression $AB + A\overline{B}$.

SOLUTION   Place a 1 in each box that represents a given expression. See Figure 27–7 and circle adjacent squares that contain terms.

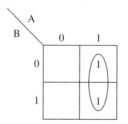

FIGURE 27–7
Karnaugh map of the
expression $AB + A\overline{B}$.

27–1–2 Three-Variable Expression

Let's consider the truth table for a three-input AND gate, as shown in Figure 27–8. In the table for Figure 27–8(b), 1s and 0s have been replaced by their respective variables. Note that for three variables there are eight possible combinations. The Karnaugh map for the three variables is shown in Figure 27–9. Each square represents the logical product of the three variables.

FIGURE 27–8
Truth table for
three-input
AND gate:
(a) using 1s and
0s; (b) using
variables.

A	B	C	ABC
0	0	0	0
0	0	1	0
0	1	0	0
0	1	1	0
1	0	0	0
1	0	1	0
1	1	0	0
1	1	1	1

(a)

A	B	C	ABC
$\overline{A}$	$\overline{B}$	$\overline{C}$	$\overline{A}\overline{B}\overline{C}$
$\overline{A}$	$\overline{B}$	C	$\overline{A}\overline{B}C$
$\overline{A}$	B	$\overline{C}$	$\overline{A}B\overline{C}$
$\overline{A}$	B	C	$\overline{A}BC$
A	$\overline{B}$	$\overline{C}$	$A\overline{B}\overline{C}$
A	$\overline{B}$	C	$A\overline{B}C$
A	B	$\overline{C}$	$AB\overline{C}$
A	B	C	ABC

(b)

The rules for simplifying a three-variable Boolean expression using a Karnaugh map are as follows:

1. A single square represents a three-variable term.
2. Two adjacent squares are combined to represent a two-variable term.
3. Four adjacent squares are combined to represent one variable.

Rule 1 for the three-variable Karnaugh map is exemplified in Figure 27–10. The terms in the squares cannot be combined. Therefore,

$$\text{map} = \overline{A}B\overline{C} + A\overline{B}\overline{C} + \overline{A}\overline{B}C$$

Figure 27–11 on the next page illustrates some possible combinations of the two adjacent squares (rule 2). Figure 27–11(a) shows how two end squares may be considered as adjacent to each other if we think of the map as a *wraparound cylinder.*

The map reading for Figure 27–11(a) is $\overline{A}\,\overline{B}\overline{C} + A\overline{B}\overline{C} = \overline{B}\,\overline{C}$, because $\overline{B}\,\overline{C}$ is common to both squares. Figure 27–11(b) shows two sets of adjacent squares. The two top-left squares read $\overline{A}\,\overline{B}\,\overline{C} + \overline{A}B\overline{C} = \overline{A}\,\overline{C}$; the two bottom-right squares read $ABC + A\overline{B}C = AC$. Therefore, the map reading for Figure 27–11(b) is

$$\text{map} = \overline{A}\,\overline{C} + AC$$

Rule 3 is illustrated by Figure 27–12 on the next page. In Figure 27–12(a), the four outer boxes may be combined because they are considered to be adjacent to each other (wraparound cylinder). The map reading for Figure 27–12(a) is $\overline{A}\,\overline{B}\overline{C} + A\overline{B}\overline{C} + \overline{A}\,\overline{B}C + A\overline{B}C = \overline{B}$.

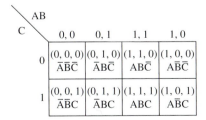

C \ AB	0, 0	0, 1	1, 1	1, 0
0	(0, 0, 0) $\overline{A}\overline{B}\overline{C}$	(0, 1, 0) $\overline{A}B\overline{C}$	(1, 1, 0) $AB\overline{C}$	(1, 0, 0) $A\overline{B}\overline{C}$
1	(0, 0, 1) $\overline{A}\overline{B}C$	(0, 1, 1) $\overline{A}BC$	(1, 1, 1) ABC	(1, 0, 1) $A\overline{B}C$

FIGURE 27–9 General form of
three-variable Karnaugh map.

C \ AB	0, 0	0, 1	1, 1	1, 0
0		①		①
1	①			

FIGURE 27–10 Single square
represents a three-variable term.

FIGURE 27–11
Two adjacent
square
combinations:
(a) two adjacent
squares rule
(wraparound
cylinder);
(b) two adjacent
squares rule.

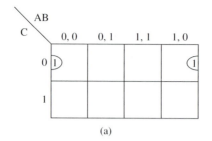
(a)

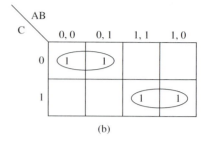
(b)

This is because $\overline{B}$ is the common variable for all the terms, which may be verified by the laws and theorems previously discussed in Chapter 26.

$\overline{A}\,\overline{B}\,\overline{C} + A\overline{B}\,\overline{C} + \overline{A}\,\overline{B}C + A\overline{B}C$	original expression
$(\overline{A}\,\overline{B}\,\overline{C} + \overline{A}\,\overline{B}C) + (A\overline{B}\,\overline{C} + A\overline{B}C)$	theorem 7b
$\overline{A}\,\overline{B}(\overline{C} + C) + A\overline{B}(\overline{C} + C)$	theorem 8a
$\overline{A}\,\overline{B}(1) + A\overline{B}(1)$	theorem 10a
$\overline{A}\,\overline{B} + A\overline{B}$	theorem 8a
$\overline{B}(\overline{A} + A)$	theorem 8a
$\overline{B}(1)$	theorem 12b
$\overline{B}$	theorem 10a

Figure 27–12(b) shows another possible combination involving four adjacent squares. The map reading for Figure 27–12(b) is

$$\overline{A}B\overline{C} + AB\overline{C} + \overline{A}BC + ABC = B$$

Note that B is the common variable in all the terms, which may be verified as follows:

$\overline{A}B\overline{C} + AB\overline{C} + \overline{A}BC + ABC$	original expression
$\overline{A}B\overline{C} + \overline{A}BC + AB\overline{C} + ABC$	theorem 7b
$\overline{A}B(\overline{C} + C) + AB(\overline{C} + C)$	theorem 8a
$\overline{A}B(1) + AB(1)$	theorem 10a
$\overline{A}B + AB$	theorem 8a

FIGURE 27–12
Four adjacent
square
combinations:
(a) four adjacent
squares
(wraparound
cylinder); (b)
four adjacent
squares.

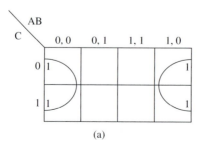
(a)

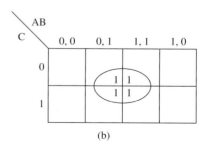
(b)

$$B(\overline{A} + A) \qquad \text{theorem 8a}$$
$$B(1) \qquad \text{theorem 12b}$$
$$B \qquad \text{theorem 10a}$$

EXAMPLE 27–3

Write the simplified expression for the Karnaugh map in Figure 27–13.

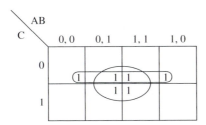

FIGURE 27–13
Karnaugh map
solution. Map =
$B + \overline{C}$.

SOLUTION   The four top squares (top row) may be combined to read $\overline{A}\,\overline{B}\,\overline{C}$ + $\overline{A}B\overline{C} + AB\overline{C} + A\overline{B}\,\overline{C} = \overline{C}$, because $\overline{C}$ is common to all the terms. The middle four squares may be combined to read $\overline{A}B\overline{C} + AB\overline{C} + \overline{A}BC + ABC = B$, because B is common to all the terms. Therefore, the resulting expression is

$$\text{map} = B + \overline{C}$$

Complex Boolean expressions may often be simplified by the construction of a Karnaugh map. Suppose we wish to simplify the following logic statement:

$$x = A\overline{B}\,\overline{C} + ABC + \overline{A}\,\overline{B}C + A\overline{B}C + \overline{A}\,\overline{B}\,\overline{C}$$

Since there are three variables, a Karnaugh map having eight squares is constructed, and 1s are placed in the square corresponding to each term of the expression, as shown in Figure 27–14. Note that the four outer squares may be combined because of the wraparound concept, resulting in

$$\overline{A}\,\overline{B}\,\overline{C} + A\overline{B}\,\overline{C} + \overline{A}\,\overline{B}C + A\overline{B}C = \overline{B}$$

The two lower-right squares may be combined to give

$$ABC + A\overline{B}C = AC$$

Therefore, the simplified expression is

$$x = \overline{B} + AC$$

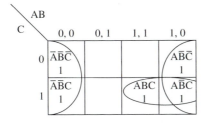

FIGURE 27–14
Karnaugh map
solution. Map =
$\overline{B} + AC$.

EXAMPLE 27–4

Write the Boolean expression that produced the Karnaugh map in Figure 27–15. Write the simplified expression.

FIGURE 27–15
Karnaugh map for the expression $\overline{A}\overline{B}\overline{C} + A\overline{B}\overline{C} + \overline{A}\overline{B}C + \overline{A}BC$ in Example 27–4.

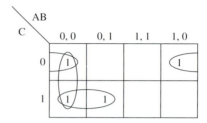

SOLUTION For the original expression,

$$\text{map} = \overline{A}\,\overline{B}\,\overline{C} + A\overline{B}\,\overline{C} + \overline{A}\,\overline{B}C + \overline{A}\,\overline{B}C$$

The squares representing $\overline{A}\,\overline{B}\,\overline{C}$ and $A\overline{B}\,\overline{C}$ are adjacent squares. $\overline{B}$ and $\overline{C}$ are common to both, so the terms simplify to $\overline{B}\,\overline{C}$. The squares representing $\overline{A}\,\overline{B}C$ and $\overline{A}BC$ are adjacent squares. $\overline{A}$ and C are common to both, so the terms simplify to $\overline{A}C$. The squares representing $\overline{A}\,\overline{B}\,\overline{C}$ and $\overline{A}\,\overline{B}C$ are adjacent squares. $\overline{A}$ and $\overline{B}$ are common to both, so the terms simplify to $\overline{A}\,\overline{B}$. The simplified expression is

$$\text{map} = \overline{A}\,\overline{B} + \overline{A}C + \overline{B}\,\overline{C}$$

EXAMPLE 27–5

Construct a Karnaugh map and simplify the expression $\overline{A}B\overline{C} + \overline{A}BC + ABC$.

FIGURE 27–16
Karnaugh map for the expression $\overline{A}B\overline{C} + \overline{A}BC + ABC$ in Example 27–5.

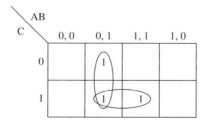

SOLUTION Figure 27–16 is the Karnaugh map. An inspection of the map shows that the squares representing $\overline{A}B\overline{C}$ and $\overline{A}BC$ are adjacent, and $\overline{A}$ and B are common to both. Then $\overline{A}B\overline{C} + \overline{A}BC$ reduces to $\overline{A}B$. The squares representing $\overline{A}BC$ and ABC are adjacent, and B and C are common to both. Then $\overline{A}BC + ABC$ reduces to BC. Therefore,

$$\overline{A}B\overline{C} + \overline{A}BC + ABC = \overline{A}B + BC$$

PRACTICE PROBLEMS 27–1

1. Write the Boolean expressions for the Karnaugh maps in Figure 27–17 and simplify the expressions.

Draw a Karnaugh map and simplify the following Boolean expressions:

2. $A\overline{B}\overline{C} + \overline{A}BC + A\overline{B}C$

3. $\overline{A}\overline{B}\overline{C} + \overline{A}BC + AB\overline{C} + A\overline{B}C$

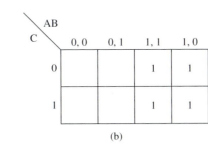

FIGURE 27–17 Karnaugh maps for problem 1, Practice Problems 27–1.

(a)

(b)

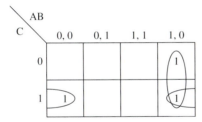

FIGURE 27–18 Karnaugh map for the expression $A\overline{B}\overline{C} + \overline{A}BC + A\overline{B}C$ in problem 2, Practice Problems 27–1.

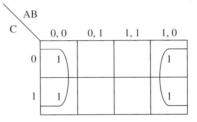

FIGURE 27–19 Karnaugh map for the expression $\overline{A}\overline{B}\overline{C} + \overline{A}BC + AB\overline{C} + A\overline{B}C$ in problem 3, Practice Problems 27–1.

SOLUTIONS

1. For Figure 27–17(a) the original expression is $\overline{A}\overline{B}\overline{C} + \overline{A}B\overline{C} + \overline{A}BC + ABC$, which reduces to $\overline{A}\overline{C} + \overline{A}B + BC$. There are three adjacent pairs and hence three terms in the answer. For Figure 27–17(b), the original expression is $AB\overline{C} + A\overline{B}\overline{C} + ABC + A\overline{B}C$, which reduces to A. There are four adjacent squares, which reduces to a single variable.

2. See Figure 27–18 for the Karnaugh map. There are two pairs of adjacent squares, so the expression reduces to $A\overline{B} + \overline{B}C$.

3. See Figure 27–19 for the Karnaugh map. There are four adjacent squares, so the expression reduces to a single variable, $\overline{B}$.

Additional practice problems are at the end of the chapter.

27-2   FOUR-VARIABLE EXPRESSION

A four-variable Karnaugh map is shown in Figure 27–20. Because there are four variables, 16 combinations are possible.

FIGURE 27–20
General form of
a four-variable
Karnaugh map.

AB

CD	0, 0	0, 1	1, 1	1, 0
0, 0	$\overline{A}\overline{B}\overline{C}\overline{D}$	$\overline{A}B\overline{C}\overline{D}$	$AB\overline{C}\overline{D}$	$A\overline{B}\overline{C}\overline{D}$
0, 1	$\overline{A}\overline{B}\overline{C}D$	$\overline{A}B\overline{C}D$	$AB\overline{C}D$	$A\overline{B}\overline{C}D$
1, 1	$\overline{A}\overline{B}CD$	$\overline{A}BCD$	$ABCD$	$A\overline{B}CD$
1, 0	$\overline{A}\overline{B}C\overline{D}$	$\overline{A}BC\overline{D}$	$ABC\overline{D}$	$A\overline{B}C\overline{D}$

The rules for a four-variable Karnaugh map are as follows:

1. A single square represents a four-variable term.
2. Two adjacent squares represent a three-variable term.
3. Four adjacent squares represent a two-variable term.
4. Eight adjacent squares represent one variable.

EXAMPLE 27–6

Simplify the map in Figure 27–21.

FIGURE 27–21
Karnaugh map
solution. Map =
A + CD +
$\overline{B}\,\overline{C}\,\overline{D}$.

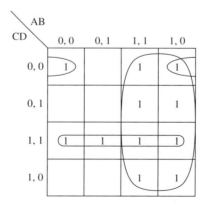

SOLUTION   Part of the problem is solved by applying rule 4 to the eight adjacent squares that are full.

$$AB\overline{C}\overline{D} + A\overline{B}\,\overline{C}\,\overline{D} + AB\overline{C}D + A\overline{B}\overline{C}D$$
$$+ ABCD + A\overline{B}CD + ABC\overline{D} + A\overline{B}C\overline{D} = A$$

A is the only variable common to all eight terms. Use rule 3 for the four adjacent squares:

$$\overline{A}\,\overline{B}CD + \overline{A}BCD + ABCD + A\overline{B}CD = CD$$

Variables C and D are common to all four terms. Use rule 2 for the two adjacent squares that are full:

$$\overline{A}\,\overline{B}\,\overline{C}\,\overline{D} + A\overline{B}\,\overline{C}\,\overline{D} = \overline{B}\,\overline{C}\,\overline{D}$$

Therefore, the map expression for Figure 27–21 is

$$\text{map} = A + CD + \overline{B}\,\overline{C}\,\overline{D}$$

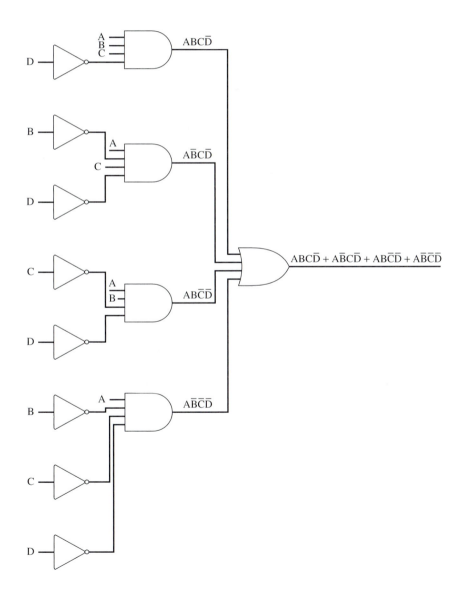

FIGURE 27–22
Logic circuit to
be simplified.

EXAMPLE 27–7

Simplify the logic circuit in Figure 27–22 using Karnaugh maps.

SOLUTION The terms are placed in the appropriate squares, as shown in Figure 27–23(a). Figure 27–23(b) shows how the two top and two bottom squares are combined by the four-adjacent-square rule. Note that $A\overline{D}$ is common to all terms. Therefore, map = $A\overline{D}$. The simplified logic circuit is shown in Figure 27–24.

FIGURE 27–23
Karnaugh map
solution for the
logic circuit in
Figure 27–22;
(a) variables
placed in
appropriate
squares;
(b) four-
adjacent-square
rule applied.

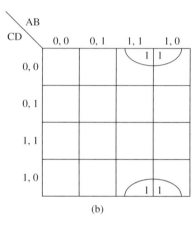

FIGURE 27–24
Simplified
circuit of Figure
27–22.

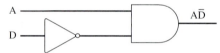

PRACTICE PROBLEMS 27–2

Use Karnaugh maps to simplify the following expressions:

1. $AB\overline{C}\overline{D} + A\overline{B}\overline{C}\overline{D} + A\overline{B}CD + \overline{A}BCD$

2. $\overline{A}\overline{B}\overline{C}\overline{D} + A\overline{B}\overline{C}\overline{D} + \overline{A}B\overline{C}\overline{D} + A B\overline{C}\overline{D} + A\overline{B}\overline{C}D + A B\overline{C}D$

3. Simplify the Boolean expression represented by the Karnaugh map in Figure 27–25.

4. Simplify the Boolean expression represented by the Karnaugh map in Figure 27–26.

SOLUTIONS

1. See Figure 27–27. Map = $A\overline{C}\overline{D} + \overline{B}CD$.
3. $\overline{A}\overline{B}\overline{C}D + \overline{A}B\overline{C}D + AB\overline{C}D + A\overline{B}\overline{C}D$
   $\qquad = \overline{C}D \qquad \text{(rule 3)}$
   $\overline{A}B\overline{C}D + \overline{A}BCD = \overline{A}BD \qquad \text{(rule 2)}$
   map = $\overline{A}BD + \overline{C}D$

2. See Figure 27–28. Map = $\overline{A}\overline{B}\overline{C} + A\overline{B}$.
4. $\overline{A}BC\overline{D} + \overline{A}B\overline{C}\overline{D} + AB\overline{C}\overline{D} + A B C\overline{D}$
   $\qquad = C\overline{D} \qquad \text{(rule 3)}$
   $\overline{A}B\overline{C}\overline{D} + \overline{A}BC\overline{D} = \overline{A}B\overline{D} \qquad \text{(rule 2)}$
   map = $\overline{A}B\overline{D} + C\overline{D}$

Additional practice problems are at the end of the chapter.

CHAPTER 27

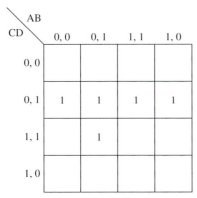

FIGURE 27–25   Karnaugh map for problem 3, Practice Problems 27–2.

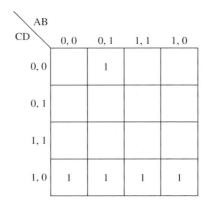

FIGURE 27–26   Karnaugh map for problem 4.

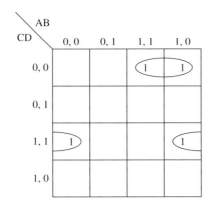

FIGURE 27–27
Karnaugh map
for problem 1.

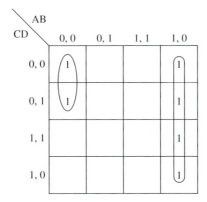

FIGURE 27–28
Karnaugh map
for problem 2.

1. Simplify $AB\overline{C} + ABC + A\overline{B}C$ using Karnaugh maps.

2. Simplify the Boolean expression represented by the Karnaugh map in Figure 27–29.

3. Simplify $\overline{A}\overline{B}\overline{C}\overline{D} + A\overline{B}\overline{C}D + A\overline{B}CD + A\overline{B}C\overline{D} + \overline{A}\overline{B}CD$ using Karnaugh maps.

4. Simplify the Boolean expression represented by the Karnaugh map in Figure 27–30.

Answers to Self-test 27–1 are at the end of the chapter.

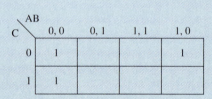

FIGURE 27–29   Karnaugh map for Self-test 27–1, problem 2.

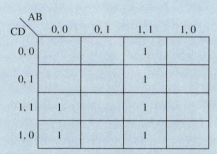

FIGURE 27–30   Karnaugh map for Self-test 27–1, problem 4.

CHAPTER 27 AT A GLANCE

PAGE	RULES	EXAMPLE
771	The rules for simplifying a two-variable Boolean expression using a Karnaugh map are as follows: **1.** Two adjacent squares are combined to represent a single variable. **2.** The resulting variable is common to both terms.	FIGURE 27–31
773	The rules for simplifying a three-variable Boolean expression using a Karnaugh map are as follows: **1.** A single square represents a three-variable term. **2.** Two adjacent squares are combined to represent a two-variable term. **3.** Four adjacent squares are combined to represent one variable.	FIGURE 27–32

$$\overline{A}\,\overline{B}\,\overline{C} + A\overline{B}\,\overline{C} + \overline{A}BC + A\overline{B}C$$
$$= \overline{B}$$
$$ABC + A\overline{B}C = AC$$
$$\text{map} = \overline{B} + AC$$

END OF CHAPTER PROBLEMS 27–1

Simplify the following Boolean expressions using Karnaugh maps.

1. $\overline{A}\overline{B} + A\overline{B}$
2. $AB + A\overline{B}$
3. $\overline{A}\,\overline{B}\overline{C} + \overline{A}B\overline{C} + \overline{A}BC$
4. $AB\overline{C} + ABC + A\overline{B}C$
5. $\overline{A}\,\overline{B}\overline{C} + \overline{A}B\overline{C} + AB\overline{C} + A\overline{B}\,\overline{C}$
6. $\overline{A}\,\overline{B}C + \overline{A}BC + ABC + A\overline{B}C$
7. $\overline{A}\,\overline{B}\overline{C} + A\overline{B}\overline{C} + \overline{A}BC + A\overline{B}C$
8. $\overline{A}\,\overline{B}\overline{C} + A\overline{B}\overline{C} + \overline{A}BC + ABC$
9. Simplify the Boolean expression represented by the Karnaugh map in Figure 27–33.
10. Simplify the Boolean expression represented by the Karnaugh map in Figure 27–34.
11. Simplify the Boolean expression represented by the Karnaugh map in Figure 27–35.
12. Simplify the Boolean expression represented by the Karnaugh map in Figure 27–36.

END OF CHAPTER PROBLEMS 27–2

Simplify the following Boolean expressions using Karnaugh maps.

1. $\overline{A}\overline{B}\overline{C}\overline{D} + \overline{A}\overline{B}\overline{C}D + AB\overline{C}D + ABCD$
2. $\overline{A}\,\overline{B}C\overline{D} + \overline{A}\overline{B}\overline{C}\overline{D} + AB\overline{C}\overline{D} + A\overline{B}C\overline{D}$
3. $\overline{A}\,\overline{B}\overline{C}\overline{D} + \overline{A}\,\overline{B}\overline{C}D + \overline{A}B\overline{C}D + AB\overline{C}D + A\overline{B}\overline{C}D$
4. $\overline{A}\,\overline{B}CD + \overline{A}B\overline{C}D + ABCD + AB\overline{C}D + \overline{A}\overline{B}C\overline{D}$
5. $\overline{A}\,\overline{B}\overline{C}\overline{D} + \overline{A}\,\overline{B}C\overline{D} + A\overline{B}C\overline{D} + A\overline{B}\overline{C}\overline{D} + \overline{A}B\overline{C}D + \overline{A}\overline{B}C\overline{D} + AB\overline{C}D + A\overline{B}C\overline{D} + ABCD$
6. $\overline{A}\,\overline{B}\overline{C}\overline{D} + \overline{A}\,\overline{B}\overline{C}D + \overline{A}\overline{B}CD + \overline{A}\overline{B}C\overline{D} + \overline{A}B\overline{C}\overline{D} + \overline{A}B\overline{C}D + \overline{A}BCD + \overline{A}BCD + AB\overline{C}\overline{D} + A\overline{B}C\overline{D}$
7. Simplify the Boolean expression represented by the Karnaugh map in Figure 27–37.
8. Simplify the Boolean expression represented by the Karnaugh map in Figure 27–38.
9. Simplify the Boolean expression represented by the Karnaugh map in Figure 27–39.
10. Simplify the Boolean expression represented by the Karnaugh map in Figure 27–40.

C \ AB	0, 0	0, 1	1, 1	1, 0
0		1	1	
1		1	1	

FIGURE 27–33 Karnaugh map for problem 9 of End of Chapter Problems 27–1.

C \ AB	0, 0	0, 1	1, 1	1, 0
0	1			1
1				1

FIGURE 27–34 Karnaugh map for problem 10 of End of Chapter Problems 27–1.

C \ AB	0, 0	0, 1	1, 1	1, 0
0	1	1	1	1
1		1		

FIGURE 27–35   Karnaugh map for problem 11 of End of Chapter Problems 27–1.

C \ AB	0, 0	0, 1	1, 1	1, 0
0		1	1	
1	1			1

FIGURE 27–36   Karnaugh map for problem 12 of End of Chapter Problems 27–1.

CD \ AB	0, 0	0, 1	1, 1	1, 0
0, 0				
0, 1	1	1	1	1
1, 1	1	1	1	1
1, 0				

FIGURE 27–37   Karnaugh map for problem 7 of End of Chapter Problems 27–2.

CD \ AB	0, 0	0, 1	1, 1	1, 0
0, 0	1	1		
0, 1	1	1		
1, 1			1	1
1, 0			1	1

FIGURE 27–38   Karnaugh map for problem 8 of End of Chapter Problems 27–2.

FIGURE 27–39 Karnaugh map for problem 9 of End of Chapter Problems 27–2.

CD \ AB	0, 0	0, 1	1, 1	1, 0
0, 0		1	1	
0, 1		1	1	
1, 1	1			1
1, 0				

FIGURE 27–40 Karnaugh map for problem 10 of End of Chapter Problems 27–2.

CD \ AB	0, 0	0, 1	1, 1	1, 0
0, 0				
0, 1	1			1
1, 1		1	1	
1, 0		1	1	

ANSWERS TO SELF-TEST

1. See Figure 27–41. Map =
   AB + AC.
2. Map = $\overline{A}\,\overline{B} + \overline{B}\,\overline{C}$.
3. See Figure 27–42. Map =
   $\overline{B}CD + A\overline{B}$.
4. Map = $\overline{A}\,\overline{B}C + AB$.

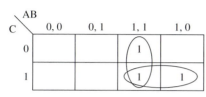

FIGURE 27–41 Karnaugh map for the expression $AB\overline{C}$ + ABC + $A\overline{B}C$.

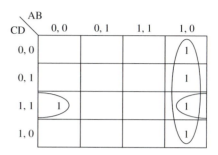

FIGURE 27–42 Karnaugh map for the expression $A\overline{B}\,\overline{C}\,\overline{D}$ + $A\overline{B}\,\overline{C}D$ + $A\overline{B}C\overline{D}$ + $A\overline{B}CD$ + $\overline{A}\,\overline{B}CD$.

Introduction to Statistics in Electronics

Chapter 28   Introduction to Statistics

Statistics is that part of math that deals with the collection, analysis, and interpretation of numerical data. In the electronics and computer industry, statistics are used to predict:

· The reliability of spacecraft electronic equipment that must operate for years with no maintenance
· The failure rates of products during the manufacturing process
· The average time it should take to repair equipment

Spacecraft computers are very reliable and expensive. Personal computers are not very reliable (they may have to be rebooted often), but they are inexpensive. Chapter 28 provides some of the fundamental concepts of statistics that must be understood in order to advance to complicated problems such as calculating the reliability of a computer system.

Introduction to Statistics

28

Introduction

Statistics are numbers that are used to analyze and interpret data in order to make decisions. Statistics is the relationship of how many to how much. How many scores were there? How often did each score appear? For example, statistics are kept on players and teams in sports. As the season progresses, teams and players are evaluated: batting and pitching averages, and team win-loss records in baseball; passing and rushing averages, and team win-loss records in football are a few of the statistics most often found in the sports section of the newspaper.

Statistical quality control and statistical process control are integral parts of all manufacturing processes. Components and assemblies that make up a complete unit—such as a television, VCR, or automobile—must be built to meet certain standards. To be competitive, units must be of high quality. At the same time, to stay in business, a company product must be profitable. In short, to be competitive a company must make a quality product in as short a time as possible. Only with strict quality control can a company maintain a quality product that is competitive.

Schools, hospitals, and other organizations use statistical data in many different ways: Determining students' grades, predicting students' chances for succeeding in a particular course of study, and predicting the success of certain operations or medicines are a few examples.

In statistics, the entire lot or group to be tested is called a *population*. When testing an entire population would be too costly, both in terms of time and money, a *sample* is taken. From the sample, conclusions are made concerning the entire population. To be a fair gauge of the entire population, the sample taken must be a *random* sample. That is, each item in the population must have an equal chance of being selected.

Suppose a company manufactures 1000 widgets per shift. Customers for these widgets realize that there will be variations in the weight and size of the widgets (there are *always* variations), so some variations are allowed. Causes of variations in manufactured goods could be temperature variations, humidity, wear on the machines making the goods, and operator error. (And, of course, whenever you measure *anything* there is always the nagging question of how accurate the measuring instrument is.)

Suppose that the customer agrees that each widget is to weigh 10 ounces ± 0.1 ounce. That means any widget that weighs between 9.90 oz and 10.1 oz is a good widget. The cost of weighing each individual widget per shift would make the price too great, so the customer accepts random sampling each shift as a way of determining the weight of the entire population. Random sampling not only tells the customer about the weight of the entire population, but it also provides the company with valuable information about its quality control. From such data, trouble areas can be spotted and corrected. For example, the analyzed data from the samples might show that one shift is turning out a higher percentage of defective parts than another shift. Such data might also alert maintenance that a particular machine is beginning to show signs of wear and needs to be overhauled or replaced. The early detection of faults, for whatever reason, is one of the most important functions of statistical process control.

One of the jobs of the technician is the analysis and interpretation of data: A circuit is built and tested; the results are recorded and compared to expected results. A resistor is measured. Was the measured value within the acceptable range? A sample of components is removed from the assembly line and tested. The test results are used to determine the quality of the entire lot.

Over the last decade, it has become more and more important that the technician be aware of data collection and the statistical methods used in the analyses and interpretation of these data.

This chapter introduces the student to some of the basic tools used in statistics and presents examples of how these tools can be used. In order to obtain valid statistical data it is often necessary to use hundreds, or even thousands, of data samples. We use artificially low numbers of data samples in this chapter's problems to make the solutions less time consuming for students. However, the mathematical procedures are the same regardless of the number of data samples.

Chapter Objectives

In this chapter you will learn how to:

1. Use frequency distribution to clarify or group data.
2. Construct a histogram.
3. Calculate the mean, median, and mode of a set of data.
4. Compute the standard deviation for a set of data.
5. Determine whether measured data fall within acceptable limits.

28-1 FREQUENCY DISTRIBUTION

In data collection, we will first arrange the data in a table. Such a table is called a *frequency distribution* table. A frequency distribution table contains the information *how much* and *how many*. The *how much* could be the magnitude of an item or score such as the measured value of a resistor or the score on a test. The *how many* is simply a tally of how many times that value appears.

To construct a frequency distribution table we determine the number of items. If there are fewer than 12 *different* items, construct a table, listing each item. Then tally the number of times each item appears.

EXAMPLE 28–1

In an electronics math quiz, the following scores were recorded: 75, 80, 60, 95, 75, 70, 90, 75, 85, 80, 70, 65, 80, 75, 70, 75, 90, 80, 65, 75. Construct a frequency distribution table.

Scores	60	65	70	75	80	85	90	95
Tally	|	||	|||	ʬ||	||||	|	||	|
	1	2	3	6	4	1	2	1

FIGURE 28–1
Frequency distribution table for Example 28–1.

SOLUTION There are eight different scores so we can construct a table with eight columns, one for each score. In the table, we need three rows: one for the scores, and two for the tally. Enter the scores in the eight columns. Tally the number of times each score appears. In the first row of the tally, we put a mark for each score as we read it. When all scores have been read, add up the tally and place the results in the second row. Check your tally to make sure you have 20 scores recorded. The frequency distribution table for Example 28–1 appears in Figure 28–1.

☞ **KEY POINT** A frequency distribution table tells us how many items are in a study and how often each item appears.

28–1–1 Histogram

A frequency distribution can be represented graphically in a bar graph called a *histogram*. In the histogram, the items or scores are plotted along the horizontal axis in equal intervals. The number of times each score appears, the frequency, is plotted along the vertical axis.

A histogram of the frequency distribution in Example 28–1 is shown in Figure 28–2 on the next page. In a histogram, the midpoint of each interval is the data value. The widths of the intervals are the same and are usually determined by the distance between adjacent midpoints. In Example 28–1, the interval width is 5 because the distance between scores is 5.

EXAMPLE 28–2

In a production run, the number of defective parts per 100 were tallied over a 20-day period. The results were: day 1—3, day 2—6, day 3—4, day 4—3, day 5—2, day 6—1, day 7—5, day 8—1, day 9—3, day 10—4, day 11—2, day 12—1, day 13—3, day 14—3, day 15—4, day 16—3, day 17—2, day 18—2, day 19—1, day 20—5. Construct a frequency distribution table and a histogram.

FIGURE 28–2
Histogram for
Example 28–1.

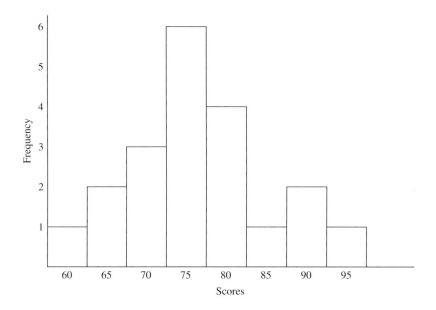

FIGURE 28–3
Frequency
distribution
table for
Example 28–2.

Defective parts/100	1	2	3	4	5	6
Tally	||||	||||	ⴼ||	|||	||	|
	4	4	6	3	2	1

FIGURE 28–4
Histogram for
Example 28–2.

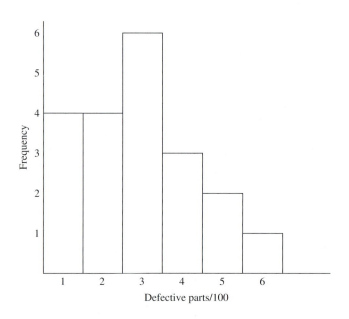

SOLUTION   There are six different values (defective parts per 100) so the frequency distribution table will have six columns. Tally how many times each number of defective parts per 100 appears. The frequency distribution table for Example 28–2 is shown in Figure 28–3. Check your tally (it should be 20) to make sure you didn't lose or add a day. The histogram for Example 28–2 is shown in Figure 28–4.

> ☞ **KEY POINT**   In a histogram, the frequency is plotted along the *y*-axis and the values are plotted along the *x*-axis.

If there are *more* than 12 different items in a data set, the table and histogram become too unwieldy. When this happens, we divide the items into a convenient number of intervals or groups. In the table, make a column for each interval; just like we did for each score. Then tally the number of data in each interval as before. In such a table: (1) intervals cannot overlap; (2) data cannot appear in more than one interval; (3) the smallest data value fits in the first interval; (4) the largest data value fits in the last interval; (5) the width of all intervals is the same.

Example 28–3 shows how to construct a table for data that are separated into intervals.

EXAMPLE 28–3

In a second quiz, the following scores were recorded: 82, 84, 76, 70, 75, 86, 58, 82, 74, 78, 86, 88, 68, 92, 84, 82, 75, 77, 66. Construct a frequency distribution table and a histogram. Determine the range and interval midpoints.

SOLUTION   There are more than 12 *different* scores, so we will divide the scores into intervals. To determine how many intervals to use, we will first determine the *range*. The range is the difference between the maximum and minimum scores. An examination of the scores shows that the highest score was 92 and the lowest score was 58.

$$\text{range} = 92 - 58 = 34$$

Choosing the number of intervals is arbitrary. We try to choose a number of intervals that provides a reasonable picture of the population distribution. With a range of 34 we could conveniently choose seven groups with an interval of 5. That is, each interval can consist of five different scores.

$$34 \div 7 = 4.86 \approx 5$$

We always round up to the next highest whole number to ensure that all data fall within the selected intervals. The actual data spread will always be one greater than the range because the range is not *inclusive*. That is, the range includes all data from the lowest number to *but not including* the highest number. In our example, the actual number of data is from 58 *through* 92, a total of 35. Therefore 7 intervals of 5 gives us an exact fit. In our table, we need seven columns—one for each interval. Tally the number of data in each interval. The frequency distribution is shown in

FIGURE 28–5

Frequency
distribution
table for
Example 28–3.

Score intervals	58–62	63–67	68–72	73–77	78–82	83–87	88–92
Tally	|	|	||	ﬄ	||||	||||	||
	1	1	2	5	4	4	2

Figure 28–5. Notice that all the scores from 58 through 62, five scores (58, 59, 60, 61, and 62), are tallied in the first interval. There is only one score (58) in this interval, so our tally is one. In the interval 63 through 67, there is only one score (66). In the interval 68 through 72, there are two scores (68 and 70). In the interval 73 through 77, there are five scores (74, 75 twice, 76, and 77). And finally in the highest interval, 88 through 92, there are two scores (88 and 92).

The histogram for Example 28–3, using intervals of 5, is shown in Figure 28–6.

Often, the number of intervals selected leaves us with more possible data values than we have. This is often referred to as an *overlap*. For example, suppose the highest score had been 90. Then the range would have been

$$\text{range} = 90 - 58 = 32$$

Again using seven intervals of five scores each, the highest interval, 88 through 92, will be wider than necessary creating an overlap of 2. Ninety-two must be the upper boundary though, because all intervals must be of the same width.

Sometimes the overlap is dealt with at the lower end. If we had chosen to do so, our boundaries would have been from 56 through 90. Sometimes the overlap is split. In this example we could have split the overlap and started the table at 57 and ended at 91. As a rule of thumb, if the overlap is less than one half the interval width, we deal with the overlap at one end or the other.

FIGURE 28–6

Histogram for
Example 28–3.

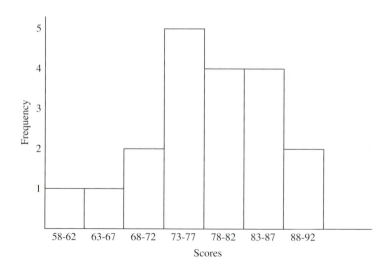

☞ **KEY POINT** The range of a set of data is the difference between the largest and smallest numbers.

☞ **KEY POINT** Select a number of intervals that will most nearly describe the spread of the population.

☞ **KEY POINT** When using intervals in a frequency distribution table: (1) intervals cannot overlap; (2) data cannot appear in more than one interval; (3) the smallest data value fits in the first interval; (4) the largest data value fits in the last interval; (5) the width of all intervals is the same.

The extremes of an interval are called the end points or boundaries. The data within an interval include all the values from the lower boundary through the upper boundary. Thus, in Example 28–3, in the interval 58 through 62, scores of 58, 59, 60, 61, and 62, a total of 5 is included. The lower boundary (end point) is 58 and the upper boundary (end point) is 62.

All scores from 63 through 67 (63, 64, 65, 66, and 67) will fall in the next group. The lower boundary is 63 and the upper boundary is 67. Finally, all values from 88 through 92 will be part of the highest interval. The lower boundary is 88 and the upper boundary is 92.

The midpoint of an interval can be found by dividing the sum of the interval boundaries by 2. Finding the midpoint of each interval will make it easier to calculate the average, which we will do in the next section.

EXAMPLE 28–4

Twenty high-gain amplifiers were constructed in the lab using components off the shelf. The design called for a voltage gain of 400 dB. The voltage gains were measured to determine the actual gain of the amplifiers. The results (in dB) were: 408, 412, 401, 380, 402, 390, 398, 405, 400, 396, 404, 401, 388, 399, 406, 402, 391, 404, 398, 403. Construct a frequency distribution table. From the table, construct a histogram. Determine the range and interval midpoints.

SOLUTION   There are more than 12 different measurements of gain, so we will divide the gains into intervals. To determine the number of intervals to use, we will first find the range. The highest gain (in dB) is 412 and the lowest gain is 380.

$$\text{Range} = 412 - 380 = 32$$

Let's use 5 intervals: $32/5 = 6.4 \simeq 7$

With 5 intervals, there will be 7 measurements in each interval. The frequency distribution is shown in Figure 28–7 on the next page. Notice that there is an overlap of 2 in the highest interval. We could have chosen to have the overlap in the lowest interval or we could have created an overlap of 1 on each end. Since the overlap was less than half the number of gains in an interval, only one end was used. We chose to use the upper end.

FIGURE 28–7
Frequency
distribution
table for
Example 28–4.

Gain in dB	380–386	387–393	394–400	401–407	408–414
Tally	|	|||	⊞	⊞||||	||
	1	3	5	9	2

The interval midpoints are 382, 387, 392, 397, 402, 407, and 412. The histogram, using intervals of 5, is shown in Figure 28–8.

FIGURE 28–8

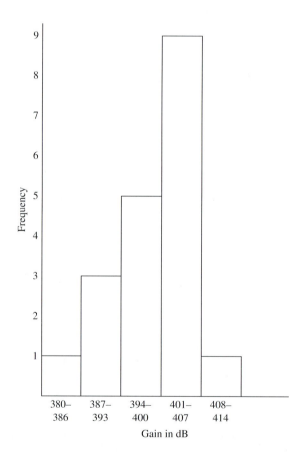

PRACTICE PROBLEMS 28–1

1. In an electronics math quiz, the following scores were recorded: 85, 60, 75, 90, 70, 80, 65, 75, 75, 85, 75, 90, 80, 85, 75, 80, 65, 75, 90, 80, 65, 95, 70, 80, 75. Construct a frequency distribution table. From the table, construct a histogram.

2. From a bin of 100 1 kΩ ±5% carbon resistors, 25 were selected to be measured in the laboratory. The results, in ohms, were: 990, 1000, 980, 1010, 1000, 1030, 1010, 1000, 990, 980, 990, 1020, 1010, 1000, 990, 1020, 990, 1010, 1000, 1010, 1020, 1010, 1020, 1010, 1040. Construct a frequency distribution table. From the table, construct a histogram.

3. In a production run, the number of defective parts per 100 were tallied over a 20-day period. The results were: day 1—6, day 2—4, day 3—3, day 4—4, day 5—2, day 6—4, day 7—3, day 8—5, day 9—4, day 10—2, day 11—1, day 12—5, day 13—4, day 14—1, day 15—3, day 16—4, day 17—5, day 18—2, day 19—4, day 20—3. Construct a frequency distribution table and a histogram.

4. In a mid-term exam, the following scores were recorded: 83, 82, 63, 91, 75, 68, 87, 80, 71, 72, 74, 85, 90, 62, 81, 77, 70, 86, 78, 73, 76, 84, 81, 74, 72. Construct a frequency distribution table and a histogram. Determine the range and the midpoints of the intervals.

5. In a production run covering a 31-day period, the following numbers of items were produced each day: 25, 51, 72, 31, 40, 44, 37, 41, 48, 43, 57, 42, 31, 28, 46, 50, 63, 37, 78, 51, 62, 61, 51,

Scores	60	65	70	75	80	85	90	95
Tally	I	III	II	IIIII II	IIIII	III	III	I
	1	3	2	7	5	3	3	1

FIGURE 28–9 Frequency distribution table for Practice Problems 28–1 #1.

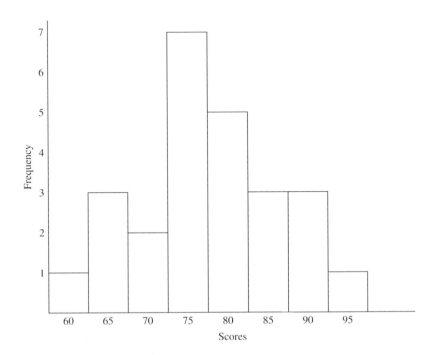

FIGURE 28–10 Histogram for Practice Problems 28–1 #1.

48, 41, 45, 34, 54, 32, 40, 79. Construct a frequency distribution table and histogram. Determine the range and interval midpoints.

6. Twenty-five high-gain amplifiers were constructed in the lab using components off the shelf. The design called for a voltage gain of 200 dB. The voltage gains were measured to determine the actual gain of the amplifiers. The results (in dB) were: 203, 201, 213, 193, 184, 200, 210, 196, 198, 191, 203, 206, 199, 204, 201, 202, 199, 197, 200, 208, 201, 198, 191, 206, 197. Construct a frequency distribution table. From the table, construct a histogram. Determine the range and interval midpoints.

SOLUTIONS

1. There are eight different scores so we can construct a table with a column for each score. Next tally the number of times each score appears. Count the scores to make sure we tallied them all. The frequency distribution table for problem 1 appears in Figure 28–9. The histogram appears in Figure 28–10.

FIGURE 28–11
Frequency distribution table for Practice Problems 28–1 #2.

Resistor values in ohms	*980*	*990*	*1000*	*1010*	*1020*	*1030*	*1040*
Tally	||	|||	|||	|||||	||||	|	|
	2	5	5	7	4	1	1

FIGURE 28–12
Histogram for Practice Problems 28–1 #2.

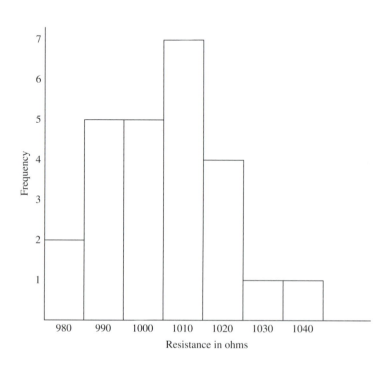

Number of defective parts/100	1	2	3	4	5	6
Tally	II	III	IIII	ⅣⅡII	III	I
	2	3	4	7	3	1

FIGURE 28–13
Frequency distribution table for Practice Problems 28–1 #3.

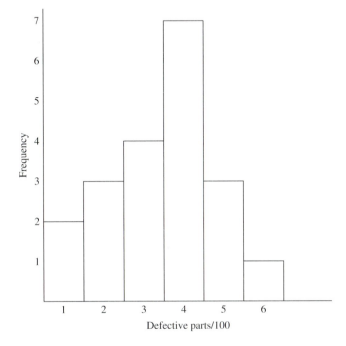

FIGURE 28–14
Histogram for Practice Problems 28–1 #3.

2. There are seven different values so we can have a separate column for each. Tally the number of resistors in each column. The frequency distribution for problem 2 is shown in Figure 28–11. The histogram for problem 2 appears in Figure 28–12. Notice in our histogram that the midpoint of each interval is the data value and each interval is 10 Ω wide.

3. There are six different items so our table will have six columns, one for each number. Tally how many times each number of defective parts per 100 appears. The frequency distribution table for Problem 3 is shown in Figure 28–13. The histogram is shown in Figure 28–14.

4. There are more than 12 *different* scores, so we will divide the grades into intervals. To determine how many intervals to use, we will first determine the *range*. An examination of the scores shows that the highest score was 91 and the lowest score was 62.

$$range = 91 - 62 = 29$$

With a range of 29 we could conveniently choose six groups with an interval of 5. $29 \div 6 = 4.83 \approx 5$. Construct the frequency distribution table. Again, we round up to ensure that all data fits. Our first interval, then, will be from 62 through 66 and our last or highest interval will be from 87

FIGURE 28–15
Frequency
distribution table
for Practice
Problems 28–1
#4.

Score intervals	62–66	67–71	72–76	77–81	82–86	87–91
Tally	||	|||	ⅢⅢ||	ⅢⅢ	ⅢⅢ	|||
	2	3	7	5	5	3

FIGURE 28–16
Histogram for
Practice
Problems 28–1
#4.

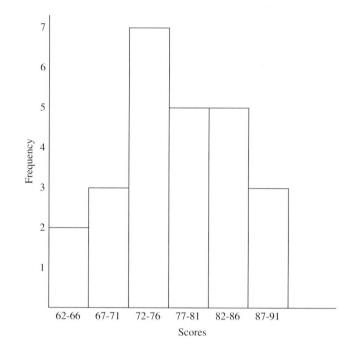

FIGURE 28–17
Frequency
distribution table
for Practice
Problems 28–1
#5.

Score intervals	25–31	32–38	39–45	46–52	53–59	60–66	67–73	74–80
Tally	|||	ⅢⅢ	ⅢⅢ|||	ⅢⅢ||	||	|||	|	||
	3	5	8	7	2	3	1	2

FIGURE 28–18
Histogram for
Practice
Problems 28–1
#5.

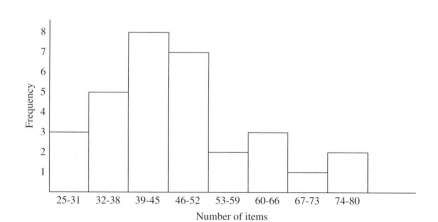

Gain in dB	184–188	189–193	194–198	199–203	204–208	209–213
Tally	I	III	‖‖	‖‖ ‖‖	IIII	II
	1	3	5	10	4	2

FIGURE 28–19
Frequency distribution table for Practice Problems 28–1 #6.

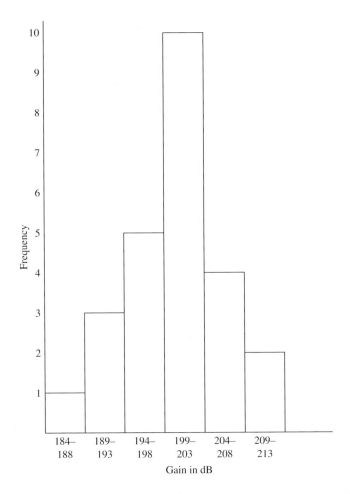

FIGURE 28–20
Histogram for Practice Problems 28–1 #6.

through 91. Tally the number of data in each interval. The frequency distribution for problem 4 is shown in Figure 28–15. The midpoints of the intervals are 64, 69, 74, 79, 84, and 89. The histogram is in Figure 28–16.

5. There are more than 12 different values so we need to determine the number of intervals to use:

range = maximum number
  − minimum number = 79 − 25 = 54

Let's use eight groups. 54 ÷ 8 = 6.75 ≈ 7. Each interval width is 7. Tally the number of items in each interval. Your groups should look like Figure 28–17. Even though the highest number of items is 79, the upper

INTRODUCTION TO STATISTICS

boundary of this interval is 80. We have dealt with the overlap (one) in the highest interval. Remember, all intervals must be of the same width. The midpoints are at the middle of each interval. The lowest interval is 25 through 31. Subtracting the lower boundary from the upper boundary:

$$31 - 25 = 6$$

$$6 \div 2 = 3$$

The midpoint of the lowest interval will be $25 + 3 = 28$. Each interval width is 7 so each midpoint will be 7 greater than the previous midpoint. The midpoints are 28, 35, 42, 49, 56, 63, 70, and 77. The histogram for problem 5 appears in Figure 28–18.

6. There are more than 12 different gains measured so we need to determine how many intervals we will use. The range is

$$213 - 184 = 29$$

Let's use 6 intervals. With 6 intervals we will have 5 different gain measurements in each interval. Using 6 intervals, your frequency distribution table should look like the one in Figure 28–19. The first interval is from 184 to 188 and the last interval is from 209 to 213. The midpoints of the intervals are 186, 191, 196, 201, 206, and 211. The histogram is shown in Figure 28–20.

Additional practice problems are at the end of the chapter.

☞ **KEY POINT** The midpoint of an interval is the number that is halfway between the upper and lower interval boundaries.

28-2 MEASURES OF CENTRAL TENDENCY

From the frequency distribution table, more specific information can be obtained. In a data distribution we usually are interested in the center of the distribution. We are interested in the *average*. There are three methods used to measure averages. These methods are referred to as measures of central tendency. These measures of central tendency are the *mean,* the *median,* and the *mode.* In a normal distribution (discussed later in the chapter) the three measures occur at the same point.

28-2-1 The Mean

The mean is the sum of all the data divided by the number of data. In the English language the word "average" is used for the mathematical "mean." A student's grades are *averaged* to determine his or her grade for the semester. A hitter's at-bats versus hits are *averaged* to determine a batting average. A quarterback's completions versus number of passes thrown are *averaged* to determine a completion average. In each case we are calculating the mean.

$$\text{mean} = \bar{x} = \frac{\Sigma x}{n} = \frac{x_1 + x_2 + x_3 + \ldots + x_n}{\text{number of data}}$$

(28–1)

where $x_1, x_2, x_3, \ldots ; x_n$ are the data. Σx (summation x) means the sum of all the values of x, the data. For example, if five test scores are 60, 65, 70, 75, and 80, what would be the average score?

$$\text{mean} = \bar{x} = \frac{\Sigma x}{n} = \frac{x_1 + x_2 + x_3 + \ldots + x_n}{\text{number of data}}$$

$$= \frac{60 + 65 + 70 + 75 + 80}{5} = \frac{350}{5} = 70$$

☞ **KEY POINT** The mean is equal to the sum of the data divided by the number of data.

28-2-2 The Median

The median is that value where half of the values are greater and half are less. That is, half the data fall below this number and half the data are above this number. If the number of data is odd, the median is the middle number. If the number of data is even, then the median is half the distance between the middle two numbers. For example, again using the test scores of 60, 65, 70, 75, and 80, the median would be 70, the middle score. If six test scores are 60, 65, 70, 75, 80, and 100, then the median would be half the distance between the two middle scores, or 72.5. This value can be found by adding the two middle scores and dividing by 2.

$$70 + 75 = 145; \ 145 \div 2 = 72.5$$

☞ **KEY POINT** The median is that value where half of the values are greater and half are less.

28-2-3 The Mode

The *mode* is the data (the numbers) that occur most frequently. For example, if six test scores are 65, 70, 70, 70, 75, and 80, the mode would be 70 because 70 is the score that occurs most often.

If the data are grouped (in intervals), then the midpoint of the interval is used in computing the mean, median, and mode.

☞ **KEY POINT** The mode is the measurement that appears most frequently in the data.

EXAMPLE 28–5

Using the data from Figure 28–21 (same as Figure 28–1), calculate the mean, median, and mode.

$$\text{mean} = \bar{x} = \frac{\Sigma x}{n} = \frac{x_1 + x_2 + x_3 + \ldots + x_n}{\text{number of data}}$$

The TI-36X SOLAR, CASIO fx-115W, and SHARP EL-506G are three calculators that include statistical functions. You may use any one of these calculators to find the mean and Σx, as well as other statistical information. See the user's manual for your calculator for the key combinations to use. In the TI-36X SOLAR, the key combination for this example is:

3rd STAT 1 60 $\Sigma +$ 65 $\Sigma +$ 65 $\Sigma +$ 70 $\Sigma +$

and so on until all 20 of the data have been entered. Notice that after $\Sigma +$ has been keyed in, the number displayed is the number of data entered, *not* the data themselves. After you key in the last data, 95, and press $\Sigma +$, the number 20 should be displayed because you have entered 20 data numbers. If the number 20 is *not* displayed, you made a mistake. After you have entered 95 $\Sigma +$ and the number 20 is displayed, both the mean and the total of the scores, Σx, are stored in the calculator. Display the mean by pressing $\bar{x}$. *Display Σx by pressing the* Σx key.

$$\text{mean} = \bar{x} = \frac{x_1 + x_2 + x_3 + \ldots + x_{20}}{\text{number of data}} = \frac{1530}{20} = 76.5$$

For the Casio the key sequence is:

MODE MODE 1   to enter the Statistical Mode.

SHIFT Scl $=$ to clear memory.

To enter data:

60 DT 65 DT DT 70 DT DT DT 75 DT ...

until all 20 values have been entered.

To display the mean: SHIFT $\bar{x}$ $=$

If you do not use a statistical calculator, simply add all the data and divide by 20, the number of data.

FIGURE 28–21
Frequency
distribution
table for
Example 28–5.

Scores	60	65	70	75	80	85	90	95
Tally	│	││	│││	₦₦I	││││	│	││	│
	1	2	3	6	4	1	2	1

$$\text{mean} = \frac{60 + (2)65 + (3)70 + (6)75 + (4)80 + 85 + (2)90 + 95}{20}$$

$$= \frac{1530}{20} = 76.5$$

The median is the score where half the data are above and half the data are below. Because the number of data is even (20), the median would be the score between the tenth and eleventh data values. Since both scores are 75, the median is 75.

An examination of the data shows that a score of 75 appears most frequently, six times. Therefore, the mode is 75.

$$\text{mean} = 76.5 \qquad \text{median} = 75 \qquad \text{mode} = 75$$

EXAMPLE 28–6

Using the data from Figure 28–22 (same as Figure 28–3), calculate the mean, median, and mode.

SOLUTION

$$\text{mean} = \bar{x} = \frac{\Sigma x}{n} = \frac{x_1 + x_2 + x_3 + \ldots + x_{20}}{20}$$

$$\bar{x} = \frac{1(4) + 2(4) + 3(6) + 4(3) + 5(2) + 6(1)}{20} = \frac{58}{20} = 2.90$$

There are 20 data values so the median falls between the tenth and eleventh values. The tenth and eleventh values are both 3, so the median is 3. Three appears the most times so the mode is also 3.

$$\text{mean} = 2.9 \qquad \text{median} = 3 \qquad \text{mode} = 3$$

Defective parts/100	1	2	3	4	5	6
Tally	││││	││││	₦₦I	│││	││	│
	4	4	6	3	2	1

FIGURE 28–22
Frequency
distribution
table for
Example 28–6.

EXAMPLE 28-7

Using the data from Figure 28–23 (same as Figure 28–5), compute the mean, median, and mode.

FIGURE 28–23

Score intervals	58–62	63–67	68–72	73–77	78–82	83–87	88–92
Tally			||	$\cancel{||||}$	||||	||||	||
	1	1	2	5	4	4	2

SOLUTION   The data are assumed to occur at the midpoint of each interval or at 60, 65, 70, 75, 80, 85, and 90. Selecting the midpoint of each interval to compute the mean, rather than using the raw scores, is the accepted method in statistics. This method introduces some error when the number of scores is small, as is the case here (see Example 28–3 for actual scores) but for larger numbers of data, which is the general case in statistics, the error would be negligible.

$$\text{mean} = \bar{x} = \frac{\Sigma x}{n} = \frac{x_1 + x_2 + x_3 + \ldots + x_{19}}{19}$$

$$\text{mean} = \frac{60 + 65 + (2)70 + (5)75 + (4)80 + (4)85 + (2)90}{19} = \frac{1480}{19}$$

$$= 77.9 \text{ rounded to three places}$$

The median would occur at the tenth score. There are nine scores above and nine scores below the tenth score. The tenth score is in the interval 78 through 82. The midpoint of the interval 78 through 82 is 80. Therefore, the median is 80.

The mode occurs in the interval 73 through 77. The midpoint of this interval is 75. Therefore, the mode is 75.

$$\text{mean} = 77.9 \qquad \text{median} = 80 \qquad \text{mode} = 75$$

EXAMPLE 28-8

Using the data from Figure 28–24 (same as Figure 28–7), compute the mean, median, and mode.

FIGURE 28–24
Frequency distribution table for Example 28–8.

Gain in dB	380–386	387–393	394–400	401–407	408–414
Tally			|||	$\cancel{||||}$	$\cancel{||||}$||||	||
	1	3	5	9	2

SOLUTION

To compute the mean we again use the interval midpoints which in this case are 383, 390, 397, 404, and 411.

$$\text{Mean} = \bar{x} = \frac{\Sigma x}{n} = \frac{x_1 + x_2 + x_3 + \ldots + x_{20}}{20} = \frac{7996}{20} = 400$$

The median would occur between the tenth and eleventh data values. Since both gain measurements fall in the interval 401 to 407, the median will be at the midpoint of that interval, which is 404. The mode also occurs in the interval 401–407, so the mode is also 404, the midpoint of that interval.

$$\text{mean} = 400 \qquad \text{median} = 404 \qquad \text{mode} = 404$$

PRACTICE PROBLEMS 28–2

1. Using the data in Figure 28–25 (same as Figure 28–9), determine the mean, median, and mode.
2. Using the data in Figure 28–26 (same as Figure 28–15), determine the mean, median, and mode.
3. Using the data in Figure 28–27 (same as Figure 28–11), determine the mean, median, and mode.
4. Using the data in Figure 28–28 (same as Figure 28–17), determine the mean, median, and mode.
5. Using the data in Figure 28–29 (same as Figure 28–13), determine the mean, median, and mode.
6. Using the data in Figure 28–30 (same as Figure 28–19), determine the mean, median, and mode.

Scores	60	65	70	75	80	85	90	95
Tally	I	III	II	NIII	NI	III	III	I
	1	3	2	7	5	3	3	1

FIGURE 28–25 Frequency distribution table for Practice Problems 28–2 #1.

Score intervals	62–66	67–71	72–76	77–81	82–86	87–91
Tally	II	III	NII	NI	NI	III
	2	3	7	5	5	3

FIGURE 28–26 Frequency distribution table for Practice Problems 28–2 #2.

FIGURE 28–27 Frequency distribution table for Practice Problems 28–2 #3.

Resistor values in ohms	***980***	***990***	***1000***	***1010***	***1020***	***1030***	***1040***
Tally	II	𝍸	𝍸	𝍸II	IIII	I	I
	2	5	5	7	4	1	1

FIGURE 28–28 Frequency distribution table for Practice Problems 28–2 #4.

Score intervals	***25–31***	***32–38***	***39–45***	***46–52***	***53–59***	***60–66***	***67–73***	***74–80***
Tally	III	𝍸	𝍸III	𝍸II	II	III	I	II
	3	5	8	7	2	3	1	2

FIGURE 28–29 Frequency distribution table for Practice Problems 28–2 #5.

Number of defective parts/100	***1***	***2***	***3***	***4***	***5***	***6***
Tally	II	III	IIII	𝍸II	III	I
	2	3	4	7	3	1

FIGURE 28–30 Frequency distribution table for Practice Problems 28–2 #6.

Gain in dB	***184–188***	***189–193***	***194–198***	***199–203***	***204–208***	***209–213***
Tally	I	III	𝍸	𝍸𝍸	IIII	II
	1	3	5	10	4	2

SOLUTIONS

1. $\text{mean} = \bar{x} = \dfrac{\Sigma x}{n}$

$$= \frac{x_1 + x_2 + x_3 + \ldots + x_{25}}{25}$$

$$\bar{x} = \frac{1940}{25} = 77.6$$

The median is the thirteenth value, which is 75. The mode is 75.

mean = 77.6 median = 75 mode = 75

2. $\text{mean} = \bar{x} = \dfrac{\Sigma x}{n}$

$$= \frac{x_1 + x_2 + x_3 + \ldots + x_{25}}{25}$$

$$\bar{x} = \frac{1935}{25} = 77.4$$

The mean is found by using the midpoints of each interval, which are 64, 69, 74, 79, 84, and 89. The median is the thirteenth value, which is in the in

3. mean $= \bar{x} = \dfrac{\Sigma x}{25}$

$ = \dfrac{x_1 + x_2 + x_3 + \ldots + x_{25}}{25}$

$ = \dfrac{25{,}130}{25} = 1005$

There are 25 different resistors measured, so the thirteenth value would be the median value, which is 1010 Ω. 1010 Ω was measured the most times so the mode is 1010 Ω.

mean = 1005   median = 1010   mode = 1010

5. mean $= \bar{x} = \dfrac{\Sigma x}{20}$

$ = \dfrac{x_1 + x_2 + x_3 + \ldots + x_{20}}{20}$

$ = \dfrac{69}{20} = 3.45$

The median would be between the tenth and eleventh values but they both occur at a value of 4, so the median is 4. Four occurs most frequently, so the mode is 4.

mean = 3.45   median = 4   mode = 4

terval 77 through 81. The midpoint of this interval, 79, is the median. The mode is the midpoint of the interval 72 through 76, or 74.

mean = 77.4   median = 79   mode = 74

4. mean $= \bar{x} = \dfrac{\Sigma x}{n}$

$ = \dfrac{x_1 + x_2 + x_3 + \ldots + x_{31}}{31}$

$\bar{x} = \dfrac{1463}{31} = 47.2$

The median would be the sixteenth value. The sixteenth value is in the interval 39 through 45, so the median will be the midpoint of that interval, which is 42. The interval 39 through 45 occurs most often, so the midpoint of that interval, 42, is the mode.

mean = 47.2   median = 42   mode = 42

6. mean $= x = \dfrac{\Sigma x}{n}$

$ = \dfrac{x_1 + x_2 + x_3 + \ldots + x_{25}}{25}$

$x = \dfrac{4995}{25} = 200$

The median would be the thirteenth value. The thirteenth value is in the interval 199–203, so the median will be the midpoint of that interval, which is 201. The interval 199–203 occurs most often, so the midpoint of that interval, 201, is the mode.

mean = 200   median = 201
mode = 201

Additional practice problems are at the end of the chapter.

1. In an electronics math quiz, the following scores were recorded: 85, 70, 70, 65, 60, 95, 70, 75, 80, 85, 90, 70, 60, 85, 75, 75, 70, 80, 85, 70. Construct a frequency distribution table and histogram. Compute the mean, median, and mode.

2. During a 25-day period in a manufacturing facility, the following numbers of items were produced: 123, 147, 159, 185, 114, 136, 144, 170, 152, 140, 138, 158, 125, 136, 166, 151, 156, 148, 142, 153, 177, 155, 133, 145, 160. Construct a frequency distribution table and histogram. Determine the range and the interval midpoints. Compute the mean, median, and mode.

28-3   STANDARD DEVIATION

All data sets will have some variability. How much the data values are different depends on many factors. The factors involved depend on what we're measuring. As mentioned before, in manufacturing processes, factors such as temperature, humidity, the precision of the manufacturing equipment, and human error account for these variations.

One measure of variability (*spread* and *dispersion* are also used to describe these differences) is the *range,* which we defined previously as the difference between the maximum and minimum values. Although the range gives us the extremes, it doesn't tell us anything about how the data are spread out between these extremes.

The most widely used measure of variability is the *standard deviation.* The standard deviation is the averaged distances of all data from the mean. The standard deviation gives statisticians a clearer picture of the spread because it relates the spread to the mean.

Statisticians use two different methods for calculating standard deviation. One is defined as the *population* standard deviation and the second is called the *sample* standard deviation. The most commonly used method is the sample standard deviation and it is the one we will use. We will use the lowercase Greek letter *sigma* (σ) to define the sample standard deviation, or simply the standard deviation. On the TI-36X calculator, that key is designated with the letters σ_{xn-1}.

The equation for determining the standard deviation is:

$$\sigma = \sigma_{xn-1} = \sqrt{\frac{\Sigma(x - \bar{x})^2}{n - 1}}$$

$(x - \bar{x})^2$ means that we are to subtract a value in the set from the mean and square it. $\dfrac{\Sigma(x - \bar{x})^2}{n}$ means that we are to find the mean of all these values. For example, suppose a set of numbers is 4, 4, 6, 7, and 8. The mean would be

$$\text{mean} = \bar{x} = \frac{\Sigma x}{n} = \frac{x_1 + x_2 + x_3 + x_4 + x_5}{5} = 5.80$$

The standard deviation would be

$$\sigma = \sqrt{\left[\frac{(4 - 5.8)^2 + (4 - 5.8)^2 + (6 - 5.8)^2 + (7 - 5.8)^2 + (8 - 5.8)^2}{5 - 1}\right]}$$

$\sigma = 1.79$

An easier way is to use the statistical calculator. When you find the mean ($\bar{x}$), you can also find σ by pressing $\boxed{\sigma_{xn-1}}$.

$\boxed{3rd}\ \boxed{STAT\ 1}\ 4\ \boxed{\Sigma+}\ 4\ \boxed{\Sigma+}\ 6\ \boxed{\Sigma+}\ 7\ \boxed{\Sigma+}\ 8\ \boxed{\Sigma+}$

$\boxed{2nd}\ \boxed{\bar{x}}$ mean $= 5.80$

$\boxed{2nd}\ \boxed{\sigma_{xn-1}}$ standard deviation $= 1.79$

EXAMPLE 28–9

Using the data in Figure 28–31, find the mean, median, mode, range, and the standard deviation.

Scores	60	65	70	75	80	85	90	95
Tally	||	||||	ЖЖ	ЖЖ|||	ЖЖ	|||	|||	|
	2	4	5	8	5	3	3	1

FIGURE 28–31
Frequency distribution table for Example 28–9.

SOLUTION You may find the mean and standard deviation, σ, using Equations 28–1 and 28–2, or you may use your statistical calculator. Follow the directions in the user's manual. Using the calculator:

$$\bar{x} = 75.8 \qquad \sigma = 9.05$$
$$\text{median} = 75 \qquad \text{mode} = 75 \qquad \text{range} = 95 - 60 = 35$$

EXAMPLE 28–10

Using the data in Figure 28–32, find the mean, median, mode, range, and the standard deviation. Again, use the interval midpoints to find the mean. The midpoints are 60, 65, 70, 75, 80, 85, and 90.

Score intervals	58–62	63–67	68–72	73–77	78–82	83–87	88–92
Tally	||	||||	ЖЖ|||	ЖЖ	||||	||	|
	2	4	8	5	4	2	1

FIGURE 28–32
Frequency distribution table for Example 28–10.

SOLUTION Using the calculator: $\bar{x} = 72.9$; $\sigma = 7.64$.

median = 70   mode = 70   range = 92 − 58 = 34

PRACTICE PROBLEMS 28–3

1. From the data set in Figure 28–33, find the mean, median, mode, range, and standard deviation.
2. From the data set in Figure 28–34, find the mean, median, mode, range, and standard deviation.
3. From the data set in Figure 28–35, find the mean, median, mode, range, and standard deviation.
4. From the data set in Figure 28–36, find the mean, median, mode, range, and standard deviation.
5. From the data set in Figure 28–37, find the mean, median, mode, range, and standard deviation.

FIGURE 28–33 Frequency distribution table for Practice Problems 28–3 #1.

Scores	*60*	*65*	*70*	*75*	*80*	*85*	*90*	*95*
Tally	II	III	卌	卌I	卌IIII	卌	IIII	II
	2	3	5	6	9	5	4	2

FIGURE 28–34 Frequency distribution table for Practice Problems 28–3 #2.

Items	*600*	*610*	*620*	*630*	*640*	*650*	*660*
Tally	I	III	卌II	卌卌	卌III	IIII	II
	1	3	7	10	8	4	2

FIGURE 28–35 Frequency distribution table for Practice Problems 28–3 #3.

Score intervals	*59–63*	*64–68*	*69–73*	*74–78*	*79–83*	*84–88*	*89–93*
Tally	III	卌I	卌卌	卌	IIII	II	I
	3	6	10	5	4	2	1

FIGURE 28–36 Frequency distribution table for Practice Problems 28–3 #4.

Score intervals	*60–64*	*65–69*	*70–74*	*75–79*	*80–84*	*85–89*	*90–94*	*95–99*
Tally	I	III	卌IIII	卌I	IIII	III	II	I
	1	3	9	6	4	3	2	1

Item intervals	200–216	217–233	234–250	251–267	268–284	285–301	302–318	319–335
Tally	l	lll	Ɲ	Ɲ Ɲ l	Ɲ llll	Ɲ l	Ɲ	lll
	1	3	5	11	9	6	5	3

FIGURE 28–37   Frequency distribution table for Practice Problems 28–3 #5.

SOLUTIONS

1. $\bar{x} = 78.1$ median $= 80$
   mode $= 80$ range $= 35$
   $\sigma = 9.20$

2. $\bar{x} = 632$ median $= 630$
   mode $= 630$ range $= 60$
   $\sigma = 14.2$

3. Midpoints are 61, 66, 71, 76, 81, 86, and 91.
   $\bar{x} = 72.8$ median $= 71$
   mode $= 71$ range $= 34$
   $\sigma = 7.59$

4. Midpoints are 62, 67, 72, 77, 82, 87, 92, and 97.
   $\bar{x} = 77.3$ median $= 77$
   mode $= 72$ range $= 39$
   $\sigma = 8.44$

5. Midpoints are 208, 225, 242, 259, 276, 293, 310, and 327.
   $\bar{x} = 272$ median $= 276$
   mode $= 259$ range $= 135$
   $\sigma = 29.3$

Additional practice problems are at the end of the chapter.

☞ **KEY POINT** The *standard deviation* is the averaged distances of all data from the mean.

In order to understand the significance of the standard deviation, let's discuss the *normal distribution* curve and then tie the two together.

THE NORMAL CURVE   28–4

When a population has a normal distribution, connecting all the midpoints in the histogram will result in a curve like the one in Figure 28–38. Such a curve is called the *normal curve.*
Characteristics of the normal curve are:

1. The curve is bell-shaped and has only one peak.
2. The shape on one side of the peak is the same as the shape on the other side.
3. There will be as many data points above the mean as there are below the mean.
4. The mean, median, and mode occur at the same point—the peak.

FIGURE 28–38
The normal
curve.

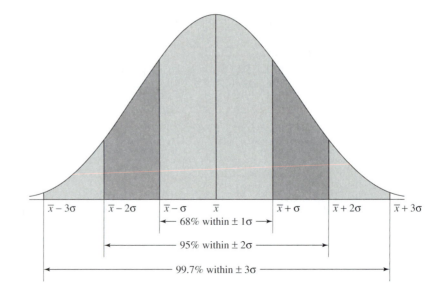

5. The mean defines the center of the curve. The standard deviation defines the spread.
6. 68% of the values fall within $\pm 1\ \sigma$; 95% of the values fall within $\pm 2\ \sigma$ and 99.7% of the values fall within $\pm 3\ \sigma$.

A small σ indicates a small spread from the mean. For example, suppose in a normal curve, the mean of a data set is 500 and σ equals 8. We know from the characteristics of the normal curve that 68% of the data values will fall between 492 and 508.

$$500 - \sigma = 500 - 8 = 492;\ 500 + \sigma = 500 + 8 = 508$$

Because the curve is symmetrical, 34% of the data fall in the group from 492 to 500 and 34% of the data fall in the group from 500 to 508.

95% of the data values will fall between $\pm 2\ \sigma$.

$$500 - 2\ \sigma = 500 - 16 = 484;\ 500 + 2\ \sigma = 500 + 16 = 516$$

Half of the 95% falls below the mean and half will be above the mean, or 47.5% of all the data lie between 484 and 500 and 47.5% of all the data lie between 500 and 516.

99.7% of the data values will fall between $\pm 3\ \sigma$.

$$500 - 3\ \sigma = 500 - 24 = 476;\ 500 + 3\ \sigma = 500 + 24 = 524$$

Half the 99.7% falls below the mean and half will be above the mean or 99.7% of all the data values fall between 476 and 524.

0.3% of all the data will be greater than $\pm 3\ \sigma$.

EXAMPLE 28–11

In a standardized test given over a period of years, 2000 students were tested. The test results yielded a normal curve. The mean was 75 and the standard deviation was 5. What was the distribution of grades?

SOLUTION   If $\sigma = 5$, then:

$$75 - 1\,\sigma = 75 - 5 = 70$$
$$75 + 1\,\sigma = 75 + 5 = 80$$

68% of all the grades fall between 70 and 80.
34% (half of 68%) fall between 70 and 75 and 34% fall between 75 and 80.

$$2000 \times 0.34 = 680$$

680 of the students' scores were between 70 and 75.
680 of the students' scores were between 75 and 80.
95% of all scores fall $\pm 2\,\sigma$ from the mean.

$$75 - 2\,\sigma = 75 - 10 = 65$$
$$75 + 2\,\sigma = 75 + 10 = 85$$

47.5% of the scores (half of 95%) fall between 65 and 75.
47.5% are in the range from 75 to 85.

$$2000 \times 0.475 = 950$$

950 of the students' scores were between 65 and 75.
950 of the students' scores were between 75 and 85. $950 + 950 = 1900$.
The normal curve shows that almost all the scores (99.7%) are $\pm 3\,\sigma$ or less from
the mean.

$$75 - 3\,\sigma = 75 - 15 = 60$$
$$75 + 3\,\sigma = 75 + 15 = 90$$
$$2000 \times 0.4985 = 997$$

997 of the students' scores were between 60 and 75.
997 of the students' scores were between 75 and 90.
Some scores were *greater than* $3\,\sigma$ from the mean. In this case, 6 (0.3% of 2000).
This area beyond $3\,\sigma$ exists because in the normal curve, there are a few scores (or
items) that are unusually large or small.

PRACTICE PROBLEMS 28–4

1. 600 amplifiers were produced in a given time period. The mean output was 25 dB with a standard deviation of 2dB. What would be a normal distribution?

2. 1500 scores were reported on a standardized test. The mean score was 70 with a standard deviation of 4. What would be the normal distribution?

1. If $\sigma = 2$, then:

   25 dB $-$ 1 $\sigma = 23$ dB
   25 dB $+$ 1 $\sigma = 27$ dB

   68% of the output of all the amplifiers are between 23 dB and 27 dB. $600 \times 0.34 = 204$. 34% or 204 amplifiers have an output between the mean (25 dB) and 23 dB. 34% or 204 amplifiers have an output between the mean and 27 dB. 95% of the output of all amplifiers are between ± 2 σ from the mean.

   25 dB $-$ 2 $\sigma = 25$ dB $-$ 4 dB $=$
   21 dB
   25 dB $+$ 2 $\sigma = 25$ dB $+$ 4 dB $=$
   29 dB
   $600 \times 0.475 = 285$
   47.5% of 600 amplifiers or 285 amplifiers will have outputs between 21 dB and 25 dB (-2σ).

   47.5% of 600 amplifiers or 285 amplifiers will have outputs between 25 dB and 29 dB ($+2$ σ). 99.7% of the output of all amplifiers are between ± 3 σ from the mean.

   25 dB $-$ 3 $\sigma = 25$ dB $-$ 6 dB $=$
   19 dB
   25 dB $+$ 3 $\sigma = 25$ dB $+$ 6 dB $=$
   31 dB
   $600 \times 0.4985 = 299$
   49.85% of the amplifiers or 299 will have outputs between 19 dB and 25 dB.

   49.85% of the amplifiers or 299 will have outputs between 25 dB and 31 dB.

   $299 + 299 = 598$. Two amplifiers will have outputs beyond ± 3 σ.

2. 1 σ above $70 = 70 + 4 = 74$. 1 σ below $70 = 70 - 4 = 66$.

   $1500 \times 34\% = 510$
   510 students scored between 66 and 70.
   510 students scored between 70 and 74.
   2 σ above $70 = 70 + 8 = 78$. 2 σ below $70 = 70 - 8 = 62$.
   $1500 \times 47.5\% = 713$
   713 students scored between 62 and 70.
   713 students scored between 70 and 78.

   3 σ above $70 = 70 + 12 = 82$. 3 σ below $70 = 70 - 12 = 58$.

   $1500 \times 49.85\% = 748$
   748 students scored between 58 and 70.
   748 students scored between 70 and 82.
   $748 + 748 = 1496$
   2 students had scores greater than 82.
   2 students had scores less than 58.

Additional practice problems are at the end of the chapter.

In the examples we have used in this chapter, our curves have been *skewed*. That is, they are different from the normal curve. In our examples and practice problems we had more data points above the mean than below the mean, or vice versa. This is usually the case when dealing with small amounts of data. What this means is that the distribution of the data will be different from the normal but all or almost all of the data will fall between ± 3 σ. A look at Example 28–5 shows that there were 20 scores. The mean was 76.5. Computing the standard deviation, we get $\sigma = 8.90$. Twelve of the scores fall between the mean and $-3\ \sigma$. Eight of the scores lie between the mean and $+3\ \sigma$. Ideally, all scores will lie within $\pm 3\ \sigma$.

In manufacturing, if the limits of the tolerances of manufactured parts are set to $\pm 3\ \sigma$, then any item that lies beyond $\pm 3\ \sigma$ is a reject. If there are parts that exceed the established limits, then steps are taken immediately to determine and correct the cause.

The normal curve is the *ideal* curve. It is the curve that is expected as the result of thousands of tests or thousands of items produced. In reality, though, most curves will be skewed. In quality control, companies continually strive to eliminate any items, parts, or processes that fall beyond the acceptable limits. These limits are often set at $\pm 3\ \sigma$.

Further discussions of statistical methods are left to courses on statistics. Our purpose here has been to acquaint the student with some of the basic concepts he or she is most likely to encounter.

SELF-TEST 28–2

1. From the data in Figure 28–39, compute the mean, median, mode, range, and standard deviation.
2. From the data in Figure 28–40, compute the mean, median, mode, range, and standard deviation.
3. In a standardized test given over a period of years, 3000 students were tested. The test results yielded a normal curve. The mean was 70 and the standard deviation was 6. What was the distribution of grades?

Scores	55	60	65	70	75	80	85	90	95
Tally	I		II	NI	NIIII	NI	IIII	III	II
	1	0	2	5	8	5	4	3	2

FIGURE 28–39   Frequency distribution table for Self-Test 28–2 #1.

Item intervals	600–640	641–681	682–722	723–763	764–804	805–845	846–886
Tally	I	II	NI NI	NI NI II	NI	II	I
	1	2	10	12	5	2	1

FIGURE 28–40   Frequency distribution table for Self-Test 28–2 #2.

CHAPTER 28 AT A GLANCE

PAGE	KEY POINTS	EXAMPLE

Assume the following data: 3, 3, 4, 4, 4, 5, 7, 7, 9.

791   *Key Point:* A frequency distribution table tells us how many items are in a study and how often each item appears.

Data	*3*	*4*	*5*	*6*	*7*	*8*	*9*
Tally	||	|||	|		||		|
	2	3	1	0	2	0	1

FIGURE 28–41

792   *Key Point:* In a histogram, the frequency is plotted along the *y*-axis and the values are plotted along the *x*-axis.

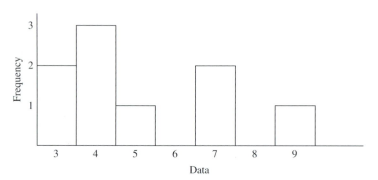

FIGURE 28–42

795   *Key Point:* The range of a set of data is the difference between the largest and smallest numbers.

range = 9 − 3 = 6

795   *Key Point:* Select a number of intervals that will most nearly describe the spread of the population.

795   *Key Point:* When using intervals in a frequency distribution table: (1) intervals cannot overlap; (2) data cannot appear in more than one interval; (3) the smallest data value fits in the first interval; (4) the largest data value fits in the last interval; (5) the width of all intervals is the same.

| 803 | *Key Point:* The *mean* is equal to the sum of the data divided by the number of data. | $\bar{x} = \dfrac{\Sigma x}{n} = \dfrac{46}{9} = 5.11$ |

| 803 | *Key Point:* The median is that value where half of the values are greater and half are less. | median $= 4$ |

| 804 | *Key Point:* The *mode* is the measurement that appears most frequently in the data. | mode $= 4$ |

| 814 | *Key Point:* The standard deviation is the averaged distances of all data from the mean. | $\sigma = \sqrt{\dfrac{\Sigma(x - \bar{x})^2}{n - 1}} = 1.97$ |

END OF CHAPTER PROBLEMS 28–1

1. In an electronics math quiz, the following scores were recorded: 85, 65, 90, 75, 60, 70, 75, 65, 80, 75, 85, 70, 75, 80, 75. Construct a frequency distribution table. From the table, construct a histogram.

2. In an electronics math quiz, the following scores were recorded: 80, 70, 75, 60, 70, 85, 70, 65, 60, 70, 80, 60, 70, 65, 75. Construct a frequency distribution table. From the table, construct a histogram.

3. In an electronics math quiz the following scores were recorded: 80, 62, 77, 98, 78, 75, 70, 60, 90, 85, 75, 83, 78, 72, 65, 82, 70, 76, 65, 80, 94, 75, 66, 73, 85. Construct a frequency distribution table. From the table, construct a histogram. Determine the range and interval midpoints.

4. In an electronics math quiz the following scores were recorded: 83, 64, 75, 72, 78, 85, 57, 73, 70, 79, 80, 76, 95, 88, 72, 81, 75, 73, 85, 74, 76, 70, 82, 65, 78, 91, 82, 80, 75, 68. Construct a frequency distribution table. From the table, construct a histogram. Determine the range and interval midpoints.

5. Twenty high-gain amplifiers were constructed in the lab using components off the shelf. The design called for a voltage gain of 400 dB. The voltage gains were measured to determine the actual gain of the amplifiers. The results (in dB) were: 409, 397, 403, 418, 413, 382, 370, 407, 392, 400, 416, 402, 394, 392, 407, 405, 384, 432, 426, 404. Construct a frequency distribution table and a histogram. Determine the range and interval midpoints.

6. Twenty high-gain amplifiers were constructed in the lab using components off the shelf. The design called for a voltage gain of 400 dB. The voltage gains were measured to determine the actual gain of the amplifiers. The results (in dB) were: 403, 402, 381, 417, 370, 397, 408, 388, 408, 415, 406, 398, 394, 411, 417, 405, 384, 434, 426, 404. Construct a frequency distribution table and histogram. Determine the range and interval midpoints.

7. In a manufacturing company, the following items were produced over a 25-day period: 312, 276, 295, 300, 325, 301, 280, 277, 290, 315, 306, 279, 320, 302, 288, 270, 260, 293, 315, 294, 340, 292, 298, 330, 310. Construct a frequency distribution table and histogram. Determine the range and interval midpoints.

8. In a manufacturing company, the following items were produced over a 25-day period: 125, 101, 113, 94, 78, 88, 94, 107, 112, 96, 102, 104, 96, 85, 89, 93, 98, 103, 119, 111, 102, 105, 98, 95, 104. Construct a frequency distribution table and histogram. Determine the range and interval midpoints.

9. From a bin of 200 2-kΩ ±5% carbon resistors, 25 were selected to be measured in the laboratory. The results were: 1.98 kΩ, 2.00 kΩ, 1.96 kΩ, 2.02 kΩ, 2.00 kΩ, 1.93 kΩ, 2.03 kΩ, 2.00 kΩ, 1.98 kΩ, 1.96 kΩ, 1.90 kΩ, 2.05 kΩ, 2.02 kΩ, 2.00 kΩ, 1.98 kΩ, 2.04 kΩ, 1.98 kΩ, 2.04 kΩ, 2.00 kΩ, 2.03 kΩ, 2.09 kΩ, 2.04 kΩ, 2.06 kΩ, 2.05 kΩ, 2.08 kΩ. Construct a frequency distribution table. From the table, construct a histogram. Determine the range and interval midpoints.

10. From a bin of 100 47-kΩ ±5% carbon resistors, 25 were selected to be measured in the laboratory. The results were: 46.5 kΩ, 45.5 kΩ, 48.7 kΩ, 47.2 kΩ, 49.8 kΩ, 47.1 kΩ, 46.7 kΩ, 47.0 kΩ, 46.8 kΩ, 45.3 kΩ, 46.8 kΩ, 49.0 kΩ, 47.3 kΩ, 46.9 kΩ, 46.2 kΩ, 47.8 kΩ, 46.0 kΩ, 48.5 kΩ, 47.0 kΩ, 48.1 kΩ, 46.8 kΩ, 44.5 kΩ, 47.6 kΩ, 47.5 kΩ, 46.5 kΩ. Construct a frequency distribution table. From the table, construct a histogram. Determine the range and interval midpoints.

11. In a production run, the number of defective parts per 100 were tallied over a 20-day period. The results were: day 1—3, day 2—4, day 3—3, day 4—2, day 5—6, day 6—4, day 7—3, day 8—5, day 9—2, day 10—1, day 11—3, day 12—5, day 13—4, day 14—3, day 15—1, day 16—3, day 17—5, day 18—2, day 19—4, day 20—3. Construct a frequency distribution table and a histogram.

12. In a production run, the number of defective parts per 100 were tallied over a 30-day period. The results were: day 1—4, day 2—3, day 3—6, day 4—3, day 5—2, day 6—5, day 7—3, day 8—5, day 9—2, day 10—3, day 11—1, day 12—4, day 13—5, day 14—2, day 15—6, day 16—3, day 17—4, day 18—5, day 19—4, day 20—3, day 21—4, day 22—5, day 23—4, day 24—5, day 25—6, day 26—2, day 27—4, day 28—3, day 29—2, day 30—4. Construct a frequency distribution table and a histogram.

END OF CHAPTER PROBLEMS 28–2

1. Using the frequency distribution table developed for problem 1 of End of Chapter Problems 28–1, find the mean, median, and mode for the data set.

2. Using the frequency distribution table developed for problem 2 of End of Chapter Problems 28–1, find the mean, median, and mode for the data set.

3. Using the frequency distribution table developed for problem 3 of End of Chapter Problems 28–1, find the mean, median, and mode for the data set.

4. Using the frequency distribution table developed for problem 4 of End of Chapter Problems 28–1, find the mean, median, and mode for the data set.

5. Using the frequency distribution table developed for problem 5 of End of Chapter Problems 28–1, find the mean, median, and mode for the data set.

6. Using the frequency distribution table developed for problem 6 of End of Chapter Problems 28–1, find the mean, median, and mode for the data set.

7. Using the frequency distribution table developed for problem 7 of End of Chapter Problems 28–1, find the mean, median, and mode for the data set.

8. Using the frequency distribution table developed for problem 8 of End of Chapter Problems 28–1, find the mean, median, and mode for the data set.

9. Using the frequency distribution table developed for problem 9 of End of Chapter Problems 28–1, find the mean, median, and mode for the data set.

10. Using the frequency distribution table developed for problem 10 of End of Chapter Problems 28–1, find the mean, median, and mode for the data set.

11. Using the frequency distribution table developed for problem 11 of End of Chapter Problems 28–1, find the mean, median, and mode for the data set.

12. Using the frequency distribution table developed for problem 12 of End of Chapter Problems 28–1, find the mean, median, and mode for the data set.

END OF CHAPTER PROBLEMS 28–3

1. From the data set in Figure 28–43, find the mean, median, mode, range, and standard deviation.

2. From the data set in Figure 28–44, find the mean, median, mode, range, and standard deviation.

3. From the data set in Figure 28–45, find the mean, median, mode, range, and standard deviation.

4. From the data set in Figure 28–46, find the mean, median, mode, range, and standard deviation.

5. From the data set in Figure 28–47, find the mean, median, mode, range, and standard deviation.

6. From the data set in Figure 28–48, find the mean, median, mode, range, and standard deviation.

7. From the data set in Figure 28–49, find the mean, median, mode, range, and standard deviation.

8. From the data set in Figure 28–50 (p. 823), find the mean, median, mode, range, and standard deviation.

9. From the data set in Figure 28–51 (p. 823), find the mean, median, mode, range, and standard deviation.

10. From the data set in Figure 28–52, (p. 823) find the mean, median, mode, range, and standard deviation.

11. From the data set in Figure 28–53, (p. 823) find the mean, median, mode, range, and standard deviation.

12. From the data set in Figure 28–54, (p. 824) find the mean, median, mode, range, and standard deviation.

FIGURE 28–43

Scores	*60*	*65*	*70*	*75*	*80*	*85*	*90*	*95*	*100*
Tally	𝍷	𝍷𝍷𝍷	𝍸	𝍸𝍷𝍷	𝍸	𝍷𝍷𝍷𝍷	𝍷𝍷	𝍷	𝍷
	1	3	5	7	5	4	2	1	1

FIGURE 28–44

Scores	*60*	*65*	*70*	*75*	*80*	*85*	*90*	*95*	*100*
Tally	𝍷	𝍷𝍷	𝍷𝍷𝍷	𝍸	𝍸 𝍸	𝍸𝍷	𝍷𝍷𝍷𝍷	𝍷𝍷	𝍷
	1	2	3	5	10	6	4	2	1

FIGURE 28–45

Scores	*55*	*60*	*65*	*70*	*75*	*80*	*85*	*90*	*95*
Tally	𝍷𝍷	𝍷	𝍷𝍷𝍷𝍷	𝍷𝍷𝍷	𝍸𝍷	𝍸 𝍸	𝍸	𝍷𝍷𝍷𝍷	𝍷𝍷
	2	1	4	3	6	10	5	4	2

FIGURE 28–46

Scores	*55*	*60*	*65*	*70*	*75*	*80*	*85*	*90*	*95*
Tally	𝍷		𝍷𝍷𝍷	𝍸𝍷	𝍸𝍷𝍷	𝍸𝍷𝍷𝍷𝍷	𝍸	𝍷𝍷	𝍷
	1	0	3	6	7	9	5	2	1

FIGURE 28–47

Score intervals	*50–56*	*57–63*	*64–70*	*71–77*	*78–84*	*85–91*	*92–98*
Tally	𝍷𝍷	𝍸	𝍸𝍷𝍷	𝍸𝍷𝍷𝍷	𝍸𝍷	𝍷𝍷𝍷𝍷	𝍷𝍷
	2	5	7	8	6	4	2

FIGURE 28–48

Score intervals	*50–56*	*57–63*	*64–70*	*71–77*	*78–84*	*85–91*	*92–98*
Tally	𝍷	𝍷𝍷	𝍸	𝍸𝍷𝍷𝍷	𝍸𝍸	𝍸𝍷𝍷	𝍷𝍷𝍷𝍷
	1	2	5	8	10	7	4

Score intervals	50–54	55–59	60–64	65–69	70–74	75–79	80–84	85–89	90–94
Tally	||	Ⅲ	ⅢⅢ	|||	||||	ⅢⅢ|	ⅢⅢ|||	|||	|
	2	5	5	3	4	6	8	3	1

FIGURE 28–49

Score intervals	50–54	55–59	60–64	65–69	70–74	75–79	80–84	85–89	90–94
Tally	|	||||	ⅢⅢ||	||||	|||	||||	ⅢⅢ ⅢⅢ	ⅢⅢ	||
	1	4	7	4	3	4	10	5	2

FIGURE 28–50

Item intervals	300–340	341–381	382–422	423–463	464–504	505–545	546–586	587–627
Tally	|	|||	ⅢⅢ	ⅢⅢ|||	ⅢⅢ ⅢⅢ||	ⅢⅢ|	||||	||
	1	3	5	8	12	6	4	2

FIGURE 28–51

Item intervals	710–760	761–811	812–862	863–913	914–964	965–1015	1016–1066	1067–1117
Tally	||	|||	ⅢⅢ||	ⅢⅢ ⅢⅢ|	ⅢⅢ||	ⅢⅢ	|||	|
	2	3	7	11	7	5	3	1

FIGURE 28–52

Item intervals	500–520	521–541	542–562	563–583	584–604	605–625	626–646
Tally	||||	ⅢⅢ|	ⅢⅢ ⅢⅢ|	ⅢⅢ||	ⅢⅢ	|||	|
	4	6	11	7	5	3	1

FIGURE 28–53

Item intervals	140–170	171–201	202–232	233–263	264–294	295–325	326–356	357–387	388–418
Tally	||	|||	||||	NI	NI NI	NII	||||	||		
	2	3	4	6	10	7	4	2	1

FIGURE 28–54

END OF CHAPTER PROBLEMS 28–4

1. 500 widgets were selected to be measured. The mean length was designed to be 40 mm. If the standard deviation was 1 mm, what would we expect a normal distribution to be?

2. 200 widgets were selected to be measured. The mean length was designed to be 200 mm. If the standard deviation was 1.5 mm, what would we expect a normal distribution to be?

3. The mean average of parts produced each day during the month of June was 600. If the standard deviation was 10, what would a normal distribution be for that 30-day period?

4. The mean average of parts produced each day during the month of June was 400. If the standard deviation was 7.5, what would a normal distribution be for that 30-day period?

5. 5000 scores were reported on a standardized test. The mean score was 73, with a standard deviation of 5.6. What would be the normal distribution?

6. 4000 scores were reported on a standardized test. The mean score was 77.5, with a standard deviation of 6.3. What would be the normal distribution?

ANSWERS TO SELF-TESTS

ANSWERS TO SELF-TEST 28–1

1. Figure 28–55 is the frequency distribution. Figure 28–56 is the histogram.   mean = 75.8
median = 75 mode = 70

2. Figure 28–57 is the frequency distribution. Figure 28–58 is the histogram. The midpoints of the intervals are 118, 127, 136, 145, 154, 163, 172, 181.   mean = 149
median = 145   mode = 154

FIGURE 28–55

Scores	60	65	70	75	80	85	90	95
Tally	||			NI	|||	||	||||				
	2	1	6	3	2	4	1	1

FIGURE 28–56

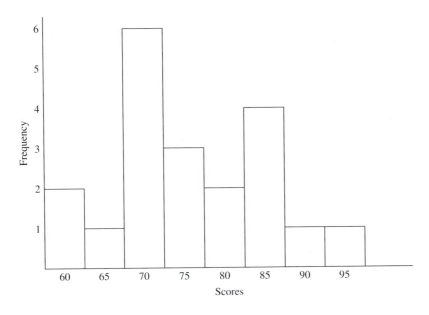

Item intervals	114–122	123–131	132–140	141–149	150–158	159–167	168–176	177–185
Tally	|	||	|||||	|||||	||||| |	|||	|	||
	1	2	5	5	6	3	1	2

FIGURE 28–57

FIGURE 28–58

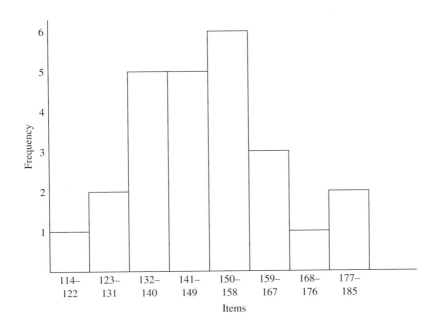

1. $\bar{x} = 77.8$ $\sigma = 9.26$ median $= 75$ mode $= 75$ range $= 40$
2. Midpoints are 620, 661, 702, 743, 784, 825, 866.
   $\bar{x} = 737$ $\sigma = 50.3$ median $= 743$ mode $= 743$ range $= 286$
3. If $\sigma = 6$, then:
   $70 - 1\,\sigma = 70 - 6 = 64.\ 70 + 1\,\sigma = 70 + 6 = 76.$
   68% of all the grades fall between 64 and 76. 34% (half of 68%) fall between 64 and 70 and 34% fall between 70 and 76.
   $3000 \times 0.34 = 1020$
   1020 of the students' scores were between 64 and 70.
   1020 of the students' scores were between 70 and 76.
   95% of all the scores fall $\pm 2\,\sigma$ from the mean.
   $70 - 2\,\sigma = 70 - 12 = 58.\ 70 + 2\,\sigma = 70 + 12 = 82.$
   47.5% of the scores (half of 95%) fall between 58 and 70.
   47.5% are in the range from 70 to 82.
   $3000 \times 0.475 = 1425$
   1425 of the students' scores were between 58 and 70.
   1425 of the students' scores were between 70 and 82. $1425 + 1425 = 2850$
   The normal curve shows that almost all the scores (99.7%) are $\pm 3\,\sigma$ or less from the mean.
   $3000 \times 0.4985 = 1496$ (rounded up from 1495.5)
   $70 - 3\,\sigma = 70 - 18 = 52.\ 70 + 3\,\sigma = 70 + 18 = 88.$
   1496 of the students' scores were between 52 and 70.
   1496 of the students' scores were between 70 and 88.
   Some scores were *greater than* $\pm 3\,\sigma$ from the mean. In this case, 9 (0.3% of 3000).

The Calculator

There are many makes and models of calculators on the market today. Many full-function calculators are available for less than $15. In addition to the basic functions that all calculators perform, you will need a calculator that includes trigonometric and logarithmic functions and displays the constant π (pi). It should also perform polar/rectangular conversions and be capable of performing basic functions in other number systems such as binary and octal. Such calculators are called *full-function* calculators. In addition to the functions mentioned above, full-function calculators perform many other functions not mentioned here. Most full-function calculators will FIX the number of digits displayed to the right of the decimal point and will display answers in scientific (SCI) and engineering (ENG) notation.

All calculators handle basic functions the same way (except for Hewlett-Packard calculators) but may require different keystrokes to change from standard notation to scientific or engineering notation and to make other more advanced computations. All full-function calculators come with an instruction manual that provides step-by-step examples and sample problems. Many full-function calculators today are solar powered so there is never a need to replace batteries. Ask your instructor and advanced students for advice as to what calculators they like for the work you will be doing.

In this section of the text, we will look at the internal circuitry of a calculator to see how it works. An understanding of the internal operations of the calculator should help you better understand how to use it.

Internal to the calculator, there are two places where data are stored (there are probably others, but let's keep it simple). These places are digital circuits called *registers*. These registers are called the *x-register* and the *y-register*. The contents of the *x*-register are displayed. You cannot see the contents of the *y*-register.

Most students I have observed prefer the TI-36X Solar (the TI-35X is the battery version) or the Casio fx-115W.

Let's multiply 3×4 using the TI-36X Solar. Here is what you do: When first using the calculator, press $\boxed{AC}$ to make sure that the calculator is cleared. Then:

1. Key in the 3.
2. Press the $\boxed{\times}$ key.
3. Key in the 4.
4. Press the $\boxed{=}$ key. *Answer:* $3 \times 4 = 12$.

3	x-register

?	y-register

FIGURE A–I

3	x-register

3	y-register

FIGURE A–2

4	x-register

3	y-register

FIGURE A–3

12	x-register

4	y-register

FIGURE A–4

When you go through these four steps, here is what happens internally in the calculator:

1. *Key in the 3.* The three is stored in the x-register and is displayed. See Figure A–1.

2. *Press the* $\boxed{\times}$ *key.* The 3 is copied into the y-register. Both the x- and y-registers now store the number 3. See Figure A–2. At the same time, the calculator is given the message "Get ready to multiply the contents of the x-register by the contents of the y-register."

3. *Key in the 4.* The contents of the x-register are replaced with the number 4. The y-register is not affected. At this point, the x-register stores 4 and the y-register stores 3. See Figure A–3.

4. *Press the* $\boxed{=}$ *key.* In step 2 we told the calculator to get ready to multiply the contents of the x-register by the contents of the y-register. Pressing the $\boxed{=}$ key says "Do it!" At this point the calculator performs the multiply function and moves the product (12) into the x-register. The old contents of the x-register (4) go to the y-register and the original contents of the y-register (3) are lost. See Figure A–4. Remember, the calculator always displays the contents of the x-register.

Let's try a different one. Let's raise 4 to the third power. $4^3 = ?$. Here is the algorithm:

1. Key in the base, which is 4. 4 is stored in the x-register and is displayed.

2. Press $\boxed{y^x}$. 4 is moved into the y-register, and the calculator is instructed to get ready to raise the contents of the y-register (which is the base, 4) to some power. This exponent will be keyed into the x-register.

3. Key in the exponent, which is 3. The contents of the x-register (4) are replaced with the number 3. Now the y-register contains the base (4), and the x-register contains the exponent (3).

4. Press the $\boxed{=}$ key. Pressing the equals key tells the calculator to raise the contents of the y-register (4) to the third power. As always, the result (64) is placed into the x-register and is displayed.

No matter what the function is, multiply, raise to a power, add, or whatever, the result is always moved to the x-register and displayed. There are various ways you can display the contents of the x-register; you can display from 1 to as many as 10 digits in floating decimal notation, or you can also display answers in scientific or engineering notation.

The Casio calculator is somewhat different. All keystrokes (numbers and math operators) are stored in a memory and the last several keystrokes are displayed in the top line of a three-line display. Up to 72 numbers and operators can be saved in memory. When the "equals" key is pressed, the calculator operates on the numbers and math operators saved in the memory and then displays the answer in the middle line of the display. The top line will now show the first few keystrokes entered. The user can now scroll through the previously entered keystrokes and change any or all numbers or math operators. When the "equals" key is pressed, a new answer, based on the edited keystrokes, is displayed. This procedure of editing and recalculating can be repeated as long as needed.

Algorithms for solving various problems appear throughout the text as problems occur for the first time.

All calculators that use *algebraic logic* follow the same rules when assigning priorities to the various mathematical operations. The algebraic hierarchy is as follows:

- Priority 1: Parentheses
- Priority 2: Reciprocals, squares and square roots, logs, trig functions
- Priority 3: Universal powers and roots
- Priority 4: Multiplication and division (same priority level)
- Priority 5: Addition and subtraction (same priority level)
- Priority 6: Equals

Our colleague, Barbara Snyder, has simplified the above rules of priority and offers the PEMDAS rule, a variation on My Dear Aunt Sally (**M**ultiply, **D**ivide, **A**dd, and **S**ubtract), to help remember the priority order: *Please Excuse My Dear Aunt Sally.*

- **P**arentheses
- **E**xponents
- **M**ultiplication and **D**ivision
- **A**ddition and **S**ubtraction

Consult your user's manual for more detailed information concerning your calculator.

Trigonometric Functions of Degrees

The following tables show the sine, cosine, tangent, and cotangent of all angles between 0 and 90 degrees. To use these tables use the following procedure: If the given angle is between 0 and 45 degrees, then locate it in the *left most* column and use the column *headings* to find the trig function you need. If the given angle is between 45 and 90 degrees, then locate it in the *right most* column and use the column *footings* to find the trig function you need.

ANGLE°	SIN	TAN	COT	COS	
0.0	.00000	.00000	∞	1.00000	**90.0**
.1	.00175	.00175	572.96	1.00000	.9
.2	.00349	.00349	286.48	0.99999	.8
.3	.00524	.00524	190.98	.99999	.7
.4	.00698	.00698	143.23	.99998	.6
.5	.00873	.00873	114.59	.99996	.5
.6	.01047	.01047	95.489	.99995	.4
.7	.01222	.01222	81.847	.99993	.3
.8	.01396	.01396	71.615	.99990	.2
.9	.01571	.01571	63.657	.99988	.1
1.0	.01745	.01746	57.290	.99985	**89.0**
.1	.01920	.01920	52.081	.99982	.9
.2	.02094	.02095	47.740	.99978	.8
.3	.02269	.02269	44.066	.99974	.7
.4	.02443	.02444	40.917	.99970	.6
.5	.02618	.02619	38.188	.99966	.5
.6	.02792	.02793	35.801	.99961	.4
.7	.02967	.02968	33.694	.99956	.3
.8	.03141	.03143	31.821	.99951	.2
.9	.03316	.03317	30.145	.99945	.1
2.0	.03490	.03492	28.636	.99939	**88.0**
.1	.03664	.03667	27.271	.99933	.9
.2	.03839	.03842	26.031	.99926	.8
.3	.04013	.04016	24.898	.99919	.7
COS	COT	TAN	SIN	ANGLE°	

ANGLE°	SIN	TAN	COT	COS	
.4	.04188	.04191	23.859	.99912	.6
.5	.04362	.04366	22.904	.99905	.5
.6	.04536	.04541	22.022	.99897	.4
.7	.04711	.04716	21.205	.99889	.3
.8	.04885	.04891	20.446	.99881	.2
.9	.05059	.05066	19.740	.99872	.1
3.0	.05234	.05241	19.081	.99863	**87.0**
.1	.05408	.05416	18.464	.99854	.9
.2	.05582	.05591	17.886	.99844	.8
.3	.05756	.05766	17.343	.99834	.7
.4	.05931	.05941	16.832	.99824	.6
.5	.06105	.06116	16.350	.99813	.5
.6	.06279	.06291	15.895	.99803	.4
.7	.06453	.06467	15.464	.99792	.3
.8	.06627	.06642	15.056	.99780	.2
.9	.06802	.06817	14.699	.99767	.1
4.0	.06976	.06993	14.301	.99756	**86.0**
.1	.07150	.07168	13.951	.99744	.9
.2	.07324	.07344	13.617	.99731	.8
.3	.07498	.07519	13.300	.99719	.7
.4	.07672	.07695	12.996	.99705	.6
.5	.07846	.07870	12.706	.99692	.5
.6	.08020	.08046	12.429	.99678	.4
.7	.08194	.08221	12.163	.99664	.3
.8	.08368	.08397	11.909	.99649	.2
.9	.08542	.08573	11.664	.99635	.1
5.0	.08716	.08749	11.430	.99619	**85.0**
.1	.08889	.08925	11.205	.99604	.9
.2	.09063	.09101	10.988	.99588	.8
.3	.09237	.09277	10.780	.99572	.7
.4	.09411	.09453	10.579	.99556	.6
.5	.09585	.09629	10.385	.99540	.5
.6	.09758	.09805	10.199	.99523	.4
.7	.09932	.09981	10.019	.99506	.3
.8	.10106	.10158	9.8448	.99488	.2
.9	.10279	.10334	9.6768	.99470	.1
6.0	.10453	.10510	9.5144	.99452	**84.0**
.1	.10626	.10687	9.3572	.99434	.9
.2	.10800	.10863	9.2052	.99415	.8
.3	.10973	.11040	9.0579	.99396	.7
.4	.11147	.11217	8.9152	.99377	.6
.5	.11320	.11394	8.7769	.99357	.5
.6	.11494	.11570	8.6427	.99337	.4
.7	.11667	.11747	8.5126	.99317	.3
	COS	COT	TAN	SIN	ANGLE°

ANGLE°	SIN	TAN	COT	COS	
.8	.11840	.11924	8.3863	.99297	.2
.9	.12014	.12101	8.2636	.99276	.1
7.0	.12187	.12278	8.1443	.99255	**83.0**
.1	.12360	.12456	8.0285	.99233	.9
.2	.12533	.12633	7.9158	.99211	.8
.3	.12706	.12810	7.8062	.99189	.7
.4	.12880	.12988	7.6996	.99167	.6
.5	.13053	.13165	7.5958	.99144	.5
.6	.13226	.13343	7.4947	.99122	.4
.7	.13399	.13521	7.3962	.99098	.3
.8	.13572	.13698	7.3002	.99075	.2
.9	.13744	.13876	7.2066	.99051	.1
8.0	.13917	.14054	7.1154	.99027	**82.0**
.1	.14090	.14232	7.0264	.99002	.9
.2	.14263	.14410	6.9395	.98978	.8
.3	.14436	.14588	6.8548	.98953	.7
.4	.14608	.14767	6.7720	.98927	.6
.5	.14781	.14945	6.6912	.98902	.5
.6	.14954	.15124	6.6122	.98876	.4
.7	.15126	.15302	6.5350	.98849	.3
.8	.15299	.15481	6.4596	.98823	.2
.9	.15471	.15660	6.3859	.98796	.1
9.0	.15643	.15838	6.3138	.98769	**81.0**
.1	.15816	.16017	6.2432	.98741	.9
.2	.15988	.16196	6.1742	.98714	.8
.3	.16160	.16376	6.1066	.98686	.7
.4	.16333	.16555	6.0405	.98657	.6
.5	.16505	.16734	5.9758	.98629	.5
.6	.16677	.16914	5.9124	.98600	.4
.7	.16849	.17093	5.8502	.98570	.3
.8	.17021	.17273	5.7894	.98541	.2
.9	.17193	.17453	5.7297	.98511	.1
10.0	.17365	.17633	5.6713	.98481	**80.0**
.1	.17537	.17813	5.6140	.98450	.9
.2	.17708	.17993	5.5578	.98420	.8
.3	.17880	.18173	5.5026	.98389	.7
.4	.18052	.18353	5.4486	.98357	.6
.5	.18224	.18534	5.3955	.98325	.5
.6	.18395	.18714	5.3435	.98294	.4
.7	.18567	.18895	5.2924	.98261	.3
.8	.18738	.19076	5.2422	.98229	.2
.9	.18910	.19257	5.1929	.98196	.1
11.0	.19081	.19438	5.1446	.98163	**79.0**
.1	.19252	.19619	5.0970	.98129	.9
	COS	COT	TAN	SIN	ANGLE°

ANGLE°	SIN	TAN	COT	COS	
.2	.19423	.19801	5.0504	.98096	.8
.3	.19595	.19982	5.0045	.98061	.7
.4	.19766	.20164	4.9594	.98027	.6
.5	.19937	.20345	4.9152	.97992	.5
.6	.20108	.20527	4.8716	.97958	.4
.7	.20279	.20709	4.8288	.97922	.3
.8	.20450	.20891	4.7867	.97887	.2
.9	.20620	.21073	4.7453	.97851	.1
12.0	.20791	.21256	4.7046	.97815	**78.0**
.1	.20962	.21438	4.6646	.97778	.9
.2	.21132	.21621	4.6252	.97742	.8
.3	.21303	.21804	4.5864	.97705	.7
.4	.21474	.21986	4.5483	.97667	.6
.5	.21644	.22169	4.5107	.97630	.5
.6	.21814	.22353	4.4747	.97592	.4
.7	.21985	.22536	4.4373	.97553	.3
.8	.22155	.22719	4.4015	.97515	.2
.9	.22325	.22903	4.3662	.97476	.1
13.0	.22495	.23087	4.3315	.97437	**77.0**
.1	.22665	.23271	4.2972	.97398	.9
.2	.22835	.23455	4.2635	.97358	.8
.3	.23005	.23639	4.2303	.97318	.7
.4	.23175	.23823	4.1976	.97278	.6
.5	.23345	.24408	4.1653	.97237	.5
.6	.23514	.24193	4.1335	.97196	.4
.7	.23684	.24377	4.1022	.97155	.3
.8	.23853	.24562	4.0713	.97113	.2
.9	.24023	.24747	4.0408	.97072	.1
14.0	.24192	.24933	4.0108	.97030	**76.0**
.1	.24362	.25118	3.9812	.96987	.9
.2	.24531	.25304	3.9520	.96945	.8
.3	.24700	.25490	3.9232	.96902	.7
.4	.24869	.25676	3.8947	.96858	.6
.5	.25038	.25862	3.8667	.96815	.5
.6	.25207	.26048	3.8391	.96771	.4
.7	.25376	.26235	3.8118	.96727	.3
.8	.25545	.26421	3.7848	.96682	.2
.9	.25713	.26608	3.7583	.96638	.1
15.0	.25882	.26792	3.7321	.96593	**75.0**
.1	.26050	.26982	3.7062	.96547	.9
.2	.26219	.27169	3.6806	.96502	.8
.3	.26387	.27357	3.6554	.96456	.7
.4	.26556	.27545	3.6305	.96410	.6
.5	.26724	.27732	3.6059	.96363	.5
	COS	COT	TAN	SIN	ANGLE°

ANGLE°	SIN	TAN	COT	COS	
.6	.26892	.27921	3.5816	.96316	.4
.7	.27060	.28109	3.5576	.96269	.3
.8	.27228	.28297	3.5339	.96222	.2
.9	.27396	.28486	3.5105	.96174	.1
16.0	.27564	.28675	3.4874	.96126	**74.0**
.1	.27731	.28864	3.4646	.96078	.9
.2	.27899	.29053	3.4420	.96029	.8
.3	.28067	.29242	3.4197	.95981	.7
.4	.28234	.29432	3.3977	.95931	.6
.5	.28402	.29621	3.3759	.95882	.5
.6	.28569	.29811	3.3544	.95832	.4
.7	.28736	.30001	3.3332	.95782	.3
.8	.28903	.30192	3.3122	.95732	.2
.9	.29070	.30382	3.2914	.95681	.1
17.0	.29237	.30573	3.2709	.95630	**73.0**
.1	.29404	.30764	3.2506	.95579	.9
.2	.29571	.30955	3.2305	.95528	.8
.3	.29737	.31147	3.2106	.95476	.7
.4	.29904	.31338	3.1910	.95424	.6
.5	.30071	.31530	3.1716	.95372	.5
.6	.30237	.31722	3.1524	.95319	.4
.7	.30403	.31914	3.1334	.95266	.3
.8	.30570	.32106	3.1146	.95213	.2
.9	.30736	.32299	3.0961	.95159	.1
18.0	.30902	.32492	3.0777	.95106	**72.0**
.1	.31068	.32685	3.0595	.95052	.9
.2	.31233	.32878	3.0415	.94997	.8
.3	.31399	.33072	3.0237	.94943	.7
.4	.31565	.33266	3.0061	.94888	.6
.5	.31730	.33460	2.9887	.94832	.5
.6	.31896	.33654	2.9714	.94777	.4
.7	.32061	.33848	2.9544	.94721	.3
.8	.32227	.34043	2.9375	.94665	.2
.9	.32392	.34238	2.9208	.94609	.1
19.0	.32557	.34433	2.9042	.94552	**71.0**
.1	.32722	.34628	2.8878	.94995	.9
.2	.32887	.34824	2.8716	.94438	.8
.3	.33051	.35020	2.8556	.94380	.7
.4	.33216	.35216	2.8397	.94322	.6
.5	.33381	.35412	2.8239	.94264	.5
.6	.33545	.35608	2.8083	.94206	.4
.7	.33710	.35805	2.7929	.94147	.3
.8	.33874	.36002	2.7776	.94088	.2
.9	.34038	.36199	2.7625	.94029	.1
COS	COT	TAN	SIN		ANGLE°

ANGLE°	SIN	TAN	COT	COS	
20.0	.34202	.36397	2.7475	.93969	**70.0**
.1	.34366	.36595	2.7326	.93909	.9
.2	.34530	.36793	2.7179	.93849	.8
.3	.34694	.36991	2.7034	.93789	.7
.4	.34857	.37190	2.6889	.93728	.6
.5	.35021	.37388	2.6746	.93667	.5
.6	.35184	.37588	2.6605	.93606	.4
.7	.35347	.37787	2.6464	.93544	.3
.8	.35511	.37986	2.6325	.93483	.2
.9	.35674	.38186	2.6187	.93420	.1
21.0	.35837	.38386	2.6051	.93358	**69.0**
.1	.36000	.38587	2.5916	.93295	.9
.2	.36162	.38787	2.5782	.93232	.8
.3	.36325	.38988	2.5649	.93169	.7
.4	.36488	.39190	2.5517	.93106	.6
.5	.36650	.39391	2.5386	.93042	.5
.6	.36812	.39593	2.5257	.92978	.4
.7	.36975	.39795	2.5129	.92913	.3
.8	.37137	.39997	2.5002	.92849	.2
.9	.37299	.40200	2.4876	.92784	.1
22.0	.37461	.40403	2.4751	.92718	**68.0**
.1	.37622	.40606	2.4627	.92653	.9
.2	.37784	.40809	2.4504	.92587	.8
.3	.37946	.41013	2.4383	.92521	.7
.4	.38107	.41217	2.4262	.92455	.6
.5	.38268	.41421	2.4142	.92388	.5
.6	.38430	.41626	2.4023	.92321	.4
.7	.38591	.41831	2.3906	.92254	.3
.8	.38752	.42036	2.3789	.92186	.2
.9	.38912	.42242	2.3673	.92119	.1
23.0	.39073	.42447	2.3559	.92050	**67.0**
.1	.39234	.42654	2.3445	.91982	.9
.2	.39394	.42860	2.3332	.91914	.8
.3	.39555	.43067	2.3220	.91845	.7
.4	.39715	.43274	2.3109	.91775	.6
.5	.39875	.43481	2.2998	.91706	.5
.6	.40035	.43689	2.2889	.91636	.4
.7	.40195	.43897	2.2781	.91566	.3
.8	.40355	.44105	2.2673	.91496	.2
.9	.40514	.44314	2.2566	.91425	.1
24.0	.40674	.44523	2.2460	.91355	**66.0**
.1	.40833	.44732	2.2355	.91283	.9
.2	.40992	.44942	2.2251	.91212	.8
.3	.41151	.45152	2.2148	.91140	.7
	COS	COT	TAN	SIN	ANGLE°

ANGLE°	SIN	TAN	COT	COS	
.4	.41310	.45362	2.2045	.91068	.6
.5	.41469	.45573	2.1943	.90996	.5
.6	.41628	.45784	2.1842	.90924	.4
.7	.41787	.45995	2.1742	.90851	.3
.8	.41945	.46206	2.1624	.90778	.2
.9	.42104	.46418	2.1543	.90704	.1
25.0	.42262	.46631	2.1445	.90631	**65.0**
.1	.42420	.46843	2.1348	.90557	.9
.2	.42578	.47056	2.1251	.90483	.8
.3	.42736	.47270	2.1155	.90408	.7
.4	.42894	.47483	2.1060	.90334	.6
.5	.43051	.47698	2.0965	.90259	.5
.6	.43209	.47912	2.0872	.90183	.4
.7	.43366	.48127	2.0778	.90108	.3
.8	.43523	.48342	2.0682	.90032	.2
.9	.43680	.48557	2.0594	.89956	.1
26.0	.43837	.48773	2.0503	.89879	**64.0**
.1	.43994	.48989	2.0413	.89803	.9
.2	.44151	.49206	2.0323	.89726	.8
.3	.44307	.49423	2.0233	.89649	.7
.4	.44464	.49640	2.0145	.89571	.6
.5	.44620	.49858	2.0057	.89493	.5
.6	.44776	.50076	1.9970	.89415	.4
.7	.44932	.50295	1.9883	.89337	.3
.8	.45088	.50514	1.9797	.89259	.2
.9	.45243	.50733	1.9711	.89180	.1
27.0	.45399	.50953	1.9626	.89101	**63.0**
.1	.45554	.51173	1.9542	.89021	.9
.2	.45710	.51393	1.9458	.88942	.8
.3	.45865	.51614	1.9375	.88862	.7
.4	.46020	.51835	1.9292	.88782	.6
.5	.46175	.52057	1.9210	.88701	.5
.6	.46330	.52279	1.9128	.88620	.4
.7	.46484	.52501	1.9047	.88539	.3
.8	.46639	.52724	1.8967	.88458	.2
.9	.46793	.52947	1.8887	.88377	.1
28.0	.46947	.53171	1.8807	.88295	**62.0**
.1	.47101	.53395	1.8728	.88213	.9
.2	.47255	.53620	1.8650	.88130	.8
.3	.47409	.53844	1.8572	.88048	.7
.4	.47562	.54070	1.8495	.87965	.6
.5	.47716	.54296	1.8418	.87882	.5
.6	.47869	.54522	1.8341	.87798	.4
.7	.48022	.54748	1.8265	.87715	.3
	COS	COT	TAN	SIN	ANGLE°

ANGLE°	SIN	TAN	COT	COS	
.8	.48175	.54975	1.8190	.87631	.2
.9	.48328	.55203	1.8115	.87546	.1
29.0	.48481	.55431	1.8040	.87462	**61.0**
.1	.48634	.55659	1.7966	.87377	.9
.2	.48786	.55888	1.7893	.87292	.8
.3	.48938	.56117	1.7820	.87207	.7
.4	.49090	.56347	1.7747	.87121	.6
.5	.49242	.56577	1.7675	.87036	.5
.6	.49394	.56808	1.7603	.86949	.4
.7	.49546	.57039	1.7532	.86863	.3
.8	.49697	.57271	1.7461	.86777	.2
.9	.49849	.57503	1.7391	.86690	.1
30.0	.50000	.57735	1.7321	.86603	**60.0**
.1	.50151	.57968	1.7251	.86515	.9
.2	.50302	.58201	1.7182	.86427	.8
.3	.50453	.58435	1.7113	.86340	.7
.4	.50603	.58670	1.7045	.86251	.6
.5	.50754	.58905	1.6977	.86163	.5
.6	.50904	.59140	1.6909	.86074	.4
.7	.51054	.59376	1.6842	.85985	.3
.8	.51204	.59612	1.6775	.85895	.2
.9	.51354	.59849	1.6709	.85806	.1
31.0	.51504	.60086	1.6643	.85717	**59.0**
.1	.51653	.60324	1.6577	.85627	.9
.2	.51803	.60562	1.6512	.85536	.8
.3	.51952	.60801	1.6447	.85446	.7
.4	.52101	.61040	1.6383	.85355	.6
.5	.52250	.61280	1.6319	.85264	.5
.6	.52399	.61520	1.6255	.85173	.4
.7	.52547	.61761	1.6191	.85081	.3
.8	.52697	.62003	1.6128	.84989	.2
.9	.52844	.62245	1.6066	.84897	.1
32.0	.52992	.62487	1.6003	.84805	**58.0**
.1	.53140	.62730	1.5941	.84712	.9
.2	.53288	.62973	1.5880	.84619	.8
.3	.53435	.63217	1.5818	.84526	.7
.4	.53583	.63462	1.5757	.84433	.6
.5	.53730	.63707	1.5697	.84339	.5
.6	.53877	.63953	1.5637	.84245	.4
.7	.54024	.64199	1.5577	.84151	.3
.8	.54171	.64446	1.5517	.84057	.2
.9	.54317	.64693	1.5458	.83962	.1
33.0	.54464	.64941	1.5399	.83867	**57.0**
.1	.54610	.65189	1.5340	.83772	.9
	COS	COT	TAN	SIN	ANGLE°

ANGLE°	SIN	TAN	COT	COS	
.2	.54756	.65438	1.5282	.83676	.8
.3	.54902	.65688	1.5224	.83581	.7
.4	.55048	.65938	1.5166	.83485	.6
.5	.55194	.66189	1.5108	.83389	.5
.6	.55339	.66440	1.5051	.83292	.4
.7	.55484	.66692	1.4994	.83195	.3
.8	.55630	.66944	1.4938	.83098	.2
.9	.55775	.67197	1.4882	.83001	.1
34.0	.55919	.67451	1.4826	.82904	**56.0**
.1	.56064	.67705	1.4770	.82806	.9
.2	.56208	.67960	1.4715	.82708	.8
.3	.56353	.68215	1.4659	.82610	.7
.4	.56497	.68471	1.4605	.82511	.6
.5	.56641	.68728	1.4550	.82413	.5
.6	.56784	.68985	1.4496	.82314	.4
.7	.56928	.69243	1.4442	.82214	.3
.8	.57071	.69502	1.4388	.82115	.2
.9	.57215	.69761	1.4335	.82015	.1
35.0	.57358	.70021	1.4281	.81915	**55.0**
.1	.57501	.70281	1.4229	.81815	.9
.2	.57643	.70542	1.4176	.81714	.8
.3	.57786	.70804	1.4124	.81614	.7
.4	.57928	.71066	1.4071	.81513	.6
.5	.58070	.71329	1.4019	.81412	.5
.6	.58212	.71593	1.3968	.81310	.4
.7	.58354	.71857	1.3916	.81208	.3
.8	.58496	.72122	1.3865	.81106	.2
.9	.58637	.72388	1.3814	.81004	.1
36.0	.58779	.72654	1.3764	.80902	**54.0**
.1	.58920	.72921	1.3713	.80799	.9
.2	.59061	.73189	1.3663	.80696	.8
.3	.59201	.73457	1.3613	.80593	.7
.4	.59342	.73726	1.3564	.80489	.6
.5	.59482	.73996	1.3514	.80386	.5
.6	.59622	.74267	1.3465	.80282	.4
.7	.59763	.74538	1.3416	.80178	.3
.8	.59902	.74810	1.3367	.80073	.2
.9	.60042	.75082	1.3319	.79968	.1
37.0	.60182	.75355	1.3270	.79864	**53.0**
.1	.60321	.75629	1.3222	.79758	.9
.2	.60460	.75904	1.3175	.79653	.8
.3	.60599	.76180	1.3127	.79547	.7
.4	.60738	.76456	1.3079	.79441	.6
.5	.60876	.76733	1.3032	.79335	.5
COS	COT	TAN	SIN	ANGLE°

ANGLE°	SIN	TAN	COT	COS	
.6	.61015	.77010	1.2985	.79229	.4
.7	.61153	.77289	1.2938	.79122	.3
.8	.61291	.77568	1.2892	.79016	.2
.9	.61429	.77848	1.2846	.78908	.1
38.0	.61566	.78129	1.2799	.78801	**52.0**
.1	.61704	.78410	1.2753	.78694	.9
.2	.61841	.78692	1.2708	.78586	.8
.3	.61978	.78975	1.2662	.78478	.7
.4	.62115	.79259	1.2617	.78369	.6
.5	.62251	.79544	1.2572	.78261	.5
.6	.62388	.79829	1.2527	.78152	.4
.7	.62524	.80115	1.2482	.78043	.3
.8	.62660	.80402	1.2437	.77934	.2
.9	.62796	.80690	1.2393	.77824	.1
39.0	.62932	.80978	1.2349	.77715	**51.0**
.1	.63068	.81268	1.2305	.77605	.9
.2	.63203	.81558	1.2261	.77494	.8
.3	.63338	.81849	1.2218	.77384	.7
.4	.63473	.82141	1.2174	.77273	.6
.5	.63608	.82434	1.2131	.77162	.5
.6	.63742	.82727	1.2088	.77051	.4
.7	.63877	.83022	1.2045	.76940	.3
.8	.64011	.83317	1.2002	.76828	.2
.9	.64145	.83613	1.1960	.76717	.1
40.0	.64279	.83910	1.1918	.76604	**50.0**
.1	.64412	.84208	1.1875	.76492	.9
.2	.64546	.84507	1.1833	.76380	.8
.3	.64679	.84806	1.1792	.76267	.7
.4	.64812	.85107	1.1750	.76154	.6
.5	.64945	.85408	1.1708	.76041	.5
.6	.65077	.85710	1.1667	.75927	.4
.7	.65210	.86014	1.1626	.75813	.3
.8	.65342	.86318	1.1585	.75700	.2
.9	.65474	.86623	1.1544	.75585	.1
41.0	.65606	.86929	1.1504	.75471	**49.0**
.1	.65738	.87236	1.1463	.75356	.9
.2	.65869	.87543	1.1423	.75241	.8
.3	.66000	.87852	1.1383	.75126	.7
.4	.66131	.88162	1.1343	.75011	.6
.5	.66262	.88473	1.1303	.74896	.5
.6	.66393	.88784	1.1263	.74780	.4
.7	.66523	.89097	1.1224	.74664	.3
.8	.66653	.89410	1.1184	.74548	.2
.9	.66783	.89725	1.1145	.74431	.1
	COS	COT	TAN	SIN	ANGLE°

ANGLE°	SIN	TAN	COT	COS	
42.0	.66913	.90040	1.1106	.74314	**48.0**
.1	.67043	.90357	1.1067	.74198	.9
.2	.67172	.90674	1.1028	.74080	.8
.3	.67301	.90993	1.0990	.73963	.7
.4	.67430	.91313	1.0951	.73846	.6
.5	.67559	.91633	1.0913	.73728	.5
.6	.67688	.91955	1.0875	.73610	.4
.7	.67816	.92277	1.0837	.73491	.3
.8	.67944	.92601	1.0799	.73373	.2
.9	.68072	.92926	1.0761	.73254	.1
43.0	.68200	.93252	1.0724	.73135	**47.0**
.1	.68327	.93578	1.0686	.73016	.9
.2	.68455	.93906	1.0649	.72897	.8
.3	.68582	.94235	1.0612	.72777	.7
.4	.68709	.94565	1.0575	.72657	.6
.5	.68835	.94896	1.0538	.72537	.5
.6	.68962	.95229	1.0501	.72417	.4
.7	.69088	.95562	1.0464	.72294	.3
.8	.69214	.95897	1.0428	.72176	.2
.9	.69340	.96232	1.0392	.72055	.1
44.0	.69466	.96569	1.0355	.71934	**46.0**
.1	.69591	.96907	1.0319	.71813	.9
.2	.69717	.97246	1.0283	.71691	.8
.3	.69842	.97586	1.0247	.71569	.7
.4	.69966	.97927	1.0212	.71447	.6
.5	.70091	.98270	1.0176	.71325	.5
.6	.70215	.98613	1.0141	.71203	.4
.7	.70339	.98958	1.0105	.71080	.3
.8	.70463	.99304	1.0070	.70957	.2
.9	.70587	.99652	1.0035	.70834	.1
45.0	.70711	1.00000	1.0000	.70711	**45.0**
	COS	COT	TAN	SIN	ANGLE°

Answers to Selected Problems

CHAPTER I

PROBLEMS 1–1

1. (a) 0   (b) 2   (c) 8000   (d) 4
3. (a) 7   (b) 0   (c) 7000   (d) 4
5. (a) 3   (b) 9   (c) 800,000   (d) 4

PROBLEMS 1–2

1. (a) 0.3   (b) 0.016   (c) 0.00278   (d) 0.1763   (e) 0.435   (f) 0.2060

3. (a) $\dfrac{7}{1000}$ (b) $\dfrac{432}{10,000}$ (c) $\dfrac{174}{1000}$ (d) $\dfrac{65}{1,000,000}$ (e) $\dfrac{16}{100,000}$ (f) $\dfrac{1234}{100,000}$

5. (a) $\dfrac{17}{1000} = 0.017$ (b) $\dfrac{4}{100} = 0.04$ (c) $\dfrac{460}{10,000} = 0.0460$

   (d) $\dfrac{27}{1,000,000} = 0.000027$ (e) $\dfrac{1780}{100,000} = 0.01780$ (f) $\dfrac{65}{1000} = 0.065$

7. (a) Thousandths   (b) Millionths   (c) Ten-thousandths
9. (a) Ten-thousandths   (b) Millionths   (c) Hundredths
11. (a) Six-thousandths   (b) One hundred forty-seven thousandths
    (c) Ninety-two hundred-thousandths   (d) Seven-millionths
    (e) Four hundred thirteen ten-thousandths   (f) One hundred one ten-thousandths

PROBLEMS 1–3

1. (a) $7\dfrac{14}{100}$ (b) $50\dfrac{2}{100}$ (c) $710\dfrac{143}{1000}$ (d) $9\dfrac{99}{1000}$ (e) $73\dfrac{653}{1000}$ (f) $207\dfrac{7834}{10,000}$

   (g) $28\dfrac{736}{100,000}$ (h) $8\dfrac{706}{10,000}$

3. (a) 5.68   (b) 25.007   (c) 7.0165   (d) 70.4   (e) 473.025   (f) 80.00743
   (g) 2475.000035   (h) 307.00008

5. (a) $93.7 = 93\dfrac{7}{10}$ (b) $30.04 = 30\dfrac{4}{100}$ (c) $11.0001 = 11\dfrac{1}{10,000}$

   (d) $905.052 = 905\dfrac{52}{1000}$ (e) $78.034 = 78\dfrac{34}{1000}$

7. (a) $273.00025 = 273\dfrac{25}{100,000}$ (b) $704.000704 = 704\dfrac{704}{1,000,000}$

(c) $2044.0504 = 2044\dfrac{504}{10,000}$ (d) $10,101.00089 = 10,101\dfrac{89}{100,000}$

(e) $90.0466 = 90\dfrac{466}{10,000}$ (f) $207.0000100 = 207\dfrac{100}{10,000,000}$

PROBLEMS 1–4

1. 20   3. 50   5. 60   7. 130   9. 870

	Tens	Hundreds
11.	270	300
13.	360	400
15.	1380	1400
17.	1410	1400
19.	8710	8700

	Tens	Hundreds	Thousands
21.	4820	4800	5000
23.	85,470	85,500	85,000
25.	78,670	78,700	79,000
27.	27,850	27,800	28,000
29.	35,490	35,500	35,000
31.	68,450	68,400	68,000
33.	73,650	73,700	74,000
35.	470,000		

PROBLEMS 1–5

	Hundredths	Tenths	Units	Tens
1.	163.78	163.8	164	160
3.	9.46	9.5	9	10
5.	88.89	88.9	89	90
7.	749.49	749.5	749	750
9.	39.28	39.3	39	40
11.	63.75	63.7	64	60
13.	478.67	478.7	479	480
15.	47.47	47.5	47	50
17.	16.55	16.5	17	20
19.	76.75	76.7	77	80

PROBLEMS 1-6

1. (a) 5 (b) 4 (c) 4 (d) 4 (e) 3 (f) 4
3. (a) 8 (b) 2   5. (a) 6 (b) 3   7. (a)   (b) 7

1. 5   3. 14   5. −6   7. 34   9. −28   11. 7   13. −14   15. 18
17. 21   19. −33

1. −42   3. 54   5. −84   7. 225   9. 930   11. −7   13. −24   15. 12
17. −32   19. 28

1. −17   3. 0   5. 17   7. 21   9. 18   11. 26   13. −21   15. 12
17. 15   19. −19

1. 16   3. 7   5. −36   7. −83   9. 4   11. 51   13. 130   15. 45
17. −144   19. 20

CHAPTER 2

1. 10^1 3. 10^3 5. 10^2

1. 1000   3. 1   5. 1,000,000

1. $\dfrac{1}{100} = 0.01$ 3. $\dfrac{1}{1000} = 0.001$ 5. $\dfrac{1}{10} = 0.1$

1. 10^{-4} 3. 10^{-6} 5. 10^{-4} 7. 10^{-3}

1. 10^{10} 3. 10^6 5. 10^4 7. 10^{-4} 9. 10^3 11. 10^{-10} 13. 10^{-12} 15. 10^{-9}

1. 10^4 3. 10^{-5} 5. 10^8 7. 10^{12} 9. 10^7 11. 10^{-3} 13. 10^{-6} 15. 10^{-7}
17. 10^3 19. 10^{-4}

1. 10^{-7} 3. 10^{-7} 5. 10^{11} 7. 10^{10} 9. 10^{-12} 11. 10^8 13. 10^{13}
15. 10^{-4} 17. 10^{-36}

PROBLEMS 2–8

1. (a) 4.75×10^2 (b) 0.0475×10^4 (c) 0.000475×10^6
3. (a) 0.659×10^2 (b) 0.00659×10^4 (c) 0.0000659×10^6
5. (a) 934×10^2 (b) 9.34×10^4 (c) 0.0934×10^6
7. (a) 47.8×10^2 (b) 0.478×10^4 (c) 0.00478×10^6
9. (a) 418×10^2 (b) 4.18×10^4 (c) 0.0418×10^6
11. (a) 0.465×10^{-2} (b) 46.5×10^{-4} (c) 4650×10^{-6}
13. (a) 0.000555×10^{-2} (b) 0.0555×10^{-4} (c) 5.55×10^{-6}
15. (a) 0.0673×10^{-2} (b) 6.73×10^{-4} (c) 673×10^{-6}
17. (a) $0.00000108 \times 10^{-2}$ (b) 0.000108×10^{-4} (c) 0.0108×10^{-6}
19. (a) 0.0000233×10^{-2} (b) 0.00233×10^{-4} (c) 0.233×10^{-6}
21. (a) 0.325×10^3 (b) $325,000 \times 10^{-3}$
23. (a) 0.722×10^3 (b) $722,000 \times 10^{-3}$
25. (a) 2.70×10^3 (b) $2,700,000 \times 10^{-3}$
27. (a) 0.180×10^3 (b) $180,000 \times 10^{-3}$
29. (a) 0.00000450×10^3 (b) 4.50×10^{-3}

PROBLEMS 2–9

1. 2.76×10^4 3. 4.78×10 5. 1.77×10^6 7. 2.73×10^2 9. 1.73×10^5
11. 7.89×10 13. 1.67×10^6 15. 5.79×10^4 17. 8.10×10^4 19. 4.71×10^5

PROBLEMS 2–10

1. 4.78×10^{-3} 3. 7.47×10^{-1} 5. 1.64×10^{-2} 7. 4.54×10^{-3} 9. 5.01×10^{-5}
11. 2.55×10^{-1} 13. 4.00×10^{-2} 15. 8.96×10^{-4} 17. 1.71×10^{-2}
19. 2.78×10^{-6}

PROBLEMS 2–11

1. 2.13×10^4 3. 1.28×10^{-1} 5. 1.05×10^4 7. 6.13
9. 2.99 11. 1.31×10^{-3} 13. 1.23×10^{10} 15. 2.18×10^{-6}
17. 6.70 19. 1.82×10^{-2} 21. 9.41×10^{-5} 23. 2.59×10^6

PROBLEMS 2–12

1. 1.10×10^4 3. 2.00×10^{-1} 5. 3.86×10^3 7. 4.82×10^5 9. 2.88×10^{-1}
11. 3.87×10^{-2} 13. 9.93×10^4 15. 4.26×10^5

PROBLEMS 2-13

Estimated answers are given, then answers rounded to three places are given.

1. $3.5 \times 10^4, 3.97 \times 10^4$ 3. $3.6 \times 10^{-5}, 3.48 \times 10^{-5}$ 5. $6.0 \times 10, 7.85 \times 10$
7. $3.3 \times 10^7, 3.18 \times 10^7$ 9. $1.0 \times 10^5, 9.90 \times 10^4$ 11. $2.0 \times 10, 2.08 \times 10$
13. $5.0 \times 10^{-2}, 4.68 \times 10^{-2}$ 15. $3.0 \times 10^2, 2.71 \times 10^2$

1. 1.09×10 3. 4.55 5. 1.48×10 7. 2.23×10 9. 1.07×10
11. 1.39 13. 1.74 15. 1.09 17. 2.53 19. 1.03×10^{-2} 21. 6.75×10^{-3}

PROBLEMS 2–15

1. 1.06×10^2 3. 1.27×10^{-4} 5. 3.26×10^{-3} 7. 1.67×10^{-3}
9. 3.91×10^{-5} 11. 5.16×10^{-6} 13. 1.10×10^{-4} 15. 2.56×10^3
17. 8.46×10 19. 3.07×10^4 21. 2.44×10^3

PROBLEMS 2–16

1. 10^6 3. 10^{10} 5. 10^{-12} 7. 10^{12} 9. 10^{-8}

PROBLEMS 2–17

1. 10^3 3. 10^{-2} 5. 10^{-2} 7. 10^5 9. 10^3

PROBLEMS 2–18

1. 6.29×10^3 3. 3.17×10^6 5. 2.26×10^{-3} 7. 7.62×10^{-5} 9. 2.03×10^9
11. 9.00×10^{10} 13. 8.76×10^{-8} 15. 8.15×10^{-9} 17. 6.75 19. 2.17

PROBLEMS 2–19

1. 9.64 3. 1.54×10 5. 8.40×10^{-2} 7. 1.02×10^{-2} 9. 5.20×10^3
11. 6.71×10^3 13. 3.04×10^{-3} 15. 5.88 17. 1.07×10^{-2} 19. 7.96×10^3
21. 1.58×10^4

CHAPTER 3

PROBLEMS 3–1

1. $56 \times 10^3 = 0.056 \times 10^6$ 3. $220 \times 10^3 = 0.22 \times 10^6$ 5. $390 \times 10^3 = 0.39 \times 10^6$
7. $180 \times 10^3 = 0.18 \times 10^6$ 9. $43 \times 10^3 = 0.043 \times 10^6$ 11. $0.22 \times 10^{-3} = 220 \times 10^{-6}$
13. $2.13 \times 10^{-3} = 2130 \times 10^{-6}$ 15. $0.0556 \times 10^{-3} = 55.6 \times 10^{-6}$
17. $0.122 \times 10^{-3} = 122 \times 10^{-6}$ 19. $0.256 \times 10^{-3} = 256 \times 10^{-6}$
21. $667 \times 10^{-9} = 667,000 \times 10^{-12}$ 23. $17.9 \times 10^{-9} = 17,900 \times 10^{-12}$
25. $6.74 \times 10^{-9} = 6740 \times 10^{-12}$ 27. $0.0177 \times 10^{-9} = 17.7 \times 10^{-12}$
29. $70 \times 10^{-9} = 70,000 \times 10^{-12}$ 31. $7.3 = 7300 \times 10^{-3} = 0.0073 \times 10^3$
33. $5.67 = 5670 \times 10^{-3} = 0.00567 \times 10^3$ 35. $0.0178 = 17.8 \times 10^{-3} = 0.0000178 \times 10^3$
37. $78.3 = 78,300 \times 10^{-3} = 0.0783 \times 10^3$ 39. $845 = 845,000 \times 10^{-3} = 0.845 \times 10^3$

PROBLEMS 3–2

1. $0.26 \text{ mA} = 260 \ \mu\text{A}$ 3. $0.632 \text{ mS} = 632 \ \mu\text{S}$ 5. $7.63 \text{ k}\Omega = 0.00763 \text{ M}\Omega$
7. $17.3 \text{ mA} = 17,300 \ \mu\text{A}$ 9. $713 \text{ k}\Omega = 0.713 \text{ M}\Omega$ 11. $5630 \text{ kHz} = 5.63 \text{ MHz}$
13. $2000 \text{ k}\Omega = 2 \text{ M}\Omega$ 15. $0.237 \text{ mS} = 237 \ \mu\text{S}$ 17. $0.3 \ \mu\text{F} = 300 \text{ nF}$
19. $0.062 \text{ mA} = 62 \ \mu\text{A}$

1. 0.800 A   3. 0.0025 S   5. 33,000 Ω   7. 470 Ω   9. 12,500 Hz   11. 0.000100 F
13. 0.00025 A   15. 0.000900 S   17. 750,000 Ω   19. 0.0300 A

1. 0.65 mA = 650 μA   3. 0.00805 mS = 8.05 μS   5. 56.2 mS = 56,200 μS
7. 0.613 mS = 613 μS   9. 7.50 kΩ = 0.00750 MΩ   11. 510 kΩ = 0.510 MΩ
13. 127 kHz = 0.127 MHz   15. 713 kΩ = 0.713 MΩ   17. 46,300 μA = 0.0463 A
19. 0.00005 μF = 50 pF   21. 3200 Ω = 0.0032 MΩ   23. 0.270 MΩ = 270,000 Ω
25. 0.000403 S = 0.403 mS   27. 1,030,000 Hz = 1030 kHz   29. 1430 pF = 0.00143 μF
31. 5.5 mS = 5500 μS   33. 0.00106 μF = 1060 pF   35. 463,000 Ω = 0.463 MΩ
37. 96.3 kHz = 0.0963 MHz   39. 7.8 mA = 7800 μA   41. 0.176 V = 176,000 μV
43. 1730 mW = 0.00173 kW   45. 25 mH = 25,000 μH   47. 0.173 S = 173,000 μS
49. 2.5 μF = 2500 nF

1. 2.04 mA = 2040 μA   3. 0.0957 mA = 95.7 μA   5. 0.167 mS = 167 μS
7. 0.0582 mS = 58.2 μS   9. 1.42 mS = 1420 μS   11. 964 Ω = 0.964 kΩ
13. 68,700 Ω = 68.7 kΩ = 0.0687 MΩ   15. 6850 Ω = 6.85 kΩ = 0.00685 MΩ
17. 115 Ω = 0.115 kΩ   19. 4690 Ω = 4.69 kΩ   21. 2.30 mS = 2300 μS
23. 0.273 mA = 273 μA   25. 70,000 Ω = 70.0 kΩ   27. 0.0417 mA = 41.7 μA
29. 9.52 V = 9520 mV   31. 4.48 V = 4480 mV   33. 84,900 Ω = 84.9 kΩ
35. 318 Ω = 0.318 kΩ   37. 0.00491 μF = 4.91 nF   39. 919 Hz = 0.919 kHz
41. 5.00×10^3 ohms = 5.00 kΩ   43. 9.99×10^3 ohms = 9.99 kΩ
45. 2.50×10^3 ohms = 2.50 kΩ   47. 571 ohms = 0.571 kΩ

1. 39.37 in   3. 15.75 in   5. 2.3 m = 90.55 in = 2.515 yd   7. 10.94 yd
9. 10,940 yd = 6.214 mi   11. 25.4 cm   13. 90.42 cm   15. 804.7 m = 0.8047 km
17. 88.51 km/h   19. 160.9 km/h   21. 1640 ft/s   23. 328.1 ft/s   25. 2.116 oz
27. 5.291 oz   29. 28.22 oz = 1.764 lb   31. 176.4 oz = 11.02 lb   33. 0.6123 kg
35. 113.4 g   37. 7.796 g

CHAPTER 4

1. $2 \cdot 3 \cdot 3$ 3. $2 \cdot 2 \cdot 11$ 5. $3 \cdot 3 \cdot 7$ 7. $2 \cdot 2 \cdot 23$ 9. $3 \cdot 7 \cdot 11$ 11. $2 \cdot 2 \cdot 2 \cdot 7$
13. $3 \cdot 7 \cdot 7$ 15. $2 \cdot 2 \cdot 3 \cdot 7$ 17. $2 \cdot 3 \cdot 5 \cdot 7$ 19. $5 \cdot 7 \cdot 13$

1. $\dfrac{1}{7}$ 3. $\dfrac{5}{8}$ 5. $\dfrac{11}{16}$ 7. $\dfrac{2}{5}$ 9. $\dfrac{21}{40}$ 11. $\dfrac{3}{5}$ 13. $\dfrac{14}{15}$ 15. $\dfrac{13}{21}$ 17. $\dfrac{49}{55}$ 19. $\dfrac{1}{3}$

21. $\dfrac{1}{2}$

PROBLEMS 4–3

1. $\dfrac{1}{6}$　　3. $\dfrac{1}{12}$　　5. $\dfrac{3}{10}$　　7. $\dfrac{5}{14}$　　9. $\dfrac{21}{32}$　　11. $\dfrac{1}{9}$　　13. $\dfrac{1}{30}$　　15. $\dfrac{2}{35}$　　17. $\dfrac{1}{15}$

19. $\dfrac{11}{75}$　　21. $\dfrac{3}{28}$

PROBLEMS 4–4

1. $\dfrac{3}{4}$　　3. $\dfrac{3}{14}$　　5. $\dfrac{5}{6}$　　7. $\dfrac{6}{7}$　　9. $\dfrac{7}{10}$　　11. $\dfrac{1}{2}$　　13. $\dfrac{14}{15}$　　15. $\dfrac{7}{10}$　　17. $\dfrac{7}{36}$

19. $\dfrac{63}{64}$　　21. $\dfrac{1}{8}$

PROBLEMS 4–5

1. $\dfrac{1}{2}$　　3. $\dfrac{5}{7}$　　5. $\dfrac{1}{2}$　　7. $\dfrac{2}{7}$　　9. $\dfrac{11}{12}$　　11. $\dfrac{5}{16}$　　13. $\dfrac{13}{24}$　　15. $-\dfrac{2}{9}$　　17. $\dfrac{9}{16}$

19. $\dfrac{11}{18}$　　21. $-\dfrac{1}{8}$

PROBLEMS 4–6

1. 30　　3. 715　　5. 1260　　7. 286　　9. 72　　11. 1575　　13. 72　　15. 260
17. 350　　19. 225

PROBLEMS 4–7

1. $\dfrac{5}{12}$　　3. $\dfrac{2}{3}$　　5. $\dfrac{1}{2}$　　7. $\dfrac{11}{16}$　　9. $-\dfrac{1}{8}$　　11. $-\dfrac{19}{48}$　　13. $\dfrac{11}{12}$　　15. $\dfrac{18}{25}$　　17. $\dfrac{3}{4}$

19. $-\dfrac{11}{56}$　　21. $\dfrac{152}{275}$　　23. $\dfrac{1}{13}$　　25. $\dfrac{11}{15}$　　27. $\dfrac{15}{64}$

PROBLEMS 4–8

1. $8\dfrac{1}{2}$　　3. $9\dfrac{1}{4}$　　5. $3\dfrac{4}{5}$　　7. $6\dfrac{1}{7}$　　9. $6\dfrac{3}{4}$　　11. $9\dfrac{4}{7}$　　13. $5\dfrac{7}{16}$　　15. $7\dfrac{5}{12}$

17. $3\dfrac{1}{32}$　　19. $2\dfrac{7}{15}$　　21. $\dfrac{21}{2}$　　23. $\dfrac{67}{16}$　　25. $\dfrac{29}{8}$　　27. $\dfrac{28}{3}$　　29. $\dfrac{31}{6}$　　31. $\dfrac{71}{8}$

33. $\dfrac{37}{3}$　　35. $\dfrac{22}{3}$　　37. $\dfrac{53}{5}$　　39. $\dfrac{61}{8}$

1. $1\dfrac{1}{4}$ 3. $1\dfrac{1}{5}$ 5. $\dfrac{5}{8}$ 7. $4\dfrac{19}{20}$ 9. $9\dfrac{3}{4}$ 11. $7\dfrac{7}{12}$ 13. 14 15. $21\dfrac{3}{8}$

17. $10\dfrac{7}{18}$ 19. $25\dfrac{2}{3}$ 21. $2\dfrac{5}{8}$ 23. $1\dfrac{3}{4}$ 25. $2\dfrac{1}{7}$ 27. $9\dfrac{3}{5}$ 29. $1\dfrac{7}{20}$ 31. $\dfrac{5}{8}$

33. $\dfrac{21}{26}$ 35. $\dfrac{14}{39}$ 37. $1\dfrac{17}{25}$ 39. $3\dfrac{1}{3}$ 41. $25\dfrac{5}{6}$ 43. $5\dfrac{1}{9}$

1. $6\dfrac{8}{15}$ 3. $5\dfrac{23}{42}$ 5. $4\dfrac{2}{15}$ 7. $8\dfrac{3}{8}$ 9. $\dfrac{21}{32}$ 11. $3\dfrac{1}{12}$ 13. $1\dfrac{7}{12}$ 15. $1\dfrac{7}{8}$

17. $2\dfrac{1}{6}$ 19. $3\dfrac{5}{12}$ 21. $8\dfrac{5}{12}$ 23. $3\dfrac{15}{16}$ 25. $5\dfrac{11}{32}$ 27. $6\dfrac{8}{15}$ 29. $9\dfrac{23}{24}$ 31. $5\dfrac{1}{8}$

33. $-1\dfrac{7}{10}$ 35. $1\dfrac{5}{24}$ W

1. $\dfrac{3}{8} + \dfrac{3}{32} = 0.469 = \dfrac{15}{32}$ 3. $\dfrac{7}{16} + \dfrac{2}{7} = 0.723 = \dfrac{81}{112}$ 5. $\dfrac{5}{8} + \dfrac{37}{64} = 1.20 = 1\dfrac{13}{64}$

$\dfrac{3}{8} - \dfrac{3}{32} = 0.281 = \dfrac{9}{32}$ $\dfrac{7}{16} - \dfrac{2}{7} = 0.152 = \dfrac{17}{112}$ $\dfrac{5}{8} - \dfrac{37}{64} = 0.0469 = \dfrac{3}{64}$

$\dfrac{3}{8} \times \dfrac{3}{32} = 0.0352 = \dfrac{9}{256}$ $\dfrac{7}{16} \times \dfrac{2}{7} = 0.125 = \dfrac{1}{8}$ $\dfrac{5}{8} \times \dfrac{37}{64} = 0.361 = \dfrac{185}{512}$

$\dfrac{3}{8} \div \dfrac{3}{32} = 4.00 = 4$ $\dfrac{7}{16} \div \dfrac{2}{7} = 1.53 = 1\dfrac{17}{32}$ $\dfrac{5}{8} \div \dfrac{37}{64} = 1.08 = 1\dfrac{3}{37}$

7. $\dfrac{3}{5} + \dfrac{3}{10} = 0.900 = \dfrac{9}{10}$ 9. $3\dfrac{3}{8} + 2\dfrac{1}{3} = 5.71 = 5\dfrac{17}{24}$ 11. $4\dfrac{2}{5} + 2\dfrac{15}{64} = 6.63 = 6\dfrac{203}{320}$

$\dfrac{3}{5} - \dfrac{3}{10} = 0.300 = \dfrac{3}{10}$ $3\dfrac{3}{8} - 2\dfrac{1}{3} = 1.04 = 1\dfrac{1}{24}$ $4\dfrac{2}{5} - 2\dfrac{15}{64} = 2.17 = 2\dfrac{53}{320}$

$\dfrac{3}{5} \times \dfrac{3}{10} = 0.180 = \dfrac{9}{50}$ $3\dfrac{3}{8} \times 2\dfrac{1}{3} = 7.88 = 7\dfrac{7}{8}$ $4\dfrac{2}{5} \times 2\dfrac{15}{64} = 9.83 = 9\dfrac{133}{160}$

$\dfrac{3}{5} \div \dfrac{3}{10} = 2.00 = 2$ $3\dfrac{3}{8} \div 2\dfrac{1}{3} = 1.45 = 1\dfrac{25}{56}$ $4\dfrac{2}{5} \div 2\dfrac{15}{64} = 1.97 = 1\dfrac{63}{65}$

13. $7\dfrac{1}{2} + 3\dfrac{5}{9} = 11.1 = 11\dfrac{1}{18}$ 15. $4\dfrac{7}{8} + 8\dfrac{7}{16} = 13.3 = 13\dfrac{5}{16}$

$7\dfrac{1}{2} - 3\dfrac{5}{9} = 3.94 = 3\dfrac{17}{18}$ $4\dfrac{7}{8} - 8\dfrac{7}{16} = -3.56 = -3\dfrac{9}{16}$

$7\dfrac{1}{2} \times 3\dfrac{5}{9} = 26.7 = 26\dfrac{2}{3}$ $4\dfrac{7}{8} \times 8\dfrac{7}{16} = 41.1 = 41\dfrac{17}{128}$

$7\dfrac{1}{2} \div 3\dfrac{5}{9} = 2.11 = 2\dfrac{7}{64}$ $4\dfrac{7}{8} \div 8\dfrac{7}{16} = 0.578 = \dfrac{26}{45}$

17. $3\dfrac{2}{3} + 6\dfrac{2}{3} = 10.3 = 10\dfrac{1}{3}$ 19. $1\dfrac{7}{12} + 3\dfrac{2}{3} = 5.25 = 5\dfrac{1}{4}$

$3\dfrac{2}{3} - 6\dfrac{2}{3} = -3.00 = -3$ $1\dfrac{7}{12} - 3\dfrac{2}{3} = -2.08 = -2\dfrac{1}{12}$

$3\dfrac{2}{3} \times 6\dfrac{2}{3} = 24.4 = 24\dfrac{4}{9}$ $1\dfrac{7}{12} \times 3\dfrac{2}{3} = 5.81 = 5\dfrac{29}{36}$

$3\dfrac{2}{3} \div 6\dfrac{2}{3} = 0.550 = \dfrac{11}{20}$ $1\dfrac{7}{12} \div 3\dfrac{2}{3} = 0.432 = \dfrac{19}{44}$

PROBLEMS 4–12

Fraction	*Decimal*	*Percent*
1. 3/8	0.375	37.5%
3. 19/400	0.0475	4.75%
5. 7/16	0.4375	43.75%
7. 1/15	0.0667	6.67%
9. 1/12	0.0833	8.33%
11. 1/3	0.333	33.3%
13. 7/10	0.700	70.0%
15. 1/20	0.0500	5.00%
17. 1/5	0.200	20.0%
19. 1/250	0.00400	0.400%
21. 7/40	0.175	17.5%
23. 11/16	0.6875	68.75%
25. 3/5	0.600	60.0%
27. 3/40	0.0750	7.50%
29. 3/20	0.150	15.0%
31. 1/200	0.00500	0.500%
33. 2 1/2	2.50	250%
35. 9/16	0.5625	56.25%
37. 5/16	0.3125	31.25%

CHAPTER 5

PROBLEMS 5–1

1. Two, binomial   3. Three, trinomial   5. Two, binomial   7. Two, binomial
9. Three, trinomial   11. The coefficients are 3, −4, and 2. The literal numbers are a, b, and c.
13. The coefficients are $\frac{2}{3}$ and 7. The literal numbers are x and y.
15. The coefficients are $\frac{5}{2}$ and −4. The literal numbers are a and b.

PROBLEMS 5–2

1. 5.29   3. 54.8   5. 256   7. 512   9. 46.7   11. 166   13. 16.0   15. 64.0
17. 256

PROBLEMS 5–3

1. 30.0   3. 150   5. 900   7. 450   9. 117   11. 0   13. 9.00
15. -6.89×10^4 17. -6.78×10^4 19. 5.86×10^{14}

PROBLEMS 5–4

1. 5.48   3. 6.32   5. 14.1   7. 32.2   9. 44.7   11. 4.47   13. 4.00
15. 21.2   17. 7.55   19. 19.6   21. 23.0   23. 15.2   25. 67.5   27. 84.4
29. 823   31. 3.66×10^3

PROBLEMS 5–5

1. 4.49 V   3. 66.6 V   5. 135 V   7. 241 V   9. 38.3 mW   11. 4.40 W
13. 7.20 W   15. 447 mW   17. 3.42 mA   19. 1.67 mA   21. 678 mA
23. 3.41 mA   25. 15.0 V   27. 9.87 V   29. 11.1 V   31. 3.16 V   33. 37.6 mW
35. 5.36 mW   37. 33.1 mW   39. 4.79 W

PROBLEMS 5–6

1. 318 Ω   3. 3.98 kΩ   5. 12.6 Ω   7. 9.73 kΩ   9. 1.63 kΩ   11. 1.45 kHz
13. 58.1 kHz   15. 1.42 kHz   17. 3.56 kHz   19. 63.7 Hz   21. 144 kΩ
23. 224 Ω   25. 38.6 kΩ   27. 8.72 kΩ   29. 7.10 MΩ

CHAPTER 6

PROBLEMS 6–1

1. $2 \cdot 2 \cdot 2 \cdot a \cdot a \cdot b \cdot b$ 3. $2 \cdot 2 \cdot 2 \cdot 3 \cdot x \cdot x \cdot y$ 5. $2 \cdot 3 \cdot 7 \cdot c \cdot c \cdot c$
7. $2 \cdot 2 \cdot 2 \cdot 11 \cdot y \cdot y \cdot z$ 9. $3 \cdot 3 \cdot 3 \cdot 5 \cdot a \cdot a \cdot c \cdot c$

PROBLEMS 6–2

1. a^3b^3 3. $a^3b^3c^3$ 5. $x^2y^3z^4$ 7. a^3b^3 9. $a^4b^2c^3$ 11. $a^4b^3c^2$ 13. $78x^2y^3$
15. $32x^3y^2z^2$ 17. $150a^2b^2c$ 19. $270x^2y^4z^3$

PROBLEMS 6–3

1. $21a^4b^3$ 3. $20a^{-4}b^{-5}$ 5. $20x^6y^3z^4$ 7. $6a^3b^{-2}c^{-1}$ 9. $60x^5y^{-4}z^4$

PROBLEMS 6–4

1. $9a^2$ 3. $8ab^2$ 5. $5a^3bc^{-1}$ 7. $\dfrac{xy}{5}$ 9. $11x^{-5}y^{-1}z^{-4}$

PROBLEMS 6–5

1. $\dfrac{b^2}{10}$ 3. $\dfrac{15}{14b^2}$ 5. $\dfrac{10a^3}{21}$ 7. $\dfrac{3x^3y^{-2}}{16}$ 9. $\dfrac{2xy^{-1}}{3}$ 11. $\dfrac{16xy^{-2}z}{3}$
13. $\dfrac{a^4b^{-2}c^{-1}}{24}$ 15. $\dfrac{x^2y^{-4}}{18}$ 17. $\dfrac{a^4b^{-4}}{3}$ 19. $\dfrac{4x^{-1}y^{-1}z^2}{3}$

1. $\dfrac{9x^{-2}}{56}$ 3. $\dfrac{2}{3x}$ 5. $18b^{-3}$ 7. $\dfrac{2abc^{-1}}{45}$ 9. 1 11. $\dfrac{10a^{-1}b^{-1}c}{9}$ 13. $\dfrac{a^3c^{-1}}{10}$

15. $\dfrac{2x^2y}{9}$ 17. $\dfrac{9a^4b^{-4}c^{-6}}{8}$ 19. $\dfrac{4x^{-3}y^{-1}z^{-1}}{5}$

1. $\dfrac{5x^{-1}}{7}$ 3. $\dfrac{4x^{-1}}{7}$ 5. $\dfrac{3a^{-1}b^{-1}}{4}$ 7. $\dfrac{a^{-1}}{2}$ 9. $\dfrac{x}{2}$ 11. $\dfrac{2x}{5}$ 13. $\dfrac{3ab}{5}$

15. $\dfrac{3ab}{5}$ 17. $\dfrac{10x}{9}$ 19. $-\dfrac{2x}{7}$ 21. $-\dfrac{5x}{24}$ 23. $\dfrac{2ab+3b}{8}$ 25. $\dfrac{7abc-3ac}{4d}$

27. $-\dfrac{2abc+3ab}{4x}$ 29. $\dfrac{6xy-4xz+3yz}{5}$

1. $\dfrac{11x}{15}$ 3. $\dfrac{13a}{6}$ 5. $-\dfrac{x}{6}$ 7. $\dfrac{x}{2}$ 9. $\dfrac{4y+5xy}{8x}$ 11. $\dfrac{9x^2+8y^2}{12xy}$ 13. $\dfrac{4by+3x}{10xy}$

15. $\dfrac{10ab^2c-15ab}{32c^2}$ 17. $\dfrac{3b}{16}$ 19. $\dfrac{44a}{105}$ 21. $\dfrac{x}{30y^2}$ 23. $\dfrac{40x^2+12y^2z-5xz^2}{30xyz}$

25. $\dfrac{4a^2c+3ab^2-9bc^2}{33abc}$ 27. $\dfrac{75y^3+24x^3y-20x^4y}{90x^2y^2}$ 29. $\dfrac{9b^4+42ab^2c-25a^3}{60a^2b^3}$

CHAPTER 7

1. 10 3. 2 5. 3 7. 9 9. 4 11. -4 13. 4 15. -9 17. 12

19. -7 21. 12 23. -10 25. 16 27. 12 29. $\dfrac{1}{10}$ 31. 8 33. $\dfrac{1}{4}$

35. -6 37. $-\dfrac{2}{7}$ 39. $-1\dfrac{1}{3}$ 41. 6 43. $1\dfrac{1}{14}$ 45. $\dfrac{3}{28}$ 47. $\dfrac{2}{3}$ 49. $1\dfrac{7}{20}$

1. 5 3. 7.35 5. 7.48 7. 1.22 9. 2.45 11. 49 13. 36 15. 190

17. $\dfrac{1}{16}$ 19. $\dfrac{1}{144}$ 21. $\dfrac{1}{64}$ 23. 81 25. $a=\dfrac{6}{x}, x=\dfrac{6}{a}$ 27. $b=\dfrac{7}{4c}, c=\dfrac{7}{4b}$

29. $R=\dfrac{E^2}{P}, E=\sqrt{PR}$ 31. $f=\dfrac{B_C}{2\pi C}, C=\dfrac{B_C}{2\pi f}$ 33. $a=\dfrac{1}{6\sqrt{b}}, b=\dfrac{1}{36a^2}$

35. $x=\dfrac{36}{25y}, y=\dfrac{36}{25x}$ 37. $f=\dfrac{X_L}{2\pi L}, L=\dfrac{X_L}{2\pi f}$ 39. $L=\dfrac{1}{(2\pi f_r)^2 C}, C=\dfrac{1}{(2\pi f_r)^2 L}$

1. $R_1 = 6.8$ kΩ   3. $G_1 = 17.9$ μS   5. $V_o = 8.25$ V   7. $R = 113$ Ω   9. $R = 1.35$ Ω
11. $R = 34.1$ kΩ   13. $E = 13.5$ V   15. $E = 136$ V   17. $E = 285$ V   19. $I = 29.7$ mA
21. $I = 13.6$ mA   23. $I = 383$ mA   25. $R = 5.00$ kΩ   27. $R = 268$ Ω
29. $R = 7.32$ Ω   31. $N_P = 1770 = 1.77 \times 10^3$ 33. $N_P = 14.1$ 35. $N_P = 79.1$
37. $Z_S = 6.94$ Ω   39. $Z_S = 745$ Ω   41. $Z_S = 1.83$ Ω   43. $N_S = 268$
45. $N_S = 141$ 47. $Z_P = 3.20$ Ω   49. $Z_P = 1.00$ kΩ   51. $C = 84.2$ nF
53. $f = 3.18$ kHz   55. $L = 128$ mH   57. $f = 500$ Hz   59. $L = 281$ mH
61. $C = 25.3$ μF   63. 6.25 Ω   65. 40 Ω   67. 5 V   69. 8.33 Ω

CHAPTER 8

PROBLEMS 8–1

1. $3a + 9$ 3. $4a - 4$ 5. $28 - 7b$ 7. $-3a + 2$ 9. $-6a - 6$ 11. $6x - 12$
13. $6x + 6xy$ 15. $6a^2 + 9ab$ 17. $9x^2 + 18x - 36$ 19. $8a^2 - 6a + 4$ 21. $a^2 - 11ab$
23. $-a^2b + 8ab$ 25. $5x^2y - 9xy$ 27. $2x^2y - 20xy$ 29. $3x^3 + 6x^2 + 3x$
31. $a^3 - 27a^2 + 42a$ 33. $a^2 + 4a - 4ab$ 35. $3x^2 + 17x - 18xy$

PROBLEMS 8–2

1. $\dfrac{44x^2}{3}$ 3. $5y^2$ 5. $8a^2 + 9a$ 7. $5b^2 - 6b$ 9. $6a^2 - 16a + 8$ 11. $2a^2 - 6a + \dfrac{9}{2}$

13. $-2x^2 + 7x$ 15. $8x^2 - 15x$ 17. $5b^2 - 6b$ 19. $\dfrac{6x^2}{5} + \dfrac{21b}{4} = \dfrac{24x^2 + 105b}{20}$

PROBLEMS 8–3

1. $a^2 + 6a + 9$ 3. $a^2 - 12a + 36$ 5. $4x^2 - 16x + 16$ 7. $16x^2 + 24x + 9$
9. $4a^2 + 12ab + 9b^2$ 11. $x^6 - 8x^3y + 16y^2$ 13. $4x^2 - 12xy + 9y^2$ 15. $a^2 + 4a + 3$
17. $y^2 - 3y - 18$ 19. $a^2 + 4a - 12$ 21. $x^2 - 5x + 4$ 23. $a^2 - 16$ 25. $4x^2 - 9$
27. $x^2 + 4xy + 3y^2$ 29. $x^2 - xy - 20y^2$ 31. $4a^2 + 5ab - 6b^2$ 33. $2a^2 - 10ab + 8b^2$
35. $6a^2 + ab - 12b^2$ 37. $6x^2 + 23xy + 20y^2$ 39. $4x^2 - 16xy + 15y^2$

PROBLEMS 8–4

1. $x^2 + 3x + 2$ 3. $3x + 1\dfrac{2}{3}$ 5. $x - 1 + \dfrac{4x}{5}$ or $1\dfrac{4}{5}x - 1$ 7. $3x + 1\dfrac{1}{5}$

9. $x - 3$ 11. $x - 7$ 13. $3x - 2$ 15. $a - 5$ 17. $2x + 3y$ 19. $3a - 4b$

21. $x - 9y$ 23. $8 + 4x$ 25. $4x - 4 + \dfrac{3}{3x + 2}$ 27. $2x + 3y - \dfrac{2xy}{4x + 5y}$

29. $4a - 2 + \dfrac{4}{a + 1}$ 31. $6a + 3 + \dfrac{1}{a - 1}$ 33. $a + b + \dfrac{2ab}{a + b}$ or $a + 3b - \dfrac{2b^2}{a + b}$

35. $6x^2 - 3x + 7 - \dfrac{3}{x + 1}$

1. $3(a + 3)$ 3. $7(b - 3)$ 5. $2xy(2x + 3)$ 7. $2a(8a + 1)$ 9. $8abc^2(3ac - 2b)$
11. $3a^2b^2(4a + 3b^2 - 2a^2b^2)$ 13. $5xy^2(2x^2 - 4x^3y^2 + 3y)$ 15. $3b^2c(5a^3c^3 - a^2bc^3 - 4)$
17. $5x^2y^2z^2(7x^2z + 5y^2z^2 - 6)$ 19. $6x^2yz^2(4xyz^2 - 7y^2 - 12)$
21. $\dfrac{7x^2z^2}{3}\left(\dfrac{z}{4} - \dfrac{2xy^2}{3}\right)$ 23. $\dfrac{6x^2y^2z}{5}\left(\dfrac{2x^2}{7} + \dfrac{4yz^2}{11} - \dfrac{3xyz^3}{5}\right)$ 25. $\dfrac{7a^2b^2c^2}{11}\left(\dfrac{3a}{2} + \dfrac{4b}{3} - \dfrac{3ac}{2}\right)$

1. $(x + 2)(x + 3)$ 3. $(x + 8)(x - 4)$ 5. $(a + 3)^2$ 7. $(a + 4)(a - 4)$
9. $(x + 7)(x - 7)$ 11. $(x - 5)(x - 3)$ 13. $(a - 7)(a - 5)$ 15. $(a - 4)(a - 7)$
17. $(y + 9)(y - 4)$ 19. $(x - 7)(x + 5)$ 21. $(x + 4)(x - 6)$ 23. $(2x + 3)(3x - 2)$
25. $(4a - 2)(2a - 4)$ 27. $(3x + 4)(3x - 2)$ 29. $(3b - 6)(2b - 2)$ 31. $(5a + 3)(2a + 5)$
33. $(2x + 3y)(x + 2y)$ 35. $(4a - 3b)(5a + 2b)$ 37. $(2x + 6y)(4x - 2y)$
39. $(3x + 3y)(4x + 2y)$

CHAPTER 9

1. $x = 1\frac{1}{5}$ 3. $a = 5\frac{1}{4}$ 5. $I = 3$ 7. $x = -12$ 9. $R = 1\frac{4}{5}$ 11. $R = -4\frac{3}{5}$
13. $R = -4\frac{3}{7}$ 15. $c = -1$ 17. $I = 6$ 19. $R = 6$ 21. $x = -\dfrac{14}{23}$ 23. $a = -\dfrac{15}{17}$
25. $x = -4$ 27. $y = -7$ 29. $G = 3\frac{4}{7}$ 31. $R = -2\frac{7}{13}$ 33. $a = \dfrac{b}{2b - 1}, b = \dfrac{a}{2a - 1}$
35. $a = \dfrac{2b}{1 + 3b}, b = \dfrac{a}{2 - 3a}$ 37. $a = \dfrac{3bx}{12b - 2x}, b = \dfrac{2ax}{12a - 3x}, x = \dfrac{12ab}{2a + 3b}$
39. $a = \dfrac{9bx}{24b + 2x}, b = \dfrac{2ax}{9x - 24a}, x = \dfrac{24ab}{9b - 2a}$ 41. $a = \dfrac{2b + 2}{b - 5}, b = \dfrac{5a + 2}{a - 2}$
43. $a = \dfrac{4b - 1}{12 - 6b}, b = \dfrac{12a + 1}{6a + 4}$ 45. $b = 1\frac{1}{2}$

1. $V_{OC} = \dfrac{V_L(R_{\text{TH}} + R_L)}{R_L}, R_L = \dfrac{V_L R_{\text{TH}}}{V_{OC} - V_L}, R_{\text{TH}} = \dfrac{V_{OC}R_L - V_L R_L}{V_L}$

3. $I_{SC} = \dfrac{I_L(G_L + G_N)}{G_L}, G_L = \dfrac{I_L G_N}{I_{SC} - I_L}, G_N = \dfrac{I_{SC}G_L - I_L G_L}{I_L}$

5. $R_o = \dfrac{R_S R_i}{R_i - R_S}, R_i = \dfrac{R_S R_o}{R_o - R_S}$ 7. $r_b = r_i - \beta r_e, r_e = \dfrac{r_i - r_b}{\beta}$

9. $R_1 = \dfrac{R_T R_2}{R_2 - R_T}, R_2 = \dfrac{R_T R_1}{R_1 - R_T}$ 11. $A = B - 1$ 13. $R_1 = \dfrac{1}{2\pi fC} - R_2$

15. $V_B = V_{CC} - I_B(R_1 + R_2), R_1 = \dfrac{V_{CC} - V_B - I_B R_2}{I_B}$

1. (a) $R_2 = 4.8\ \text{k}\Omega$ (b) $R_1 = 773\ \Omega$ 3. (a) $I_T = 12\ \text{mA}$ (b) $G_2 = 1.79\ \text{mS}$
5. (a) $h_{oe} = 74.3\ \mu\text{S}$ (b) $R_L = 20.8\ \text{k}\Omega$ 7. $R_1 = 3\ \text{k}\Omega$ 9. $Z_2 = 220\ \text{k}\Omega$
11. $R_2 = 3.94\ \text{k}\Omega$ 13. $R_B = 550\ \text{k}\Omega$ 15. $R_1 = 9.34\ \text{k}\Omega$

1. $x = 4, x = -4$ 3. $x = -4, x = -2$ 5. $x = 3, x = -6$ 7. $x = 2, x = 4$
9. $x = 8, x = -7$ 11. $x = 8, x = -5$ 13. $x = 4, x = 9$ 15. $x = 6, x = -3$
17. $x = -7, x = -3$ 19. $x = 1, x = 2$ 21. $x = -1.55, x = -6.45$
23. $x = -4.73, x = -1.27$ 25. $x = -4.87, x = 2.87$ 27. $x = 15.9, x = -0.941$
29. $x = 7.53, x = -0.531$ 31. $x = 0.333, x = -1$ 33. $x = 3.27, x = -0.766$
35. $x = -2, x = -1$ 37. $x = -1.77, x = -0.566$ 39. $x = 5.81, x = -1.15$

CHAPTER 10

1. $I = 9\ \text{A}$ 3. $I_3 = 3\ \text{mA}$ 5. $I_2 = 45\ \mu\text{A}$ 7. $I_1 = 960\ \text{mA}$
9. $I_T = 25\ \text{mA}, I_3 = 15\ \text{mA}$ 11. $I_1 = 800\ \text{mA}, I_2 = 800\ \text{mA}$ 13. $I_2 = 200\ \mu\text{A}, I_3 = 200\ \mu\text{A}$
15. $I_T = 1.4\ \text{A}, I_3 = 800\ \text{mA}, I_4 = 600\ \text{mA}$ 17. $I_T = 3.2\ \text{mA}, I_1 = 3.2\ \text{mA}, I_2 = 500\ \mu\text{A}$
19. $I_1 = 1.7\ \text{mA}, I_3 = 3.8\ \text{mA}$ 21. $I_T = 14.2\ \text{mA}, I_2 = 4.8\ \text{mA}$
23. $I_1 = 30\ \text{mA}, I_3 = 10\ \text{mA}, I_4 = 10\ \text{mA}, I_6 = 70\ \text{mA}$

1. $V_2 = 18\ \text{V}$ 3. $E = 13\ \text{V}$ 5. $V_3 = 11\ \text{V}$ 7. $E = 18\ \text{V}$ 9. $V_1 = 14\ \text{V}$
11. $V_1 = 12\ \text{V}, V_2 = 8\ \text{V}$ 13. $V_2 = 13\ \text{V}, V_3 = 13\ \text{V}$ 15. $E = 17\ \text{V}, V_2 = 5\ \text{V}$
17. $V_1 = 3\ \text{V}, V_4 = 9\ \text{V}$ 19. $V_1 = 1.7\ \text{V}, V_3 = 800\ \text{mV}$ 21. $E = 60\ \text{V}, V_3 = 10\ \text{V}$
23. $V_2 = 25\ \text{V}, V_4 = 17\ \text{V}, V_6 = 50\ \text{V}$ 25. $E = 25\ \text{V}, V_2 = 12\ \text{V}, V_3 = 5.5\ \text{V}$

1. (a) 10 V (b) 5 V (c) 25 V (d) 30 V (e) −5 V
3. (a) 12 V (b) 43 V (c) −43 V (d) −35 V (e) −8 V
5. (a) 11 V (b) 5 V (c) 19 V (d) −13 V
7. (a) $V_1 = 34\ \text{V}$ (b) $V_2 = 16\ \text{V}$ (c) 6 V (d) −50 V

CHAPTER 11

1. (a) $R_T = 45\ \text{k}\Omega$ (b) $R_T = 1.95\ \text{k}\Omega$ (c) $R_T = 16.8\ \text{k}\Omega$ (d) $R_T = 300\ \text{k}\Omega$
3. (a) $R_2 = 1.8\ \text{k}\Omega$ (b) $R_2 = 39\ \text{k}\Omega$
5. (a) $R_T = 2.29\ \text{k}\Omega$ (b) $R_T = 130\ \text{k}\Omega$ (c) $R_T = 297\ \Omega$ (d) $R_T = 1.36\ \text{M}\Omega$
7. (a) $R_3 = 5.6\ \text{k}\Omega$ (b) $R_3 = 82\ \text{k}\Omega$
9. (a) $G_T = 150\ \mu\text{S}, R_T = 6.67\ \text{k}\Omega$ (b) $G_T = 1.2\ \text{mS}, R_T = 831\ \Omega$
   (c) $G_T = 76.8\ \mu\text{S}, R_T = 13.0\ \text{k}\Omega$ (d) $G_T = 20\ \mu\text{S}, R_T = 50.0\ \text{k}\Omega$

11. (a) $G_T = 685$ µS, $R_T = 1.46$ kΩ (b) $G_T = 31.9$ µS, $R_T = 31.3$ kΩ
(c) $G_T = 2.39$ mS, $R_T = 419$ Ω (d) $G_T = 26.7$ µS, $R_T = 37.4$ kΩ
13. (a) $R_2 = 4.70$ kΩ (b) $R_2 = 199$ Ω

PROBLEMS 11–2

1. (a) $R_T = 1.01$ kΩ (b) $R_T = 18.3$ kΩ (c) $R_T = 36.6$ kΩ (d) $R_T = 735$ kΩ
3. (a) $R_T = 5.08$ kΩ (b) $R_T = 657$ Ω (c) $R_T = 25$ kΩ (d) $R_T = 755$ Ω
5. (a) $R_T = 23.3$ kΩ (b) $R_T = 324$ Ω (c) $R_T = 685$ kΩ (d) $R_T = 7.98$ kΩ
7. 1.11 kΩ   9. 95.1 kΩ

PROBLEMS 11–3

1. (a) $R_T = 950$ Ω, $I = 9.47$ mA, $V_1 = 2.56$ V, $V_2 = 6.44$ V
(b) $R_T = 4.5$ kΩ, $I = 2.67$ mA, $V_1 = 3.2$ V, $V_2 = 8.80$ V
(c) $R_T = 340$ kΩ, $I = 118$ µA, $V_1 = 14.1$ V, $V_2 = 25.9$ V
(d) $R_T = 13.3$ kΩ, $I = 752$ µA, $V_1 = 7.52$ V, $V_2 = 2.48$ V
3. (a) $R_T = 1.43$ kΩ, $I = 7.00$ mA, $V_1 = 7.00$ V, $R_2 = 429$ Ω
(b) $R_T = 16.5$ kΩ, $I = 2.43$ mA, $V_1 = 16.5$ V, $R_2 = 9.68$ kΩ
(c) $R_T = 41.0$ kΩ, $I = 366$ µA, $V_1 = 3$ V, $R_2 = 32.8$ kΩ
(d) $R_T = 311$ kΩ, $I = 321$ µA, $V_1 = 70.7$ V, $R_2 = 91.2$ kΩ
5. (a) $R_1 = 60.9$ kΩ, $R_T = 93.9$ kΩ, $V_1 = 13.0$ V, $V_2 = 7.03$ V
(b) $R_1 = 15$ kΩ, $R_T = 25$ kΩ, $V_1 = 15$ V, $V_2 = 10$ V
(c) $R_1 = 560$ Ω, $R_T = 1.56$ kΩ, $V_1 = 4.31$ V, $V_2 = 7.69$ V
(d) $R_1 = 270$ kΩ, $R_T = 450$ kΩ, $V_1 = 5.40$ V, $V_2 = 3.60$ V
7. (a) $V_2 = 225$ mV, $R_1 = 82.7$ Ω, $R_T = 3.08$ kΩ, $E = 231$ mV
(b) $V_2 = 24.0$ V, $R_1 = 7.15$ kΩ, $R_T = 19.2$ kΩ, $E = 38.3$ V
(c) $V_2 = 5.50$ V, $R_1 = 27$ kΩ, $R_T = 60$ kΩ, $E = 10$ V
(d) $V_2 = 1.22$ V, $R_1 = 120$ Ω, $R_T = 202$ Ω, $E = 3$ V
9. (a) $R_T = 61$ kΩ, $I = 148$ µA, $V_1 = 4.87$ V, $V_2 = 1.48$ V, $V_3 = 2.66$ V
(b) $R_T = 388$ kΩ, $I = 103$ µA, $V_1 = 22.7$ V, $V_2 = 10.3$ V, $V_3 = 7.01$ V
(c) $R_T = 2.58$ kΩ, $I = 9.69$ mA, $V_1 = 7.95$ V, $V_2 = 11.6$ V, $V_3 = 5.43$ V
(d) $R_T = 8.60$ kΩ, $I = 1.74$ mA, $V_1 = 3.14$ V, $V_2 = 7.50$ V, $V_3 = 4.36$ V
11. (a) $V_2 = 10$ V, $V_3 = 10$ V, $R_1 = 5$ kΩ, $R_3 = 10$ kΩ, $R_T = 25$ kΩ
(b) $V_2 = 7.05$ V, $V_3 = 6.25$ V, $R_1 = 44.7$ kΩ, $R_3 = 41.7$ kΩ, $R_T = 133$ kΩ
(c) $V_2 = 13.6$ V, $V_3 = 36.5$ V, $R_1 = 18$ kΩ, $R_3 = 22$ kΩ, $R_T = 48.2$ kΩ
(d) $V_2 = 1.41$ V, $V_3 = 1.16$ V, $R_1 = 27.9$ Ω, $R_3 = 74.6$ Ω, $R_T = 194$ Ω
13. (a) $E = 36$ V, $R_2 = 2.43$ kΩ, $R_3 = 4.87$ kΩ, $V_1 = 14.1$ V, $V_3 = 14.6$ V
(b) $E = 11.7$ V, $R_2 = 13.7$ kΩ, $R_3 = 14.5$ kΩ, $V_1 = 4.05$ V, $V_3 = 3.91$ V
(c) $E = 30$ V, $R_2 = 100$ kΩ, $R_3 = 1.50$ kΩ, $V_1 = 13.3$ V, $V_3 = 10$ V
(d) $E = 2$ V, $R_2 = 300$ Ω, $R_3 = 99.6$ Ω, $V_1 = 1.08$ V, $V_3 = 229$ mV
15. $I = 0.500$ mA, $V_1 = 2.50$ V, $V_2 = 0.500$ V
17. $I = 3.00$ mA, $V_1 = V_2 = V_3 = 3.00$ V, $R_1 = R_2 = R_3 = 1.00$ kΩ

PROBLEMS 11–4

1. (a) $G_T = 5.49$ mS, $R_T = 182$ Ω, $V = 547$ mV, $I_1 = 2.02$ mA, $I_2 = 976$ µA
(b) $G_T = 40.3$ µS, $R_T = 24.8$ kΩ, $V = 17.3$ V, $I_1 = 255$ µA, $I_2 = 445$ µA

(c) $G_T = 8.25$ mS, $R_T = 121$ Ω, $V = 4.56$ V, $I_1 = 20.7$ mA, $I_2 = 16.9$ mA

(d) $G_T = 48.2$ μS, $R_T = 20.8$ kΩ, $V = 2.08$ V, $I_1 = 37.1$ μA, $I_2 = 62.9$ μA

3. (a) $G_T = 3.68$ mS, $R_T = 272$ Ω, $V = 6.80$ V, $I_1 = 10$ mA, $R_2 = 453$ Ω

(b) $G_T = 694$ μS, $R_T = 1.44$ kΩ, $V = 7.20$ V, $I_1 = 4.00$ mA, $R_2 = 7.20$ kΩ

(c) $G_T = 391$ μS, $R_T = 2.56$ kΩ, $V = 5.88$ V, $I_1 = 1.25$ mA, $R_2 = 5.6$ kΩ

(d) $G_T = 361$ μS, $R_T = 2.77$ kΩ, $V = 83.2$ V, $I_1 = 17.7$ mA, $R_2 = 6.76$ kΩ

5. (a) $G_T = 500$ μS, $R_T = 2$ kΩ, $I_1 = 4.26$ mA, $I_2 = 5.74$ mA, $R_2 = 3.48$ kΩ

(b) $G_T = 14$ μS, $R_T = 71.4$ kΩ, $I_1 = 139$ μA, $I_2 = 211$ μA, $R_2 = 118$ kΩ

(c) $G_T = 2.05$ mS, $R_T = 487$ Ω, $I_1 = 11.9$ mA, $I_2 = 8.12$ mA, $R_2 = 1.20$ kΩ

(d) $G_T = 41.2$ μS, $R_T = 24.3$ kΩ, $I_1 = 1.13$ mA, $I_2 = 870$ μA, $R_2 = 55.9$ kΩ

7. (a) $G_T = 94.2$ μS, $R_T = 10.6$ kΩ, $I_T = 1.13$ mA, $I_2 = 600$ μA, $R_1 = 22.6$ kΩ

(b) $G_T = 267$ μS, $R_T = 3.75$ kΩ, $I_T = 1.33$ mA, $I_2 = 333$ μA, $R_1 = 5.00$ kΩ

(c) $G_T = 155$ μS, $R_T = 6.44$ kΩ, $I_T = 4.8$ mA, $I_2 = 3.09$ mA, $R_1 = 18.1$ kΩ

(d) $G_T = 133$ μS, $R_T = 7.53$ kΩ, $I_T = 287$ μA, $I_2 = 23.7$ μA, $R_1 = 8.21$ kΩ

9. (a) $G_T = 2.20$ mS, $R_T = 454$ Ω, $V = 785$ mV, $I_1 = 291$ μA, $I_2 = 654$ μA, $I_3 = 785$ μA

(b) $G_T = 470$ μS, $R_T = 2.13$ kΩ, $V = 1.06$ V, $I_1 = 131$ μA, $I_2 = 142$ μA, $I_3 = 227$ μA

(c) $G_T = 89.0$ μS, $R_T = 11.2$ kΩ, $V = 5.62$ V, $I_1 = 255$ μA, $I_2 = 144$ μA, $I_3 = 100$ μA

(d) $G_T = 1.18$ mS, $R_T = 846$ Ω, $V = 1.69$ V, $I_1 = 627$ μA, $I_2 = 434$ μA, $I_3 = 942$ μA

11. (a) $G_T = 317$ μS, $R_T = 3.15$ kΩ, $R_1 = 18$ kΩ, $R_3 = 5.6$ kΩ, $I_2 = 1.05$ mA, $I_3 = 2.25$ mA

(b) $G_T = 15.0$ μS, $R_T = 66.7$ kΩ, $R_1 = 267$ kΩ, $R_3 = 133$ kΩ, $I_2 = 74.1$ μA, $I_3 = 151$ μA

(c) $G_T = 12.8$ μS, $R_T = 78.0$ kΩ, $R_1 = 678$ kΩ, $R_3 = 99.9$ kΩ, $I_2 = 31.2$ μA, $I_3 = 234$ μA

(d) $G_T = 4.36$ mS, $R_T = 230$ Ω, $R_1 = 914$ Ω, $R_3 = 561$ Ω, $I_2 = 20.3$ mA, $I_3 = 24.6$ mA

13. (a) $I_1 = 3.20$ mA, $I_3 = 8.8$ mA, $I_T = 15$ mA, $R_1 = 4.69$ kΩ, $R_2 = 5$ kΩ, $R_3 = 1.7$ kΩ, $R_T = 1$ kΩ

(b) $I_1 = 60.0$ mA, $I_3 = 120$ mA, $I_T = 240$ mA, $R_1 = 500$ Ω, $R_2 = 500$ Ω, $R_3 = 250$ Ω, $R_T = 125$ Ω

(c) $I_1 = 6.09$ mA, $I_3 = 3.04$ mA, $I_T = 20.0$ mA, $R_1 = 1$ kΩ, $R_2 = 559$ kΩ, $R_3 = 2.0$ kΩ, $R_T = 304$ Ω

(d) $I_1 = 70.4$ μA, $I_3 = 26.4$ μA, $I_T = 150$ μA, $R_1 = 75.2$ kΩ, $R_2 = 99.8$ kΩ, $R_3 = 201$ kΩ, $R_T = 35.3$ kΩ

15. $E = 24.0$ V, $I_T = 8.00$ mA, $R_2 = 4.00$ kΩ

17. $V_1 = V_2 = V_3 = 90.0$ V, $I_1 = 90.0$ mA, $I_2 = 54.0$ mA, $I_3 = 36.0$ mA, $R_2 = 1.67$ kΩ, $R_3 = 2.50$ kΩ, $R_T = 500$ Ω

PROBLEMS 11–5

1. (a) $R_T = 29.4$ kΩ, $I_T = 510$ μA, $I_1 = I_T = 510$ μA, $I_2 = 300$ μA, $I_3 = 211$ μA, $V_1 = 5.10$ V, $V_2 = V_3 = 9.90$ V

(b) $R_T = 487$ Ω, $I_T = 18.5$ mA, $I_1 = 18.5$ mA, $I_2 = 10.1$ mA, $I_3 = 8.34$ mA, $V_1 = 3.33$ V, $V_2 = V_3 = 5.67$ V

3. (a) $V_1 = 6$ V, $V_2 = V_3 = 4$ V, $I_2 = 85.1$ μA, $I_3 = 415$ μA, $R_3 = 9.64$ kΩ, $R_T = 20$ kΩ

(b) $R_1 = 4.05$ kΩ, $R_T = 20$ kΩ, $V_1 = 2.43$ V, $V_2 = 9.57$ V, $V_3 = 9.57$ V, $I_2 = 245$ μA, $I_3 = 355$ μA

5. (a) $R_T = 11.8$ kΩ, $I_T = 5.10$ mA, $I_1 = I_2 = 1.76$ mA, $I_3 = 3.33$ mA, $V_1 = 21.2$ V, $V_2 = 38.8$ V, $V_3 = 60$ V

(b) $R_T = 19.9$ kΩ, $I_T = 151$ μA, $I_1 = 87.0$ μA, $I_2 = 87.2$ μA, $I_3 = 63.8$ μA, $V_1 = 652$ mV, $V_2 = 2.35$ V, $V_3 = 3$ V

7. (a) $R_T = 1.22$ kΩ, $I_T = 32.8$ mA, $I_4 = 8.51$ mA, $I_1 = 24.3$ mA, $I_2 = 15.6$ mA, $I_3 = 8.69$ mA, $V_1 = 16.5$ V, $V_2 = V_3 = 23.5$ V, $V_4 = 40$ V
(b) $R_T = 263$ kΩ, $I_T = 114$ μA, $V_1 = 10.5$ V, $V_2 = 19.5$ V, $V_3 = 19.5$ V, $V_4 = 30$ V, $I_1 = 70.1$ μA, $I_2 = 41.5$ μA, $I_3 = 28.7$ μA, $I_4 = 44.1$ μA

9. (a) $R_T = 8.96$ kΩ, $I_T = 2.79$ mA, $I_1 = 2.79$ mA, $I_2 = 1.11$ mA, $I_3 = 756$ μA, $I_4 = 924$ μA, $V_1 = 8.37$ V, $V_2 = V_3 = V_4 = 16.6$ V
(b) $R_T = 100$ kΩ, $R_1 = 60$ kΩ, $R_2 = 75$ kΩ, $I_2 = 133$ μA, $I_3 = 66.7$ μA, $I_4 = 50$ μA, $I_1 = 250$ μA

11. (a) $R_T = 61.2$ kΩ, $I_T = 653$ μA, $I_1 = I_4 = I_T = 653$ μA, $I_2 = 441$ μA, $I_3 = 213$ μA, $V_1 = 6.53$ V, $V_2 = V_3 = 11.9$ V, $V_4 = 21.6$ V
(b) $R_T = 578$ kΩ, $I_T = I_1 = I_4 = 26.0$ μA, $I_2 = 15.3$ μA, $I_3 = 10.6$ μA, $V_1 = 3.89$ V, $V_2 = V_3 = 7.21$ V, $V_4 = 3.89$ V

13. $R_T = 399$ Ω, $I_T = 30.1$ mA, $I_2 = 19.1$ mA, $I_3 = 11.0$ mA, $V_1 = 3.01$ V, $V_2 = V_3 = 8.99$ V

15. $I_1 = I_2 = 7.50$ mA, $I_3 = I_4 = 15.0$ mA, $I_T = 22.5$ mA, $V_1 = V_2 = E/2 = 22.5$ V, $V_3 = 15.0$ V, $V_4 = 30.0$ V, $I_T = 22.5$ mA, $R_T = 2.00$ kΩ

CHAPTER 12

PROBLEMS 12–1

1. $V_1 = 41.4$ V, $V_2 = 16.4$ V, $V_3 = 33.6$ V, $I_1 = 18.8$ mA, $I_2 = 9.13$ mA, $I_3 = 28.0$ mA
3. $V_1 = 36.8$ V, $V_2 = 45.0$ V, $V_3 = 18.3$ V, $V_4 = 21.7$ V, $I_1 = 13.6$ mA, $I_2 = 13.6$ mA, $I_3 = 18.3$ mA, $I_4 = 4.63$ mA
5. $V_1 = 0.992$ V, $V_2 = 9.99$ V, $V_3 = 8.01$ V, $I_1 = 0.301$ mA, $I_2 = 2.13$ mA, $I_3 = 2.43$ mA
7. $V_1 = 14.8$ V, $V_2 = 10.2$ V, $V_3 = 170$ mV, $I_1 = 21.8$ mA, $I_2 = 21.6$ mA, $I_3 = 170$ μA
9. $V_1 = 14.2$ V, $V_2 = 7.72$ V, $V_3 = 18.1$ V, $V_4 = 3.12$ V, $I_1 = 6.44$ mA, $I_2 = 6.44$ mA, $I_3 = 5.49$ mA, $I_4 = 945$ μA

PROBLEMS 12–2

1. $V_{OC} = 20.4$ V, $R_{TH} = 15$ kΩ, $I_L = 818$ μA, $V_L = 8.18$ V
3. $V_{OC} = 12$ V, $R_{TH} = 1$ kΩ, $I_L = 6.86$ mA, $V_L = 5.14$ V
5. $V_{OC} = 25.4$ V, $R_{TH} = 16.2$ kΩ, $I_L = 812$ μA, $V_L = 12.2$ V

PROBLEMS 12–3

1. $V_{OC} = 23.5$ V, $R_{TH} = 848$ Ω, $I_L = 7.18$ mA, $V_L = 18.0$ V
3. $V_{OC} = 66.4$ V, $R_{TH} = 2.64$ kΩ, $I_L = 14.3$ mA, $V_L = 28.6$ V
5. $V_{OC} = 1.51$ V, $R_{TH} = 3.78$ kΩ, $I_L = 315$ μA, $V_L = 0.315$ V
7. $V_{OC} = 2.53$ V, $R_{TH} = 5.94$ kΩ, $I_B = 43.8$ μA, $I_C = 3.50$ mA, $V_C = 13.0$ V

PROBLEMS 12–4

1. (a) $I_{SC} = 1.36$ mA, $G_N = 66.7$ μS, $V_L = 8.18$ V, $I_L = 818$ μA
(b) $I_{SC} = 2.78$ mA, $G_N = 85.9$ μS, $V_L = 24.9$ V, $I_L = 639$ μA
3. (a) $I_{SC} = 12.0$ mA, $G_N = 1$ mS, $V_L = 5.14$ V, $I_L = 6.86$ mA
(b) $I_{SC} = 495$ μA, $G_N = 49.5$ μS, $V_L = 3.31$ V, $I_L = 331$ μA
5. (a) $I_{SC} = 1.56$ mA, $G_N = 61.6$ μS, $V_L = 12.2$ V, $I_L = 812$ μA
(b) $I_{SC} = 1.52$ mA, $G_N = 299$ μS, $V_L = 1.90$ V, $I_L = 949$ μA

CHAPTER 13

1. x-intercept $= 4, 0$
   y-intercept $= 0, 4$
3. x-intercept $= 3, 0$
   y-intercept $= 0, -6$
5. x-intercept $= 3, 0$
   y-intercept $= 0, 1$
7. x-intercept $= 4, 0$
   y-intercept $= 0, 2.67$
9. x-intercept $= 6, 0$
   y-intercept $= 0, 1.5$
11. x-intercept $= -6, 0$
    y-intercept $= 0, -8$
13. x-intercept $= 7, 0$
    y-intercept $= 0, -3.5$
15. x-intercept $= -2, 0$
    y-intercept $= 0, 3$
17. x-intercept $= 0.714, 0$
    y-intercept $= 0, -2.5$
19. x-intercept $= 0.583, 0$
    y-intercept $= 0, 3$

1. x-intercept $= 3, 0$; y-intercept $= 0, 6$; slope $= -2$

3. x-intercept $= -1, 0$; y-intercept $= 0, \dfrac{5}{2}$; slope $= \dfrac{5}{2}$

5. x-intercept $= 5, 0$; y-intercept $= 0, 3$; slope $= -\dfrac{3}{5}$

7. x-intercept $= -7, 0$; y-intercept $= 0, 2$; slope $= \dfrac{2}{7}$

9. x-intercept $= 8, 0$; y-intercept $= 0, -6$; slope $= \dfrac{3}{4}$

11. x-intercept $= 2, 0$; y-intercept $= 0, -10$; slope $= 5$
13. x-intercept $= -3/2, 0$; y-intercept $= 0, -3$; slope $= -2$

1. $y = -\dfrac{1}{3}x + 1$; slope $= -\dfrac{1}{3}$; y-intercept $= 1$
3. $y = -\dfrac{3}{4}x + 3$; slope $= -\dfrac{3}{4}$; y-intercept $= 3$

5. $y = \dfrac{1}{4}x - 3$; slope $= \dfrac{1}{4}$; y-intercept $= -3$
7. $y = \dfrac{2}{5}x + 2$; slope $= \dfrac{2}{5}$; y-intercept $= 2$

9. $y = \dfrac{3}{7}x - 3$; slope $= \dfrac{3}{7}$; y-intercept $= -3$
11. $y = -\dfrac{1}{3}x + 2$; slope $= -\dfrac{1}{3}$; y-intercept $= 2$

13. $y = \dfrac{2}{3}x - \dfrac{8}{3}$; slope $= \dfrac{2}{3}$; y-intercept $= -\dfrac{8}{3}$
15. $y = \dfrac{1}{2}x - 3$; slope $= \dfrac{1}{2}$; y-intercept $= -3$

17. $y = \dfrac{3}{2}x - 4$; slope $= \dfrac{3}{2}$; y-intercept $= -4$
19. $y = -3x - 9$; slope $= -3$; y-intercept $= -9$

1. $R_{\text{dc}} \approx 1.67 \text{ k}\Omega$, $r_{\text{ac}} \approx 1.14 \text{ k}\Omega$
3. $\Delta I_f \approx 6.2 \text{ mA}$, slope $= 31 \text{ mS}$, $r_d = 32.3 \text{ }\Omega$
5. (a) $V_C \approx 50 \text{ V}$, $V_R \approx 50 \text{ V}$
   (b) $V_C \approx 94 \text{ V}$, $V_R \approx 6 \text{ V}$

7. (a) $V_C \approx 15.6$ V, $V_R \approx 4.4$ V   (b) $V_C \approx 19.6$ V, $V_R \approx 0.4$ V
9 (a) ≈ 1.7 TC   (b) ≈ 0.5 TC
11. Slope = 133 μS, $R = 7.50$ kΩ

CHAPTER 14

PROBLEMS 14–2

1. $x = 3, y = 2$ 3. $x = 2, y = -4$ 5. $x = 2, y = 1$ 7. $x = -2, y = -1$
9. $x = 4, y = 3$ 11. $x = 4.57, y = 1.71$ 13. $x = 1.03, y = 0.147$

PROBLEMS 14–3

1. $x = 1.60, y = 1.20$ 3. $x = 1.14, y = 0.786$ 5. $x = 1.31, y = -0.517$
7. $x = 5.00, y = 5.00$ 9. $x = -2.83, y = -1.17$

PROBLEMS 14–4

1. $x = 1.60, y = 1.20$ 3. $x = 1.14, y = 0.786$ 5. $x = 1.31, y = -0.517$
7. $x = 5.00, y = 5.00$ 9. $x = -2.83, y = -1.17$

PROBLEMS 14–5

1. $x = 1$ 3. $x = -21.3$ 5. $x = 3.80$
   $y = 2$ $y = 20$ $y = 33.2$
   $z = 3$ $z = 16.3$ $z = -25.0$

PROBLEMS 14–6

1. $I_1 = 1.76$ mA
   $I_2 = 1.17$ mA
   $I_3 = 587$ μA
   $V_1 = 8.27$ V
   $V_2 = 11.7$ V
   $V_3 = 11.7$ V

PROBLEMS 14–7

1. $I_1 = 12.7$ mA   3. $I_1 = 4.63$ mA   5. $I_1 = 94.8$ mA   7. $I_1 = 4$ mA
   $I_2 = 9.08$ mA $I_2 = 220$ μA $I_2 = 92.5$ mA $I_2 = 2.79$ mA
   $I_3 = 21.8$ mA $I_3 = 4.44$ mA $I_3 = 2.35$ mA $I_3 = 1.2$ mA
   $V_1 = 38.2$ V $V_1 = 55.6$ V $V_1 = 9.48$ V $V_1 = 10.8$ V
   $V_2 = 18.2$ V $V_2 = 4.44$ V $V_2 = 30.5$ V $V_2 = 9.2$ V
   $V_3 = 21.8$ V $V_3 = 44.4$ V $V_3 = 517$ mV $V_3 = 1.2$ V

CHAPTER 15

PROBLEMS 15–2

1. $-4 - j4$ 3. $-4 - j8$ 5. $3.1 - j4.5$ 7. $11 + j7$ 9. $1.9 - j4.2$ 11. $-1 - j7$
13. $10 - j6$ 15. $0 + j0$ 17. $6 + j0$ 19. $0 - j6$

PROBLEMS 15–3

1. $15 + j27$ 3. $25 + j25$ 5. $27 - j8$ 7. $30 + j10$ 9. $20 + j2$
11. $20 - j8$ 13. $10 - j6$ 15. $17 - j10$ 17. $-10 - j12$ 19. $-21 - j20$
21. $5 + j3$ 23. $-1 - j5$ 25. $1 + j20$ 27. $20 + j50$ 29. $12 - j10$
31. $4 - j24$ 33. $-2 + j0$ 35. $3 - j2$ 37. $2 + j2$ 39. $-9 + j0$

PROBLEMS 15–4

1. $19 - j4$ 3. $13 + j0$ 5. $7 - j24$ 7. $-12 - j18$ 9. $14 + j38$
11. $33 - j13$ 13. $1 - j1$ 15. $0.8 + j1.6$ 17. $0.310 + j0.276$ 19. $-1.04 + j0.280$
21. $1.02 - j0.780$ 23. $0.780 + j0.0244$

CHAPTER 16

PROBLEMS 16–1

1. (a) $\phi = 55°$ (b) Side R is greater   (c) Side R (d) Side R 3. (a) 4.83 kΩ
   (b) 487 Ω (c) 98.9 kΩ 5. (a) 2.99 mS   (b) 800 μS   (c) 425 μS

PROBLEMS 16–2

1. $Z = 26.1$; angle ϕ is greater.   3. $R = 245$ 5. $X = 68.1$
7. $Z = 976$ Ω; angle θ is greater.   9. $Z = 731$ kΩ; angle θ is greater.
11. $Z = 84.5$ kΩ; angle ϕ is greater.   13. $X = 6.80$ kΩ; angle θ is greater.
15. $X = 150$ kΩ; angle θ is greater.   17. $R = 27$ kΩ; angle θ is greater.
19. $R = 90.5$ Ω; angle ϕ is greater.   21. $Y = 832$ μS; angle θ is greater.
23. $Y = 4.37$ mS; angle ϕ is greater.   25. $B = 21.3$ μS; angle ϕ is greater.
27. $B = 4.54$ mS; angle ϕ is greater.   29. $G = 556$ μS; angle ϕ is greater.
31. $G = 13.3$ mS; angle θ is greater.

PROBLEMS 16–3

1. (a) $\sin \theta = \dfrac{X}{Z}$, $\cos \theta = \dfrac{R}{Z}$, $\tan \theta = \dfrac{X}{R}$ (b) $\sin \phi = \dfrac{R}{Z}$, $\cos \phi = \dfrac{X}{Z}$, $\tan \phi = \dfrac{R}{X}$

1. 0.45865   3. 0.89571   5. 0.20279   7. 0.91566   9. 2.3906   11. 0.84806

1. 23.4°   3. 60.5°   5. 56.2°   7. 38.7°   9. 55.8°

1. $\theta = 32°$ 3. $\theta = 55°$ 5. $a = 11.1$ 7. $R = 42.9\ k\Omega$ 9. $R = 1.58\ k\Omega$
   $\phi = 58°$ $b = 140$ $b = 13.3$ $Z = 46.4\ k\Omega$ $Z = 1.71\ k\Omega$
   $c = 47.2$ $c = 244$

11. $Z = 11.5\ k\Omega$ 13. $Z = 33.4\ k\Omega$ 15. $X = 11.5\ k\Omega$ 17. $X = 207\ k\Omega$
    $\theta = 60.8°$ $\theta = 36.1°$ $Z = 23.1\ k\Omega$ $Z = 340\ k\Omega$

19. $R = 15.3\ k\Omega$ 21. $R = 599\ \Omega$ 23. $X = 18.4\ k\Omega$ 25. $X = 1.81\ k\Omega$
    $X = 37.0\ k\Omega$ $X = 508\ \Omega$ $\theta = 56.9°$ $\theta = 45.1°$

27. $R = 33.7\ k\Omega$ 29. $R = 22.0\ k\Omega$ 31. $B = 28.0\ \mu S$ 33. $B = 290\ \mu S$
    $\theta = 27.4°$ $\theta = 34.8°$ $Y = 66.2\ \mu S$ $Y = 319\ \mu S$

35. $\theta = 34.9°$ 37. $\theta = 53.7°$ 39. $G = 185\ mS$ 41. $G = 6.40\ mS$
    $Y = 15.2\ mS$ $Y = 21.5\ \mu S$ $Y = 189\ mS$ $Y = 7.20\ mS$

43. $\theta = 47.2°$ 45. $\theta = 45.1°$ 47. $G = 59.4\ \mu S$ 49. $G = 3.77\ mS$
    $B = 6.86\ mS$ $B = 2.34\ mS$ $B = 37.1\ \mu S$ $B = 5.38\ mS$

51. $\theta = 40.7°$ 53. $\theta = 32.1°$
    $G = 8.72\ mS$ $G = 271\ \mu S$

CHAPTER 17

1. angle $B = 60°$ 3. angle $A = 124.9°$ 5. angle $B = 43.51°$ or   angle $B = 136.5°$
   side $a = 28.53$ angle $B = 20.13°$ angle $C = 108.5°$ or   angle $C = 15.52°$
   side $b = 32.26$ side $a = 143.0$ side $c = 30.30$ or   side $c = 8.549$

7. angle $C = 80°$ 9. angle $A = 121.2°$ or   angle $A = 14.82°$
   side $b = 18.79$ angle $B = 36.82°$ or   angle $B = 143.2°$
   side $c = 19.70$ side $a = 228.4$ or   side $a = 68.28$

11. angle $B = 57.7°$ or   angle $B = 122.3°$ 13. angle $A = 25°$ 15. angle $B = 32.66°$
    angle $C = 97.3°$ or   angle $C = 32.7°$ side $a = 30.34$ angle $C = 52.34°$
    side $c = 23.47$ or   side $c = 12.78$ side $c = 69.35$ side $c = 95.37$

17. angle $A = 100.4°$ or   angle $A = 11.5°$ 19. angle $B = 10°$ 21. angle $B = 40°$
    angle $B = 45.55°$ or   angle $B = 134.5°$ side $a = 16.22$ side $a = 186.2$
    side $a = 8.266$ or   side $a = 1.68$ side $b = 3.438$ side $b = 239.4$

23. angle $A = 102.3°$ or   angle $A = 21.7°$ 25. angle $A = 29.31°$ 27. angle $A = 63°$
    angle $C = 49.72°$ or   angle $C = 130°$ angle $C = 50.70°$ side $a = 15.21$
    side $a = 83.25$ or   side $a = 31.5$ side $a = 174.0$ side $c = 6.393$

29. angle $A = 140°$ 31. angle $A = 61.26°$ 33. angle $A = 24.74°$
    side $a = 178.5$ angle $C = 48.74°$ angle $C = 35.26°$
    side $c = 95$ side $a = 233.3$ side $a = 72.47$

35. angle A = 83.02° or   angle A = 30.98°
     angle C = 63.98° or   angle C = 116.0°
     side a = 36.45   or   side a = 18.9

37. angle B = 38°
     side a = 80.76
     side b = 67.98

39. angle B = 60°
     side a = 5.87
     side b = 5.266

PROBLEMS 17–2

1. angle A = 30.80°
   angle B = 108.0°
   angle C = 41.17°

3. angle A = 108.2°
   angle B = 22.33°
   angle C = 49.47°

5. angle B = 35.1°
   angle C = 49.9°
   side a = 104

7. angle B = 19.21°
   angle C = 80.79°
   side a = 35.92

9. angle A = 45.33°
   angle C = 34.67°
   side b = 692.4

11. angle A = 121.9°
   angle B = 39.57°
   angle C = 18.57°

13. angle A = 60°
   angle B = 81.79°
   angle C = 38.21°

15. angle A = 57.02°
   angle C = 93.0°
   side b = 5.007

17. angle A = 28.10°
   angle C = 46.90°
   side b = 82.02

19. angle B = 52.88°
   angle C = 32.12°
   side a = 749.6

21. angle B = 23.41°
   angle C = 36.59°
   side a = 21.79

23. angle A = 48.19°
   angle B = 48.19°
   angle C = 83.62°

25. angle A = 59.75°
   angle B = 45.35°
   angle C = 74.90°

27. angle A = 54.03°
   angle B = 91.12°
   angle C = 34.84°

29. angle A = 36.69°
   angle B = 71.66°
   angle C = 71.66°

CHAPTER 18

PROBLEMS 18–1

1. 1.67 ms   3. 833 μs   5. 568 ns   7. 2.50 kHz   9. 752 Hz   11. 40.0 kHz
13. 0.436 rad   15. 1.57 rad   17. 3.14 rad   19. 1.31 rad   21. 10.5 rad
23. 8.37°   25. 28.6°   27. 90°   29. 100°   31. 573°

PROBLEMS 18–2

1. (a) 377 rad/s, 21,600°/s   (b) 942 rad/s, 54,000°/s   (c) 1260 rad/s, 7.20×10^4 degrees/s
  (d) 1880 rad/s, 1.08×10^5 degrees/s   (e) 3770 rad/s, 2.16×10^5 degrees/s
  (f) 5030 rad/s, 2.88×10^5 degrees/s   (g) 7850 rad/s, 4.50×10^5 degrees/s
3. (a) 1.91 Hz   (b) 7.96 Hz   (c) 63.7 Hz   (d) 143 Hz   (e) 398 Hz   (f) 796 Hz   (g) 1.11 kHz
5. (a) 27.8 Hz   (b) 41.7 Hz   (c) 139 Hz   (d) 333 Hz   (e) 556 Hz   (f) 2.04 kHz   (g) 4.58 kHz
7. (a) 0.314 rad, 18.0°   (b) 0.785 rad, 45.0°   (c) 3.14 rad, 180°   (d) 8.80 rad, 504°
9. (a) 1.60 rad, 91.8°   (b) 4.01 rad, 230°   (c) 16.0 rad, 918°   (d) 44.9 rad, 2570°
11. (a) 14.3 kHz   (b) 1.03 kHz   (c) 9.26 kHz   (d) 62.2 kHz
13. (a) 175 Hz   (b) 1.72 kHz   (c) 61.1 kHz   (d) 1.67 MHz
15. (a) 31.8 μs   (b) 3.98 μs   (c) 151 μs   (d) 11.1 μs   (e) 50 μs   (f) 167 μs
17. (a) 10.6 μs   (b) 1.33 μs   (c) 50.2 μs   (d) 3.70 μs   (e) 16.7 μs   (f) 55.6 μs

PROBLEMS 18–3

1. (a) 3.91 V   (b) 9.57 V   (c) 17.7 V   (d) −9.57 V   (e) −25 V   (f) −9.57 V
3. (a) 8.59 V   (b) 10.9 V   (c) −6.75 V   (d) −9.37 V   (e) −11.2 V   (f) −12.0 V
5. (a) 700 μA   (b) 0   (c) −700 μA   (d) 0   (e) 700 μA   (f) 0
7. (a) 86.7 μA   (b) 173 μA   (c) 260 μA   (d) 346 μA   (e) 432 μA   (f) 518 μA

1. (a) $65 \text{ V} = 91.9 \text{ V}_{pk} = 184 \text{ V}_{p\text{-}p}$ (b) $110 \text{ V} = 156 \text{ V}_{pk} = 311 \text{ V}_{p\text{-}p}$
3. (a) $220 \text{ V} = 311 \text{ V}_{pk} = 622 \text{ V}_{p\text{-}p}$ (b) $440 \text{ V} = 622 \text{ V}_{pk} = 1.24 \text{ kV}_{p\text{-}p}$
5. (a) $600 \text{ } \mu\text{A} = 849 \text{ } \mu\text{A}_{pk} = 1.70 \text{ mA}_{p\text{-}p}$ (b) $1.2 \text{ mA} = 1.70 \text{ mA}_{pk} = 3.39 \text{ mA}_{p\text{-}p}$
7. (a) $37.8 \text{ mA} = 53.5 \text{ mA}_{pk} = 107 \text{ mA}_{p\text{-}p}$ (b) $3 \text{ A} = 4.24 \text{ A}_{pk} = 8.49 \text{ A}_{p\text{-}p}$
9. (a) $350 \text{ } \mu\text{V}_{pk} = 700 \text{ } \mu\text{V}_{p\text{-}p} = 247 \text{ } \mu\text{V}$ (b) $400 \text{ mV}_{pk} = 800 \text{ mV}_{p\text{-}p} = 283 \text{ mV}$
11. (a) $80 \text{ V}_{pk} = 160 \text{ V}_{p\text{-}p} = 56.6 \text{ V}$ (b) $200 \text{ V}_{pk} = 400 \text{ V}_{p\text{-}p} = 141 \text{ V}$
13. (a) $200 \text{ } \mu\text{A}_{pk} = 400 \text{ } \mu\text{A}_{p\text{-}p} = 141 \text{ } \mu\text{A}$ (b) $9.45 \text{ mA}_{pk} = 18.9 \text{ mA}_{p\text{-}p} = 6.68 \text{ mA}$
15. (a) $500 \text{ mA}_{pk} = 1.00 \text{ A}_{p\text{-}p} = 354 \text{ mA}$ (b) $1.75 \text{ A}_{pk} = 3.50 \text{ A}_{p\text{-}p} = 1.24 \text{ A}$
17. (a) $320 \text{ } \mu\text{V}_{p\text{-}p} = 160 \text{ } \mu\text{V}_{pk} = 113 \text{ } \mu\text{V}$ (b) $70 \text{ mV}_{p\text{-}p} = 35 \text{ mV}_{pk} = 24.7 \text{ mV}$
19. (a) $17 \text{ V}_{p\text{-}p} = 8.50 \text{ V}_{pk} = 6.01 \text{ V}$ (b) $283 \text{ V}_{p\text{-}p} = 142 \text{ V}_{pk} = 100 \text{ V}$
21. (a) $50 \text{ } \mu\text{A}_{p\text{-}p} = 25 \text{ } \mu\text{A}_{pk} = 17.7 \text{ } \mu\text{A}$ (b) $700 \text{ } \mu\text{A}_{p\text{-}p} = 350 \text{ } \mu\text{A}_{pk} = 247 \text{ } \mu\text{A}$
23. (a) $9.38 \text{ mA}_{p\text{-}p} = 4.69 \text{ mA}_{pk} = 3.32 \text{ mA}$ (b) $650 \text{ mA}_{p\text{-}p} = 325 \text{ mA}_{pk} = 230 \text{ mA}$

1. (a) $i = 2.35 \text{ mA}, e = 8.75 \text{ V}$ (b) $i = 3.99 \text{ mA}, e = 9.31 \text{ V}$ (c) $i = 0, e = -4.23 \text{ V}$
   (d) $i = -3.8 \text{ mA}, e = -7.31 \text{ V}$
3. (a) $i = 23.6 \text{ mA}, e = 16.6 \text{ V}$ (b) $i = 20.5 \text{ mA}, e = 28.3 \text{ V}$ (c) $i = 12.4 \text{ mA}, e = 25.6 \text{ V}$
   (d) $i = -11.0 \text{ mA}, e = -24.8 \text{ V}$
5. (a) $i = 140 \text{ } \mu\text{A}, e = -53 \text{ V}$ (b) $i = 485 \text{ } \mu\text{A}, e = -1.40 \text{ V}$ (c) $i = 185 \text{ } \mu\text{A}, e = 76.5 \text{ V}$
   (d) $i = -600 \text{ } \mu\text{A}, e = -45.9 \text{ V}$
7. (a) $i = 21.9 \text{ mA}, e = 23.5 \text{ V}$ (b) $i = 24.7 \text{ mA}, e = 55.8 \text{ V}$ (c) $i = -12.2 \text{ mA}, e = 12.2 \text{ V}$
   (d) $i = -3.31 \text{ mA}, e = 34.8 \text{ V}$

CHAPTER 19

1. $E = 28.3 \text{ V}, \theta = -45°$ 3. $E = 14.4 \text{ V}, \theta = -33.7°$ 5. $V_C = -7.73 \text{ V}, \theta = -50.6°$
7. $V_C = -34 \text{ V}, \theta = -58.3°$ 9. $E = 19.6 \text{ V}, V_C = -12.6 \text{ V}$ 11. $E = 6.07 \text{ V}, V_C = -2.56 \text{ V}$
13. $V_R = 14.6 \text{ V}, V_C = -10.5 \text{ V}$ 15. $V_R = 62.4 \text{ V}, V_C = -31.8 \text{ V}$
17. $V_R = 22.4 \text{ V}, \theta = -41.8°$ 19. $V_R = 30 \text{ V}, \theta = -53.1°$ 21. $E = 10.4 \text{ V}, V_R = 8.80 \text{ V}$
23. $E = 31.9 \text{ V}, V_R = 10.9 \text{ V}$

1. $X_C = 6.37 \text{ k}\Omega, Z = 11.9 \text{ k}\Omega, \theta = -32.5°$ 3. $X_C = 796 \text{ }\Omega, Z = 1.06 \text{ k}\Omega, \theta = -48.7°$
5. $X_C = 3.18 \text{ k}\Omega, Z = 3.76 \text{ k}\Omega, \theta = -57.9°$ 7. $X_C = 20.5 \text{ k}\Omega, Z = 33.9 \text{ k}\Omega, \theta = -37.3°$
9. $X_C = 692 \text{ }\Omega, Z = 3.37 \text{ k}\Omega, \theta = -11.8°$
11. $X_C = 1.33 \text{ k}\Omega, Z = 2.4 \text{ k}\Omega, \theta = -33.6°, V_R = 25 \text{ V}, V_C = -16.6 \text{ V}, I = 12.5 \text{ mA}$
13. $X_C = 15.9 \text{ k}\Omega, Z = 18.8 \text{ k}\Omega, \theta = -57.9°, V_R = 5.32 \text{ V}, V_C = -8.47 \text{ V}, I = 532 \text{ } \mu\text{A}$
15. $X_C = 5.31 \text{ k}\Omega, Z = 6.25 \text{ k}\Omega, \theta = -58.1°, V_R = 15.8 \text{ V}, V_C = -25.5 \text{ V}, I = 4.8 \text{ mA}$
17. $X_C = 49 \text{ k}\Omega, Z = 50 \text{ k}\Omega, \theta = -78.5°, V_R = 4 \text{ V}, V_C = -19.6 \text{ V}, I = 400 \text{ } \mu\text{A}$
19. $X_C = 398 \text{ }\Omega, Z = 616 \text{ }\Omega, \theta = -40.3°, V_R = 76.3 \text{ V}, V_C = -64.6 \text{ V}, I = 162 \text{ mA}$
21. $\theta = -46.9°, X_C = 8.76 \text{ k}\Omega, f = 90.8 \text{ kHz}, V_R = 27.3 \text{ V}, V_C = -29.2 \text{ V}, I = 3.33 \text{ mA}$
23. $\theta = -51.3°, X_C = 937 \text{ }\Omega, f = 17 \text{ kHz}, V_R = 9.38 \text{ V}, V_C = -11.7 \text{ V}, I = 12.5 \text{ mA}$
25. $\theta = -39.5°, X_C = 2.23 \text{ k}\Omega, f = 2.38 \text{ MHz}, V_R = 23.1 \text{ V}, V_C = -19.1 \text{ V}, I = 8.57 \text{ mA}$

27. $\theta = -29.6°$, $X_C = 31.8$ kΩ, $f = 200$ Hz, $V_R = 6.52$ V, $V_C = -3.70$ V, $I = 116$ μA
29. $\theta = -44.7°$, $X_C = 802$ Ω, $f = 1988$ Hz, $V_R = 8.53$ V, $V_C = -8.44$ V, $I = 10.5$ mA

PROBLEMS 19–3

1. $E = 12.8$ V, $\theta = 51.3°$ 3. $E = 24.4$ V, $\theta = 55°$ 5. $E = 7.52$ V, $V_L = 3.41$ V
7. $E = 71.3$ V, $V_L = 65.8$ V   9. $V_L = 8.62$ V, $\theta = 45.9°$ 11. $V_L = 41.1$ V, $\theta = 43.3°$
13. $V_R = 26$ V, $\theta = 30°$ 15. $V_R = 13.7$ V, $\theta = 36°$ 17. $E = 3.4$ V, $V_R = 1.6$ V
19. $E = 31$ V, $V_R = 26.8$ V   21. $V_R = 4.77$ V, $V_L = 7.63$ V   23. $V_R = 35.4$ V, $V_L = 35.4$ V

PROBLEMS 19–4

1. $X_L = 3.77$ kΩ, $Z = 5.42$ kΩ, $\theta = 44°$ 3. $X_L = 377$ Ω, $Z = 626$ Ω, $\theta = 37.0°$
5. $X_L = 13.4$ kΩ, $Z = 30$ kΩ, $\theta = 26.3°$ 7. $X_L = 1.26$ kΩ, $Z = 2.20$ kΩ, $\theta = 34.9°$
9. $X_L = 785$ Ω, $Z = 1.27$ kΩ, $\theta = 38.1°$, $V_R = 11.8$ V, $V_L = 9.27$ V, $I = 11.8$ mA
11. $X_L = 471$ Ω, $Z = 559$ Ω, $\theta = 57.5°$, $V_R = 26.9$ V, $V_L = 42.2$ V, $I = 89.5$ mA
13. $X_L = 19.9$ kΩ, $Z = 33.6$ kΩ, $\theta = 36.4°$, $V_R = 16.1$ V, $V_L = 11.9$ V, $I = 596$ mA
15. $X_L = 78.5$ Ω, $Z = 127$ Ω, $\theta = 38.1°$, $V_R = 9.44$ V, $V_L = 7.41$ V, $I = 94.4$ mA
17. $X_L = 3.73$ kΩ, $\theta = 38.4°$, $f = 913$ Hz, $E = 12.1$ V, $V_R = 9.45$ V, $I = 2.01$ mA
19. $X_L = 26.5$ kΩ, $\theta = 55.8°$, $f = 56.1$ kHz, $E = 4.48$ V, $V_R = 2.52$ V, $I = 140$ μA
21. $X_L = 28.1$ kΩ, $\theta = 52.0°$, $f = 5.59$ kHz, $E = 21.6$ V, $V_R = 13.3$ V, $I = 605$ μA
23. $X_L = 1.6$ kΩ, $\theta = 64.9°$, $f = 2.55$ kHz, $E = 29.8$ V, $V_R = 12.6$ V, $I = 16.8$ mA
25. $X_L = 8.65$ kΩ, $L = 320$ mH, $\theta = 51.8°$, $V_R = 4.72$ V, $E = 7.63$ V, $I = 694$ μA
27. $X_L = 4.27$ kΩ, $L = 90.5$ mH, $\theta = 62.7°$, $V_R = 2.32$ V, $E = 5.06$ V, $I = 1.05$ mA
29. $X_L = 68.4$ kΩ, $L = 1.98$ H, $\theta = 55.5°$, $V_R = 13.7$ V, $E = 24.3$ V, $I = 292$ μA
31. $X_L = 5.51$ kΩ, $L = 73.1$ mH, $\theta = 54.7°$, $V_R = 6.02$ V, $E = 10.4$ V, $I = 1.54$ mA

PROBLEMS 19–5

1. $Z_{polar} = 30.9 \, \underline{/48°}$ kΩ
   $Z_{rect} = 20.7$ kΩ $+ j23$ kΩ
3. $Z_{polar} = 952 \, \underline{/-51°}$ Ω
   $Z_{rect} = 599$ Ω $- j740$ Ω
5. $Z_{polar} = 857 \, \underline{/-17°}$ Ω
   $Z_{rect} = 820$ Ω $- j251$ Ω
7. $Z_{polar} = 6.35 \, \underline{/-45°}$ kΩ
   $Z_{rect} = 4.49$ kΩ $- j4.49$ kΩ
9. $Z_{polar} = 50 \, \underline{/-36.9°}$ kΩ
   $Z_{rect} = 40$ kΩ $- j30$ kΩ
11. $Z_{polar} = 216 \, \underline{/33.7°}$ kΩ
   $Z_{rect} = 180$ kΩ $+ j120$ kΩ
13. $Z_{polar} = 4.6 \, \underline{/-40°}$ kΩ
   $Z_{rect} = 3.49$ kΩ $- j3$ kΩ
15. $Z_{polar} = 5.35 \, \underline{/44.5°}$ kΩ
   $Z_{rect} = 3.82$ kΩ $+ j3.75$ kΩ
17. $Z_{polar} = 20.6 \, \underline{/60.9°}$ kΩ
   $Z_{rect} = 10$ kΩ $+ j18$ kΩ
19. $Z_{polar} = 2.31 \, \underline{/-38.9°}$ kΩ
   $Z_{rect} = 1.8$ kΩ $- j1.45$ kΩ
21. $E_{polar} = 16.2 \, \underline{/51.8°}$ V
   $E_{rect} = 10$ V $+ j12.7$ V
23. $E_{polar} = 7.60 \, \underline{/-55.3°}$ V
   $E_{rect} = 4.32$ V $- j6.25$ V
25. $E_{polar} = 12 \, \underline{/-28.4°}$ V
   $E_{rect} = 10.6$ V $- j5.7$ V
27. $E_{polar} = 15 \, \underline{/34.5°}$ V
   $E_{rect} = 12.4$ V $+ j8.5$ V
29. $E_{polar} = 15 \, \underline{/-40°}$ V
   $E_{rect} = 11.5$ V $- j9.64$ V
31. $E_{polar} = 14.0 \, \underline{/34.6°}$ V
   $E_{rect} = 11.5$ V $+ j7.93$ V
33. $E_{polar} = 136 \, \underline{/-28°}$ V
   $E_{rect} = 120$ V $- j64$ V
35. $E_{polar} = 5.00 \, \underline{/58.7°}$ V
   $E_{rect} = 2.60$ V $+ j4.28$ V
37. $E_{polar} = 30 \, \underline{/32.2°}$ V
   $E_{rect} = 25.4$ V $+ j16$ V
39. $E_{polar} = 10 \, \underline{/-36.9°}$ V
   $E_{rect} = 8$ V $- j6$ V

PROBLEMS 19–6

1. $4.64 \, \underline{/47.2°}$ kΩ   3. $25.9 \, \underline{/-25.8°}$ kΩ   5. $8.99 \, \underline{/-7.39°}$ kΩ
7. $82.6 \, \underline{/1.88°}$ kΩ   9. $420 \, \underline{/30°}$ kΩ   11. $12.8 \, \underline{/-11°}$ kΩ

1. $Z_{\text{rect}} = 1.76 \text{ k}\Omega + j3 \text{ k}\Omega$
   $Z_{\text{polar}} = 3.48 \underline{/59.6°} \text{ k}\Omega$
7. $Z_{\text{rect}} = 200 \Omega - j281 \Omega$
   $Z_{\text{polar}} = 345 \underline{/-54.5°} \Omega$
   $V_R = 14.5 \text{ V}$
   $V_C = -65.9 \text{ V}; I = 72.5 \text{ mA}$
   $V_L = 45.5 \text{ V}$; in the ESC, $R = 200 \Omega$ and $C = 1.13 \mu\text{F}$

3. $Z_{\text{rect}} = 1.15 \text{ k}\Omega + j1.3 \text{ k}\Omega$
   $Z_{\text{polar}} = 1.74 \underline{/48.5°} \text{ k}\Omega$

5. $R = 3.40 \text{ k}\Omega$ and $L = 624 \text{ mH}$

1. $f_r = 2.6 \text{ kHz}$ (a) $X_L = 12.2 \text{ k}\Omega$, $X_C = 12.2 \text{ k}\Omega$, $Z_{\text{polar}} = 470 \underline{/0°} \Omega$, $I = 21.3 \text{ mA}$, $V_R = 10 \text{ V}$, $V_L = 261 \text{ V}$, $V_C = -261 \text{ V}$. The ESC consists of 470Ω of resistance. (b) $X_L = 10.8 \text{ k}\Omega$, $X_C = 13.8 \text{ k}\Omega$, $Z_{\text{polar}} = 3.04 \underline{/-81.1°} \text{ k}\Omega$, $I = 3.29 \text{ mA}$, $V_R = 1.55 \text{ V}$, $V_L = 35.5 \text{ V}$, $V_C = -45.4 \text{ V}$. In the ESC $R = 470 \Omega$ and $C = 23.1 \text{ nF}$. (c) $X_L = 12.7 \text{ k}\Omega$, $X_C = 11.8 \text{ k}\Omega$, $Z_{\text{polar}} = 1.10 \underline{/63.3°} \text{ k}\Omega$, $I = 9.56 \text{ mA}$, $V_R = 4.49 \text{ V}$, $V_L = 122 \text{ V}$, $V_C = -113 \text{ V}$. In the ESC: $R = 470 \Omega$ and $L = 55.1 \text{ mH}$.

3. $f_r = 225 \text{ Hz}$ (a) $X_L = 1.41 \text{ k}\Omega$, $X_C = 1.41 \text{ k}\Omega$, $Z_{\text{polar}} = 300 \underline{/0°}$, $I = 83.3 \text{ mA}$, $V_R = 25 \text{ V}$, $V_L = 118 \text{ V}$, $V_C = -118 \text{ V}$. The ESC consists of 300Ω of resistance. (b) $X_L = 1.26 \text{ k}\Omega$, $X_C = 1.59 \text{ k}\Omega$, $Z_{\text{polar}} = 450 \underline{/-48.1°} \Omega$, $I = 55.6 \text{ mA}$, $V_R = 16.8 \text{ V}$, $V_L = 69.9 \text{ V}$, $V_C = 88.5 \text{ V}$. The ESC consists of 300Ω of resistance and $2.38 \mu\text{F}$ of capacitance. (c) $X_L = 1.88 \text{ k}\Omega$, $X_C = 1.06 \text{ k}\Omega$, $Z_{\text{polar}} = 872 \underline{/69.9°} \Omega$, $I = 28.7 \text{ mA}$, $V_R = 8.60 \text{ V}$, $V_L = 53.9 \text{ V}$, $V_C = -30.4 \text{ V}$. The ESC consists of 300Ω of resistance and 437 mH of inductance.

CHAPTER 20

1. $I_{\text{polar}} = 40 \underline{/54°} \text{ mA}$
   $I_{\text{rect}} = 23.5 \text{ mA} + j32.4 \text{ mA}$
7. $I_{\text{polar}} = 442 \underline{/37.6°} \mu\text{A}$
   $I_{\text{rect}} = 350 \mu\text{A} + j270 \mu\text{A}$
13. $I_{\text{polar}} = 1.19 \underline{/33°} \text{ mA}$
   $I_{\text{rect}} = 1 \text{ mA} + j649 \mu\text{A}$
19. $I_{\text{polar}} = 47.6 \underline{/71°} \text{ mA}$
   $I_{\text{rect}} = 15.5 \text{ mA} + j45 \text{ mA}$
25. $I_{\text{polar}} = 12 \underline{/51.3°} \text{ mA}$
   $I_{\text{rect}} = 7.5 \text{ mA} + j9.37 \text{ mA}$
31. $I_{\text{polar}} = 43.0 \underline{/35.5°} \text{ mA}$
   $I_{\text{rect}} = 35.0 \text{ mA} + j25.0 \text{ mA}$

3. $I_{\text{polar}} = 2.75 \underline{/67.4°} \text{ mA}$
   $I_{\text{rect}} = 1.06 \text{ mA} + j2.54 \text{ mA}$
9. $I_{\text{polar}} = 8.51 \underline{/55.9°} \text{ mA}$
   $I_{\text{rect}} = 4.77 \text{ mA} + j7.05 \text{ mA}$
15. $I_{\text{polar}} = 683 \underline{/20.5°} \mu\text{A}$
   $I_{\text{rect}} = 640 \mu\text{A} + j239 \mu\text{A}$
21. $I_{\text{polar}} = 6.31 \underline{/59.5°} \text{ mA}$
   $I_{\text{rect}} = 3.20 \text{ mA} + j5.44 \text{ mA}$
27. $I_{\text{polar}} = 1 \underline{/49.5°} \text{ mA}$
   $I_{\text{rect}} = 650 \mu\text{A} + j760 \mu\text{A}$
33. $I_{\text{polar}} = 400 \underline{/30.0°} \mu\text{A}$
   $I_{\text{rect}} = 346 \mu\text{A} + j200 \mu\text{A}$

5. $I_{\text{polar}} = 175 \underline{/22°} \mu\text{A}$
   $I_{\text{rect}} = 162 \mu\text{A} + j65.6 \mu\text{A}$
11. $I_{\text{polar}} = 141 \underline{/45°} \mu\text{A}$
   $I_{\text{rect}} = 100 \mu\text{A} + j100 \mu\text{A}$
17. $I_{\text{polar}} = 53.9 \underline{/62°} \text{ mA}$
   $I_{\text{rect}} = 25.3 \text{ mA} + j47.6 \text{ mA}$
23. $I_{\text{polar}} = 333 \underline{/55.7°} \mu\text{A}$
   $I_{\text{rect}} = 188 \mu\text{A} + j275 \mu\text{A}$
29. $I_{\text{polar}} = 3.15 \underline{/36.0°} \text{ mA}$
   $I_{\text{rect}} = 2.55 \text{ mA} + j1.85 \text{ mA}$
35. $I_{\text{polar}} = 6.08 \underline{/35.7°} \text{ mA}$
   $I_{\text{rect}} = 4.94 \text{ mA} + j3.55 \text{ mA}$

1. $Y_{\text{rect}} = 303 \mu\text{S} + j377 \mu\text{S}$
   $Y_{\text{polar}} = 484 \underline{/51.2°} \mu\text{S}$
   $I_{\text{rect}} = 2.51 \text{ mA} + j3.12 \text{ mA}$
   $V = 8.27 \text{ V}$

3. $Y_{\text{rect}} = 1.33 \text{ mS} + j1.88 \text{ mS}$
   $Y_{\text{polar}} = 2.31 \underline{/54.7°} \text{ mS}$
   $I_{\text{rect}} = 577 \mu\text{A} + j816 \mu\text{A}$
   $V = 433 \text{ mV}$

5. $Y_{\text{rect}} = 147 \mu\text{S} + j94.2 \mu\text{S}$
   $Y_{\text{polar}} = 175 \underline{/32.7°} \mu\text{S}$
   $I_{\text{rect}} = 16.8 \text{ mA} + j10.8 \text{ mA}$
   $V = 115 \text{ V}$

7. $Y_{rect} = 100 \ \mu S + j188 \ \mu S$
$Y_{polar} = 213 \ \underline{/62.1°} \ \mu S$
$I_{rect} = 4.69 \ mA + j8.83 \ mA$
$V = 46.9 \ V$

9. $Y_{rect} = 66.7 \ \mu S + j66 \ \mu S$
$Y_{polar} = 93.8 \ \underline{/44.7°} \ \mu S$
$I_{rect} = 1.42 \ mA + j1.41 \ mA$
$V = 21.3 \ V$

11. $R = 5.96 \ k\Omega$
$f = 1.27 \ kHz$
$I_T = 4.67 \ mA$
$V = 17.9 \ V$

PROBLEMS 20–3

1. $I_{polar} = 3.8 \ \underline{/-30°} \ mA$
$I_{rect} = 3.29 \ mA - j1.9 \ mA$
7. $I_{polar} = 14.4 \ \underline{/-33.7°} \ mA$
$I_{rect} = 12 \ mA - j8 \ mA$
13. $I_{polar} = 2.28 \ \underline{/-28.5°} \ mA$
$I_{rect} = 2 \ mA - j1.09 \ mA$
19. $I_{polar} = 1.03 \ \underline{/-45°} \ mA$
$I_{rect} = 730 \ \mu A - j730 \ \mu A$
25. $I_{polar} = 27 \ \underline{/-47.8°} \ mA$
$I_{rect} = 18.1 \ mA - j20 \ mA$
31. $I_{polar} = 6.70 \ \underline{/-45.5°} \ mA$
$I_{rect} = 4.70 \ mA - j4.77 \ mA$

3. $I_{polar} = 210 \ \underline{/-39°} \ \mu A$
$I_{rect} = 163 \ \mu A - j132 \ \mu A$
9. $I_{polar} = 50 \ \underline{/-53.1°} \ mA$
$I_{rect} = 30 \ mA - j40 \ mA$
15. $I_{polar} = 4.45 \ \underline{/-73°} \ mA$
$I_{rect} = 1.3 \ mA - j4.25 \ mA$
21. $I_{polar} = 14 \ \underline{/-40°} \ mA$
$I_{rect} = 10.7 \ mA - j9 \ mA$
27. $I_{polar} = 300 \ \underline{/-30°} \ \mu A$
$I_{rect} = 260 \ \mu A - j150 \ \mu A$
33. $I_{polar} = 1.35 \ \underline{/-52.3°} \ mA$
$I_{rect} = 825 \ \mu A - j1.07 \ mA$

5. $I_{polar} = 1.34 \ \underline{/-23.4°} \ mA$
$I_{rect} = 1.23 \ mA - j532 \ \mu A$
11. $I_{polar} = 234 \ \underline{/-50.2°} \ \mu A$
$I_{rect} = 150 \ \mu A - j180 \ \mu A$
17. $I_{polar} = 42.4 \ \underline{/-27.4°} \ \mu A$
$I_{rect} = 37.6 \ \mu A - j19.5 \ \mu A$
23. $I_{polar} = 690 \ \underline{/-43.5°} \ \mu A$
$I_{rect} = 501 \ \mu A - j475 \ \mu A$
29. $I_{polar} = 1.83 \ \underline{/-30.4°} \ mA$
$I_{rect} = 1.58 \ mA - j925 \ \mu A$
35. $I_{polar} = 100 \ \underline{/-42.4°} \ mA$
$I_{rect} = 73.8 \ mA - j67.5 \ mA$

PROBLEMS 20–4

1. $Y_{polar} = 63.9 \ \underline{/-38.5°} \ \mu S$
$Y_{rect} = 50 \ \mu S - j39.8 \ \mu S$
$I_{polar} = 40 \ \underline{/-38.5°} \ mA$
$I_{rect} = 31.3 \ mA - j24.9 \ mA$
7. $Y_{polar} = 48.8 \ \underline{/-40.8°} \ \mu S$
$Y_{rect} = 37.0 \ \mu S - j31.8 \ \mu S$
$I_{polar} = 400 \ \underline{/-40.7°} \ \mu A$
$I_{rect} = 303 \ \mu A - j261 \ \mu A$

3. $Y_{polar} = 166 \ \underline{/-53°} \ \mu S$
$Y_{rect} = 100 \ \mu S - j133 \ \mu S$
$I_{polar} = 200 \ \underline{/-53°} \ mA$
$I_{rect} = 120 \ mA - j160 \ mA$
9. $Y_{polar} = 1.49 \ \underline{/-26.4°} \ mS$
$Y_{rect} = 1.33 \ mS - j663 \ \mu S$
$I_{polar} = 30 \ \underline{/-26.4°} \ mA$
$I_{rect} = 26.9 \ mA - j13.4 \ mA$

5. $Y_{polar} = 266 \ \underline{/-36.8°} \ \mu S$
$Y_{rect} = 213 \ \mu S - j159 \ \mu S$
$I_{polar} = 100 \ \underline{/-36.8°} \ \mu A$
$I_{rect} = 80.1 \ \mu A - j59.9 \ \mu A$

PROBLEMS 20–5

1. $269 \ \underline{/-27°} \ \Omega$ 3. $158 \ \underline{/40°} \ \mu S$ 5. $657 \ \underline{/44.1°} \ \Omega$ 7. $8.55 \ \underline{/-27.8°} \ k\Omega$
9. $2.59 \ \underline{/-38.2°} \ k\Omega$

PROBLEMS 20–6

1. $I_R = 1.21 \ mA$, $I_C = 1.59 \ mA$, $Y_{polar} = 110 \ \underline{/52.8°} \ \mu S$, $Z_{polar} = 9.06 \ \underline{/-52.8°} \ k\Omega$, $V = 18.1 \ V$.
In the ESC $R = 5.47 \ k\Omega$ and $C = 630 \ pF$.
3. $I_R = 24.1 \ mA$, $I_C = 32.0 \ mA$, $Y_{polar} = 3.54 \ \underline{/53.0°} \ mS$, $Z_{polar} = 283 \ \underline{/-53.0°} \ \Omega$, $V = 11.3 \ V$.
In the ESC $R = 170 \ \Omega$ and $C = 235 \ nF$.
5. $I_R = 192 \ \mu A$, $I_C = 160 \ \mu A$, $Y_{polar} = 27.7 \ \underline{/-39.7°} \ \mu S$, $Z_{polar} = 36.1 \ \underline{/39.7°} \ k\Omega$, $V = 9.04 \ V$.
In the ESC $R = 27.8 \ k\Omega$ and $L = 24.5 \ mH$.
7. $I_R = 4.05 \ mA$, $I_C = 2.93 \ mA$, $Y_{polar} = 1.81 \ \underline{/-35.8°} \ mS$, $Z_{polar} = 551 \ \underline{/35.8°} \ \Omega$, $V = 2.76 \ V$.
In the ESC $R = 447 \ \Omega$ and $L = 10.3 \ mH$.

9. $I_R = 10.4$ mA, $I_C = 4.88$ mA, $I_L = -22.0$ mA, $Y_{polar} = 193 \,/-58.8°\,$ μS, $Z_{polar} = 5.18 \,/58.8°\,$ kΩ, $V = 104$ V. In the ESC $R = 2.68$ kΩ and $L = 28.2$ mH.

11. $I_R = 257$ μA, $I_C = 4.95$ mA, $I_L = -4.64$ mA, $Y_{polar} = 228 \,/49.9°\,$ μS, $Z_{polar} = 4.38 \,/-49.9°\,$ kΩ, $V = 1.75$ V. In the ESC $R = 28.2$ kΩ and $C = 3.17$ nF.

PROBLEMS 20–7

1. $Y_{rect} = 621$ μS $+ j726$ μS, $C = 28.9$ nF

3. $Y_{rect} = 37.4$ μS $+ j33.1$ μS, $C = 43.9$ pF

5. $Y_{rect} = 289$ μS $- j219$ μS, $L = 275$ mH

7. $Y_{rect} = 39.2$ μS $- j32.8$ μS, $L = 485$ mH

9. $Z_{rect} = 3.15$ kΩ $- j1.55$ kΩ, $C = 10.3$ nF

11. $Z_{rect} = 216$ Ω $- j361$ Ω, $C = 136$ nF

13. $Z_{rect} = 52.6$ Ω $+ j28.5$ Ω, $L = 45.3$ mH

15. $Z_{rect} = 249$ Ω $+ j1.48$ kΩ, $L = 29.2$ mH

PROBLEMS 20–8

1. $Z = 2.00$ kΩ $+ j19.9$ kΩ

3. $Z = 7$ kΩ $- j1$ kΩ

5. $Z = 16.9$ kΩ $- j26.3$ kΩ

7. $Z = 8.61$ kΩ $- j9.21$ kΩ

9. $R = 1.13$ kΩ and $L = 528$ mH. The phase angle is 71.2°.

11. $R = 8.49$ kΩ, $L = 895$ mH, $\theta = 73.2°$

13. $R = 23.8$ kΩ, $C = 8.45$ nF, $\theta = -38.3°$

15. $R = 6.18$ kΩ, $C = 6.76$ nF, $\theta = -29.6°$

CHAPTER 21

PROBLEMS 21–1

1. $I_{max} = 31.9$ mA, $P_{max} = 479$ mW, $V_{o(max)} = 15$ V. At f_{co}, $X_C = 470$ Ω, $I = 22.6$ mA, $P = 239$ mW, $V_o = 10.6$ V.

3. $I_{max} = 53.6$ μA, $P_{max} = 16.1$ μW, $V_{o(max)} = 300$ mV. At f_{co}, $X_C = 5.6$ kΩ, $I = 37.9$ μA, $P = 8.04$ μW, $V_o = 212$ mV.

5. $f_{co} = 265$ Hz. At f_{co}, $I = 11.8$ mA, $V_o = 14.1$ V, $P = 167$ mW.

7. $f_{co} = 362$ Hz. At f_{co}, $I = 12.9$ mA, $V_o = 28.3$ V, $P = 364$ mW.

PROBLEMS 21–2

1. $V_{OC} = 6.47$ V, $R_{TH} = 2.16$ kΩ, $f_{co} = 147$ Hz, $V_o = 2.11$ V

3. $V_{OC} = 20$ V, $R_{TH} = 2$ kΩ, $f_{co} = 39.8$ Hz, $V_o = 14.1$ V

5. $V_{OC} = 10$ V, $R_{TH} = 5$ kΩ, $f_{co} = 31.8$ Hz, V_o at $f_{co} = 3.25$ V

7. $V_{OC} = 5.66$ V, $R_{TH} = 4.08$ kΩ, $f_{co} = 39$ Hz

9. $I_{SC} = 31.9$ mA, $G_N = 2.96$ mS, $f_{co} = 9.43$ kHz, V_o at $f_{co} = 7.62$ V

11. $I_{SC} = 5$ mA, $G_N = 891$ μS, $f_{co} = 142$ kHz, V_o at $f_{co} = 3.97$ V

PROBLEMS 21–3

1. $f_1 = 169$ Hz, $f_2 = 183$ kHz, V_o at $f_m = 5.74$ V, V_o at $f_{co} = 4.06$ V

3. $f_1 = 15.0$ Hz, $f_2 = 128$ kHz, V_o at $f_m = 5.66$ V, V_o at $f_{co} = 4.00$ V

5. $f_1 = 21.7$ Hz, $f_2 = 5.66$ kHz, V_o at $f_m = 1.28$ V, V_o at $f_{co} = 904$ mV

7. $f_1 = 692$ Hz, $f_2 = 1.35$ MHz, V_o at $f_m = 2.14$ V, V_o at $f_{co} = 1.51$ V

9. $f_1 = 43.9$ Hz, $f_2 = 87.9$ kHz

11. $f_1 = 10.3$ Hz, $f_2 = 241$ kHz

13. $BW = 80$ Hz, $f_1 = 1960$ Hz, $f_2 = 2.04$ kHz. At f_r, $V_R = 20$ V, $I = 20$ mA, $P = 400$ mW. At f_1 and f_2, $V_R = 14.1$ V, $I = 14.1$ mA, $P = 200$ mW.

15. $BW = 393$ Hz, $f_1 = 5.30$ kHz, $f_2 = 5.70$ kHz. At f_r, $V_R = 5$ V, $I = 41.7$ mA, $P = 208$ mW. At f_1 and f_2, $V_R = 3.54$ V, $I = 29.5$ mA, $P = 104$ mW.

17. $f_r = 1.01$ kHz, $Q = 14.1$, $BW = 71.6$ Hz, $f_1 = 971$ Hz, $f_2 = 1.04$ kHz

19. $f_r = 3.56$ kHz, $Q = 5.96$, $BW = 597$ Hz, $f_1 = 3.26$ kHz, $f_2 = 3.86$ kHz

CHAPTER 22

1. $\log 1000 = 3$, $10^3 = 1000$ 3. $\log 1 = 0$, $10^0 = 1$

5. $\log 10{,}000{,}000 = 7$, $10^7 = 10{,}000{,}000$ 7. $\log 0.1 = -1$, $10^{-1} = 0.1$

9. $\log 0.000001 = -6$, $10^{-6} = 0.000001$

1. $\log 3.42 = 0.534$, $10^{0.534} = 3.42$ 3. $\log 4.05 = 0.607$, $10^{0.607} = 4.05$

5. $\log 6.44 = 0.809$, $10^{0.809} = 6.44$ 7. $\log 9.45 = 0.975$, $10^{0.975} = 9.45$

9. $\log 2.50 = 0.398$, $10^{0.398} = 2.50$

1. $\log 14 = 1.146$, $10^{1.146} = 14$ 3. $\log 743 = 2.871$, $10^{2.871} = 743$

5. $\log 1070 = 3.029$, $10^{3.029} = 1070$ 7. $\log 43.4 = 1.638$, $10^{1.638} = 43.4$

9. $\log 10{,}000 = 4$, $10^4 = 10{,}000$ 11. $\log 23{,}400 = 4.369$, $10^{4.369} = 23{,}400$

13. $\log 47{,}700 = 4.679$, $10^{4.679} = 47{,}700$ 15. $\log 14.6 = 1.164$, $10^{1.164} = 14.6$

17. $\log 37{,}000 = 4.568$, $10^{4.568} = 37{,}000$ 19. $\log 2{,}000{,}000 = 6.301$, $10^{6.301} = 2{,}000{,}000$

1. $\log 0.00377 = -2.424$, $10^{-2.424} = 0.00377$ 3. $\log 0.037 = -1.432$, $10^{-1.432} = 0.037$

5. $\log 0.746 = -0.1273$, $10^{-0.1273} = 0.746$ 7. $\log 0.407 = -0.39$, $10^{-0.39} = 0.407$

1. $\log (27 \times 56) = \log 27 + \log 56 = 1.431 + 1.748 = 3.18$, $10^{3.18} = 1.51 \times 10^3$

3. $\log (107 \times 243{,}000) = \log 107 + \log 243{,}000 = 2.029 + 5.386 = 7.415$, $10^{7.415} = 2.60 \times 10^7$

5. $\log (43 \times 0.143) = \log 43 + \log 0.143 = 1.633 + (-0.8447) = 0.7888$, $10^{0.7888} = 6.15$

7. $\log (0.043 \times 0.0106) = \log 0.043 + \log 0.0106 = -1.367 + (-1.975) = -3.341$, $10^{-3.341} = 4.56 \times 10^{-4}$

9. $\log (0.0000176 \times 0.000357) = \log 0.0000176 + \log 0.000357 = -4.754 + (-3.447) = -8.202$, $10^{-8.202} = 6.28 \times 10^{-9}$

11. $\log (256 \times 0.00246) = \log 256 + \log 0.00246 = 2.408 + (-2.609) = -0.2008$, $10^{-0.2008} = 0.630$

13. $\log (78 \times 0.473) = \log 78 + \log 0.473 = 1.892 + (-0.3251) = 1.567$, $10^{1.567} = 36.9$

15. $\log (4730 \times 0.000506) = \log 4730 + \log 0.000506 = 3.675 + (-3.296) = 0.3790,\ 10^{0.3790} =$ 2.39
17. $\log (760 \times 7360) = \log 760 + \log 7360 = 2.881 + 3.867 = 6.748,\ 10^{6.748} = 5.59 \times 10^{6}$
19. $\log (2560 \times 0.00457) = \log 2560 + \log 0.00457 = 3.408 + (-2.340) = 1.068,\ 10^{1.068} = 11.7$

PROBLEMS 22–6

1. $\log \dfrac{1760}{43} = \log 1760 - \log 43 = 3.246 - 1.633 = 1.612,\ 10^{1.612} = 40.9$

3. $\log \dfrac{65}{467} = \log 65 - \log 467 = 1.813 - 2.669 = -0.856,\ 10^{-0.856} = 0.139$

5. $\log \dfrac{0.00273}{172} = \log 0.00273 - \log 172 = -2.564 - 2.236 = -4.799,\ 10^{-4.799} =$ 1.59×10^{-5}

7. $\log \dfrac{4300}{0.15} = \log 4300 - \log 0.15 = 3.633 - (-0.824) = 4.457,\ 10^{4.457} = 2.87 \times 10^{4}$

9. $\log \dfrac{0.0675}{0.00103} = \log 0.0675 - \log 0.00103 = -1.171 - (-2.987) = 1.816,\ 10^{1.816} = 65.5$

11. $\log \dfrac{0.00467}{73.2} = \log 0.00467 - \log 73.2 = -2.331 - 1.865 = -4.196,\ 10^{-4.196} = 6.38 \times 10^{-5}$

13. $\log \dfrac{0.147}{0.0432} = \log 0.147 - \log 0.0432 = -0.8327 - (-1.365) = 0.5318,\ 10^{0.5318} = 3.41$

15. $\log \dfrac{4730}{0.443} = \log 4730 - \log 0.443 = 3.675 - (-0.3536) = 4.029,\ 10^{4.029} = 1.07 \times 10^{4}$

17. $\log \dfrac{0.445}{635} = \log 0.445 - \log 635 = -0.3516 - 2.803 = -3.154,\ 10^{-3.154} = 7.01 \times 10^{-4}$

19. $\log \dfrac{0.00275}{4.35} = \log 0.00275 - \log 4.35 = -2.561 - 0.6385 = -3.199,\ 10^{-3.199} = 6.32 \times 10^{-4}$

PROBLEMS 22–7

1. $\log 3.4^{3} = 3 \log 3.4 = 3 \times 0.5315 = 1.594,\ 10^{1.594} = 39.3$
3. $\log 6^{-1.3} = -1.3 \log 6 = -1.3 \times 0.7782 = -1.012,\ 10^{-1.012} = 9.74 \times 10^{-2}$
5. $\log 0.43^{2.5} = 2.5 \log 0.43 = 2.5 \times -0.3665 = -0.9163,\ 10^{-0.9163} = 0.121$
7. $\log 6^{4.3} = 4.3 \log 6 = 4.3 \times 0.7782 = 3.346,\ 10^{3.346} = 2.22 \times 10^{3}$
9. $\log 0.27^{-4.4} = -4.4 \log 0.27 = -4.4 \times -0.5686 = 2.502,\ 10^{2.502} = 3.18 \times 10^{2}$
11. $\log 8.7^{6.3} = 6.3 \log 8.7 = 6.3 \times 0.9395 = 5.919,\ 10^{5.919} = 8.30 \times 10^{5}$
13. $\log 3.4^{3.2} = 3.2 \log 3.4 = 3.2 \times 0.5315 = 1.701,\ 10^{1.701} = 50.2$
15. $\log 6.35^{-2.3} = -2.3 \log 6.35 = -2.3 \times 0.8028 = -1.846,\ 10^{-1.846} = 1.42 \times 10^{-2}$
17. $\log 0.045^{3.6} = 3.6 \log 0.045 = 3.6 \times (-1.347) = -4.848,\ 10^{-4.848} = 1.42 \times 10^{-5}$
19. $\log 0.00735^{-2.5} = -2.5 \log 0.00735 = -2.5 \times (-2.134) = 5.334,\ 10^{5.334} = 2.16 \times 10^{5}$

Logarithmic Form	Exponential Form	Logarithmic Form	Exponential Form
1. $\ln 2 = 0.6931$	$e^{0.6931} = 2$	17. $\ln 8.43 = 2.132$	$e^{2.132} = 8.43$
3. $\ln 0.932 = -0.07042$	$e^{-0.07042} = 0.932$	19. $\ln 0.346 = -1.061$	$e^{-1.061} = 0.346$
5. $\ln 60 = 4.094$	$e^{4.094} = 60$	21. $\ln 3.67 = 1.30$	$e^{1.30} = 3.67$
7. $\ln 0.073 = -2.617$	$e^{-2.617} = 0.073$	23. $\ln 0.107 = -2.235$	$e^{-2.235} = 0.107$
9. $\ln 247 = 5.509$	$e^{5.509} = 247$	25. $\ln 176 = 5.17$	$e^{5.17} = 176$
11. $\ln 0.00146 = -6.529$	$e^{-6.529} = 0.00146$	27. $\ln 0.293 = -1.228$	$e^{-1.228} = 0.293$
13. $\ln 1100 = 7.003$	$e^{7.003} = 1100$	29. $\ln 873 = 6.772$	$e^{6.772} = 873$
15. $\ln 0.00027 = -8.217$	$e^{-8.217} = 0.00027$		

31. 10.4   33. 43.4   35. 70.1   37. 200   39. 4.71   41. 1.54   43. 0.996
45. 1.16   47. 0.730   49. 0.868

CHAPTER 23

1. $10^x = 340$ 3. $10^x = 14$ 5. $10^x = 7.4$ 7. $10^y = 0.014$ 9. $10^y = 0.734$
11. $10^y = -0.00178$ 13. $x = 10^{-2.43}$ 15. $x = 10^{-3.643}$ 17. $x = 10^{-0.143}$
19. $y = 10^{1.44}$ 21. $y = 10^{2.83}$ 23. $y = 10^{4.14}$ 25. $\log y = 3.6$ 27. $\log y = 4.73$
29. $\log y = 2.73$ 31. $a = \log 732$ 33. $a = \log 47.3$ 35. $a = \log 276$
37. $y = \log 0.043$ 39. $y = \log 0.177$ 41. $y = \log 0.00178$ 43. $\log x = -3.4$
45. $\log x = -2.76$ 47. $\log x = -1.17$

1. 2.29×10^7 3. 1.00   5. 3.78×10^{-1} 7. 0.992   9. 2.66   11. −1.76
13. −2.60   15. 4.67   17. 1.45×10^4 19. 1.08   21. 2.51×10^{-4}
23. 6.71×10^{-1} 25. 2.87   27. −2.15   29. 2.78×10^3 31. -5.33×10^{-1}
33. 7.91   35. 1.00×10^3 37. 2.29×10^7 39. 1.46   41. 5.85×10^{-1}

1. −0.355   3. 1.17   5. 2.10×10^3 7. 1.00×10^{-4} 9. 5.56   11. 1.57
13. 1×10^8 15. 2.00×10^5 17. 2.85   19. 1.27   21. 5.00×10^{-6}
23. $y = 1.80 \times 10^5$ 25. 2.35   27. 0.316   29. $y = 0.422$

1. 2.40   3. 2.71   5. 5.30   7. −7.82   9. −2.71   11. 14.9   13. 3.83
15. 1.83×10^{-2}

1. 0.981   3. 0.470   5. 13.2   7. 31.6   9. 2.75   11. 3.64   13. 2.61   15. 2.89
17. 0.950   19. 0.472   21. 0.549

CHAPTER 24

1. 60 dB   3. $A_p = 43$ dB   5. $A_p = 44$ dB   7. 3.79 mW   9. $P_{in} = 1$ W
11. $P_{in} = 3.15$ mW   13. 1.58 W   15. $P_{out} = 3.15$ W   17. $P_o = 316$ mW   19. 100 mW
21. $P_o = 1.58$ W   23. $P_o = 100$ W   25. 88.6 dB   27. $A_v = 49.5$ dB   29. $A_v = 106$ dB
31. 500 mV   33. $V_{out} = 27.3$ mV   35. $V_{out} = 103$ V   37. 126 μV   39. $V_{in} = 2.67$ mV
41. $V_{in} = 60$ μV   43. 706 mV   45. $V_{in} = 2.00$ V   47. $V_{in} = 7.06$ V   49. $V_{in} = 3.16$ V

PROBLEMS 24–3

1. $A_{VM} = -3.9$ dB; $f_1 = 339$ Hz; $f_2 = 9.17$ kHz   3. $A_{VM} = -2.02$ dB; $f_1 = 140$ Hz;
$f_2 = 8.54$ kHz

5. $A_{VM} = -7.96$ dB; $f_1 = 145$ Hz; $f_2 = 4.02$ kHz   7. At 40 Hz, $A_V = 15.9$ dB. At 25 kHz,
$A_V = 15.3$ dB.

9. At 40 Hz, $A_V = 3.01$ dB. At 12 kHz, $A_V = 0$ dB   11. $A_{VM} = -4.56$ dB. At 30 Hz,
$A_V = -18.0$ dB. At 50 kHz, $A_V = -10.6$ dB.

PROBLEMS 24–4

1. $V_C = 34.7$ V, $V_R = 65.3$ V   3. 2.47 ms   5. 145 μs   7. 5.41 ms   9. 9.79 kΩ
11. 9.53 nF

CHAPTER 25

PROBLEMS 25–1

1. 7   3. 10   5. 13   7. 23   9. 38   11. 56   13. 101   15. 119
17. 240   19. 195   21. 101_2 23. 1100_2 25. 10101_2
27. 101101_2 29. 111011_2 31. 1000100_2 33. 1100000_2 35. 10000111_2
37. 11010010_2

PROBLEMS 25–2

1. 12_{10} 3. 63_{10} 5. 190_{10} 7. 2174_{10} 9. 5068_{10} 11. 24_8 13. 120_8
15. 550_8 17. 2611_8 19. 7315_8

PROBLEMS 25–3

1. 26_{10} 3. 76_{10} 5. 512_{10} 7. 4522_{10} 9. $41{,}003_{10}$ 11. 16_{16} 13. 61_{16}
15. 200_{16} 17. $A8C_{16}$ 19. $17BB_{16}$

PROBLEMS 25–4

1. $312_8 = CA_{16}$ 3. $233_8 = 9B_{16}$ 5. $201_8 = 81_{16}$ 7. $6311_8 = CC9_{16}$
9. $5303_8 = AC3_{16}$ 11. 1111_2 13. 100100_2 15. 10000000_2 17. 101110000_2
19. 1000000000_2 21. 1100_2 23. 100100_2 25. 1001111_2 27. 11100011_2
29. 110011010_2

1. $75_8 = 111101_2 = 61_{10} = 3D_{16}$ 3. $140_8 = 1100000_2 = 96_{10} = 60_{16}$
5. $2A_{16} = 101010_2 = 52_8 = 42_{10}$ 7. $1A7_{16} = 110100111_2 = 647_8 = 423_{10}$
9. $100110_2 = 46_8 = 38_{10} = 26_{16}$ 11. $111000101_2 = 705_8 = 453_{10} = 1C5_{16}$
13. $10_{10} = 1010_2 = 12_8 = A_{16}$ 15. $290_{10} = 100100010_2 = 442_8 = 122_{16}$

PROBLEMS 25–6

1. 357_8 3. 775_8 5. 776_8 7. 1423_8 9. 1107_8

PROBLEMS 25–7

1. ED_{16} 3. $CA5_{16}$ 5. $D56_{16}$ 7. $179D_{16}$ 9. $1DA6C_{16}$

PROBLEMS 25–8

1. 10101_2 3. 110011_2 5. 101001_2 7. 1100010_2 9. 1101001_2

PROBLEMS 25–9

1. 45_8 3. 42_8 5. 245_8 7. 337_8 9. 1515_8

PROBLEMS 25–10

1. $1A_{16}$ 3. $3E1_{16}$ 5. $5F1_{16}$ 7. AFD_{16} 9. FDC_{16}

PROBLEMS 25–11

1. 101_2 3. 10_2 5. 10_2 7. 11_2 9. 10110_2

PROBLEMS 25–12

1. 0110_2 3. 010011_2 5. 01101_2 7. 100011_2 9. 01011_2 11. 1000011_2

PROBLEMS 25–13

1. 14_8 3. 143_8 5. $5D_{16}$ 7. $3E2_{16}$ 9. 36_8 11. 227_8 13. 16_{16} 15. $1A3_{16}$

CHAPTER 26

PROBLEMS 26–1

1.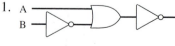

3.

5.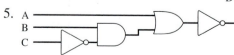

7.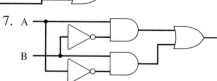

9.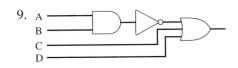

11.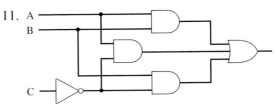

13.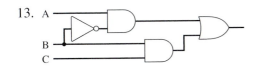

15.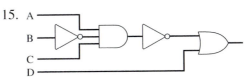

17.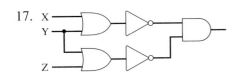

19.

21. $A\overline{B} + \overline{B}C$ 23. $\overline{\overline{X}YZ}$ 25. $\overline{\overline{A}\,\overline{B}} + C$ 27. $(A + B)(\overline{A + C})D$ 29. $Z(X + Y)$

31. $\overline{\overline{AB}\,(A + B)}$ 33. $\overline{X + \overline{Y}} \cdot \overline{\overline{Y} + Z}$ 35. $\overline{\overline{A + B}(B + C)(C + D)}$ 37. $\overline{XY + \overline{YZ}}$

39. $A\overline{B} + \overline{B}C$

PROBLEMS 26–2

1. (a) $A((A + B)B)$ (b) $AB(A + B) = AAB + ABB = AB + AB = AB$

(c) A ──┐
         │)── X
      B ──┘

(d)

A	B	X
0	0	0
0	1	0
1	0	0
1	1	1

3. (a) $(\overline{A} + B + C) + (\overline{A} + B) + C$ (b) $\overline{A} + B + C + \overline{A} + B + C = \overline{A} + B + C$

(c)

(d)

A	B	C	X
0	0	0	1
0	0	1	1
0	1	0	1
0	1	1	1
1	0	0	0
1	0	1	1
1	1	0	1
1	1	1	1

5. (a) $(AB)(\overline{B} + C)C$ (b) $ABC(\overline{B} + C) = AB\overline{B}C + ABCC = 0 + ABC = ABC$

(c)

(d)

A	B	C	X
0	0	0	0
0	0	1	0
0	1	0	0
0	1	1	0
1	0	0	0
1	0	1	0
1	1	0	0
1	1	1	1

1. (a) $\overline{(A + B)} + \overline{C} = \overline{(A + B)C}$ (b)

(c)

A	B	C	X
0	0	0	1
0	0	1	1
0	1	0	1
0	1	1	0
1	0	0	1
1	0	1	0
1	1	0	1
1	1	1	0

3. (a) $(A + B)\overline{(A + C)}D = ((A + B) \ (\overline{A}\,\overline{C}))D = A\overline{A}\,\overline{C}D + \overline{A}B\overline{C}D = \overline{A}B\overline{C}D$

(b)

(c)

A	B	C	D	X
0	0	0	0	0
0	0	0	1	0
0	0	1	0	0
0	0	1	1	0
0	1	0	0	0
0	1	0	1	1
0	1	1	0	0
0	1	1	1	0
1	0	0	0	0
1	0	0	1	0
1	0	1	0	0
1	0	1	1	0
1	1	0	0	0
1	1	0	1	0
1	1	1	0	0
1	1	1	1	0

5. (a) $\overline{A\overline{B} + \overline{B}C} = \overline{\overline{B}(A + C)} = B + \overline{(A + C)}$ (b)

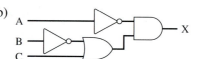

(c)

A	B	C	X
0	0	0	1
0	0	1	0
0	1	0	1
0	1	1	1
1	0	0	0
1	0	1	0
1	1	0	1
1	1	1	1

7. (a) $\overline{A + B\overline{C}} = \overline{A}\,\overline{(B\overline{C})} = \overline{A}(\overline{B} + C)$ (b)

(c)

A	B	C	X
0	0	0	1
0	0	1	1
0	1	0	0
0	1	1	1
1	0	0	0
1	0	1	0
1	1	0	0
1	1	1	0

9. (a) $\overline{(B+C)}(C+D) = (\overline{B}\,\overline{C})(C+D) = \overline{B}\,\overline{C}C + \overline{B}\,\overline{C}D = \overline{B}\,\overline{C}D$ (b)

(c)

B	C	D	X
0	0	0	0
0	0	1	1
0	1	0	0
0	1	1	0
1	0	0	0
1	0	1	0
1	1	0	0
1	1	1	0

11. (a) $\overline{\overline{AB} + C\overline{D}} = \overline{(\overline{AB})}\,\overline{(C\overline{D})} = (AB)(\overline{C}+D)$ (b)

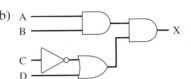

(c)

A	B	C	D	X
0	0	0	0	0
0	0	0	1	0
0	0	1	0	0
0	0	1	1	0
0	1	0	0	0
0	1	0	1	0
0	1	1	0	0
0	1	1	1	0
1	0	0	0	0
1	0	0	1	0
1	0	1	0	0
1	0	1	1	0
1	1	0	0	1
1	1	0	1	1
1	1	1	0	0
1	1	1	1	1

1. (PARK) · (key) = (start)   3. (Power) · (ON) · (OS) = (start)
5. (Down arrow) + (page down) + ((scroll bar) · (left mouse button)) = (scroll down)

CHAPTER 27

PROBLEMS 27–I

1. $\overline{B}$ 3. $\overline{A}C + \overline{A}B$ 5. $\overline{C}$ 7. $B\overline{C} + \overline{B}C$ 9. B   11. $\overline{C} + \overline{A}B$

PROBLEMS 27–2

1. $\overline{AB}\overline{C} + ABD$ 3. $\overline{C}D + \overline{A}\,\overline{B}\,\overline{C}$ 5. $ABC + \overline{D}$ 7. D   9. $B\overline{C} + \overline{B}CD$

CHAPTER 28

PROBLEMS 28–I

1. Table C–28A is the frequency distribution table. Figure C–28A is the histogram.
3. Table C–28B is the frequency distribution table. Figure C–28B is the histogram. Range = 98 − 60 = 38. Eight groups of five starting at 60 and extending to 99 are used. Interval midpoints are 62, 67, 72, 77, 82, 87, 92, and 97.

Scores	60	65	70	75	80	85	90
Tally	|	||	||	ⅣⅠ	||	||	|
	1	2	2	5	2	2	1

TABLE C–28A Frequency distribution table for problem I of End of Chapter Problems 28–I.

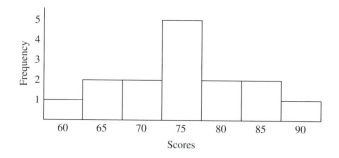

FIGURE C–28A Histogram for problem I of End of Chapter Problems 28–I.

Scores	60-64	65-69	70-74	75-79	80-84	85-89	90-94	95-99
Tally	||	|||	||||	ⅣⅠⅠ	||||	||	||	|
	2	3	4	7	4	2	2	1

TABLE C–28B Frequency distribution table for problem 3 of End of Chapter Problems 28–I.

FIGURE C–28B
Histogram for
problem 3 of
End of Chapter
Problems 28–1.

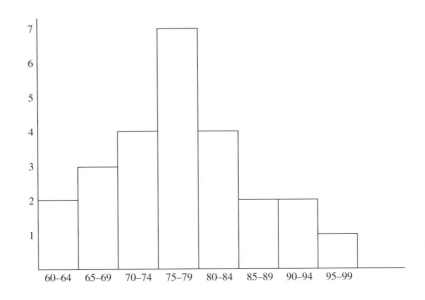

5. Range = 432 dB − 370 dB = 62. Seven groups of 9-dB intervals from 370 dB to 433 dB are used. Interval midpoints (in dB) are 374, 383, 392, 401, 410, 419, and 428. Table C–28C is the frequency distribution table and Figure C–28C is the histogram.
7. Range = 340 − 260 = 80 items. Nine groups of 9 items ranging from 260 to 341 are used. Interval midpoints are 264, 273, 282, 291, 300, 309, 318, 327, and 336. Table C–28D is the frequency distribution table. Figure C–28D is the histogram.

TABLE C–28C
Frequency
distribution
table for
problem 5 of
End of Chapter
Problems 28–1.

Gain	370–378	379–387	388–396	397–405	406–414	415–423	424–432
Tally			||	|||	|||	||||	||	||
	1	2	3	6	4	2	2

FIGURE C–28C
Histogram for
problem 5 of
End of Chapter
Problems 28–1.

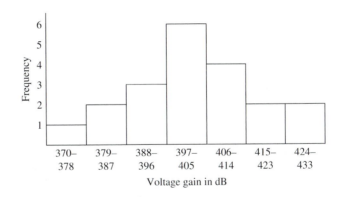

Voltage gain in dB

Items produced	260–268	269–277	278–286	287–295	296–304	305–313	314–322	323–331	332–340
Tally	|	|||	||	⨍||| |	||||	|||	|||	||	|
	1	3	2	6	4	3	3	2	1

TABLE C–28D   Frequency distribution table for problem 7 of End of Chapter Problems 28–1.

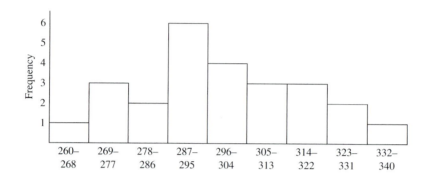

FIGURE C–28D
Histogram for problem 7 of End of Chapter Problems 28–1.

9. Range = 2090 Ω − 1900 Ω = 190 Ω. Eight groups of 25 Ω ranging from 1900 Ω to 2099 Ω are used. Interval midpoints (in ohms) are 1912, 1937, 1962, 1987, 2012, 2037, 2062, and 2087. Table C–28E is the frequency distribution table and Figure C–28E is the histogram.

11. Table C–28F is the frequency distribution table and Figure C–28F is the histogram.

Resistor values	1900–1924	1925–1949	1950–1974	1975–1999	2000–2024	2025–2049	2050–2074	2075–2099
Tally	|	|	||	||||	⨍||| ||	⨍|||	|||	||
	1	1	2	4	7	5	3	2

TABLE C–28E   Frequency distribution table for problem 9 of End of Chapter Problems 28–1.

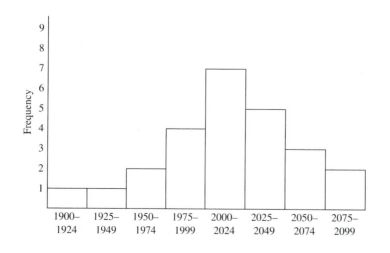

FIGURE C–28E
Histogram for problem 9 of End of Chapter Problems 28–1.

TABLE C–28F
Frequency distribution table for problem 11 of End of Chapter Problems 28–1.

Defective parts/100	1	2	3	4	5	6
Tally	II	III	NⅢI	IIII	III	I
	2	3	7	4	3	1

FIGURE C–28F
Histogram for problem 11 of End of Chapter Problems 28–1.

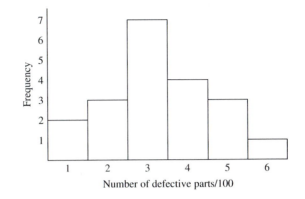

PROBLEMS 28–2

1. Mean = 75.0; Median = 75; Mode = 75   3. Mean = 77.4; Median = 77; Mode = 77
5. Mean = 40.3; Median = 40.1; Mode = 40.1   7. Mean = 299; Median = 300; Mode = 291
9. Mean = 2014; Median = 2012; Mode = 2012   11. Mean = 3.30; Median = 3; Mode = 3

PROBLEMS 28–3

1. Mean = 77.4; Median = 75; Mode = 75; σ = 9.41. Range = 40
3. Mean = 77.4; Median = 80; Mode = 80; σ = 10.2. Range = 40
5. Mean = 73.4; Median = 74; Mode = 74; σ = 11.2. Range = 48
7. Mean = 71.9; Median = 72; Mode = 82; σ = 11.2. Range = 44
9. Mean = 473; Median = 484; Mode = 484; σ = 67.4. Range = 327
11. Mean = 561; Median = 552; Mode = 552; σ = 32.3. Range = 146

PROBLEMS 28–4

1. 340 (68%) are between 39 mm and 41 mm long.
   475 (95%) are between 38 mm and 42 mm long.
   499 (99.7%) are between 37 mm and 43 mm long.
   1 will be greater than 3σ from the mean.
3. During 20 days (68%), between 590 and 610 parts will be produced per day.
   During 28 days (95% rounded down to a full day) between 580 and 620 parts will be produced each day.
   Between 570 and 630 parts will be produced each day.
5. 3400 of the scores will fall between 67.4 and 78.6.
   4750 of the scores will fall between 61.8 and 84.2.
   4985 of the scores will fall between 56.2 and 90.8.
   15 of the scores will be less than 56.2 or greater than 90.8.

Index